THE
CARBON
FOOTPRINT
HANDBOOK

THE
CARBON
FOOTPRINT
HANDBOOK

Edited by
Subramanian Senthilkannan Muthu

CRC Press
Taylor & Francis Group
Boca Raton London New York

CRC Press is an imprint of the
Taylor & Francis Group, an **informa** business

First published in paperback 2024

First published 2016
by CRC Press
2385 NW Executive Center Drive, Suite 320, Boca Raton FL 33431

and by CRC Press
4 Park Square, Milton Park, Abingdon, Oxon, OX14 4RN

First issued in hardback 2019

CRC Press is an imprint of Taylor & Francis Group, LLC

© 2016, 2019, 2024 Taylor & Francis Group, LLC

Library of Congress Cataloging-in-Publication Data

The carbon footprint handbook / Subramanian Senthilkannan Muthu, editor.
    pages cm
  "A CRC title."
  Includes bibliographical references and index.
  ISBN 978-1-4822-6222-3
  1. Atmospheric carbon dioxide. 2. Greenhouse gas mitigation. 3. Environmental protection. 4. Climate change mitigation. I. Muthu, Subramanian Senthilkannan, editor.

TJ163.3.C385 2016
363.738'74--dc23   2015008644

ISBN: 978-1-4822-6222-3 (hbk)
ISBN: 978-1-03-291898-3 (pbk)
ISBN: 978-0-429-16049-3 (ebk)

DOI: 10.1201/b18929

Visit the Taylor & Francis Web site at
http://www.taylorandfrancis.com

and the CRC Press Web site at
http://www.crcpress.com

Contents

Preface ix
Editor xi
Contributors xiii

SECTION I   Methodological Aspects of Carbon Footprint

Chapter 1   The Science of Carbon Footprint Assessment 3

T.V. Ramachandra and Durga Madhab Mahapatra

Chapter 2   Challenges and Merits of Choosing Alternative Functional Units 45

Benjamin C. McLellan

Chapter 3   Methodology for Carbon Footprint Calculation in Crop and Livestock Production 61

Kun Cheng, Ming Yan, Genxing Pan, Ting Luo, and Qian Yue

Chapter 4   End of Life Scenarios and the Carbon Footprint of Wood Cladding 85

Andreja Kutnar and Callum Hill

Chapter 5   Carbon Footprints and Greenhouse Gas Emission Savings of Alternative Synthetic Biofuels 101

Diego Iribarren, Jens F. Peters, Ana Susmozas, Pedro L. Cruz, and Javier Dufour

Chapter 6   Issues in Making Food Production GHG Efficient: Challenges before Carbon Footprinting 125

Divya Pandey and Madhoolika Agrawal

Chapter 7   Modeling the Carbon Footprint of Wood-Based Products and Buildings 143

Ambrose Dodoo, Leif Gustavsson, and Roger Sathre

Chapter 8   Applications of Carbon Footprint in Urban Planning and Geography 163

Taehyun Kim

SECTION II   Modeling Aspects of Carbon Footprint

Chapter 9   Quantifying Spatial–Temporal Variability of Carbon Stocks and Fluxes in Urban Soils: From Local Monitoring to Regional Modeling 185

V.I. Vasenev, J.J. Stoorvogel, N.D. Ananyeva, K.V. Ivashchenko, D.A. Sarzhanov, A.S. Epikhina, I.I. Vasenev, and R. Valentini

Chapter 10 Urban Carbon Footprint Evaluation of a Central Chinese City: The Case of
Zhengzhou City 223

Lipeng Hou and Rongqin Zhao

Chapter 11 Carbon Footprint Estimation from a Building Sector in India 239

Venu Shree, Varun Goel, and Himanshu Nautiyal

Chapter 12 The Carbon Footprint of Dwelling Construction in Spain 259

*Jaime Solís-Guzmán, Patricia González-Vallejo,
Alejandro Martínez-Rocamora, and Madelyn Marrero*

Chapter 13 Carbon Footprint: Calculations and Sensitivity Analysis for Cow Milk
Produced in Flanders, a Belgian Region 281

*Ray Jacobsen, Valerie Vandermeulen, Guido Vanhuylenbroeck, and
Xavier Gellynck*

Chapter 14 Digitizing the Assessment of Embodied Energy and Carbon Footprint
of Buildings Using Emerging Building Information Modeling 303

F.H. Abanda, A.H. Oti, and J.H.M. Tah

SECTION III   Carbon Footprint Assessment—Case Studies

Chapter 15 Product Carbon Footprint: Case Study of a Critical Electronic Part
(Subassembly of a Product) 333

Winco K.C. Yung and Subramanian Senthilkannan Muthu

Chapter 16 GHG Emissions from Municipal Wastewater Treatment in Latin America 351

*Leonor Patricia Güereca-Hernandez, María Guadalupe Paredes-Figueroa,
and Adalberto Noyola-Robles*

Chapter 17 Carbon Footprint of the Operation and Products of a Restaurant: A Study
and Alternative Perspectives 369

*Benjamin C. McLellan, Yoshiki Tanaka, Thi Thu Huyen Dinh,
Huong Long Dinh, Piradee Jusakulvijit, Faizi Ashley Taro Freemantle, and
Aibek Hakimov*

Chapter 18 Cultivation of Microalgae: Implications for the Carbon Footprint of
Aquaculture and Agriculture Industries 389

Kirsten Heimann, Samuel Cires, and Obulisamy P. Karthikeyan

Chapter 19 Carbon Footprint of Agricultural Products 431

Bidisha Chakrabarti, S. Naresh Kumar, and H. Pathak

Chapter 20 The Carbon Footprint of Sugar Production in Eastern Batangas, Philippines 451

Teodoro C. Mendoza, Rex B. Demafelis, Anna Elaine D. Matanguihan,
Justine Allen S. Malabuyoc, Richard V. Magadia Jr., Amabelle A. Pector,
Klarenz A. Hourani, Lavinia Marie A. Manaig, and Jovita L. Movillon

Chapter 21 A Two-Phase Carbon Footprint Management Framework: A Case Study
on the Rockwool Supply Chain 473

Eirini Aivazidou, Christos Keramydas, Agorasti Toka, Dimitrios Vlachos,
and Eleftherios Iakovou

Chapter 22 Product Carbon Footprint Estimation of a Ton of Paper: Case Study of a Paper
Production Unit in West Bengal, India 487

Debrupa Chakraborty

Chapter 23 Product Carbon Footprint Assessment of a Personal Electronic Product: Case
Study of an Electronic Scale 503

Winco K.C. Yung and Subramanian Senthilkannan Muthu

Index 523

Preface

It is emergent to act toward reducing the vulnerable impacts of climate change. Climate change or carbon footprint is no longer a new term to our society. Interest in quantifying and reducing the carbon footprint of products, processes, organizations, and even individuals has been raised, and the topic of reducing our carbon footprint is becoming more familiar and popular. Recognizing the importance of working toward mitigating the impact of climate change, companies, governments, and researchers, students, and private individuals alike are striving to reduce the carbon footprint within the possible boundaries.

Almost every industrial sector and its corresponding organizations are keenly working toward the quantification of carbon footprint, mitigation plans, and development of carbon labels pertaining to that particular sector. For most organizations, reduction of carbon emissions within a specified time frame is at the top of their agendas. Many countries have introduced stringent regulations to curb carbon emissions, and many countries are planning to implement the same. To gain sufficient knowledge and to keep abreast with the latest developments on our carbon footprint is pivotal for everyone involved in the sustainability and to anyone interested in protecting our planet from the impact of climate change. Hence, this handbook is composed of 23 scientific chapters under three sections.

The first section deals with methodological aspects of the carbon footprint. This overall topic is divided into eight interesting chapters, namely: The science of carbon footprint assessment; Challenges and merits of choosing alternative functional units; Methodology for CF calculation in crop and livestock production; End-of-life scenarios and the carbon footprint of wood cladding; Carbon footprints and greenhouse gas emission savings of alternative synthetic biofuels; Issues in making food production GHG efficient; Modeling the carbon footprint of wood-based products and buildings; and Applications of carbon footprint in urban planning and geography.

The crux of the carbon footprint and its results lie in the credibility of the models and results of an assessment. Modeling aspects of the carbon footprint are important, and hence a separate section composed of six chapters is dedicated to dealing with this issue. Chapters include Carbon footprint: Calculations and sensitivity analysis for milk produced in Flanders, a Belgian region; Digitizing the assessment of embodied energy and carbon footprint of buildings using emerging building information modeling; The carbon footprint of dwelling construction in Spain; Carbon footprint estimation from a building sector in India; Urban carbon footprint evaluation of a central Chinese city: The case study of Zhengzhou city; and Quantifying spatial–temporal variability of carbon stocks and fluxes in urban soils.

Case studies are valuable for illustrating the applicability and results of any concept. Hence, a separate section is provided to include many interesting case studies that offer bountiful information to the readers and practitioners of carbon footprint. Nine chapters from different perspectives have been chosen to discuss the case studies on carbon footprint under this section. Chapters include Product carbon footprint: Case study of a critical electronic part (subassembly of a product); GHG emissions from municipal wastewater treatment in Latin America; Carbon footprint of the operation and products of a restaurant; Cultivation of Microalgae: Implications for the carbon footprint of aquaculture and agriculture industries; Carbon footprint of agricultural products; The carbon footprint of sugar production in eastern Batangas, Philippines; A two-phase carbon footprint management framework: A case study on the rockwool supply chain; Product carbon footprint estimation of a ton of paper: Case study of a paper production unit in West Bengal, India, and Product carbon footprint assessment of a personal electronic product: Case study of an electronic scale.

All the very important aspects including even minute information pertaining to carbon footprint assessment and modeling have been collated for this handbook and presented in the 23 chapters examining the detailed aspects of the carbon footprint. I would like to take this opportunity to thank all the contributors of these 23 chapters for their sincere efforts and for so successfully enriching this handbook with the technical content in their chapters. I am confident that readers and practitioners of carbon footprint will certainly benefit from this handbook, which highlights the essential details pertaining to carbon footprint assessment. This handbook will undoubtedly become an important reference for the researchers and students, industrialists, climate change, and sustainability professional carbon footprint practitioners.

Editor

Dr. Subramanian Senthilkannan Muthu is currently working for SGS as a global sustainability consultant based in Hong Kong. He earned his diploma, bachelors, and masters degrees in textile technology from premier institutes of India. He was awarded a doctorate from the Institute of Textiles and Clothing of the Hong Kong Polytechnic University for his dissertation entitled "Eco-functional Assessment of Grocery Shopping Bags." He also has more than 7 years of industrial experience in textile manufacturing, testing, and sustainability evaluation of textiles and clothing materials. Dr. Subramanian has conducted many carbon footprint assessments of various products produced from various countries across the globe in his current position with SGS HK. He has conducted carbon footprint training along with many invited presentations at various conferences across the globe on the subject of the carbon footprint. An outstanding student of his discipline, Dr, Subramanian received numerous awards and medals including many gold medals in his field of study. He has more than 75 academic publications in various textiles and environmental journals to his credit. Additionally, he has 2 patents, 10 book chapters, and around 21 books and numerous conference publications. He serves as an editor, editorial board member, and reviewer for many international peer-reviewed journals of textiles and environmental science disciplines. He is the editor-in-chief of *Textiles and Clothing Sustainability*, an open access journal of Springer. He is also the series editor of two books for Springer, namely *Textile Science and Clothing Technology* and *Environmental Footprints and Eco-design*.

Contributors

F.H. Abanda
Oxford Institute for Sustainable Development
Department of Real Estate and Construction
Faculty of Technology, Design and
    Environment
Oxford Brookes University
Oxford, United Kingdom

Madhoolika Agrawal
Department of Botany
Banaras Hindu University
Varanasi, India

Eirini Aivazidou
Department of Mechanical Engineering
Aristotle University of Thessaloniki
Thessaloniki, Greece

N.D. Ananyeva
Institute of Physico-chemical and Biological
    Problems in Soil Science, RAS
Moscow Region, Russia

Bidisha Chakrabarti
Centre for Environment Science and Climate
    Resilient Agriculture
Indian Agricultural Research Institute
New Delhi, India

Debrupa Chakraborty
Department of Commerce
Netaji Nagar College
Kolkata, India

Kun Cheng
Institute of Resource, Ecosystem and
    Environment of Agriculture
Nanjing Agricultural University
Nanjing, People's Republic of China

Samuel Cires
College of Marine and Environmental Sciences
and
Centre for Sustainable Fisheries and Aquaculture
James Cook University
Queensland, Australia

Pedro L. Cruz
Systems Analysis Unit
Instituto IMDEA Energía
Móstoles, Spain

Rex B. Demafelis
College of Engineering and Agro-Industrial
    Technology
University of the Philippines Los Baños
Laguna, Philippines

Thi Thu Huyen Dinh
Graduate School of Energy Science
Kyoto University
Kyoto, Japan

Huong Long Dinh
Graduate School of Energy Science
Kyoto University
Kyoto, Japan

Ambrose Dodoo
Sustainable Built Environment Group
Department of Built Environment and Energy
    Technology
Linnaeus University
Växjö, Sweden

Javier Dufour
Systems Analysis Unit
Instituto IMDEA Energía
and
Department of Chemical and Energy
    Technology
ESCET
Rey Juan Carlos University
Móstoles, Spain

A.S. Epikhina
Laboratory of Agroecological Monitoring,
    Ecosystem Modeling and Prediction
Russian State Agricultural University
Moscow, Russia

Faizi Ashley Taro Freemantle
Graduate School of Energy Science
Kyoto University
Kyoto, Japan

Xavier Gellynck
Department of Agricultural
    Economics
Ghent University
Ghent, Belgium

Varun Goel
Department of Mechanical Engineering
National Institute of Technology
    Hamirpur
Hamirpur, India

Patricia González-Vallejo
Department of Building Construction II
School of Building Engineering
University of Seville
Seville, Spain

Leonor Patricia Güereca-Hernandez
Instituto de Ingeniería
Universidad Nacional Autónoma
    de México
Ciudad Universitaria
Distrito Federal, México

Leif Gustavsson
Sustainable Built Environment Group
Department of Built Environment and Energy
    Technology
Linnaeus University
Växjö, Sweden

Aibek Hakimov
Graduate School of Energy Science
Kyoto University
Kyoto, Japan

Kirsten Heimann
College of Marine and Environmental Sciences
and
Centre for Sustainable Fisheries and
    Aquaculture
James Cook University
Queensland, Australia

Callum Hill
Norsk Institutt for Skog og Landskap
Ås, Norway
and
JCH Industrial Ecology Limited
Bangor, United Kingdom

Lipeng Hou
School of Resources and Environment
North China University of Water Resources
    and Electric Power
Zhengzhou, China

Klarenz A. Hourani
College of Engineering and Agro-Industrial
    Technology
University of the Philippines Los Baños
Laguna, Philippines

Eleftherios Iakovou
Department of Mechanical Engineering
Aristotle University of Thessaloniki
Thessaloniki, Greece

Diego Iribarren
Systems Analysis Unit
Instituto IMDEA Energía
Móstoles, Spain

K.V. Ivashchenko
Institute of Physico-chemical and Biological
    Problems in Soil Science, RAS
Moscow Region, Russia

Ray Jacobsen
Department of Agricultural Economics
Ghent University
Ghent, Belgium

Piradee Jusakulvijit
Graduate School of Energy Science
Kyoto University
Kyoto, Japan

Obulisamy P. Karthikeyan
College of Marine and Environmental
    Sciences
and
Centre for Sustainable Fisheries and
    Aquaculture
James Cook University
Queensland, Australia

Christos Keramydas
Department of Mechanical Engineering
Aristotle University of Thessaloniki
Thessaloniki, Greece

Taehyun Kim
Korea Environmental Information Center
Korea Environment Institute
Sejong-si, Republic of Korea

S. Naresh Kumar
Centre for Environment Science and Climate
    Resilient Agriculture
Indian Agricultural Research Institute
New Delhi, India

Andreja Kutnar
Andrej Marušič Institut
University of Primorska
Koper, Slovenia

Ting Luo
Institute of Resource, Ecosystem
    and Environment of Agriculture
Nanjing Agricultural University
Nanjing, People's Republic of China

Richard V. Magadia Jr.
College of Engineering and Agro-Industrial
    Technology
University of the Philippines Los Baños
Laguna, Philippines

Durga Madhab Mahapatra
Energy and Wetlands Research Group
Center for Ecological Sciences [CES]
Karnataka, India

Justine Allen S. Malabuyoc
College of Engineering and Agro-Industrial
    Technology
University of the Philippines Los Baños
Laguna, Philippines

Lavinia Marie A. Manaig
College of Engineering and Agro-Industrial
    Technology
University of the Philippines Los Baños
Laguna, Philippines

Madelyn Marrero
Department of Building Construction II
School of Building Engineering
University of Seville
Seville, Spain

Alejandro Martínez-Rocamora
Department of Building Construction II
School of Building Engineering
University of Seville
Seville, Spain

Anna Elaine D. Matanguihan
College of Engineering and Agro-Industrial
    Technology
University of the Philippines Los Baños
Laguna, Philippines

Benjamin C. McLellan
Graduate School of Energy Science
Kyoto University
Kyoto, Japan

Teodoro C. Mendoza
College of Agriculture
University of the Philippines Los Baños
Laguna, Philippines

Jovita L. Movillon
College of Engineering and Agro-Industrial
    Technology
University of the Philippines Los Baños
Laguna, Philippines

Subramanian Senthilkannan Muthu
Global Sustainability Services
SGS (HK) Limited
Kwai Chung, N.T. Hong Kong

Himanshu Nautiyal
Department of Mechanical Engineering
THDC Institute of Hydropower Engineering
    and Technology
Uttarakhand, India

Adalberto Noyola-Robles
Instituto de Ingeniería
Universidad Nacional Autónoma de México
Ciudad Universitaria
Distrito Federal, México

A.H. Oti
Oxford Institute for Sustainable Development
Department of Real Estate and Construction
Oxford Brookes University
Oxford, United Kingdom

Genxing Pan
Institute of Resource, Ecosystem and
    Environment of Agriculture
Nanjing Agricultural University
Nanjing, People's Republic of China

Divya Pandey
Department of Botany
Banaras Hindu University
Varanasi, India

María Guadalupe Paredes-Figueroa
Instituto de Ingeniería
Universidad Nacional Autónoma de México
Ciudad Universitaria
Distrito Federal, México

H. Pathak
Centre for Environment Science and Climate
    Resilient Agriculture
Indian Agricultural Research Institute
New Delhi, India

Amabelle A. Pector
College of Engineering and Agro-Industrial
    Technology
University of the Philippines Los Baños
Laguna, Philippines

Jens F. Peters
Systems Analysis Unit
Instituto IMDEA Energía
Móstoles, Spain

T.V. Ramachandra
Energy and Wetlands Research Group
Center for Ecological Sciences
Centre for Sustainable Technologies (astra)
Centre for Infrastructure, Sustainable
    Transportation and Urban Planning
Indian Institute of Science
Karnataka, India

D.A. Sarzhanov
Laboratory of Agroecological Monitoring,
    Ecosystem Modeling and Prediction
Russian State Agricultural University
Moscow, Russia

Roger Sathre
Sustainable Built Environment Group
Department of Built Environment and Energy
    Technology
Linnaeus University
Växjö, Sweden

Venu Shree
Department of Architecture
National Institute of Technology Hamirpur
Hamirpur, India

Jaime Solís-Guzmán
Department of Building Construction II
School of Building Engineering
University of Seville
Seville, Spain

J.J. Stoorvogel
Soil Geography and Landscape Group
Wageningen University
Wageningen, the Netherlands

Ana Susmozas
Systems Analysis Unit
Instituto IMDEA Energía
Móstoles, Spain

J.H.M. Tah
Oxford Institute for Sustainable
    Development
Department of Real Estate and
    Construction
Oxford Brookes University
Oxford, United Kingdom

Yoshiki Tanaka
Graduate School of Energy Science
Kyoto University
Kyoto, Japan

Agorasti Toka
Department of Mechanical Engineering
Aristotle University of Thessaloniki
Thessaloniki, Greece

R. Valentini
Laboratory of Agroecological Monitoring,
  Ecosystem Modeling and Prediction
Russian State Agricultural University
Moscow, Russia
and
Tuscia University
Viterbo, Italy

Valerie Vandermeulen
Department of Agricultural Economics
Ghent University
Ghent, Belgium

Guido Vanhuylenbroeck
Department of Agricultural Economics
Ghent University
Ghent, Belgium

I.I. Vasenev
Laboratory of Agroecological Monitoring,
  Ecosystem Modeling and Prediction
Russian State Agricultural University
Moscow, Russia

V.I. Vasenev
Soil Geography and Landscape Group
Wageningen University
Wageningen, the Netherlands
and
Laboratory of Agroecological Monitoring,
  Ecosystem Modeling and Prediction
Russian State Agricultural University
and
Landscape Architecture and Design Group
Peoples' Friendship University of Russia
Moscow, Russia

Dimitrios Vlachos
Department of Mechanical Engineering
Aristotle University of Thessaloniki
Thessaloniki, Greece

Ming Yan
Institute of Resource, Ecosystem and
  Environment of Agriculture
Nanjing Agricultural University
Nanjing, People's Republic of China

Qian Yue
Institute of Resource, Ecosystem and
  Environment of Agriculture
Nanjing Agricultural University
Nanjing, People's Republic of China

Winco K.C. Yung
Department of Industrial and Systems
  Engineering
The Hong Kong Polytechnic University
Hung Hom, Hong Kong

Rongqin Zhao
School of Resources and Environment
North China University of Water Resources
  and Electric Power
Zhengzhou, China

Section I

Methodological Aspects of Carbon Footprint

1 The Science of Carbon Footprint Assessment

T.V. Ramachandra and Durga Madhab Mahapatra

CONTENTS

1.1 The Science of Carbon Footprint 4
    1.1.1 Importance and Need for Assessment 4
    1.1.2 Definition of C Footprint: A Brief Review 6
    1.1.3 Issues Related to Quantification: Methodological Issues 7
    1.1.4 GHG Emissions from Wastewater Sector 10
1.2 C Footprint of Municipal Wastewater 11
    1.2.1 GHG Emissions for Wastewater Treatment Plant at Bengaluru 11
    1.2.2 Quantification of GHG Emissions in Treatment Plants 13
1.3 Carbon Sequestration and Biofuel Prospects 16
    1.3.1 Materials and Methods 17
        1.3.1.1 Study Area 17
        1.3.1.2 Sampling: Algal Screening, Selection, and Densities.......................18
        1.3.1.3 Characterization of the Growth Environment and Water Quality.............18
        1.3.1.4 Harvesting of Algal Biomass 18
        1.3.1.5 Monitoring the Growth 18
        1.3.1.6 Spectral Signature and Biochemical Composition Analysis by ATR-FTIR.....19
    1.3.2 Lipid Extraction and Analysis 19
        1.3.2.1 Lipid Extraction 19
        1.3.2.2 Fatty Acid Composition Using GC–MS 19
    1.3.3 Scope for Biofuel as a Viable Energy Source in Cities of Karnataka 20
        1.3.3.1 Distribution, Morphological Features, and Cellular Characteristics of
                L. ovum 20
    1.3.4 Nutrient Requirements and Growth Conditions 20
        1.3.4.1 Nutrient Concentrations and *L. ovum* Growth at
                Wastewater-Treatment Units 20
        1.3.4.2 Monitoring of Algal Growth...22
        1.3.4.3 Lipid Composition by GC–MS Analysis 24
        1.3.4.4 C Sequestration in Wastewater Algae...26
    1.3.5 Viability of Algae-Based Biofuel as an Energy Source in Karnataka 27
1.4 Integrated Wetlands Ecosystem to Mitigate Carbon Emissions...................................32
    1.4.1 Wetlands/Algae Pond as Wastewater Treatment Systems 32
    1.4.2 Integrated Wetland System..33
        1.4.2.1 Nutrients (Nitrates and Phosphates)..33
        1.4.2.2 BOD and COD..35
    1.4.3 Integrated Wastewater Management System..35
    1.4.4 Integrated Wetlands Ecosystem: Sustainable Model to Mitigate GHG Emissions 37
        1.4.4.1 Functional Aspects of the Integrated Wetland Systems 37
Acknowledgments...39
References...39

1.1   THE SCIENCE OF CARBON FOOTPRINT

Carbon footprint refers to the amount of greenhouse gases (GHGs) produced due to human activities, measured in units of carbon dioxide. "Carbon footprint" assessment has gained importance in recent years and has been used as a benchmark for quantitation of GHG emissions in context of climate dynamics aided by anthropogenic, unregulated resource usage and consumption practices (Wiedmann and Minx 2007; East 2008; Finkbeiner 2009; Peters 2010). Carbon footprint (C footprint) is a subset of "ecological footprint" (Wackernagel and Rees 1996), which implies the land resources needed to immobilize/sequester the total GHG as CO_2 generated by anthropogenic activity. C footprint estimations have become more prevalent with towering GHG emissions, leading to global warming with rapid changes in the climate (East 2008). C footprint is a measure of the exclusive, total amount of CO_2 emissions that is directly and indirectly released by an activity or is accumulated over the life stages of a product (Wiedmann and Minx 2007). This is comparable to life-cycle impact assessment and is an indicator of global warming potential (GWP) and global sustainability. Although there has been a great deal of environmental deterioration, awareness on quantifying the magnitude of C footprint is scarce. However, the measurement of C footprint requires clarity on (i) inclusion of indirect emissions required for upstream production process, (ii) direct and onsite emission of the production process, (iii) accounting for all stages of the life cycle of a process in terms of services and goods, and (iv) systems boundary and quantification approaches.

The science of C footprint is frequently discussed as the well-known term carbon footprint, (Kleiner 2007) but variant terminologies to C footprint are climate footprint, GHG footprint, embodied carbon, and carbon flows (Courchene and Allan 2008; Edgar and Peters 2009). The term "carbon footprint" comprised of two words, the first of which, carbon, is the lifeline of all living organisms and constitutes the most dominant biopolymer in every single living organism. On an average, living organisms comprise of ~40–50% of C on a dry weight basis. Carbon forms the most abundant biopolymers globally and comprises the bulk of any living organic matter. For every organism, food mostly consists of C in some form; for example, the food source of human beings comes from plants and animals. The plants capture/immobilize CO_2 from the atmosphere using solar energy to combine it with water in order to create sugars from which they build their cytoskeleton (photosynthesis).

The term "footprint" explains the measurement or the impression of C and is represented as global hectares (Brown et al. 2009). However, the C footprint can have a variety of quantifying scales as to determine the impact or the stress/quantity of C emission (tons)/quantity normalized over CO_2 as CO_2 equivalents (tons of CO_2-eq.) as in the case of GHG potential/area for land measurements. For wastewater sector, it is attributed to the sum of all emissions associated with the collection, treatment, and ultimate disposal of wastewater. Significant sources of these emissions include the indirect emissions from the purchase of electricity, direct emissions resulting from the treatment process, fugitive emissions from the waste itself, and transportation-related emissions.

1.1.1   IMPORTANCE AND NEED FOR ASSESSMENT

The last two centuries have witnessed a soaring increase in atmospheric concentrations of GHGs. Human activities such as industry, agriculture, deforestation, waste disposal, and, especially, unprecedented use of fossil fuel have been producing increasing amounts of GHGs. For example, the concentrations of CO_2 increased from approximately 280 parts per million by volume (ppmv) in preindustrial age to 372 ppmv in 2001 and have continued to increase by about 0.5% per year (IPCC 2007), whereas current CH_4 atmospheric concentration is going up at a rate of 0.02 ppmv per year. Similarly, there has been a substantial increase in sources of N_2O by anthropogenic activities of about 40%–50% over preindustrial levels. Owing to this rapid increase in GHG concentrations in the atmosphere, due to human-induced activities, the entire humanity is under the threat of global warming (rise in global temperature) and associated climate change and its far-reaching consequences.

The magnitude of impact of GHGs differs and is based on the radiative forcing together with the mean time span of the existence of GHG gas in the atmosphere, which is jointly expressed as the GWP and mostly expressed as carbon dioxide equivalent (CO_2-eq). The most significant contributors to global warming are the six Kyoto gases and, to a lesser extent, the chlorofluoro carbons (highlighted during the Montreal protocol; IPCC 2006, 2007). The largest share of the GHG is contributed by CO_2 (~60%) followed by methane and nitrous oxide and the largest GHG-producing sector being the energy supply (~26%) (IPCC 2007). The global anthropogenic GHG emissions and the GHG emissions by various sectors (IPCC 2007) have been elucidated in Figure 1.1.

Anthropogenic GHG emissions grew at an average annual rate of 2.2% during 2000–2010, compared with a growth rate of 1.3% per year in the period 1970–2000. CO_2 emissions from fossil fuels and industrial processes contributed 65% of global GHG emissions in 2010. The increase in anthropogenic GHG emissions comes from energy supply (47%), industry (30%), transport (11%), and building (3%) sectors. Globally, population and economic growth continue to be the most important drivers of increase in CO_2 emissions from fossil-fuel combustion (IPCC 2014).

(a)

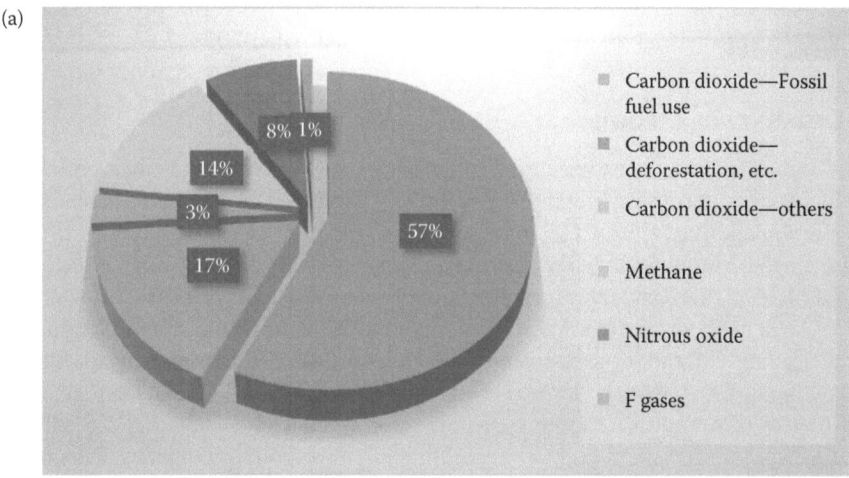

(b)

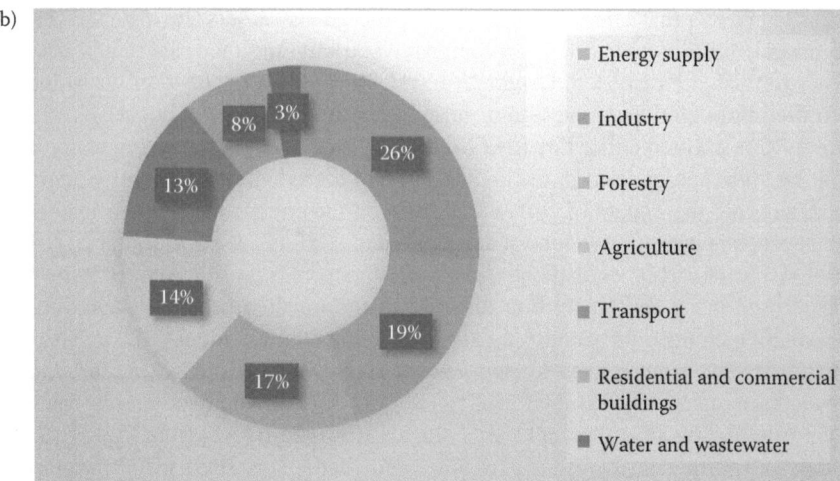

FIGURE 1.1   (a) Global anthropogenic GHG emissions. (b) GHG emissions by sector. (From IPCC. 2007. Climate change 2007: Synthesis report: Contribution of working groups I, II and III to the fourth assessment report. Intergovernmental Panel on Climate Change (IPCC).)

The greenhouse effect of these gases is estimated by the GWP that depends on radiative forcing and the timeframe of consideration (usually taken as 100 years). The GWP factors for a 100-year horizon for (a) CO_2 is 1 (base/reference), (b) CH_4 is 21, and (c) N_2O is 298 (IPCC 2007). In other words, over a time period of 100 years, 1 ton of CH_4 will have a warming effect equivalent to 25 tons of CO_2 (IPCC 2006).

Increase in the emission levels has led to 0.74°C rise in the mean global temperature (IPCC 2007), which in turn can lead to global warming resulting in glaciers melting followed by increase in sea level and submersion of coastal and low-lying areas (Kerr 2007). The imbalance in nature is evident through the recent episodes of cyclones, tsunamis, and extreme weather conditions. The C footprint is accounted as a quantitative expression of the GHG emissions (Carbon Trust 2006) from various activities that aid in devising tools/methods for emission management and figuring out appropriate mitigation measures. Today, it has become imperative for a detailed GHG inventory on a global scale, and specific methods have to be devised for C abatement to reduce and evaluate C footprint and GHG emission. The C footprinting exercise will enable quantification of emissions, tracing important sources of emissions for which appropriate management and emission reduction techniques can be developed along with increasing the efficiency of the process. Stringent legislative measures should be followed for accounting C footprint and identification of C footprint reduction methods for regions and organizations (Courchene and Allan 2008).

1.1.2 DEFINITION OF C FOOTPRINT: A BRIEF REVIEW

The term "carbon footprint" has been used to measure the GHG emissions that a particular product or service generates during its lifetime and is mainly expressed as CO_2 equivalents (CO_2-eq.) that comprises of emissions of CO_2, CH_4, and N_2O. The C footprint can be a subset of life-cycle analysis (LCA), emphasizing only the GWP part (Weidema et al. 2008; Finkbeiner 2009). However, the C footprinting calculations and representation have been inconsistent and divergent, and the utility and potential of such methods/strategies have been questioned. Most of the recent literatures have dealt with either sectoral C footprint assessments or region-specific C footprint analysis. The C footprint terminology has also been widely observed and reported as a measure of GHG expressed as CO_2-eq.

According to EPA (2010), C footprint is the total amount of GHGs that are emitted into the atmosphere each year by a person, family, building, organization, or company. This includes GHG emissions from fuel that an individual burns directly, such as by heating a home or riding in a car. It also includes GHGs that come from producing the goods or services that the individual uses, including emissions from power plants that make electricity and factories that make products and landfills where trash are disposed. Grubb (2007) defines it as a measure of the amount of carbon dioxide emitted through the combustion of fossil fuels. In the case of a business organization, it is the amount of CO_2 emitted either directly or indirectly as a result of its everyday operations. It also reflects on the fossil energy used in the manufacture of a product or commodity upon reaching the market. Considering the impact of human activities on the environment in terms of the amount of GHGs produced, it is measured in tons of carbon dioxide (ETAP 2007). C footprint reflects the total emission of GHGs in carbon equivalents from a product across its lifecycle from the production of raw material used in its manufacture to disposal of the finished product (excluding in-use emissions). It is also a technique for identifying and measuring the individual GHG emissions from each activity within a supply-chain process step and the framework for attributing these to each output product. According to the Global Footprint Network, 2007, "the demand on bio-capacity required to sequester (through photosynthesis) carbon dioxide (CO_2) emissions from fossil fuel combustion" (GFN 2007). The parliamentary office of Science and Technology (POST 2006) reports "A 'carbon footprint' is the total amount of CO_2 and other greenhouse gases, emitted over the full life cycle of a process or product. It is expressed as grams of CO_2 equivalent per kilowatt hour of generation (gCO_2-eq./kW h), which accounts for the different global warming effects of other greenhouse gases."

1.1.3 Issues Related to Quantification: Methodological Issues

i. Greenhouse gas records and measurements

a. *GHGs*: The choice of the GHGs depends on the nature, requirement, and the guidelines of the sector/activities. For example, in the case of wastewater, the most important gas emitted is CO_2 as an outcome of bacterial metabolism, and along with it are the emissions of CH_4 and N_2O. In the case of any industry, for example, coal-based power plant CO_2 becomes the most important GHG compared to CH_4 and other gases found in traces. Many earlier studies account only CO_2 emissions for measurements of C footprint (Wiedmann and Minx 2007; Craeynest and Streatfeild 2008) and six major GHG emissions (Bokowski et al. 2007; Energetics 2007; Garg and Dornfeld 2008; Matthews et al. 2008a,b; Peters 2010).

b. *Systems boundary*: The systems boundary delineates the targeted region and considers the activities within the boundary. This depends on the organizational and operational boundaries. The organizational boundary can be ascribed as the boundary of the organization on the economic and business grounds and its associated activities under the study. The operational boundary involves the direct and indirect emissions (WRI/WBCSD 2004; Carbon Trust 2007a,b; BSI 2008).

The operational boundary considers (i) direct emissions (on-site) and (ii) indirect emission sources (embodied emissions—consumption of purchased power). All indirect emissions—difficult to quantify, for example, in the case of wastewater direct emissions, sources include all direct GHG emissions, except the direct CO_2 biogenic emissions (fixed biologically during treatment). Indirect emissions are ascribed to the consumption of purchased energy, that is, acquired electricity, heating, cooling, and so on, and such emissions are an outcome of activities within the organizational boundary but are emitted from sources controlled or owned by some other organization, as central electricity supply, and so on. This constitutes one of the largest sources of emissions for wastewater pumping and treatment facilities due to highenergy, intensive pumping process (Figure 1.2). Thus, wastewater treatment units have to ensure energy-efficient initiatives to reduce indirect GHG emissions.

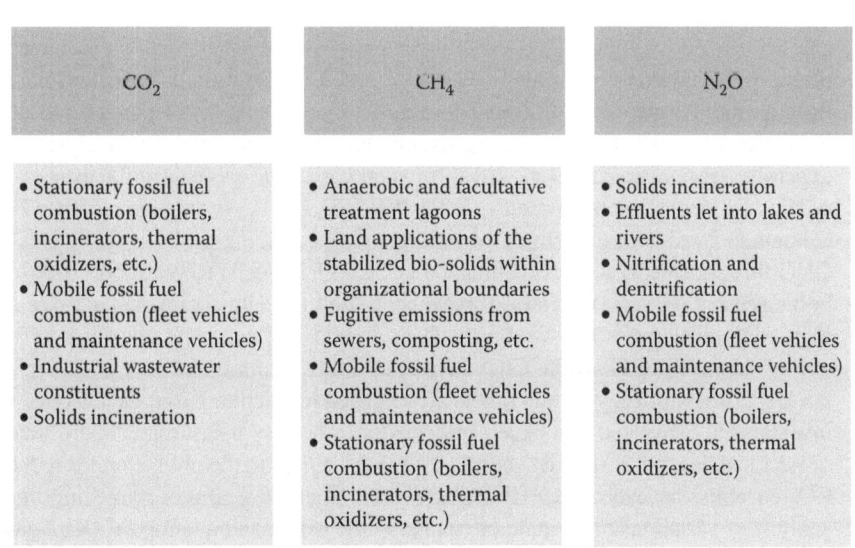

FIGURE 1.2 Method of accounting indirect emissions.

Other indirect emission sources are:
a. Supply-chain GHG emissions as upstream/downstream transport of chemicals, materials, and fuels and upstream production
b. Bio-solids reuse, including land application or other methods that are outside the organizational boundary
c. Landfilling of bio-solids
d. Emissions from services contracted with outside vendors as bio-solids hauling fuel emissions or contracted wastewater treatment by adjacent municipalities
e. Emissions from employee commuting and business travel

All these are difficult to quantify and require additional guidelines for facilitating inventories. Whatever C footprint studies conducted to date have emphasized and accounted for the quantity of GHG removal, but gross sequestration of GHGs has not been addressed (Peters 2010). Therefore, C sequestration becomes vital in all such calculations and is to be incorporated in C footprint measurements.

c. *Compilation of GHG emission data*: Emissions have been quantified either directly through on-site real-time measurements or through estimations based on the emission factors and various empirical models. The emission factors or the model-based data generation are the most used means for acquiring data. These emissions are calculated from specific emission factors using the data on fuel consumption, efficiency of the process, energy, and other activities resulting in emissions (CO_2, in particular). Emission factors for various sectors and industrial processes and land-uses are available in PAS 2050 (IPCC 2006, 2007). Country-specific emission factors developed in many countries as in national laboratories under UNFCC, US EPA (WRI/WBCSD 2004; IPCC 2007). However, the verification of these is highly required at varied organizational, operational, and geographical contexts. The region-specific emission factors and models have been precisely developed (WRI/WBCSD 2004; IPCC 2006). In cases of fugitive emissions, the direct measurements are done through specialized GHG-capture techniques such as optical, biochemical, IR sensors, and quantitative measurements through IR spectroscopy and gas chromatography (USCCTP 2005; Berg et al. 2006). Many other advanced techniques include flux towers (Velasco et al. 2005) and cavity ring-down spectrometers for measurements in air (Kelly et al. 2009). Apart from the primary data collection techniques, a wide variety of secondary data sources are accessible at the global scale in recent times. On a much larger scale, a CO_2 emission database from different countries is now obtainable with various national-level GHG inventories with adequate data on fuel and energy usage, IEA (International Energy Agency), and UNDP (United Nations Development Programme) (Brown et al. 2009). These direct methods are most precise and accurate, but the cost incurred for the data acquisition through experiments and analyses are quite expensive (Ramachandra et al. 2013). In such cases, the secondary databases are economic and yield acceptable results (WRI/WBCSD 2004). As an example, the GHG calculation tool provides guidelines for setting and customizing the tools for calculating the GHG flux specific to sector/organization (USCCTP 2005; WRI/WBCSD 2006). Space-borne sensors (optical, microwave) have been used for data acquisition and are coupled with GHG inventories that would result in a vivid and complete coverage for acquisition of data (USCCTP 2005; Chambers et al. 2007; Ramachandra et al. 2013). NASA, US Department of Energy, and others are engaged in satellite-based data acquisition and inventorying GHG and associated information. The flux measurements are taken with respect to a base year (starting year) as 1990 with regard to the obligation for reduction of CO_2-eq. emission levels under UNFCCC (2008). The base year is of prime importance for analysis of variations/magnitude of change in the extent and quantity of GHG gases and often needs to have reproducibility and reliability (Carbon Trust 2007b; Brewer 2008a,b).

d. *Calculation of C footprint*: This is done for all GHG gas quantities/flux as CO_2-eq. considering emission factors (WRI/WBCSD 2004; Wiedmann and Minx 2007; BSI 2008). There are two approaches followed for quantification of C footprint: (a) bottom-up, based on process analysis (PA), and (b) top-down, considering environment input–output analysis. The bottom-up approach inherits the understanding of environmental impacts of individual product/region/category from the source to the sink. This approach has disadvantage for identification of appropriate system boundary and is based only on-site and most first-order impacts (Lenzen 2001). This approach becomes far fetching with increase in the systems area such as sectors/governments, and so on. In contrast, the top-down or the environmental input–output (EIO) analysis (Wiedmann et al. 2006) is most commonly applied in economics and provides a majority of the economic activities at sectoral scales. This approach can result in a robust economic set up as the boundary. Input–output approach is largely carried out for sectoral C footprinting assessment and might not be suitable for small operations/processes. Hybrid approach integrates PA with the EIO approach (Suh et al. 2004; Heijungs and Suh 2006).

e. *C footprints assessment of a product*: C footprint assessment of a product is done either through bottom-up, based on PA, or top-down, based on EIO analysis considering the full life-cycle impacts through LCA, (Life Cycle Assessment).

   Bottom-up method involves a PA considering the environmental impacts of individual products from cradle to grave. However, this will have a problem of a system boundary only on-site, that emphasizes the need for identification of appropriate system boundaries. PA-based LCAs run into difficulties for larger entities such as government, households, or particular industrial sectors.

   Top-down approach involving EIO analysis provides a picture of all economic activities at the sector level. This with consistent environmental data aids in a comprehensive and robust way taking into account all impacts of the whole economic system as boundary. Application of this approach to assess microsystems such as products or processes is limited, as it assumes homogeneity of prices, outputs, and their carbon emissions at the sector level. Advantage of input–output-based approaches is a much smaller requirement of time and manpower once the model is in place.

   The best option for a comprehensive and robust analysis is to combine PA and input–output methods. Such an approach allows one to preserve the detail and accuracy of bottom-up approaches in lower-order stages, whereas higher-order requirements are covered by the input–output part of the model.

f. *C footprint standards for products*: Carbon footprint of a product involves measuring, managing, and communicating GHG emissions related to the product's goods and services. This is based on an LCA from a global warming perspective. This helps in enhancing market reputation, engaging with suppliers, clients, and other stakeholders, and more importantly this constitutes a first step toward a more comprehensive environmental assessment. Three commonly used C footprint standards applicable to the product are: PAS 2050, GHG Protocol, and ISO 14067 (Kulkarni and Ramachandra 2009; http://www.pre-sustainability.com/product-carbon-footprint-standards-which-standard-to-choose).

   PAS 2050 has been developed by the British Standards (BSI) and came into effect in October 2008. PAS 2050 has already been applied by many companies worldwide.

   The GHG Protocol product standard was developed by the WRI/WBCSD and tested by 60 companies in a road test in 2010. The GHG Protocol product standard was launched in October 2011.

ISO DS 14067 specifies principles, requirements, and guidelines for the quantification and communication of the C footprint of a product (CFP), based on International Standards on LCA (ISO 14040 and ISO 14044) for quantification and on environmental labels and declarations (ISO 14020, ISO 14024, and ISO 14025).

All three standards developed on existing LCA methods established through ISO 14040 and ISO 14044 provide requirements and guidelines on the decisions to be made when conducting a C footprint study. Decisions involve LCA issues, such as goal and scope definition, data collection strategies, and reporting. Moreover, these standards provide requirements on specific issues relevant for C footprints, including land-use change, carbon uptake, biogenic carbon emissions, soil carbon change, and green electricity.

1.1.4 GHG Emissions from Wastewater Sector

GHGs produced during wastewater treatment are CO_2, CH_4, and N_2O. Globally, wastewater is the fifth largest emitter of CH_4 and sixth largest emitter of N_2O that contributes ~10% of the total CH_4 emissions and 3% of N_2O emissions, respectively. CH_4 and N_2O emissions are expected to grow by ~20% and 13% by 2020 (US EPA 2006). GHGs such as CO_2 are emitted due to (a) treatment process and (b) electricity consumption (Ramachandra et al. 2014a). The next section discusses GHGs due to wastewater treatment based on field experiments. Bengaluru city generates ~1200 million liters per day (MLD) of domestic wastewater that gets either partially treated or untreated and joins receiving surface waters at various locations in the city. This has contributed toward C footprint.

The GHG emissions of the select treatment plants were estimated based on conventional treatment technologies and the energy units consumed during the operation of the treatment plants. The direct GHG emissions accounted for CO_2 generated through organic matter degradation (aerated tanks-aerobic process), CH_4 generated from the clarifiers and also from the basins, N_2O emissions from the effluent discharge into surface water bodies and the diesel generator used in the treatment units. Indirect emissions were calculated from offsite generation of electric power that is consumed at wastewater treatment plant, which is explained in Section 1.2.

Wastewaters generated from domestic households are a source of nutrients contributing to growth and development of algae and consequent lipid extraction. The CO_2 sequestration in algae is accomplished by (a) direct sequestration, which comprises the capture of CO_2 from wastewaters before its emission to atmosphere and subsequent storage and (ii) indirect sequestration, based on the capture of CO_2 that is already in the atmosphere, through photosynthesis. Carbon present in wastewater can be either inorganic C (carbonic acid/bicarbonates/carbonates) or organic C (dissolved organic acids, etc.). Many wastewater algae capable of consuming both C forms are called mixotrophic algae and are extremely beneficial in rapid C fixation and reduction of wastewater C footprint. Often, the bulk of the wastewater C fixed in the algal biomass is stored in the form of lipids (neutral) that can be potentially used as feedstock for biofuel as biodiesel and gasoline. The sequestration aspect of microalgae is discussed in Section 1.3.

Increased and unprecedented population growth has resulted in enormous stress on potable water from a daily consumption point of view and also with regard to increased wastewater generated by the city. Land-use analyses show 584% growth in built-up area during the last four decades with the decline of vegetation by 66% and water bodies by 74%. Apart from these, rapid urbanization in recent times has led to mammoth wastewater generation. Untreated or partially treated wastewaters are fed to surface water, leading to GHG emissions. The sustained inflow of untreated or partially treated sewage to wetlands leads to the enrichment of nutrients such as carbon (C), nitrogen (N), and so on. Installation of conventional treatment plants has helped in the removal of dissolved and suspended biological matter, while the effluents are still nutrient-rich and need further treatment and stabilization (Ramachandra et al. 2014a,b).

Section 1.4 presents the possibility of mitigating GHG emissions from wastewater sector through integrated wetlands ecosystem. Integration with wetlands (consisting of typha beds and algal pond) would help in the complete removal of nutrients in a cost-effective way. However, this requires regular maintenance of harvesting macrophytes and algae (from algal ponds). Harvested algae would have energy value, which could be used for biofuel production. The joint activity of algae and macrophytes in the wetland systems helps in the removal of 77% chemical oxygen demand (COD), ~90% biochemical oxygen demand (BOD), ~33% NO_3−N, and ~75% PO_4^{3-}−P. Implementation of integrated wetland system helps in treating the water and more importantly in the mitigation of GHG emissions from wastewater sector.

1.2   C FOOTPRINT OF MUNICIPAL WASTEWATER

Indirect emissions due to electricity consumption (fossil fuel-based energy: coal, gas, oil) in the treatment plants have been considered along with GHG production during the wastewater treatment process. Coal-based power generation has been considered, which is the most common in India. The GHG emitted from treatment plants depends on the technology. The wastewater treated in the various treatment plants in the city are also sources of CH_4. N_2O is formed during the denitrification processes, while CO_2 results from both aerobic and anaerobic processes. GHGs such as CO_2 are emitted due to (a) treatment process and (b) electricity consumption. During an anaerobic process, the BOD of wastewater is either incorporated into biomass or is transformed into CO_2 and CH_4. A small quantity of this biomass is further converted into CO_2 and CH_4 via endogenous respiration. Other emission sources of carbon dioxide are sludge digesters (if operational). But, during aerobic treatment, CO_2 is produced through the breakdown of organic matter in the activated sludge process and some through the primary clarifiers. CH_4 can be produced from wastewater directly and/or from wastewater sludge during anaerobic digestion. The magnitude of methane emission depends on the nature of environment (redox potential), temperature, and more importantly on the amount of degradable organic matter in wastewater. N_2O gas emanates as a decomposable product of nitrate, urea, and proteins. Centralized wastewater treatment systems comprise a variety of unit processes, for example, ponds, lagoons, and advanced treatment technology that might lead to N removal and discharge to receiving waters. Both nitrification and denitrification occur, to some extent, during the treatment process. N_2O is an intermediate product of both processes and is more often associated with denitrification.

1.2.1   GHG EMISSIONS FOR WASTEWATER TREATMENT PLANT AT BENGALURU

Bengaluru city generates ~1200 MLD of domestic wastewater that gets either partially treated or untreated and joins receiving surface waters at various locations in the city. The locations of the treatment plants are depicted in Figure 1.3 and the details regarding the installed capacity and the technology used are provided in Table 1.1. Most of the treatment plants follow an extended aeration or activated sludge process. The estimation of C has been performed based on a general mechanically aerated process set up as shown in Figure 1.4.

Methods: The quantification of C footprint has been done through field experiments (during 2012–2013) and also through GHG protocol (as per IPCC 1996; 2006). Preparation of GHG inventory involves:

a. *Organizational boundaries:* This includes the wastewater treatment plants and the electricity supply to run the treatment plants.
b. *Operational boundaries:* This deals with calculation of emissions associated with the operation/working of the treatment plants.
   Measuring direct GHG emissions: Here CO_2, CH_4, and N_2O concentrations are quantified for the operational wastewater treatment plant. CO_2 emissions as an outcome of

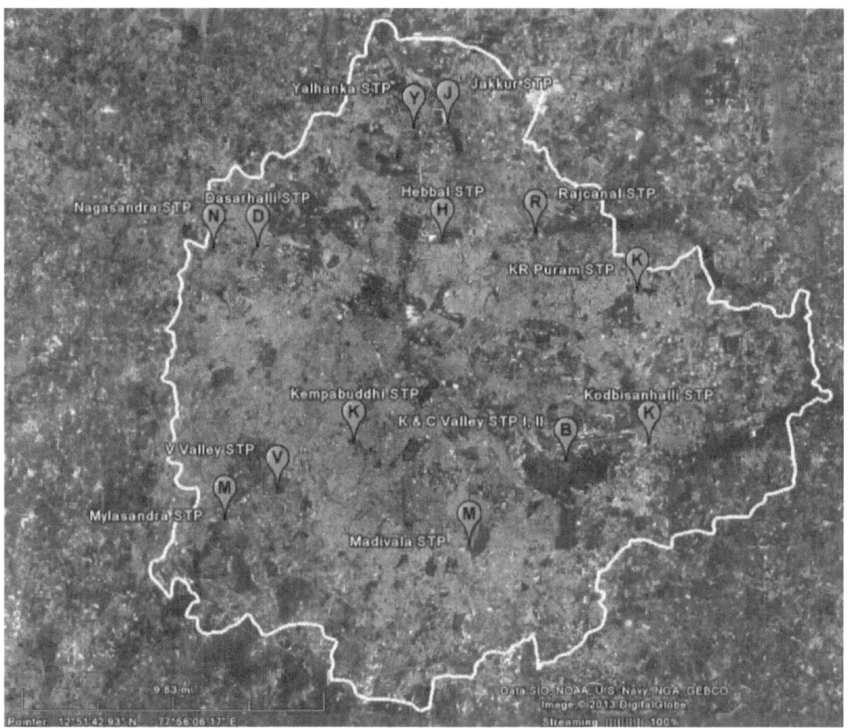

FIGURE 1.3   Greater Bengaluru and the locations of the existing wastewater treatment plants.

TABLE 1.1

Wastewater Treatment Plants with Their Installed Capacities and the Various Treatment Technologies Used in Bengaluru

S. No.	Location	Capacity in MLD	Treatment Facility
Bengaluru—CWSS I, II, III Stages			
1	Vrishabhavati Valley	180	Secondary: trickling filters
2	K & C Valley I	163	Secondary: activated sludge process
3	Hebbal Valley	60	Secondary: activated sludge process
4	Madivala	4	Secondary: UASB + oxidation ponds + constructed wetlands
5	Kempambudhi	1	Secondary: extended aeration
6	Yelahanka	10	Activated sludge process + filtration + chlorination tertiary
Bengaluru—CWSS IV Stage, Phase I			
7	Mylasandra	75	Secondary: extended aeration
8	Nagasandra	20	Secondary: extended aeration
9	Jakkur	10	Secondary: UASB + extended aeration
10	K.R. Puram	20	Secondary: UASB + extended aeration
11	Kadabeesananahalli	50	Secondary: extended aeration
12	K & C Valley II	30	Secondary: extended aeration
13	K & C Valley III	55	Secondary: CMAS
14	Rajacanal	40	Secondary: extended aeration
	Total	**718**	

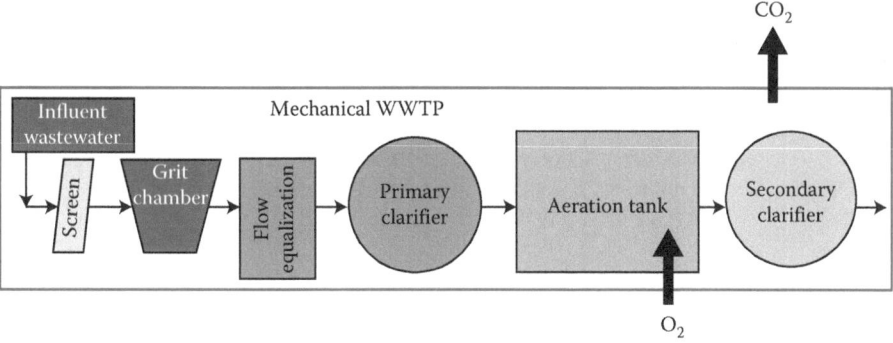

FIGURE 1.4 A schematic representation of a mechanical wastewater treatment plant.

microbial C transformation during treatment have not been taken into account based on IPCC Guidelines as they are biogenic and not from fossil sources.

Indirect GHG emissions: Here emissions occurring due to import of electricity and steam/gas are accounted and taken into consideration.

c. Estimation of emissions over time: The GHG emissions have been calculated for 12 months (to account for seasonal variations, during 2012–2013).

d. Quantification of GHG emissions: This was performed according to IPCC Guidelines for National Greenhouse Gas Inventories (1996; 2006) for quantification of GHG emissions from wastewater treatment plants.

1.2.2 QUANTIFICATION OF GHG EMISSIONS IN TREATMENT PLANTS

The GHG emissions for the various treatment plants were estimated based on conventional treatment technologies and the energy units consumed during the operation of the treatment plants considering the volume of water (functional unit). The direct GHG emissions were accounted for CO_2 generated through organic matter degradation (aerated tanks-aerobic process), CH_4 generated from the clarifiers and also from the basins, and N_2O emissions from the effluent discharge into surface water bodies and the diesel generator used in the treatment units. Indirect emissions were calculated from offsite generation of electric power that is consumed at wastewater treatment plant.

The estimation of the wastewater C footprint of the present wastewater treatment plants in Bengaluru is presented in Tables 1.2 through 1.9. The results show the highest C footprint through emissions by electricity usage, that is, ~38.4 ktons CO_2-eq. The total emission is ~41.53 ktons CO_2-eq., CH_4 being produced in treatment plants should be captured and used for cogeneration of electricity or can be used as a fuel on site. Although magnitude of N_2O generated was

TABLE 1.2
Estimation of Organically Degradable Material in Domestic Wastewater

	I	II	III	IV
City	Population (P)	Degradable Organic Component (BOD) (kg BOD/cap/year)	Correction Factor for Industrial BOD Discharged into Sewers	Organically Degradable Material in Wastewater (TOW) (kg BOD/year) IV = I × II × III
Bengaluru	8,499,399	12.41	1.25	132,803,109.4

TABLE 1.3

Estimation of Methane Emission Factor for Domestic Wastewater

	I	II	III
Type of Treatment or Discharge	Maximum Methane-Producing Capacity (BO) (kg CH$_4$/kg BOD)	Methane Correction Factor for the Treatment System (MCF$_j$)	Emission Factor (EF$_j$) (kg CH$_4$/kg BOD) III = I × II
Aerobic	0.6	0.05	0.03

TABLE 1.4

Estimation of Methane Emissions from Domestic Wastewater

	I	II	III	IV	V	VI	VII	VIII	IX	X
Income Group (High, Medium, and Low)	Fraction of Population Income Group (U$_j$) Fraction	Degree of Utilization (Ti$_j$) Fraction	Emission Factor (EF$_j$) (kg CH$_4$/kg BOD)	Organically Degradable Material in Wastewater (TOW) (kg BOD/year)	Sludge Removed (S) (kg BOD/year)	Methane Recovered and Flared (R) (kg CH$_4$/year)	Net Methane Emissions (CH$_4$) (kg CH$_4$/year) VII = [(I × II × III) × (IV − V)] − VI	GWP for CH$_4$ (IPCC 2007)	Total CO$_2$-eq. kg CO$_2$-eq./year	Total CO$_2$-eq. tCO$_2$-eq./year
HI	0.3	0.07	0.03	132,803,109.4	0	0				
MI	0.5	0.03								
LI	0.2	0.1								
Avg.	0.333	0.066								
Total							86,774	25	2,168,350	2168.35

TABLE 1.5

Estimation of Nitrogen in Effluent

I	II	III	IV	V	VI	VII
Population (P)	Per Capita Protein Consumption (Protein) kg/person/year	Fraction of Nitrogen in Protein (Fnpr) kg N/kg protein	Fraction of Nonconsumption Protein (Fnon-con)	Fraction of Industrial and Commercial Codischarged Protein (Find-com)	Nitrogen Removed with Sludge (Default is Zero) (N sludge) kg	Total Nitrogen in Effluent (N effluent) VII = [(I × II × III × IV × V)] − VI
8,499,399	0.056	0.16	1.4	1.25	0	133,270.58

TABLE 1.6
Estimation of Emission Factor and Emissions of Indirect N_2O Emissions from Wastewater

I	II	III	IV	V	VI	VII	VIII
Nitrogen in Effluent (kg N/year)	Emission Factor (kg N_2O-N/kg N)	Conversion Factor of kg N_2O-N into kg N_2O 44/28[a]	Emissions from Wastewater Plants (Default as Zero) kg N_2O/year	Total N_2O Emissions kg N_2O/year	GWP for N_2O (IPCC 2007)	Total CO_2-eq. kg CO_2-eq./ year	Total CO_2-eq. tCO_2-eq./year
133,271	0.0005	1.57	0	104.62	298	31,176.76	31.18

[a] Indicates the division of the molar mass of the gas i.e. N_2O by the amount of di-nitrogen. This is found to be [44 (MM of N_2O)/28 (MM of N2) = 1.57].

TABLE 1.7
Emissions from Diesel Generator Set during Treatment Process

Total Diesel Consumption (L/year)	Emission Factor (tCO_2-eq./L) of Diesel	Total CO_2-eq. (tCO_2-eq./year)
364,000	0.00255	928.2

TABLE 1.8
Electricity Consumption[a] of the Plant

Total Electricity (MWH)	Emission Factor (0.91 tCO_2-eq.)	Total Scope 2 Emissions CO_2-eq. (tCO_2eq./year)
42,200	0.91 tCO_2/MwH	38,402 tCO_2-eq.

[a] The electricity consumption includes the power required for (a) pump station, (b) clarifiers, (c) aerators, (d) control room, and (e) sludge pump.

TABLE 1.9
Total Emissions of the Treatment Plants[a]

Emission Scopes	Source	CO_2-eq. Emissions
Scope I	CH_4	2168.35 tCO_2-eq.
	N_2O	31.18 tCO_2-eq.
	Generator (fuel)	928.2 tCO_2-eq.
Scope II	Electricity used	38,402 tCO_2-eq.
Scope III	Not attempted	
Total		**41,529.73 tCO_2-eq.**

[a] Biogenic CO_2—36,135 tCO_2 per year.

comparatively much less than other components due to low nitrification, the CO_2 generated during the treatment process is huge. Considering the entire wastewater volume of Bengaluru region as 1200 MLD, the total organic carbon (TOC) in city wastewaters is about 180 tons/day. Therefore, the net CO_2 generated in the system is about 99 tons/day. This accounts to ~36,135 tCO_2 per annum. Considering the biogenic CO_2 for the footprinting, the total C footprint of Bengaluru region is ~77,665 tCO_2-eq./year.

1.3   CARBON SEQUESTRATION AND BIOFUEL PROSPECTS

Wastewater treatment is still a challenge in India due to the enormous load (1,70,000 MLD). At the same time, escalating fuel prices and nonavailability of indigenous oil reserves have necessitated viable oil options for automotive transportation, energy generation, and so on. The resident algal species such as euglenoides occurring naturally in the wastewater lagoons due to their heterotrophic nature are ideal for wastewater treatment (and mitigation of carbon emissions) as well as lipid production.

The CO_2 sequestration in algae is accomplished as (a) direct sequestration, which comprises the capture of CO_2 from the wastewaters before its emission to atmosphere, and its subsequent storage and (ii) indirect sequestration, based on the capture of CO_2 that is already in the atmosphere, through photosynthesis. Carbon present in wastewater can be either inorganic C (carbonic acid/bicarbonates/carbonates) or organic C (dissolved organic acids, etc.). Many wastewater algae are capable of consuming both C forms and are called mixotrophic algae and are extremely beneficial in rapid C fixation and reduction of wastewater C footprint. Often, bulk of the wastewater C fixed in the algal biomass is stored in the form of lipids (neutral) that can be potentially used as feedstocks for biofuel as biodiesel and gasoline.

Soaring global oil demands and perishing stock of finite oil reserves have created enormous interest for a clean, green, and alternative energy sources such as algal biofuels (Ramachandra et al. 2009; Ackom 2010). Among the major organisms considered for oil, bacteria accumulate lipid up to 80% of dry weight (Gouda et al. 2008), yeast up to 70% (Li et al. 2007; Easterling et al. 2009; Hu et al. 2009), algae up to 50%–75% (Chisti 2007), and fungi up to 40% of dry weight (Seraphim et al. 2004). Among these, algae are promising due to their higher growth rates in natural conditions (sunlight and atmospheric CO_2) with minimal feed requirements (Chanakya et al. 2012; 2013). Species such as euglenoides have higher lipid generation (Coleman et al. 1988) with continuous growth and typical mixotrophic mode of nutrition (Yamane et al. 2001). Microalgae contain high amount of lipids, including triglycerides (Borowitzka 2010), suitable for production of biodiesel. Microalgae species *Botryococcus braunii* have a very high content of hydrocarbons similar to crude oil of oil deposits (Metzger and Largeau 2005). These microalgae thrive well in tropical climates in any biosystems like open ponds, eutrophic lakes, wastewater lagoons, algal ponds, and so on. Wastewaters generated from domestic households are a source of nutrients contributing to growth and development of algae and consequent lipid extraction. Lagoons are also cost-effective wastewater treatment options adopted in most part of the country, paving the path for a sustained energy generation.

Neochloris oleoabundans (Li et al. 2008), *Chlorella prototheecoides* (Xu et al. 2006), and *Chlorella vulgaris* (Liang et al. 2009; Widjaja et al. 2009) are some reported high lipids accumulating strains in controlled conditions. However, studies on the resident algal population/species in natural environments are scarce. In this context, investigations on euglenoid *Lepocinclis ovum* focusing on the growth environments, cell characteristics in relation to the lipid content, composition of lipid, and quantum of lipid production have been done for the first time. Growth and consequent oil accumulation under natural conditions deviate significantly from that of photobioreactors with the favorable experimental conditions. In the natural conditions, *L. ovum* competes for dominance with other algal species and predators under hostile conditions. Algae under such stressed conditions appear to have higher lipid accumulation and prove to be a viable candidate for biofuel generation. Generation of a large quantity of wastewater and subsequent treatment through lagoons (for economic reasons) provide an opportunity to harvest microalgae for biofuel generation without additional requirements of land, water, or nutrients. However, there are challenges related to mass cultivation and downstream processing in realizing this vision.

Lipid generation at mass scale from the algae grown in wastewater requires understanding of algal species assemblages, enhancement of biomass accumulation of viable species, enumeration, quantification, and lipid profiling. Algal growth has been monitored through the measurement of cell densities with direct cell count, optical density (OD) and estimation of chlorophyll, volatile solids, and so on. Factors such as nutrient uptake, cell development, cell division, increment in biomass, and lipid accumulation help evaluate cellular responses to local environment conditions. Cell composition and percentage fraction with substrate accumulation are best analyzed through GC–MS for lipid quantitation and composition analysis (Akotoa et al. 2005). Objectives of the study are to evaluate the scope of euglenoides—*L. ovum* predominantly grown in wastewater lagoons for mass biofuel production. This involves

   i. Assessment of growth of euglenoides—*L. ovum* collected from lagoons fed with municipal wastewater;

   ii. Investigation of lipid content and the composition of lipids; and

   iii. Potential of algae-based biofuel in meeting the regional energy demand.

1.3.1 MATERIALS AND METHODS

1.3.1.1 Study Area

Mysore lies between 12°9′ and 11°6′ latitude and 77°7′ and 76°4′ longitude in Karnataka, with the elevation of 650 m above mean sea level with a population of 1.2 million (in 2001). The temperature ranges from 14°C to 34°C and the monsoon showers starts from the end of May through October. The city is blessed with rivers like Cauvery for domestic water consumptions. Part of the city is dependent on groundwater for various purposes.

The wastewater generated in Mysore city due to gravity finds its way into three valleys—northern region leading Kesere valley, the southeastern to the Dalvai tank valley, and southwest to the Malalvadi lake valley. The northern drainage system leads to 30.0 MLD sewage treatment plant (STP; powered by aerators) at Kesare Village, Mysore. The southwestern drainage flow connects to the Vidyaranyapuram STP (67.65 MLD), which uses organic microbe-mix inoculum's (OS-1 and -2) for treatment before entering Dalvai kere. Further south is the STP at Rayankere (60 MLD), H. D. Kote Road. The study area (Figure 1.5) is essentially the catchment slope of the feeder to Dalvai lake. It lies between 12.273681° and 12.270031° latitude and 76.650737° and 76.655947° longitude at the foothills of Chamundi hills. The STP is also surrounded by numerous solid waste dumps. The effluent of the STP flows through the Dalvai kere and then traverses about 20 km before reaching Kabini river. The flow is considered laminar, the residence time of wastewater in the facultative ponds is 11.8 days and in maturation ponds is 2.5 days. The lagoon at Vidyaranyapuram consists of facultative (2, each of 5.15 ha spatial extent and depth) and maturation ponds (2 of 2.5 ha each).

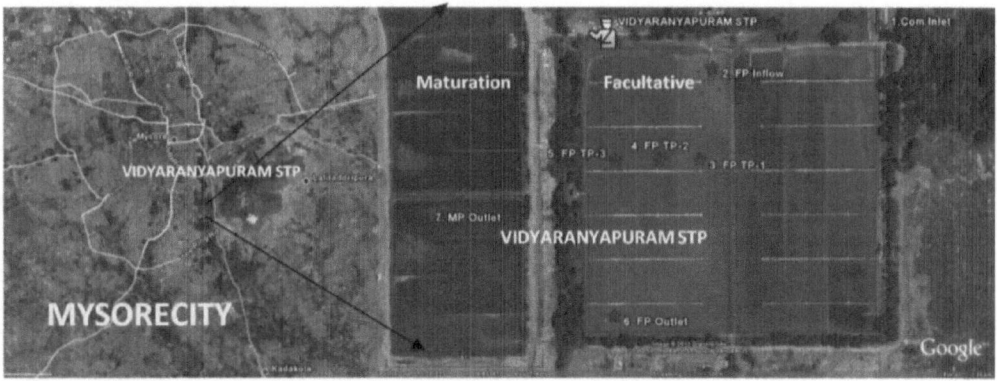

FIGURE 1.5   Sampling site—Vidyaranyapuram STP.

1.3.1.2   Sampling: Algal Screening, Selection, and Densities

For the biomass growth studies, the wastewater samples collected from the STP were filtered and used in the growth experiments. Algal samples were collected from anaerobic and aerobic regions of the facultative ponds as well as various locations of the inflows, outflows, and maturation ponds. The abundance of various species was calculated. The algal samples were identified using standard keys (Prescott 1964) based on their external appearance, color, morphological features, size, habitat, orientation of chloroplast, cellular structure and pigments, and so on. The wastewater samples collected were concentrated by centrifuging 15 ml volume. Algae were enumerated using three replicates of 20 µL of the concentrated sample through microscopic observations. Samples collected from the field were weighed for total dry weight. Quantification for unit area/volume was done taking 10-L volume for microalgae.

1.3.1.3   Characterization of the Growth Environment and Water Quality

Water samples (1 L) were collected from the sampling locations between 9 and 12 a.m. Sampling was always from the sunny side of the banks/platform so as to avoid shaded areas. pH, water temperature, electrical conductivity, total dissolved solids (TDSs), salinity, dissolved oxygen (DO), dissolved free carbon dioxide (free CO_2), and turbidity were measured on-site using the standard methods. In addition, visual clarity/transparency of the lake/STP wastewater was measured with Secchi disk. Water samples collected were analyzed for the various estimations of carbon, nitrogen and its various forms, and phosphorus according to standard methods (APHA 1998).

1.3.1.4   Harvesting of Algal Biomass

Algal biomass was collected from the wastewater ponds with buckets of 2.5-L capacity. The higher settling rate of *L. ovum* helped concentrate the algae, which were then transferred to the laboratory and carefully washed. After microscopic analysis, the samples were washed thrice thoroughly with deionized water and were concentrated by centrifuging at 5000 rpm for 20 min. The pellet was scrapped carefully using spatula and was exposed to drying at room temperature. Samples were preserved through refrigeration till further use.

1.3.1.5   Monitoring the Growth

Conducive growth environments were created for the cultivation with adequate sunlight and provisions for gaseous exchanges to stimulate the growth of the algae *L. ovum* collected and isolated from the wastewater ponds. The isolated algae *L. ovum* were grown, with the modified Chu's media with the composition: $Ca(NO_3)_2$ (20 g/L), K_2HPO_4 (2.5 g/L), $MgSO_4 \cdot 7H_2O$ (12.5 g/L), Na_2CO_3 (12.5 g/L), Na_2SiO_3 (10 g/L), $FeCl_3$ (0.4 g/L), and trace metal solution H_3BO_3 (2.48 g/L), $MnSO_4 \cdot H_2O$ (1.47 g/L), $ZnSO_4 \cdot 7H_2O$ (0.23 g/L), $CuSO_4 \cdot 5H_2O$ (0.1 g/L), $(NH_4)_6Mo_7O_{24} \cdot 4H_2O$ (0.07 g/L), $Co(NO_3)_2 \cdot 6H_2O$ (0.14 g/L), and finally the vitamin mix thiamine, HCl (Vit. B_1) (50 mg/L), biotin (Vit. H) (0.5 g/L), and cyanocobalamin (Vit. B_{12}) (0.5 g/L). A control was also prepared without any inoculations.

The wastewater collected from the influent of the lagoon was centrifuged, filtered, and sterilized for the growth studies. The initial pH was adjusted to 7.3 and the media was sterilized for 30 min before the growth experiments. The experiment was performed in conical flasks, each containing 100 mL of the wastewater media. The algal growth, that is, biomass growth, was measured every 24 h with a spectrophotometer (HACH, DR-2800) at 658 nm, using deionized water as the blank control. Probable relationship of biomass with the OD is given by

$$\text{Dry weight (g/L)} = 0.9712 \times OD_{658} \quad (r = 0.9973)$$

Approximately 10 mL of each sample was used and measured in 2.5-cm cuvettes. The conical flasks were agitated by hand prior to measurement in order to prevent settling of the algae. These experiments were performed in the batch mode for 9 days, as the retention time of the wastewater facultative pond in the sewage ponds at Mysore was about 9 days. During this period, algal species were analyzed by OD measurements and also by direct cell counting method for obtaining the cell densities through conventional microscopy. These parameters were kept constant during the growth periods. Environmental scanning electron microscope (ESEM) Quanta was used for imaging of *L. ovum* cells after fixation with glutaraldehyde.

1.3.1.6 Spectral Signature and Biochemical Composition Analysis by ATR-FTIR

Fourier transform-infrared (FTIR) spectroscopy was used to study the macromolecular composition of algal biomass during the initial lag phase, the intermediate exponential phase, and final stationary phases of the *L. ovum* cell culture. Attenuated total reflectance (ATR-FTIR) spectra were collected on a Bruker Alpha Spec Instrument. The dried algal cells were pressed against the diamond cell prior to scanning. The finely powdered cellular extracts from *L. ovum* sp. were observed for their functionalities in the spectrogram. The spectra were collected in the Mid IR range (128 scans) from 4000 to 800 cm^{-1} (at a spectral resolution of 2 cm^{-1}) and data were analyzed using Origin Pro 8 SR0, v8.0724 (B724) with an initial base line correction and scaled up to amide I_{max}. The peaks with each of the spectral curves were carefully fitted and calculated. The changes in the carbohydrate, protein, lipids, and phosphates of the algal cells were determined from the variations in the intensities of specific bands of the FTIR spectrograms (Giordano et al. 2001; Murdock and Wetzel 2009).

1.3.2 Lipid Extraction and Analysis

1.3.2.1 Lipid Extraction

The algal cells were lyzed by sonication using an ultrasonic bath (frequency 35 kHz) for 45 min after concentration. The lipids were extracted from the solvent mixture as per modified Bligh and Dyer's (1959) method. One gram of dry algal biomass in triplicates was mixed with a solvent mixture of chloroform and methanol in a ratio of 1:2 (v/v). The organic chloroform layer was separated and was evaporated using a rotary evaporator fixed to a water bath maintained at 60°C. The various lipid classes were separated by thin-layer chromatography (TLC). The solvent system used for elution of lipids was a combination of petroleum ether:diethyl ether:acetic acid in a ratio of 70:30:1 (v/v) (Pal et al. 2011). Bands were visualized after staining the TLC plates with iodine vapors. The triacylglyceride (TAG) layer was immediately and carefully scrapped out and was processed for total lipid extraction. The fatty acids were analyzed by methylation of lipid samples using boron trifluoride–methanol (BF_3–MeOH), where BF_3–MeOH converts fatty acids into their methyl esters in a couple of minutes. The extracted samples were heated at 60°C for 15 min followed by cooling in ice bath for 5 min and then 1 ml of water and hexane were added, respectively. After settling, the top hexane layer was removed and washed using anhydrous sodium sulfate for further purification. The samples prepared were analyzed through the gas chromatography column. The methyl esters formed were assessed via gas chromatography (GC) and identified using mass spectroscopy (MS) on the basis of their retention time and abundance.

1.3.2.2 Fatty Acid Composition Using GC–MS

Fatty acid composition was analyzed through gas chromatography (Agilent Technologies 7890C, GC System) using detection by mass spectrometry (Agilent Technologies 5975C insert MSD with Triple-Axis Detector). The injection and detector temperatures were maintained at 250°C and 280°C, respectively (ASTM D 2800). One microliter of the sample was injected into the column, whose initial temperature was maintained at 40°C. After 1 min, the oven temperature was raised to 150°C at a ramp rate of 10°C min^{-1}. The oven temperature was then raised to 230°C at a ramp rate of 3°C min^{-1} and finally it was raised to 300°C at a ramp rate of 10°C min^{-1} and this temperature was maintained for 2 min. The methylated sample was loaded onto the silica column, with helium gas as carrier in splitless mode. The total run time was calculated to be 47.667 min. Fatty acids were identified by comparing the retention time obtained to that of known standards.

1.3.3 Scope for Biofuel as a Viable Energy Source in Cities of Karnataka

Estimates indicate that 40,700–53,200 L of oil/ha (max: 3,54,000 L/ha/year) of biofuel can be produced from algae from different regions having varying insolation capacities (Chisti 2007; Weyer et al. 2010). The density of the oil is about 918 g/L and is similar to the soybean oil (Weyer et al. 2010). Yield of conversion of algal crude lipid to biodiesel is about 70%–90% (Amin 2009) and a maximum of 98% (Chisti 2007). The heating value of the algal oil is about 42 GJ/ton (Amin 2009). Energy demand for regional transportation sector in Karnataka state has been compiled from the published government reports (State of Environment Report 2003).

1.3.3.1 Distribution, Morphological Features, and Cellular Characteristics of *L. ovum*

L. ovum indicative of higher organic loads (Affan et al. 2005) are unicellular flagellates belonging to the class Euglenozoa. During the study periods, *L. ovum* species were found at a lower N:P ratio ranging from 3.6 to 5.2, at a considerable organic load and at slightly acidic pH.

The body of the cell is more or less oval, and sometimes cylindrical and chloroplasts are numerous and discoidal in shape. Many storage bodies, that is, paramylum, were observed which were large (6–8 µm) and ring-shaped. The cells were 80–100 µm size in dimensions. These characteristics are provided in Table 1.10. These species normally found at lagoons, wastewater-treatment lagoons, ditches, swamps, tanks and reservoirs, small ponds, shallow sea bays, acidic lakes (Wollman et al. 2000), brackish waters (John et al. 2002), and also in well-aerated systems with moderate organic loads (Subakov-Simi et al. 2008). In the present study, these species were predominant in wastewater regions with higher organic regime lagoons.

1.3.4 Nutrient Requirements and Growth Conditions

1.3.4.1 Nutrient Concentrations and *L. ovum* Growth at Wastewater-Treatment Units

A slightly lower pH value during the sampling period were conducive for the predominance of *L. ovum*. However, with an increase in pH, the algal species richness generally increases, while diversity and abundance are low at lower pH (Wollmann et al. 2000).

TABLE 1.10
L. ovum Cellular Characteristics in Wastewaters

Cell Characteristics	Values
Cell count (C)	8.74×10^5 cells/mL
Cell shape: average ratio of length to width of individual cells	Round: diameter 80–100 µm
Cell volume: the real volume of a million cells, in mm³	33,493 µm³
Packed cell volume (Vc): centrifuged volume in mL of the cells contained in 1 L of the sample after centrifugation at 5000 rpm	2.5 mL
Cell weight: dry weight of the cells in µg per million cells	0.48 mg/million cells
Cell density: dry weight of the cells in mg/m³ of real volume	640×10^3 mg/m³
Cell index: the packed volume per million cells (Vc/C)	2.86×10^{-3} m³
Cell color	Pea green to bottle green
Cell types (stages)	Round, ovoid (mature)
Cell storage product	Starch: paramylon bodies (ring-shaped)
Bio-volume: volume of cells compared to volume of water they are occupied in 1 µm³	0.03 µm³
C (%): content in g/100 g of the cell dry weight	34.99
N (%): content in g/100 g of the cell dry weight	3.98
P (%): content in g/100 g of the cell dry weight	0.94

Their intensive growth of *L. ovum* is indicative of its ability to grow at lower pH with higher organic loads. The very high count is due to lower washout rate, with the longer retention period of the facultative ponds (10 days). In addition, under higher organic loads, some *Euglena* sp. (e.g., *Euglena gracilis*) through heterotrophy influences H^+ concentration, resulting in a marked decrease in water pH (Yamane et al. 2001). The lower DO levels have related to more of oxygen-demanding materials such as algal biomass, dissolved organic matter (DOM), ammonia, and sediment (Sanchez et al. 2007; Mahapatra et al. 2011a,b,c). Euglenoides can thrive in hypoxic, anoxic, and anaerobic environments as they are facultative anaerobic genera (Castro-Guerrero et al. 2005) and decrease in the hypoxic conditions (Moreno-Sanchez et al. 2000). Higher anoxygenic photosynthesis and high algal biomass for mixotropic culture of *euglenoides* reported here are similar to the earlier reports (Rosenberg 1963; Yamane et al. 2001; Ruiz et al. 2004). It is observed that the euglenoides were present in both anoxic and hypoxic conditions, and this could be a result of the switchover of the mode of metabolism. Similar cases have also been found in earlier studies (Tittel et al. 2003).

The physicochemical parameter of the growth environment of *L. ovum* in wastewaters is provided in Table 1.11. Among the various habitat parameters analyzed, the pH and DO were lower. The concentration of soluble orthophosphates reached 8.73 mg/L. Phosphorus, which regulates/limits the algal growth, is one of the most important nutrients. Higher growth of *L. ovum* species should involve higher uptake rates of phosphorus that should decrease the P concentration. But because of anoxic conditions, the sediment-bound P might have been released (Donnelly et al. 1998) that would be available to the euglenoides observed in the present study. Higher concentrations of N species as ammonium–N play a major role in the proliferation of euglenoid species (Duttagupta et al.

TABLE 1.11
Physicochemical Parameters of the Growth Environment of *L. ovum*

Parameters	Mean	±St. Dev.
pH	7.03	0.20
Temperature (°C)	24.60	2.68
Electrical conductivity (µS/cm)	1117.40	277.70
TDSs (mg/L)	649.33	156.02
Total suspended solids (mg/L)	588.00	56.00
Turbidity (NTU)	213.21	43.44
DO (mg/L)	0.22	0.35
Free CO_2 (mg/L)	38.59	12.88
COD (mg/L)	374.50	24.48
BOD (mg/L)	262.58	33.29
Nitrates (mg/L)	0.43	0.08
Ammonia (mg/L)	58.00	14.00
TKN (mg/L)	77.00	12.00
Phosphates (mg/L)	8.73	1.26
Total phosphates (mg/L)	12.20	4.60
Alkalinity (mg/L)	370.00	77.90
Total hardness (mg/L)	287.00	97.66
Calcium (mg/L)	68.00	15.00
Magnesium (mg/L)	26.00	6.60
Chlorides (mg/L)	176.16	116.35
SAR	18.75	3.79
TOC (mg/L)	102	5.9
Sodium (mg/L)	274.02	42.04
Potassium (mg/L)	26.75	13.02
ORP (mV)	−141.33	66.48

2004). A number of studies conducted earlier reveals that water bodies with high levels of organic compounds, for example, fish or sugar factory ponds (Lungu and Obuh 2004), municipal and dairy wastewaters (Bernal et al. 2008), and industrial wastewaters (Veeresh et al. 2010), had high population densities of euglenoids. The lower N:P (3.6–5.2) suggests P accumulation in the systems and is indicative of N-limiting conditions.

Euglenoides are well known for their sustained heterotrophic growth in the dark conditions (Ogbonna and Tanakaah 1998). It is observed that the regions with higher nutrient concentrations had higher euglenoid cell densities. Electrolyte concentrations during the higher growth periods were moderate (Table 1.11). *L. ovum* occurrence at a very high salinity (i.e., >3000 mg/L) and higher alkaline conditions (Wen and Zhi-Hui 1999) has been reported. This suggests the variability of the habitat and environmental conditions favorable for *L. ovum* in lagoons with varying conditions, which help in the nutrient recovery while accumulating lipid. Peptones are necessary as nitrogen sources in the darkness (Lwoff 1932) and studies have revealed that the euglenoides like *E. anabaena* and *E. deses* grow with inorganic N and *E. gracilis* can selectively take both inorganic and organic sources as N source (Hall and Schoenborn 1938; Schoenborn 1942). In this study, the euglenoides were found in the anoxic regions and are anticipated to feed on peptone type of protein, that is, N sources from the wastewaters.

1.3.4.2 Monitoring of Algal Growth

Algal growth studies were monitored through OD measurements at 658 nm. Growth of biomass (OD) was measured daily with the spectrophotometer. OD values showed that there was a considerable increase in the algal density with time (Figure 1.6). The maximum algal growth was found to be 470 mg/L at the end of the 7th day. *L. ovum* species grown in the wastewater media was under the lag period for the initial 3 days (due to the change in the growth environment), which picked up from the 4th day onwards, evident from a rapid growth of cell mass and an increased cell density due to better acclimatization and nutrient uptake. In contrary, algae grown in Chu media have higher lag period and the biomass increased during the 5th and 6th days. Similar studies using Chu media report maximum growth of 400 mg/L for *Botryococcus braunii* after the 11th day in batch systems (Sawayama et al. 1994). The growths of the algae were highest during the 7th day (3.67×10^5 cells/mL—OD 0.475), which started decreasing during the 8th day. Studies with *Chlamydomonas reinhardtii* grown in the influents, center, and effluents of municipal wastewaters showed higher biomass productivities of 2 g/day (OD 2.5) with the peak cell concentrations of 5.4×10^7 cells/mL in flask cultures with the cells reaching their exponential stage at the 7th day (Kong et al. 2010). Algal cultures grown in Chu media show the highest densities during the 6th day after which the densities started to decrease steeply, which could be attributed to lack of organic carbon in the medium. Studies have also reported reduced cell densities of *Haematococcus pluvialis* cultured in Chu media compared to the other media's (Fabregas et al. 2000).

Comparison of cell densities of algae grown in Chu and wastewater with the control is given in Figure 1.7. During the initial periods after the cell inoculations, the algal cells increased faster in

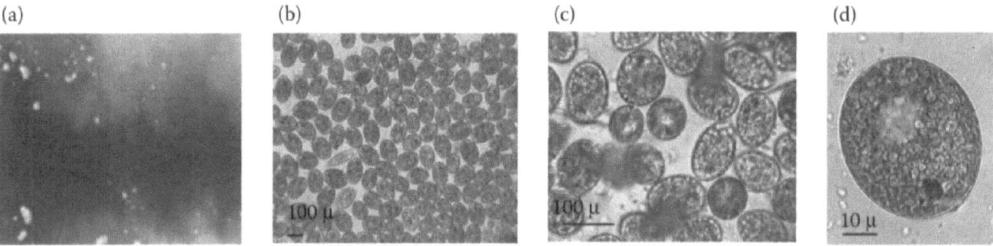

FIGURE 1.6 (a) *L. ovum* proliferation in the facultative ponds, (b) shape, (c) size, and (d) densities of *L. ovum* (at various levels of magnification).

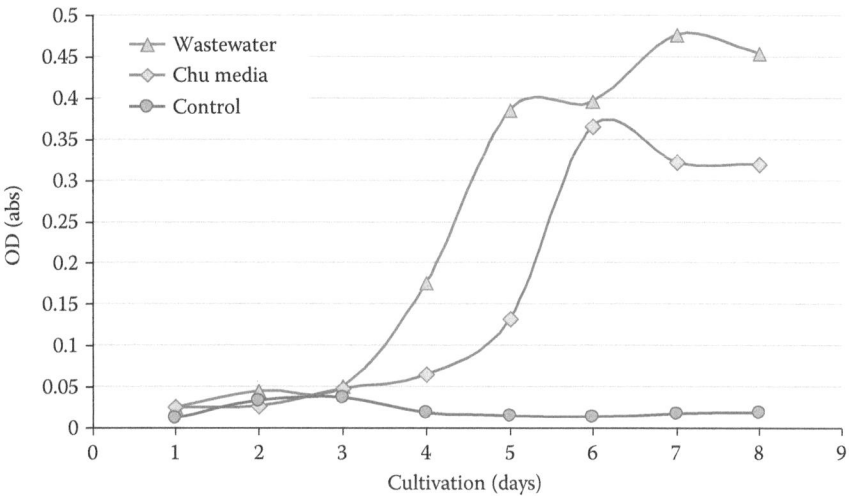

FIGURE 1.7 Comparative assessment of cell densities in diverse media.

the Chu media due to changes in the environment and nutrient conditions. During the lag phases, the cells were in clumps (i.e., attached to each other) due to mucilaginous secretions under stress conditions. Cell densities were higher on 6th and 7th days, with a higher biomass index due to high density of matured population.

The FTIR spectra of *L. ovum* biomass at initial (lag phase), intermediate (exponential phase), and final (stationary phase) stages of the culture are illustrated in Figure 1.8. The region 3300–3000 cm⁻¹ is characteristic for C–H stretching vibrations of C≡C, C=C, and some aromatic hydrocarbons; on the other hand, the regions between 3000 and 2700 are assigned for C–H stretching vibrations of >CH₂, –CH₃, C–H, and CHO functional groups (Giordano et al. 2001; Sigee et al. 2002;

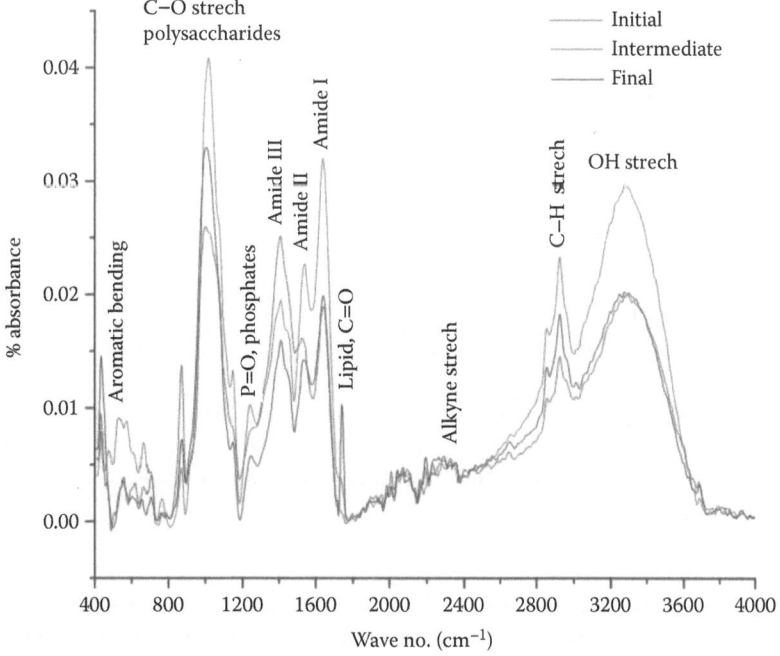

FIGURE 1.8 FTIR spectra of various stages of the cell growth.

Dumas and Miller 2003). The bands present between 3600 and 3300 cm^{-1} are due to the presence of unsaturates (olephinic C–H stretching vibrations). The band present at ~1640 cm^{-1} indicates the presence of C=O of lipids/esters (Sigee et al. 2002; Stehfest et al. 2005). The bands between 1700 and 1500 cm^{-1} indicate proteins. As depicted in Figure 1.8, peaks at ~1640 cm^{-1} are characteristics of amide I bands (Dumas and Miller 2003), and are due to C=O stretching vibrations of peptide bonds (Backmann et al. 1996). These bands provide essential information regarding the secondary structure of proteins (Fischer et al. 2006). The peaks at ~1540 cm^{-1} are for amide II bands, attributed to N–H bending vibrations in proteins (Fischer et al. 2006). The clusters of peaks at ~1357–1423 cm^{-1} are attributed to the amide III bands. The bands present at ~1245 cm^{-1} are for the phosphates and polyphosphates present in the algal cells (Stehfest et al. 2005). The wave numbers from 1200 to 900 cm^{-1} indicate stretching vibrations of polysaccharides due to C–C, C–O, C–O–C stretch (Yee et al. 2004) and rocking vibrations of CH_3 and CH_2. The presence of bands at ~2925 and 2855 cm^{-1} is indicative of asymmetric and symmetric CH_2 starching in lipid.

During the culture experiment, the initial lag phase showed lower absorbances in the polysaccharide region (1001 cm^{-1}) and amides I (1640 cm^{-1}) and II (1527 cm^{-1}) regions due to constraints in acclimatization, and the absorbance intensities are illustrated in Table 1.12 and depicted in Figure 1.8. During this stage, the cells just begin to grow and multiply. A very small peak was observed in the band corresponding to lipids (1732 cm^{-1}) in the culture systems. The observations made during the exponential stages of the growth showed higher absorbances for polysaccharides, proteins, especially amide I and II, phosphates, and a slight increase in the lipid content as provided in Figure 1.8 and Table 1.12. However, during the stationary phase, that is, final stages, a distinct lipid band (~1740 cm^{-1}) was observed indicating lipid accumulation. The comparison of band intensities shows a conversion of carbon from carbohydrates and proteins into lipids. This is evident from the decreased intensities of polysaccharides and proteins. The major reasons for lipids accumulation at the later stages of growth are N and P limitations. This is in agreement with other studies (Sigee et al. 2002; Stehfest et al. 2005; Dean et al. 2007). There was not much of a change in amide I bands when the lag phases were compared with the final stages of algae during the growth. The protein content significantly decreased from the exponential stages to the final starvation stages indicating N stress that could be a possible reason for lipid accumulation. Marked changes in the intensities were observed at the various stages of cell culture indicating variable nutrient uptake, assimilation, and accumulation. The SEM image of *L. ovum* is depicted in Figure 1.9.

1.3.4.3 Lipid Composition by GC–MS Analysis

Total lipid content of *L. ovum* is 18.4 ± 1.04%, which is relatively lower compared to the earlier reports with euglenoides (Constantopolous and Bloch 1967). Algal species grown in batch mode in the wastewater has a total lipid content (of oil content):

a. 31% in *Scenedesmus obliquus* grown in secondary-treated municipal wastewaters (Martinez et al. 2000)
b. 25.25% in *C. reinhardtii* when cultured in all stages of municipal wastewaters (Kong et al. 2010)
c. 18% in *Botryococcus braunii* in municipal wastewaters (Orpez et al. 2009)
d. 9% in mixed population *Scenedesmus* sp. followed by *Micractinium* sp., *Chlorella* sp., and *Actinastrum* sp. in primary-treated wastewaters (Woertz et al. 2009)

Compared to these, the lipid content of *Chlorella* sp. grown in a semicontinuous mode in wastewaters was 42% with an effective COD, NH_3–N, and P removal (Feng et al. 2011). Fatty acid methyl esters (FAME) were separated according to the retention times in the column. The proportion of each fatty acid was calculated based on the peak area (area under the curve) and are provided in Table 1.13, which illustrates the composition of the fatty acid (saturates and unsaturates). The total saturated fatty acids (68.1%), monounsaturates (19.34%), polyunsaturates (12.96%), and fatty

TABLE 1.12
Band Assignments for Infrared Spectra

No.	Initial Lag Phase		Intermediate Exponential Phase		Final Stationary Phase			Main Groups	Band Assignments
	Intensities	Wave Number (cm⁻¹)	Intensities	Wave Number (cm⁻¹)	Intensities	Wave Number (cm⁻¹)	Wave Number Range (cm⁻¹)		
1	0.01997	3295	0.02972	3275	0.02027	3274	3029–3639	OH	Water v(O–H) stretching
								Amide A	Protein v(N–H) stretching (amide A)
2	0.01418	2922	0.02329	2923	0.01831	2925	2809–3012	C–H	Lipid–carbohydrate Mainly v_{as}(CH$_2$) and v_s(CH$_2$) stretching
3	0.00101	1732	0.00368	1732	0.01032	1740	1763–1712	Lipids	Fatty acids v(C=O) stretching of esters
4	0.01896	1640	0.03198	1639	0.01988	1640	1583–1709	Amide I	Protein amide I band: Mainly v(C=O) stretching
5	0.01613	1527	0.02267	1540	0.01421	1535	1481–1585	Amide II	Protein amide II band: Mainly δ(N–H) bending and v(C–N) stretching
6	0.02511	1408	0.01952	1409	0.01596	1411	1357–1423	Amide III	Protein δ_{as}(CH$_2$) and δ_{as}(CH$_3$) bending of methyl, lipid δ_s(N(CH$_3$)$_3$) bending of methyl
7	0.00824	1256	0.01027	1242	0.00579	1245	1190–1350	Phosphates	Nucleic acid (other phosphate-containing compounds as phospholipids) v_{as}(>P=O) stretching of phosphodiesters
8	0.02596	1001	0.04088	1017	0.03295	1006	980–1072	Saccharides	Carbohydrate v(C–O–C) of polysaccharides

Source: Adapted from Giordano M et al. 2001. *J Phycol* 37:271–279; Sigee DC et al. 2002. *Eur J Phycol* 37:19–26; Benning LG et al. 2004. *Geochim Cosmochim Acta* 68:729–741; Dean AP et al. 2010. *Bioresour Technol* 101:4499–4507; Dean AP, Martin MC, Sigee DC. 2007. *Phycologia* 46:151–159; Stehfest K, Toepel J, Wilhelm C. 2005. *Plant Physiol Biochem* 43:717–726.

Note: v_{as}, asymmetric stretch; v_s, symmetric stretch; δ_{as}, asymmetric deformation; δ_s, symmetric deformation.

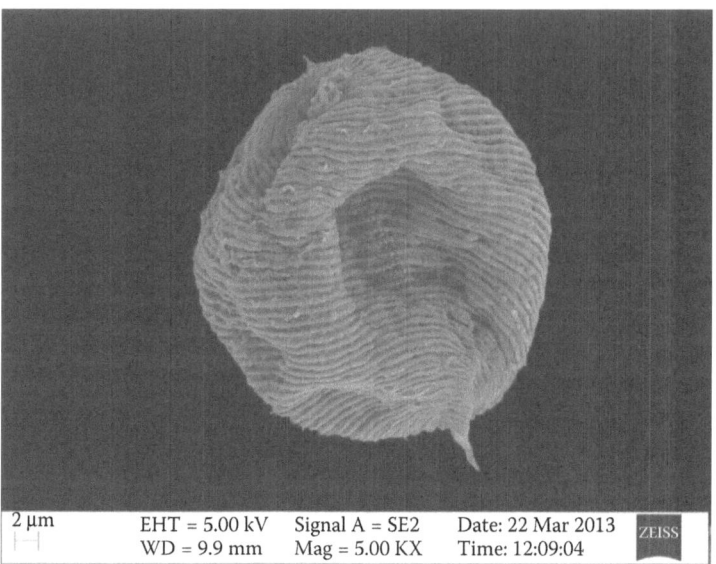

| 2 μm | EHT = 5.00 kV | Signal A = SE2 | Date: 22 Mar 2013 | ZEISS |
| | WD = 9.9 mm | Mag = 5.00 KX | Time: 12:09:04 | |

FIGURE 1.9 SEM image of *L. ovum*.

acids (72.94%) reflect properties of biodiesel. Important fatty acids from biofuel perspective of *L. ovum* using the ultrasonication techniques are given in Figure 1.10. The present analysis showed 15 different types of fatty acids. The fatty acid composition has higher percentage of palmitic acid [C16:0] 30.32% > oleic acid [C18:1(9)] 12.94% > stearic acid [C18:0] 12.04% > methyl tetradecanoate [C14:0] 10.38% > pentadecanoic acid [C15:0] 8.97% > linoleic acid [C18:2(9,12)] 8.44% > linolenic acid [C18:3(9,12,15)] 4.52%. These results are in agreement with earlier studies (Knothe 2008), with palmitic acid (C16:0), stearic acid (C18:0), oleic acid (C18:1), and linoleic acid (C18:2) are the most common fatty acids for providing a reasonable balance of fuel properties. The chromatogram derived from the gas chromatography and mass spectrogram (GC and MS) analyses of the fatty acids for *L. ovum* is elucidated in Figure 1.11, which highlights the peaks of the fatty-acid methyl esters.

These results indicate that *L. ovum* constitute a better candidate for biodiesel production, due to higher content of requisite fatty acids (68.1% saturates: Table 1.13) comparable to biodiesel properties. Similar findings were also reported earlier (Rasoul-Amini et al. 2009), and also algal oils having high polyunsaturated fatty acids with four or more double bonds (Damiani et al. 2010). Higher content of saturated fatty acids provides an excellent cetane number (better ignition quality) and oxidative stability to biodiesel (Chinnasamy et al. 2010). The present study reveals that appropriate optimization of euglenoid microhabitat, development, growth conditions, and lipid composition to enhance the net productivity from wastewater systems would ensure a sustained energy supply while treating wastewater at decentralized levels which foster human welfare.

1.3.4.4   C Sequestration in Wastewater Algae

The average biomass productivity found in the Mysore pond systems ranged from 10 to 20 g/m^2/day and considering the surface BOD loading 400–900 kg/ha/day, the CO_2 sequestration potential in the native *L. ovum* species was calculated to be 1.23–1.48 tons CO_2-eq./ha/day providing an annual C sequestration of ~490 tons CO_2-eq./ha and from this ~10 tons CO_2-eq./ha/year is the C sequestered and transformed to lipids that can be used effectively as a biofuel. This unique way of C sequestration and transformation into valorizable algal biomass with ~20% lipids paves path for new avenues for energy sustainability with adequate reduction of wastewater C footprint. The algal communities also create aerobic environment, thus abating the formation of methane which is a potential GHG.

TABLE 1.13

Fatty Acid Profile of *L. ovum* through GC–MS Analyses

CN:U	FAME	Peak Area	Corr. Area	Corr. %max	Retention Time	%Comp.
C 10:0	Decanoic acid, methyl ester	22,363	60,004	0.31	14.137	0.09
C 11:0	Undecanoic acid, methyl ester	35,981	87,061	0.45	16.561	0.14
C 12:0	Dodecanoic acid, methyl ester	186,635	527,647	2.73	19.01	0.83
C 13:0	Tridecanoic acid, methyl ester	611,028	2,006,749	10.37	22.054	3.14
C 14:0	Methyl tetradecanoate	1,795,343	6,625,155	34.24	25.291	10.38
C 15:0	Pentadecanoic acid, methyl ester	1,438,245	5,578,805	28.83	28.537	8.97
C 16:0	Hexadecanoic acid, methyl ester	622,964	2,439,440	12.61	31.038	30.32
C 16:1(7)	7-Hexadecenoic acid, methyl ester	54,675	213,817	1.11	31.791	0.34
C 16:1(9)	9-Hexadecenoic acid, methyl ester	4,012,960	9,349,604	0.00	31.791	3.82
C 17:0	Heptadecanoic acid, methyl ester	217,340	1,333,190	6.89	34.836	2.09
C 17:1(10)	*cis*-10-Heptadecenoic acid, methyl ester	372,136	1,428,305	7.38	34.069	2.24
C 18:0	Octadecanoic acid, methyl ester	1,028,784	4,427,325	22.88	36.861	12.05
C 18:1(9)	9-Octadecenoic acid, methyl ester	1,875,793	7,746,598	40.03	37.217	12.94
C 18:2(9,12)	9,12-Octadecadienoic acid, methyl ester	249,646	958,305	4.95	37.478	8.44
C 18:3(9,12,15)	9,12,15-Octadecatrienoic acid, methyl ester	705,164	2,881,726	14.89	37.071	4.52
			Saturated fatty acids (saturates)			68.01
			Monoenoic fatty acids (mono-unsaturated fatty acids)			19.34
			Polyenoic fatty acids (poly-unsaturated fatty acids)			12.96
			C16–C18 (fatty acids important from biodiesel perspective)			72.94
			Total lipid content			18.48

1.3.5   Viability of Algae-Based Biofuel as an Energy Source in Karnataka

The current investigations reveal that resident algal populations growing in wastewater systems are not only aiding in treating wastewaters, but also ensure reliable substrate for biofuel generation in Mysore being one of the rapidly growing cities in Karnataka generates about 250 MLD of sewage (Table 1.14). Three treatment plants with the installed capacity of 150 MLD are in operation. The conventional wastewater treatment plants face the problem due to the lack of assured supply of

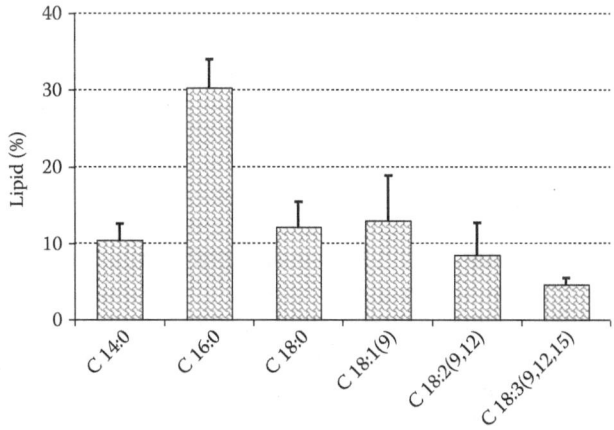

FIGURE 1.10   Comparative account of the major fatty acid composition in *L. ovum*.

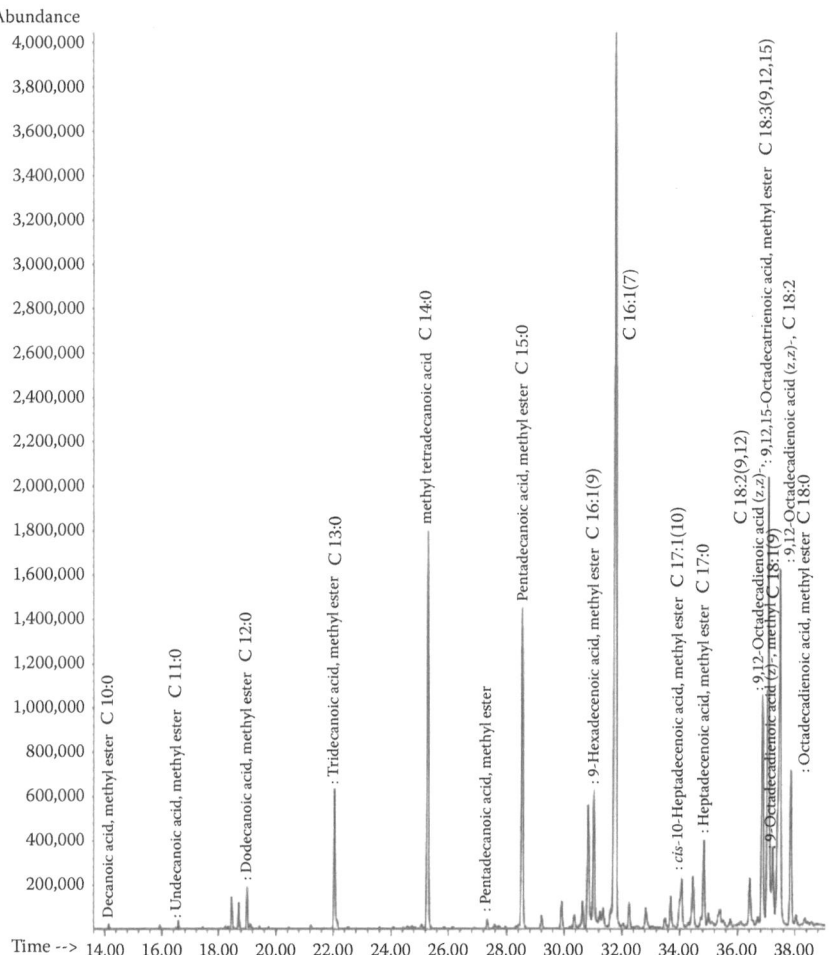

FIGURE 1.11 Chromatogram depicting peaks of major fatty acids in *L. ovum*.

electricity and also under capacity to handle the increasing load. However, algal population present at various unit processes of the treatment plants, that is, inflows, facultative ponds, maturation ponds, and the outflows, are aiding in the nutrient uptake or in the treatment. Very high densities of *L. ovum* were observed in the facultative ponds at a very high surface BOD loading 400–900 kg/ha/day with a retention time of 10–12 days. The area required to treat 100 MLD of wastewater is about 20 ha.

On an average, 640 mg of algal dry weight were derived from 1 L of the wastewater and the cells take roughly 7–8 days to reach the exponential growth stage with 4.4×10^5 cells/mL. At such cellular densities, the biomass productivities were about 250–300 tons/ha/year and about 18% of the dried algal biomass constitutes lipid which is about 40–50 tons/ha/year of lipid from *L. ovum*. Various studies conducted earlier have estimated the regional or local estimates of the potential of algal biofuel. Studies reveal that after allowing for maximum photorespiration the maximum yields of carbohydrate-type molecules are about 410 tons/ha/year at tropics (Williams and Laurens 2010), 182 tons/ha/year through culturing diatoms *Phaeodactylum tricornutum* in bioreactors (Fernandez et al. 1998), 60 tons/ha/year in raceway ponds at subtropics with *Pleurochrysis* sp. (Moheimani and Borowitzka 2007), 175 tons/ha/year bioreactor raceway pond systems (Chisti 2007), 50–300 tons/ha/year in closed photobioreactors (Sheehan et al. 1988), 22.87–137.25 tons/ha/year at about

TABLE 1.14
District-Wise Analysis of the Algal Biofuel Potential from Wastewaters

Districts	Total Geographic Area (Ha)	Population (Census Data 2011)	Per Capita Water Use (L) in the Region per Day	Wastewater Generation (MLD)	Total Wasteland (Ha)	Area Req. for Algal Pond (Ha)	Waste Land % to Total Geog. Area	Land Req. for Algal Ponds % to Waste Land	Algal Lipid Potential (Tons/Year)
Bagalkote	657,500	1,890,826	226,899,120	158.83	69,932	79.41	10.64	0.11	3964.38
Bengaluru Urban	219,000	9,588,910	1,150,669,200	805.47	6525	402.73	2.98	6.17	20,104.49
Belagavi	1,341,500	4,778,439	573,412,680	401.39	143,000	200.69	10.66	0.14	10,018.67
Bellary	841,900	2,532,383	303,885,960	212.72	109,120	106.36	12.96	0.10	5309.50
Bidar	544,800	1,700,018	204,002,160	142.80	36,127	71.40	6.63	0.20	3564.33
Bijapur	1,049,400	2,175,102	261,012,240	182.71	32,996	91.35	3.14	0.28	4560.41
Chamarajnagar	568,500	1,020,962	122,515,440	85.76	17,150	42.88	3.02	0.25	2140.59
Chikamagalur	720,100	1,137,753	136,530,360	95.57	60,676	47.79	8.43	0.08	2385.46
Chitradurga	844,000	1,660,378	199,245,360	139.47	119,096	69.74	14.11	0.06	3481.21
Davanagere	596,600	1,946,905	233,628,600	163.54	50,342	81.77	8.44	0.16	4081.96
Dharwad	423,000	1,846,993	221,639,160	155.15	14,508	77.57	3.43	0.53	3872.48
Gadag	465,700	1,065,235	127,828,200	89.48	37,237	44.74	8.00	0.12	2233.41
Kalaburagi	1,622,400	2,564,892	307,787,040	215.45	78,401	107.73	4.83	0.14	5377.66
Hassan	681,400	1,776,221	213,146,520	149.20	31,595	74.60	4.64	0.24	3724.10
Haveri	485,100	1,598,506	191,820,720	134.27	29,147	67.14	6.01	0.23	3351.49
Kodagu	410,200	554,762	66,571,440	46.60	8368	23.30	2.04	0.28	1163.14
Kolar	822,300	1,540,231	184,827,720	129.38	59,536	64.69	7.24	0.11	3229.31
Koppala	718,900	1,391,292	166,955,040	116.87	43,740	58.43	6.08	0.13	2917.04
Mandya	496,100	1,808,680	217,041,600	151.93	39,077	75.96	7.88	0.19	3792.15
Mangalore	484,300	2,083,625	250,035,000	175.02	20,889	87.51	4.31	0.42	4368.61
Mysore	626,900	2,994,744	359,369,280	251.56	20,430	125.78	3.26	0.62	6278.90
Raichur	682,800	1,924,773	230,972,760	161.68	74,544	80.84	10.92	0.11	4035.56
Shivamogga	846,500	1,755,512	210,661,440	147.46	39,178	73.73	4.63	0.19	3680.68
Tumakuru	1,059,800	2,681,449	321,773,880	225.24	99,718	112.62	9.41	0.11	5622.03
Udupi	359,800	1,177,908	141,348,960	98.94	25,710	49.47	7.15	0.19	2469.65
Uttara Kannada	1,029,100	1,436,847	172,421,640	120.70	57,639	60.35	5.60	0.10	3012.55
Bengaluru Rural	250,000	987,257	118,470,840	82.93	19,000	41.46	7.60	0.22	2069.92
Ramanagaram	331,500	1,082,739	129,928,680	90.95	20,000	45.48	6.03	0.23	2270.11
Total	1,917,9100	5,870,3342	7,044,401,040	4931.08	1,363,681	2465.54			123,079.77

30%–70% oil on a dry weight basis (Chisti 2007), 12.6 tons/ha/year for tropical countries, that are comparatively much lower than the optimal estimates (Rodolfi et al. 2008). In a recent study, the optimal yields of algal lipids at the tropics were estimated to be as low as 36–45 tons/ha/year (Weyer et al. 2010). The resident algal populations (*L. ovum*) isolated from the facultative lagoons with high organic loads help in wastewater treatment and also in biofuel generation. Replication of these systems in all urbanizing towns would aid as low-cost option for wastewater treatment while meeting the energy demand at local levels.

For effective algal growth and year-round productivity, the ponds have to be shallow (1 m) to reduce the shading effects caused by the higher cell densities. This requires 2.5 times more area than what is currently being practiced as area required for the facultative lagoons. These types of shallow systems will treat wastewaters and at the same time generate ample algal biomass for deriving energy. At such high productivities of 250 tons/ha/year, and assuming that entire wastewater treatment and consequent lipid generation would meet the transport and irrigation (diesel pump) energy requirement in rural area, the algal biofuel proves to be a sustainable option. Taking the per capita generation of wastewater as 80% of the total water consumption the daily generation of wastewater is 96 L/day/capita. The wastewater has to travel a certain distance to reach such community/regional algal ponds. As most of the advanced cities of India are connected to underground drainage (UGD), the chances of wastewater infiltration are low but pilferages cannot be ignored at the source points—on the way to the treatment units and while joining receiving waters. In the case of small cities, most of the wastewater generated is adequately allowed to infiltrate in pits and soak ways (Parkinson et al. 2008). The degree of infiltration varies based on the nature of soil types. The infiltration is higher with sand and silt type of soils (150 L/m^2/day).

During this travel, approximately 10% of the wastewaters are lost mainly by infiltration and to a smaller extent through evaporation. Accounting for all these losses, the wastewater generated in Karnataka state is around 4931 MLD with the land requirement of 2465 ha for implementing algae-based wastewater lagoons and subsequent lipid recovery. The total wasteland area in Karnataka is about 1.37 million hectares.

District-wise total area required for integrated algal pond-type treatment cum lipid recovery systems are given in Table 1.14 for Karnataka, a federal state in India. The percentage of land required for algal lipid production (only from wastewaters) is less than 1% of the total wastelands area required for catering the entire wastewaters of Karnataka with simultaneous biofuel generation except for Bengaluru, where the area required for high-detention algal ponds is about 6% of the wasteland. The district-wise algal lipid potential derived from wastewaters is elucidated in Figure 1.12. This illustrates that Bengaluru, Mysore, Belagavi, Kalaburagi, and Tumakuru have higher lipid potential (5000–20,000 tons/year; Figure 1.12). The petrol and diesel consumption was 5,09,918 and 22,98,370 tons (in 2003–2004), respectively (State of the Environment Report 2003). Estimates show that fuel oil consumption for the transport sectors in the state at present is estimated to be around 30,00,000 tons/year. To meet the energy requirement for the transport sector in the state, only 5% of the total wasteland (60,000 ha) area is required to build high-detention algal ponds, which paves a way for sustainable biofuel generation and wastewater treatment.

The present studies on resident algal population in the wastewater treatment ponds revealed the dominance of an euglenoid *L. ovum* through mixotrophy (both autotrophy and heterotrophy) at anoxic waters with higher organic loads and ammonium-N concentrations, where other algae fail to grow. The species isolated and grown in Chu media showed lower growth rates compared to the wastewaters that attained maximum biomass productivity at the 7th day. The total lipid content was found to be 18.6%. The fatty acid composition showed 68% saturates, 18% monounsaturates, and 13% polyunsaturates. The essential fatty acids from the biodiesel perspective were 76% showing the algae as potential qualifier for a biodiesel feedstock. The fatty acid profile of *L. ovum* demonstrated that palmitic acid (16:0), stearic acid (18:0), linoleic acid (18:1), methyl tetradecanoic acid (14:1), oleic acid (18:1), and linolenic acid (18:3) are major fatty acids of *L. ovum*, implying the potentials of microalgal lipids for biodiesel.

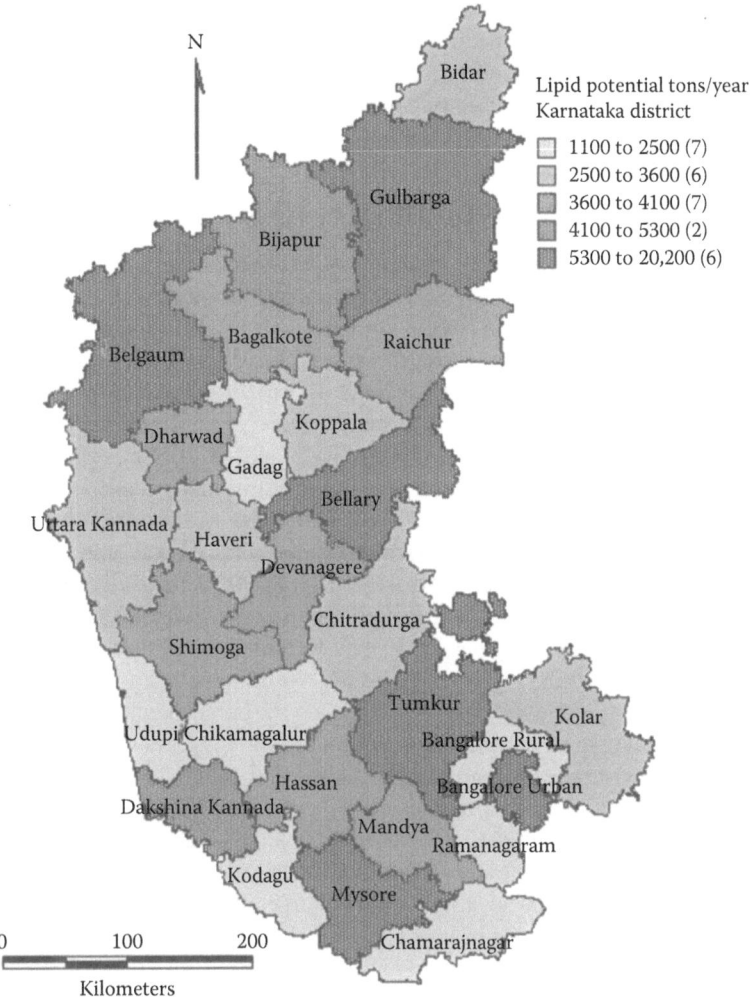

N

Bidar

Lipid potential tons/year
Karnataka district

☐ 1100 to 2500 (7)
☐ 2500 to 3600 (6)
☐ 3600 to 4100 (7)
☐ 4100 to 5300 (2)
☐ 5300 to 20,200 (6)

Gulbarga

Bijapur

Bagalkote   Raichur

Belgaum

Dharwad   Koppala

Gadag

Uttara Kannada   Bellary

Haveri

Devanagere

Chitradurga

Shimoga

Tumkur   Kolar

Udupi Chikamagalur   Bangalore Rural

Hassan   Bangalore Urban

Dakshina Kannada

Mandya

Ramanagaram

Kodagu   Mysore

0   100   200   Chamarajnagar

Kilometers

FIGURE 1.12   District-wise algal lipid potential in tons/year for Karnataka state.

The evaluation of the resident wastewater algae for the regional energy demand for transport showed positive results. With the current wastewater generation rate, about 26.78 tons/year of lipid can be generated from 1 MLD of wastewaters. The land required for the cultivation of wastewater-grown algae is calculated to be 0.5 ha/MLD. The lipid potential calculated at the maximum growth conditions was found to vary between 40 and 50 tons/ha/year due to seasonal variabilities. The biomass productivities can be further optimized in wastewater ponds to establish a sustainable biomass production process. This resident wastewater algae *L. ovum* can be grown in the wastewaters of the cities in Karnataka and simultaneously treat the wastewater and generate biofuel to meet the regional energy requirement.

The estimate of the potential of algae to replace vehicle fuels in the Karnataka state shows that an area corresponding to at least 5% of the total wastelands would be required to satisfy current fuel oil demand in the transport corridor. However, future fuel demand is most likely to be met from a combination of algal substrates at the end of each stage of processing by algal bio-refinery approach rather than only wastewaters as a solution.

The potentials of the existing resident wastewater algae, which otherwise cause a nuisance by their prolific outbreaks in eutrophied freshwater, can be harvested as resources. The large quantum of terrestrial nutrients (sewage) getting into these waters are a sustainable nutrient feed for the algae for its growth and development. Further investigations have to be carried out to better understand the wastewater ecology, nutrient dependencies, mode of nutrition, and the growth microenvironment of the promising microalgae *L. ovum* and consequently optimizing the growth rates and lipid accumulation of these resident algae for integrated wastewater treatment and sustainable bioenergy generation.

1.4 INTEGRATED WETLANDS ECOSYSTEM TO MITIGATE CARBON EMISSIONS

Increased and unprecedented population growth has resulted in enormous stress on potable water from a daily consumption point of view and also in regard to increased wastewater generated by the city. Unplanned growth has led to radical land-use conversion of forests, surface water bodies, and so on with the irretrievable loss of land prospects. Land-use analyses show 584% growth in built-up area during the last four decades with the decline of vegetation by 66% and water bodies by 74%. Analyses of the temporal data reveal an increase in urban built-up area of 342.83% (during 1973–1992), 129.56% (during 1992–1999), 106.7% (1999–2002), 114.51% (2002–2006), and 126.19% (during 2006–2010) (Ramachandra et al. 2012; Ramachandra and Uttam Kumar, 2008).

Rapid urbanization in recent times has led to the mammoth wastewater generation. Untreated or partially treated wastewaters are fed to surface water that finds its way into groundwater sources. The sustained inflow of untreated or partially treated sewage to wetlands leads to the enrichment of nutrients such as carbon (C), nitrogen (N), and phosphorus (P), evident from the algae bloom and profuse growth of macrophytes. This has led to the contamination of existing water resources with pathogens and nutrients resulting in algal bloom due to eutrophic status of surface water, thereby contaminating the nearby groundwater sources affecting the human health. On the other hand, macrophytes grow profusely in this nutrient-rich environment and progressively cover the entire surface of the water body hindering the passage of sunlight and diffusion of gases to the underlying water layers. Absence of sunlight in these parts affects algal growth and photosynthetic O_2 generation and critically depletes the DO concentration and hence affects the local biota.

Treatment and disposal of wastewater generated in the neighborhood constitute key environmental challenges faced in urban localities due to burgeoning population in the recent decade. Nutrient-laden wastewater generated in municipalities is either untreated or partially treated and is directly fed into the nearby water bodies regularly, resulting in nutrient enrichment and algal blooms. Conventional wastewater-treatment options are energy- and capital-intensive apart from their inability to remove nutrient completely. In this backdrop, algal processes are beneficial and remove nutrients with carbon sequestration and resultant biomass production. Algae grow rapidly and uptake nutrients (C, N, and P) available in the wastewater (GOI 2008; Mahapatra et al. 2013a,b; Sharachchandra Lele et al. 2013) and hence are useful in nutrient remediation. Treatment of sewage and letting into wetlands would help in further treatment (removal of N, P, and heavy metals). This also prevents contamination of groundwater resources. Thus, wetlands provide a cost-effective option to handle sewage generated in the community and also help in addressing the water crisis in the region.

Microalgae and native macrophytes of the wetlands help in the treatment due to abilities to uptake nutrients and heavy metals. Techniques have been developed for exploiting the algae's fast growth and nutrient removal capacity (Karin 2006). The nutrient removal is basically an effect of assimilation of nutrients as the algae grow. Also, nutrient stripping happens due to high pH induced by the algae as in ammonia volatilization, phosphorus precipitation, and so on.

1.4.1 WETLANDS/ALGAE POND AS WASTEWATER TREATMENT SYSTEMS

Wetlands aid in water purification (nutrient, heavy metal, and xenobiotics removal) and flood control through physical, chemical, and biological processes. When sewage is released into an environment containing macrophytes and algae, a series of actions takes place. Through contact with biofilms, plant roots and rhizomes processes like nitrification, ammonification, and plant uptake will decrease the nutrient level (nitrate and phosphates) in wastewater (Garcia et al. 2010). Algae-based lagoons treat wastewater by natural oxidative processes. Various zones in lagoons function equivalent to cascaded anaerobic lagoon, facultative aerated lagoons followed by maturation ponds (Mahapatra et al. 2013b). Microbes aid in the removal of nutrients and are influenced by wind, sunlight, and other factors (Mahapatra et al. 2011b,c, 2013b).

The conventional wastewater treatment systems (sewage treatment systems) are expensive and require input of external energy sources (e.g., electricity, organic carbon) and chemical additives. These treatment systems generate concentrated waste streams necessitating environmentally sound disposal. There is an urgent need to develop innovative, environment-friendly, and cost-effective approaches for treating sewage generated in the community every day. Untreated sewage leads to the neighborhood contamination of land and water resources (groundwater). An easy way to check the sewage contamination is to test the level of nutrients (nitrates and phosphates). Nitrate is a substance that develops from organic waste. Algae convert nitrate into organic compounds (proteins) through photosynthesis in the presence of sunlight. Algae can exhibit growth rates that are higher than other plants due to their extraordinarily efficient light and nutrient utilization. By taking advantage of rapid availability of nutrient-enriched water, high solar intensity, and favorable microclimate for algal growth, higher densities of algae can be grown continuously that provides ample biomass and at the same time treats wastewater within a short period of time.

Algal bacterial symbiosis is very effective in these tropical conditions. Algae, the primary producers, generate O_2 (during photosynthesis) which aid in the efficient oxidation of organic matter with the help of the chemoorganotrophic bacteria. The type and diversity of the algae grown are potential indicators of treatment process (Mahapatra and Ramachandra 2013; Ramachandra et al. 2012; Mahapatra et al. 2013a,b; Mahapatra et al. 2014). And bacterial system disintegrates and degrades the organic matter, providing the algae with an enriched supply of CO_2, minerals, and nutrients.

The focus of the current investigation is to assess the efficacy of wetlands in Jakkur lake system. This has been done through water quality assessment (physico-chemical analysis) at various stages of the integrated wetland system consisting of STP (10 MLD), wetlands (with macrophytes), algal pond, and Jakkur lake (Figure 1.13). Nitrate and phosphate levels were monitored at various stages of wetland ecosystem.

Jakkur lake (Figure 1.13) situated at 13°04′N and 77°36′E, northeast of Bengaluru. Ten MLD STP is functional in this locality. Partially treated water is let into Jakkur lake through wetlands (consisting of emergent macrophytes and algae). Water samples were collected (Figure 1.13) from inlet (S6), outlets (S1, S2, and S3), middle (S4, S5, and S9), and at treatment plant outlets (S6 and S7) totaling nine locations. The treated water from the treatment plant passes through the wetlands to Jakkur lake.

1.4.2 Integrated Wetland System

Integrated wetland system at Jakkur consists of (i) treatment plant (treats sewage partially before letting to wetlands, (ii) constructed wetlands consisting of macrophytes, (iii) algal pond, and (iv) lake (Figure 1.13). Jakkur lake with wetlands is manmade and constructed about 200 years ago to meet the domestic and irrigation water requirement of Jakkur village located about 100 m southwest in the downstream of the lake (Figure 1.13).

1.4.2.1 Nutrients (Nitrates and Phosphates)

Nutrients essentially comprise of various forms of N and P that readily mineralize (inorganic mineral ions) to enable uptake by microbes and plants. Accumulation of nitrates and inorganic P induces changes in water quality and affects its integrity leading to higher net productivity. Nitrates in excess amounts together with phosphates accelerate aquatic plant growth in surface water causing rapid

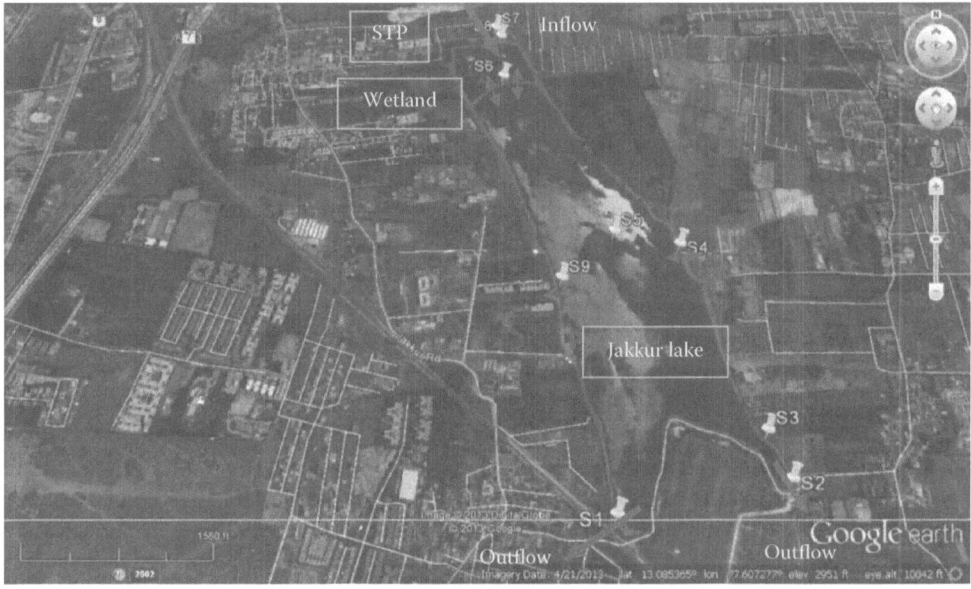

FIGURE 1.13   Water sampling locations in wetland system.

oxygen depletion or eutrophication in the water. Nitrates at high concentrations (10 mg/L or higher) in surface and groundwater used for human consumption are particularly toxic to young children affecting the oxygen-carrying capacity of red blood cells (RBCs) causing cyanosis (methemoglobinemia). In the current study nitrate values ranged from 0.2 to 0.38 mg/L and phosphate values ranged from 0.09 to 1.29 mg/L. The nitrate and phosphate values are higher at the wetland inlets and significantly reduced after the passage through wetlands and algal pond as elucidated in Figure 1.14.

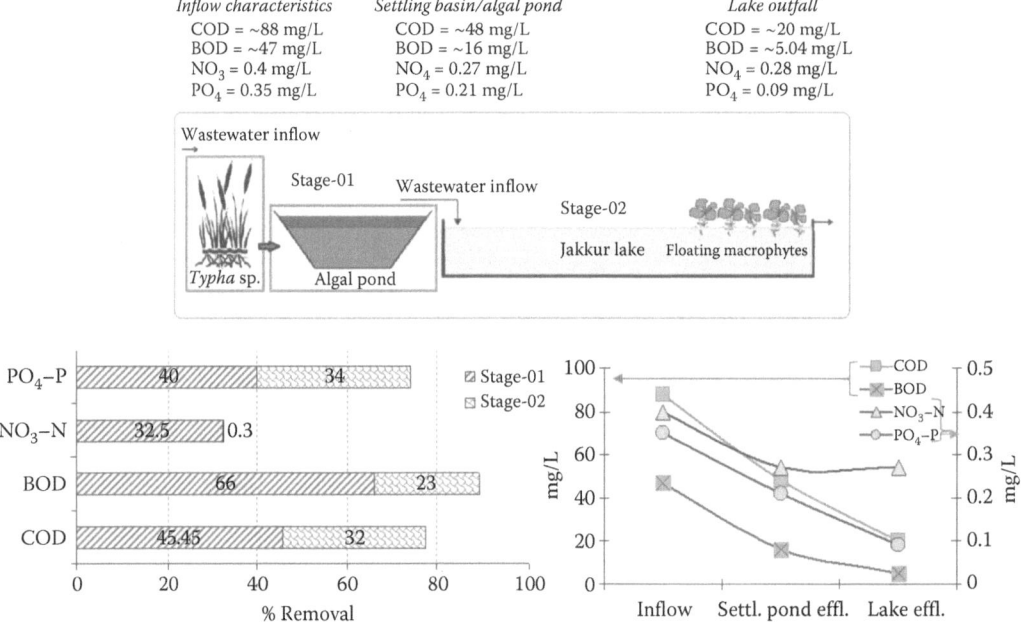

FIGURE 1.14   Integrated wastewater management system.

1.4.2.2 BOD and COD

BOD and COD are important parameters that indicate the presence of organic content. BOD is the amount of oxygen required by bacteria while stabilizing decomposable organic matter under aerobic conditions. It is required to assess the pollution of surface and groundwater where contamination occurs due to disposal of domestic and industrial effluents. COD determines the oxygen required for chemical oxidation of most organic matter and oxidizable inorganic substances with the help of strong chemical oxidant. In conjunction with the BOD, the COD test is helpful in indicating toxic conditions and the presence of biologically resistant organic substances (Sawyer and McCarty 1978). In the current study, the BOD values ranged from 17 to 128 mg/L. There was a reduction of 66% in BOD after the algal pond and 23% removal in the water which flows out of the lake. The COD values ranged from 16 to 161 mg/L. The COD reduced is by 45% in the algae pond and 32% in the lake as shown in Figure 1.14.

1.4.3 INTEGRATED WASTEWATER MANAGEMENT SYSTEM

The treatment of domestic sewage in natural systems such as constructed wetlands and lagoons is being practiced in developing nations. Significant advantages are its simple construction and operation and economic viability (Mahapatra et al. 2011a). Lagoon systems are associated with a high growth rate of beneficial phytoplankton that are caused by the influence of light and the continuous nutrient inflow. Algal growth contributes toward the treatment of wastewater by transforming dissolved nutrients into particle aggregates (biomass). Algal retention in the lagoon helps in the treatment, which has to be harvested at regular intervals to ensure effective treatment. Wetlands consisting of reed-bed and algal pond help in the removal of nutrients (Mahapatra et al. 2013a,b).

The emergent macrophytes (such as *Typha*) act as a filter in removing suspended matter and avoiding anaerobic conditions by the root zone oxidation, and the dissolved nutrients would be taken up by the lagoon algae. This type of treatment helps in augmenting the existing treatment system in complete removal of nutrients and bacteria. The combination of wetlands (with macrophyte assemblages), algal lagoon, and a sustained harvesting of algae and macrophytes would provide complete solution to wastewater treatment systems with minimal maintenance. Integrated wetland system at Jakkur provides an opportunity to assess the efficacy of treatment apart from providing insights for replicating similar systems to address the impending water scarcity in the rapidly urbanizing Bengaluru.

The treatment plant (1.6 ha) with an installed capacity of 10 MLD comprises of an Upflow Anaerobic Sludge Blanket Reactor (UASB) with an extended aeration system for sewage treatment. The treatment effluent then gets into wetlands (settling basin) of spatial extent ~4.63 ha consisting of diverse macrophytes such as *Typha* sp., *Cyperus* sp., *Ludwigia* sp., *Alternanthera* sp., *Eichhornia* sp., and so on in the shallow region (with an area of ~1.8 ha) followed by deeper algal basin (covering an area of about 2.8 ha). This being the significant functional component with macrophytes and algae jointly helps in the nutrient removal and wastewater remediation. The water from the settling basin flow passes through three sluices of which only the middle one was functional in the low flow conditions during the investigation. This water flows into Jakkur lake that spans over 45 ha. There were notably less occurrence of floating macrophytes, except near the outfalls (~0.5 ha) due to blockage of the outflow channels by solid wastes and debris. These macrophytes are being managed by local fishermen. Water in the Jakkur lake is clear with acceptable phytoplankton densities and abundant diversity indicating a healthy trophic status.

The nutrient analysis shows (illustrated in Figure 1.14) that treatment happens due to immergent macrophytes of the wetlands and algae, which removes ~45% COD, ~66% BOD, ~33% NO_3–N, and ~40% PO_4^{3-}–P. Jakkur lake treats the water and acts as the final level of treatment in the stage 2 that removes ~32% COD, ~23% BOD, ~0.3% NO_3–N, and ~34% PO_4^{3-}–P. The synergestic mechanism of STPs followed by wetlands helps in the complete removal of nutrients to acceptable levels according to CPCB norms.

Jakkur STP has been reported to treat only 6 MLD of sewage that is drawn from Yelahanka town. Yet untreated sewer channels were observed carrying voluminous wastewater into the Jakkur system alongside the treatment plant effluents. The major nutrient removal and polishing is done by the manmade wetland and the lake. This wetland comprises of emergent macrophytes such as *Typha angustata*, and so on and thus provides a key role of oxygenation in soil subsystems through root zone oxidation and entrapment of necessary nutrients that otherwise would cause an algal bloom in the lake. The algal species in this manmade wetland region (Figure 1.15) primarily comprised of members of chlorophyceae followed by cyanophyceae, euglenophyceae, and bacillariophyceae (Figure 1.16). The relative abundances are provided in the pie diagrams below.

Similarly, macrophytes play an important role in the effluent stabilization. The distribution of the macrophytes in the wetland area as well as at the outfalls of the lake is provided in Figure 1.17. *T. angustata* species were dominating (54%) in the wetland area followed by *Alternanthera philoxeroides* (28%). However, even though the macrophyte population was scarce in the lake, but still amongst them *Eichhornia crassipes* (84%) were dominating (Figure 1.18), which were only restricted to the outlet reaches provisioned by the net intervenetion to aid fishing and restrict floating macrophyte growth in the core fishing area.

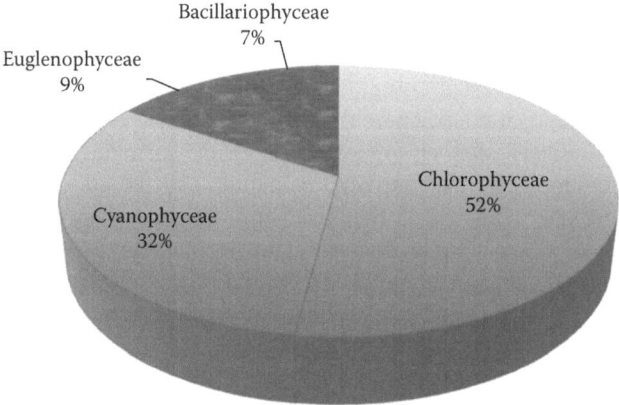

FIGURE 1.15   Composition of algae in man-made wetland system.

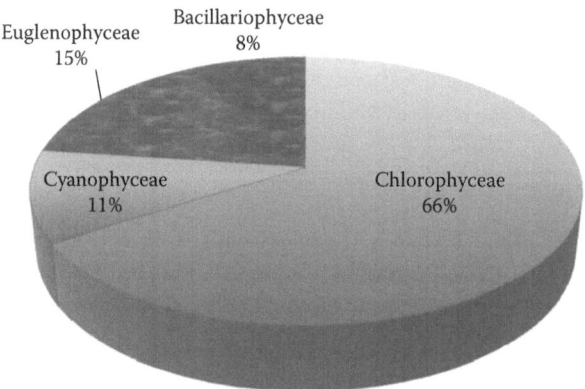

FIGURE 1.16   Composition of algae in Jakkur lake.

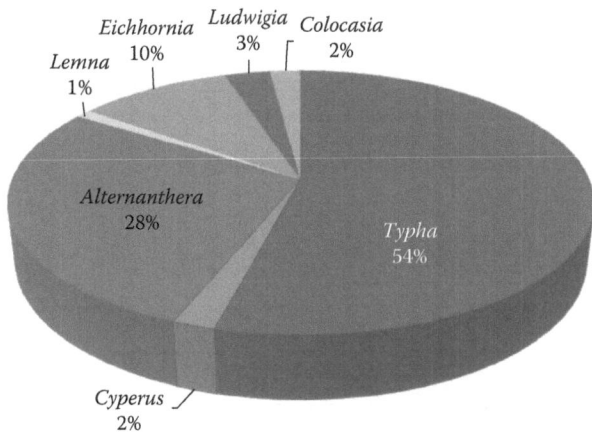

FIGURE 1.17 Composition of macrophytes in man-made wetland system.

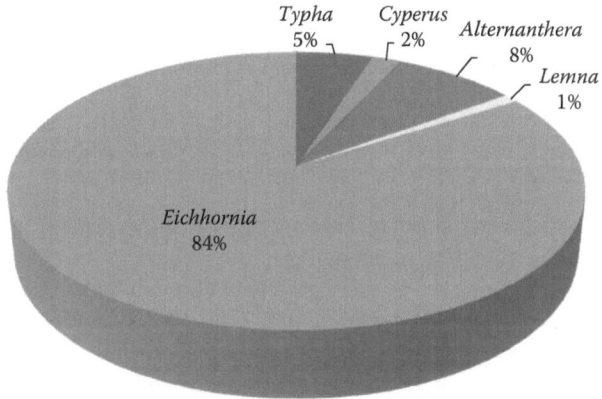

FIGURE 1.18 Composition of macrophytes in Jakkur lake.

1.4.4 INTEGRATED WETLANDS ECOSYSTEM: SUSTAINABLE MODEL TO MITIGATE GHG EMISSIONS

Performance assessment of an integrated wetland ecosystem at Jakkur provides vital insights toward mitigating water crisis in Bengaluru. An integrated system is outlined in Figure 1.19. The STP followed by wetland systems would help in treating water for sustainable recycle and reuse.

1.4.4.1 Functional Aspects of the Integrated Wetland Systems

Sewage treatment plant (STP): The purpose of sewage treatment is to remove contaminants (carbon and solids) from sewage to produce an environmentally safe water. The treatment based on physical, chemical, and biological processes includes three stages—primary, secondary, and tertiary. Primary treatment entails holding the sewage temporarily in a settling basin to separate solids and floatable. The settled and floating materials are filtered before discharging the remaining liquid for secondary treatment to remove dissolved and suspended biological matter. The effluents of STP are still nutrient-rich and need further treatment and stabilization for further water utilities in the vicinity.

Integration with wetlands (consisting of typha beds and algal pond) would help in the complete removal of nutrients in a cost-effective way. A nominal residence time (~5 days) would help in the

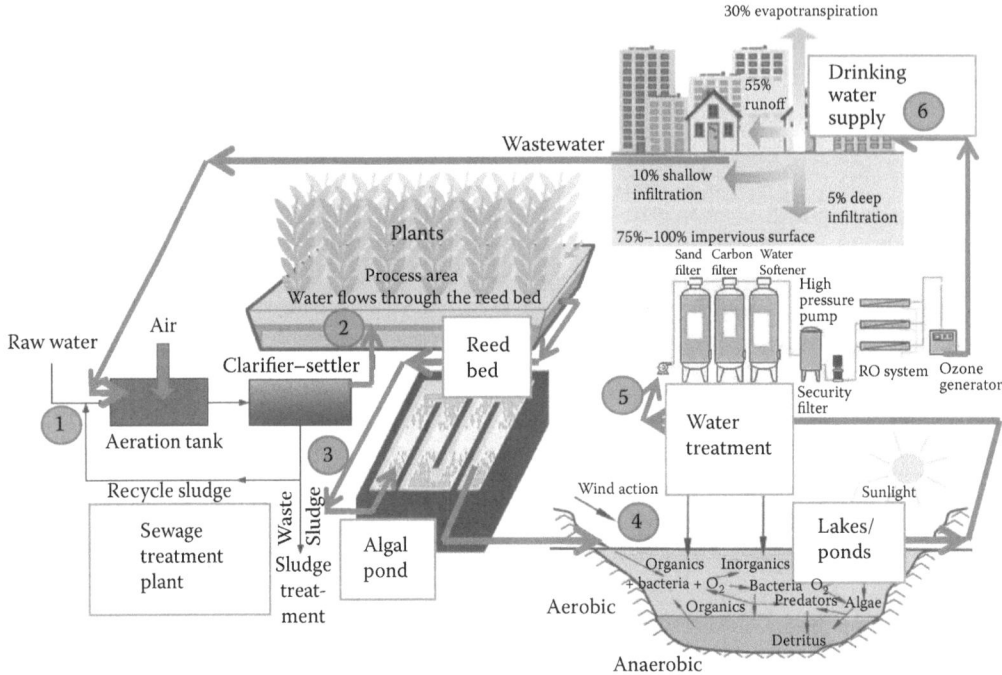

FIGURE 1.19 Integrated wetland system for managing water and wastewater.

removal of pathogens apart from nutrients. However, this requires regular maintenance of harvesting macrophytes and algae (from algal ponds). Harvested algae would have energy value, which could be used for biofuel production. The joint activity of algae and macrophytes in the wetland systems helps in the removal of 77% COD, ~90% BOD, ~33% NO_3–N, and ~75% PO_4^{3-}–P. In the case of integrated wetland systems, algae play a major role in C footprint reduction by actively uptaking the C in dissolved form (carbonic acid, bicarbonates, and carbonates) and also have the ability to take dissolved organic carbon in the dark and thus have great potential to immobilize C (up to 80% of the total input C) in the biomass.

Pilot-scale experiment in the laboratory has revealed nutrient removal of algae are 86, 90, 89, 70, and 76% for TOC, TN, Amm.–N, TP, and OP, respectively, and lipid content varied from 18 to 28.5% of dry algal biomass. Biomass productivity is of ~122 mg/L/day and lipid productivity of ~32 mg/L/day. Gas chromatography and mass spectrometry (GC–MS) analyses of the FAME showed a higher content of desirable fatty acids (biofuel properties) with major contributions from saturates such as palmitic acid [C16:0; ~40%], stearic acid [C18:0; ~34%], followed by unsaturates such as oleic acid [C18:1(9); ~10%] and linoleic acid [C18:2(9,12); ~5%]. The decomposition of algal biomass and reactor residues with calorific exothermic heat content of 123.4 J/g provides the scope for further energy derivation (Mahapatra et al. 2014).

The study reveals that surface water bodies (lakes, ponds, tanks, etc.) in Bengaluru are subjected to high nutrient loads due to the sustained inflow of untreated or partially treated sewage, altering physicochemical and biological integrity of water bodies. The treated water from STP in Jakkur still contains nutrients as primary and secondary treatments do not completely remove nutrients. However, passage of STP effluents through wetlands (consisting of emergent macrophytes and algal pond) ensures removal of nutrients to an extent ensuring potability of water. This study investigates the water quality at different stages in the integrated wetland system. The physicochemical and biological parameters were monitored as water enters the algal pond (wetland) from the STP, outlet of wetlands and at the inlet, middle, and outlets of Jakkur lake. The nutrient analysis highlights nutrient removal by wetlands due to macrophytes and algae, which removes

77% COD, ~90% BOD, ~33% NO_3–N, and ~75% PO_4^{3-}–P. The first stage comprising of emergent vegetatation and algal pond removes ~45% COD, ~66% BOD, ~33% NO_3–N, and ~40% PO_4^{3-}–P. Jakkur lake as a second stage treats the water and acts as the final level of treatment and removes ~32% COD, ~23% BOD, ~0.3% NO_3–N, and ~34% PO_4^{3-}–P. The combination of all the stages leads to a complete removal of nutrients to acceptable levels according to CPCB norms. This study provided vital insights toward an environmentally sound option of managing wastewater, while mitigating GHG emissions.

ACKNOWLEDGMENTS

We are grateful to the Department of Biotechnology (DBT), the Ministry of Environment and Forests, Government of India, and the Indian Institute of Science for the infrastructure and financial support. We thank the central facilities at the Department of Biochemistry and SID and Innovation Centre for fatty acid composition analysis through GC–MS and ATR-FTIR analyses. We are grateful to the Institute Nano Initiative (INI) for providing facilities for scanning electron microscopy (SEM) imaging.

REFERENCES

Ackom EK. 2010. Sustainability standards for Canada's bioethanol industry. *Biofuels* 1:237–241.

Affan A, Jewel SA, Haque M, Khan S, Lee JB. 2005. Seasonal cycle of phytoplankton in aquaculture ponds in Bangladesh. *Algae* 20:43–52.

Akotoa L, Pel R, Irtha H, Udo A, Brinkmana T, Vreuls RJJ. 2005. Automated GC–MS analysis of raw biological samples: Application to fatty acid profiling of aquatic micro-organisms. *J Anal Appl Pyrol* 73:69–75.

Amin S. 2009. Review on biofuel oil and gas production processes from micro-algae. *Energy Convers Manage* 50:1834–1840.

APHA, 1998. *Standard Methods for the Examination of Water and Wastewater*, 20th edn. American Waterworks Association, Washington, DC, USA.

Backmann J, Schultz C, Fabian H, Hahn U, Saenger W, Naumann D. 1996. Thermally induced hydrogen exchange processes in small proteins as seen by FTIR spectroscopy. *Proteins* 24:379–387.

Benning LG, Phoenix VR, Yee N, Tobin MJ. 2004. Molecular characterization of cyanobacterial silification using synchrotron infrared micro-spectroscopy. *Geochim Cosmochim Acta* 68:729–741.

Berg W, Brunsch R, Hellebrand HJ, Kern J. 2006. Methodology for measuring gaseous emissions from agricultural buildings, manure, and soil surfaces. In *Workshop on Agricultural Air Quality*, June 5–8, 2006, pp. 233–241.

Bernal CB, Vazquez G, Quintal IB, Bussy AL. 2008. Microalgal dynamics in batch reactors for municipal wastewater treatment containing dairy sewage water. *Water Air Soil Pollut* 190:259–270.

Bligh EG, Dyer WJ. 1959. A rapid method of lipid extraction and purification. *Can J Biochem Physiol* 37:911–917.

Bokowski G, White D, Pacifico A, Talbot S, DuBelko A, Phipps A. 2007. Towards campus climate neutrality: Simon Fraser University's carbon footprint. Simon Fraser University.

Borowitzka MA. 2010. Algae oils for biofuels: Chemistry, physiology, and production. In: Cohen Z, Ratledge C (eds), *Single Cell Oils. Microbial and Algal Oils*. AOCS Press, Urbana, pp. 271–289.

Brewer RS. 2008a. Literature review on carbon footprint collection and analysis. http://csdl.ics.hawaii.edu/techreports/09-05/09-05.pdf (accessed on August 12, 2014).

Brewer RS. 2008b. Carbon metric collection and analysis with the personal environmental tracker. In *Proceedings of the UbiComp 2008 Workshop on Ubiquitous Sustainability: Citizen Science and Activism*, September 21–24, 2008, Seoul.

Brown MA, Southworth F, Sarzynski A. 2009. The geography of metropolitan carbon footprints. *Policy Soc* 27:285–304.

BSI. 2008. Publicly available specification 2050. Specification for the assessment of the life cycle greenhouse gas emissions of goods and services. British Standards.

Carbon Trust. 2006. Carbon footprints in the supply chain: The next step for business. Report Number CTC616, November 2006. The Carbon Trust, London, UK. http://www.carbontrust.co.uk (accessed on August 12, 2014).

Carbon Trust. 2007a. *Carbon Footprint Measurement Methodology*, Version 1.1. The Carbon Trust, London, UK. http://www.carbontrust.co.uk (accessed on August 12, 2014).

Carbon Trust. 2007b. Carbon footprinting. An introduction for organizations. http://www.carbontrust.co.uk/publications/publicationdetail.htm?productid=CTV033 (accessed on August 12, 2014).

Castro-Guerrero NA, Chavez R, Moreno-Sanchez R. 2005. Physiological role of rhodoquinone in *Euglena gracilis* mitochondria. *Biochem Biophys Acta* 1710:113–121.

Chambers JQ, Fisher JI, Zeng H, Chapman EL, Baker DB, Hurtt GC. 2007. Hurricane Katrina's carbon footprint on U.S. gulf coast forests. *Science* 318:1107.

Chanakya HN, Mahapatra DM, Sarada R, Abitha R. 2013. Algal biofuel production and mitigation potential in India. *Mitig Adapt Strateg Global Change* 18:113–136.

Chanakya HN, Mahapatra DM, Sarada R, Chauhan VS, Abitha R. 2012. Sustainability of large-scale algal biofuel production in India. *J Indian Inst Sci* 92(1):63–98.

Chinnasamy S, Bhatnagar A, Hunt RW, Das KC. 2010. Microalgae cultivation in a wastewater dominated by carpet mill effluents for biofuel applications. *Bioresour Technol* 101:3097–3105.

Chisti Y. 2007. Biodiesel from microalgae. *Biotechnol Adv* 25:294–306.

Coleman LW, Rosen BH, Schwartzbach SD. 1988. Environmental control of carbohydrate and lipid synthesis in Euglena. *Plant Cell Physiol* 29:423–432.

Constantopolous G, Bloch K. 1967. Effect of light intensity on the lipid composition of *Euglena gracilis. J Biol Chem* 242:3538–3542.

Courchene TJ, Allan JR. 2008. Climate change: The case of carbon tariff/tax. *Policy Options* 3:59–64.

Craeynest L, Streatfeild D. 2008. The World Bank and its carbon footprint: Why the World Bank is still far from being an environment bank. World Wildlife Fund, 35 pp.

Damiani MC, Popovich CA, Constenla D, Leonardi PI. 2010. Lipid analysis in *Haematococcus pluvialis* to assess its potential use as a biodiesel feedstock. *Bioresour Technol* 101:3801–3807.

Dean AP, Estrada B, Sigee DC, Pittman JK. 2010. Using FTIR spectroscopy for rapid determination of lipid accumulation in response to nitrogen limitation in freshwater microalgae. *Bioresour Technol* 101:4499–4507.

Dean AP, Sigee DC. 2006. Molecular heterogeneity in *Aphanizomenon flosaquae* and *Anabaena flosaquae* (Cyanophyta): A synchrotron-based Fourier-transform infrared study of lake micro populations. *Eur J Phycol* 41:201–212.

Dean AP, Martin MC, Sigee DC. 2007. Resolution of codominant phytoplankton species in a eutrophic lake using synchrotron based Fourier transform infrared spectroscopy. *Phycologia* 46:151–159.

Department of Mines and Geology. 2011. Groundwater hydrology and groundwater quality in and around Bangalore city.

Donnelly TH, Barnes CJ, Wasson RJ, Murray AS, Short DL. 1998. Catchment phosphorus sources and algal blooms—An interpretative review. Technical Report 18/98. CSIRO Land and Water, Canberra.

Dumas P, Miller L. 2003. The use of synchrotron infrared microspectroscopy in biological and biomedical investigations. *Vib Spectrosc* 32:3–21.

Duttagupta S, Gupta S, Gupta A. 2004. Euglenoid blooms in the floodplain wetlands of Barak Valley, Assam, North-Eastern India. *J Environ Biol* 25:369–373.

East AJ. 2008. What is a carbon footprint? An overview of definitions and methodologies. In *Vegetable Industry Carbon Footprint Scoping Study—Discussion Papers and Workshop*, September 26, 2008. Horticulture Australia Limited, Sydney.

Easterling ER, French WT, Hernandez R, Lich M. 2009. The effect of glycerol as a sole and secondary substrate on the growth and fatty acid composition of *Rhodotorula glutinis. Bioresour Technol* 100:356–361.

Edgar GH, Peters GP. 2009. Carbon footprint of nations: A global, trade linked analysis. *Environ Sci Technol* 43:6414–6420.

Energetics. 2007. The reality of carbon neutrality. http://www.energetics.com.au/file?node_id=21228.

EPA (Environmental Protection Agency). 2010. 2010 U.S. Greenhouse Gas Inventory Report: Inventory of U.S. Green-house Gas Emissions and Sinks, 1990–2008. EPA-430-R-10-006.

ETAP. 2007. The Carbon Trust Helps UK Businesses Reduce their Environmental Impact, Press Release. http://ec.europa.eu/environment/etap/pdfs/jan07_carbon_trust_initiative.pdf (accessed on August 12, 2014).

Fabregas J, Domínguez A, Regueiro M, Maseda A, Otero A. 2000. Optimization of culture medium for the continuous cultivation of the microalga *Haematococcus pluvialis. Appl Microbiol Biotechnol* 53:530–535.

Feng Y, Li C, Zhang D. 2011. Lipid production of *Chlorella vulgaris* cultured in artificial wastewater medium. *Bioresour Technol* 102:101–105.

Fernandez AFG, Camacho GF, Perez SJA, Sevilla FJM, Grima ME. 1998. Modelling of biomass productivity in tubular photobioreactors for microalgal cultures: Effects of dilution rate, tube diameter and solar irradiance. *Biotechnol Bioeng* 58:605–616.

Finkbeiner M. 2009. Carbon footprinting—Opportunities and threats. *Int J Life Cycle Assess* 14:91–94.

Fischer G, Braun S, Thissen R, Dott W. 2006. FT-IR spectroscopy as a tool for rapid identification and intraspecies characterization of airborne filamentous fungi. *J Microbiol Methods* 64:63–77.

Garcia J, Rousseau DPL, Morato J, Lesage E, Matamoros V, Bayona JM. 2010. Contaminant removal processes in subsurface-flow constructed wetlands: A review. *Crit Rev Environ Sci Technol* 40(7):561–661.

Garg S, Dornfeld D. 2008. An indigenous application for estimating carbon footprint of academia library based on life cycle assessment. Laboratory for Manufacturing and Sustainability, University of California, Berkeley, CA. http://escholarship.org/uc/item/8zp825mq.

GFN. 2007. Ecological Footprint Glossary. Global Footprint Network, Oakland, CA. http://www.footprintnetwork.org/gfn_sub.php?content=glossary (accessed on August 12, 2014).

Giordano M, Kansiz M, Heraud P, Beardall J, Wood B, McNaughton D. 2001. Fourier transform infrared spectroscopy as a novel tool to investigate changes in intracellular macromolecular pools in the marine alga *Chaetoceros muellerii* (Bacillariophyceae). *J Phycol* 37:271–279.

Gouda MK, Omar SH, Aouad LM. 2008. Single cell oil production by *Gordonia* sp. DG using agro-industrial wastes. *World J Microbiol Biotechnol* 24:1703–1711.

GOI (Government of India) 2008. Groundwater information booklet. Bangalore Urban District, Karnataka. http://cgwb.gov.in/district_profile/karnataka/bangalore_urban_brochure.pdf (accessed on July 5, 2014).

Grubb E. 2007. *Meeting the Carbon Challenge: The Role of Commercial Real Estate Owners*. Users & Managers, Chicago.

Hall RP, Schoenborn HW. 1938. Studies on the question of autotrophic nutrition in *Chlorogonium euchlorum*, *Euglena anabaena* and *Euglena deses*. *Arch Protistenk* 90:259–271.

Heijungs R, Suh S. 2006. Reformulation of matrix-based LCI: From product balance to process balance. *J Clean Prod* 14(1):47–51.

http://www.pre-sustainability.com/product-carbon-footprint-standards-which-standard-to-choose (accessed on November 30, 2014).

Hu C, Zhao X, Zhao J, Wu S, Zhao ZK. 2009. Effects of biomass hydrolysis by-products on oleaginous yeast *Rhodosporidium toruloides*. *Bioresour Technol* 100:4843–4847.

IPCC. 1996. *Revised IPCC Guidelines for National Green-house Gas Inventories*. IPCC/OECD/IEA, Bracknell.

IPCC. 2006. *National Greenhouse Gas Inventories: Land Use, Land Use Change and Forestry*. Hayama, Japan.

IPCC. 2007. Climate change 2007: Synthesis report: Contribution of working groups I, II and III to the fourth assessment report. Intergovernmental Panel on Climate Change (IPCC).

IPCC. 2014. https://www.dieselnet.com/news/2014/04ipcc.php (accessed on November 30, 2014)

John DM, Whitton BA, Brook, AJ. 2002. *The Freshwater Alga Flora of the British Isles. An Identification Guide to Freshwater and Terrestrial Algae*. British Phycological Society, Natural History Museum, Cambridge University Press, London.

Karin L. 2006. Wastewater treatment with microalgae—A literature review. *Vatten* 62:31–38.

Kelly LM, Shepson PB, Strim BP, Karion A, Sweeney C, Gurney KR. 2009. Aircraft-based measurements of the carbon footprint of Indianapolis. *Environ Sci Technol* 43:7816–7823.

Kerr AR. 2007. How urgent is climate change? *Science* 318:1230–1231.

Kleiner K. 2007. The corporate race to cut carbon. *Nature* 3:40–43.

Knothe G. 2008. "Designer" biodiesel: Optimizing fatty ester composition to improve fuel properties. *Energy and Fuels* 22:1358–1364.

Kong Q, Li L, Martinez B, Chen P, Ruan R. 2010. Culture of microalgae *Chlamydomonas reinhardti* in wastewater for biomass feedstock production. *Appl Biochem Biotechnol* 160:9–18.

Kulkarni V, Ramachandra TV. 2009. *Environment Management*. TERI Press, New Delhi, 442 pp.

Lenzen M. 2001. Errors in conventional and input–output based lifecycle inventories. *J Ind Ecol* 4(4):127–148.

Li Y, Horsman M, Wang B, Wu N, Lan CQ. 2008. Effects of nitrogen sources on cell growth and lipid accumulation of green alga *Neochloris oleoabundans*. *Appl Microbiol Biotechnol* 8:629–636.

Li Y, Zhao Z, Bai F. 2007. High-density cultivation of oleaginous yeast *Rhodosporidium toruloides* Y4 in fed-batch culture. *Enzyme Microbial Technol* 41:312–317.

Liang Y, Sarkany N, Cui Y. 2009. Biomass and lipid productivities of *Chlorella vulgaris* under autotrophic, heterotrophic and mixotrophic growth conditions. *Biotechnol Lett* 31:1043–1049.

Lungu AI, Obuh PA. 2004. Euglenophyta of the wastewater treatment facilities of the sugar industry (Moldova). *Int J Algae* 6:158–170.

Lwoff A. 1932. *Recherches biochimiques sur la nutrition des Protozoaires.* Mono Inst Pasteur Paris, Masson, p. 158.

Mahapatra DM, Chanakya HN, Ramachandra TV. 2011a. Assessment of treatment capabilities of Varthur Lake, Bangalore, India. *Int J Environ Technol Manage* 14:84–102.

Mahapatra DM, Chanakya HN, Ramachandra TV. 2011b. C:N ratio of sediments in a sewage fed Urban Lake. *Int J Geol* 3:86–92.

Mahapatra DM, Chanakya HN, Ramachandra TV. 2011c. Role of macrophytes in a sewage fed urban lake. *Inst Integr Omics Appl Biotechnol* 2:1–9.

Mahapatra DM, Chanakya HN, Ramachandra TV. 2013a. *Euglena* sp. as a suitable source of lipids for potential use as biofuel and sustainable wastewater treatment. *J Appl Phycol* 25:855–865.

Mahapatra DM, Chanakya HN, Ramachandra TV. 2013b. Treatment efficacy of algae based sewage treatment plants. *Environ Monit Assess* 185:7145–7164.

Mahapatra DM, Chanakya HN, Ramachandra TV. 2014. Bioremediation and lipid synthesis of myxotrophic algal consortia in municipal wastewater. *Bioresour Technol* 168:142–150.

Mahapatra DM, Ramachandra TV. 2013. Algal biofuel: Bountiful lipid from *Chlorococcum* sp. proliferating in municipal wastewater. *Curr Sci* 105:47–55.

Martinez ME, Sanchez S, Jimenez JM, El Yousfi F, Munoz L. 2000. Nitrogen and phosphorus removal from urban wastewater by the microalga *Scenedesmus obliquus. Bioresour Technol* 73:263–272.

Matthews SC, Hendrickson CT, Weber CL. 2008a. Estimating carbon footprints with input output models. In *International Input Output Meeting on Managing the Environment,* July 9–11, 2008, Seville.

Matthews SC, Hendrickson CT, Weber CL. 2008b. The importance of carbon footprint estimation boundaries. *Environ Sci Technol* 42(16):5839–5842.

Metzger P, Largeau C. 2005. *Botryococcus braunii*: A rich source for hydrocarbons and related ether lipids. *Appl Microbiol Biotechnol* 6:486–496.

Moheimani NR, Borowitzka MA. 2007. Limits to productivity of the alga *Pleurochrysis carterae* (Haptophyta) grown in outdoor raceway ponds. *J Appl Phycol* 18:703–712.

Moreno-Sanchez R, Covian R, Jasso-Chavez R, Rodriguez-Enriquez S, Pacheco-Moises F, Torres-Marquez ME. 2000. Oxidative phosphorylation supported by an alternative respiratory pathway in mitochondria from Euglena. *Biochim Biophys Acta* 1457:200–210.

Murdock JN, Wetzel DL. 2009. FT-IR microspectroscopy enhances biological and ecological analysis of algae. *Appl Spectrosc Rev* 44:335–361.

Ogbonna JC, Tanakaah H. 1998. Cyclic autotrophic heterotrophic cultivation of photosynthetic cells: A method of achieving continuous cell growth under light/dark cycles. *Bioresour Technol* 65:65–72.

Orpez R, Martinez ME, Hodaifa G, El Yousfi F, Jbari N, Sanchez S. 2009. Growth of the microalga *Botryococcus braunii* in secondarily treated sewage. *Desal* 246:625–230.

Pal D, Khozin-Goldberg I, Cohen Z, Boussiba S. 2011. The effect of light, salinity and nitrogen availability on lipid production by *Nannochloropsis* sp. *Appl Microbiol Biotechnol* 90:1429–1441.

Parkinson J, Tayler K, Colin J, Nema A. 2008. *A Guide to Decisionmaking: Technology Options for Urban Sanitation in India.* Government of India, September 2008, pp. 42–46.

Peters GP. 2010. Carbon footprints and embodied carbon at multiple scales. *Curr Opin Environ Sust.* doi: 10.1016/j.cosust.2010.05.004.

POST. 2006. Carbon footprint of electricity generation. POSTnote 268, October 2006. Parliamentary Office of Science and Technology, London, UK. http://www.parliament.uk/documents/upload/postpn268.pdf (accessed on August 12, 2014).

Prescott, G.W. 1964. How to know the fresh-water algae: An illustrated key for identifying the more common freshwater algae to genus, with hundreds of species named and pictured and with numerous aids for the study. Dubuque, Iowa, W.C. Brown Co. 272 pp.

Ramachandra TV, Mahapatra DM, Karthick B, Gordon R. 2009. Milking diatoms for sustainable energy: Biochemical engineering vs gasoline secreting diatom solar panels. *Ind Eng Chem Res* 48:8769–8788.

Ramachandra TV, Mahapatra DM, Samantray S, Joshi NV. 2013. Biofuel from urban wastewater: Scope and challenges. *Renew Sustain Energy Rev* 21:767–777.

Ramachandra TV, Shwetmala K, Dania TM. 2014b. Carbon footprint of solid waste sector in Greater Bangalore, India, In: S.S. Muthu (ed.), *Assessment of Carbon Footprint in Different Industrial Sectors, Vol. 1, EcoProduction.* Springer Science, Singapore, pp. 265–292. doi: 10.1007/978-981-4560-41-2_11.

Ramachandra TV, Sreejith K, Bharath HA. 2014a. Sector-wise assessment of carbon footprint across major cities in India, In S.S. Muthu (ed.), *Assessment of Carbon Footprint in Different Industrial Sectors, Vol. 2, Eco-Production.* Springer Science, Singapore, pp. 207–228. doi: 10.1007/978-981-4585-75-0_8.

Ramachandra TV, Sudarshan B, Mahapatra DM, Gautham Krishnadas. 2012. Impact of indiscriminate disposal of untreated effluents from thermal power plant on water resources. *Ind J Environ Prot* 32(9):705–718.

Ramachandra TV, Uttam Kumar. 2008. Wetlands of Greater Bangalore, India: Automatic Delineation through Pattern Classifiers. *Electron Green J* 26:11.

Rasoul-Amini S, Ghasemi Y, Morowvat MH, Mohagheghzadeh A. 2009. PCR amplification of 18S rRNA, single cell protein production and fatty acid evaluation of some naturally isolated microalgae. *Food Chem* 116:129–136.

Rodolfi L, Zittelli GC, Bassi N, Padovani G, Biondi N, Bonini G. 2008. Microalgae for oil: Strain selection, induction of lipid synthesis and outdoor mass cultivation in a low-cost photo-bioreactor. *Biotechnol Bioeng* 102:100–112.

Rosenberg A. 1963. A comparison of lipid patterns in photosynthesizing and non-photosynthesizing cells of *Euglena gracilis. Biochemistry* 2:1148–1154.

Ruiz LB, Rocchetta I, dos Santos Ferreira V, Conforti VTD. 2004. Isolation, culture and characterization of a new strain of *Euglena gracilis. Phycol Res* 52:168–174.

Sanchez E, Colmenarejo MF, Vicente J, Rubio A, Garcia MG, Travieso L, Borja R. 2007. Use of the water quality index and dissolved oxygen deficit as simple indicators of watersheds pollution. *Ecol Indicators* 7:315–328.

Sawayama S, Inoue S, Yokoyama S. 1994. Continuous culture of hydrocarbon-rich microalga *Botryococcus braunii* in secondarily treated sewage. *Appl Microbiol Biotechnol* 41:729–731.

Sawyer CN, McCarty PL. 1978. *Chemistry for Environmental Engineering.* 3rd edn. McGraw-Hill Book Company, New York.

Schoenborn HW. 1942. Studies on the nutritional requirements of *Euglena gracilis* in darkness. *Physiol Zool* 15:325–332.

Seraphim P, Michael K, George A. 2004. Single cell oil (SCO) production by *Mortierella isabellina* grown on high-sugar content media. *Bioresour Technol* 95:287–291.

Sharachchandra L, Srinivasan V, Jamwal P, Thomas BK, Eswar M, Zuhail TM. 2013. Water management in Arkavathy basin: A situation analysis. Environment and Development. Discussion Paper No.1. Bengaluru: Ashoka Trust for Research in Ecology and the Environment.

Sheehan J, Dunahay T, Benemann J, Roessler P. 1998. Look back at the U.S. Department of Energy's Aquatic Species Program: Biodiesel from algae. Close-Out Report. NREL Report No. TP-580-24190.

Sigee DC, Dean A, Levado E, Tobin MJ. 2002. Fourier-transform infrared spectroscopy of *Pediastrum duplex*: Characterization of a micro-population isolated from a eutrophic lake. *Eur J Phycol* 37:19–26.

State of the environment Report 2003, Karnataka. http://parisara.kar.nic.in/PDF/ENERGY.pdf.

Stehfest K, Toepel J, Wilhelm C. 2005. The application of micro-FTIR spectroscopy to analyse nutrient stress-related changes in biomass composition of phytoplankton algae. *Plant Physiol Biochem* 43:717–726.

Subakov-Simi G, Karad Zi V, Krizmani J, Cvijan M, Maljevi E 2008. Euglenophyta of the Danube River in Serbia. *Arch Biol Sci Belg* 60:159–162.

Suh S, Lenzen M, Treloar GJ, Hondo H, Horvath A, Huppes G, Jolliet O et al. 2004. System boundary selection in life-cycle inventories using hybrid approaches. *Environ Sci Technol* 38(3):657–664.

Tittel J, Bissinger V, Zippel B, Gaedke U, Bell E, Lorke A, Kamjunke N. 2003. Mixotrophs combine resource use to out-compete specialists: Implications for aquatic food webs. *PNAS* 100:12776–12781.

UNFCCC. 2008. Offsetting the carbon footprint for Poznan. http://unfccc.int/press/news_room/newsletter/items/4603.php (accessed on August 12, 2014).

USCCTP. 2005. Technology options for near and long term future. http://www.climatetechnology.gov/library/2005/tech-options/index.htm (accessed on August 12, 2014).

USEPA. 2006. Global anthropogenic non-CO_2 greenhouse gas emissions: 1990–2020. June 2006 revised. Washington, DC: US Environmental Protection Agency.

Veeresh M, Veeresh AV, Basvaraj D, Basaling H, Hosetti B. 2010. Dynamics of industrial waste stabilization pond treatment process. *Environ Monit Assess* 169:55–65.

Velasco E, Pressley S, Allwine E, Westberg H, Lamb B. 2005. Measurements of CO_2 fluxes from the Mexico City urban landscape. *Atmos Environ* 39(38):7433–7446.

Wackernagel M, Rees WE. 1996. *Our Ecological Footprint: Reducing Human Impact on the Earth.* New Society Publishers, Gabriola Island, BC.

Weidema BP, Thrane M, Christensen P, Schmidt J Lokke S. 2008. Carbon footprint: A catalyst for life cycle assessment. *J Ind Ecol* 12(1):3–6.

Wen Z, Zhi-Hui H. 1999. Biological and ecological features of inland saline waters in North Hebei, China. *Int J Salt Lake Res* 8:267–285.

Weyer, KM, Bush DR, Darzins A, Willson BB. 2009. Theoretical maximum algal oil production. *Bio Energ Res* 3(2):204–213.

Weyer KM, Bush DR, Darzins A, Willson BD. 2010. Theoretical maximum algal oil production. *Bioene Res* 3:204–213.

Widjaja A, Chien C, Ju Y. 2009. Study of increasing lipid production from freshwater microalgae *Chlorella vulgaris*. *J Taiwan Inst Chem Eng* 40:13–20.

Wiedmann T, Minx J, Barrett J, Wackernagel M. 2006. Allocating ecological footprints to final consumption categories with input-output analysis. *Ecol Econ* 56(1):28–48.

Wiedmann T, Minx J. 2007. A definition of carbon footprint. ISAUK Research Report 07-01, Durham.

Williams PJB, Laurens LML. 2010. Microalgae as biodiesel and biomass feedstocks: Review and analysis of the biochemistry, energetics and economics. *Energy Environ Sci* 3:554–590.

Woertz I, Feffer A, Lundquist T, Nelson Y. 2009. Algae grown on dairy and municipal wastewater for simultaneous nutrient removal and lipid production for biofuel feedstock. *J Environ Eng* 135:1115–1122.

Wollmann K, Deneke R, Nixdorf B, Packroff G. 2000. Dynamics of planktonic food webs in three mining lakes across a pH gradient (pH 2–4). *Hydrobiology* 433:3–14.

WRI/WBCSD. 2004. The greenhouse gas protocol: A corporate accounting and reporting standard revised edition. World Business Council for Sustainable Development and World Resource Institute, Geneva.

WRI/WBCSD. 2006. The greenhouse gas protocol: Designing a customized greenhouse gas calculation tool. World Business Council for Sustainable Development and World Resource Institute, Geneva.

Xu H, Miao X, Wu Q. 2006. High quality biodiesel production from a microalga *Chlorella protothecoides* by heterotrophic growth in fermenters. *J Biotechnol* 126:499–507.

Yamane Y, Utsunomiya T, Watanabe M, Sasaki K. 2001. Biomass production in mixotrophic culture of *Euglena gracilis* under acidic condition and its growth energetics. *Biotechnol Lett* 23:1223–1228.

Yee N, Benning LG, Phoenix VR, Ferris FG. 2004. Characterization of metal cyanobacteria sorption reactions: A combined macroscopic and infrared spectroscopic investigation. *Environ Sci Technol* 38:775–782.

2 Challenges and Merits of Choosing Alternative Functional Units

Benjamin C. McLellan

CONTENTS

2.1 Introduction 45
    2.1.1 Definitions...46
    2.1.2 Literature Review 46
2.2 Impacts of the FU 47
    2.2.1 Case Study 1: Coffee 47
    2.2.2 Case Study 2: Cement and Concrete 49
2.3 FUs for Policy and Decision Making 51
    2.3.1 Policy Making...51
        2.3.1.1 The Target Level: Organization as an FU 52
        2.3.1.2 The Level of Inducements—Actions and Accounts as FUs 52
    2.3.2 Consumer Products...53
        2.3.2.1 Single-Serve or Single-Item Consumable Products...................54
        2.3.2.2 Multiserve or Multi-Item Consumable Products 54
        2.3.2.3 Durable Products 55
    2.3.3 Design and Industrial Operations...56
    2.3.4 Some FUs Applied in Industry 56
    2.3.5 Inappropriate Assignation of the FU 57
2.4 Conclusions...58
References...58

2.1 INTRODUCTION

The functional unit (FU) is a fundamental component of any life-cycle assessment (LCA) or carbon footprint (CF), highlighted and described in both CF and LCA standards as one of the first decisions to be made in the scoping of a study. The FU can be critical in determining the relative rank and subsequent selection of alternative products based on a CF. The importance of the FU from the point of view of assessment can sometimes be understated, particularly given the many other areas of a CF study that need to be considered in order to obtain good, accurate, and ultimately useful study results. In a purely academic sense, LCA studies in modern academic literature are generally theoretically sound in the selection of FUs, but the argument can be made that some of the FUs selected are not ideal with regard to final selection.

Given the broad coverage of LCA and CF studies, it is difficult to give a generalizable procedure to enable the selection of the most appropriate FU for each case—arguably the standards cannot accomplish this alone, and hence the development of product category rules (PCRs) and sectoral guidance documents to enable greater specific detail relevant to particular product systems and industrial sectors. Likewise, this chapter does not seek to cover every situation, but to illustrate with a number of typical categories of example.

This chapter aims to put the FU into context, to give examples of how it can affect outcomes and to provide some guidance and commentary toward selection of appropriate FU for consumer or policy decision making. The most important aspect of this chapter is to illustrate the importance of integrating both the product system under assessment and the perspective of the end user who must base their decision upon the outcome of the study.

2.1.1 DEFINITIONS

Two major standards are available, free of cost, for carbon footprinting: the British Standards Institute's Publicly Available Standard (PAS 2050) (BSI 2011), and the World Business Council for Sustainable Development and World Resources Institute's "The Greenhouse Gas Protocol" (WBCSD and WRI 2013a). Both offer good guidance in the calculation of product carbon footprints (PCFs) and general guidance, as well as sector-specific guidance in some cases, that can assist in selecting appropriate scope of study including the function unit. There is also an International Organization for Standardization (ISO) technical standard (ISO/TS 14067), which has been developed to complement existing standards on environmental management and LCA, to fill the need posed by the expansion of carbon footprinting activity—as a subset of LCA (Standardization IOF 2013).

PAS 2050 section offers the following definition of the FU: "quantified performance of a product system for use as a reference unit" (PAS 2050) (BSI 2011). As a general statement, this is not particularly useful until more explanation is given. Some further discussion in the text gives the following guidance: "Where a product is commonly available on a variable unit size basis, the calculation of GHG emissions shall be proportional to the unit size (e.g., per kilogram ...)" (BSI 2011). The further notes specify that where preferred units have otherwise been set (in related guidance), these should be used. Most relevant to this chapter, they note that the FU may differ according to the purpose of the assessment activity, which is why this chapter looks further into a variety of examples, hoping to expand some guidance in this area.

In the GHG Protocol, this is slightly differently phrased, but with similar meaning: "The unit of analysis is defined as the performance characteristics and services delivered by the product being studied" (WBCSD and WRI 2013a). They also offer the guidance that the FU "typically includes the function (service) a product fulfils, the duration of the service life (amount of time needed to fulfil the function), and the expected quality level." Further, they distinguish between the FU and a reference flow, which is when the final product is not clear, and therefore the life-cycle PCF cannot be fully calculated. The GHG Protocol gives good guidance on the selection of an appropriate FU for companies and is a suggested further reference for readers interested in this area.

The ISO 14067 standard uses the same definition as PAS 2050, then goes on to explain that: "Comparisons between systems shall be made on the basis of the same function(s), quantified by the same functional unit(s) in the form of their reference flows" (Standardization IOF 2013). Overall, the GHG Protocol offers a clearer explanation of the FU for the general user and is thus preferred in some contexts (such as education or wider industry application), while PAS 2050 and ISO/TS 14067 are equally compiled in a higher tone of language, sufficient for experienced users but not for a broader audience.

In general, it is important that the FU should be readily delineated, first through physically measurable or countable quantities and thereafter by specifications that quantitatively or qualitatively define the quality and type of service being provided by the FU.

2.1.2 LITERATURE REVIEW

Since the early days of LCA development globally, the consideration of FUs has been seen as important, particularly as standardization was being undertaken. The need for cross-comparability of studies emerged rapidly as an issue that required a standard consideration of all aspects of an LCA study and the FU was widely discussed (see, e.g., the review of the early discussions provided by other authors; Cooper 2003). These discussions continue for specific and general FUs (Frijia et al. 2011) and are relevant to carbon footprinting as a subset of LCA.

Significant effort has been spent examining and specifying appropriate FUs for particular product systems, such as degreasing in the metal products industry in Germany (Finkbeiner et al. 1997) which highlights at least 10 alternatives and nutritional value and yields as alternative FUs in horticultural production (among five alternatives) (Martínez-Blanco et al. 2011). In the assessment of buildings, it was highlighted that by choosing a subsystem definition of the FU, vastly different results on the importance of operational and constructional energy can be identified (Frijia et al. 2011). It has further been explained that in the case of bioenergy pathways, the choice of preferred pathway can be changed by changing the FU (Choudhary et al. 2014).

Among the key issues described in the selection of the FU are lifetime, performance, and system dependency (Cooper 2003). It was confirmed that the selection of an appropriate FU had not been standardized effectively, and the authors offered a useful guide consisting of four steps for specifying the FU and reference flows, of which the second point specifies the FU (description paraphrased here; for full details, see Cooper 2003): specification of the functional unit—FU should include (a) the magnitude of the service, (b) the duration of the service including the product's lifespan, and (c) the expected level of quality. Methods for determining various economically and environmentally focused FUs for manufacturing systems are described by Plehn et al. (2012). Particularly relevant to the context of decision making described in this chapter is that they incorporate stakeholder opinion in the selection of certain FUs.

Decision making using LCA as a tool has been quite widely discussed; however, in general, the topic of the FU is not the focus of these studies. Rather, such studies discuss the inclusion of LCA as a tool to assist decision making (Stewart 1999; Seppälä et al. 2001; Dante et al. 2002; Benetto et al. 2007). The approach in the current study parallels this to some extent, although not aiming to design the framework in which to seat the CF, but rather to make the CF a more useful component of decision making, by reflecting on the necessary configuration of the FU that makes most sense to the final end user. The following sections will outline why the selection of the FU is important and how appropriate FUs might be selected in order to fit with the needs of decision making.

2.2 IMPACTS OF THE FU

The selection of an FU that is appropriate to the end user is a topic best demonstrated using a number of examples. This section provides two examples to demonstrate the importance of selecting an appropriate FU. The first case study is based on the assessment of the CF of coffee products, as described in detail by Hassard et al. (2014), whereas the second is drawn from a study of the comparative CFs of geopolymer cements and ordinary Portland cement (OPC) (McLellan et al. 2011). Section 2.3 discusses the general considerations for alternative product types and decision maker categories and provides a further illustration of the potential impacts of choosing the wrong or less-than-optimal FU.

2.2.1 Case Study 1: Coffee

In the first case study, the aim of the assessment was to identify the PCF of alternative coffee products available on or around Kyoto University's main campus. The assessment described the results on the basis of three alternative FUs: (a) per volume of the product; (b) per gram of roasted coffee utilized; and (c) per serve (where each product was considered as a unit regardless of the total volume or mass of coffee utilized). Each of these FUs thereby produced alternative information (as shown in Figure 2.1).

A quick analysis of these graphs indicates that the ranking of each product is different according to the FU chosen. None of these FUs can be argued to be theoretically incorrect. However, the different FUs give different information and therefore are more or less appropriate in answering

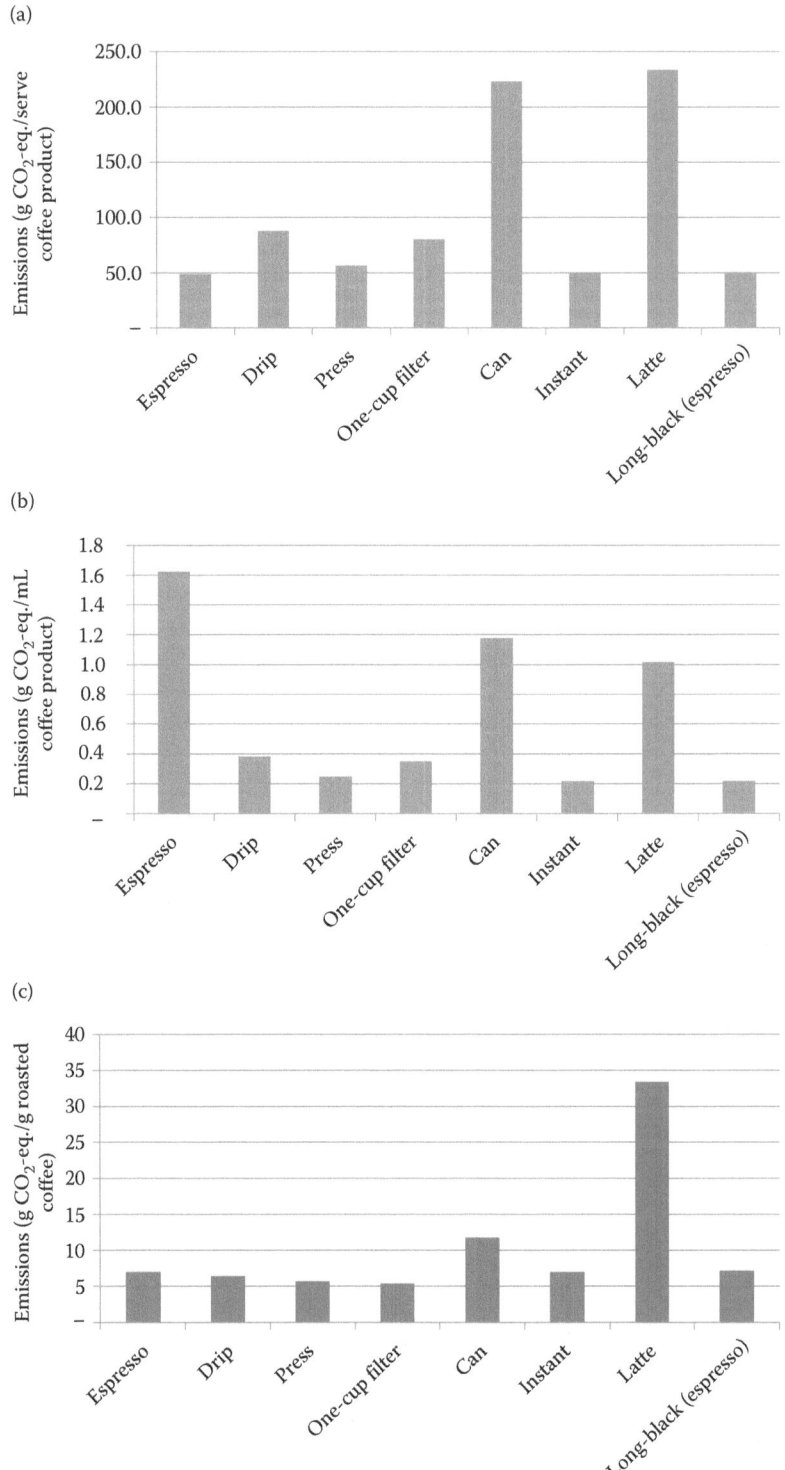

FIGURE 2.1 Overall PCF across alternative coffee products and FUs: (a) per serve, (b) per mL of coffee product, and (c) per gram of roast coffee.

alternative questions about the products and the CF thereof. In case (a), the PCF on a volumetric unit can be used by a consumer or a product designer wanting to have the lowest footprint from a set or standardized volume. This may be the case if a company uses set-sized cups into which alternative products can be poured. For the case of a mass of roasted material (b), the information could be utilized to identify the best way to process a set amount of coffee through to a low-impact final product. In this case, the addition of milk could be a differentiating element that would otherwise not be considered if this FU was chosen from the start. Finally, case (c), which was the primary FU in the original study, enables a consumer with a variety of alternative volumes and coffee-intensities of products to be able to choose the one with the lowest PCF overall. This is a logical PCF for labeling a consumer product, but could also be disregarded in many cases due to the discrepancy in mass and volume of the products.

The comparison between the final and the first two of these FUs also brings us the important and persistent argument of whether specific (in the sense of per mass or per volume) emissions are better to be reported, or conversely, whether absolute figures are better. With regard to the overall picture of an individual or industry's carbon budget, the absolute figure gives us important information that addresses what is ultimately important in the global warming argument—what the total greenhouse gas emissions are. In the specific case, as is often reported in industries keen to both grow in production and be seen to be reducing their environmental impact, the emissions per unit may be lower due to efficiency improvements, but the overall figure may still grow if production output is raised in the process.

2.2.2 CASE STUDY 2: CEMENT AND CONCRETE

The second case study discusses the life cycle of cement and concrete—globally one of the most significant minerals with regard to emissions (McLellan et al. 2012). The production chain from input materials through to concrete is shown in Figure 2.2. Limestone and a number of minor components are mixed, calcined, and ground to produce clinker (FU1). The clinker is mixed with a number of dry admixtures, which may or may not include supplementary cementitious materials (SCMs) such as fly ash and ground granulated blast furnace slag. This product is known as cement (FU2); in its purer form, OPC. The cement is not ready to bind material until it is wet, hence the addition of water to create the binder—this is an intermediary not always considered, but is useful in the comparison with the geopolymer, in which the binder is almost always wet to begin with (thus

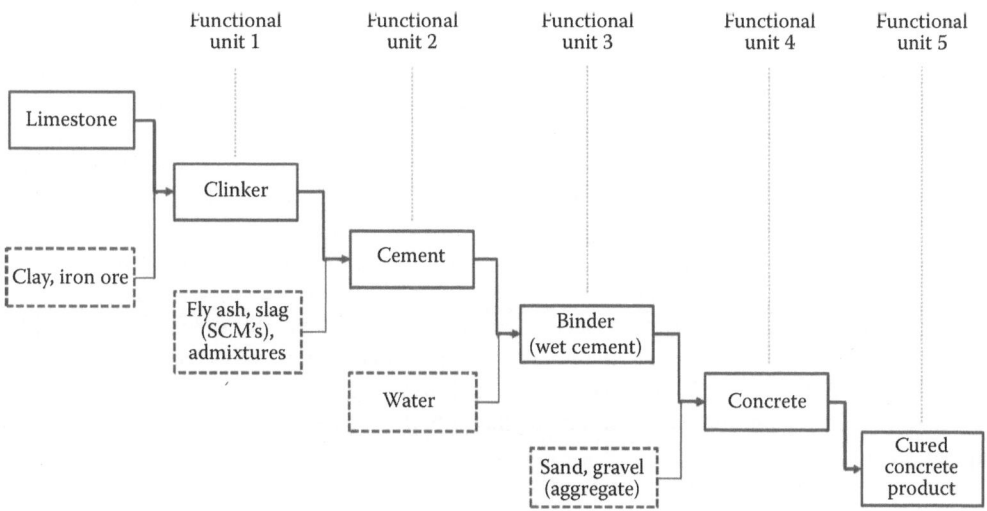

FIGURE 2.2 General concrete production cycle showing alternative FUs.

enabling the inclusion of the added water component of the wet cement binder to be closer to a like-for-like comparison). The wet binder is FU3, which is then mixed with large quantities of aggregate (sand and gravel) to produce a concrete mixture which, on a wet volume basis, can be FU4. Wet concrete mixes are somewhat predictable in their final properties, if given proper treatment, so the FU at this point is already quite well specified. However, if we wish to be more specific, we can utilize a fifth functional unit FU5, specifying the specific strength properties and potentially specifying the shape, dimensions, and other properties of a final cured concrete product (beam, slab, statue, etc.). Curing processes can vary, so there is some value in differentiating at the final stage to look at the cured product. Within these specific FUs, the method of defining (on a mass, volume, absolute product basis) can vary, so a few specific examples are shown here.

In the second case study, three of these alternative FUs are utilized: (a) binder (the geopolymer or OPC paste ready to be mixed) (FU3), (b) 1 ton of concrete (FU4 or FU5), and (c) 1 cm^3 of concrete (FU4 or FU5). By way of further explanation for those unfamiliar with concrete, regular concrete is made up of a mixture of sand, crushed rock or gravel (aggregate), cement, water, and a series of admixtures which aid in promoting certain properties in the concrete, such as fast-setting, workability of the wet concrete, higher strength, and so on. The overwhelming majority of concrete is aggregate material (around 80% by mass), whereas the important binding components (the binder) only make up a small percentage of the overall mass. With regard to the final product, and without consideration of some of the important properties attributable to admixtures, the most important performance characteristic is generally the compressive strength.

A comparison of alternative binders could be made either on a mass versus mass basis—where 1 ton or 1 kg of a binder is the FU—or on the equivalent mass required to obtain the same or equivalent properties in the final product (same compressive strength). (In some cases, it might require much more of one binder or the other to achieve the same functionality in the final concrete.) When cement production is being considered, the sectoral guidance specifies the emissions with regard to the production rate of cement at a plant (tons per year). The emissions are calculated based on the cement production rate, the ratio of clinker to cement, and the fraction of $CaCO_3$ in the final product. These same characteristics could be applied to the selection of an FU for comparing cements as a dry binder, but it is more useful perhaps to identify the FU as a ton of cement, which may include varying levels of admixtures and SCMs such as fly ash or slag, but still give equivalent performance to unblended cement (O'Brien et al. 2009). It should be noted that this type of FU may also be deemed to be a "reference flow," where the FU refers to an intermediate product.

When considering the overall final product, the comparison of equivalent strength concrete products is generally undertaken using the industry standard FU of 1 cm^3 of concrete (assuming the compressive strength to be equal). It is important to note that this FU has an important practical linkage to industrial application—the volume of concrete required is readily measured and important to estimate for construction. In traditional applications of concrete, it is not the mass of concrete that is considered important, but the strength and volume. The validity of this FU is also evident when we consider that the aggregate density is likely to play the largest role in determining the mass of concrete (due to its large proportion in the mix), whereas the importance of aggregate density to the concrete properties is not nearly so significant.

Though not necessarily included explicitly in the FU, and more likely in the definition of the scope of a CF assessment, the geographical location of delivery of the product (relative to the feedstock materials) is an important aspect to consider. This is typically defined as being FOB (free on board) at the port of export or import, or at the "plant gate." In the second case study, described in McLellan et al. (2011), this point is made very clear, as the inclusion of transportation impacts makes a significant difference to the potential outcomes and interpretation of the footprint results.

The two case studies shown in this section have illustrated the importance of selecting an appropriate FU. They have shown that the results of a comparison can vary significantly—particularly with regard to best and worst performing options—when an alternative FU is specified. Importantly, these studies shown that the selection of FUs that make sense from the end user perspective is vitally important and that the industry or specified standard, though sometimes sensible (case study 2), may at other times be inappropriate in cases where it does not represent the real choice available to the user (case study 1).

2.3   FUs FOR POLICY AND DECISION MAKING

The case studies in the previous section make it clear that FUs need to be selected carefully to purvey information accurately and in the most useful form for the end user or audience. This section examines how the FU can be selected appropriately for the purpose of decision making by consumers and in policy making.

The important question to be asked in this chapter then is "How do we select a functional unit that is 'fit-for-purpose'?" In order to answer this question, we must first consider what the different requirements of each of the key end user groups might be, and from there determine what FU might provide for these requirements appropriately. In this chapter, we will consider three alternative mainstream purposes for undertaking carbon footprinting and the needs of the relevant stakeholders. The three purposes to be considered are:

1. Policy making
2. Consumer products
3. Design and industrial operations

This list of purposes is by no means complete, but FUs for most other purposes could be readily adapted from these. Moreover, the discussion presented here is deliberately general in nature, as there are many fields and subsectors in which these broad categories are relevant, and it is anticipated that the FU should be tuned more specifically when the actual application is known.

Importantly, the connection of the end user to the FU requires a number of key elements that should be included in the selection and specification process. Firstly, it is important that the FU be able to be specified, and preferably quantified, in a manner that the end user, the study manager, and the final audience of any report or action can readily understand and come to a consensus on. Secondly, the FU performance characteristics should be specified at a suitable level, so as to enable comparison on the basis of service delivery without undue over-specification which may lead to deliberate elimination of alternatives or excessive focus on nonmaterial aspects of the life cycle with regard to the CF (e.g., if two cars offer the same functionality apart from the specific shade of yellow that they are available in, then the FU does not necessarily need to include the color). On the other hand, the FU must include essential functions that are required for equitable comparison (e.g., the number of passengers that can legally travel in the car at one time). With these general aspects in place, the FU should then be adjusted to fit the purpose of the assessment or decision being made, with care being taken that this does not result in important losses of information or transparency.

2.3.1   POLICY MAKING

The first area of application is policy making. This is considered in the sense of organizations and governments needing to make broad-coverage policies—most likely in this case to be referring to the reduction of the organizational CF. Such policies almost inevitably refer to targets of overall emissions and penalties or benefits that will be applied to assist in policy-fulfillment through enforced or encouraged behavioral change. This therefore leads to the consideration of two key types of FU—the FU at the target level and the FU at the level of the operations, flows, or products to which penalties and benefits are applied (and which ultimately should sum to the overall target measure).

2.3.1.1   The Target Level: Organization as an FU

The FU at the organizational level will typically be the organization as a whole, including its respective parts and functions, but generally represented by a "black" or "grey" box model in which the fundamental constituents are aggregated so as to allow meaningful representation and data management—in the political arena, this overall FU may be the municipality, state or nation, and so on, with neighborhoods, economic subsectors, or statistical districts as potential operational levels within the FU. The selection of this FU typically leads to an absolute overall CF that can be compared with other countries, cities, states, companies on the basis of total impact. Such a comparison of absolutes will often be criticized on the basis of not presenting a fair comparison or compensation for the differences in the size and make-up of the units being compared (population, geography, economy, etc.). This is one of the sticking points of global negotiations on climate change, which has in part led to the various Annexes and discrepancies in national targets under the international agreements on climate change (Kyoto Protocol) (UNFCCC 2006).

An often-utilized alternative for the absolute CF representation of the FU then is a specific value, representing the ratio of two elements considered to be important in leveling the comparison of competing units. This is most commonly expressed at the national level in indicators such as the emissions per unit of gross domestic product ($tCO_{2-eq.}/\$GDP$) or emissions per capita ($tCO_{2-eq.}/$person). Although still representing the CF of the FU, these relative indicators are not the FU *per se*. It is important to mention here however that as the argument for either relative or absolute reporting can be made, the selection of one type may leave room for reporting of only favorable trends. (For example, a large growth in GDP can decrease the specific emissions per unit GDP while increasing the total value. Inversely, the absolute emissions could be stable while GDP decreases, thus hiding a decrease in efficiency.) In standard sustainability reporting frameworks such as the Global Reporting Initiative (GRI) (GRI 2006), organizations are encouraged to report on both an absolute and relative basis. This approach gives appropriate data for both types of comparison to be made, while also reducing the likelihood of adverse trends being hidden by reporting on only one of these bases.

2.3.1.2   The Level of Inducements—Actions and Accounts as FUs

The component flows and units within the organization should add ultimately to give the organizational level FU (in the absolute case). However, it is often these flows and subunits which become the focus of legislation to regulate performance through the attribution of inducements (penalties or benefits) that accrue to either improvement or failure to comply. It is important from the policy perspective that the FU at which this performance is measured is appropriately considered.

Some examples of these units may be seen in the various carbon pricing schemes and their compensation schemes globally. For example, "households" become a minimum unit for many policies—households are compensated (whether equitably or not is another argument) on the basis of being a household, although this is often disaggregated by the character of the FU (i.e., how many people in the household, whether they own or rent their home, etc.). However, for most policy targets, the FU of the household does not give an appropriate unit with regard to the penalization and requirement of behavior change. In this case, it is often most appropriate to consider the actual physical flows that are the cause of the national CF, and the typical surrogate for these which is the price of supply or the account.

Like the FU of the household, the account (and for the sake of example, the electricity bill will be used) can be considered as an FU and penalties or inducements applied to it. A carbon tax per account, however, is likely to result in inequitable distribution of either inducement, as the users of that account are variable. However, applying to the unit of electricity (the physical or the financial) is a unit that enables scalable—and therefore hopefully more equitable (on a user-pays system) than the account FU. The flow in this case is also a useful unit for attempting to induce behavior change, as it is readily measurable and the consumer can take steps to improve their performance in order to best obtain the benefits or avoid the punishment attributable to the policy.

2.3.2 Consumer Products

Consumer products and consumer decision making are the focus of much of the carbon footprinting activity in the world. Whether government-prescribed, industry market-advantage-driven, or consumer-led, CF labeling schemes and other environmental information activities are largely focused on providing consumers with information upon which to make their purchasing choices. Consumer products vary widely across the many areas of life and are often designed with market placement in mind, rather than attempting to provide uniformity of FU characteristics by which the consumer can select. This has implications for FU selection, in that care must be taken to select according to both the specific product and the market in which it is sold or compared. In such business-to-consumer contexts, the onus is on the business to identify and understand the needs of the market and to balance the PCF expression and therefore the FU with the other needs of the consumer. This may in some cases imply a need to change the basis upon which other data (e.g., nutritional data) is expressed, or vice versa, to make the PCF align with existing information supplied.

As with the example of coffee earlier, it is important in such environmental-labeling schemes to enable the consumer to make an informed choice based on the function unit that is presented to them in the real world. In most schemes, this means presenting the overall CF of the product or pack as sold on the shelf, as well as a suggested serving or standardized unit. With durable products, the FU may also need to be clarified with a lifetime or testing procedure performance figure, such as is common with motor vehicles. Three types of consumer products will therefore be considered in this section: consumables (single and multiple item or serve) and durables.

Before entering into a detailed discussion of the FUs for each of these consumer product types, it is useful to consider some of the barriers and challenges to using and promoting a PCF as a tool for consumer decision making by consumers.

One of the foremost challenges in encouraging consumers to change behavior or product selection based on the PCF is the impact of competing priorities. For example, in Table 2.1, the relative pair-wise comparison of three priorities in the selection of coffee products is shown, namely, cost, taste, and the environment. The percentage indicates the percentage of respondents prioritizing the cost or taste of a coffee product in comparison with the environmental impact of that product.

It is apparent from Table 2.1 that the majority of respondents considered taste and cost to be more important in the selection of a coffee product than the environment. Table 2.2 further supports this case, indicating that the environmental choice of alternative coffee products is far outranked by other issues. In the same study, only 29% of the respondents indicated that knowing the PCF of the product would influence them to change to an alternative product, with 37% unsure and 22% sure that they would not. This indicates one of the major hurdles that needs to be overcome—other priorities are considered higher than the PCF in product choice (at least in the food and beverage sector in Kyoto—other sectors and other countries would likely have different priority levels). However, there is a large proportion of uncertain consumers who may be able to be swayed to change their preference if a convincing or attractive environmental argument can be presented.

TABLE 2.1

Priorities of Coffee Drinkers (Showing the Majority Response for Each Comparison) (66 Respondents at Kyoto University, July 2014)

	Percentage of Respondents		
Criteria	Cost or Taste	Neutral	Environment
Prioritize cost over environment	52	21	27
Prioritize taste over environment	48	27	24

TABLE 2.2

Criteria Ranked by the Respondents (Given Up to Three Choices—Many Chose Just One) (66 Respondents at Kyoto University, July 2014)

Criteria	Ranking of the Criteria (No. of Responses in Each Rank)		
	1st	2nd	3rd
I like the taste	36	3	8
It is convenient (to obtain)	22	16	4
It is fast	3	7	5
It costs the least	2	6	10
It is the best for my health	2	3	3
I think it is the best environmental choice	1	1	4
My friends drink the same	1	1	3
I like the atmosphere where I drink it	0	4	5

2.3.2.1   Single-Serve or Single-Item Consumable Products

In the case of single-serve or single-item consumables, it may appear that the FU is relatively straight forward to select. Certainly, if there is only one item, then the obvious FU must be that single item in the form in which it is available to the consumer. However, as shown with the coffee example above, it is often the case that two alternative products are offered in different sizes, packaging alternatives, hot or cold, and so on. In this event, it is often useful to adopt alternative FUs or to specify the FU with a variety of criteria to add further information to the study and allow comparison. Considering the underlying tenet of selecting the FU that enables the consumer to choose based on the product offerings in front of them however, in the case of a single product such additional FUs will either be confusing or of little consequence, as the minimal purchase is a single item, and its usage, consumption or wastage is at the will of the consumer. The FU of a standard cup of coffee or a standard volume of wine can be interesting, but if the available size is not standard, then the absolute PCF for the product at hand is most useful. All other FUs become academic in this sense.

2.3.2.2   Multiserve or Multi-Item Consumable Products

In the case of multiserve or multi-item consumables, there is more leeway to select appropriate FUs. With regard to consumer choice in the food industry, this is already established for individual producers—offering serving size and respective vitamin and energy content figures is commonplace, although the serving size is not always standardized across like products. In such cases, it is useful to offer an expression of the FU as a volumetric or mass-based unit—preferably one that can be readily compared (e.g., 100 not 147 mL). In all cases, it is useful to have alternative FUs—on both the multipack (minimal consumer purchase) level, and as an expression of the PCF attributed or distributed to the individual items or serve. In the course of calculating the PCF, it is both likely and important (especially where packaging and transportation logistics influence the result) that the FU for the study be initially the product or package as a whole, with the total PCF then being divided into an appropriate serve, mass, or volume-based FU.

There are three likely configurations of multiserve products: packs consisting of a number of discrete serves which are relatively indivisible and uniform (e.g., individually wrapped candy bars or pencils in a larger pack); discrete serves which may be naturally variable or nonuniform (e.g., apples in a crate); or, relatively continuous materials that can be readily distributed into serves or amounts of entirely flexible quantity (e.g., milk in a carton). In the first case, we can treat the individual serves or items as single-serve products (except that we must take into account the additional packing associated with containing the multiserve item). In the third case (continuous materials), the FU may be expressed as a standard serve or might be better expressed over an easily calculable volume as described above. The value of a volume or mass expression to the consumer is most useful if all the competing products are expressed in the same unit, and the consumer then requires less effort to calculate and compare alternatives (this is made difficult with nonstandard serving expressions). In the third case, the FU of a serving size may become more difficult. If a crate contains a uniform 5 kg of nonuniformly sized apples, for example, then to express the FU as one apple requires at least some level of probabilistic distribution of the PCF among the potential apple sizes that could be expected in a sample (and perhaps expressing as an "average apple"). Distributing the total PCF among all the apples on the assumption of uniformity or the average number in the box would certainly lead to error in the attribution. Alternatively, expressing the FU as 1 or 10 g of apple is generally only as useful as the overall PCF for the full crate—the consumer is not likely to be weighing their apples individually before they eat them. It may of course be useful for comparison across alternative producers or alternative products that could be purchased instead (e.g., apples vs. oranges). So, in the case of discrete, nonuniform multi-item packs, it may be of most use to retain the FU as the multi-item pack, but the choice must be made with due consideration to the specifics of the product.

2.3.2.3 Durable Products

Durable products can largely be considered in a similar manner to single-item consumables—typically they are indivisible, and the PCF must be based largely on this. However, with durable products, the FU must represent the service provided by the product, not just the item itself. It is important to note that with durable products, the lifetime, usage rate, maintenance, and the end-of-life treatment of the product must be adequately considered in specifying the FU and scope.

When the FU is being specified, considering the lifetime of a product is important, although alternative products may have different lifetimes, which requires some careful consideration as in practice it is not reasonable to apply the same lifetime if products are known to deteriorate at different rates. Likewise, the usage rate will affect the lifetime in many cases, and, in conjunction with the rate of maintenance, will have a large impact on the PCF attributed to the subsequent service of a durable product. Maintenance considerations should also factor-in the difference in lifetime of different parts of a product, some of which may need replacing before the effective lifetime of the product is completed.

Particularly, when it comes to multifunction items, the allocation of services or distinguishing between single-function and multifunction mode should be considered. For example, the treatment of a multifunction printer for comparison with a fax machine should consider carefully how to allocate the PCF, and whether in fact the FU should not be focused on the machine itself, but on the levelized lifetime emissions per transferred page. The number of multifunction items has been growing significantly—and there is much convergence between, for instance, smartphones, laptop computers, cameras, and GPS systems (among other electronics)—which makes this concept of greater importance. Arguments have been made that "the functional unit requires equality of all benefits of the compared systems" (Fleischer and Schmidt 1996)—which is allowed for in the ISO standards (see below), but not required if the exclusion of particular system functions can be adequately explained for a specific study. In the cited example of natural raw materials, the authors identified that without accounting for all of the functions of a forest (not just as a source of physical materials—wood) that the optimal environmental decision may not be made (Fleischer and Schmidt 1996). The ISO14067 states that "If additional functions of any of the systems are not taken into account in the comparison of FUs, then these omissions shall be explained and documented. As an alternative approach, systems associated with the delivery of these functions may be added to the boundary of the other system to make the systems more comparable" (Standardization IOF 2013).

Thus, in the realm of durable products, it is most useful for the consumer to be given an FU that expresses the PCF with regard to the service expected to be delivered. For example, the PCF of a vehicle, comparing a two-seater sports car with a family sedan, may not be seen as a comparable FU—as a product itself (not considering the usage of the item, but solely the manufacturing), then the FU of "one car" may be applicable, but with regard to the service provided, the FU of "one person-kilometer" is perhaps more appropriate and fair. It may be argued of course that the consumer is not likely to be considering these two items—that they are in different markets—but even within the same market, the fuel efficiency is a clear example of the use of a service-based FU for comparison.

2.3.3 Design and Industrial Operations

Design—both of products and of processes or operations—has been increasingly considering the environmental and broader sustainability footprint, and how to incorporate these into the design process itself (McLellan et al. 2009). In particular, LCA has been applied as one tool in this endeavor, and the PCF can be considered as a subset of LCA. When considered from the point of view of the designer of products or processes, we will consider the inputs that they require, rather than considering the product output explicitly. The reasoning behind this is that the designer wants to be able to accurately, rapidly, and cheaply understand the impact of their design choices on the PCF of the product that they are designing. Furthermore, considering the FU of the final product leads to a consideration of the provided service or the durable product produced as per the above considerations.

For example, in the design of a chair, the designer may wish to use a variety of materials and forms, and these may change more based on esthetics of the final design than on the inherent properties of the materials themselves. However, with a known set of alternative materials to fill the final design properties, the PCF of those materials may come into play. There are some excellent analyses and a broad literature (both academic and applied) on material selection (Ashby 2012). In the case of the chair design, the designer is most likely to seek to know the PCF of a part (e.g., a screw) or a certain volume or mass of material (e.g., steel). A database of PCFs of parts may be readily available and these will largely be in the category of the durable product above. Equally, or sometimes more useful for the designer is the FU that describes a PCF per unit mass or volume, as this can be applied to various shapes and sizes of alternative designs with relative ease. This is also the case with other supply chains—representative of many business-to-business operations, in which components or materials may be supplied rather than the final product.

With regard to industrial operations, the design process is relatively similar. As products become larger (to the extent of the whole of an industrial operation, for instance), the importance of including the other material properties in the FU of a part or material becomes greater. For example, comparing concrete beams with steel beams for the construction of a factory requires that the steel and concrete present the same strength and durability fundamentally, so the FU (as with the concrete example in Section 2.2) must represent the service provided. The designer can calculate the PCF for plant components on their own if given an appropriate database of materials, but in many cases it would be more useful to have the component PCF in place.

2.3.4 Some FUs Applied in Industry

In association with general guidelines for carbon footprinting, which offer a broad guidance on the selection and specification of FUs, some of the sectoral guidance documents for specific industries supply useful FU examples. For example, in the draft ICT sector guidance from the GHG Protocol, the following examples are given for software and hardware:

"Completing a definable task (human or machine driven) via the use of software, where the task can be defined by its input, processing and output commands. For example this unit could be used to assess 'the energy consumption of using a word processing program for 1 hour'" (WBCSD and WRI 2013b).

"Typically, the functional unit needs to clearly define the magnitude of the ICT hardware's duty or service, the duration of its duty or service life under assessment, and the expected level of quality" (WBCSD and WRI 2013b). They give the example of a wireless router, as described in Table 2.3.

TABLE 2.3
Example FU Specification (after WBCSD and WRI 2013b)

Item	Magnitude	Duration	Quality
Wireless router	Wireless data connection with 2 antennas Data routing at 2.4 GHz 4 Ethernet ports each at 10/100	Mbps 5-year lifetime	Wireless data transfer specification per IEEE 802.11n

For the case of software, the service element of the FU is clearly important, as the software itself does not accrue a CF except by virtue of the media in which it is delivered and the utilization of the computer itself.

In other industries, PCRs may specify an appropriate FU. Many PCRs are being developed, particularly, in the move toward environmental labeling. Considering the end user's decision-making context, these are not always detailed and specific enough, particularly when they are applied to general product categories such as base metals (mass or area) or "wild caught fish" (1 kg of fish), which may often be considered as intermediate products (perhaps more useful to supply chain management than end users).

2.3.5 INAPPROPRIATE ASSIGNATION OF THE FU

It is readily perceivable that inappropriate or nonideal FUs can be selected, as shown in the example of cement and concrete, multiple FUs are possible, and the unit appropriate to the end user is not necessarily the one that is chosen to study. The implications of choosing an inappropriate FU can range from the CF being difficult to apply through to the study results being rendered entirely useless. In some cases, the inappropriate FU can cause poor decision making to ensue. Let us examine a few clear examples.

For the first example, consider the context of a "well-to-wheel" CF for alternatively fuelled vehicles—hydrogen fuel cell vehicles versus standard gasoline vehicles. The FU to be chosen in this case must represent the appropriate service—the delivery of a given number of passengers, over a given distance or range, possibly within a given timeframe, over a predetermined life of vehicle. An appropriate FU, having taken these into account, might then come to a reference unit of passenger-kilometers (standard use within transport industry). If the initial FU does not adequately account for the variation in cars (hydrogen vehicles may not have the same range or space for passengers), then applying a unit of vehicle-kilometers (as in some studies) will not give us an accurate outcome to compare and make a decision upon. (If a two-person hydrogen vehicle with a range of 300 km and a lifetime of 5 years is compared with a four-person gasoline vehicle with a range of 600 km and a lifetime of 10 years, then apart from all other factors, two hydrogen vehicles are needed to deliver the same number of people and they must be replaced once in the period, leading to ultimately four hydrogen vehicles being necessary to equate to one gasoline vehicle. On the other hand, if we assume the average passenger numbers per car trip in countries such as Australia might be 1–2 people, and the average trip is below 300 km, then the unit of vehicle kilometers may be appropriate.)

Returning to the example of coffee described in Section 2.2, the consumer standing ready to purchase a coffee product is not likely to be concerned with the volumetric or mass-based FU, rather considering only the overall per-purchasable-product (PPP) CF as relevant. If the CF is only provided on one of the other bases, then the time required to calculate the PPP-CF will be too long for most consumers—in short, the information and its reporting will have been wasted. Using a volumetric basis could also be misconstrued by companies trying to upscale the size of a can (from 190 to 200 mL, for example), and they may utilize the FU as a direct multiplier, not taking into account the change in mass of can material (in which case, the result will be an incorrect CF).

In the case of a multiservice or multifunction product, taking only the functions that are the focus of a study for the comparison (and leaving out noncore or additional functions) can lead to an undervaluing of a multifunction product against a single-function product. This was argued earlier on with regard to natural materials, but the alternatives of justifying the omission or adding the alternative functions to a single-function product have limitations to their validity and their applicability.

2.4 CONCLUSIONS

This chapter has sought to offer a reminder of the importance of choosing an appropriate FU when undertaking a carbon footprinting study, not just from the point of view of theoretical consistency, but also from the perspective of the final PCF audience. Decision making using the results of a PCF study of alternatives requires an FU that encapsulates the essential function of the products at a scale that cannot be further diminished without losing function. Selection of the FU has its challenges, particularly when multiproduct or multiservice products are considered, and the importance of fundamentally achieving comparability must remain the primary consideration. However, when the decision maker is considered as well as the theoretical validity of the FU, the context of such comparability is expanded. Moreover, although some FUs might fulfill this for certain purposes, for other purposes, an alternative FU—even one that is derived from the same data set—might be more appropriate.

The chapter has outlined the general requirements for selecting FUs to meet the decision making needs of three types of decision-making processes:

1. Policy making
2. Consumer products
3. Design and industrial operations

These decision-making strands can also be considered from the perspective of alternative user-types, for example, policy making has both legislative or operationalizing users as well as the final consumer on which it impacts. Design and industrial operations may also be considered from the perspective of the designer or the client. Although ideally a consultative process would be used to identify the most appropriate FU and the most useful expression of the results, even an explicit consideration of the end user requirements by the CF developer should improve the usability of the results.

The importance of making CF studies fit-for-purpose is vital in order to enable uptake in decision making. Particularly, in the context of competing priorities (time, money, convenience) and competing information sources, the end user is likely to disregard information that is provided on an inconvenient study basis. Also of vital importance to making the best environmental choices is that the FU accurately reflect the products for comparison.

REFERENCES

Ashby MF. 2012. *Materials and the Environment: Eco-Informed Material Choice*, 2nd edition. Oxford: Elsevier, 628 pp.

Benetto E, Tiruta-Barna L, Perrodin Y. 2007. Combining lifecycle and risk assessments of mineral waste reuse scenarios for decision making support. *Environmental Impact Assessment Review* 27(3):266–285.

BSI. 2011. *Publicly Available Specification: PAS 2050: Specification for the Assessment of the Life Cycle Greenhouse Gas Emissions of Goods and Services*. London: British Standards Institute, 38 pp.

Choudhary S et al. 2014. Reference and functional unit can change bioenergy pathway choices. *The International Journal of Life Cycle Assessment* 19(4):796–805.

Cooper J. 2003. Specifying functional units and reference flows for comparable alternatives. *The International Journal of Life Cycle Assessment* 8(6):337–349.

Dante RC et al. 2002. Life cycle analysis of hydrogen fuel: A methodology for a strategic approach of decision making. *International Journal of Hydrogen Energy* 27(2):131–133.

Finkbeiner M, Hoffmann E, Kreisel G. 1997. Environmental auditing: The functional unit in the life cycle inventory analysis of degreasing processes in the metal-processing industry. *Environmental Management* 21(4):635–642.

Fleischer G, Schmidt W-P. 1996. Functional unit for systems using natural raw materials. *The International Journal of Life Cycle Assessment* 1(1):23–27.

Frijia S, Guhathakurta S, Williams E. 2011. Functional unit, technological dynamics, and scaling properties for the life cycle energy of residences. *Environmental Science Technology* 46(3):1782–1788.

GRI. 2006. *Sustainability Reporting Guidelines*. Amsterdam, the Netherlands: Global Reporting Initiative.

Hassard HA et al. 2014. Product carbon footprint and energy analysis of alternative coffee products in Japan. *Journal of Cleaner Production* 73(0):310–321.

Martínez-Blanco J et al. 2011. Comparing nutritional value and yield as functional units in the environmental assessment of horticultural production with organic or mineral fertilization. *The International Journal of Life Cycle Assessment* 16(1):12–26.

McLellan BC et al. 2009. Incorporating sustainable development in the design of mineral processing operations—Review and analysis of current approaches. *Journal of Cleaner Production* 17(16):1414–1425.

McLellan BC et al. 2011. Costs and carbon emissions for geopolymer pastes in comparison to ordinary Portland cement. *Journal of Cleaner Production* 19(9–10):1080–1090.

McLellan BC et al. 2012. Renewable energy in the minerals industry: A review of global potential. *Journal of Cleaner Production* 32(0):32–44.

O'Brien K, Ménaché J, O'Moore L. 2009. Impact of fly ash content and fly ash transportation distance on embodied greenhouse gas emissions and water consumption in concrete. *The International Journal of Life Cycle Assessment* 14(7):621–629.

Plehn J et al. 2012. A method for determining a functional unit to measure environmental performance in manufacturing systems. *CIRP Annals—Manufacturing Technology* 61(1):415–418.

Seppälä J, Basson L, Norris GA. 2001. Decision analysis frameworks for life-cycle impact assessment. *Journal of Industrial Ecology* 5(4):45–68.

Standardization IOF. 2013. *Greenhouse Gases—Carbon Footprint of Products—Requirements and Guidelines for Quantification and Communication*. Geneva, Switzerland: International Organization for Standardization.

Stewart M. 1999. Environmental life cycle considerations for design related decision making in minerals processing. In *Department of Chemical Engineering*. Cape Town: University of Cape Town.

UNFCCC. 2006. *United Nations Framework Convention on Climate Change*. Available from http://unfccc. int/resource/docs/convkp/kpeng.pdf (accessed May 2, 2006).

WBCSD and WRI. 2013a. The Greenhouse Gas Protocol. In: *A Corporate Accounting and Reporting Standard*. Conches-Geneva, Switzerland: World Resources Institute and World Business Council for Sustainable Development.

WBCSD and WRI. 2013b. *The Greenhouse Gas Protocol: ICT Sector Guidance (Draft)*. Conches-Geneva, Switzerland: World Resources Institute and World Business Council for Sustainable Development.

3 Methodology for Carbon Footprint Calculation in Crop and Livestock Production

Kun Cheng, Ming Yan, Genxing Pan, Ting Luo, and Qian Yue

CONTENTS

3.1 Introduction 62
3.2 GHG Emission Sources 63
    3.2.1 System Boundaries 63
    3.2.2 GHG Emission Sources in Crop Production 63
        3.2.2.1 Direct Emission 63
        3.2.2.2 Indirect Emission 64
    3.2.3 GHG Emission Sources in Livestock Production 64
        3.2.3.1 Direct Emissions 64
        3.2.3.2 Indirect Emission 65
3.3 Carbon Footprint Calculation Methods 65
    3.3.1 Crop Production 65
        3.3.1.1 Direct Emissions 65
        3.3.1.2 Indirect Emissions 66
        3.3.1.3 CF Assessment 66
    3.3.2 Livestock Production 67
        3.3.2.1 Direct Emission 67
        3.3.2.2 Indirect Emissions 68
        3.3.2.3 Assessment of CF of Livestock Production 68
3.4 Data Sources 69
    3.4.1 Emission Factors 69
        3.4.1.1 Crop Production 69
        3.4.1.2 Livestock Production 71
    3.4.2 Activity Data 74
        3.4.2.1 Statistical Data 74
        3.4.2.2 Field Survey Data 74
3.5 Case Studies 75
    3.5.1 CF of Main Grain Crop Production in Shandong Province, China ... 75
        3.5.1.1 Scope and Objective 75
        3.5.1.2 Data Source 76
        3.5.1.3 CF Calculation 76
        3.5.1.4 Results 77
        3.5.1.5 Sensitivity Analysis 78
        3.5.1.6 Conclusions 79
    3.5.2 CF of Milk Production Based on a Site Survey in Sichuan Province, China 79
        3.5.2.1 Scope and Objective 79
        3.5.2.2 Data Source 79
        3.5.2.3 CF Calculation 79

        3.5.2.4 Results 81
        3.5.2.5 Sensitivity Analysis 81
        3.5.2.6 Conclusions 82
    3.5.3 Limitations and Recommendations 82
Acknowledgments 83
References 83

3.1 INTRODUCTION

The Fifth Assessment Report of IPCC (2013) disclosed that the global mean temperature increased by 0.85°C from 1880 to 2012. Specifically, the last three decades have been warmer at the Earth's surface than any preceding decade since 1850. Global greenhouse gas (GHG) emissions due to human activities have grown rapidly by 54% between 1990 and 2011 (IPCC 2013). Rapidly increasing anthropogenic GHG emissions are making a significant contribution to global climate change.

As a primary food producer, GHG emissions from agriculture contribute by 13% to the global GHG emissions and are likely to increase in the coming decades (Smith and Gregory 2013). Fifty-six percent of the global anthropogenic non-CO_2 emissions was derived from agriculture (U.S. EPA 2012). In addition, food production and consumption accounted for almost one-fifth of total anthropogenic GHG emissions globally (Smith et al. 2014). However, indirect emissions in the process of crop production relevant to the manufacture of agricultural chemicals and farm mechanical operations also made a significant contribution to global GHG emissions.

Livestock production is already known to contribute to GHG emissions. As reported by the Food and Agriculture Organization of the United Nations (FAO), GHG emissions induced by livestock production were estimated to be 7516 million metric tons per year of CO_2 equivalents (CO_2-eq.), accounting for 18% of annual worldwide GHG emissions. Livestock production emitted 103 million tons of methane in 2004 through enteric fermentation and manure management, equivalent to 2369 million tons of CO_2-eq. In China, a large livestock producer, GHG emissions from animal enteric fermentation and manure management have been estimated at 445 Tg CO_2-eq., accounting for 45.7% of the nation's total agricultural emissions in 2005 (NDRC 2012). However, the life cycle and supply chain of domesticated animals raised for food have been vastly underestimated as a source of GHGs and in fact account for at least half of all human-caused GHGs. The above-mentioned would easily qualify livestock for a hard look indeed in the search for ways to address climate change.

The population explosion and subsequently the growing demand for resources intensify the food and energy crises and also attract increasing attentions (Steinfeld et al. 2006; FAO 2009; Godfray et al. 2010). The production of crop and livestock has been operated with maximum resource input available to reach a maximum production, though the total cropland area has been decreasing and GHGs have been emitted increasingly since the end of twentieth century (Huang 2013).

Concerns about GHG emissions reduction to mitigate climate change and food security have recently inspired the assessment of the carbon footprint (CF) for various activities and products (Hertwich and Peters, 2009). CF, defined as a measure of the total amount of carbon dioxide emissions that is directly and indirectly caused by an activity or is emitted over the life cycles of a product, is analyzed to measure the climate change impact of products and activities in terms of the amount of GHGs emitted in a whole life cycle.

Growing interest in GHG mitigation in agriculture has provided a strong enticement to assess the CF of different agricultural production. Calculation of CF in agriculture identifies the contributions of agricultural production to climate change and the components of emission sources. Meanwhile, determining a CF of a certain agricultural product is useful for identifying how low-carbon economy could be implemented to abate climate change. In this chapter, a methodology is introduced to clarify how to calculate CF of crop and livestock production according to the CF definition and life-cycle assessment method. Then an assessment approach is supplied to evaluate the CFs under different scenarios and detect climate change mitigation strategies.

3.2 GHG EMISSION SOURCES

3.2.1 System Boundaries

The system boundary for CF calculation is the farm gate of agricultural production. For crop production, the system boundary is the land boundary of studied cropland. However, the system boundary for livestock production is the farm gate of studied livestock such as dairy, sheep, pig, and so on. All agricultural inputs are traced back to production and raw material extraction and all major GHG emissions (CH_4, N_2O, and fossil CO_2) associated with inputs and farm management are accounted for.

3.2.2 GHG Emission Sources in Crop Production

CF of crop production was generally assessed by taking into account all the GHG emissions caused by or associated with material used, and farm machine operated and irrigation and drainage power exhausted for crop production in a crop life cycle (Lal et al. 2004; Hillier et al. 2009; Cheng et al. 2011). During the full life-cycle analysis of crop production, the total carbon emissions were estimated both of the direct and indirect emissions within the farm gate from crop sowing to grain harvest under a single cropping system. Hence, identifying GHG emission sources during agricultural production is the first step to conduct a CF research.

3.2.2.1 Direct Emission

3.2.2.1.1 Nitrous Oxide Emissions from Soil

Nitrous oxide (N_2O) could absorb terrestrial thermal radiation and thus contributes to global warming of the atmosphere. On a mass basis, N_2O is approximately 265 times more potent than CO_2 in a 100-year time horizon (IPCC 2013). N_2O could be generated in the processes of nitrification and denitrification in soil. In most soils, an increase in available N could enhance the nitrification and denitrification rates which then increase the production of N_2O. Available N could be supplied through human-induced N additions or change of management practices. Approximately, 58% of global N_2O emissions were from agriculture in 2005 (Smith et al. 2007). Therefore, N_2O emissions from soil could be identified as a source of CF in crop production.

3.2.2.1.2 Methane Emissions from Flooded Rice Paddy

Atmospheric concentration of methane (CH_4) has increased by 1.5 times compared with preindustrial times, accounting for 18% of the global warming potential (GWP), which ranks CH_4 as having the second highest radiative forcing of the long-lived GHGs. Rice paddies, which are characterized by relatively high organic carbon levels and prolonged anaerobic conditions during rice growth, are one of the major anthropogenic sources of CH_4 accounting for almost 20% of agricultural CH_4 emission. Carbon input and anaerobic condition are the important factors to generate CH_4. Soil type, temperature, and rice cultivar could also affect CH_4 emissions. The two main pathways for CH_4 emissions from soil to the atmosphere are transfer through rice plants and ebullition in the water. CH_4 emission from flooded rice paddy is one of the important emission sources when we calculate CF in rice production.

3.2.2.1.3 Organic Carbon Stock Changes in Soil

According to IPCC (2007), the net flux of CO_2 in agriculture is estimated to be approximately balanced though large annual CO_2 exchanges occurring between the atmosphere and agricultural lands. Soil carbon sequestration, with an estimated 89% contribution to the technical potential, is one of the most important pathways to achieve GHG mitigation potential. Although CO_2 emits much from soil each year, carbon input could not be ignored in the calculation of CF. In general, organic carbon stock change in soil is a thoughtful way to reflect the net CO_2 flux in cropland. Actually, carbon stock change could not be measured in 1 year and there is also not an appropriate method to accurately calculate SOC stock change in 1 year. However, we usually assess CF for carbon production in a given year. Given these, SOC stock change is not taken into account CF calculation in this methodology.

3.2.2.1.4 *CO₂ Emissions from Machine Operation*

Farm machinery is used for spraying and tillage, harvesting, strapping, and transportation in the process of crop production. Energy, such as gasoline and diesel oil, is necessary for operating farm machinery. In this process, there is an amount of CO_2 emitted by energy consumed. CO_2 emission from machine operation should be taken into account CF calculation.

3.2.2.2 Indirect Emission

3.2.2.2.1 *GHG Emissions from Manufacturing of Agricultural Inputs*

Inputs of fertilizer, pesticide, herbicides, and plastic film are necessary to supply nutrient, protect plant health, and prevent water loss for crop production. However, there is a huge CO_2 emission when these agricultural materials are produced in the factories. These emissions occur outside the farm gate but are induced by crop production. Given these, the indirect emissions from manufacturing of agricultural materials should be thought over when assessing CF for crop production.

3.2.2.2.2 *GHG Emissions by Irrigation*

In most agricultural regions of the world, irrigation is necessary for water sully in plant growth except for rainfed cropland. Energy use for irrigation is also one of the main GHG emission sources in farming process. The main components of energy use associated with irrigation are related to processes which apply water to field by lifting, conveying or pressurizing it, and these processes could be powered by diesel or electricity (Wang et al. 2012). The emissions of GHG reasonably occur when diesel is used and electricity is generated.

3.2.3 GHG Emission Sources in Livestock Production

CF of livestock and poultry production could be generally assessed by taking into account all the GHG emissions caused by or associated with material used, farm management, and power exhausted for livestock and poultry production.

3.2.3.1 Direct Emissions

3.2.3.1.1 *Manure Treatment*

Both CH_4 and N_2O emissions occur from livestock manure management systems. Manure management account for 7% of total non-CO_2 emissions in agriculture (Smith et al. 2007). CH_4 could be emitted when confined manure management operations are conducted. However, N_2O emissions vary significantly between different manure treatment types and can also result in indirect emissions due to other forms of nitrogen loss from the system.

3.2.3.1.2 *Enteric Fermentation*

According to the FAO, 37% of human-induced methane comes from livestock production. Ruminant animals have a rumen which could produce CH_4 when microbial fermentation takes place. Cattle are an important CH_4 source in many countries due to their large population and high CH_4 emission rate by their ruminant digestive system. So emissions of CH_4 from enteric fermentation should be taken into account when CF of ruminant livestock production is estimated.

3.2.3.1.3 *Farm Management*

Machinery operation is needed especially in aggregated farm management. Machinery operation could exhaust energy such as gasoline and diesel oil which induce CO_2 emissions.

3.2.3.2 Indirect Emission

3.2.3.2.1 *CO$_2$ Emissions from Manufacturing of Agricultural Inputs*

Agricultural input includes forage, pesticides, detergents, medicines, and so on. GHG emissions occur by production and transport of these inputs. Specifically, the GHG emissions from forage input include crop cultivation, forage production, and forage transportation.

3.2.3.2.2 *CO$_2$ Emissions from Electricity Use*

Electricity is also used in livestock farms, and CO$_2$ emissions occur when electricity is generated. This is an indirect emission source in livestock production.

3.3 CARBON FOOTPRINT CALCULATION METHODS

3.3.1 CROP PRODUCTION

CF of crop production could be estimated by summarizing all the GHG emissions that is directly and indirectly caused by or associated with material used, farm management, and power exhausted over the life stages of crop production.

3.3.1.1 Direct Emissions

3.3.1.1.1 *Nitrous Oxide Emissions from Soil*

The direct N$_2$O emissions (E_{N_2O}, kg CE ha^{-1}) induced by fertilizer N input could be estimated using the following equation:

$$E_{N_2O} = N \times EF_{N_2O} \times \frac{44}{28} \times 265 \tag{3.1}$$

where N represents the amount of chemical fertilizer-N application (kg N), EF_{N_2O} is the emission factor of N$_2$O emission induced by N fertilizer application (kg N$_2$O–N kg^{-1} N fertilizer), 44/28 is the molecular weight of N$_2$ in relation to N$_2$O, 298 is the net GWP of N$_2$O in a 100-year horizon, and 12/44 is the molecular weight of CO$_2$ in relation to CE. According to IPCC (2006), EF_{N_2O} is estimated as 0.01 for dry cropland and 0.003 for rice paddy.

3.3.1.1.2 *Methane Emissions from Flooded Rice Paddy*

Seasonal CH$_4$ emissions from rice paddies should be estimated for rice CF calculation. Direct CH$_4$ emissions are estimated using the IPCC's methodology (IPCC 2006) with the equation:

$$E_{CH_4} = EF_d \times t \times A \times 28 \tag{3.2}$$

$$EF_d = EF_c \cdot SF_w \cdot SF_p \cdot SF_o \cdot SF_{s,r} \tag{3.3}$$

$$SF_o = \left(1 + \sum_i ROA_i \cdot CFOA_i\right)^{0.59} \tag{3.4}$$

where E_{CH_4} represents the GHG emissions (kg CO$_2$-eq.) from CH$_4$ emitted from rice paddy in a single season, EF_d is a daily emission factor (kg CH$_4$/ha/day), t is rice growing period (days), A is area of rice paddy (ha), and 28 is the relative molecular warming forcing of CH$_4$ in a 100-year horizon (IPCC 2013), EF_c is the emission factor for continuously flooded fields without organic amendments, SF_w represents scaling factor to account for the differences in water regime during the rice growing period, SF_p represents scaling factor to account for the differences in water regime in the preseason before the rice growing season, SF_o represents scaling factor should vary for both type and amount of organic amendment applied (Equation 3.4), $SF_{s,r}$ represents scaling factor for soil type, rice cultivar, and so on, if these factors are available. ROA_i is the application rate of organic amendment i, in dry weight for straw and fresh weight for others (kg/ha), and $CFOA_i$ is the conversion factor for organic amendment i.

3.3.1.1.3 CO_2 Emissions from Machine Operation

CO_2 emissions from machine operation could be calculated according to the amount of fuel, net caloric value of a given fuel and emission factor of this fuel, as shown in Equation 3.5:

$$E_M = \sum_i (EF_i \cdot W_i \cdot NCV_i) \tag{3.5}$$

where E_M is the CO_2 emissions from machine operation (kg CO_2-eq.), W_i is the amount of fuel i consumed (t or L), NCV_i represents net caloric value for fuel i (GJ/kg or L), EF_i represents emission factor for fuel i (kg CO_2-eq./GJ).

3.3.1.2 Indirect Emissions

3.3.1.2.1 GHG Emissions from Manufacturing of Agricultural Inputs

GHG emissions induced by manufacturing of individual materials used for crop production, protection and management including fertilizer, pesticides, and plastic film, which are calculated using Equation 3.6:

$$E_{AI} = \sum AI \times EF_{AI} \tag{3.6}$$

where E_M is GHG emissions from manufacture of individual materials such as fertilizer, pesticide, agricultural film, and so on (kg CO_2-eq.), AI denotes the amount of a kind of agricultural inputs in kg ha^{-1}, and EF is the emission factor of manufacturing a unit of the input material (kg CO_2-eq. kg^{-1}).

3.3.1.2.2 GHG Emissions by Irrigation

The energy use for irrigation, which would be one of the main GHG emission sources in farm operations, is also calculated. The main components of energy use associated with irrigation are related to pumping water to field, being generally powered either by diesel or by electricity (Wang et al. 2012). In general, GHG emissions induced by irrigation (E_{IRRI}, kg CE ha^{-1}) could be calculated using the approach developed by Wang et al. (2012):

$$E_{IRRI} = IR_{ij} \times EF_j \tag{3.7}$$

where IR_{ij} represents the amounts of water irrigated for crop i in region j in m^3 and EF_j is the emission factor of irrigation for region j in kg CO_2-eq. m^{-3}.

3.3.1.3 CF Assessment

Overall, the total CF of grain crop in a single cropping season in terms of land used (CF_A, kg CO_2-eq./ha) was assessed by summarizing all of the individual GHG emissions mentioned above:

$$CF_A = \frac{E_{N_2O} + E_{CH_4} + E_M + E_{AI} + E_{IRRI}}{A} \tag{3.8}$$

where A is the total area of studied field (ha).

With the estimated CF_A, CF in terms of grain production (CF_Y, kg CO_2-eq./kg grain produced) (GHG intensity in other words) was evaluated using Equation 3.9:

$$CF_Y = \frac{CF_A}{Y} \tag{3.9}$$

where Y denotes grain yield of a given crop (kg/ha).

3.3.2 LIVESTOCK PRODUCTION

CF of livestock production was generally assessed by taking into account all the GHG emissions caused by or associated with material used, farm management, and energy exhausted for livestock production. A life-cycle assessment (LCA) and input–output analysis was used to describe the total GHG emissions by livestock production. CF calculated was expressed in carbon dioxide equivalent (CO_2-eq.) per unit livestock production.

3.3.2.1 Direct Emission

3.3.2.1.1 CH_4 Emissions from Enteric Fermentation

CH_4 is produced in herbivores as a by-product of enteric fermentation, a digestive process by which carbohydrates are broken down by microorganisms into simple molecules for absorption into the bloodstream (IPCC 2006). The CH_4 emissions from enteric fermentation are estimated using Equation 3.10:

$$E_{EF} = H \times EF_{EF} \times 28 \tag{3.10}$$

where E_{EF} is the CH_4 emissions from enteric fermentation (kg CO_2-eq.), H denotes the number of ruminant head, EF_{EF} is the emission factors for enteric fermentation (kg CH_4/head/a), and 28 is the net GWP of CH_4 in a 100-year horizon.

3.3.2.1.2 N_2O and CH_4 Emissions by Manure Treatment

Livestock production can result in both CH_4 and N_2O emissions from livestock manure management systems, which are significant GHG sources in the agricultural sector. CH_4 and N_2O emissions during manure treatment (E_M, kg CO_2-eq.) are estimated using Equations 3.11 through 3.13:

$$E_M = H \times EF_{CH_4} \times 28 + (N_2O_{D(mm)} + N_2O_{G(mm)}) \times 265 \tag{3.11}$$

$$N_2O_{D(mm)} = \left[\sum_S \left[\sum_T (N_T \times Nex_{(T)} \times MS_{(T,S)}) \right] \times EF_{3(S)} \right] \times \frac{44}{28} \tag{3.12}$$

$$N_2O_{G(mm)} = \left[\sum_S \left[\sum_T (N_{(T)} \times Nex_{(T)} \times MS_{(T,S)}) \times \left(\frac{Frac_{GasMS}}{100} \right)_{(T,S)} \right] \times EF_4 \right] \times \frac{44}{28} \tag{3.13}$$

where H denotes the number of head (head), EF_{CH_4} is the CH_4 emission factor (kg CH_4/head/a), $N_2O_{D(mm)}$ the direct emissions of nitrous oxide under manure management system (kg N_2O/a), $N_2O_{G(mm)}$ the indirect nitrous oxide emissions of manure management system (kg N_2O/a), the direct and indirect N_2O emission from manure treatment are calculated using the second method recommended by IPCC (2006), 298 is the net GWP of N_2O in a 100-year horizon, $N_{(T)}$ is the number of head of livestock species/category T in the given region, $Nex_{(T)}$ is the annual average N excretion per head of species/category T in the country (kg N/animal/year), $MS_{(T,S)}$ is the fraction of total annual nitrogen excretion for each livestock species/category T that is managed in manure management system S in the given region (dimensionless), $EF_{3(S)}$ is the emission factor for direct N_2O emissions from manure management system S in the country (kg N_2O-N/kg N) in manure management system S, $Frac_{GasMS}$ is the percent of managed manure nitrogen for livestock category T that volatilizes as NH_3 and NO_x in the manure management system S (%), S is manure management system, T is species/category of livestock, EF_4 is the emission factor for N_2O emissions from atmospheric deposition of nitrogen on soils and water surfaces, kg N_2O-N (kg NH_3–N + NO_x–N volatilized)$^{-1}$, default value is 0.01 kg N_2O-N (kg NH_3–N + NO_x–N volatilized)$^{-1}$, 44/28 is conversion of (N_2O–N) (mm) emissions to N_2O (mm) emissions.

3.3.2.1.3 Energy Use in Farm Management

The GHG emissions from energy use induced by farm management (CF_M, kg CO_2-eq.) include fuel and electricity consumed in the process of farm management. These emissions could be calculated by Equations 3.14 and 3.15:

$$E_F = \sum_i (EF_i \cdot W_i \cdot NCV_i) \qquad (3.14)$$

$$E_E = E \times EF_E \qquad (3.15)$$

where E_F is CO_2 emissions from fuel use (kg CO_2-eq.), W_i is the amount of fuel i consumed (t or L), NCV_i represents net caloric value for fuel i (GJ/kg or L), EF_i represents emission factor for fuel i (kg CO_2-eq./GJ). E is the amount of electricity used in the life-cycle analysis of livestock production (kw h) and EF_E represents emission factor for electricity generation (kg CO_2-eq./kw h).

3.3.2.2 Indirect Emissions

3.3.2.2.1 GHG Emissions from Manufacture of Forage

According to the field survey conducted by previous study, the types of forage include mixed forage, self-made forage, concentrates, and green forage. The GHG emissions from manufacture of forage include crop cultivation, forage processing, and forage transportation. In general, these emissions could be estimated using the following equation:

$$E_{Forge} = F_C \times EF_C + F_P \times EF_P + E_T \qquad (3.16)$$

where E_{Forge} is the GHG emissions by forge input, F_C is the amount of crop production used for forge manufacture (kg), EF_C represents the emission factor for a given crop production (kg CO_2-eq./kg production); F_P means the amount of forge used in livestock production (kg), EF_P denotes emission factor in forage processing (kg CO_2-eq./kg forge produced). E_T is the GHG emissions by forage transportation calculated by multiplying amount of fuel consumption by emission factor of fuel.

3.3.2.3 Assessment of CF of Livestock Production

Overall, total CF of livestock production could be assessed by summarizing all of individual GHG emissions mentioned above:

$$CF = E_{EF} + E_M + E_F + E_E + E_{Forage} \qquad (3.17)$$

With the estimated CF, CF in terms of livestock and poultry production (CF_P, kg CO_2-eq./kg production) (GHG intensity in other words) was calculated using Equation 3.18:

$$CF_P = \frac{CF}{P} \tag{3.18}$$

where P denotes the production of a given livestock (kg).

3.4 DATA SOURCES

3.4.1 EMISSION FACTORS

3.4.1.1 Crop Production

Assessment of CF in crop production considers various emission factors including direct N_2O emission by N input, CH_4 emission from rice paddy, CO_2 emission by fuel combustion, carbon dioxide respired by an adult, manufacture of agricultural inputs, and energy use in irrigation.

1. *Emission factors of direct N_2O emission by N input*
   Direct N_2O emission factors have been developed by IPCC (2006) and some studies over the world. Some emission factors developed are listed in Table 3.1. However, the region-specific emission factors should be used in CF assessment in these regions if available.
2. *Emission factors of direct CH_4 emission from rice paddy*
   All the emission and scaling factors that relate to calculate CH_4 emissions from rice paddy are listed in Tables 3.2 through 3.5 according to IPCC methodology (IPCC 2006). Region-specific emission factors should be used in CF assessment in these regions if available.

TABLE 3.1
Emission Factor of N Fertilizer-Induced N_2O Emissions

Abbreviation	Emission Factor	Condition	Region	Literature
EF_{N_2O}	0.01 tN_2O–N/t fertilizer-N	Mineral soil	Global	IPCC (2006)
	0.003 tN_2O–N/t fertilizer-N	Rice paddy	Global	IPCC (2006)
	0.0002 tN_2O–N/t fertilizer-N	F[a] in rice paddy	China	Zou et al. (2007)
	0.0042 tN_2O–N/t fertilizer-N	F-D-F[a] in rice paddy	China	Zou et al. (2007)
	0.0073 tN_2O–N/t fertilizer-N	F-D-IF-M[a] in rice paddy	China	Zou et al. (2007)
	$0.0186P^{[b]}$ tN_2O–N/t fertilizer-N	Mineral soil	China	Lu et al. (2006)

[a] F: flood; D: drainage; IF: intermittent flood; M: moist but nonwaterlogged by intermittent irrigation.
[b] P: annual precipitation, m.

TABLE 3.2
Default CH_4 Baseline Emission Factor

Abbreviation	Emission Factor[a] (kg CH_4 ha^{-1} day^{-1})
EF_c	1.30

Source: Adapted from Yan XY et al. 2005. *Global Change Biology* 11:1131–1141.

[a] It was assumed that no flooding occurred for less than 180 days prior to rice cultivation, and that the fields were continuously flooded without organic amendments during rice cultivation.

TABLE 3.3

Default CH$_4$ Emission Scaling Factors for Water Regimes during the Rice Season

		Scaling Factor (SF_w)	
	Water Regime	Aggregated Case	Disaggregated Case
	Upland[a]	0	0
Irrigated[b]	Continuously flooded	0.78	1
	Intermittently flooded—single aeration		0.6
	Intermittently flooded—multiple aeration		0.52
Rainfed and	Regular rainfed (50 cm)	0.27	0.28
deep water[c]	Drought prone		0.25
	Deep water (more than 50 cm for a significant period of time)		0.31

Source: Adapted from Yan XY et al. 2005. *Global Change Biology* 11:1131–1141.

[a] Fields are never flooded for a significant period of time.

[b] Fields are flooded for a significant period of time and water regime is fully controlled.

   • *Continuously flooded:* Fields have standing water throughout the rice growing season and may only dry out for harvest (end-season drainage).

   • *Intermittently flooded:* Fields have at least one aeration period of more than 3 days during the cropping season.

      • *Single aeration:* Fields have a single aeration during the cropping season at any growth stage (except for end-season drainage).

      • *Multiple aeration:* Fields have more than one aeration period during the cropping season (except for end-season drainage).

[c] Fields are flooded for a significant period of time and water regime depends solely on precipitation.

TABLE 3.4

Default CH$_4$ Emission Scaling Factors for Water Regimes before the Cultivation Period

	Scaling Factor (SF_p)	
Water Regime Prior to Rice Cultivation	Aggregated Case	Disaggregated Case
Non flooded preseason <180 days		1
Non flooded preseason >180 days	1.22	0.68
Flooded preseason (>30 days)[a]		1.9

Source: Adapted from Yan XY et al. 2005. *Global Change Biology* 11:1131–1141.

[a] Short preseason flooding periods of less than 30 days are not considered in selection of SF_p.

TABLE 3.5

Default Conversion Factor for Different Types of Organic Amendment

Organic Amendment	Conversion Factor (CFOA)
Straw incorporated shortly (<30 days) before cultivation[a]	1
Straw incorporated long (>30 days) before cultivation[a]	0.29
Compost	0.05
Farm yard manure	0.14
Green manure	0.50

Source: Adapted from Yan XY et al. 2005. *Global Change Biology* 11:1131–1141.

[a] Straw application means that straw is incorporated into the soil; it does not include case that straw just placed on the soil surface nor that straw was burnt on the field.

TABLE 3.6
Default Net Calorific Values (NCVs) of Main Kinds of Fuel

Abbreviation	Fuel Type	Net Calorific Value (TJ/Gg)
NCV_i	Motor gasoline	44.3
	Aviation gasoline	44.3
	Jet gasoline	44.3
	Gas/diesel oil	43.0
	Biogasoline	27.0
	Biodiesels	27.0

Source: Adapted from IPCC. 2006. *2006 IPCC Guidelines for National Greenhouse Gas Inventories*. National Greenhouse Gas Inventories Programme. Tokyo, Japan: IGES.

3. *Emission factors of fuel combustion*
   Net caloric values and emission factors for main kinds of fuel are shown in Tables 3.6 and 3.7 according to IPCC (2006).
4. *Carbon dioxide respired by an adult*
   Carbon dioxide respired by an adult is 0.9 kg CO_2-eq. day^{-1} $person^{-1}$ (Yang 1996).
5. *Emission factors of manufacture of agricultural inputs*
   Emission factors of manufacture of agricultural inputs vary largely among different regions and techniques of production. However, there are limited studies focused on development emission factors of these. Herewith, this chapter lists some emission factors by previous studies (Table 3.8). Specific emission factor development is an essential research field in future.
6. *Emission factor of irrigation*
   Emission factors of irrigation have been developed by researches for some regions (Table 3.9). For example, emission factors of irrigation in each province of China have been developed, and the details of this calculator were also presented in the study of Wang et al. (2012). However, region-specific emission factors for irrigation should be developed in further studies for the precise of CF calculation.

3.4.1.2   Livestock Production

1. *CH_4 emission factor of enteric fermentation*
   The amount of CH_4 that is released depends on the type of digestive tract, age, and weight of the animal, and the quality and quantity of the feed consumed (IPCC 2006). The emission factor could be found in Tables 3.10a and 3.10b according to IPCC (2006).

TABLE 3.7
Default CO_2 Emission Factors for Main Kinds of Fuel Combustion

Abbreviation	Fuel Type	Default CO_2 Emission Factor (kg CO_2/TJ)
EF_i	Motor gasoline	69,300
	Aviation gasoline	70,000
	Jet gasoline	70,000
	Gas/diesel oil	74,100
	Biogasoline	70,800
	Biodiesels	70,800

Source: Adapted from IPCC. 2006. *2006 IPCC Guidelines for National Greenhouse Gas Inventories*. National Greenhouse Gas Inventories Programme. Tokyo, Japan: IGES.

TABLE 3.8
Emission Factors for Manufacturing Fertilizer, Pesticide, and Plastic Film

Abbreviation	Emission Source	Emission Factor	Region	Literature
EF_{AI}	N fertilizer	6.38 kg CO_2-eq. kg^{-1} N	China	Lu et al. (2008)
	N fertilizer	10.85 kg CO_2-eq. kg^{-1} N	Scotland, UK	Hillier et al. (2009)
	N fertilizer	4.95 kg CO_2-eq. kg^{-1} N	United States	Dubey and Lal (2009)
	P fertilizer	0.73 kg CO_2-eq. kg^{-1} P_2O_5	United States	Dubey and Lal (2009)
	K fertilizer	0.55 kg CO_2-eq. kg^{-1} K_2O	United States	Dubey and Lal (2009)
	Farmyard manure	0.55 kg CO_2-eq. kg^{-1}	Scotland, UK	Hillier et al. (2009)
	Insecticides	17.05 kg CO_2-eq. kg^{-1}	United States	Dubey and Lal (2009)
	Insecticides	1.32 kg CO_2-eq. kg^{-1}	Scotland, UK	Hillier et al. (2009)
	Herbicides	26.22 kg CO_2-eq. kg^{-1}	United States	Dubey and Lal (2009)
	Herbicides	23.1 kg CO_2-eq. kg^{-1}	Scotland, UK	Hillier et al. (2009)
	Fungicide/nematicide	11.59 kg CO_2-eq. kg^{-1}	Scotland, UK	Hillier et al. (2009)
	Film	18.99 kg CO_2-eq. kg^{-1}	China	Cheng et al. (2011)

TABLE 3.9
Emission Factors for Irrigation

Abbreviation	Emission Factor	Region	Literature
EF_j	91.67 kg CO_2-eq. ha^{-1}	Punjab, India	Dubey and Lal (2009)
	0.49 kg CO_2-eq. m^{-3}	Hebei province, China	Wang et al. (2012)
	0.30 kg CO_2-eq. m^{-3}	Inner Mongolia, China	Wang et al. (2012)
	0.21 kg CO_2-eq. m^{-3}	Liaoning province, China	Wang et al. (2012)
	0.50 kg CO_2-eq. m^{-3}	Gansu province, China	Wang et al. (2012)
	0.40 kg CO_2-eq. m^{-3}	Hunan province, China	Wang et al. (2012)

TABLE 3.10a
Emission Factors for Enteric Fermentation

Abbreviation	Livestock	Developed Countries (kg CH_4 $head^{-1}$ $year^{-1}$)	Developing Countries (kg CH_4 $head^{-1}$ $year^{-1}$)	Liveweight (kg)
EF_{EF}	Buffalo	55	55	300
	Sheep	8	5	65 (developed countries); 45 kg (developing countries)
	Goats	5	5	40
	Camels	46	46	570
	Horses	18	18	550
	Mules and Asses	10	10	245
	Deer	20	20	120
	Alpacas	8	8	65
	Swine	1.5	1.0	

Source: Adapted from IPCC. 2006. 2006 *IPCC Guidelines for National Greenhouse Gas Inventories*. National Greenhouse Gas Inventories Programme. Tokyo, Japan: IGES.

TABLE 3.10b
Enteric Fermentation Emission Factors for Cattle

Region	Cattle Category	Emission Factors (kg CH$_4$/head/year)	Comments
North America	Dairy	128	Average milk production of 8400 kg head^{-1} year^{-1}
	Other Cattle	53	Includes beef cows, bulls, calves, growing steers/heifers, and feedlot cattle
Western Europe	Dairy	117	Average milk production of 6000 kg head^{-1} year^{-1}
	Other Cattle	57	Includes bulls, calves, and growing steers/heifers
Eastern Europe	Dairy	99	Average milk production of 2550 kg head^{-1} year^{-1}
	Other Cattle	58	Includes beef cows, bulls, and young
Oceania	Dairy	90	Average milk production of 2200 kg head^{-1} year^{-1}
	Other Cattle	60	Includes beef cows, bulls, and young
Latin America	Dairy	72	Average milk production of 800 kg head^{-1} year^{-1}
	Other Cattle	56	Includes beef cows, bulls, and young
Asia	Dairy	68	Average milk production of 1650 kg head^{-1} year^{-1}
	Other Cattle	47	Includes multi-purpose cows, bulls, and young
Africa and the Middle East	Dairy	46	Average milk production of 475 kg head^{-1} year^{-1}
	Other Cattle	31	Includes multi-purpose cows, bulls, and young
Indian Subcontinent	Dairy	58	Average milk production of 900 kg head^{-1} year^{-1}
	Other Cattle	27	Includes cows, bulls, and young. Young comprise a large portion of the population

Source: Adapted from IPCC. 2006. 2006 *IPCC Guidelines for National Greenhouse Gas Inventories.* National Greenhouse Gas Inventories Programme. Tokyo, Japan: IGES.

2. *Emission factors of manure management*

The main factors affecting CH$_4$ emissions are manure production and the portion of the manure that decomposes anaerobically. Direct N$_2$O emissions occur via combined nitrification and denitrification of nitrogen contained in the manure. The CH$_4$ emission factors could be found in Annex 10A.2 in Chapter 10 of IPCC (2006), whereas the N$_2$O emission factors could be obtained in Tables 10.19 and 10.20 in Chapter 10 of IPCC (2006).

3. *Emission factors of fuel and electricity consumed*

Net caloric values and emission factors for main kinds of fuel are shown in Tables 3.6 and 3.7 according to IPCC (2006). Emission factor of electricity generation is shown in Table 3.11. However, region-specific emission factors should be used when CFs of livestock production are calculated in these region if available.

4. *Carbon dioxide respired by an adult*

Carbon dioxide respired by an adult is 0.9 kg CO$_2$-eq. day^{-1} person^{-1} (Yang 1996).

5. *Emission factors of manufacture of forge*

According to the description above, GHG emissions from manufacture of forage include crop cultivation, forage processing, and forage transportation. Emissions from transportation could be calculated by the emission factors supplied in Tables 3.6 and 3.7. Emission factors of crop cultivation actually are the CF of crop production used in forge production, which could be estimated by the approach supplied by this chapter or just obtained from

TABLE 3.11
Emission Factors for Electricity Generation

Abbreviation	Region	Emission Factors (kg CO_2-eq./kW h)	Reference
EF_E	Global	0.538	CER (2007)
	North China	1.0302	The NDRC climate
	Northeast China	1.1120	division (2013)
	East China	0.8100	
	Middle China	0.9779	
	Northwest China	0.9720	
	South China	0.9223	

previous studies. There are limited studies focused on the development of emission factors by forge processing, which should be conducted in future studies.

3.4.2 Activity Data

3.4.2.1 Statistical Data

There are national or regional statistical yearbooks in the government of each country and region. Annually, being one of the most important sections, agriculture production data should be involved in these yearbooks. Therefore, input data for CF calculation could be collected from various yearbooks that reported agricultural production statistical data. The input data requested by CF calculation of crop production include crop yields, cultivated area, as well as the amounts of various agricultural inputs including fertilizer, pesticide, agricultural plastic film, irrigated water, diesel use during crop production. However, the input data, including scale of livestock production, yield of livestock, manure treatment method, and amounts of manure, as well as the amount of individual agricultural inputs consisting of forge, fuel, and electricity, should be collected for livestock production CF calculation. The input data requested is listed in Tables 3.12 and 3.13 for crop and livestock production, respectively.

3.4.2.2 Field Survey Data

Field survey data are also an important source for CF study. Data collection could be performed by face-to-face interview with farmers during crop and livestock harvest each year. Data describing material and energy inputs for crop and livestock production in a single crop production cycle or a whole year should be recorded to create a database. During the survey, a number of farms should be randomly selected for the field questionnaire survey according to local scare. The data for a single cropping season inquired with the interview include: (a) locations of farms and contacts of farmers; (b) amounts of fertilizers (N, P, K) and pesticides used; (c) machinery operation for spraying, tillage,

TABLE 3.12
Data Request for CF Calculation in Crop Production

No.	Location	Cultivated Area (ha)	Yield (kg)	N Fertilizer (kg N/ha)	P Fertilizer (kg P_2O_5/ ha)	K Fertilizer (kg K_2O/ ha)	Other Fertilizer (kg/ha)	Pesticide (kg/ha)	Film (kg/ha)	Irrigation (m³/ha)	Fuel (L or kg/ha)
Region 1											
Region 2											
Region 3											
…											

TABLE 3.13

Data Request for CF Calculation in Livestock Production

No.	Location	Scale (Heads)	Yield (kg/year)	Forge (kg/head)	Electricity (kW h)	Fuel (L or kg)	Manure Treatment Method	Manure (kg/year)
Region 1								
Region 2								
Region 3								
...								

TABLE 3.14

Data Surveyed for CF Calculation in Crop Production

No.	Location	Cultivated Area (ha)	Yield (kg)	N Fertilizer (kg N)	P Fertilizer (kg P_2O_5)	K Fertilizer (kg K_2O)	Other Fertilizer (kg)	Pesticide (kg)	Film (kg)	Irrigation (m³)	Fuel (L or kg)
Site 1											
Site 2											
Site 3											
...											

TABLE 3.15

Data Surveyed for CF Calculation in Livestock Production

No.	Location	Scale (Heads)	Yield (kg/year)	Forge (kg/year)	Electricity (kW h/year)	Fuel (L or kg/year)	Manure Treatment Method	Manure (kg/year)
Site 1								
Site 2								
Site 3								
...								

harvesting, and transportation as well as for irrigation; (d) cropping system and water regime management for rice paddy; and (e) farm size and the total grain produced. Data obtained for CF calculation of livestock production include: (a) location of a farm and the contacts of farmers; (b) scale of a farm, (c) amounts of forage, fuel, and electricity; (d) manure treatment method and amounts of manure, (e) annual yield of livestock production. The requested data for the field survey of crop and livestock production are listed in Tables 3.14 and 3.15, respectively.

3.5 CASE STUDIES

3.5.1 CF of Main Grain Crop Production in Shandong Province, China

3.5.1.1 Scope and Objective

Wheat and corn are the major grain crops, accounting for almost 64% of total arable land area in China (FAOSTAT 2014). Shandong Province is one of the most important grain crop producers in China. The yields of wheat and maize in Shandong province accounted for 18 and 10% of total wheat and maize, respectively, in China. In order to meet the food demand by the increasing population, production of these crops has been achieved with maximum available input of resources to reach a maximum yield. In addition, these two grain crops could have significant contributions to agricultural

GHG emissions. Given that determining the CF of crop production is useful for identifying how low-carbon economy could be implemented to abate climate change, this case study was aimed to quantify the CFs of the two grain crops in Shandong province. Special attention has been given to the contribution of individual inputs to total CF in crop production. In addition, sensitivity analysis was conducted to assess the sensitivity of the total CF to different factors considered in this study.

3.5.1.2 Data Source

Input data for CF calculation includes crop grain yields, the amounts of various agricultural inputs including fertilizer, pesticide, agricultural plastic film, irrigated water, and diesel in 2012. The original data were available from "Compilation of the national agricultural costs and returns" and "Yearbook of China water resources." The data collected for CF calculation was shown in Table 3.16.

3.5.1.3 CF Calculation

The "farm gate" (up to harvest) was considered as the boundary of this study. Therefore, the direct emissions from soil N_2O and CH_4 emissions, machine operation, and indirect emissions from manufacturing of agricultural inputs and irrigation in wheat and maize production were considered for CF calculation up to the "farm gate" (up to harvest).

3.5.1.3.1 Direct Emissions

3.5.1.3.1.1 Nitrous Oxide Emissions from Soil   For maize field,

$$E_{N_2O} = N \times EF_{N_2O} \times \frac{44}{28} \times 265 = 170.69 \times 0.0186 \times 0.75 \times \frac{44}{28} \times 265 = 991.57 \quad (\text{kg CO}_2\text{-eq./ha})$$

For wheat field,

$$E_{N_2O} = 197.55 \times 0.0186 \times 0.75 \times \frac{44}{28} \times 265 = 1147.6 \quad (\text{kg CO}_2\text{-eq./ha})$$

3.5.1.3.1.2 Methane Emissions from Flooded Rice Paddy   There is no CH_4 emission in maize and wheat field, therefore, $E_{CH_4} = 0$.

3.5.1.3.1.3 CO₂ Emissions from Machine Operation   For maize field,

$$E_M = \sum_i (EF_i \cdot W_i \cdot NCV_i) = 74{,}100 \times 131.40 \times 43 \times 10^{-6} = 418.68 \quad (\text{kg CO}_2\text{-eq./ha})$$

For wheat field,

$$E_M = 74{,}100 \times 207 \times 43 \times 10^{-6} = 659.56 \quad (\text{kg CO}_2\text{-eq./ha})$$

TABLE 3.16
Original Data for CF Calculation of Grain Crop Production

Crop	N Fertilizer (kg/ha)	P Fertilizer (kg/ha)	K Fertilizer (kg/ha)	Film (kg/ha)	Pesticide (kg/ha)	Diesel (kg/ha)	Irrigation (m³/ha)	Yield (kg/ha)
Maize	170.69	86.41	48.75	0	9.39	131.40	945	7515.75
Wheat	197.55	115.50	60.45	0	7.81	207.00	2400	6546.75

3.5.1.3.2   Indirect Emissions

3.5.1.3.2.1   GHG Emissions from Manufacturing of Agricultural Inputs   For maize production,

$$E_{AI} = \sum AI \cdot EF_{AI} = 170.69 \times 6.38 + 86.41 \times 0.73 + 48.75 \times 0.55 + 9.39 \times 17.05$$
$$= 1338.99 \quad (kg\,CO_2\text{-eq./ha})$$

For wheat production,

$$E_{AI} = 197.55 \times 6.38 + 115.5 \times 0.73 + 60.45 \times 0.55 + 7.81 \times 17.05 = 1511.09 \quad (kg\,CO_2\text{-eq./ha})$$

3.5.1.3.2.2   GHG Emissions by Irrigation   For maize production,

$$E_{IRRI} = IR_{ij} \times EF_j = 945 \times 0.26 = 245.7 \quad (kg\,CO_2\text{-eq./ha})$$

For wheat production,

$$E_{IRRI} = IR_{ij} \times EF_j = 2400 \times 0.26 = 624 \quad (kg\,CO_2\text{-eq./ha})$$

3.5.1.3.3   CF Assessment

For maize production,

$$CF_A = \frac{E_{N_2O} + E_{CH_4} + E_M + E_{Labor} + E_{AI} + E_{IRRI}}{A} = \frac{991.57 + 0 + 418.68 + 1338.99 + 245.7}{1}$$
$$= 2994.94 \quad (kg\,CO_2\text{-eq./ha})$$

$$CF_Y = \frac{CF_A}{Y} = \frac{2994.94}{7515.75} = 0.40 \quad (kg\,CO_2\text{-eq./kg yield})$$

For wheat production,

$$CF_A = \frac{1147.6 + 0 + 659.56 + 1511.09 + 624}{1} = 3942.25 \quad (kg\,CO_2\text{-eq./ha})$$

$$CF_Y = \frac{3942.25}{6546.75} = 0.60 \quad (kg\,CO_2\text{-eq./kg yield})$$

3.5.1.4   Results

The CFs of maize and wheat production were calculated based on the CF methodology provided by this chapter. As shown in Figure 3.1, wheat production had the higher CFs being 3942.25 kg CO_2-eq./ha and 0.6 kg CO_2-eq./kg yield than maize production with the CFs of 2994.94 kg CO_2-eq./ha and 0.4 kg CO_2-eq./kg yield.

Proportions of different inputs in CF were calculated to assess the contribution of each emission source to total CF (Figure 3.2). Both for maize and wheat production, most of the CF (70 and 61%, respectively) was derived from N fertilizer application including N_2O emissions from soil and GHG emissions from manufacturing of N fertilizer. Machine operation turned to be the second biggest

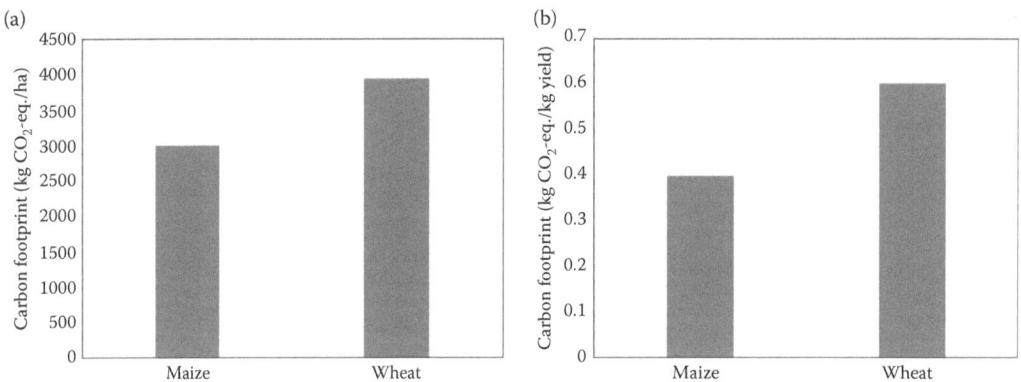

FIGURE 3.1 CFs in terms of land use (a) and grain yield (b) in maize and wheat production in Shandong, China.

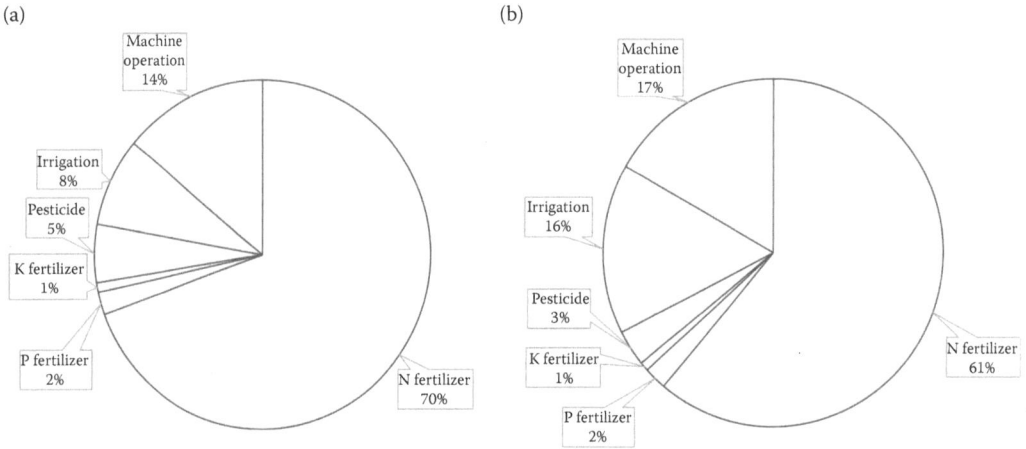

FIGURE 3.2 Contributions of agricultural inputs to total CF of maize (a) and wheat production (b).

contributor by 14% in maize production. However, machine operation and irrigation made the similar contribution (17 and 16%, respectively) to total CF in wheat production, which is different from maize production with a contribution of only 8% in irrigation.

3.5.1.5 Sensitivity Analysis

The sensitivity of the total CF to different inputs and grain yield, including fertilizer, irrigation, pesticide, machine operation, and grain yield, was evaluated in this study. As shown in Table 3.17, the calculated CF responded differently to changes in the different parameters, which were increased and decreased by 10% in our evaluation. The most sensitive parameter was grain yield (9.09 and −9.09% for 10% decrement and increment, respectively) and N fertilizer input (6.95 and −6.95% for 10% decrement and increment in maize production, and 6.11 and −6.11% for 10% decrement and increment in wheat production, respectively) for both maize and wheat production. The sensitivity of K fertilizer had a much smaller influence on CF in our evaluation. Sensitivity analysis gave us a suggestion that the reliability of data sources of these inputs is a key factor to obtain the accurate CF results. These also indicated that the increase in grain production and N fertilizer use efficiency could be a strategy to reduce CF in grain crop production.

TABLE 3.17
Sensitivity Analysis of Various Agricultural Inputs and Grain Yield to Total CF (in %)

Item	Maize		Wheat	
	−10%	10%	−10%	10%
N fertilizer	−6.95	6.95	−6.11	6.11
P fertilizer	−0.21	0.21	−0.21	0.21
K fertilizer	−0.09	0.09	−0.08	0.08
Pesticide	−0.53	0.53	−0.34	0.34
Irrigation	−0.82	0.82	−1.58	1.58
Machine operation	−1.40	1.40	−1.67	1.67
Yield	9.09	−9.09	9.09	−9.09

3.5.1.6 Conclusions

The production of wheat and maize in Shandong province of China shows CFs of 0.6 and 0.4 kg CO_2-eq./kg grain, respectively, mainly due to N fertilizer inputs and machine operation. This chapter highlights opportunities for GHG mitigation by improving N fertilizer use efficiency and increasing grain yield in crop production of Shandong province of China.

3.5.2 CF OF MILK PRODUCTION BASED ON A SITE SURVEY IN SICHUAN PROVINCE, CHINA

3.5.2.1 Scope and Objective

China's livestock production ranks first in the world, and GHG emissions from animal enteric fermentation and manure management have been estimated at 445 Tg CO_2-eq., accounting for 45.7% of the nation's total agricultural emissions in 2005 (NDRC 2012). Thus, livestock production in China could play a key role in global climate change. To identify the contributions of livestock production to climate change and the key mitigation options, quantifying and assessing the CF in China's livestock and poultry production is urgently required. According to statistical data, Sichuan province is one of the largest livestock producer in China with the output of pork, poultry, eggs, and milk being 4.964, 0.93, 1.464, and 0.7118 million tons, respectively, which could emit much GHG to atmosphere. Therefore, Sichuan province was chosen to assess the CF of milk production in this case study. The main purposes of the study are to quantify the CFs of the milk production and to identify the contributions of individual GHG sources to total CF.

3.5.2.2 Data Source

This study involved a field survey of Hongya county, Sichuan province (103.37°E, 29.91°N) that is one of the major livestock production regions in China (Hu et al. 2010). Data collection was performed by face-to-face interviews with farmers. The obtained data included: (I) number of livestock head, (II) amounts of forage, fuel, and electricity, (III) manure treatment method and amounts of manure, and (IV) annual yield of livestock production (Table 3.18).

3.5.2.3 CF Calculation

The "farm gate" (up to harvest) was considered as the boundary of this study. Therefore, the direct emissions from enteric fermentation, manure treatment and machine operation, and indirect emissions from manufacturing of forage in milk production were considered for CF calculation up to the "farm gate" (up to harvest).

TABLE 3.18
Original Data for CF Calculation of Milk Production

Items	Data	Items	Data
Adult cow	280 heads	Cattle manure for hay production	24,841 kg
Bred cattle (>1 year)	12 heads	Electricity for irrigation	125 kw h
Calf (<1 year)	0	Gasoline for forage transport	80.416 kg
Hay	5610.78 t	Gasoline during farm management	321.66 kg
Concentrated forage	688.92 t	Electricity	23,061 kw h
Green forage	425.492 t	Milk production	1588.44 t

3.5.2.3.1 Direct Emission

3.5.2.3.1.1 CH$_4$ Emissions from Enteric Fermentation In this calculation, enteric fermentation factors could be obtained from a previous study in China as shown in Table 3.19.

Hence, CH$_4$ emissions from enteric fermentation could be calculated as follow:

$$E_{EF} = H \times EF_{EF} \times 28 = 280 \times 86.2 \times 28 + 12 \times 61.2 \times 28 = 696,371 \quad (\text{kg CO}_2\text{-eq.})$$

3.5.2.3.1.2 N$_2$O and CH$_4$ Emissions by Manure Treatment All the manure in this farm was directly applied into soil for forge cultivation. Therefore, this emission was 0 kg CO$_2$-eq. for this farm. The N$_2$O emissions by manure input are calculated in forge cultivation part.

3.5.2.3.1.3 Energy Use in Farm Management

$$E_F = \sum_i (EF_i \cdot W_i \cdot NCV_i) = 44.3 \times 69,300 \times 321.66 \times 10^{-6} = 987.49 \quad (\text{kg CO}_2\text{-eq.})$$

$$E_E = E \times EF_E = 23,061 \times 0.9223 = 21,269.16 \quad (\text{kg CO}_2\text{-eq.})$$

3.5.2.3.2 Indirect Emissions

3.5.2.3.2.1 GHG Emissions from Manufacture of Forage As shown in Table 3.17, the forage used in this farm includes hay, concentrated forage, and green forage. The emission factors of concentrated forage and green forage production are 0.57 and 0.98 kg CO$_2$-eq./kg forage, respectively. However, hay was produced by itself. The GHG emissions in the process of hay production include N$_2$O emissions by manure input and irrigation, which could be calculated according to CF methodology for crop production supplied by this chapter.

$$E_{N_2O} = N \times EF_{N_2O} \times \frac{44}{28} \times 265 = (24,841 \times 0.062) \times 0.02 \times \frac{44}{28} \times 265 = 12,827.18 \quad (\text{kg CO}_2\text{-eq.})$$

$$E_E = E \times EF_E = 125 \times 0.9223 = 115.29 \quad (\text{kg CO}_2\text{-eq.})$$

TABLE 3.19
Emission Factors for Enteric Fermentation of Cow

Item	Average Weight (kg)	Emission Factor (kg CH$_4$/head/a)
Adult cow	500	86.2
Bred cattle (>1 year)	400	61.2
Calf (<1 year)	180	39.5

Total GHG emissions by hay production are:

$$E_{Forge\_hay} = 12,827 + 115.29 = 12,942.29 \quad (\text{kg CO}_2\text{-eq.})$$

GHG emissions by all the forage input are also calculated:

$$E_{Forge} = F_C \times EF_C + F_P \times EF_P + E_T = (688,920 \times 0.57 + 425,492 \times 0.98 + 12,942.29)$$
$$+ 0 + 80.416 \times 69,300 \times 321.66 \times 10^{-6} = 824,401.4 \quad (\text{kg CO}_2\text{-eq.})$$

3.5.2.3.3 Assessment of CF of Milk Production

$$CF = E_{EF} + E_M + E_F + E_E + E_{labor} + E_{Forage} = 696,371 + 0 + 987.49 + 21,269.16 + 814,401.4$$
$$= 1,533,029.05 \quad (\text{kg CO}_2\text{-eq.})$$

CF of milk production in terms of production is calculated below:

$$CF_P = \frac{CF}{P} = \frac{1,533,029.05}{1,588,440} = 0.97 \quad (\text{kg CO}_2\text{-eq./kg milk produced})$$

3.5.2.4 Results

According to this calculation, the CF of milk production in this farm is 0.97 kg CO_2-eq./kg milk produced. Proportions of GHG emissions by different inputs in CF were calculated to assess the contribution of each emission source to the total CF (Figure 3.3). For milk production, forage input contributed the most GHG emission in the total CF (53%), and the contributions of enteric fermentation were 45%. Emissions in the process of farm management made a contribution as small as 2%.

3.5.2.5 Sensitivity Analysis

The sensitivity of the total CF to different inputs and milk yield was further evaluated in this study, including CH_4 emissions from enteric fermentation, energy use in farm management, GHG

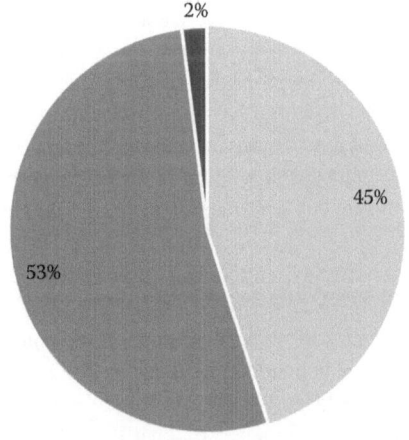

2%

45%

53%

■ Enteric fementation ■ Forage input   ■ Farm management

FIGURE 3.3 Contributions of agricultural inputs to total CF of milk production.

TABLE 3.20

Sensitivity Analysis of Various Agricultural Inputs and Milk Yield to Total CF in Milk Production (in %)

Item	−10%	10%
CH_4 emissions from enteric fermentation	−4.51	4.51
Energy use in farm management	−0.14	0.14
GHG emissions from manufacture of forage	−5.34	5.34
Yield	9.09	−9.09

emissions from manufacture of forage and milk yield, respectively. As shown in Table 3.20, the calculated CF responded differently to changes in the different parameters. The most sensitive parameters were milk yield (9.09 and −9.09% for 10% decrement and increment, respectively), GHG emissions from manufacture of forage (−5.34 and 5.34% for 10% decrement and increment, respectively), and CH_4 emissions from enteric fermentation (−4.51 and 4.51% for 10% decrement and increment, respectively). The sensitivity of energy use in farm management had a much smaller influence on CF in our evaluation. Sensitivity analysis gave us a suggestion that the reliability of data sources of these inputs is a key factor to obtain the accurate CF results. These also indicated that the increase of milk production and forage use efficiency and decrease of CH_4 emissions from enteric fermentation could be a strategy to reduce CF in milk production.

3.5.2.6 Conclusions

This case study quantified the CF of milk production in Sichuan province of China using data from a questionnaire survey. GHG emissions from enteric fermentation and forage use contributed significantly to the total CF.

3.5.3 Limitations and Recommendations

As a basic assessment of CF of crop and livestock production, there would be still an issue of uncertainty. As an estimation for regional CF of crop production, the direct N_2O emission from N fertilizer use was estimated using default value by IPCC (2006). In addition, emissions of potassium and phosphorus from fertilizers and from use of pesticides were estimated using those reported by Dubey and Lal (2009). Other factors such as CO_2 emission from N fertilizer and plastic film manufacturing, as well as irrigation energy consumption were estimated from synthesizing data available from Chinese literature though limited. The selection of emission factors may cause some uncertainties in the CF calculation. Hence, the development of region-specific emission factors is one of key assignments for further studies. There could be some uncertainties induced by data collection, especially by field survey. In the second case study, we chose a farm in Sichuan province as a sample to calculate the CF of milk production. However, there may be some biases or errors for the data obtained from field survey due to the survey approaches and the knowledge of farmers. Therefore, the reasonable approach of field survey such as the design of survey forms and the selection of farm samples is very important for further studies, and increasing the number of samples may lower the uncertainties of the estimates.

We have not considered the impact of soil carbon balance on farm gate emissions in the first case study, because there were limited data concerned on soil carbon balance in a given crop production in Shandong province. The soil forms a large C pool, and there is thus scope for large amounts of C to be gained or lost from soils as a consequence of farming practice. For example, Lal (2004) reported management of repeated disturbance has turned many arable soils into C sources. But, some studies showed that the C sequestration in soil had not changed in conventional till, while reducing tillage or no till would lead to C sequestration in the soil (Smith et al. 2008). Soil carbon balance should be taken into account when CFs are estimated under different agricultural management related to carbon balance in further studies.

ACKNOWLEDGMENTS

The two case studies were conducted by Ting Luo, Ming Yan, and Qian Yue.

REFERENCES

Cheng K, Pan G, Smith P, Luo T, Li L, Zheng J, Zhang X, Han X, Yan M. 2011. Carbon footprint of China's crop production—An estimation using agro-statistics data over 1993–2007. *Agriculture Ecosystems and Environment* 142:231–237.

Commission for Energy Regulation (CER). 2007. Fuel mix and CO_2 emissions factors disclosure 2007. Reference CER/08/195. www.cer.ie.

Dubey A, Lal R. 2009. Carbon footprint and sustainability of agricultural production systems in Punjab, India, and Ohio, USA. *Journal of Crop Improvement* 23:332–350.

FAO. 2009. *The State of Food and Agriculture-Livestock in the Balance*. Rome, Italy: Food and Agriculture Organization, 180 pp.

Godfray HCJ, Beddington JR, Crute IR, Haddad L, Wawrence D, Muir J, Pretty J, Robinson S, Thomas SM, Toulmin C. 2010. Food security: The challenge of feeding 9 billion people. *Science* 327:812–818.

Hertwich EG, Peters GP. 2009. Carbon footprint of nations: A global, trade-linked analysis. *Environment Science and Technology* 43:6414–6420.

Hillier J, Hawes C, Squire G, Hilton A, Wale S, Smith P. 2009. The carbon footprints of food crop production. *International Journal of Agricultural Sustainability* 7:107–118.

IPCC. 2006. In: Eggleston HS, Buendia L, Miwa K, Ngara T, Tanabe K (eds), *2006 IPCC Guidelines for National Greenhouse Gas Inventories*. National Greenhouse Gas Inventories Programme. Tokyo, Japan: IGES.

IPCC. 2007. Climate Change 2007: Synthesis Report. In: Solomon S, Qin D, Manning M, Chen Z, Marquis M, Averyt KB, Tignor M, Miller HL (eds). *Contribution of Working Groups I, II and III to the Fourth Assessment Report of the Intergovernmental Panel on Climate Change*. Cambridge, UK and New York: Cambridge University Press, pp. 95–212.

IPCC. 2013. Summary for policymakers. In: Stocker TF, Qin D, Plattner G-K et al. (eds), *Climate Change 2013: The Physical Science Basis*. Contribution of Working Group I to the Fifth Assessment Report of the Intergovernmental Panel on Climate Change. Cambridge, UK and New York: Cambridge University Press.

Lal R. 2004. Carbon emission from farm operations. *Environment International* 30:981–990.

Lu F, Wang XK, HAN B. 2008. Assessment on the availability of nitrogen fertilization in improving carbon sequestration potential of China's cropland soil. *Chinese Journal of Applied Ecology* 19(10):2239–2250 (in Chinese with English summary).

National Development and Reform Commission of China (NDRC). 2012. The People's Republic of China second national communication on climate change. URL: http://nc.ccchina.gov.cn/WebSite/NationalCCC/UpFile/File115.pdf (accessed June 5, 2012).

Smith P, Gregory PJ. 2013. Climate change and sustainable food production. *Proceedings of the Nutrition Society* 72(1):21–28.

Smith P, Martino D, Cai ZC et al. 2008. Greenhouse gas mitigation in agriculture. *Philosophical Transactions of the Royal Society of London Series B* 363:789–813.

Smith P, Martino D, Cai Z et al. 2014. Agriculture, forestry and other land use (AFOLU). In: Metz B, Davidson OR, Bosch PR, Dave R, Meyer LA (eds), *Climate Change 2014: Mitigation*. Contribution of Working Group III to the Fifth Assessment Report of the Intergovernmental Panel on Climate Change. Cambridge, UK and New York: Cambridge University Press.

Steinfeld D, Gerber P, Wassenaar T, Caste LV, Rosales M, de Haan C. 2006. Livestock's long shadow. Environmental issues and options. Rome, Italy: food and agriculture organization of the UN. http://www.virtualcentre.org/on/library/key.pub/long shad/AO701EOO.pdf

The NDRC Climate Division. 2013. *Baseline Emission Factors for Regional Power Grids in China*.

U.S. EPA. 2012. Global anthropogenic non-CO_2 greenhouse gas emissions: 1990–2030. Washington, DC. http://www.epa.gov/climatechange/Downloads/EPAactivities/EPA_Global_NonCO2_Projections_Dec2012.pdf

Wang J, Rothausen SGSA, Conway D, Zhang L, Xiong W, Holman IP, Li Y. 2012. China's water–energy nexus: Greenhouse-gas emissions from groundwater use for agriculture. *Environmental Research Letters* 7. doi:10.1088/1748-9326/7/1/014035.

Yan XY, Yagi K, Akiyama H, Akimoto H. 2005. Statistical analysis of the major variables controlling methane emission from rice fields. *Global Change Biology* 11:1131–1141.

Yang SH. 1996. The research of City Trees effects of carbon and oxygen balance. *City Environment and Ecology* 9(001):37–39 (in Chinese).

Zou JW, Huang Y, Zheng XH, Wang YS. 2007. Quantifying direct N_2O emissions in paddy fields during rice growing season in mainland China: Dependence on water regime. *Atmospheric Environment* 41:8030–8042.

4 End of Life Scenarios and the Carbon Footprint of Wood Cladding

Andreja Kutnar and Callum Hill

CONTENTS

4.1 Introduction 85
4.2 Role of Wood and Wood Products in the Bioeconomy 87
    4.2.1 Wood as Building Material...87
        4.2.1.1 External Wooden Claddings 87
4.3 Life-Cycle Assessment 89
    4.3.1 Methodologies for Calculating Carbon Footprint of Bio-Based Materials 91
    4.3.2 LCA of Wood Products 92
        4.3.2.1 Carbon Storage in Wood Products..93
    4.3.3 Cascade Use of Wood..95
4.4 Wood in Sustainable Buildings...96
4.5 Conclusions...97
Acknowledgment 98
References..98

4.1 INTRODUCTION

Climate change is a global issue, which must be addressed all over the world. Many of the world's political and economic decisions are increasingly determined by resource and energy scarcity and by climate change. Under these circumstances, a balance must be achieved between economics, ecology, and social welfare (sustainability). In Europe, several strategic documents, directives, and initiatives were launched with the aim of developing a low-carbon economy in the European Union (EU). Among these documents are the EU Sustainable Development Strategy (European Commission 2009), which was published in 2006, and reviewed in 2009, the EU Roadmap 2050 (European Parliament Council 2008; European Commission 2011), and the recycling society directive (Directive 2008/98/EC; European Parliament Council 2008). Another important document is the Europe 2020 Strategy for smart, sustainable, and inclusive growth, where one of the headlines relates to climate and energy. Furthermore, in 2012, the European Commission launched a strategy entitled "Innovating for Sustainable Growth: A Bioeconomy for Europe" (COM 2012, 60 final). The Commission states that Europe needs to radically change its approach regarding production, consumption, processing, storage, recycling, and disposal of biological resources in order to cope with increasing global population, rapid depletion of many resources, increasing environmental pressures, and climate change. Furthermore, the European Commission adopted the Communication "Towards a circular economy: a zero waste programme for Europe" (COM 2014, 0398 final) to establish a common and coherent EU framework to promote the circular economy. The deliverables of a circular economy could be: boosting recycling and preventing the loss of valuable materials; creating jobs and economic growth; showing how new business models, eco-design, and industrial symbiosis can move us toward zero waste; and reducing greenhouse emissions and environmental impacts.

The forest-based sector is a key pillar of Europe's bioeconomy. Using wood products can contribute to significant CO_2 saving in terms of greenhouse gas (GHG) emissions, embodied energy, and energy efficiency (Hill 2011). The goals of sustainable development are to increase economic efficiency, protect and restore ecological systems, and improve human well-being, or a combination of all the three. Working toward these goals is expected to lead to new concepts, products, and processes, optimizing multiple use and reuse of forest-based resources. Life-cycle analysis, industrial ecology, and the cradle-to-cradle concepts will be used as key tools in sustainable economic development. These tools will also lead to new business opportunities through innovative products optimized for end use requirements and sustainable resource utilization.

The European Commission has published the new EU forest strategy for forests and the forest-based sector (COM 2013, 659 final). One action within the new EU forest strategy is dedicated to innovation and research and to rural development. Among other aspects, the EU forest strategy underlines the importance of building with wood and the importance of considering innovation and research activities in the whole forestry–wood chain.

Among other mitigation strategies, the Intergovernmental Panel on Climate Change (IPCC) in its Fourth Assessment Report (WG3 Chapter 9) lists "increasing off-site carbon stocks in wood products and enhancing product and fuel substitution using forest-derived biomass to substitute products with high fossil fuel requirements, and increasing the use of biomass energy to substitute fossil fuels" as an option (IPCC 2007). Sustainable production of timber, fiber, or energy from sustainably managed forests is considered to provide the best long-term mitigation benefit. However, in the Special Report of the IPCC "Renewable Energy Sources and Climate Change Mitigation" (IPCC 2012), the conclusion of Chapter 2 emphasizes that in order to achieve the high potential deployment levels of biomass for energy, increases in the competing demand for fiber must be moderate, land must be properly managed, and forestry yields must increase substantially. Expansion of bioenergy in the absence of good governance and close monitoring of land use carries the risk of low GHG benefits. Furthermore, conflicting demands for virgin fiber between the wood-based composites sector and the biomass energy sector are likely and could undermine efforts to mitigate GHG emissions. However, burning virgin fiber may not be the best option when compared with multiple use and long-life products, which are only incinerated at their end of their service life. The correct choice requires appropriate analysis.

The aim of this chapter is to demonstrate the role wood products can play in mitigating climate change, with emphasis on proper handling of wood products at the end of their service life. The use of timber in products has many environmental advantages: it is a renewable resource that can be produced sustainably, it contains sequestered atmospheric carbon, and it can be incinerated at the end of product life to recover energy. It is essential that the forest sector is placed at the center of the developing bio-based economy and has a strong voice within Europe. The value of the forest for mankind and the environment is irrefutable, and the value of the multitude of products made of wood is of great importance, socially, economically, and environmentally. Over the last 50 years, wood, sawn timber, in particular, has largely disappeared from many technological applications, diminishing its contribution to sustainability in the one area where it could be most significant: as a substitute for energy-intensive materials (e.g., in the built environment) (Hill and Norton 2014a). However, there is currently a resurgence of interest in timber products due to the environmental benefits they provide, a phenomenon that other industrial sectors are well aware of. Several recent studies dealing with environmental impact assessment of timber processing are highlighted in this chapter. Furthermore, further research activities based on the holistic approach taking into account total life cycle, cascade use of wood, and objective environmental impact assessment are discussed. Only with these activities, wood and wood products will be recognized as the key material in European and global bio-based economy.

4.2 ROLE OF WOOD AND WOOD PRODUCTS IN THE BIOECONOMY

Wood is a natural, renewable, reusable, and recyclable raw material. It can play a major role in mini-mizing negative effects on the climate and environment, when it is sourced from sustainably man-aged forests. Forest biomass is currently the most important source of renewable energy and now accounts for around half of the EU's total renewable energy consumption (COM 2013, 659 final). Wood products can contribute to climate change mitigation: they act as a carbon pool during their service life, as they withdraw CO_2 from its natural cycle, and they can substitute for more energy-intense products after their service life, as they can substitute for fossil fuels if they are incinerated (Werner et al. 2006).

Leturcq (2014) studied the effective ways of using wood production from the perspective of cli-mate change mitigation. Two major opposing concepts were compared: carbon sequestration and biomass carbon neutrality. The latter advocates that woody resources are regarded as "renewable" and their use as an energy source "neutral" with respect to the greenhouse effect. And the other concept calls for increasing the carbon stock in forests, in wood products and other long-term wood storage. Comparison of the carbon footprint of heat generation between wood and other fuels has revealed that the intrinsic carbon emission factor for wood is the highest among all fuels in common use. Furthermore, the concept of wood carbon neutrality neglects the possibility of carbon storage in long-life wood products (e.g., through reuse or in the built environment). Also, Kim and Song (2014) compared life-cycle assessments (LCAs) of two wood waste-recycling systems, production of particleboard, and energy recovery and concluded that the particleboard production scenario has a greater environmental benefit in reducing GHG emissions.

4.2.1 WOOD AS BUILDING MATERIAL

Wood is a commonly used construction material in many parts of the world because of its reason-able cost, ease of work, attractive appearance, and adequate life if protected from moisture and insects. Many wood species have the necessary physical characteristics to perform well in build-ing construction. Pajchrowski et al. (2014) summarized the qualities of wood important for its use as a building material. They range from raw material renewability to carbon balance benefits and embodied energy to its mechanical, thermal, and physical properties. However, wood properties vary among species, between trees of the same species, and between pieces from the same tree. This leads to variability in the performance of wood, which is one of its inherent deficiencies as a material (Dinwoodie 2000). To overcome this deficiency, a broad range of wood-based composites, which minimize the variability of physical, mechanical, and thermal properties of solid wood, has been developed. Wood composites are a family of materials, which contain wood either in whole or fiber form as the basic constituent (Bodig and Jayne 1982).

Wood is a durable material, which has to be accompanied with appropriate building applica-tions and design. The natural durability of wood has been proved by the multitude of buildings that have stood for centuries. However, while wood's natural attributes make it a sustainable building material, they also make wood vulnerable to decay and wood-destroying insects. Therefore, proper design, installation, and detailing are critical to ensure long-term durability. When wood is used in exposed applications, or in areas where it is subjected to moisture and insects, it must be protected with mechanical barriers, coatings, and, in some instances, preservative treatments. Several emerg-ing environmentally friendly processes of wood modification (chemical, thermal, and impregna-tion/polymerization) have been developed, which can improve the intrinsic properties of wood, and provide desired form and functionality.

4.2.1.1 External Wooden Claddings

One wood product category commonly subjected to wood modification is external wood claddings. These products require dimensional stability and durability, as well as decay resistance. There are various wood modification methods, all of which permanently alter the internal components (Figure 4.1). These include methods like heat treatment, acetylation, and polymer grafting (Hill 2006).

Wood modification by heat treatment has been generally accepted as a possible way of improving some characteristics of wood (Esteves and Pereira 2009). In different heat-treatment processes (exposing wood to temperatures between approximately 150°C and 220°C), the main purpose is to achieve new material properties, such as increased biological durability, enhanced dimension stability, and the possibility of controllable color changes. At present, primarily five different commercial treatments have emerged: the Finnish Thermowood, the Dutch Plato Wood, the French Retification and Bois Perdure processes, and the Oil Heat Treatment (Militz 2002; Homan and Jorissen 2004; Esteves and Pereira 2009). New heat treatment processes are also emerging in other countries, such as Denmark (WTT) and Austria (Huber Holz) (Esteves and Pereira 2009). The main differences between these methods are based on the materials used (e.g., wood species, fresh or dried wood, moisture content, dimensions), process conditions applied (e.g., one or two process stages, wet or dry processing, steaming, heating medium, oxygen or nitrogen as sheltering gas, heating and cooling speed) and the equipment necessary for treatment (e.g., process vessel, kiln).

In the French Retification process (Retified wood), developed by École des Mines de Saint-Etienne, wood of approximately 12% moisture content (MC) is heated slowly in a specific kiln up to 210–240°C in a nitrogen atmosphere with less than 2% oxygen. Another process developed in France is Le Bois Perdure (BCI-MBS), which is a two-step process. Fresh wood is first dried in the kiln and then heated to 240°C in a steam atmosphere (steam generated from moisture of the timber).

In Finland, ThermoWood has been developed at the Finnish Research Center VTT together with the Finnish industry. It consists of three phases: preliminary heating stage at 100°C, kiln drying process between 100°C and 150°C, and a slow temperature-heating period to the actual treatment temperature (150°C–240°C), which is held for 0.5–4 h.

In Germany, the OHT-Process (oil-heat treatment, Menz Holz) has been developed based on a thermal treatment in a hot bath of oil. The treatment is performed in a closed vessel with hot oil circulating around the wood, which excludes oxygen from the wood during treatment. The heating medium is crude vegetable oil, such as rape seed, linseed oil, or sunflower oil. Temperatures between 180°C and 220°C are used for a 2–4 h period, excluding heating up and cooling down, and depend on the dimensions of the wood (Rapp and Sailer 2001).

The heat treatment method developed in the Netherlands is the PLATO (Providing Lasting Advanced Timber Option) technology (Boonstra et al. 1998). This technology consists of a two-stage treatment under relatively mild conditions, temperatures below 200°C, and combines successively a hydrothermolysis step with a dry curing step. It uses the presence of abundant moisture

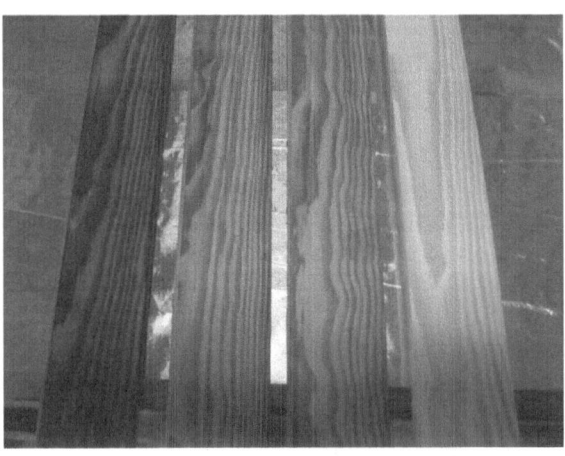

FIGURE 4.1 Color changes in Scots pine wood due to thermal processing.

in the woody cell wall during the hydrothermolysis, which provokes an increased reactivity of the cell-wall components under relatively low temperature. In the first step (hydrothermolysis) of the process, green or air-dried wood is treated at temperatures typically between 160°C and 190°C (4–5 h) under increased pressure (super atmospheric pressure). A conventional wood drying process (3–5 days) is used to dry the treated wood to a low MC (ca. 10%). In the second step, the dry intermediate product is heated again to temperatures between 170°C and 190°C (14–16 h). The heating medium can be steam or heated air.

The service life of thermally modified wood differs among processes. The Finnish ThermoWood used as uncoated ThermoWood pine exterior cladding has a service life of 30 years. The producer delivered this calculation based on the quality of the workmanship at the time of installation and the exposure of cladding to the weather including location, elevation, and design. Furthermore, the producer reports that if ThermoWood pine cladding is coated over the life of the product, its service life can be as high as 60 years. However, it is necessary that the coating is maintained in line with the coating manufacturer recommendations over the service life of the product.

An example of wood modification with acetylation is Accoya® Wood Cladding, which is produced with the nontoxic acetylation process of radiate pine wood. Accoya® Wood has outstanding dimensional stability and good durability properties. It is resistant to a wide range of insects, including termites, and forms an effective barrier to any attack. Additionally, it is resistant to decay. Uncoated Accoya® Decking and Cladding are maintenance-free and meet durability Class 1 (EN 350-1, 1994). The expected lifespan is at least 60 years when more than 20 cm above the ground.

Another modified wood used for cladding is Kebony Cladding, which is produced by forming stable, locked-in furan polymers in the wood cell walls. The process is based on impregnation with furfuryl alcohol, which is produced from agricultural crop waste. Impregnation is followed by heating, which induces in situ polymerization of the furfuryl alcohol. It should be noted that a plant-derived waste product is used to give enhanced strength and durability to wood. The Kebony cladding has high durability, low life-cycle costs, and long lifespan. Kebony's durability against rot decay conforms to Durability Class 1–2, which is the most durable category of the European norm EN 350. The producer states that the Kebony claddings are practically maintenance-free.

Manufacturers of these products have, to some extent, considered the environmental impacts related to their products and some also obtained environmental products declarations (EPDs). However, a more detailed consideration reveals that the global environmental impact of timber modification processing and further uses of the modified wood products are not yet included in the development of processes and products. In fact, their environmental impacts are, to a large extent, still unknown, which must change to meet the demands of increasingly conscientious business and consumer markets desiring to make environmentally responsible decisions regarding the goods and services they purchase. Furthermore, due to the EU's on-going transition to a recycling society (Directive 2008/98/EC), modified timber product design should include efficient product reuse, recycling, and end-of-life use (cradle to cradle) to more effectively contribute to a low-carbon economy. Consequently, the need has emerged to develop methodologies that allow for informed purchasing decisions to be made regarding environmental impacts, which is discussed in detail bellow.

4.3 LIFE-CYCLE ASSESSMENT

LCA is a rational, quantified approach to determining specific environmental impacts of a product or system through its entire life cycle. As solutions are sought to reduce the impacts of buildings, LCA is seen as an objective measure for comparing building designs. LCAs clearly have an important role to play in assessing the sustainability of green buildings and are a valuable tool in decision making.

Studies found LCA to have started in 1960s (Hunt et al. 1992). From the time when LCA analysis was developed until today, numerous methodologies to classify, characterize, and normalize environmental effects were developed. The most common, for example, CML 2 (2000), IPCC

GHG emissions, Ecopoints 97, and Eco-indicator 99 (PRé Consultants 2010) are focused on the following indicators: acidification, eutrophication, thinning the ozone layer, various types of eco-toxicity, air contaminations, usage of resources, and GHG emissions. At first, LCA analysis was mostly focused on environmental effects like acidification and eutrophication, whereas in more recent years mostly on GHG emissions, which are also called carbon footprints. The carbon footprint (or climate declaration) is expressed in terms of the amount of emitted carbon dioxide or its equivalent of other GHGs.

The LCA methodology involves four steps (ISO 14040 2006; Puettmann et al. 2010). The goal and scope definition step spells out the purpose of the study and its breadth and depth. The second step, life-cycle inventory (LCI) quantifies the environmental inputs and outputs associated with a product over its entire life cycle. Inventory analysis entails quantifying the inventory flows for a product system. Inventory flows include inputs of water, energy, and raw materials, and releases to air, land, and water. The next LCA step, impact assessment (LCIA), characterizes these inventory flows in relation to a set of environmental impacts as identified in the LCI. Finally, the interpretation step combines environmental impact in accordance with the goals of the study.

For a product, the life cycle starts with procuring the raw material, primary processing, secondary processing or manufacturing, packaging, shipping and handling, installation, in-use energy consumption, maintenance, and end-of-life strategies. LCA is performed for various stages of a products life. For example, cradle-to-gate refers to LCA from raw material stage to the point it is shipped from the manufacturer. Similarly, cradle-to-grave involves LCA of all stages of the product or the material, starting from raw material procurement to end-of-life strategies. For wood products, the life cycle generally starts with extraction of raw resources from the natural environment or recovery of materials from a previous use. The raw resources are then manufactured into useable products. The finished products are shipped to a site, consuming energy in the process. During the service life of the product, it may consume energy based on its use (e.g., energy used to maintain the product). Over time, renovations or retrofitting may be performed on the products, which may require additional materials and energy. Finally, the product is removed/demolished and its materials disposed of either as construction waste or recycled for reuse. Each of these steps consumes energy and materials and produces waste. The purpose of LCA is to quantify how a product or system affects the environment during each phase of its life. Examples of parameters that may be quantified include: energy consumption, resource use, GHG production, solid waste generation, and pollution generation. Another option is to perform LCA from cradle-to-cradle, which means that the analysis includes efficient product reuse, recycling, and end-of-life use.

With the objective to develop a framework that allows for comparability of environmental performance between products, ISO 14025 (2009) was introduced. This describes the procedures required in order to produce Type III EPD. This is based on the principle of developing product category rules (PCRs), which specify how the information from an LCA is to be used to produce the EPD. A PCR will typically specify what the functional unit is for the product. Within the framework of ISO 14025, it is only necessary for the production phase (cradle to gate) of the lifecycle to be included in the EPD. It is also possible to include other lifecycle stages, such as the in-service stage and the end-of-life stage, but this is not compulsory.

European EPDs until now have been on the PCR for "wood materials," which was released by the German Institute for Construction and Environment (Institut Bauen und Umwelt e.V.), in November 2009. The PCR determines five impacts on the environment: global warming potential [global warming (GWP100)], acidification potential, eutrophication potential, smog potential (photochemical oxidation), and ozone depletion potential (ozone layer depletion). As an example, in Table 4.1, the results of a cradle-to-gate LCA analysis of 1 m^3 of planed kiln-dried softwood sawn-wood calculated by CML 2 baseline 2000 V2.05 are shown. The emission factors were taken from database Ecoinvent 3 (2013). This PCR will now be superceded by the standard EN 16485 (2014), which was recently published.

TABLE 4.1
LCA Analysis of 1 m³ of Planed Kiln-Dried Softwood Sawnwood

Impact Category	Unit	Sawnwood, Softwood, Kiln-Dried, Planed
Acidification	kg SO_2-eq.	−0.261
Eutrophication	kg PO_4^--eq.	0.057
Global warming (GWP100)	kg CO_2-eq.	−211
Ozone layer depletion (ODP)	kg CFC-11-eq.	−7.67E−07
Photochemical oxidation	kg C_2H_4-eq.	−0.456

4.3.1 Methodologies for Calculating Carbon Footprint of Bio-Based Materials

Different methodologies can be used to calculate the carbon footprint of wood products. Some of them, like climate declaration, do not account for biogenic CO_2. When using these methodologies, disregarding carbon storage in the assessment of the carbon footprint of wood products may bias the comparison with competing products that do not store biogenic carbon (Garcia and Freire 2014). Furthermore, the embedded carbon should be reported when presenting cradle-to-gate results to avoid misleading comparisons with other products because the embodied carbon at the gate may be released during the use or end-of-life phases (Garcia and Freire 2014).

Pawelzik et al. (2013) compared seven LCA methodologies and emphasized how they account for biogenic carbon storage. The methodologies compared were from the French Environment and Energy Management Agency (ADEME), the European Commission's Lead Market Initiative, the GHG Protocol Initiative, ISO/TS 14067, the International Reference Life Cycle Data (ILCD) Handbook, PAS 2050, and process/material carbon footprint. The authors concluded that the lack of standardized approaches to accounting for the biogenic carbon storage in bio-based materials presents a key challenge to LCA calculations. In order to properly account for the benefits of storing atmospheric carbon in long-life products, it is necessary to develop internationally agreed accounting procedures. In particular, it is necessary to determine the effect of the carbon-storage duration.

Hill et al. (2014) reviewed the methodologies for evaluating the atmospheric carbon stored in products. The ILCD Handbook, published by the European Commission Joint Research Centre (Institute for Environment and Sustainability), considers a 100-year assessment period and recommends that fossil and biogenic carbon releases (as carbon dioxide and methane) are differentiated. Carbon emissions associated with land-use changes and from biomass associated with virgin forests are treated as fossil carbon. In contrast, emissions associated with plantation forests are inventoried as biogenic carbon. Uptake of atmospheric carbon dioxide is inventoried as "resources from air." A methodology is given for accounting for the removal and storage of atmospheric carbon dioxide. The uptake of atmospheric carbon dioxide is inventoried as "Carbon Dioxide—Resources from Air" and the emissions as "Carbon Dioxide (Biogenic)—Emissions to Air." These two flows then cancel each other out. Meanwhile, the issue of the storage in the product is calculated by declaring a correction flow for delayed emission of the carbon dioxide and giving it a value of 0.01× the CO_2 equivalent mass stored per year. The same method is used to calculate the storage of fossil carbon in a long-life product, except that there is no consideration given to the category "Carbon Dioxide—Resources from Air." Thus, there is a net effect of the release of the fossil-derived CO_2 at the end of life, but the compensatory effect of the delayed emission of the fossil carbon is accounted for.

The British Standards Institution developed Publically Available Specification 2050 (PAS 2050 2011). This standard uses a 100-year assessment period for the storage of atmospheric carbon dioxide as the basis for calculation. However, although a temporal weighting factor was included in the

TABLE 4.2

Fossil CO_2-eq., Biogenic CO_2-eq., CO_2-eq. from Land Transformation, and CO_2 Uptake of 1 m^3 of Planed Kiln-Dried Softwood Sawnwood

	[kg CO_2 eq.]
Fossil CO_2-eq.	−186
Biogenic CO_2-eq.	440
CO_2-eq. from land transformation	1.28
CO_2 uptake	−1229

original (2008) version of this specification, for the 2011 version, it was no longer a requirement and was removed to an annex (E) "Calculation of the weighted average impact of delayed emissions arising from the use and final disposal phases of products."

In the prestandard (prEN 16485) describing PCRs for wood and wood-based products in construction, calculations were included for reporting sequestered carbon in long-life timber products which were based upon the PAS 2050 methodology. However, in the published (2014) version of EN 16485, these temporal calculation methods were no longer present.

The GHG Protocol method uses the characterization factors per substance that are identical to the IPCC 2007 GWP (100a) method. The only difference is that carbon uptake and biogenic carbon emissions are included in this method and that a distinction is made between: fossil-based carbon (carbon originating from fossil fuels), biogenic carbon (carbon originating from biogenic sources such as plants and trees), carbon from land transformation (direct impacts), and carbon uptake (CO_2 that is stored in plants and trees as they grow). In Table 4.2, the carbon footprint of 1 m^3 of planed kiln-dried softwood sawnwood was calculated with GHG Protocol in Simapro 8.0 using the emission factors from database Ecoinvent 3 (2013). The results could be different, if different scenarios for biogenic carbon storage and release would be taken into consideration (Kutnar and Hill 2014). The results of net carbon footprint of sawnwood from old grown forest would be different from net carbon footprint of sawnwood from plantation forest.

4.3.2 LCA of Wood Products

Several studies have dealt with LCA of forests and primary wood products, but most of these tend to be cradle to gate, producing an information deficit across the entire value chain. Reasons for limited studies dealing with cradle to grave and with cradle to cradle lay in lack of data about the use phase related to use-phase maintenance, repair, refurbishment, and replacement, as well as data related to deconstruction, demolition, and waste processing, and data related to reuse, recovery, and recycling. It is important that future studies consider the whole life cycle, including the end of life and recycling of wood products. Obtaining the listed data can be a challenge, since we do not have control over it. Furthermore, the service life of different wooden products should be determined in standardized way, which would objectively support the LCA in the whole life cycle.

The number of LCA studies in the wood sector is relatively limited, geographically distributed, uses a variety of databases, and impacts assessment protocols (Kutnar and Hill 2014). Comparison between different production processes is not possible given the availability of information, different system boundaries, and assessment criteria. A comparison of different production methods using common calculation rules is clearly required. Furthermore, very limited research has been performed on environmental impact assessments of wood modification processes and products, precluding modified wood as a material of choice in the Green Guide Specifications. It is essential that research of timber processing and the resultant products place more emphasis on the interactive assessment of processes parameters, developed product properties, and environmental impact, including recycling and disposal options at the end of the service life. Intelligent concepts for reuse and cycling of valuable materials at the end of a single product life could reduce the amount of waste destined for landfills or down-cycling and align well with the cradle-to-cradle concept. With new and innovative production technologies, reduced overall energy consumption, increased recycling of wood products, and reuse and refining of side-streams, the sector can become leader on the path to achieving resource efficiency, a low-carbon economy, and sustainability.

Werner and Richter (2007) reviewed the results of international research on the environmental impact of the life cycle of wood products used in the building sector compared with functionally equivalent products from other materials. The study concluded that fossil fuel consumption, potential contributions to the greenhouse effect, and quantities of solid waste tend to be minor for wood products compared with competing products. Furthermore, impregnated wood products tend to be more critical than comparative products with respect to toxicological effects and/or photo-generated smog depending on the type of preservative. Bolin and Smith (2011a) compared environmental impacts related to borate-treated lumber, commonly used as structural lumber in the United States, with the primary alternative product, galvanized steel framing, with a cradle-to-grave LCA. The cradle-to-grave life-cycle impacts of borate-treated lumber framing were approximately four times less for fossil-fuel use, 1.8 times less for GHGs, 83 times less for water use, 3.5 times less for acidification, 2.5 times less for ecological impact, 2.8 times less for smog formation, and 3.3 times less for eutrophication than those for galvanized steel framing. The same authors, Bolin and Smith (2011b), also performed a cradle-to-grave LCA of alkaline copper quaternary (ACQ)-treated lumber used for decking and to determine how the impacts compare to the primary alternative product, wood plastic composite (WPC) decking. The study assumed that the ACQ decking has a service life of 10 years and that they are demolished and disposed in a solid waste landfill after the end of use. Similarly, the LCI of WPC was assumed to be 10 years. In both, compared decking materials, no maintenance, like chemical cleaners and refurbishing, was included in the LCI. The results of the cradle-to-grave LCA showed that ACQ-treated lumber impacts were 14 times less for fossil-fuel use, almost three times less for GHG emissions, potential smog emissions, and water use, four times less for acidification, and almost half for ecological toxicity than those for WPC decking. Impacts were approximately equal for eutrophication. Hill and Norton (2014b) discussed the environmental impact of the wood modification process in relation to life extension of the material. A comparison among different wood modification treatments was made. By determination of carbon neutrality, they identify the point at which the benefits of life extension compensate for the increased environmental impact associated with the modification. The effect of increased maintenance intervals of modified wood products could be a powerful argument in favor of their use.

Kutnar and Hill (2014) used a cradle-to-gate analysis to present the carbon footprint of 14 different primary wood products. The largest source of emissions for all sawn timber products, is removing the timber from the forest, while for kiln-dried sawn timber the drying process follows closely behind. For fiber composites (MDF and HDF), the extra energy required to convert the raw material into fibers, in addition to the energy required to apply pressure and heat to the products is responsible for the bulk of the emissions from these products. The adhesives used in particleboard, plywood, and OSB are responsible for the largest fraction of emissions from these products. This is especially significant considering the low total volume they represent in the final products. Glulam emissions derive mostly from the harvest and initial production of the softwood, but also from the extra energy required to apply pressure and set the adhesives used. Altering the system boundaries would yield different results. Furthermore, the results would have been different if the carbon footprint calculation accounted for carbon sequestration of wood, or considered the use of recycled wood products and other similar issues pertinent to LCA. Additionally, the results would have also differed if the full life cycle of products was considered (e.g., cradle to grave, cradle to cradle).

4.3.2.1 Carbon Storage in Wood Products

Trees capture atmospheric carbon dioxide via photosynthesis and a proportion of this sequestered carbon is stored in the aboveground woody biomass. Wood is composed of three main biopolymers (cellulose, hemicellulose, and lignin) and to a first approximation the elementary composition can be assigned a stoichiometric ratio of CH_2O. This means that atmospheric carbon comprises a minimum of 40% of the dry wood mass (increasing somewhat with increasing lignin content). Each ton of dry wood therefore equates to the removal of approximately 1.5 tons of atmospheric carbon dioxide (the ratio of the molecular weight of CO_2 compared to CH_2O—44/30). The net benefit of this ability to store atmospheric carbon depends on the length of time before the material is subsequently oxidized and the carbon released back to the atmosphere (Kutnar and Hill 2014). In all situations where carbon flows and stocks are considered, it is essential that a distinction be made between biogenic and fossil carbon sources. Even with biogenic carbon, it is important to differentiate between carbon that is held in long-term storage (such as in an old-growth forest) and that derived from newer managed or plantation forests. Kutnar and Hill (2014) discussed different scenarios of accounting for carbon storage in wood products. When an old growth forest is burnt and the land cleared for alternative use, fossil carbon is released to the atmosphere (carbon stored in old growth forest is treated the same as subterranean fossil carbon). The carbon content was previously held in long-term (historical) storage. Thus, although technically this is biogenic carbon, it represents carbon that would have been in storage prior to the industrial revolution was part of the natural biogenic cycle and can be considered equivalent to fossil carbon. The concentration of this "fossil" carbon in the atmosphere gradually decreases after the release (the Bern cycle) as it is removed by sequestration in oceanic and terrestrial sinks. The second scenario is when a new forest plantation is established and the trees are allowed to grow for 50 years before harvesting and restocking. Carbon is removed from the atmosphere as the atmospheric carbon dioxide is photosynthetically bound in the biomass. The overall result is a benefit because atmospheric carbon dioxide has been sequestered. If the forest biomass is subsequently burnt for energy recovery after 50 years, then the aboveground biomass is oxidized and the accumulated atmospheric carbon is lost. The overall result is still a benefit in terms of carbon sequestration. This is because there has been removal of atmospheric carbon dioxide during the 100-year period of consideration. When the aboveground biomass is subsequently burnt, this results in the return of atmospheric carbon dioxide. This only applies because a new forest was created. However, burning virgin woody biomass cannot seriously be considered an effective mitigation strategy. Far better is one in which the calorific value of the biomass is utilized and substituted for a fossil-fuel alternative. The benefit then arises not only from the storage of atmospheric carbon in the growing biomass, but additionally from the avoided emission of the fossil carbon. The third scenario is when the biogenic carbon embedded in the plantation forest is stored in timber products for 50 years, before it is used to generate energy. In this way, three benefits are realized. During the growth phase of the forest, carbon dioxide is sequestered due to the incremental growth of the trees. After harvesting, the carbon continues to be stored in the timber products. It is only at the end of the life that this stored carbon is released into the atmosphere. However, if the wood is burnt with energy recovery, then there is also the benefit of the avoided emission of the fossil carbon. An even better option is to cascade the wood material down the product value chain through several life cycles before final incineration with energy recovery.

Levasseur et al. (2013) demonstrated the importance of considering biogenic carbon and the timing of GHG emissions using an illustrative case study performed on a fictitious wooden chair for four end-of-life scenarios (incineration, landfilling, refurbishment, and incineration with energy recovery) with five different methods. The authors considered a traditional LCA approach that does not account for biogenic carbon, a traditional LCA approach modified to account for biogenic carbon uptake and emissions, the application of the PAS 2050 carbon footprint specification, the application of the ILCD Handbook method, and the dynamic LCA approach. The authors concluded that a comparative LCA could change depending on the chosen time horizon for the analysis. In order to have transparent choices, it is suggested that the dynamic LCA approach is used. Additionally, Pawelzik et al. (2013) discussed accounting for biogenic carbon storage in LCA and recommended it should be accounted for depending on a product specific life cycle and the duration of carbon storage.

4.3.3 Cascade Use of Wood

Another aspect that has to be considered is cascade use of wood. Efficient resource use is the core concept of cascading, which is a sequential use of a certain resource for different purposes (Höglmeier et al. 2013). This means that the same unit of a resource is used for multiple high-grade material applications followed by a final use for energy generation (Figure 4.2). Intelligent concepts for reuse and recycling of valuable materials at the end of single product life will reduce the amount of waste to be landfilled. Reuse of engineered solid wood products from large structures is a valuable source of already graded and dried wood for recycled construction wood in the timber frame housing industry. At the end of their first product life, the majority of wood-based products provide material qualities and quantities that would allow a further use in solid or stepwise disintegrated form (as strands, particles, or fibers). Despite this, recovered wood is currently mainly burned in waste incineration plants, or—with higher power efficiencies—in special wood combustion facilities. However, this ignores the preferred option to hold the bio-based, carbon-storing wooden materials in a maximum quality level by reuse in solid form and subsequent recycling of reclaimed wood in as many steps of a material cascade as possible. In the wood product sector, the potential of efficient material cascading of engineered solid wood products is largely underdeveloped.

Höglmeier et al. (2013) discussed the potential for cascading of recovered wood from building deconstruction. The study focused on recovered wood from the building sector in large dimensions and without contamination that should first be used to produce timber of smaller dimensions, such as lamellas, which can be chipped after their service life and used in the next cascading step as particles or fibers, and finally as energy. The study concluded that in southeast Germany, 45% of recovered wood from building deconstruction is potentially suitable for use as raw material for particle or fiberboard production. Furthermore, 26% of the recovered wood is suitable for reuse and 27% could be channeled into other high-value secondary applications. The results showed that by utilization of recovered wood in a cascade could increase the time span of carbon storage in the wood products and consequently delay its contribution to the greenhouse effects.

The amount of carbon dioxide released into the atmosphere while wood is burned and decomposed is the same as that absorbed by the tree during growth. Still, incineration of wood products at end of life provides various environmental benefits. The use of woody biomass as feedstock for biofuels production avoids the food versus fuel debate, which makes it more attractive from the environmental perspective (Wang 2005). However, Rivela et al. (2006a,b) applied a multicriterial

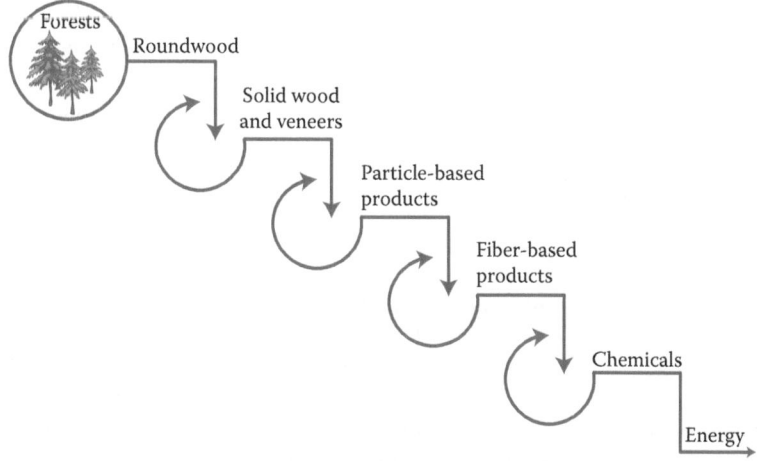

FIGURE 4.2 Cascading use of wood. (Adapted after Höglmeier K, Weber-Blaschke G, Richter K. 2013. *Resources Conservation and Recycling* 78:81–91.)

approach in order to define the most adequate use of wood wastes. The study concluded that based on the environmental, economical, and social considerations, the use of forest residues in particleboard manufacture is more sustainable than their use as fuel. The results of LCA study might differ due to the type and management of raw materials, conversion technologies, end-use technologies, system boundaries, and reference energy systems with which the bioenergy chain is compared (Hall and Scrase 1998). Furthermore, Kim and Song (2014), by applying the temporary biomass carbon storage and using the LCA methodology based on ISO 14040 and ISO 14044, demonstrated that particleboard production from wood wastes produces 428 kg CO_2-eq. less compared to particleboard from fresh wood and that energy production using wood wastes is 154 kg CO_2-eq. less compared to that of the combined heat- and power-generation process. The results clearly show that cascading through several life cycles prior to incineration is a better option.

4.4 WOOD IN SUSTAINABLE BUILDINGS

The environmental properties of wood—and other construction materials—are currently entering building codes. However, the limited availability of emissions data and its poor integration to real-life decision making within the construction sector have kept construction industries from using environmental arguments for material choices. For buildings, the life cycle generally starts with extraction of raw resources from the natural environment or recovery of materials from a previous use. The raw resources are then manufactured into useable products, such as steel, concrete, and so on. The finished products are then shipped to the site, consuming energy in the process. On the site, the products are assembled into a building. The building consumes energy throughout its service life. In due course of time, renovation or retrofit is performed on the building, which uses more materials and energy. Finally, the building is removed/demolished and its materials disposed of either as construction waste or recycled for reuse. Each of these steps consumes energy and materials and produces waste. The purpose of the LCA quantifies how a building product or system affects the environment during each phase of its life. As sustainability becomes a dominant aspect of building development, the environmental impact of building materials should be included during planning with special attention paid to the life cycle and embodied energy of the materials used. In the last few decades, several studies with the aim to reduce the environmental impact of buildings were performed. Similarly, new methods of assessing a building's sustainability are under development (Ding 2008; Haapio and Viitaniemi 2008; Sinha and Kutnar 2012), as well as different green building rating systems to assess how well a building project meets a specified set of sustainability criteria (Sinha and Knowles 2014).

A building uses most of its energy during its service life, which is about 90% of the total life-cycle energy (Citherlet and Defaux 2007; Newsham et al. 2009). Therefore, the adoption of a life-cycle approach, where not only current energy concerns are accounted for, but also long-term energy, environmental, and social impacts, should be taken into account and integrated into design. In sustainable design, durability is increasingly being included on priority lists under the assumption that designing for longevity is an environmental imperative. However, this is unsupported in the absence of LCA and accurate lifespan predictions. In the worst case, designing for longevity can lead to design choices that are well intentioned but, in fact, yield poor environmental results. Rather than attempt to predict the future and design permanent structures with an infinite lifespan, design for easy adaptation and material recovery should be considered.

Building performance assessment is complex, as it has to respond to multiple criteria, like energy consumption, acoustical performance, thermal occupant comfort, indoor air quality, and many other issues. An assessment requires the use of predictive models that involve numerous design and physical parameters as their inputs (Hopfe et al. 2013). Kuittinen et al. (2013) calculated the energy efficiency and carbon efficiency for eight wood-framed buildings from different European regions (Austria, Finland, Germany, Italy, and Sweden). Using real buildings as an example, the study showed how the use of wood affects the carbon footprint and primary energy demand of buildings and concluded that country-specific data should be used in calculations. Furthermore, Carre (2011) performed cradle-to-grave analysis to compare environmental indicators of five different constructions of typical single-story Australian homes with the same energy performance. Building materials, construction, operation, maintenance, and end-of-life management phases were included. The results showed that global warming indicators are remarkably similar between the construction types. Variation between construction types is minimal for most indicators due to the dominance of the operational aspect of the building life cycle. Furthermore, a comparison of life-cycle impacts excluding operation showed that timber-based construction tends to have lower global warming impacts than alternatives. In general, construction types incorporating timber tend to have lower global warming, resource use, and embodied energy outcomes.

Using wood in the built environment has several benefits besides neutral carbon balance and the possibility of recovering energy. These include other stages of a building's life cycle, such as: transport (lower weight of shipping), building site (shorter construction time and lower energy and water use, as well as lower amount of building waste), and demolition (lower electric energy consumption) (Pajchrowski et al. 2014). On the other hand, using wood as a building material results in more frequent conservation and renovation activities during the use phase of a building's life cycle, which each carries its own environmental consequences due to material production and generated waste (Pajchrowski et al. 2014). However, Kuittinen et al. (2013) delivered arguments why wood and wood-based products have a huge potential on carbon-efficient construction over the whole life cycle of a building. The carbon-efficient construction is due to carbon-neutral source of renewable raw material, local sourcing, low embodied energy, eco-efficient technical performance, zero waste, and carbon storage. In the use phase, they argue that, because of the flexibility, ease of renovation and refurbishment, as well as the limited construction waste produced, wood and wood-based products have high material substitution potential with high energy recovery. Furthermore, even fire and acoustic performance can be superior to competing materials when proper design choices are made. Also, the thermal insulation properties of wood are leading to CO_2 savings in terms of embodied energy and in-use energy. In the end-of-life stage, the contribution to carbon-efficient construction is due to its easy dismantling and limited waste as it can be recycled and/or recovered as energy. Additionally, cascading wood multiplies material substitution and energy substitution effects, which increases availability of wood as a raw material and is associated with the same beneficial climate mitigation effects as the direct use of wood from forests.

4.5 CONCLUSIONS

Wood has a negative carbon footprint because of the carbon dioxide fixed by the original living tree. The emissions associated with harvesting, transporting, and processing sawnwood products are small compared with the total amount of carbon stored in the wood. This means that even when energy used for harvesting, transport, and processing are taken into account, sawnwood still has a negative footprint. The processes of wood modification require additional energy inputs to process raw materials, manufacturing byproducts, and recycled wood into the desired form, as well as additives to form the composite matrices, which considerably increases the carbon footprint of these wood products. Therefore, it is essential that research of timber processing and the resultant products place more emphasis on the interactive assessment of processes parameters, developed product properties, and environmental impact, including recycling and disposal options at the end of the service life toward up-cycling after their service life based on the cradle-to-cradle concept. Intelligent concepts for reuse and cycling of valuable materials at the end of a single product life could reduce the amount of waste destined for landfills or down-cycling. With new and innovative production technologies, reduced overall energy consumption, increased recycling of wood products, and reuse and refining of side streams, the sector can become leader on the path to achieving the European Commission's ambitious target of 80% reduction in CO_2 emissions by 2050.

ACKNOWLEDGMENT

Andreja Kutnar would like to acknowledge the Slovenian Research Agency for financial support within the frame of the project Z4-5520.

REFERENCES

Bodig J, Jayne B. 1982. *Mechanics of Wood and Wood Composites*. New York: Van Nostrand Reinhold Company, p. 712.

Bolin CA, Smith ST. 2011a. Life cycle assessment of borate-treated lumber with comparison to galvanized steel framing. *Journal of Cleaner Production* 19:630–639.

Bolin CA, Smith ST. 2011b. Life cycle assessment of ACQ-treated lumber with comparison to wood plastic composite decking. *Journal of Cleaner Production* 19: 620–629.

Boonstra M, Tjeerdsma BF, Groeneveld HAC. 1998. Thermal modification of non-durable wood species. 1. The PLATO technology: Thermal modification of wood. International Research Group on Wood Preservation. Document No. IRG/WP/998-40123.

Carre A. 2011. A comparative life cycle assessment of alternative constructions of a typical Australian house design. Project report PNA147-0809. Melbourne: Forest and Wood Products Australia.

Citherlet S, Defaux T. 2007. Energy and environmental comparison of three variants of a single family house during its whole life span. *Building and Environment* 42:591–598.

COM 2012. 60 final 2012. Innovating for sustainable growth: A bioeconomy for Europe; published on 13 February 2012. http://ec.europa.eu/research/bioeconomy/pdf/201202_innovating_sustainable_growth_en.pdf (accessed September 28, 2014).

COM 2013. 659 final 2013. A new EU Forest Strategy: For forests and the forest-based sector; published on September 20, 2013. http://ec.europa.eu/agriculture/forest/strategy/communication_en.pdf (accessed September 28, 2014).

COM 2014. 0398 final 2014. Towards a circular economy: A zero waste programme for Europe; published on July 2, 2014. http://eur-lex.europa.eu/legal-content/EN/NOT/?uri=CELEX:52014DC0398 (accessed September 28, 2014).

Ding CKC. 2008. Sustainable construction: The role of environmental assessment tools. *Journal of Environmental Management* 86:451–464.

Dinwoodie JM. 2000. *Timber: Its Nature and Behaviour*. Second edition, London, New York: EFN Spon, 257 pp.

Ecoinvent 3.0. 2013. Swiss Centre for Life Cycle Inventories, Dübendorf, Switzerland.

EN 350-1: 1994. 1994. Durability of wood and wood-based products—Natural durability of solid wood: Guide to the principles of testing and classification of the natural durability of wood.

EN 16485. 2014. Round and sawn timber. Environmental product declarations. Product category rules for wood and wood-based products for use in construction.

Esteves MB, Pereira H. 2009. Wood modification by heat treatment: A review. *Bioresources* 4:370–404.

European Commission. 2009. Mainstreaming sustainable development into EU policies: 2009 Review of the European Union Strategy for Sustainable Development. Communication. Brussels: European Commission European Commission. http://eur-lex.europa.eu/LexUriServ/LexUriServ.do?uri=CELEX:52009DC0400:EN:NOT (accessed September 15, 2013).

European Commission. 2011. A roadmap for moving to a competitive low carbon economy in 2050. Communication. Brussels: European Commission European Commission. http://eur-lex.europa.eu/LexUriServ/LexUriServ.do?uri=CELEX:52011DC0112:EN:NOT (accessed September 15, 2013).

European Parliament Council. 2008. Directive 2008/98/EC of the European Parliament and of Brussels: European Parliament European Parliament. http://eur-lex.europa.eu/LexUriServ/LexUriServ.do?uri=CELEX:32008L0098:EN:NOT (accessed September 15, 2013).

Garcia R, Freire F. 2014. Carbon footprint of particleboard: A comparison between ISO/TS 14067, GHG Protocol, PAS 2050 and Climate Declaration. *Journal of Cleaner Production* 66:199–209.

Hall DO, Scrase JI. 1998. Will biomass be the environmentally friendly fuel of the future? *Biomass & Bioenergy* 15(4–5):357–367.

Haapio A, Viitaniemi P. 2008. A critical review of building environmental assessment tools. *Environmental Impact Assessment Review* 28:469–482.

Hill C. 2006. *Wood Modification: Chemical, Physical and Other Processes*. Chichester, UK: John Wiley & Sons.

Hill C. 2011. *An Introduction to Sustainable Resource Use*. London, UK: Taylor & Francis.

Hill C, Norton A. 2014a. Environmental standards and embedded carbon in the built environment. Chapter 10. In: Schmitz-Hoffmann C, Schmidt M, Hansmann B, Palekhov D (eds), *Voluntary Standard Systems, A Contribution to Sustainable Development*. Heidelberg: Springer.

Hill C, Norton A. 2014b. The environmental impacts associated with wood modification balanced by the benefits of life extension. *European Wood Conference on Wood Modification*, Lizboa, March 10–12, 2014, 8 pp.

Hill C, Norton A, Kutnar A. 2014. Environmental impacts of wood composites and legislative obligations. Chapter 12. In: Ansell M (ed.), *Wood Composites—From Nanocellulose to Superstructures*. Cambridge, UK: Woodhead.

Homan WJ, Jorissen AJM. 2004. Wood modification developments. *Heron* 49:361–386.

Hopfe CJ, Augenbroe GA, Hensen LM. 2013. Multi-criteria decision making under uncertainty in building performance assessment. *Building and Environment* 69:81–90.

Höglmeier K, Weber-Blaschke G, Richter K. 2013. Potentials for cascading of recovered wood from building deconstruction—A case study for south-east Germany. *Resources Conservation and Recycling* 78:81–91.

Hunt R, Sellers J, Franklin W. 1992. Resource and environmental profile analysis: A life cycle environmental assessment for products and procedures. *Environmental Impact Assessment Review* 12:245–269.

IPCC. 2007. *Intergovernmental Panel on Climate Change. Fourth Assessment Report*. Cambridge: Cambridge University Press.

IPCC. 2012. *Intergovernmental Panel on Climate Change, Renewable Energy Sources and Climate Change Mitigation, Special Report*. Cambridge: Cambridge University Press.

ISO 14040. 1997. Environmental management—Life cycle assessment—Principles and framework. Standard. Geneva, Switzerland: International Standards Organization.

ISO 14044. 2006. Environmental management—Life cycle assessment—Requirements and guidelines. Standard. Geneva, Switzerland: International Standards Organization.

ISO 14025. 2009. Environmental Labels and declarations—Type III environmental declarations—Principles and procedures. Standard. Geneva: International Standards Organization.

ISO/TS 14067. 2013. Greenhouse gases—Carbon footprint of products—Requirements and guidelines for quantification and communication. Technical Specifications. Geneva: International Standards Organization.

Kim MH, Song HB. 2014. Analysis of the global warming potential for wood waste recycling systems. *Journal of Cleaner Production* 69:199–207.

Kuittinen M, Ludvig A, Weiss G. 2013. *Wood in Carbon Efficient Construction: Tools, Methods and Application*. CEI-Bois, Brussels, 163 pp.

Kutnar A, Hill C. 2014. Assessment of carbon footprinting in the wood industry. In: Muthu SS (ed.), *Assessment of Carbon Footprint in Different Industrial Sectors, Vol. 2 (EcoProduction)*. Singapore: Springer, pp. 135–172.

Levasseur A, Lesage P, Margni M, Samson R. 2013. Biogenic carbon and temporary storage addressed with dynamic life cycle assessment. *Journal of Industrial Ecology* 17(1):117–128.

Leturcq P. 2014. Wood preservation (carbon sequestration) or wood burning (fossil-fuel substitution), which is better for mitigating climate change? *Annals of Forest Science* 71:117–124.

Militz H. 2002. Heat treatment technologies in Europe: Scientific background and technological state-of-art. Int. Res. Group Wood Preservation. Document No. IRG/WP 02-40241.

Newsham GR, Mancini S, Birt BJ. 2009. Do LEED-certified buildings save energy? Yes, but …. *Energy and Buildings* 41:897–905.

Pajchrowski G, Noskowiak A, Lewandowska A, Strykowski W. 2014. Wood as a building material in the light of environmental assessment of full life cycle of four buildings. *Construction and Building Materials* 52:428–436.

PAS 2050. 2011. Specification for the assessment of the life cycle greenhouse gas emissions of goods and services.

Pawelzik P, Carus M, Hotchkiss J, Narayan R, Selke S, Wellisch M, Weiss M, Wicke B, Patel MK. 2013. Critical aspects in the life cycle assessment (LCA) of bio-based materials—Reviewing methodologies and deriving recommendations. *Resources, Conservation and Recycling* 73:211–228.

PRé Consultants. 2010. Impact assessment methods. http://www.pre.nl/simapro/impact_assessment_methods. htm#EP97.

Rapp AO, Sailer M. 2001. Oil heat treatment of wood in Germany—State of the art. Review on heat treatments of wood. In: Rapp AO (ed.), Proceedings of the special seminar held in Antibes, France, on 9 February 2001, Forestry and Forestry Products, France. *COST Action* E22:43–60.

Rivela B, Hospido A, Moreira T, Feijoo G. 2006a. Life cycle inventory of particleboard: A case study in the wood sector. *The International Journal of Life Cycle Assessment* 11(2):106–113.

Rivela B, Moreira M, Muñoz I, Rieradevall J, Feijoo G. 2006b. Life cycle assessment of wood wastes: A case study of ephemeral architecture. *Science of the Total Environment* 357(1–3):1–11.

Sinha A, Kutnar A. 2012. Carbon footprint versus performance of aluminum, plastic, and wood window frames from cradle to gate. *Buildings* 2:542–553.

Sinha A, Knowles C. 2014. Green building and the global forest sector. In: Hansen E, Panwar R, Vlosky R (eds), *The Global Forest Sector: Changes, Practices, and Prospects*. Boca Raton, FL: CRC Press, pp. 261–280.

Wang M. 2005. Energy and greenhouse gas emissions impacts of fuel ethanol. Center for Transportation Research Energy System Division, Argonne National Laboratory. NGCA Renewable Fuels Forum, The National Press Club, August 23, 2005.

Werner F, Taverna R, Hofer P, Richter K. 2006. Greenhouse gas dynamics of an increased use of wood in buildings in Switzerland. *Climate Change* 74:319–347.

Werner F, Richter K. 2007. Wood building products in comparative LCA. A literature review. *The International Journal of Life Cycle Assessment* 12(7):470–479.

5 Carbon Footprints and Greenhouse Gas Emission Savings of Alternative Synthetic Biofuels

*Diego Iribarren, Jens F. Peters, Ana Susmozas,
Pedro L. Cruz, and Javier Dufour*

CONTENTS

5.1   Introduction 101
5.2   Production of Synthetic Biofuels 102
    5.2.1   Fast Pyrolysis and Hydro-upgrading 103
    5.2.2   Indirect Gasification and FT Synthesis 104
5.3   CF Framework 106
    5.3.1   Goal and Scope 106
        5.3.1.1   Pyrolysis-Based System (PYR-HT) 108
        5.3.1.2   Gasification-Based System (GAS-FT) 108
    5.3.2   Data Acquisition 109
        5.3.2.1   Agriculture 109
        5.3.2.2   PYR-HT Production Process 109
        5.3.2.3   GAS-FT Production Process 111
        5.3.2.4   Biofuel Use 112
5.4   Results and Discussion 114
    5.4.1   Carbon Footprints and GHG Emission Savings 114
    5.4.2   Carbon Footprint as a Single Indicator 116
        5.4.2.1   A Thermodynamic Perspective 116
        5.4.2.2   An Environmental Perspective 119
5.5   Conclusions 120
Acknowledgment 120
References 120

5.1   INTRODUCTION

Increasing greenhouse gas (GHG) concentrations in the atmosphere and the corresponding increase in global average temperature have raised concerns about the consequences of man-made climate change. As the energy sector is one of the main sources of anthropogenic GHG emissions, policy measures for mitigating climate change focus on increasing the share of renewable energies in this sector. In this sense, the most relevant policy framework within the European context is the Renewable Energy Directive (RED) (European Union 2009), which sets a target GHG reduction of 20% until 2020. Furthermore, a 20% contribution of renewable energy to the transport sector is aspired. This is of special importance, because transport is the energy sector with the greatest dependence on fossil fuels.

Because the electrification of individual motor transport is associated with major societal changes, policy makers rely on biofuels in order to achieve the 20% target of the RED (European Union 2009). Nevertheless, conventional (i.e., first generation) biofuels made from oil and sugar or starch crops show relatively low GHG savings, compete for feedstocks with the food sector, and have potentially high environmental impacts in other categories such as acidification and eutrophication (Zah et al. 2007; Fargione et al. 2008; Campbell and Doswald 2009; Fischer et al. 2009; Banse et al. 2011; Hein and Leemans 2012). Second-generation biofuels are considered to be the solution for avoiding these drawbacks as they are produced from waste or lignocellulosic biomass feedstocks (European Environment Agency 2013a). In this respect, the recently agreed proposal for amendment of the RED strongly promotes advanced biofuels from these types of feedstock (European Union 2014). In particular, lignocellulosic energy crops (short-rotation plantations) show potentially high yields at low agricultural inputs as well as positive impacts on soil carbon and biodiversity (Abrahamson et al. 1998; Heller et al. 2003; Rödl 2008), potentially resulting in a better life-cycle performance than conventional energy crops (Perttu 1998; Dallemand et al. 2008).

In order to assess the environmental performance of biofuels in a comprehensive way, a wide range of potential environmental impacts has to be considered from the agricultural stage to the final use phase of the biofuels. Life-cycle assessment (LCA) is the methodology of choice for this purpose and is often used for assessing biofuel systems (Cherubini and Strømman 2011; Malça and Freire 2011; Iribarren et al. 2012; Muench and Guenther 2013). It is a standardized methodology that evaluates the environmental impacts of a product system along its whole life cycle for a thorough set of impact categories (International Organization for Standardization 2006a,b). In addition to this cradle-to-grave scope (also called well-to-wheels analysis), alternative scopes are often used when assessing biofuel systems. These alternative scopes usually limit the LCA study to fuel production (cradle-to-gate or well-to-tank analysis), omitting vehicle operation (i.e., the tank-to-wheels phase). Nevertheless, the well-to-wheels scope is the recommended perspective when evaluating the life-cycle performance of alternative fuel system options (Brinkman et al. 2005).

Given the relevance of the global warming impact category, life-cycle schemes limited to the evaluation of GHG emissions have been developed. In this regard, carbon footprinting (CF) specifications such as PAS 2050:2011 (British Standards Institution 2011) and ISO/TS 14067:2013 (International Organization for Standardization 2013) have arisen to quantify the life-cycle GHG emissions of product systems in terms of CO_2 equivalents, with relevant differences between the different evaluation alternatives. As environmental concerns and policies regarding the energy sector are mainly focused on climate change, the calculation of carbon footprints plays a significant role in the field of energy systems analysis (Iribarren and Dufour 2014). RED guidelines provide a basis for the CF of energy products (European Union 2009). In this chapter, the carbon footprints of synthetic biofuels produced through the thermochemical conversion of biomass from short-rotation plantations are assessed and discussed. In line with recent studies in this field (Iribarren et al. 2013a; Peters et al. 2015a), this chapter attempts to overcome usual concerns on the source of inventory data (Weinberg and Kaltschmitt 2013), resource availability (Wang et al. 2013; Dang et al. 2014), and limited environmental discussion (van Vliet et al. 2009; Swain et al. 2011; Hsu 2012; Han et al. 2013).

Section 5.2 presents the thermochemical pathways considered herein for the production of synthetic biofuels. A more detailed definition of the specific systems subject to assessment is included in Section 5.3, in which the CF framework is set and data inventories are provided. Section 5.4 deals with the main results from the study (i.e., carbon footprints and GHG emission savings) and discusses the suitability of carbon footprints as single performance indicators. Finally, Section 5.5 summarizes the main conclusions drawn from this chapter.

5.2 PRODUCTION OF SYNTHETIC BIOFUELS

High expectations are put on second-generation biofuels as a sustainable solution for further increasing the share of renewable energy in the transport sector. Nevertheless, complex processing

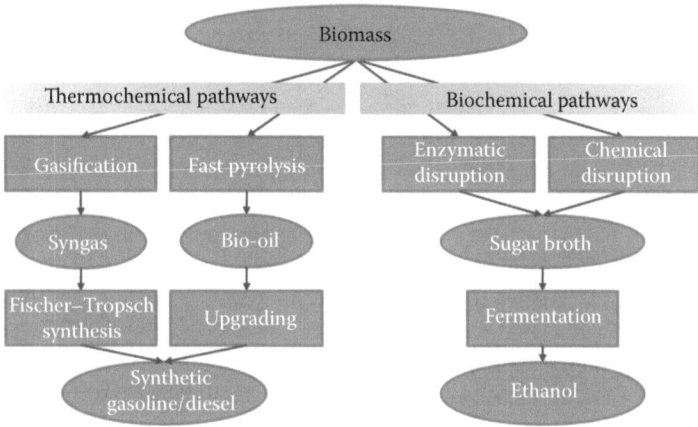

FIGURE 5.1   Main pathways for the production of second-generation biofuels from lignocellulosic biomass.

technologies are required for converting lignocellulosic biomass into liquid fuels suitable for modern internal combustion engines, and the corresponding processes are still in development. Two main types of pathways are generally considered for the production of second-generation biofuels (Figure 5.1): thermochemical and biochemical pathways (Huber et al. 2006; Anex et al. 2010; Venderbosch et al. 2010).

In contrast to typical biochemical pathways, thermochemical routes are able to process the whole biomass into fuel. Common biochemical pathways convert only the cellulose and hemicellulose fraction of the biomass into fermentable sugars, leaving the lignin fraction as a residue, thereby limiting the maximum fuel yields. Furthermore, these biochemical pathways produce ethanol, a fuel that can be mixed with conventional fuels without significant modifications in vehicle engines but only in limited percentages. Therefore, the main advantage of thermochemical pathways is not only the higher theoretical conversion efficiency, but also the properties of the final fuel product. Synthetic fuels of high quality, suitable for direct blending with fossil fuels in any share as well as for exclusive use in vehicle engines (drop-in fuels), can be obtained. On the other hand, thermochemical pathways are technically challenging and no large-scale or commercial plants using these technologies exist to date. The two principal thermochemical routes for the production of synthetic biofuels are (i) biomass fast pyrolysis followed by hydro-upgrading of the obtained bio-oil and (ii) biomass gasification followed by Fischer–Tropsch (FT) synthesis from biosyngas. Two typical configurations of the corresponding processing plants are described in Sections 5.2.1 and 5.2.2.

5.2.1   Fast Pyrolysis and Hydro-upgrading

Biomass pyrolysis is the thermal decomposition of biomass in the absence of oxygen. When using lignocellulosic biomass as feedstock, this process yields gas, char, and a tarry liquid called bio-oil, which is the product of interest for the production of biofuels. Maximum yields of bio-oil are achieved under fast pyrolysis conditions at reactor temperatures of around 500°C and very short residence times (Venderbosch and Prins 2010; Bridgwater 2013; Nachenius et al. 2013). Depending on the biomass feedstock, liquid yields can make up between 60 and 75% of the products.

Bio-oil is a tarry liquid of low quality. It is acidic and abrasive, tends to polymerization and phase separation when stored for a long time, and has a significant water content (10%–30%, depending on the feedstock moisture) (Bridgwater et al. 1999; Bridgwater 2002). Owing to its poor properties, it is not suitable for use in high-end machinery such as modern vehicle engines or turbines, and an upgrading process is required for converting bio-oil into synthetic fuels (Chiaramonti et al. 2007; Venderbosch and Prins 2010; Meier et al. 2013).

Among the possibilities for bio-oil upgrading, hydrotreating is considered to be the most promising pathway (Venderbosch et al. 2010; Bridgwater 2012; Xiu and Shahbazi 2012). In this technology, bio-oil reacts under catalyst influence with hydrogen, rejecting the oxygen contained in the bio-oil in the form of water. The hydrotreating product (a synthetic hydrocarbon mixture) can be distilled to provide drop-in fuels for vehicle engines (Elliott 2007; Wildschut, 2009). The main drawbacks are the high amount of high-pressure hydrogen required by the process and the technical problems caused by catalyst coking and deactivation in the hydrotreating reactors. Figure 5.2 shows a schematic flowsheet of a biorefinery plant producing synthetic fuels via the hydroprocessing of bio-oil from fast pyrolysis of lignocellulosic biomass.

The biomass feedstock is dried and ground in a pretreatment phase according to the requirements of the pyrolysis reactor. In the reactor, biomass is converted into the three principal pyrolysis products: gas, char, and bio-oil. A hot cyclone separates the solid char particles from the hot vapors that leave the reactor and are then quench-cooled using recirculated bio-oil. A water-cooled condenser further reduces the temperature of the product stream, thus maximizing liquid recovery. Quick cooling of the hot vapors is crucial for maximizing liquid yields, since "freezing" the pyrolysis reactions avoids secondary decomposition reactions that decrease the bio-oil yields. The liquid obtained in this way constitutes the bio-oil product, whereas the noncondensable gases are used for their processing in the steam reformer. The char fraction recovered by the cyclone is sent to the char combustor, where it is burnt completely for generating the process heat required by the pyrolysis reactor and the biomass dryer. A hot-sand cycle transfers the combustion heat to the pyrolysis reactor, whereas the dryer directly uses the hot exhaust gases from the combustor (direct-contact dryer).

The bio-oil is then converted into an almost oxygen-free hydrocarbon via hydrotreating. Owing to the thermal instability of the bio-oil, a two-stage process is required, with a first reactor stabilizing the bio-oil under moderate temperatures (around $270°C$) and a second reactor deeply deoxygenizing the intermediate product under higher temperatures (around $350°C$). The hydrotreating reactions are strongly exothermic and reactor cooling is required, generating intermediate-pressure steam for external use (by-product). The obtained product is dewatered and the liquid organic fraction refined to gasoline and diesel products. This is carried out using two distillation columns, with the first one separating the gasoline fraction and the second one obtaining the diesel-range product. A heavy residue is obtained from the distillation process; it is further processed in a hydrocracking reactor to a shorter-chain hydrocarbon, thereby increasing the fuel yield.

The hydrotreating and hydrocracking reactors require significant amounts of pressurized hydrogen. This is produced in a steam-reforming process using the noncondensable gases from the upgrading process, the gases from the pyrolysis section, and a small share of additional natural gas. Steam reacts at high temperatures with the hydrocarbons contained in the gas streams, yielding mainly hydrogen, CO, and CO_2. A pressure-swing-adsorption (PSA) unit separates the produced high purity hydrogen, which is then pressurized and fed to the hydroprocessing reactors. The off-gases from the PSA are burnt in the combustion chamber of the steam reforming reactor, providing the heat required for the endothermic steam-reforming reaction.

5.2.2 Indirect Gasification and FT Synthesis

Biomass gasification is the conversion of biomass into a gas fuel by partial oxidation in the presence of a gasification agent such as air, oxygen, and/or steam. This gas fuel is called syngas (or biosyngas) and is mainly made up of carbon monoxide, hydrogen, methane, steam, and traces of undesirable compounds such as tars and dust (McKendry 2002). This process take place at high temperature ($500°C–1400°C$), whereas the operating pressure varies from atmospheric pressure to 33 bar depending on the plant scale and the final application of the syngas product (Ciferno and Marano 2002).

Several factors (e.g., biomass composition, gasification technology, and gasifying agent) determine the composition of the biosyngas. The most common gasifying agent is air because it is inexpensive. However, owing to the dilution effect of nitrogen, the syngas produced presents a relatively

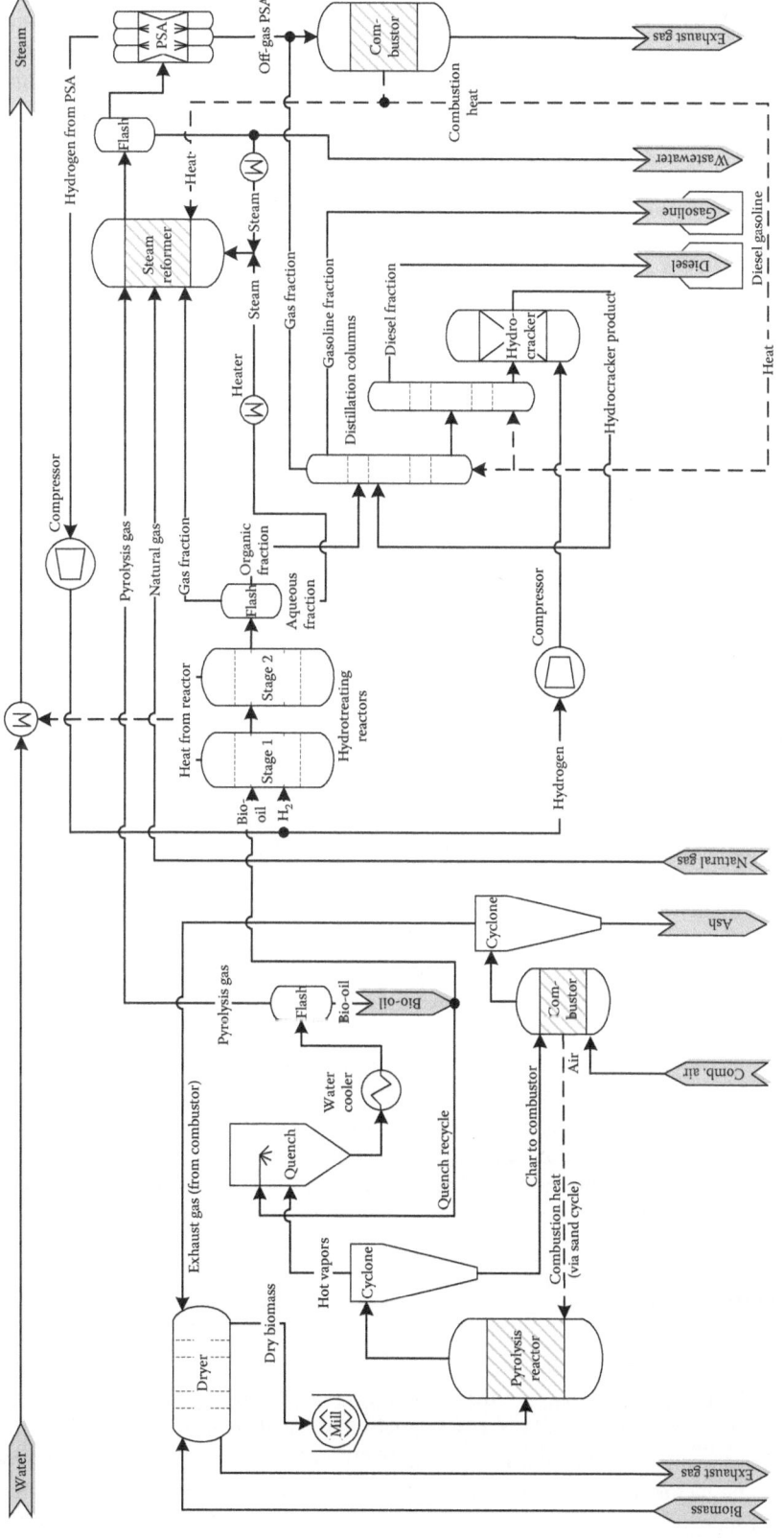

FIGURE 5.2 Simplified diagram of a biorefinery plant based on fast pyrolysis and hydro-upgrading.

low higher heating value (3.7–6.4 MJ m^{-3} stp) when compared with syngas produced using oxygen and/or steam (9.2–16.5 MJ m^{-3} stp) (Li et al. 2004). Fixed bed, fluidized bed, and indirect gasifiers are the main types of reactors used (McKendry 2002). In particular, indirect gasifiers are being widely studied since they produce a hydrogen-rich gas with a suitable heating value (Mann 1995; Pfeifer et al. 2004; Shen et al. 2008; Iribarren et al. 2014).

Syngas is a versatile fuel gas that can be used to obtain several energy products such as electricity, heat, hydrogen, and synthetic fuels (Huber et al. 2006). The latter are produced via the FT synthesis, which involves the reaction between CO and H$_2$ in the presence of a catalyst (usually Fe- or Co-based catalyst) to produce liquid hydrocarbons according to the following simplified reaction: $nCO + (n + m/2)H_2 \rightarrow C_nH_m + nH_2O$. The FT process is carried out at temperatures between 150°C and 300°C and pressures ranging from 10 to 40 bar. The products obtained through the FT synthesis are hydrocarbons of different chain length, mainly paraffins and olefins, which can be used to obtain gasoline and diesel by means of a refining process similar to that used in the petrochemical industry. This pathway based on biomass gasification and FT synthesis has attracted much interest as a potential system to coproduce biofuels and electricity when coupled with combined-cycle strategies (Larson et al. 2009; Liu et al. 2011; Iribarren et al. 2013a). Figure 5.3 shows a schematic flowsheet of a biorefinery plant coproducing synthetic fuels and electricity from a gasification-based biosyngas feedstock via FT synthesis coupled with a combined-cycle process.

In the plant, the lignocellulosic biomass is dried and milled before being fed into an indirect gasifier. The latter is a low-pressure char indirect gasifier designed by Battelle Columbus Laboratory, operating at 870°C and 1.6 bar. This gasifier consists of two fluidized-bed reactors: the gasifier itself (in which biomass and steam react producing syngas and char), and a combustor where the char is burnt in order to supply the heat needed for the gasification process. The heat generated in the combustor is transferred to the gasifier by the bed material, which circulates through both reactors (Spath et al. 2005). The exhaust gas from the combustor is used to dry the biomass feedstock before gasification.

The produced syngas is introduced in a tar-reforming reactor, where tars and light hydrocarbons are converted into CO and H$_2$. Afterward, the syngas undergoes a cleaning process that includes scrubbing, removal of sulfur compounds via the LO-CAT® process, and acid gas removal via absorption with amines (Spath et al. 2005; Swanson et al. 2010; Susmozas et al. 2013).

The clean syngas has a H$_2$/CO molar ratio of ca. 2.5. Because the molar ratio required for the FT reaction is approximately 2, a small amount of H$_2$ is separated from the stream in a PSA unit, with part of this hydrogen being used in the refining area. The obtained syngas is introduced in a slurry FT reactor that operates at 200°C and 25 bar and uses a Co catalyst. Two streams are obtained from the FT reactor: a liquid stream made up of wax (C$_{20+}$) and a gas stream consisting of C$_1$–C$_{20}$ hydrocarbons and unconverted syngas. The latter is cooled down to separate the C$_1$–C$_4$ fraction and the unconverted syngas (tail gas) from the C$_{5+}$ fraction. The C$_{5+}$ stream and the wax stream are sent to the refining area, where gasoline, diesel, and fuel gas (C$_1$–C$_4$) are produced.

Finally, the tail gas and the fuel gas are sent to the power-generation section, which consists of a combined cycle with a gas turbine, a heat-recovery steam generator, and a steam turbine with three pressure levels (87/31/2.4 bar). This combined cycle supplies the electricity demanded by the whole plant while also generating an electricity surplus (Liu et al. 2011; Iribarren et al. 2013b).

5.3 CF FRAMEWORK

5.3.1 GOAL AND SCOPE

The study aims to calculate and compare the carbon footprints of the synthetic biofuels produced through the two thermochemical pathways previously described in Section 5.2. LCA standards (International Organization for Standardization 2006a,b) are followed. This study facilitates the identification of the technology with the potentially most favorable global-warming performance from a well-to-wheels perspective. Moreover, comparing the carbon footprints of these synthetic

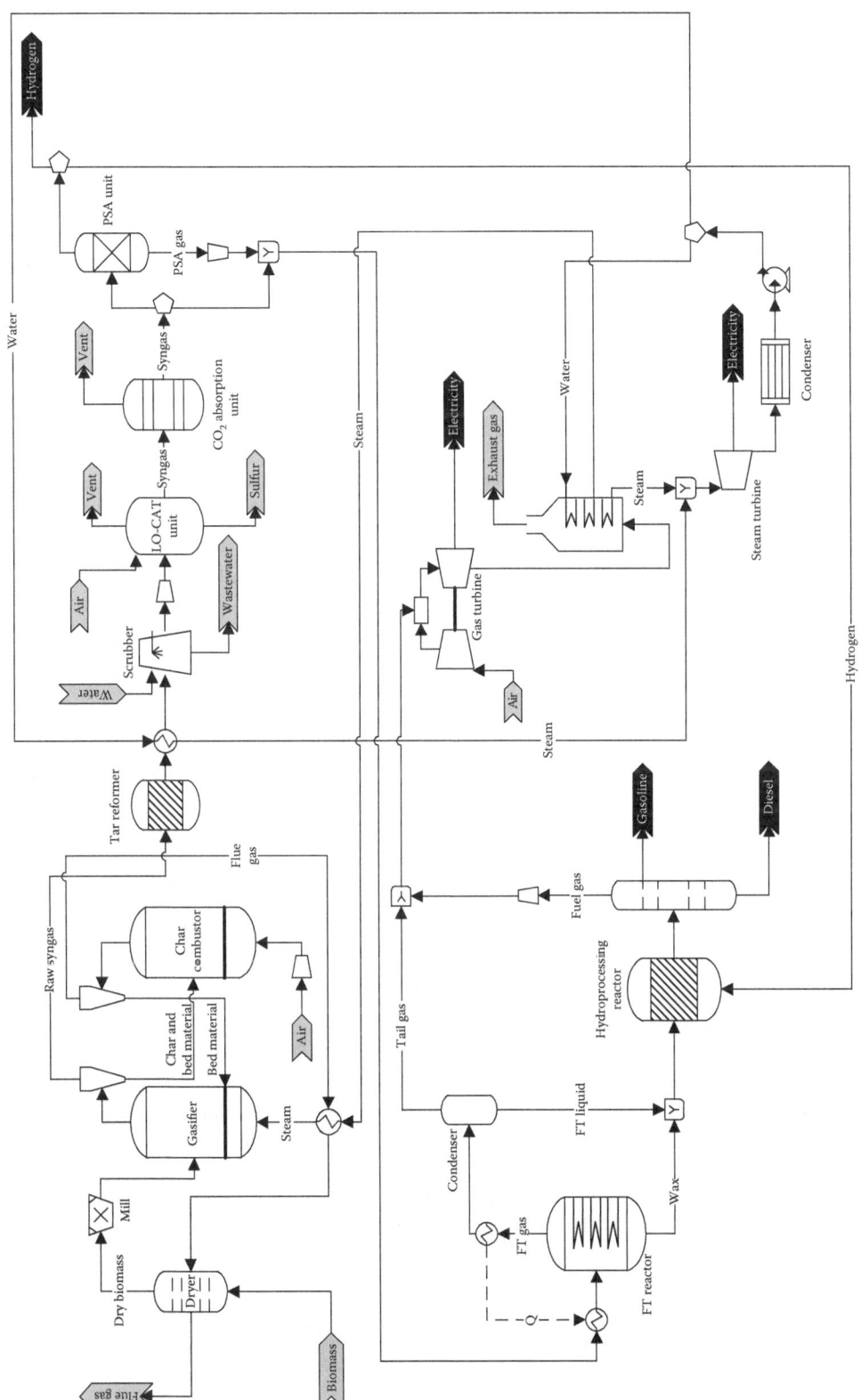

FIGURE 5.3 Simplified diagram of a biorefinery plant based on indirect gasification and FT synthesis.

biofuels with those of equivalent conventional fossil fuels leads to the calculation of the GHG emission savings linked to these synfuels according to the RED (European Union 2009). The common functional unit (FU) used for the assessment is 1 MJ of potential energy supplied by the mix of synthetic biofuels (gasoline and diesel) according to their lower heating value. The carbon footprints of the different fuels are calculated using the characterization factors reported by the Intergovernmental Panel on Climate Change (Myhre et al. 2013). Capital goods are not included in the assessment (British Standards Institution 2011). A detailed description of the boundaries of the assessed systems is given in Sections 5.3.1.1 and 5.3.1.2, whereas details on data acquisition are provided in Section 5.3.2.

5.3.1.1 Pyrolysis-Based System (PYR-HT)

The first system (PYR-HT) involves the use of synthetic biofuels (viz. gasoline and diesel) produced via the hydroprocessing of bio-oil from fast pyrolysis of poplar biomass. The system's boundaries set for the assessment cover the whole process chain from the agricultural phase to the combustion of the fuels in vehicle engines. A block diagram of the system and its boundaries is shown in Figure 5.4.

The pyrolysis section produces, apart from bio-oil, char and gas. The gas is fed to the steam reformer in the upgrading section, whereas the char is completely burnt in the char combustor providing the process heat required by the pyrolysis process. Therefore, no by-products arise from the pyrolysis section.

On the other hand, the upgrading section not only produces synthetic gasoline and diesel, but also processes steam from the cooling of the hydrotreating reactors. This steam is not consumed within the biorefinery and constitutes a by-product of the system. Since the three products are valuable outputs of the process (Peters et al. 2015a), the environmental impacts have to be distributed between them. This allocation is carried out herein by assigning the environmental burdens to the different products according to their energy content. This is in line with the RED and facilitates the calculation of the GHG emission savings associated with the synthetic biofuels (European Union 2009). According to the energy content of the products as obtained from the process simulation (Section 5.3.2.2), 49.0% of the environmental impacts caused by the production process are allocated to the gasoline product, 46.0% to diesel, and 5.0% to the steam by-product.

5.3.1.2 Gasification-Based System (GAS-FT)

The second system (GAS-FT) deals with the use of synthetic biofuels (gasoline and diesel) produced from a gasification-based biosyngas feedstock via FT synthesis coupled with a combined-cycle process. The scope of this system is equivalent to that of the PYR-HT system, covering the chain from poplar plantation to fuel use. Moreover, the composition of the synthetic fuels obtained from both systems does not differ significantly and the same emission factors are therefore used regarding fuel use in vehicle engines. A block diagram of the GAS-FT system and its boundaries is presented in Figure 5.5. As can be observed in Figure 5.5, the biorefinery is energy self-sufficient.

The production section of the GAS-FT system yields not only gasoline and diesel, but also surplus hydrogen and electricity as co-products. In order to distribute the life-cycle inventory data and the environmental burdens between the different products, an allocation approach is used. According to the energy content of the products, 17.0% of the environmental impacts caused by the production process are allocated to gasoline, 44.5% to diesel, 16.9% to hydrogen, and 21.6% to electricity.

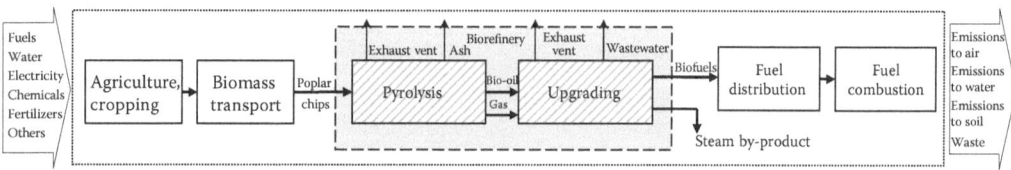

FIGURE 5.4 Block diagram of the PYR-HT system.

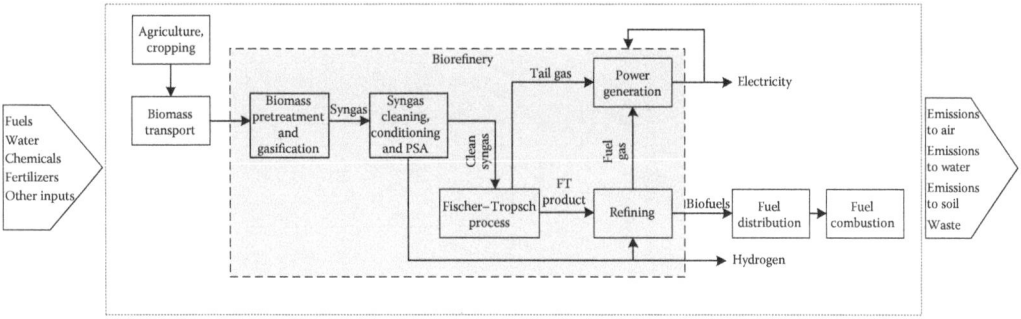

FIGURE 5.5 Block diagram of the GAS-FT system.

5.3.2 Data Acquisition

The assessed biofuel processes are still in development or in laboratory phase. Hence, operational data from real plants are not available and the evaluation of this type of process is usually based on process simulation (Jones et al. 2009; Anex et al. 2010; Swanson et al. 2010; Wright et al. 2010; Susmozas et al. 2013). In this chapter, detailed Aspen Plus® simulations of the conversion plants are used as a source of inventory data for the CF study. Additionally, the ecoinvent® database and specific literature data are used for the inventory of background processes such as transport and the production of chemicals (Frischknecht et al. 2007). A complete description of data sources and data acquisition is provided in Sections 5.3.2.1 through 5.3.2.4 for the key phases included in each system.

5.3.2.1 Agriculture

The inventory data for the agricultural phase are based on literature data. Special attention is paid to an extensive data gathering in this stage, since the agricultural phase is usually identified as a key source of environmental impacts in bioenergy systems (Peters et al. 2015b). The same biomass feedstock is used in the PYR-HT and GAS-FT systems: hybrid poplar from short-rotation plantations in central Spain (Gasol et al. 2009a,b; Sevigne et al. 2011; Peters et al. 2015a,b). These plantations are placed on fallow or abandoned agricultural land, providing biomass potential without the need for conversion of natural lands or crop displacement. The change of land use from abandoned cropland to irrigated short-rotation plantation leads to an increase in carbon stock in the soil. This is estimated according to the guidelines from the Intergovernmental Panel on Climate Change (Eggleston et al. 2006; Batjes 2010).

Spain shows a high bioenergy potential with significant shares of unused cropland (European Environment Agency 2013a; European Union 2013), but the climatic conditions lead to require irrigation in order to attain high crop yields. This is taken into account by estimating the irrigation water requirements based on literature data (Gasol et al. 2009b; Sevigne et al. 2011) adjusted to the local conditions of the assumed location.

The transport distance for the biomass feedstock from the plantation to the processing plant (capacity of 192,500 t y^{-1} of dry biomass feedstock for both PYR-HT and GAS-FT) is also estimated based on literature data (Gasol et al. 2009b), resulting an average distance of 94.4 km.

5.3.2.2 PYR-HT Production Process

The Aspen Plus® simulation of the PYR-HT production process includes all processing steps taking place at the biorefinery plant based on the fast pyrolysis of poplar biomass and the hydro-upgrading of the obtained bio-oil (Peters et al. 2015a). A simplified flowsheet of this simulation is shown in Figure 5.6.

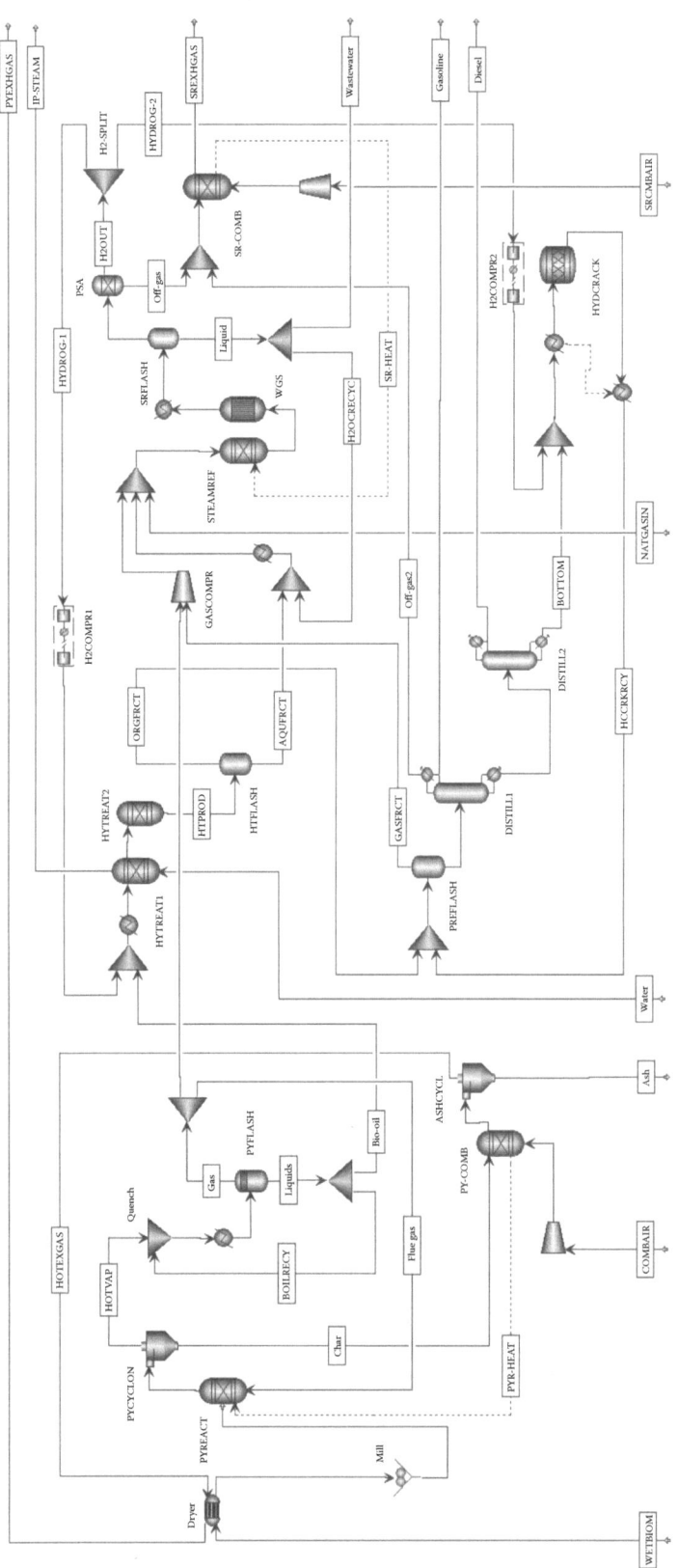

FIGURE 5.6 Aspen Plus® flowsheet of the PYR-HT biorefinery.

A kinetic reaction model is used for the calculation of the pyrolysis products. This model simulates the pyrolysis reaction as a function of the biomass composition and reactor conditions. It models the pyrolysis products with a high level of detail (33 model compounds) and shows high correlation with existing literature and experimental data (Peters et al. 2014, 2015a). The pyrolysis reactor is a circulating fluidized-bed reactor with a vapor residence time of 1.5 s and a reactor temperature of 520°C. The obtained yields are 68.8% bio-oil, 14.4% gas, and 16.8% char. The char fraction is burnt completely in the char combustor of the pyrolysis section for generating the process heat required by the reactor and the biomass dryer.

The bio-oil product is fed to the hydrotreating reactors, with the first-stage reactor operating at 270°C and 170 bar, and the second-stage reactor at 350°C and 140 bar. Both reactors use a commercial CoMo catalyst. A linear regression model is used for the simulation of the hydrotreating reactions (Peters et al. 2015a). This model iteratively adjusts the composition of a typical product distribution for hydrotreating reactions of bio-oil until fitting the composition of the specific bio-oil obtained from the selected feedstock. The hydrocarbon product is modeled with a high level of detail using 52 model compounds, thus allowing a realistic simulation of the distillation columns.

The operational parameters of the distillation columns are dynamically adjusted in order to produce gasoline and diesel fractions with distillation curves fulfilling the corresponding specifications for fossil fuels (Spanish Ministry of Industry, Tourism and Trade 2010). An atmospheric column with eight stages is required for the first distillation and a column with nine stages operating under vacuum for the second one.

For the hydrocracking reactor (operating at 90 bar and 430°C), a set of hypothetical reactions is implemented, giving a product composition according to typical values for hydrocracking reactions published in the literature (Elliott 2007; Elliott et al. 2009).

The gas streams from the pyrolysis and hydroprocessing sections, together with natural gas, undergo steam reforming at 950°C and 50 bar with a steam-to-carbon ratio of 4.5. The amount of natural gas fed to the reforming process is adjusted dynamically in order to satisfy the hydrogen demand of the hydrotreating and hydrocracking processes. The reactor calculates the reaction products by Gibbs free energy minimization under the given reactor conditions. A subsequent water gas shift reactor, operating at 350°C, increases the hydrogen yield further. This block also uses Gibbs free energy minimization for calculating the products of the water gas shift reaction. The gas stream is separated into an off-gas stream and a hydrogen product stream in a PSA unit, recovering 90% of the hydrogen at very high purity. The combustion chamber of the steam-reforming reactor is modeled as a separate component for simulation purposes, using Gibbs free energy minimization for the calculation of the reaction products. The off-gases of the process are burnt in this chamber at 1000°C. The main inventory data obtained from the simulation of the PYR-HT process are gathered in Table 5.1.

5.3.2.3   GAS-FT Production Process

The GAS-FT production process is simulated in Aspen Plus® in order to obtain key inventory data (Cruz et al. 2014). Figure 5.7 shows the simulation diagram of the process under study. In the simulation, poplar feedstock is defined as a nonconventional component by specifying its proximate and ultimate analyses (the same as for the PYR-HT simulation).

The low-pressure indirect gasifier consists of two reactors: the gasifier itself and the char combustor. The former is modeled using two Aspen Plus® blocks: an RYield block where poplar biomass is decomposed into its constituent elements, and an RGibbs block in which these elements and steam react to produce syngas and char (Susmozas et al. 2013). This RGibbs block uses the Gibbs free energy minimization method along with correlations based on experimental data in order to determine the composition of the syngas output. The char combustor is simulated by means of an RGibbs block. In order to simplify the simulation, the bed material that circulates through the gasifier and the char combustor is simulated as a heat stream.

The syngas produced in the gasifier is introduced in a tar reformer, which is simulated with an RGibbs block. In this block, the chemical equilibrium is restricted by implementing the reforming

TABLE 5.1

Main Inventory Data of the PYR-HT Production Process

Inputs

From the Technosphere

Poplar chips (kg)	0.17
Poplar transport (t km)	3.22×10^{-2}
Natural gas (MJ)	8.27×10^{-2}
Electricity (kW h)	4.13×10^{-2}
Cooling water (kg)	5.86
Catalyst (kg)	6.21×10^{-6}

Outputs

Products

Gasoline (MJ)	0.54
Diesel (MJ)	0.46
Steam (kW h)	1.52×10^{-2}

Waste to Treatment

Ash to landfill (kg)	2.25×10^{-3}
Catalyst to landfill (kg)	6.21×10^{-6}
Wastewater to treatment plant (kg)	7.23×10^{-4}

Emissions to the Air

CO_2 (kg)	8.21×10^{-2}
CO (kg)	7.34×10^{-9}
NO_2 (kg)	2.10×10^{-6}
NO (kg)	3.55×10^{-5}
SO_2 (kg)	3.34×10^{-5}
Particulates	3.02×10^{-6}

Emissions to the Water

Wastewater (kg)	1.16×10^{-2}

reactions and their conversion (Susmozas et al. 2013). The syngas from the tar reformer is conditioned and a small amount of hydrogen is separated through a PSA unit so that an H_2/CO molar ratio of approximately 2 is obtained. This PSA unit is modeled with a separator block assuming 85% recovery.

The conditioned syngas is fed to an FT slurry reactor. This unit is modeled with an RStoic block along with an Aspen Plus® calculator block computing the hydrocarbon distribution according to the Anderson–Schulz–Flory (ASF) distribution (Spyrakis et al. 2009; Swanson et al. 2010). The product from the FT reactor is assumed to consist only of paraffins. The refining area is simulated based on literature data (Iribarren et al. 2013a; Swanson et al. 2010).

Finally, the gas turbine of the power-generation section is simulated taking into account an isentropic efficiency of 80% and a mechanical efficiency of 95% (Susmozas et al. 2013). The heat recovery steam generator is simulated with a series of heat exchangers representing the different areas of the unit (economizer, evaporator, and superheater). The three pressure levels of the steam turbine are modeled with a series of three turbine blocks. The main inventory data obtained from the simulation of the GAS-FT process are presented in Table 5.2.

5.3.2.4 Biofuel Use

The use of the synthetic biofuels (gasoline and diesel) in passenger cars is included within the well-to-wheels assessment. This use phase includes the combustion emissions from fuel use in passenger cars,

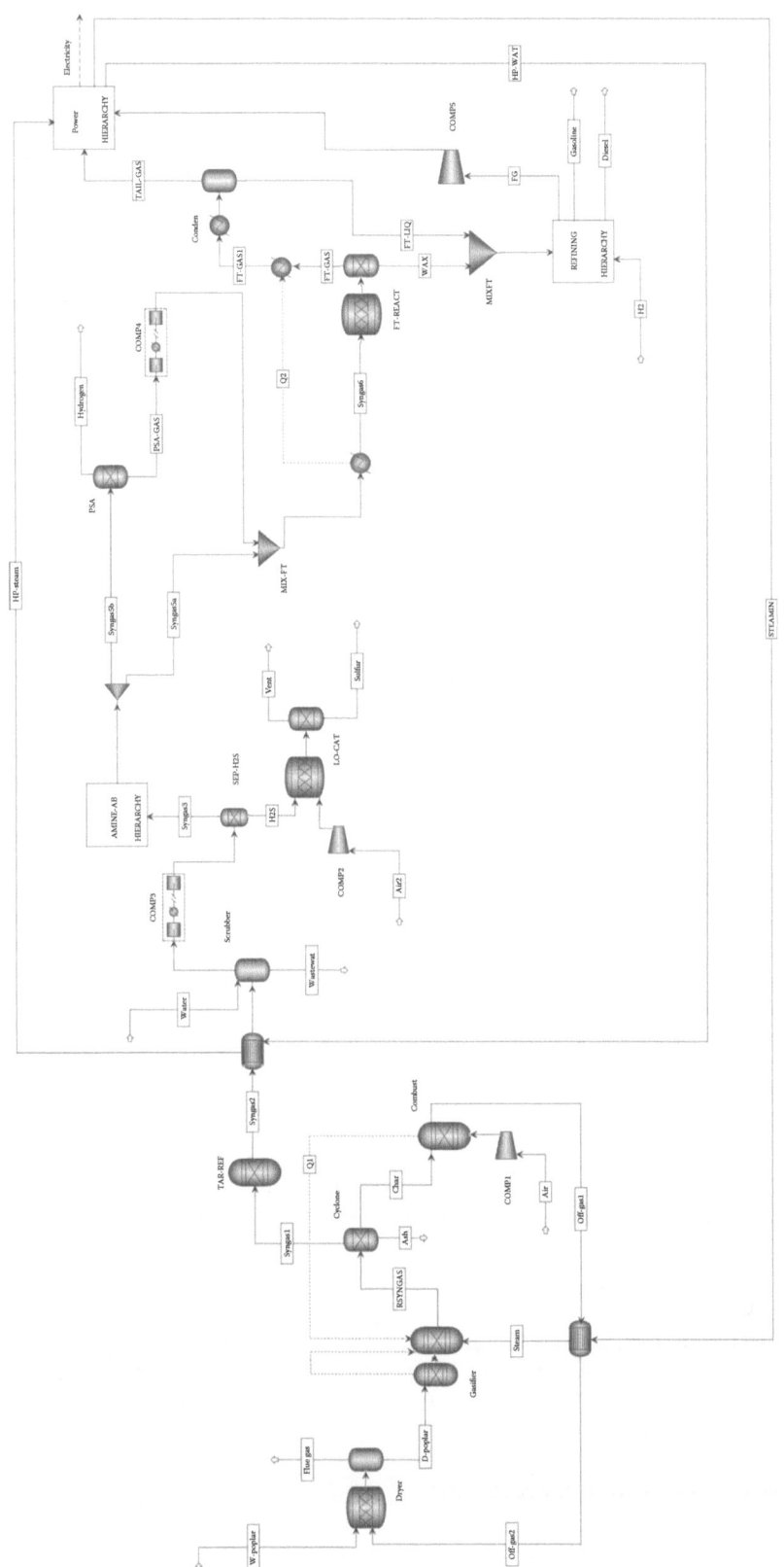

FIGURE 5.7   Aspen Plus® flowsheet of the GAS-FT biorefinery.

TABLE 5.2

Main Inventory Data of the GAS-FT Production Process

Inputs

From the Technosphere

Poplar chips (kg)	0.64
Poplar transport (t km)	0.12
Olivine (kg)	9.57×10^{-3}
Monoethanolamine (kg)	2.70×10^{-2}
Co catalyst (kg)	7.78×10^{-6}
Water (kg)	6.45

From the Environment

Air (kg)	2.17

Outputs

Products

Gasoline (MJ)	0.29
Diesel (MJ)	0.71
Hydrogen (kg)	2.36×10^{-3}
Electricity (kW h)	0.10

Waste to Treatment

Ash to landfill (kg)	8.64×10^{-3}
Catalysts to landfill (kg)	1.12×10^{-4}
Sulfur to landfill (kg)	9.21×10^{-5}
Wastewater to treatment plant (kg)	5.92

Emissions to the Air

O_2 (kg)	0.23
N_2 (kg)	1.66
H_2O (kg)	0.82
CO_2 (kg)	0.48
NO_2 (kg)	1.94×10^{-5}

Emissions to the Water

Wastewater (kg)	0.14

which are based on the European Environment Agency (2013b). In this respect, the same emission factors are used for the synthetic biofuels and the conventional fossil fuels as reported in the EMEP/EEA air pollutant emission inventory guidebook (European Environment Agency 2013b). It should be noted that, as previously stated in Section 5.3.1, capital goods are not included in the assessment.

5.4 RESULTS AND DISCUSSION

In this section, the carbon footprints of the synthetic biofuels are calculated and the processes with the highest contribution to the GHG emissions are identified. These carbon footprints are compared both to one another and to those of conventional fossil gasoline and diesel. Furthermore, the use of carbon footprints as single indicators is discussed from different perspectives.

5.4.1 CARBON FOOTPRINTS AND GHG EMISSION SAVINGS

The carbon footprints of the synthetic fuels from the two bioenergy systems are calculated by implementing the life-cycle inventories in SimaPro 8 (Goedkoop et al. 2013) and applying the latest characterization factors provided by the Intergovernmental Panel on Climate Change (Myhre et al. 2013). The well-to-wheels carbon footprints (per MJ of synfuel) obtained for the synthetic gasoline biofuels are 32.23 g CO_2-eq. MJ^{-1} (GAS-FT) and 38.44 g CO_2-eq. MJ^{-1} (PYR-HT). The corresponding values for the diesel biofuels are 30.91 g CO_2-eq. MJ^{-1} (GAS-FT) and 37.22 g CO_2-eq. MJ^{-1} (PYR-HT). Thus, the CF results for the biofuel production mixes of the two systems are 31.29 g CO_2-eq. MJ^{-1} (GAS-FT) and 37.88 g CO_2-eq. MJ^{-1} (PYR-HT). Hence, the synthetic biofuels of the GAS-FT system show a better life-cycle performance in terms of GHG emissions than those of the PYR-HT system.

Applying a system perspective, Figure 5.8 shows the contribution of the key process steps to the carbon footprints of the two bioenergy systems (values per FU, i.e., 1 MJ of the synthetic biofuel mix). For GAS-FT, significant GHG emissions arise from the gasification section ("Thermoch. conv." in Figure 5.8), followed by the gas cleaning, combustion, and power-generation process steps. In contrast to PYR-HT, the upgrading process of the GAS-FT system (i.e., FT synthesis) shows low GHG emissions. Transport processes contribute only minor shares to the overall carbon footprint.

As can be observed in Figure 5.8, the GHG emissions from the processing of the biomass to synthetic fuel are generally higher for GAS-FT in the different process steps. Nevertheless, since the GAS-FT system consumes much more biomass per energy output than the PYR-HT system, a higher amount of CO_2 is absorbed during biomass growth. Furthermore, the analysis of the agricultural phase reveals that the favorable net C balance of GAS-FT is mainly due to land-use change. The conversion of fallow agricultural land into short-rotation plantations causes an increase in soil carbon. This favors the system with higher land demand per energy output because more land has to be converted leading to a higher carbon stock increase. Without taking into account this effect, the GAS-FT system would actually result in a higher carbon footprint than the PYR-HT system. Therefore, the obtained results should be taken with care since fallow land is finite and a biofuel system generally ought to maximize the fuel yield from the available land area.

When comparing the synthetic biofuels with their fossil-fuel equivalents, significantly lower carbon footprints are obtained for the bioenergy products (Figure 5.9). Compared to the carbon footprints of 89.61 g CO_2-eq. MJ^{-1} for fossil gasoline and 83.04 g CO_2-eq. MJ^{-1} for fossil diesel (Jungbluth 2007; European Environment Agency 2013b), GHG emission savings of 64.04% (gasoline) and 62.77% (diesel) are achieved by the GAS-FT synfuels, thus fulfilling the target set in the RED (from January 1, 2018, the GHG emission saving from the use of biofuels shall be at least 60%

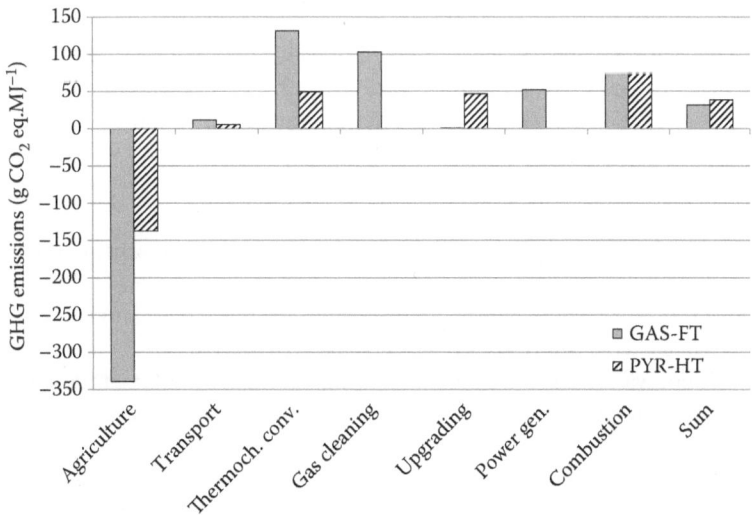

FIGURE 5.8 Contribution of the process steps to the carbon footprint of the GAS-FT and PYR-HY bioenergy systems (values per FU: 1 MJ).

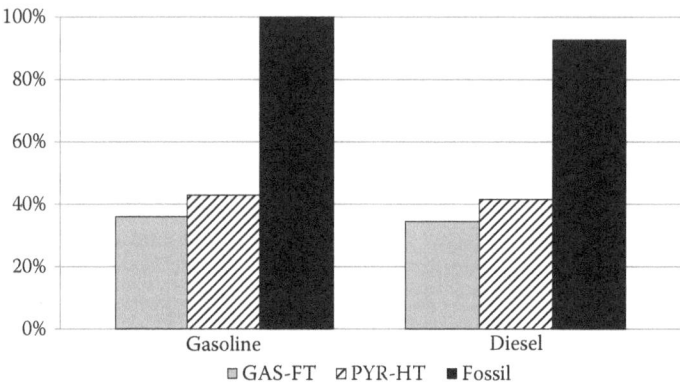

FIGURE 5.9 Comparison of the carbon footprints of synthetic biofuels and conventional fossil fuels.

for biofuels and bioliquids produced in installations in which production started on or after January 1, 2017) (European Union 2009). On the other hand, the PYR-HT synfuels account for GHG emission savings of 57.10% (gasoline) and 55.18% (diesel) and do not reach the RED goal.

Overall, the evaluated synthetic biofuels show significantly high GHG emission savings, with GAS-FT exceeding the 60% emission reduction relative to fossil fuels, and PYR-HT coming close. Hence, these bioenergy products can be considered to be attractive alternatives to conventional fossil fuels (and first-generation biofuels) from a life-cycle global warming perspective.

5.4.2 Carbon Footprint as a Single Indicator

Currently, reduction in GHG emissions is the paramount environmental driver for policy decisions on renewable energy. However, renewable energy (especially bioenergy) is often associated with low energy efficiencies and major impacts in other environmental categories (Zah et al. 2007; Peters et al. 2015b). For instance, agricultural activity may cause significant impacts in biodiversity and eutrophication, whereas increasing land demand puts pressure on natural land. Within this context, the limitation of the assessment to one impact category (global warming) can be questioned (Iribarren and Dufour 2014).

In this study, the GAS-FT system shows better CF results than the PYR-HT at the expense of a higher biomass demand. In this respect, high GHG savings should ideally not be attained by comprimising the performance of the system in other categories. In this section, further aspects of the biofuel systems are assessed and the findings contrasted with the previous CF results.

5.4.2.1 A Thermodynamic Perspective

The technical aspects of a system determine its feasibility, while inherently affecting its economic and environmental performance. Technical aspects can be assessed from a thermodynamic perspective, which facilitates decision making. In particular, the thermodynamic behavior of a system can be evaluated using exergy analysis (Tsatsaronis and Cziesla 2004a).

The exergy concept was conceived in the nineteenth century by Sadi Carnot, but it was applied to industrial systems in the twentieth century. Exergy is defined as the maximum theoretical useful work obtained when a system is brought into thermodynamic equilibrium with the environment by means of processes in which the system interacts only with this environment (Sciubba and Wall 2007). The environment is defined as the surroundings of the system; it is assumed to be a large system at equilibrium in which state variables (temperature, pressure) and the chemical potentials of the components remain constant. Therefore, the environment is free of irreversibilities and its exergy is zero.

The total exergy of a system is calculated as the sum of its physical ($\dot{E}^{PH}$), kinetic ($\dot{E}^{KIN}$), potential ($\dot{E}^{POT}$), and chemical ($\dot{E}^{CH}$) exergies. Considering systems at rest with respect to the environment, both kinetic and potential components are zero, and therefore the exergy is calculated as the sum of the chemical and physical components (Tsatsaronis and Cziesla 2002, 2004a).

The physical exergy depends on temperature and pressure. It is generally defined by Equation 5.1, where e, h, and s denote specific exergy, enthalpy, and entropy, respectively. The subscript 0 denotes reference state (25°C and 1 bar) (Bejan et al. 1996):

$$e^{PH} = (h - h_0) - T_0(s - s_0) \tag{5.1}$$

The chemical exergy is based on the difference in the composition between the system and the environment. Thus, the chemical composition of the environment has to be defined. Since the natural environment is not in equilibrium, an exergy-reference environment needs to be modeled (Tsatsaronis and Cziesla 2002). Chemical exergies of different compounds can be retrieved from the literature in order to facilitate the calculation of exergy values (Kotas 1995; Szargut 2011).

Thermodynamic processes are governed by the laws of mass and energy conservation. However, exergy is not generally conserved, but destroyed because of irreversibility within a system. Furthermore, exergy is lost when the energy associated with a material or energy stream is released to the environment (Tsatsaronis and Cziesla 2002). In this sense, the thermodynamic inefficiencies in an energy conversion system are related to exergy destruction and loss. An exergy analysis leads to identify the system components with high exergy destruction as well as the processes behind this performance (Tsatsaronis and Cziesla 2004b).

When performing an exergy analysis, the exergy balance is carried out at the level of the system, subsystems, and components. The exergy balance of each component depends on the type of component, with most common components having well-established exergy balances in the literature (Kotas 1995; Bejan et al. 1996; Tsatsaronis and Cziesla 2002; Szargut 2011). For each component i, the exergy balance is shown in Equation 5.2, where $\dot{E}_F$ is the exergy of the fuel (resources expended to generate the product), $\dot{E}_P$ is the exergy of the product (desired output), and $\dot{E}_D$ is the exergy destruction:

$$\dot{E}_{F,i} = \dot{E}_{P,i} + \dot{E}_{D,i} \tag{5.2}$$

When evaluating an overall system, the exergy loss to the surroundings ($\dot{E}_L$) has also to be taken into account, as shown in Equation 5.3:

$$\dot{E}_{F,i} = \dot{E}_{P,i} + \dot{E}_{D,i} + \dot{E}_{L,i} \tag{5.3}$$

The exergetic efficiency (ε) of a component is defined as the ratio between the product and fuel exergies (Equation 5.4):

$$\varepsilon_i = \dot{E}_{P,i} \cdot \dot{E}_{F,i}^{-1} \tag{5.4}$$

Exergy destruction rates ($\dot{E}_D$) and exergetic efficiencies (ε) characterize the thermodynamic performance of a system. In this section, the exergy analysis of the PYR-HT and GAS-FT production processes is carried out. All data required for the analysis come from the simulation of the bioenergy plants in Aspen Plus®.

Table 5.3 summarizes the results from the exergy analysis. The PYR-HT production process shows a significantly higher exergetic efficiency score than GAS-FT. Although the same biomass feedstock is used and similar products and subprocesses are involved in both bioenergy systems, the conversion pathway of each system is highly different, which clearly affects the exergy results.

TABLE 5.3
Exergy Results of the PYR-HT and GAS-FT Production Processes

	PYR-HT	GAS-FT
Exergy of fuel (MW)	169.18	159.81
Exergy of product (MW)	100.23	39.50
Exergy loss (MW)	5.17	44.07
Exergy destruction (MW)	63.78	76.23
Exergetic efficiency (%)	59.24	24.72

The PYR-HT production process includes a conversion step (pyrolysis) followed by a refining step (hydro-upgrading), whereas GAS-FT comprises two conversion steps (biomass gasification and FT synthesis) plus a power-generation step (combined cycle). Each conversion step implies thermodynamic irreversibility, which means a decrease in efficiency. In other words, coupling sub-processes negatively affects the thermodynamic performance of a process. For instance, in the GAS-FT case study, gasification, FT synthesis, and power generation show individual efficiencies of 63.97, 90.26, and 64.99%, respectively. These values highly determine the relatively low overall efficiency of the GAS-FT production process (24.72%; Table 5.3). Moreover, the exergy loss in GAS-FT is significantly high, mainly due to the hydrocarbon loss in the cleaning step as well as to the emission of hot flue gases. This entails a reduction in the exergy of product, thus making the exergy results less favorable.

Regarding the exergy of the fuel (Table 5.3), PYR-HT consumes more resources (+9.37 MW) than GAS-FT, which is a self-sufficient process in terms of electricity and heat. If irreversibility associated with the production of electricity is taken into account, the overall efficiency of PYR-HT would decrease, whereas the exergy destruction would increase.

Regarding the exergy of the product, Figure 5.10 shows the contribution of each product to the overall result. Gasoline and diesel together make up 98.18% of the overall exergy of the product in PYR-HT, whereas this percentage drops to 68.67% in GAS-FT. The coproduction of electricity and hydrogen in the GAS-FT process implies a reduction in biofuel yields. In GAS-FT, 12.25 and 4.72% of the exergy of the fuel are converted into diesel and gasoline, respectively, whereas these percentages are 28.14% (diesel) and 30.02% (gasoline) in PYR-HT.

The results derived from the exergy analysis indicate possible conflict with previous CF results. Thus, while GAS-FT could be favored in terms of life-cycle GHG emissions, PYR-HT could be considered a better option from a thermodynamic perspective (e.g., in terms of exergetic efficiency).

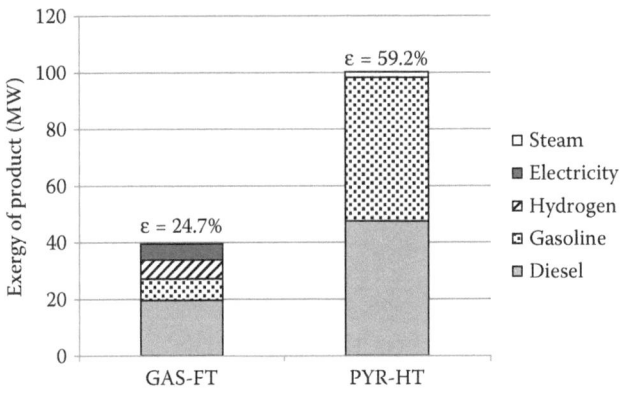

FIGURE 5.10 Contribution of each product to the overall exergy of the product.

5.4.2.2   An Environmental Perspective

In order to evaluate the results of the assessment under a broader environmental perspective, other potential impacts (apart from global warming) are evaluated using the CML method (Guinée et al. 2001): abiotic depletion potential (ADP), acidification potential (AP), eutrophication potential (EP), and land competition (LC). The LCA framework remains the same as in the CF study. Table 5.4 presents the new environmental characterization results, whereas Figure 5.11 shows the comparison of the life-cycle environmental profiles of the synthetic biofuels and the conventional fossil fuels. Inventory data for the LCA of conventional fossil fuels are based on the ecoinvent database for fuel production (Jungbluth 2007) and on the European Environment Agency (2013b) for fuel use.

The synthetic biofuels show significantly higher environmental impacts than the fossil fuels in three of the assessed categories (LC, EP, and AP). This is closely linked to the agricultural activity

TABLE 5.4

Additional Environmental Characterization Results for the Evaluated Fuels (Values per MJ of Fuel)

		Gasoline			Diesel		
		GAS-FT	PYR-HT	Fossil	GAS-FT	PYR-HT	Fossil
ADP	mg Sb-eq.	349.68	330.51	575.59	348.49	330.55	540.82
AP	mg SO_2-eq.	410.57	392.24	173.83	409.17	392.28	127.76
EP	mg PO_4^{3-}-eq.	185.10	95.40	17.33	184.47	95.41	14.41
LC	cm^2 a	1410.71	574.32	0.10	1405.88	574.38	0.06

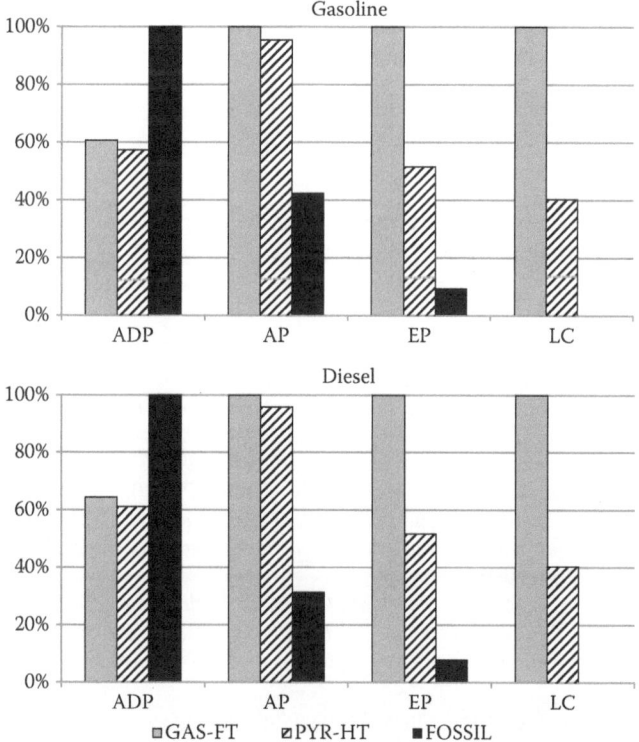

FIGURE 5.11   Relative environmental impacts in comparison with the fossil-fuel equivalents.

for biomass production. Although a high-yielding and low-input crop is used (short-rotation poplar), significant impacts arise from fertilization and electricity and fuel consumption during agricultural activity.

When comparing the two synthetic biofuels, the main discrepancy is found in LC. In this respect, the GAS-FT system shows significantly lower energy output per unit of biomass processed than the PYR-HT system, leading to more than twice the land demand. The high biomass demand is also the reason for the increased impacts of the GAS-FT biofuels in all remaining assessed categories.

Although the GAS-FT biofuels show better carbon footprints than PYR-HT biofuels, they generally score worse when considering other environmental impact categories. Furthermore, even though high GHG emission savings are associated with these synthetic biofuels, they perform worse than fossil fuels in other categories such as acidification and eutrophication. Hence, the findings from the CF study are likely to be contradictory to those from a wider assessment in terms of environmental impacts.

5.5 CONCLUSIONS

Synthetic biofuels produced via thermochemical routes have the potential to attain high GHG emission savings when compared with conventional fossil fuels. In this sense, synthetic fuels from biomass gasification followed by FT synthesis achieve GHG savings above 60%, whereas those from biomass pyrolysis followed by bio-oil hydrotreating achieve slightly lower savings. Nevertheless, the favorable carbon footprints obtained for the gasification-based synthetic biofuels (compared with pyrolysis-based biofuels) are due to the high biomass consumption of the system in combination with the carbon stock increase for short-rotation plantations on fallow land.

Including additional criteria to the assessment (e.g., thermodynamic and environmental indicators) facilitates a more sensible decision-making process. For instance, under a broader environmental perspective, pyrolysis-based synthetic biofuels are generally identified as better alternatives to conventional fossil fuels than synfuels based on biomass gasification. This is also supported by the findings from the thermodynamic assessment, which reports better results for the pyrolysis-based process.

ACKNOWLEDGMENT

This research was partly supported by the Spanish Ministry of Economy and Competitiveness (ENE2011-29643-C02-01 and IPT-2012-0219-120000).

REFERENCES

Abrahamson LP, Robison DJ, Volk TA et al. 1998. Sustainability and environmental issues associated with willow bioenergy development in New York (USA). *Biomass and Bioenergy* 15:17–22.

Anex RP, Aden A, Kazi FK et al. 2010. Techno-economic comparison of biomass-to-transportation fuels via pyrolysis, gasification, and biochemical pathways. *Fuel* 89:S29–S35.

Banse M, van Meijl H, Tabeau A, Woltjer G, Hellmann F, Verburg PH. 2011. Impact of EU biofuel policies on world agricultural production and land use. *Biomass and Bioenergy* 35:2385–2390.

Batjes NH. 2010. *IPCC Default Soil Classes Derived from the Harmonized World Soil Database*. Wageningen: ISRIC-World Soil Information.

Bejan A, Tsatsaronis G, Moran M. 1996. *Thermal Design and Optimization*. New York: John Wiley & Sons.

Bridgwater AV. 2002. *Fast Pyrolysis of Biomass: A Handbook*, Vol. 2. Newbury: CPL Press.

Bridgwater AV. 2012. Upgrading biomass fast pyrolysis liquids. *Environmental Progress & Sustainable Energy* 31:261–268.

Bridgwater AV. 2013. Fast pyrolysis of biomass for the production of liquids. In: Rosendahl L (ed.), *Biomass Combustion Science, Technology and Engineering*. Cambridge: Woodhead Publishing, pp. 130–171.

Bridgwater AV, Czernik S, Diebold J et al. 1999. *Fast Pyrolysis of Biomass: A Handbook*. Newbury: CPL Press.

Brinkman N, Wang M, Weber T, Darlington T. 2005. *Well-to-Wheels Analysis of Advanced Fuel/Vehicle Systems—A North American Study of Energy Use, Greenhouse Gas Emissions, and Criteria Pollutant Emissions.* Lemont: Argonne National Laboratory.

British Standards Institution. 2011. *PAS 2050:2011 Specification for the Assessment of the Life Cycle Greenhouse Gas Emissions of Goods and Services.* London: BSI.

Campbell A, Doswald N. 2009. *The Impacts of Biofuel Production on Biodiversity: A Review of the Current Literature.* Cambridge: UNEP-WCMC.

Cherubini F, Strømman AH. 2011. Life cycle assessment of bioenergy systems: State of the art and future challenges. *Bioresource Technology* 102:437–451.

Chiaramonti D, Oasmaa A, Solantausta Y. 2007. Power generation using fast pyrolysis liquids from biomass. *Renewable and Sustainable Energy Reviews* 11:1056–1086.

Ciferno JP, Marano JJ. 2002. *Benchmarking Biomass Gasification Technologies for Fuels, Chemicals and Hydrogen Production.* North Aurora: National Energy Technology Laboratory.

Cruz PL, Susmozas A, Iribarren D, Dufour J. 2014. Simulation of alternative systems for the co-production of biofuels and electricity from lignocellulosic biomass. In: *ANQUE ICCE BIOTEC—Science, the Key for a Better Life—Madrid 2014—Abstracts*, 143. Madrid: ANQUE.

Dallemand JF, Petersen JE, Karp A. 2008. *Short Rotation Forestry, Short Rotation Coppice and Perennial Grasses in the European Union: Agro-environmental Aspects, Present Use and Perspectives.* Luxembourg: European Communities.

Dang Q, Yu C, Luo Z. 2014. Environmental life cycle assessment of bio-fuel production via fast pyrolysis of corn stover and hydroprocessing. *Fuel* 131:36–42.

Eggleston HS, Buendia L, Miwa K, Ngara T, Tanabe K. 2006. *2006 IPCC Guidelines for National Greenhouse Gas Inventories.* Hayama: The Intergovernmental Panel on Climate Change.

Elliott DC. 2007. Historical developments in hydroprocessing bio-oils. *Energy & Fuels* 21:1792–1815.

Elliott DC, Hart TR, Neuenschwander GG, Rotness LJ, Zacher AH. 2009. Catalytic hydroprocessing of biomass fast pyrolysis bio-oil to produce hydrocarbon products. *Environmental Progress & Sustainable Energy* 28:441–449.

European Environment Agency. 2013a. *EU Bioenergy Potential from a Resource-Efficiency Perspective.* Luxembourg: EEA.

European Environment Agency. 2013b. *EMEP/EEA Air Pollutant Emission Inventory Guidebook 2013.* Luxembourg: EEA.

European Union. 2009. *Directive 2009/28/EC of the European Parliament and of the Council of 23 April 2009 on the promotion of the use of energy from renewable sources and amending and subsequently repealing Directives 2001/77/EC and 2003/30/EC.* Strasbourg: Official Journal of the European Union.

European Union. 2013. *Eurostat Pocketbooks—Agriculture, Forestry and Fishery Statistics.* Luxembourg: European Union.

European Union. 2014. *Proposal for a Directive of the European Parliament and of the Council amending Directive 98/70/EC relating to the quality of petrol and diesel fuels and amending Directive 2009/28/EC on the promotion of the use of energy from renewable sources—Political agreement.* Brussels: Council of the European Union.

Fargione J, Hill J, Tilman D, Polasky S, Hawthorne P. 2008. Land clearing and the biofuel carbon debt. *Science* 319:1235–1238.

Fischer G, Hizsnyik E, Prieler S, Shah M, van Velthuizen H. 2009. *Biofuels and Food Security.* Vienna: OFID.

Frischknecht R, Jungbluth N, Althaus HJ et al. 2007. *Overview and methodology.* Ecoinvent Report No. 1. Dübendorf: Swiss Centre for Life Cycle Inventories.

Gasol CM, Gabarrell X, Anton A et al. 2009a. LCA of poplar bioenergy system compared with *Brassica carinata* energy crop and natural gas in regional scenario. *Biomass and Bioenergy* 33:119–129.

Gasol CM, Martínez S, Rigola M et al. 2009b. Feasibility assessment of poplar bioenergy systems in the Southern Europe. *Renewable and Sustainable Energy Reviews* 13:801–812.

Goedkoop M, Oele M, Leijting J, Ponsioen T, Meijer E. 2013. *Introduction to LCA with SimaPro.* Amersfoort: PRé Consultants.

Guinée JB, Gorrée M, Heijungs R et al. 2001. *Life Cycle Assessment—An Operational Guide to the ISO Standards.* Leiden: Institute of Environmental Sciences.

Han J, Elgowainy A, Dunn JB, Wang MQ. 2013. Life cycle analysis of fuel production from fast pyrolysis of biomass. *Bioresource Technology* 133:421–428.

Hein L, Leemans R. 2012. The impact of first-generation biofuels on the depletion of the global phosphorus reserve. *Ambio* 41:341–349.

Heller M, Keoleian GA, Volk TA. 2003. Life cycle assessment of a willow bioenergy cropping system. *Biomass and Bioenergy* 25:147–165.

Hsu DD. 2012. Life cycle assessment of gasoline and diesel produced via fast pyrolysis and hydroprocessing. *Biomass and Bioenergy* 45:41–47.

Huber GW, Iborra S, Corma A. 2006. Synthesis of transportation fuels from biomass: chemistry, catalysts, and engineering. *Chemical Reviews* 106:4044–4098.

International Organization for Standardization. 2006a. *ISO 14040:2006 Environmental management—Life Cycle Assessment—Principles and framework*. Geneva: ISO.

International Organization for Standardization. 2006b. *ISO 14044:2006 Environmental management—Life Cycle Assessment—Requirements and guidelines*. Geneva: ISO.

International Organization for Standardization. 2013. *ISO/TS 14067:2013 Greenhouse gases—Carbon footprint of products—Requirements and guidelines for quantification and communication*. Geneva: ISO.

Iribarren D, Dufour J. 2014. Carbon footprint as a single indicator in energy systems: The case of biofuels and CO_2 capture technologies. In: Muthu SS (ed.), *Assessment of Carbon Footprint in Different Industrial Sectors*, Vol. 2. Singapore: Springer, pp. 81–104.

Iribarren D, Peters JF, Dufour J. 2012. Life cycle assessment of transportation fuels from biomass pyrolysis. *Fuel* 97:812–821.

Iribarren D, Susmozas A, Dufour J. 2013a. Life-cycle assessment of Fischer–Tropsch products from biosyngas. *Renewable Energy* 59:229–236.

Iribarren D, Susmozas A, Dufour J. 2013b. Enhancing the life-cycle environmental performance of an energy system for the coproduction of synthetic biofuels and electricity. In: Chalmers University of Technology, and The Swedish Life Cycle Center (eds), *Perspectives on Managing Life Cycles—Proceedings of the 6th International Conference on Life Cycle Management*. Gothenburg: Chalmers University of Technology, pp. 828–831.

Iribarren D, Susmozas A, Petrakopoulou F, Dufour J. 2014. Environmental and exergetic evaluation of hydrogen production via lignocellulosic biomass gasification. *Journal of Cleaner Production* 69:165–175.

Jones SB, Valkenburg C, Walton CW et al. 2009. *Production of gasoline and diesel from biomass via fast pyrolysis, hydrotreating and hydrocracking: A design case*. Richland: Pacific Northwest National Laboratory.

Jungbluth N. 2007. *Erdöl, ecoinvent report No. 6-IV*. Dübendorf: Swiss Centre for Life Cycle Inventories.

Kotas TJ. 1995. *The Exergy Method of Thermal Plant Analysis*. Malabar: Krieger Publishing Company.

Larson ED, Jin H, Celik FE. 2009. Large-scale gasification-based coproduction of fuels and electricity from switchgrass. *Biofuels, Bioproducts and Biorefining* 3:174–194.

Li XT, Grace JR, Lim CJ, Watkinson, Chen HP, Kim JR. 2004. Biomass gasification in a circulating fluidized bed. *Biomass and Bioenergy* 26:171–193.

Liu G, Larson ED, Williams RH, Kreutz TG, Guo X. 2011. Making Fischer–Tropsch fuels and electricity from coal and biomass: Performance and cost analysis. *Energy & Fuels* 25:415–437.

Malça J, Freire F. 2011. Life-cycle studies of biodiesel in Europe: A review addressing the variability of results and modeling issues. *Renewable and Sustainable Energy Reviews* 15:338–351.

Mann MK. 1995. *Technical and economic assessment of producing hydrogen by reforming syngas from the Battelle indirectly heated biomass gasifier*. Golden: National Renewable Energy Laboratory.

McKendry P. 2002. Energy production from biomass (part 3): Gasification technologies. *Bioresource Technology* 83:55–63.

Meier D, van de Beld B, Bridgwater, AV, Elliott DC, Oasmaa A, Preto F. 2013. State-of-the-art of fast pyrolysis in IEA bioenergy member countries. *Renewable and Sustainable Energy Reviews* 20:619–641.

Muench S, Guenther E. 2013. A systematic review of bioenergy life cycle assessments. *Applied Energy* 112:257–273.

Myhre G, Shindell D, Bréon FM et al. 2013. Anthropogenic and natural radiative forcing. In: Stocker TF, Qin D, Plattner GK et al. (eds), *Climate Change 2013: The Physical Science Basis—Contribution of Working Group I to the Fifth Assessment Report of the Intergovernmental Panel on Climate Change*. Cambridge: Cambridge University Press, pp. 659–740.

Nachenius RW, Ronsse F, Venderbosch RH, Prins W. 2013. Biomass pyrolysis. In: Murzin DY (ed.), *Advances in Chemical Engineering—Chemical Engineering for Renewables Conversion*. San Diego: Academic Press, pp. 75–139.

Perttu KL. 1998. Environmental justification for short-rotation forestry in Sweden. *Biomass and Bioenergy* 15:1–6.

Peters JF, Banks SW, Susmozas A, Dufour J. 2014. Experimental validation of a predictive pyrolysis model in Aspen Plus®. In: Hoffmann C, Baxter D, Maniatis K, Grassi A, Helm P (eds), *Papers of the 22nd European Biomass Conference*. Florence: ETA-Florence Renewable Energies, pp. 1160–1163.

Peters JF, Iribarren D, Dufour J. 2015a. Simulation and life cycle assessment of biofuel production via fast pyrolysis and hydroupgrading. *Fuel* 139:441–456.

Peters JF, Iribarren D, Dufour J. 2015b. Life cycle assessment of pyrolysis oil applications. *Biomass Conversion and Biorefinery* 5:1–19.

Pfeifer C, Rauch R, Hofbauer H. 2004. In-bed catalytic tar reduction in a dual fluidized bed biomass steam gasifier. *Industrial Engineering Chemistry Research* 43:1634–1640.

Rödl A. 2008. *Ökobilanzierung der Holzproduktion im Kurzumtrieb*. Hamburg: Universität Hamburg.

Sciubba E, Wall G. 2007. A brief commented history of exergy from the beginnings to 2004. *International Journal of Thermodynamics* 10:1–26.

Sevigne E, Gasol CM, Brun F et al. 2011. Water and energy consumption of *Populus* spp. bioenergy systems: A case study in Southern Europe. *Renewable and Sustainable Energy Reviews* 15:1133–1140.

Shen L, Gao Y, Xiao J. 2008. Simulation of hydrogen production from biomass gasification in interconnected fluidized beds. *Biomass and Bioenergy* 32:120–127.

Spanish Ministry of Industry, Tourism and Trade. 2010. *Royal Decree 1088/2010*. Madrid: Boletín Oficial del Estado.

Spath P, Aden A, Eggeman T, Ringer M, Wallace B, Jechura J. 2005. *Biomass to hydrogen production detailed design and economics utilizing the Battelle Columbus Laboratory indirectly-heated gasifier*. Golden: National Renewable Energy Laboratory.

Spyrakis S, Panopoulos KD, Kakaras E. 2009. Synthesis, modeling and exergy analysis of atmospheric air blown biomass gasification for Fischer–Tropsch process. *International Journal of Thermodynamics* 12:187–192.

Susmozas A, Iribarren D, Dufour J. 2013. Life-cycle performance of indirect biomass gasification as a green alternative to steam methane reforming for hydrogen production. *International Journal of Hydrogen Energy* 38:9961–9972.

Swain PK, Das LM, Naik SN. 2011. Biomass to liquid: A prospective challenge to research and development in 21st century. *Renewable and Sustainable Energy Reviews* 15:4917–4933.

Swanson RM, Satrio JA, Brown RC, Platon A, Hsu DD. 2010. *Techno-economic analysis of biofuels production based on gasification*. Golden: National Renewable Energy Laboratory.

Szargut J. 2011. Appendix: Standard chemical exergy. In: Bakshi BR, Gutowski TG, Sekulić DP (eds), *Thermodynamics and the Destruction of Resources*. New York: Cambridge University Press, 489 pp.

Tsatsaronis G, Cziesla F. 2002. Thermoeconomics. In: Meyers RA (ed.), *Encyclopedia of Physical Science and Technology*. New York: Academic Press, pp. 659–680.

Tsatsaronis G, Cziesla F. 2004a. Exergy and thermodynamic analysis. In: Frangopoulos CA (ed.), *Exergy, Energy System Analysis, and Optimization*. Oxford: Eolss Publishers, pp. 34–45.

Tsatsaronis G, Cziesla F. 2004b. Exergy analysis of simple processes. In: Frangopoulos CA (ed.), *Exergy, Energy System Analysis, and Optimization*. Oxford: Eolss Publishers, pp. 79–107.

van Vliet OPR, Faaij APC, Turkenburg WC. 2009. Fischer–Tropsch diesel production in a well-to-wheel perspective: A carbon, energy flow and cost analysis. *Energy Conversion and Management* 50:855–876.

Venderbosch RH, Ardiyanti AR, Wildschut J, Oasmaa A, Heeres HJ. 2010. Stabilization of biomass-derived pyrolysis oils. *Journal of Chemical Technology and Biotechnology* 85:674–686.

Venderbosch RH, Prins W. 2010. Fast pyrolysis technology development. *Biofuels, Bioproducts and Biorefining* 4:178–208.

Wang B, Gebreslassie BH, You F. 2013. Sustainable design and synthesis of hydrocarbon biorefinery via gasification pathway: Integrated life cycle assessment and technoeconomic analysis with multiobjective superstructure optimization. *Computers and Chemical Engineering* 52:55–76.

Weinberg J, Kaltschmitt M. 2013. Life cycle assessment of mobility options using wood based fuels— Comparison of selected environmental effects and costs. *Bioresource Technology* 150:420–428.

Wildschut J. 2009. *Pyrolysis Oil Upgrading to Transportation Fuels by Catalytic Hydrotreatment*. Groningen: Rijksuniversiteit Groningen.

Wright MM, Satrio JA, Brown RC, Daugaard DE, Hsu DD. 2010. *Techno-economic analysis of biomass fast pyrolysis to transportation fuels*. Golden: National Renewable Energy Laboratory.

Xiu S, Shahbazi A. 2012. Bio-oil production and upgrading research: A review. *Renewable and Sustainable Energy Reviews* 16:4406–4414.

Zah R, Böni H, Gauch M, Hischier R, Lehmann M, Wäger P. 2007. *Life Cycle Assessment of Energy Products: Environmental Assessment of Biofuels*. Bern: EMPA.

6 Issues in Making Food Production GHG Efficient

Challenges before Carbon Footprinting

Divya Pandey and Madhoolika Agrawal

CONTENTS

6.1 Introduction 125
6.2 Food-Producing Systems and GHG Emissions 126
    6.2.1 Crop Cultivation 126
    6.2.2 Livestock Rearing and Dairy 127
    6.2.3 Food Miles 127
6.3 Scope of Reducing GHG Emissions from Croplands and Agricultural Soils as Carbon
    Offsetting Option 128
    6.3.1 Reducing Emissions from Crop Cultivation and Livestock Rearing 129
    6.3.2 Improving Food-Processing Efficiency 129
    6.3.3 Wise Packaging and Transport 130
6.4 Carbon Footprinting as a Quantitative Representation of GHG Efficiency and Its
    Application to Food Production 131
    6.4.1 Counting C Sequestration 132
    6.4.2 Incorporating Land-Use Change 133
    6.4.3 The Case of Biofuels 133
6.5 Case Studies 134
    6.5.1 Crop Cultivation 134
    6.5.2 Dairy and Meat Production 134
6.6 Carbon Labeling and Policy Implications 136
6.7 Recommendations for Further Research Directions 137
6.8 Conclusions 137
Acknowledgments 137
References 138

6.1 INTRODUCTION

Food and nutrition security is the most fundamental right to every living being, though in the present times, the concept of food consumption in human race is much more than just being an essence for living. The wide differences in food choices, varieties in tastes, and association of food with sociocultural practices in many cases have led to the evolution of a diverse and huge food market with remarkably different and complex trails of processing systems. The global population is expected to rise to 9 billion by 2050 which will increase the food demand to almost double (Foley et al. 2011). Even in the today's world, nearly 1 billion people are undernourished, which suggests that food systems are under immense pressure to produce sufficiently and ensure that it reaches everyone in future. A critical analysis of global agriculture reveals that agricultural productivity increased by 20% in the last two decades, and it seems unrealistic to double the production by 2050. This means that more land would need to be created for agriculture which may lead to deforestation and hence environmental externalities. Such environmental externalities may also include climate change, which eventually will cause a great uncertainty to food-producing systems. It is therefore proposed that food production systems must increase productivity with efficient consumption and lowest possible environmental costs, mainly greenhouse gas (GHG) emissions. This chapter attempts to present an overview of carbon footprints (CFs) of food production systems and identification of approaches having lowest CF. The concept of CF calculation and its application to agriculture have been discussed in detail by Pandey and Agrawal (2014). In this review, detailed discussion of challenges before application of CFs to food production along with case studies is provided.

6.2 FOOD-PRODUCING SYSTEMS AND GHG EMISSIONS

At the global level, food production is the largest GHG emitter after energy production. According to the fifth assessment report of IPCC, Agriculture, Forestry and Other Land Uses (AFOLU) represent 20%–24% of total anthropogenic emissions (IPCC 2013). Though emissions from AFOLU show a decreasing trend over the decade, the emissions associated with croplands and cattle rearing continue to increase. Annual GHG emissions from agriculture, including croplands and livestock rearing, are estimated to be between 5.38 $GtCO_2$-eq. in 2011 compared with 5.0 $GtCO_2$-eq. yr^{-1} during 2000–2010 (Tubiello et al. 2014). As a positive sign, the GHG intensity of agricultural products decreased significantly during 1990–2010, even though additional mitigation activities are urgently required in the absence of which future emissions may increase by 30% by 2050.

The food sector is complex because there are various levels of raw material production, processing, trade, and consumption. Many times, it becomes very critical to differentiate between the relevant system boundaries for estimating emissions. Cultivation systems, livestock management, food-processing units, consumption, and wastage have high uncertainties with respect to GHG intensities. The entire supply system can be divided into a few basic phases presented in Figure 6.1, all of which individually are significant emission sources. For simplicity and ease of calculation, most of the studies focus separately on crop cultivation, transportation, and food processing and marketing. However, if we talk about C footprinting of food products, it must cover all of these activities in one boundary. Here, we discuss GHG emissions from different stages briefly.

6.2.1 CROP CULTIVATION

During crop cultivation or pasture management alone, there are a number of inputs and activities that emit GHGs directly or indirectly. GHG emissions from the "cradle-to-farm-gate" in the case of croplands represent about 20%–40% of the total CF of food products (Barber et al. 2011), whereas for horticultural products, they fall between 15 and 25% (Mithraratne et al. 2010). Most notable emissions from croplands are the soil-borne or those associated with fertilizer and manure

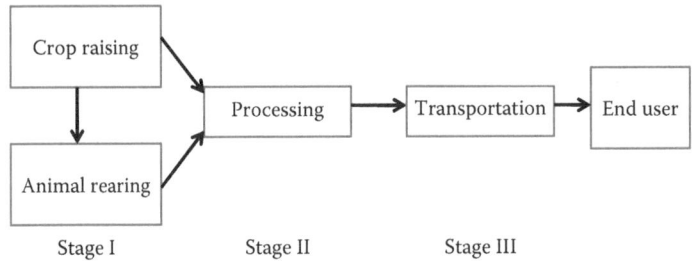

FIGURE 6.1 Important stages in food production and supply chain.

management and operational energy. Fertilizers alone are responsible for as much as 40% of the CF of crop cultivation at the farm stage (Ingwersen 2012), while a large amount of energy is devoted to irrigation, land management, harvesting, and grain processing (Pandey et al. 2013). The contribution of these activities depends on crop, available resources, and prevalent practices. Among different crops, some are well known for high emissions. Rice fields are identified as the dominant source of CH_4, whereas N_2O is emitted mainly from N-fertilized croplands. Water-intensive rice fields are not only responsible for a large amount of energy for irrigation, but the crop-growing conditions favor a high degree of methanogenesis and it is estimated that rice fields emit 32–44 Tg CH_4 yr^{-1} (Le Mer and Roger 2001). Asia contributes up to 90% of these emissions with more than 50% from China and India alone (Yan and Crookes 2009). During rice cultivation, mulching and organic manure applications also increase emissions of CH_4 (Ma et al. 2007; Qin et al. 2010a), whereas midseason drainage can potentially cut the emissions (Zou et al. 2005). On the other hand, upland soils like those under maize and wheat cultivation act as net CH_4 sinks (Robertson et al. 2000) or even as sources (Chatskikh et al. 2008; Ma et al. 2013) depending on the soil type, crop management practices, and environmental conditions. As a widely recognized effect, application of mineral N increases the emissions of N_2O.

6.2.2 LIVESTOCK REARING AND DAIRY

According to FAO, the livestock supply chain is responsible for emissions of 7.1 $GtCO_2$-eq. annually, which is approximately 14.5% of all anthropogenic GHG emissions. The main sources of emissions associated with livestock-based food production include livestock feed production and processing, which account for nearly 45% of the total enteric fermentation (39%) and manure decomposition (10%). About 6% of emissions is attributable to the processing and transportation of animal food products. It is estimated that milk production alone from farm to sale represents approximately 2.7%–4.0% of global anthropogenic emissions, and about 80%–90% of this emission takes place before the farm gate (Gerber et al. 2010). Emissions and hence CFs differ greatly among regions, dairy sizes, and management. According to an EPA estimate (2011), in the European Union share of dairy cattle rearing to total annual emissions is 1.2%, whereas in the United States, it is 0.55%. The highest CFs are estimated to be left by sub-Saharan African dairies with an average of 7.5 CO_2-eq. kg^{-1} whereas South and West Asia, North Africa, and Central and South America leave CFs between 3.0 and 5.0 kg CO_2-eq. kg^{-1} (FAO 2010). Table 6.1 presents an overview of CFs of milk and beef production as calculated in different studies.

Hagemann et al. (2011) quantified and compared the GHG emissions associated with bovine milk production from 38 countries following the method proposed by "International Farm Comparison Network" and found that 70%–95% of total emissions took place from enteric fermentation and manure. These figures appear small, but it is due to the fact that other sectors of emissions are very large compared with these. But, if we look at the countries with large cattle population or area under pastors including India, Brazil, New Zealand, and Ireland, the dairy sector becomes very important (FAO 2010). In such cases, animal-derived products may contribute to 70%–90% of the total agricultural emissions related with animal rearing in addition to those occurring from pastureland (Ledgard et al. 2009, 2011).

6.2.3 FOOD MILES

Transportation is an energy-intensive affair, and when it comes to food products, the term "food miles" is used to express the distance traveled by the food from its point of production to the point of sale (SAFE Alliance 1994). Food miles may hold the highest share in the total CF, which can be as high as 55% in the case of imported products (Svanes and Aronsson 2013) and its increase by five times can increase GHG emissions by 120% (Point 2008). In some countries, food miles are very high. For example, in the UK, food transport accounts for nearly 19 $MtCO_2$ emissions (Smith 2005).

TABLE 6.1

CFs of Milk and Beef Production in Different Case Studies

Product	CF	Region	Reference
Milk	1.06 kg CO_2-eq. kg^{-1}	Denmark	Kristensen et al. (2011)
	0.96–1.08 kg CO_2-eq. kg^{-1}	Sweden	Cederberg and Flysjö (2004), Cederberg et al. (2009)
	1.03 kg CO_2-eq. kg^{-1}	Sweden	Cederberg and Stadig (2003)
	1.06 kg CO_2-eq. l^{-1}	UK	Williams et al. (2006)
	1.50 kg CO_2-eq. kg^{-1}	Ireland	Casey and Holden (2005)
	0.64–0.85 kg CO_2-eq. kg^{-1}	Germany	Hirschfeld et al. (2008)
	1.0–1.3 kg CO_2-eq. kg^{-1}	Germany	Haas et al. (2001)
	1.26 kg CO_2-eq. kg^{-1}	Northern Germany	Hagemann et al. (2011)
	0.99 kg CO_2-eq. kg^{-1}	California	Hagemann et al. (2011)
	2.88 kg CO_2-eq. kg^{-1}	Brazil	Hagemann et al. (2011)
	5.05 kg CO_2-eq. kg^{-1}	Bangladesh	Hagemann et al. (2011)
	2.34 kg CO_2-eq. kg^{-1}	Cameroon	Hagemann et al. (2011)
	1.58 kg CO_2-eq. kg^{-1}	New Zealand	Hagemann et al. (2011)
	0.72 kg CO_2-eq. kg^{-1}	New Zealand	Basset-Mens et al. (2005)
	0.91 kg CO_2-eq. kg^{-1}	New Zealand	Beukes et al. (2010)
	1.4–1.5 kg CO_2-eq. kg^{-1}	The Netherlands	Thomassen et al. (2008)
	1.57 kg CO_2-eq. kg^{-1}	Tehran	Daneshi et al. (2014)
Beef	20.1 kg CO_2-eq. kg^{-1}	Ireland	Casey and Holden (2006)
	15.6 kg CO_2-eq. kg^{-1}	Sweden	Cederberg and Stadig (2003)
	22.6 kg CO_2-eq. kg^{-1}	Japan	Ogino et al. (2004)
	26.0–34.0 kg CO_2-eq. kg^{-1}	New Zealand	White et al. (2010)
	26.9–34.9 kg CO_2-eq. kg^{-1}	United States	Pelletier et al. (2010)

It may always not be true in all the cases. Weber and Matthews (2008) found that in the United States, on an average, 11% of the CF of food comes along the food miles.

6.3 SCOPE OF REDUCING GHG EMISSIONS FROM CROPLANDS AND AGRICULTURAL SOILS AS CARBON OFFSETTING OPTION

The biggest scope of improving GHG efficiency lies in the activities which are the biggest emitters and have been discussed in previous sections. It requires technical and behavioral changes supported by suitable policies. While food processing and marketing can be made very much GHG-efficient, controlling emissions at farm scale is not easy because reductions are sensitive to the region, climatic and environmental conditions, as well as social and economic status of the farmers. Additionally, many emissions are just unavoidable. Nevertheless, adoption of best practices based on available knowledge along with continuous monitoring must be promoted. On the basis of current understanding, these are the basic keys for reducing GHG emissions from food production: reducing soil-borne emissions, improving energy efficiency at farm, processing transportation, and strengthening C sequestration.

IPCC (2013) estimated that the economic potential for mitigating GHG emissions from the agriculture sector is 0.5–10.6 GtCO_2-eq. annually. The potential practices target better cultivation techniques to reduce CH_4 and N_2O emissions from croplands by modifying tillage, fertilizer and water management, better cultivars, and effective management of crop residues. Adopting these, it is estimated that global agriculture can mitigate 5500–6000 MtCO_2-eq. yr^{-1} (Smith et al. 2008). Additionally, minimizing food wastage alone can reduce 0.6–6.0 GtCO_2-eq. annually, whereas change in dietary habit toward more vegetarian can offset 0.7–7.3 GtCO_2-eq. annually.

6.3.1 Reducing Emissions from Crop Cultivation and Livestock Rearing

There are a number of methods that can help managing soil-borne emissions. Since such emissions are the result of microbial activities, changing microbial processes has been evaluated by scientists under diverse conditions and crops. Chemical inhibition of methanogenesis, nitrification, and denitrifiction has been successfully tested in experimental farms. Though they can help cutting the emissions significantly, they are not applicable on large-scale due to cost intensiveness and likelihood of unforeseen negative impacts on environment or crop quality. The most feasible methods are changing farming practices and improving yields. As an example, diversified cropping systems are found to be more GHG-efficient compared with the prevailing monoculture systems. Gan et al. (2011a) demonstrated that cropping of oilseed, pulse, and cereal crops in particular sequences could reduce the CFs of durum wheat monoculture by 22%. Similarly, coffee agro-forestry plantation system could sequester CO_2 at four times faster rate than the coffee monoculture (Hergoualc'h et al. 2012). Conservation tillage management is another effective and rather popular way of reducing GHG emissions and facilitating C sequestarion in cultivated soils under different crops and cropping systems in different geographical regions (Arshad et al. 1990; Bayer et al. 2000).

The majority of long-term measurements indicated that organic manure application increased C stock in the soil when compared with mineral fertilizer use in different cropping practices, the range of C sequestration offered was from 70 to 551 kg C ha^{-1} (Mandal et al. 2007, 2008; Majoomdar et al. 2008).

Livestock feed and enteric fermentation are the two main emission sources. While emissions from enteric fermentation are unavoidable, they can be controlled significantly by improved feeding mixtures to an extent as high as 38% (Peters 2010). Philippe et al. (2012) observed that a straw flow feeding system reduced N_2O by 55%, CH_4 by 46%, and CO_2 by 10%, when compared with straw-based deep litter. Among the grain feedlot and grassland pastoral systems for beef raising, Peters (2010) observed that grain-fed systems leave lower CF. Such observations are, however, highly dependent on the region, livestock breeds, and management as evident by the similar study of Subak (1999) but with opposite conclusion.

In livestock breeding strategy, longer calving intervals and higher culling rates lead to overall increase in emissions (Beukes et al. 2010; O'Brien et al. 2010). Crosson et al. (2010) found that increase in the live weight can reduce the GHG intensity of beef production 5%–6%.

6.3.2 Improving Food-Processing Efficiency

The food-processing industry makes use of a number of inputs and hence their CFs depend largely on the CFs of the inputs they use. Hence, increasing energy efficiency in all the sectors is important to reduce CFs of all other sectors. Here, we discuss the processes associated with food processing exclusively. Common methods of increasing energy efficiency in agriculture relates to improved farm machineries and irrigation, whereas during the processing phase, process intensification and integration can substantially cut the emissions arising from food processing units (Xu et al. 2014).

Intensification of processes targets optimizing productivity, which tends to reduce GHG intensity of the final product. This principle is found effective for processes like mixing, drying, emulsification, crystallization, and so on (Reay 2008). Process integration, on the other hand, integrates different processes; hence many processes can run together, or the by-product of one process can act as a resource for the other. This is particularly true for processes where heating or cooling is required. Careful control and optimization of all heating or cooling processes together substantially reduce the energy consumption and hence GHG emissions (Masanet et al. 2008; Friedler 2010). Tuan et al. (2012) observed a saving of 22.6% CO_2-eq. by modifying vapor recompression technique for evaporator. Heat recovery also increases energy efficiency of processes such as drying, evaporation and concentration, frying, pasteurization and sterilization, dehydration, and peeling (Masanet et al. 2008).

Just like energy, proper water management can be very important for some food processes and hence water reuse and efficiency becomes important (Lim and Park 2008; Griffiths-Sattenspiel and Wilson 2009).

Operating temperatures ranging from freezing to baking are crucial for the food industry and optimal temperature controls essentially save energy. Many researchers have been working on this aspect, mostly through numerical computational fluid dynamics—for example, for bakeries (Ghani and Farid 2007), heating of glass-jarred liquid food (Lespinard and Mascheroni 2012), optimization of design for refrigeration management, sterilization and cooling of orange juice (Wang et al. 2011), and cooling of packed strawberries (Ferrua and Singh 2011).

Replacement or augmentation of conventional energy sources with the renewable is also a practical and well-demonstrated tool in a number of food-processing units. Solar energy is being used for processes like defrosting, pasteurization, sterilization, and drying (Schnitzer et al. 2007; Pazheri et al. 2012). Bio-fuel is also a potential candidate to address energy demand of food-processing units, since most of them can generate energy just by recycling their organic wastes to obtain ethanol or methanol (Escobar et al. 2009) or as a substitute of coal (Intini et al. 2012).

All of the above methods target avoiding emissions, but technology is now available to capture and convert CO_2 into chemically useful forms before its emission into the atmosphere. Such techniques, at the moment, are costly to most of the small-scale industries and often reduce thermal efficiencies (Rao and Rubin 2002; Tan et al. 2009). Common techniques for industrial CO_2 capture include adsorption, absorption, membrane separation, hydrate-based separation, chemical-looping combustion, and mineral carbonation techniques (Yang et al. 2008; Kwon et al. 2011).

6.3.3   WISE PACKAGING AND TRANSPORT

For food items, packaging must ensure hygiene, preserve freshness, and at the same time it is an important part of marketing. Different packaging options primarily depend on the item, size, distance of transportation, and shelf life. Accordingly, different options and techniques have different environmental costs, and various researches provide a glimpse. Interestingly, it has been discovered that recyclable materials are not always more climate-friendly. For example, De Monte et al. (2005) analyzed different packaging systems for retail coffee and concluded that polylaminate bags had lower CFs than metallic cans. They also found that bigger pack sizes had lower CF per kg of coffee. Similarly, between glass and plastic packaging for baby foods, Humbert et al. (2009) found that plastic pot packaging reduced the CF by 28%–31% compared with the glass jar system. On the other hand, Poovarodom et al. (2012) analyzed that the retort cup system for tuna products could reduce the CF by 10% compared with metal can packaging. Reducing the packaging material, for example, by avoiding multiple layers, can significantly cut GHG emissions (Pimentel et al. 2008).

For reducing the emissions related with transportation, cutting the transportation distance and increasing it more energy-efficient are the only solutions. For cutting the food miles, making locally grown food system more sustainable by proper storage and marketable is essential (Van Hauwermeiren et al. 2007). It also requires that local food market can address consumer demands for quality food (Murdoch et al. 2000; Cowell and Parkinson 2003). Food miles become very long in the case of international trade, but in many countries (mostly developed), domestic food transportation also becomes long. However, in some cases, local grown foods may be more GHG-intensive. For example, field-grown lettuces from Spain have lower CF even if transported to UK when compared with glasshouse-raised lettuce in UK itself (Hospido et al. 2009). With advancement in plant-raising techniques, some experiments have been made with rooftop greenhouses systems, where roof tops of buildings can be used for growing crops under hydroponic culture. Sanye-Mengual et al. (2012) analyzed the impact of such change for cultivating tomatoes in Spain and calculated a reduction of 0.4 kg CO_2-eq. kg^{-1}.

6.4 CARBON FOOTPRINTING AS A QUANTITATIVE REPRESENTATION OF GHG EFFICIENCY AND ITS APPLICATION TO FOOD PRODUCTION

CF is a very popular mode of GHG expression for widely different entities and refers to the quantity of GHGs expressed in terms of CO_2-eq. emitted into the atmosphere by an individual, organization, process, product, or event from within a specified boundary. Due to its ease of conveying GHG intensity to general public, the number of CF calculations applied to different products is increasing, though there are still numerous ambiguities and challenges before harnessing its potential application. To facilitate uniform calculations, standard methods have evolved, but they address mainly products with well-defined processes. The standard methods of product carbon footprinting cover food products, however addressing the cultivation of crops and fodder becomes challenging under these methods due to large uncertainties associated with quantification of emissions. Common resources for product C footprinting are discussed below:

a. *GHG protocol of World Resource Institute (WRI)/World Business Council on Sustainable Development (WBCSD)*: It has two parts: (1) A Product Life Cycle Accounting and Reporting Standard and (2) Corporate Accounting and Reporting Standard: These guidelines provide general calculation tools and deal with quantification of GHG reductions resulting due to adoption of mitigation methods in its Project protocol. It forms the basis for most GHG accounting guidelines including ISO 14064 (parts 1 and 2) (WRI/WBCSD 2004, 2005).

b. *ISO 14067*: Based on standards on life-cycle assessment (ISO 14040, 14044), quantification, and on environmental labels and declarations (ISO 14020, 14024, and 14025) for communication, it guides the principles, requirements, quantification, and communication of the CF of a product. The standard of LCA it is based upon specifies determination of boundaries, quantification of GHG emissions, and removal (ISO 2006a, 2006b, 2013).

c. *Publicly Available Specifications-2050 (PAS 2050) of British Standard Institution (BSI)*: It specifies the requirements for assessing the life-cycle GHG emissions of goods and services (BSI 2008).

d. *IPCC guidelines for National GHG inventories*: All anthropogenic sources of GHG emissions are classified into four sectors—energy, industrial process and product use (IPPU), agriculture, forestry, and other land use (AFOLU), and waste. 2006 guidelines are an updated version of earlier 1996 guidelines. All countries signatory to UNFCCC are committed to prepare, update, and communicate their national GHG emissions/removal inventories following these guidelines.

All of these are applicable to product carbon footprinting and hence food-processing units, however, IPCC (2006) addresses crop cultivation part. These developed guidelines and standards direct accounting for the GHGs coupled with the complete "life-cycle assessment" or "cradle to grave analyses." To simplify this procedure, all the activities during the life stages are divided into three tiers according to the order of GHG emissions as below:

$Tier_1$: direct, that is, onsite emissions
$Tier_2$: emissions embodied in purchased energy
$Tier_3$: all indirect emissions not covered under $tier_2$ such as those associated with transport of purchased and sold goods, business travels, waste disposal, and so on.

Direct or onsite emissions refer to the activities that emit GHG right at the site of the activity. Most prominent processes representing this tier include fossil fuel burning such as farm machines, burners, boilers, and so on. For cultivation part, soil-borne GHG emissions also contribute to $tier_1$. Indirect or embodied emissions refer to the emissions that do not take place at the site of the activity, but are carried along the inputs, which are being used. For example, the packaging material purchased for processed food has been responsible for certain emissions during its manufacturing and transportation. Such emissions are attributable to the food-packaging process as indirect emissions. The tier$_2$ and tier$_3$ both comprise of indirect emissions but tier$_2$ deals specifically with energy since purchased energy in the form of electricity or heat is an important component of nearly all the industrial processes and grouping them together makes the calculations relatively easy.

While most of the CF studies and standards focus only on emissions, a significant level of C sequestration makes it important to also consider GHG removals. Estimation of C sequestration is, however, not easy particularly when it comes to the case where it needs to be compared with emissions. The main calculation challenge is to devise a method where two processes with very different temporal scales can be compared. This is so because sequestration is a rather slow process, but emissions are very fast (Pandey et al. 2011). Another challenge is how to measure C sequestration rates reliably, and assessing the degree of permanence of soil C stock is also a matter of debate (Lal 2004). Most of the cultivated soils are C-depleted and hence net sequestrants, but sequestration of C in soil does not go on indefinitely, but rather in a sigmoidal curve fashion attaining equilibrium after a more or less linear active phase (Lal 2004). On these lines, depleted soils of tropics and subtropics offer significantly large sequestration potential, but owing to harsh environmental conditions, it becomes questionable if the sequestered C will be irreversibly lost under unfavorable conditions (Lal et al. 2004; Powlson 2011).

6.4.1 Counting C Sequestration

Significant efforts have been made to develop methods for counting C sequestration in CF calculation, including temporary storage and even release of sequestered carbon over time, but no single robust method has yet been developed. The GHG Protocol and PAS 2050 have been revised (WRI/WBCSD 2011) and now they address temporary and long-term CO_2 removals, but their assumptions and approaches are different from each other. IPCC Good Practice Guidance for Land Use, Land Use Change and Forestry recommends the annual stock change method for both short- and long-term C removals (Penman et al. 2003). Other methods have also been developed and are commonly used. Some of these are briefly described below:

a. The conventional LCA method employs constant characterization factor throughout the life cycle of a product and hence fixed GWP. In this method, only long-term sequestration is meaningful, and both biogenic emissions and C sequestration can be counted. Net GHG emissions for bio-products are calculated as the balance of CO_2 taken up during biomass growth, and the CO_2 released to the atmosphere over the complete life cycle of the product. In case CO_2 sequestration balances or exceeds the emissions, the product is considered as C-neutral, irrespective of the difference in timing of uptake and release. However, the time period of assessment can be fixed if the objective requires counting of emissions only within a defined assessment period. Such an approach may fail to cover significant emission or removals that may occur shortly after the assessment period, but is suitable for long-time horizons.

b. The Moura-Costa method dealing with sequestration and temporary storage of carbon (Moura-Costa and Wilson 2000) can be used to calculate an equivalence factor between kg CO_2-eq. and kg CO_2-year, which is then used for accounting sequestration and temporary storage for the number of years CO_2 is removed and kept out of the atmosphere. This approach also applies on a fixed assessment period. But under this approach, starting and ending points of the assessment period are not fixed. This means that the impact of an emission is estimated for the chosen time length, irrespective of the point of time when emission took place during the life cycle.

c. The Lashof method (Fearnside 2002) assumes that CO_2 withdrawn from the atmosphere is equivalent to delayed emission from fossil fuels, and hence it can calculate credits for C sequestration and temporary removal, but it cannot exceed 100% credit.

d. The PAS 2050 (BSI 2008, 2011) method applies to different approaches for short- and long-term storage times. PAS 2050 considers short-term C storage in the same way as delaying an emission from the time of product manufacture and up to 100 years. In such cases, linear approximation of the Lashof method is used, while for longer storage times, the average amount of carbon stored over 100 years is estimated. Emissions that would occur after 100 years are ignored. In the revised PAS 2050 (BSI 2011), carbon footprinting should not count the C storage of less than 100 years.

e. The European Commission's ILCD Handbook method (European Commission 2010) does not include temporary C removals during CF calculation unless the objective of calculation strictly requires. CO_2 removals which take place for less than 100 years are credited on annual basis. Emissions or loss of sequestered C which may occur after 100,000 years are not considered in CF calculations.

f. The dynamic LCA approach (Levasseur et al. 2010) has provisions to distribute GHG emissions temporally over the life cycle. Hence, the radiative forcing due to the activity can be calculated for any at any point of time. This approach fixes the time horizon at the beginning of the accounting period.

The GHG Protocol (WRI and WBCSD 2011) also assumes that only long-term C sequestration, that is, for more than 100 years, should be considered in CF calculations. An indicator called "GWPbio" has been developed to assess the climate change impact of biogenic CO_2 emissions specifically in biomass (Cherubini et al. 2011).

Brandão et al. (2013) critically reviewed the available methods for assessing carbon sequestration, temporary storage, and release of biogenic carbon in carbon footprinting, but different assumptions, applications, and merits–demerits left them inconclusive over identification of the most appropriate method. They nevertheless could conclude that choice of time horizon for assessing climate change impacts significantly affected the estimation of benefits of temporary carbon removals.

6.4.2 INCORPORATING LAND-USE CHANGE

In addition to C sequestration, some experts suggest that agriculture is a major cause of land-use changes and hence in order to represent real CF, effects of land-use changes must be taken into account. Many researchers conducted studies to assess the contribution of land-use changes in CFs. Hermansen and Nguyen (2012) estimated that pig meat produced from two systems without land-use impacts had CF of 3.5 and 3.2 kg CO_2-eq. kg^{-1}, respectively. But when potential land-use change related with pig feed was considered, the CF increased to 7.1 and 5.2 kg CO_2-eq. kg^{-1}, respectively, thus indicating that land-use change was responsible to nearly half of the CF. Cederberg et al. (2011) also calculated that conversion of forest to pasture in the Legal Amazon Region of Brazil, which increased the CF of beef production to more than 700 kg CO_2-eq. kg^{-1}. Similarly, deforestation increased the CF of pineapple production by 10-folds (Ingwersen 2012). Though the contribution of land-use changes is substantial, they are usually not considered while calculating CFs due to methodological issues and large uncertainties.

6.4.3 THE CASE OF BIOFUELS

Biofuels appear as a new challenge for GHG estimation studies since their role in reducing GHG emissions is under debate, mostly due to their land-use change implications. Therefore, if biofuels should be included in GHG assessments, there arise some important questions. To address the land-use change issue due to biofuel production, the concept of "ecosystem carbon payback time"

(ECPT) has been proposed. For evaluating effects of substitution of conventional energy source by bio-fuels, it becomes important to the number of years required for avoiding fossil-fuel emissions from bio-fuels to compensate for losses in ecosystem carbon stocks during land conversion. It is found that bio-fuel crop cultivation into degraded or cultivated land could provide almost immediate carbon savings. Land fuel versus fuel debate is critical in many parts of the world. Thus, cultivation of biofuel crops on degraded land only seems to be the most accepted practice.

6.5 CASE STUDIES

6.5.1 Crop Cultivation

Though a majority of C footprinting cases have been applied to consumer products, over the past few years, food production covering crop cultivation, animal products, food processing, and marketing are being considered and their numbers are rising. Pandey and Agrawal (2014) presented in detail the CFs of crop cultivation and case studies in detail, presented here in Table 6.2. They found that CFs of crop cultivation range widely and can even assume negative values under certain management practices, indicating net C sequestration or GHG sink. For example, wheat cultivation may act as CH_4 sink and has left the CF of -8.11 kg CO_2-eq. kg^{-1} (Pandey et al. 2013). Fodders are also important crops. Mogensen et al. (2014) calculated CFs of different cattle fodders at the farm gate and observed that grass pallets had the highest CF of 1.19 kg CO_2-eq. kg^{-1}, whereas the fodder obtained as the by-product of other crops like wheat or barley straw, CFs ranged from 0.43 to 0.68 kg CO_2-eq. kg^{-1}.

Regarding final food products, available studies indicate that animal-derived products leave bigger CF compared with vegetarian diet in a particular region (European Commission 2006; Williams et al. 2006; Sim et al. 2007; Pathak et al. 2010), and its value depends significantly on the methods of cultivation, processing, and transportation (Hospido et al. 2009; Stoessel et al. 2012). Among animal-derived food, dairy products are essential components of diet all over the world, at the same time it is very diverse, culturally important, and perhaps the most common processed food in the market. There have been a special attention in a number of studies focusing on different products, process techniques, and system boundaries (Berlin 2002; Høgaas Eide 2002; Flysjö 2011; Sheane et al. 2011; van Middelaar et al. 2011).

6.5.2 Dairy and Meat Production

Due to diversity in products and processing techniques, special methods have been developed for dairy sector. The most common of these besides PAS 2050 and GHG protocol include "Consequential modeling based on following ISO 14040/44" and average attributional modeling (for Sweden and United Kingdom). The International Dairy Federation (IDF) also has developed a methodology with common life-cycle assessment applicable across the global dairy industry sector. It recommends a hybrid of consequential and attributional modeling approaches (IDF 2010). Modeling approaches are important not only for calculating CF, but also in predicting changes in CFs associated with changing demands, technological approaches, as well as cobenefits. For example, milk may be the principal product for dairy industry, yet it may be associated with changes in other animal-based products like meat.

Using the allocation factors developed on the basis of studies conducted in Australian dairies, Flysjö et al. (2014) attempted to develop a model for calculating CFs of dairy products of the food production company, "Arla Foods." For the boundary, "farm to customer," CFs of different dairy products ranged from 1.1 to 1.2 kg CO_2-eq. kg^{-1} of fresh product, the lowest being for butter and butter blend, whereas the highest was for milk powder-based products. Dalgaard et al. (2014) also attempted to develop an "Arla model" for milk to be able to use it flexibly with guidelines recommended in different countries.

TABLE 6.2

CFs of Some Agricultural Systems and Cultivation Practices

Agricultural System (Crop)	Region	Activity and Tiers	CF (kg CO_2-e ha^{-1})[a]	Reference
1. Canola-mustard (sowing to farm gate)	Canada	Tiers 1 and 2: Soil-borne N_2O, land preparation, pesticide and fertilizer spray, harvester Tier$_3$: Fertilizers and pesticides (factory to farm)	0.548–0.966	Gan et al. (2011a)
2. Durum wheat (sowing to farm gate)	Canada	Tiers 1 and 2: Soil-borne N_2O, land preparation, pesticide and fertilizer spray, harvester Tier$_3$: Fertilizers and pesticides (factory to farm)	0.383–0.533	Gan et al. (2011c)
3. Barley (sowing to farm gate)	Canada	Tiers 1 and 2: Soil-borne N_2O, land preparation, pesticide and fertilizer spray, harvester Tier$_3$: Fertilizers and pesticides (factory to farm)	0.252–0.456	Gan et al. (2011b)
4. Spring wheat (sowing to farm gate)	Canada	Tiers 1 and 2: Soil-borne N_2O, land preparation, pesticide and fertilizer spray, harvester, changes in soil C Tier$_3$: Fertilizers and pesticides (factory to farm)	0.357–0.140	Gan et al. (2012)
5. Potato (farm to table)	Sweden	Tiers 1 and 2: Soil-borne N_2O, soil-borne CO_2, changes in soil C electricity	0.12	Roos et al. (2010)
6. Rice (sowing to farm gate)	Italy	Soil-borne CH_4 and N_2O, fertilizer, farm machines (land preparation, harvester, grain processing), irrigation, pesticide and fertilizer spray, harvester	2.90	Blengini and Busto (2009)
7. Biofuels (ethanol, methyl ester, FT diesel) (cultivation to power generation)	Europe, Canada, SE Asia, Brazil, USA	Tiers 1 and 2: No demarcation between tiers Soil borne CH_4, N_2O, CO_2, change in soil C, power generation Tier$_3$: Fertilizers and pesticides (factory to farm)	In g CO_2-eq. MJ^{-1} 17–140	Hoefnagels et al. (2010)
8. 57 farms with different organic, conventional and integrated farming (sowing to farm gate)	Scotland	Tiers 1 and 2: Soil-borne N_2O; land preparation; harvesting Fertilizer, pesticide spray	In $\times 10^2$ kg CO_2-eq. ha^{-1} Organic farming: 7.49 Conventional farming: 16.06 Integrated farming: 12.38 Leguminous crops: 4.50 Potato: 19.44 Cereals: 11.16–15.69	Hillier et al. (2012)
9. Rice cultivation under conventional and no tillage practices	India	Tiers 1, 2 and 3 Soil-borne CH_4, N_2O, CO_2, change in soil C, farm machines, irrigation, fertilizers, pesticides and grain processing.	Conventional tillage: 1.76×10^5 No tillage: 9.82×10^4	Pandey et al. (2013a)

(Continued)

TABLE 6.2 (*Continued*)
CFs of Some Agricultural Systems and Cultivation Practices

Agricultural System (Crop)	Region	Activity and Tiers	CF (kg CO_2-e ha^{-1})[a]	Reference
10. Wheat cultivation under conventional and no tillage practices	India	Tiers 1, 2 and 3 Soil-borne CH_4, N_2O, CO_2, change in soil C, farm machines, irrigation, fertilizers, pesticides and grain processing	Conventional tillage: 2.64 × 10 No tillage: 8.33 × 10	Pandey et al. (2013b)

Source: Adapted and modified from Springer Science+Business Media: *Assessment of Carbon Footprint in Different Industrial Sectors, Vol. 1, EcoProduction,* Carbon Footprint Estimation in the Agriculture Sector, 2014, I: 25–47, Pandey D, Agrawal M, Singapore.

[a] Units of CFs are kg CO_2-eq. kg^{-1} unless stated otherwise, EF, emission factor.

6.6 CARBON LABELING AND POLICY IMPLICATIONS

Growing consumer interest in "environment-friendly products" is motivating manufacturers to adopt more efficient production techniques and highlight their contribution to society and environment. Therefore, a wide range of environmental and social labels have been introduced to reflect environmental or social impacts of a product in simplified and easy-to-understand manner. "Carbon label" or "CF label" is one of the most common environmental labels which reflect the GHG intensity of the product. It has already been introduced in many countries as pilot projects including United Kingdom, Switzerland, and Japan, where it is being monitored for further modification so as to expand on larger area (Vandenbergh and Cohen 2010; Vandenbergh et al. 2011). ISO 14020, 14024, and 14025 provide standards for development of environmental labels, though many organizations often come up with their own indicators, many of which have same meaning and compete within similar products.

Usefulness of C labeling of products, however, has been debated. On the one hand, it appears to create a sense of environmental consciousness among consumers. On the other hand, for lack of a single robust method and guidelines for publishing C labels, manufacturers can easily make the CF much less perceivable than the real values (Schmidt 2009). For increasing the credibility, third-party assessments have been promoted. Nevertheless, growing political commitments to reduce GHG emissions, C labeling may also become a part of global food trade policy. Analyses of international food export from developing countries by Edwards-Jones et al. (2009) concluded that it is very unlikely that CFs will be the dominating factor in import policy by developed countries like United Kingdom, but it may become an essential component.

Since C labeling has huge support from the consumer side as is also evident in surveys conducted by different organizations (Tukker et al. 2006; L.E.K. Consulting LLP 2007), consumer psychological implications need to be considered in recommending any standard of C labeling guidelines if it is to be used as an effective consumer-facing policy tool (Upham et al. 2011). Definitely, consumers welcome any additional information they can get about the available choices in the market, it is not easy to judge the realistic impact of introducing C labeling on all the products.

A survey made by Upham et al. (2011) in the United Kingdom indicated that most consumers preferred low C-intensive products if their prices did not differ much, but most of them were unable to comprehend how much climate benefit their purchase decision actually made.

Though manufacturers always have an incentive to maintain and increase their customer base by disclosing their products' CF, issues related to CF calculation also arise here. Hence, the need for a single standard is urgently felt. Moreover, it is being recognized that the carbon labels must explain themselves instead of just revealing a figure (Schmidt 2009).

6.7 RECOMMENDATIONS FOR FURTHER RESEARCH DIRECTIONS

Major gaps and challenges before calculation of CFs have already been discussed in previous sections. However, there are certain issues that need extensive research. Counting emissions related with trade is an important challenge. Since products travel large distances and often cross international boundaries, allocation of emission between source and destination points is often confusing. Though some studies have taken care of these issues (Hertwich 2005), the opinions do not converge and are debated. Another important research direction is to apply CF calculations to different products and processes. Because of very high diversity of this sector, more case studies are required to identify major methodological challenges and to estimate process-specific emission factors. Extensive research is also required to understand the roles of consumer behavior, social responsibilities, and legislative mechanisms in regulating the intended use of CFs and carbon labels. It would be important to undertake the cost–benefit analyses of carbon labeling of products by the food processing and supply chains.

6.8 CONCLUSIONS

Application of C footprinting to food production and processing systems is growing rapidly and CF of a wide variety of products has been calculated ranging from cereals, dairy products, and processed and packaged foods. The estimates of CFs vary depending on the region, processing techniques, as well as the activities covered and methodology adopted. Though there is no single method which can individually address the CF calculations of food products, standards for product C footprinting and agriculture have been developed which together cover the complete life cycle of food products addressing the production, processing, and supply chain. Yet, there are issues that need to be addressed so as to ensure uniformity and relevance. Because of the broad definition of CF, it is being argued that land-use changes, C sequestration, and avoidance of emissions due to alternative energy sources must be considered in calculating CF for food products, but practical difficulties in counting all these effects with an acceptable level of uncertainty pose a big question. Nevertheless, methods for these aspects are being developed and modified, but it will take long time to devise a standard methodology, which can present a real representation of GHG intensity for food production. But even with the present methods, C footprint labeling has been welcomed in many countries by policy makers, and it is now considered as an important variable in price and trade decisions. Since consumers are now responding to C labels, it is important to monitor the presentation of such labels in order to protect consumer rights. Despite challenges, the developments in the concept, methods, and representation of CF and CF labels have been very fast and continued. Though there is a long way to go with CF methods and labels as a potential indicator of GHG efficiency, these may have far reaching implications on environment and economics.

ACKNOWLEDGMENTS

The authors acknowledge University Grants Commission (UGC), New Delhi, for financial support in the form of research project. Divya Pandey is thankful to the Council for Scientific and Industrial Research (CSIR), New Delhi, and the National Academy of Sciences (NASI), Allahabad, India, for research fellowships. She also expresses her sincere thanks to Prof. Manju Sharma, Distinguished Women Scientist Chair, NASI, Allahabad for research guidance.

REFERENCES

Arshad MA, Schnitzer M, Angers DA, Ripmeester DA. 1990. Effects of till vs no-till on the quality of soil organic matter. *Soil Biology and Biochemistry* 22:595–599.

Barber A, Pellow G, Marber M. 2011. Carbon footprint of New Zealand arable production—Wheat, maize silage, maize grain and ryegrass seed. MAF Technical Paper No. 2011/97. AgriLink, New Zealand: Foundation for Arable Research and Ministry of Agriculture and Forestry, 68 pp.

Bayer C, Martin-Neto L, Mielniczuk J, Ceretta CA. 2000. Effect of no-till cropping systems on soil organic matter in a sandy clay loam Acrisol from Southern Brazil monitored by electron spin resonance and nuclear magnetic resonance. *Soil & Tillage Research* 53:95–104.

Berlin J. 2002. Environmental life cycle assessment (LCA) of Swedish semi-hard cheese. *International Dairy Journal* 12:939–953.

Beukes PC, Gregorini P, Romera AJ, Levy G, Waghorn GC. 2010. Improving production efficiency as a strategy to mitigate greenhouse gas emissions on pastoral dairy farms in New Zealand. *Agriculture, Ecosystem & Environment* 136:358–365.

Blengini GA, Busto M. 2009. The life cycle of rice: LCA of alternative agri-food chain management systems in Vercelli (Italy). *Journal of Environmental Management* 90:1512–1522.

Brandão M, Levasseur A, Kirschbaum MUF, Weidema BP, Cowie AL, Jørgensen et al. 2013. Key issues and options in accounting for carbon sequestration and temporary storage in life cycle assessment and carbon footprinting. *The International Journal of Life Cycle Assessment* 18(1):230–240.

BSI. 2008. PAS 2050:2008 Specification for the assessment of the life cycle greenhouse gas emissions of goods and services. London: British Standards Institution.

BSI. 2011. PAS 2050:2011 Specification for the assessment of the life cycle greenhouse gas emissions of goods and services. London: British Standards Institution.

Cederberg C, Flysjö A. 2004. *Life Cycle Inventory of 23 Dairy Farms in South-Western Sweden.* Gothenburg, Sweden: Swedish Institute for Food and Biotechnology, SIK-Rapport.

Cederberg C, Persson UM, Neovius K, Molander S, Clift R. 2011. Including carbon emissions from deforestation in the carbon footprint of Brazilian beef. *Environmental Science & Technology* 45(5):1773–1779.

Cederberg C, Sonesson U, Henriksson M, Sund V, Davis J. 2009. Greenhouse gas emissions from Swedish production of meat, milk and eggs: 1990 and 2005. SIK Report 793. Gothenburg, Sweden: The Swedish Institute of Food and Biotechnology.

Cederberg C, Stadig M. 2003. System expansion and allocation in life cycle assessment of milk and beef production. *The International Journal of Life Cycle Assessment* 8:350–356.

Chatskikh D, Olesen JE, Hansen EM, Elsgaard L, Petersen BM. 2008. Effects of reduced tillage on net greenhouse gas fluxes from loamy sand soil under winter crops in Denmark. *Agriculture, Ecosystem & Environment* 128:117–126.

Cherubini F, Peters GP, Berntsen T, Stromman AH, Hertwich E. 2011. CO_2 emissions from biomass combustion for bioenergy: Atmospheric decay and contribution to global warming. *GCB Bioenergy* 3(5):413–426.

Crosson P, Foley PA, Shalloo L, O'Brien D, Kenny DA. 2010. Greenhouse gas emissions from Irish beef and dairy production systems. Advances in animal biosciences. *Food, Feed, Energy and Fibre from Land—A Vision for 2020 Proceedings of the British Society of Animal Science and the Agricultural Research Forum.* Cambridge, UK: Cambridge University Press, 350 pp.

Dalgaard R, Schmidt J, Flysjö A. 2014. Generic model for calculating carbon footprint of milk using four different LCA modelling approaches. *Journal of Cleaner Production.* DOI: http://dx.doi.org/10.1016/j.jclepro.2014.01.025.

Daneshi A, Esmaili-Sari A, Daneshi M, Baumann H. 2014. Greenhouse gas emissions of packaged fluid milk production in Tehran. *Journal of Cleaner Production* 80:150–158.

De Monte M, Padoano E, Pozzetto D. 2005. Alternative coffee packaging: An analysis from a life cycle point of view. *Journal of Food Engineering* 66(4):405–411.

Edwards-Jones G, Plassmann K, York EH, Hounsome B, Jones DL, Milà i Canals L. 2009. Vulnerability of exporting nations to the development of a carbon label in the United Kingdom. *Environmental Science & Policy* 12(4):479–490.

Escobar JC, Lora ES, Venturini OJ, Yáñez EE, Castillo EF, Almazan O. 2009. Biofuels: Environment, technology and food security. *Renewable & Sustainable Energy Reviews* 13(6):1275–1287.

European Commission. 2010. *International Reference Life Cycle Data System (ILCD) Handbook—General Guide for Life Cycle Assessment—detailed Guidance.* Joint Research Centre, Institute for Environment and Sustainability. Luxembourg: Publications Office of the European Union.

FAO (Food and Agriculture Organization). 2010. *Greenhouse Gas Emissions from the Dairy Sector.* Rome, Italy: A Life Cycle Assessment.

Fearnside PM. 2002. Why a 100-year time horizon should be used for global warming mitigation calculations. *Mitigation and Adaptation for Strategies for Global Change* 7(1):19–30.

Ferrua MJ, Singh RP. 2011. Improved airflow method and packaging system for forced-air cooling of strawberries. *International Journal of Refrigeration* 34(4):1162–1173.

Flysjö A. 2011. Potential for improving the carbon footprint of butter and blend products. *Journal of Dairy Science* 94:5833–5841.

Flysjö A, Thrane M, Hermansen JE. 2014. Method to assess the carbon footprint at product level in the dairy industry. *International Dairy Journal* 34:86–92.

Foley JA, Ramankutty N, Brauman KA, Cassidy ES, Gerber JS, Johnston M, Mueller et al. 2011. Solutions for a cultivated planet. *Nature* 478.7369:337–342.

Friedler F. 2010. Process integration, modelling and optimisation for energy saving and pollution reduction. *Applied Thermal Engineering* 30(16):2270–2280.

Gan YT, Liang BC, May W, Malhi SS, Niu J, Wang X. 2012. Carbon footprint of spring barley in relation to preceding oilseeds and N fertilization. *The International Journal of Life Cycle Assessment* 17:635–645.

Gan YT, Liang BC, Wang XY, McConkey BG. 2011b. Lowering carbon footprint of durum wheat through diversifying cropping systems. *Field Crops Research* 122:199–206.

Gan YT, Liang L, Huang G, Malhi SS, Brandt SA, Katepa-Mupondwa F. 2011a. Carbon footprint of canola and mustard is a function of the rate of N fertilizer. *The International Journal of Life Cycle Assessment* 17:58–68.

Gerber P, Vellinga T, Opio C, Henderson B, Steinfeld H. 2010. Greenhouse gas emissions from the dairy sector, a life cycle assessment. Rome, Italy: FAO, Food and Agriculture Organisation of the United Nations, Animal Production and Health Division, 11 pp.

Griffiths-Sattenspiel B, Wilson W. 2009. *The Carbon Footprint of Water.* River Network. Portland, U.S.A.

Haas G, Wetterich F, Köpke U. 2001. Comparing intensive, extensified and organic grassland farming in southern Germany by process life cycle assessment. *Agriculture, Ecosystem & Environment* 83:43–53.

Hagemann M, Hemme T, Ndambi A, Alqaisi O, Sultana N. 2011. Benchmarking of greenhouse gas emissions of bovine milk production systems for 38 countries. *Animal Feed Science and Technology* 166:46–58.

Hergoualc'h K, Blanchart E, Skiba U, Hénault C, Harmand JM. 2012. Changes in carbon stock and greenhouse gas balance in a coffee (*Coffea arabica*) monoculture versus an agroforestry system with *Inga densiflora*, in Costa Rica. *Agriculture, Ecosystems & Environment* 148(1):102–110.

Hermansen JE, Nguyen TLT. 2012. Life cycle assessment in the agri food chain. In: Boye J, Arcand Y (eds), *Green Technologies in Food Production and Processing.* Springer Science & Business Media, New York, U.S.A., 704 pp.

Hertwich EG. 2005. Lifecycle approaches to sustainable consumption: A critical review. *Environmental Science and Technology* 39:4673–4684.

Hillier J, Brentrup F, Wattenbach M, Walter C, Garcia-Suarez T, Mila-i-Canals L et al. 2012. Which cropland greenhouse gas mitigation options give the greatest benefits in different world regions? Climate and soil-specific predictions from integrated empirical models. *Global Change Biology* 18:1880–1894.

Hoefnagels R, Smeets E, Faaij A. 2010. Greenhouse gas footprints of different biofuel production systems. *Renewable & Sustainable Energy Reviews* 14:1661–1694.

Høgaas Eide M. 2002. Life cycle assessment of industrial milk production. *The International Journal of Life Cycle Assessment* 7:115–126.

Hospido A, Milà i Canals L, McLaren S, Truninger M, Edwards-Jones G, Clift R. 2009. The role of seasonality in lettuce consumption: A case study of environmental and social aspects. *The International Journal of Life Cycle Assessment* 14(5):381–391.

Humbert S, Rossi V, Margni M, Jolliet O, Loerincik Y. 2009. Life cycle assessment of two baby food packaging alternatives: Glass jars vs. plastic pots. *The International Journal of Life Cycle Assessment* 14(2):95–106.

IDF. 2010. A common carbon footprint for dairy, the IDF guide to standard life cycle assessment methodology for the dairy industry. Bulletin of the International Dairy Federation 445/2010. Brussels, Belgium: International Dairy Federation, pp. 21–23.

Ingwersen WW. 2012. Life cycle assessment of fresh pineapple from Costa Rica. *Journal of Cleaner Production* 35:152–163.

Intini F, Kühtz S, Rospi G. 2012. Life cycle assessment (LCA) of an energy recovery plant in the olive oil industries. *International Journal of Energy and Environment* 3(4):541–552.

IPCC. 2013. Climate Change 2013: Contribution of working Group I to the Fifth Assessment Report of the Intergovernmental Panel on Climate Change T.F. Stocker, D. Qin, G.-K. Plattner, M. Tignor, S.K. Allen, J. Boschung (Eds.), et al., Climate change 2013: The physical science basis, Cambridge University Press, Cambridge and New York (2013).

ISO. 2006a. ISO 14064-1:2006. Greenhouse gases, Part 1: Specification with guidance at the organization level for quantification and reporting of greenhouse gas emissions and removals. http://www.iso.org (assessed December 9, 2008).

ISO. 2006b. ISO 14064-2:2006. Greenhouse gases, Part 2: Specification with guidance at the project level for quantification, monitoring and reporting of greenhouse gas emission reductions or removal enhancements. http://www.iso.org (assessed December 5, 2009).

ISO. 2013. ISO/TS 14067: Greenhouse gases—Carbon footprint of products—Requirements and guidelines for quantification and communication (assessed 15 June, 2013).

Kristensen T, Mogensen L, Trydeman Knudsen M, Hermansen JE. 2011. Effect of production system and farming strategy on greenhouse gas emissions from commercial dairy farms in a life cycle approach including effect of different allocation methods. *Livestock Science* 140:136–148.

Kwon S, Fan M, DaCosta HFM, Russell AG. 2011. Factors affecting the direct mineralization of CO_2 with olivine. *Journal of Environmental Sciences* 23(8):1233–1239.

Lal R. 2004. Carbon emission from farm operations. *Environment International* 30:981–990.

Le Mer J, Roger P. 2001. Production, oxidation, emission and consumption of methane by soils: a review. *European Journal of Soil Biology* 37:25–50.

Ledgard SF, Lieffering M, Coup D, O'Brien B. 2011. Carbon footprinting of New Zealand lamb from an exporting nation's perspective. *Animal Frontiers* 1:27–32.

Ledgard SF, McDevitt J, Boyes M, Lieffering M, Kemp R. 2009. Greenhouse gas footprint of lamb meat: Methodology report. Report to MAF. Hamilton: AgResearch, 36 pp.

L.E.K. Consulting LLP. 2007. *The L.E.K.* Consulting carbon footprint report 2007: Carbon footprints and the evolution of brand–consumer relationships. L.E.K. Consulting Research Insights, I. London, L.E.K. LLB.

Lespinard AR, Mascheroni RH. 2012. Influence of the geometry aspect of jars on the heat transfer and flow pattern during sterilization of liquid foods. *Journal of Food Process Engineering* 35(5):751–762.

Levasseur A, Lesage P, Margni M, Deschênes L, Samson R. 2010. Considering time in LCA: Dynamic LCA and its application to global warming impact assessments. *Environmental Science & Technology* 44:3169–3174.

Lim SR, Park JM. 2008. Cooperative water network system to reduce carbon footprint. *Environmental Science & Technology* 42(16):6230–6236.

Ma J, Li XL, Xu H, Han Y, Cai ZC, Yagi K. 2007. Effects of nitrogen fertilizer and wheat straw application on CH_4 and N_2O emissions from a paddy rice field. *Australian Journal of Soil Research* 45:359–367.

Ma YC, Kong XW, Yang B, Zhang XL, Yan XY, Yang JC, Xiong ZQ. 2013. Net global warming potential and greenhouse gas intensity of annual rice–wheat rotations with integrated soil–crop system management. *Agriculture, Ecosystems & Environment* 164:209–219.

Mandal B, Majumder B, Bandyopadhyay PK, Hazra GC, Gangopadhyay A, Samantaray RN, Mishra A, Chaudhury J, Saha MN, Kundu S. 2007. The potential of cropping systems and soil amendments for carbon sequestration in soils under long-term experiments in subtropical India. *Global Change Biology* 13:357–369.

Masanet E, Worrell E, Graus W, Galitsky C. 2008. Energy efficiency improvement and cost saving opportunities for the fruit and vegetable processing industry. *Energy Star Guide*. An Energy Star Guide for Energy and Plant Managers. Lawrence Berkeley National Laboratory, Berkeley.

Mithraratne N, Barber A, McLaren SJ. 2010. Carbon footprinting for the kiwifruit supply chain—Report on methodology and scoping study. Landcare Research Contract Report LC0708/156 (revised edition). Prepared for Ministry of Agriculture and Forestry, 77 pp. http://www.landcareresearch.co.nz/publications/researchpubs/Kiwifruit_Methodology_Report_2010.pdf (assessed November 18, 2012).

Mogensen L, Kristensen T, Lan TL, Nguyen T, Trydeman MT, Hermansen JE. 2014. Method for calculating carbon footprint of cattle feeds—Including contribution from soil carbon changes and use of cattle manure. *Journal of Cleaner Production* 73:40–51.

Moura-Costa P, Wilson C. 2000. An equivalence factor between CO_2 avoided emissions and sequestration— Description and applications in forestry. *Mitigation and Adaptation for Strategies for Global Change* 5:51–60.

O'Brien D, Shalloo L, Grainger C, Buckley F, Horan B, Wallace M. 2010. The influence of strain of Holstein– Friesian cow and feeding system on greenhouse gas emissions from pastoral dairy farms. *Journal of Dairy Science* 93:3390–3402.

Pandey D, Agrawal M. 2014. Carbon Footprint Estimation in the Agriculture Sector. In: Muthu SS (ed), *Assessment of Carbon Footprint in Different Industrial Sectors, Vol. 1, EcoProduction*. Singapore: Springer Science+Business Media. I: 25–47.

Pandey D, Agrawal M, Bohra JS. 2013. Impact of four tillage permutations in rice–wheat system on GHG performance of wheat cultivation through carbon footprinting. *Ecological Engineering* 60:261–270.

Pathak H, Jain N, Bhatia A, Patel J, Aggarwal PK. 2010. Carbon footprints of Indian food items. *Agriculure, Ecosystems & Environment* 139:66–73.

Pazheri F, Kaneesamkandi ZM, Othman M. 2012. Bagasse Saving and Emission Reduction in power dispatch at sugar factory by co-generation and solar energy. In: *2012 IEEE International Power Engineering and Optimization Conference (PEOCO2012)*, Melaka, pp. 407–410.

Pelletier, Nathan, Pirog, Rich, Rasmussen, Rebecca. 2010. Comparative life cycle environmental impacts of three beef production strategies in the Upper Midwestern United States. *Agricultural Systems* 103:380–389.

Penman J, Gytarsky M, Hiraishi T, Krug T, Kruger D, Pipatti R, Buendia L et al. 2003. Good practice guidance for land use, land-use change and forestry. IPCC National Greenhouse Gas Inventories Programme and Institute for Global Environmental Strategies (IGES). Kanagawa, Japan: Intergovernmental Panel on Climate Change.

Peters GP. 2010 Carbon footprints and embodied carbon at multiple scales. *Current Opinion in Environmental Sustainability* 2:245–250.

Pimentel D, Williamson S, Alexander CE, Gonzalez-Pagan O, Kontak C, Mulkey SE. 2008. Reducing energy inputs in the US food system. *Human Ecology* 36(4):459–471.

Point E. 2008. Life cycle environmental impacts of wine production and consumption in Nova Scotia, Canada. Master's thesis, Dalhousie University, Halifax.

Poovarodom N, Ponnak C, Manatphrom N. 2012. Comparative carbon footprint of packaging systems for tuna products. *Packaging Technology and Science* 25(5):249–257.

Qin J, Hu F, Li D, Li H, Lu J, Yu R. 2010a. The effect of mulching, tillage and rotation on yield in non-flooded compared with flooded rice production. *Journal of Agronomy and Crop Science* 196:397–406.

Rao AB, Rubin ES. 2002. A technical, economic, and environmental assessment of amine-based CO_2 capture technology for power plant greenhouse gas control. *Environmental Science & Technology* 36(20):4467–4475.

Reay D. 2008. The role of process intensification in cutting greenhouse gas emissions. *Applied Thermal Engineering* 28(16):2011–2019.

Robertson GP, Paul EA, Harwood RR. 2000. Greenhouse gases in intensive agriculture: Contributions of individual gases to radiative forcing of the atmosphere. *Science* 289:1922–1925.

Roos E, Sundberg C, Hansson PA. 2010. Uncertainties in the carbon footprint of food products: A case study on table potatoes. *The International Journal of Life Cycle Assessment* 15:478–488.

Schmidt HJ. 2009. Carbon footprinting, labelling and life cycle assessment. *The International Journal of Life Cycle Assessment* 14:S6–S9.

Schnitzer H, Brunner C, Gwehenberger G. 2007. Minimizing greenhouse gas emissions through the application of solar thermal energy in industrial processes. *Journal of Cleaner Production* 15(13):1271–1286.

Sheane R, Lewis K, Hall P, Holmes-Ling P, Kerr A, Stewart K et al. 2011. Identifying opportunities to reduce the carbon footprint associated with the Scottish dairy supply chain e Main report. Edinburgh, UK: Scottish Government.

Sim S, Barry M, Clift R, Cowell SJ. 2007. The relative importance of transport in determining an appropriate sustainability strategy for food sourcing. *The International Journal of Life Cycle Assessment* 12(6):422–431.

Smith A. 2005. *The Validity of Food Miles as an Indicator of Sustainable Development*. London: UK DEFRA, pp. 1–117.

Smith P, Martino D, Cai Z, Gwary D, Janzen H, Kumar P, McCarl B et al. 2008. Greenhouse gas mitigation in agriculture. *Philosophical Transactions of the Royal Society B* 363:789–813.

Subak S. 1999. Global environmental costs of beef production. *Ecological Economics* 30:79–91.

Svanes E, Aronsson AKS. 2013. Carbon footprint of a Cavendish banana supply chain. *The International Journal of Life Cycle Assessment* 18:1450–1464.

Tan RR, Sum Ng DK, Yee Foo DC. 2009. Pinch analysis approach to carbon-constrained planning for sustainable power generation. *Journal of Cleaner Production* 17(10):940–944.

Thomassen MA, van Calker KJ, Smits MCJ, Iepema GL, de Boer IJM. 2007. Life cycle assessment of conventional and organic milk production in the Netherlands. *Agricultural Systems* 96:95–107.

Tuan CI, Yeh YL, Chen CJ, Chen TC. 2012. Performance assessment with Pinch technology and integrated heat pumps for vaporized concentration processing. *Journal of the Taiwan Institute of Chemical Engineers* 43(2):226–234.

Tubiello FM, Salvatore M, Córdor Golec RD, Ferrara A, Rossi S, Biancalani R, Federici S, Jacobs H, Flammini A. 2014. Agriculture, forestry and other land use emissions by sources and removals by sinks. Food and Agriculture Organization. Statistics Division Working Paper Series ESS/14-02. Italy: FAO, 75 pp.

Tukker A, Huppes G, Guinée J, Heijungs R, Koning A, Oers et al. 2006. Environmental Impact of Products (EIPRO). Analysis of the life cycle environmental impacts related to the final consumption of the EU-25. Technical Report Series EUR 22284 EN. European Commission, Joint Research Centre, Institute for Prospective Technological Studies, pp. 1–136.

Upham P, Dendler L, Bleda M. 2011. Carbon labelling of grocery products: Public perceptions and potential emissions reductions. *Journal of Cleaner Production* 19 348–355.

Vandenbergh MP, Cohen MA. 2010. Climate change governance: Boundaries and leakage. *NYU Environmental Law Journal* 18:221–292.

Vandenbergh MP, Dietz T, Stern PC. 2011. Time to try carbon labelling. *Nature Climate Change* 1:4–6.

Van Hauwermeiren A, Coene H, Engelen G, Mathijs E. 2007. Energy lifecycle inputs in food systems: A comparison of local versus mainstream cases. *Journal of Environmental Policy & Planning* 9(1):31–51.

van Middelaar C, Berentsen P, Dolman M, de Boer I. 2011. Eco-efficiency in the production of Dutch semi-hard cheese. *Livestock Science* 139:91–99.

Wang JF, Xie J, Tao LR, Wang YH, Yang XY. 2011. Cooling process after HTST orange juice sterilization optimization based on CFD. *Advanced Materials Research* 201:2727–2730.

Weber C, Matthews HS. 2008. Food-miles and the relative climate impacts of food choices in the United States. *Environmental Science & Technology* 42:3508–3513.

White TA, Snow VO, King WMcG. 2010. Intensification of New Zealand beef farming systems. *Agricultural Systems* 103:21–35.

Williams A, Audsley E, Sanders DL. 2006. Determining the environmental burdens and resource use in the production of agricultural and horticultural commodities. Main Report. Defra Research Project IS0205. Bedford, UK: Cranfield University and Defra, 75 pp.

WRI, WBCSD. 2004. *The Greenhouse Gas Protocol. A Corporate Accounting and Reporting Standard* (revised edition). Washington, DC, USA/Conches-Geneva, Switzerland: World Resources Institute and World Business Council for Sustainable Development.

WRI, WBCSD. 2011. Product life cycle accounting and reporting standard. World Resources Institute and World Business Council for Sustainable Development, Washington.

Xu Z, Sun DW, Zeng XA, Liu D, Pu H. 2014. Research developments in methods to reduce the carbon footprint of the food system: A review. *Critical Reviews in Food Science and Nutrition*. DOI: 10.1080/10408398.2013.821593.

Yan X, Crookes RJ. 2009. Reduction potentials of energy demand and GHG emissions in China's road transport sector. *Energy Policy* 37: 658–668.

Yang H, Xu Z, Fan M, Gupta R, Slimane RB, Bland AE, Wright I. 2008. Progress in carbon dioxide separation and capture: A review. *Journal of Environmental Science* 20(1):14–27.

Zou JW, Huang Y, Jiang JY, Zheng XH, Sass RL. 2005. A 3-year field measurement of methane and nitrous oxide emissions from rice paddies in China: Effects of water regime, crop residue, and fertilizer application. *Global Biogeochemical Cycles* 19:GB2021.

7 Modeling the Carbon Footprint of Wood-Based Products and Buildings

Ambrose Dodoo, Leif Gustavsson, and Roger Sathre

CONTENTS

7.1 Introduction 143
7.2 GHG Assessment Standards 145
7.3 Climate Implications of Wood Products and Buildings 145
7.4 Definition of Functional Unit 146
7.5 Indicators for Carbon Footprint Characterization 146
7.6 Setting of System Boundaries 147
    7.6.1 Activity-Related System Boundaries 148
        7.6.1.1 Production Stage 148
        7.6.1.2 Operation Stage and Service Life 149
        7.6.1.3 Maintenance and Renovation 150
        7.6.1.4 End-of-Life Stage 150
    7.6.2 Time-Related System Boundaries 152
    7.6.3 Space-Related System Boundaries 153
7.7 Accounting for Electricity Production 154
7.8 Treatment of Allocation 155
7.9 Key Recommendations and Outlooks 156
7.10 Summary 157
References 157

7.1 INTRODUCTION

Efforts to reduce greenhouse gas (GHG) emissions and thereby mitigate global climate change are receiving increasing attention in many countries today. There is growing recognition that the current trends in energy supply and demand are not consistent with the goals of sustainable development. Of the global primary energy supply of 549 EJ in 2011, fossil fuels constituted 82%, and biofuels, nuclear, and hydro accounted for about 10, 5, and 2%, respectively (IEA 2013a). Fossil-fuel combustion is the major anthropogenic source of GHG emissions (IPCC 2013). A less significant share of anthropogenic CO_2 emission is also connected to non-energy related activities including land-use practices and industrial process reactions. Fossil-fuel combustion and industrial process reactions accounted for 78% of the global total GHG emission increase between 1970 and 2010 (IPCC 2014). Figure 7.1 shows a breakdown of the global total primary energy supply (TPES) and associated CO_2 emission by fuel type in 2011 (IEA 2013b). Major studies suggest that fossil fuels are very likely to account for a significant share of future primary energy use, even if effective measures are implemented to promote resource efficiency and sustainable energy systems in the global community (IPCC 2000a,b; IEA 2011). There is growing interest in strategies to reduce fossil-fuel use, thereby creating a resource-efficient built environment with low-carbon footprint.

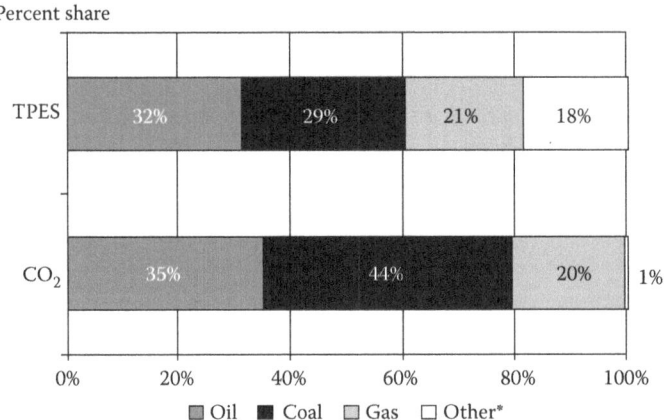

Percent share

FIGURE 7.1 Breakdown of the global TPES (total primary energy supply) by fuel type and associated CO_2 emission in 2011. Other* includes combustible renewable/waste, nuclear, hydro, geothermal, solar, wind, and tide. (From IEA [International Energy Agency]. 2013b. *IEA statistics. CO_2 Emissions from Fuel Combustion: Highlights*, 2013 Edition ©. www.iea.org [September 28, 2014].)

The building sector worldwide accounts for 40% of the total energy use and for 30% of the total GHG emission (UNEP 2007) and offers significant opportunities to mitigate climate change (IPCC 2007). In this context, wood-based products and materials can play an important role in the transition to a low-carbon and a resource-efficient society, serving as energy resource as well as physical and structural material in diverse applications, for example, in buildings. Several recent studies have documented the potential for reducing energy use and carbon footprint of the built environment by use of sustainably produced wood-based materials in place of other materials (IPCC 2007; Upton et al. 2008; Dodoo and Gustavsson 2013). Less energy input is needed to manufacture wood products compared with alternative materials (Koch 1992; Buchanan and Honey 1994; Gustavsson and Sathre 2006; Sathre and O'Connor 2010). Wood-based building materials mainly use biomass residues for processing energy (e.g., kiln-drying) and have lower carbon and primary energy balances than alternative materials (Gustavsson et al. 2006). The storage of carbon in wood materials and the increased availability of forest and woody by-products for energy purposes are other dynamics by which the use of wood-based material affects climate (Gustavsson et al. 2006; Lippke et al. 2011). Significant quantities of biomass residues are produced from the wood product chain and can be used instead of fossil fuels (Figure 7.2). Using wood-based material instead of fossil-fuel-intensive

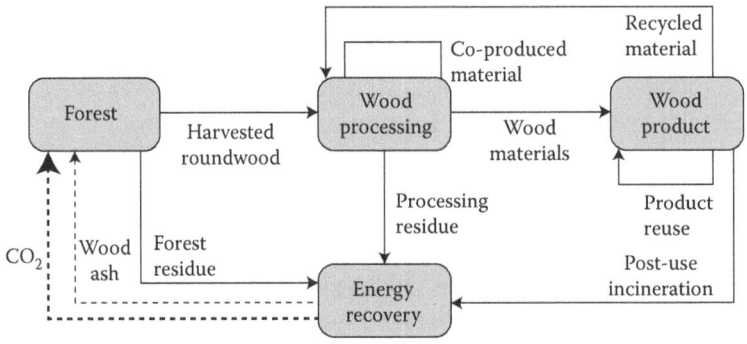

FIGURE 7.2 Schematic diagram of system-wide integrated material flows of wood products. (Reprinted from In: *Handbook of Recycling: State-of-the-Art for Practitioners, Analysts, And Scientists*, ISBN 978-0-12-396459-5, Dodoo A, Gustavsson L, Sathre R, Lumber. Chapter 12, Copyright 2014a, with permission from Elsevier.)

materials provides permanent and cumulative reduction in CO_2 emission, whereas sequestration of biological carbon is typically temporary (Schlamadinger and Marland 1996).

Generally, carbon footprint analysis can be conducted using bottom-up and top-down models. Bottom-up models begin with detailed disaggregated information for a product system and then generate aggregate system behavior to characterize the relationship between the individual components of the system (Sathre 2007). Bottom-up models provide specific information about the individual processes and systems studied, allowing detailed comparison of alternatives. Top-down models begin with the aggregate information for a system and then proceed to disaggregate this to characterize the components (Lenzen and Treloar 2002; Fulton et al. 2011). In recent years, hybrid models which integrate aspects of both bottom-up and top-down models are becoming increasingly common for analyses of life cycle and carbon footprint impacts.

A comprehensive analysis of the climate impacts caused by products and buildings requires a system-wide life-cycle perspective. Robust GHG assessment methodologies and tools can play an important role in making informed decisions from a climate mitigation perspective. In this chapter, we present methodological issues to accurately characterize the carbon footprint over the full life cycle of wood-based products and buildings.

7.2 GHG ASSESSMENT STANDARDS

Various assessment standards and tools have been developed to explore climate implications of the built environment. Life-cycle assessment (LCA) is a commonly used tool for evaluation of environmental implications of products during their life-cycle stages. ISO 14040 and 14044 (2006a,b) provide general framework and guidelines for LCA. These standards suggest that an LCA study should include all stages and impacts throughout the life cycle of a product. LCA identifies and quantifies the environmental impacts associated with the flows of energy and materials in a system. LCA includes a several impact categories, for example, acidification, global warming potential (GWP), eutrophication, ozone depletion, human toxicity, and abiotic resource depletion. Carbon footprint refers to the total net CO_2 and other GHG emitted to the atmosphere during the life cycle of a product or material (Wiedmann and Minx 2008). Thus, carbon footprint analysis is similar to the GWP impact category of LCA. The International Organization for Standardization's technical specification (ISO/TS) 14067:2013 provides general principles and guidelines for quantification of carbon footprint of a product. ISO/TS 14067:2013 is based on the ISO 14040 series of standards and suggests that a scientific approach should be used in quantifying the carbon footprint of products, focusing on relevance, completeness, consistency, accuracy, and transparency. Other standards and frameworks increasingly referred in carbon footprint studies are the Publicly Available Specification (PAS) 2050:2011 and the GHG Protocol (2011). The PAS 2050:2011 builds on the principles ISO 14040 series of standards and was developed by the British Standards Institution to provide framework methodology for analysis of GHG impacts of goods and services from a life-cycle perspective. The GHG Protocol was developed for accounting of GHG emissions of products, with the support of the World Resources Institute and World Business Council for Sustainable Development. In general, existing standards provide broad guidelines regarding analytical approaches, but more specific methods are required for detailed analysis of carbon footprint of product systems and for practical application.

7.3 CLIMATE IMPLICATIONS OF WOOD PRODUCTS AND BUILDINGS

Assessing the carbon footprint of wood-based products and building systems is more complex than that of many other building products and systems (IEA 2001; Gustavsson and Sathre 2011). One issue is the relatively much longer time frame involved when dealing with wood-based products, spanning from the time of seed germination through to forest growth, use and disposal of the product. Also, a variety of useful products can be obtained at different points along the wood product value chain (Figure 7.2). Furthermore, buildings are complex systems comprising multiple components and have relatively long lifespan. The design and construction conditions of buildings are typically heterogeneous, making each building unique (IEA 2001). It is important that these dynamics are considered in carbon footprint analysis of wood-based products and building systems. Essential methodological aspects that need to be addressed to quantify the carbon footprint of wood-based products and buildings include functional unit definition; indicators for carbon footprint characterization; setting of system boundaries in terms of activities, time, and place; treatment of allocation; and accounting of impacts of energy supply systems.

The potential use of wood building products and systems instead of non-wood alternatives raises two important questions. The first is: What would have taken place without the wood use, that is, how would the reference system have performed? The second is: How will the wood utilization system perform? In principle, marginal changes will occur in both systems, and a consequential-based approach is required to appropriately characterize the effect of these marginal changes. A study by Plevin et al. (2014) explored the challenges involved in estimating benefits of climate change mitigation measures. They emphasized the use of a consequential-based approach as the use of attributional-based approach for assessment of impacts of climate mitigation measures can be misleading.

7.4 DEFINITION OF FUNCTIONAL UNIT

Definition of an appropriate functional unit is important in carbon footprint analysis. Functional unit refers to the unit of analysis and provides a reference to which the inputs and outputs of a product system are related. The functional unit should be "consistent with the goal and scope of the study," and a "study shall clearly specify the functions of the system being studied" (ISO/TS 14067:2013). Different functional units have been used in carbon analyses of buildings (Gustavsson and Sathre 2011). These units include, for example, 1 m^2 of a building's gross or usable floor area, total gross or usable floor area, and a complete building. Functional units based on material mass, volume, or isolated structural characteristics of building components are insufficient for informed decision making. Functioning of different materials cannot always be directly compared, as materials may often fulfill more than one function. Equal amounts of different materials may not fulfill the same function and thus provide different services. For example, 1 kg of lumber may not fulfill the same function as the same quantity of steel. Therefore, the functional unit should reflect the complex interactions between multiple components and functions of a system. In carbon footprint analysis of building and construction systems, this could be achieved by considering the complete building (Gustavsson and Sathre 2011).

7.5 INDICATORS FOR CARBON FOOTPRINT CHARACTERIZATION

Various definitions and evaluation indicators for carbon footprint analysis are found in literature (see, e.g., Matthews et al. 2008; Wiedmann and Minx 2008; Weidema et al. 2008; Hertwich and Peters 2009; Wright et al. 2011; Wu 2011). In some carbon footprint studies, only CO_2 flows are analyzed, whereas in other studies, further GHG flows besides CO_2 are included. Wiedmann and Minx (2008) argued that carbon footprint analysis should include only direct and indirect CO_2 emissions linked to a product system, as the phrase suggests. They suggested that the phrase "climate footprint" may be appropriate if other GHGs beyond CO_2 flows are considered. In an exploration of a standard definition for carbon footprint, Wright et al. (2011) argued that accounting for only CO_2 flows would lead to omission of almost one-quarter of GHGs. The authors indicated that this may lead to a significant gap in the management of the omitted climate destabilization gases. They suggested that carbon footprint analysis should encompass the flows of CO_2 and CH_4 in a system because it is relatively straightforward to account for these GHGs, whereas inclusion of all GHGs may be very time-consuming and expensive. In the context of climate change and completeness, ISO/TS 14067:2013 states that "all GHG emissions and removals that provide a significant contribution" in a product system should be included in carbon footprint analysis. Thus, all relevant GHGs (e.g., CO_2, CH_4, N_2O) in systems should be considered if their climate impact is significant. In this regard, the carbon footprint of different GHGs is quantified in CO_2-equivalent, as the multiplication of the mass of the GHGs and their respective GWP factor over a given time horizon. The ISO/TS 14067:2013 suggests a 100-year time horizon for an analysis and provides GWP factors for different GHGs for this period. Another metric, cumulative radiative forcing (CRF) allows more sophisticated analysis of time-dependent climate forcing effects of complex forestry systems (Sathre et al. 2013).

7.6 SETTING OF SYSTEM BOUNDARIES

The way that a system boundary is defined is crucial to the accuracy of a carbon footprint analysis (Matthews et al. 2008). ISO/TS 14067:2013 indicates that an analysis shall "consider all stages of the life cycle of a product when assessing the [carbon footprint], from raw material acquisition to final disposal" and "include all GHG sources and sinks together with carbon storage that provide a significant contribution to the assessment of GHG emissions and removals arising from the whole or partial system being studied." Carbon footprint analyses with cradle-to-grave or cradle-to-gate perspective can be found in life-cycle literature. Cradle-to-grave system boundaries track the flows and impacts of a system from the stage of raw material extraction through to use and post-use stages. Cradle-to-gate system boundaries consider the flows and impacts up to the factory exit gate and thus omit the use and post-use stages. Narrow system boundaries are incapable of establishing the full climate impact of wood products, as use of wood products involves material and energy flows in different economic sectors (Sathre 2007; Gustavsson and Sathre 2011). To obtain robust results, system boundaries of carbon footprint analysis must be broad enough to include all significant climatic impacts. For buildings, all significant life-cycle GHG flows (Figure 7.3), including from production, operation, and end-of-life stages, need to be considered in carbon footprint analysis. Establishment of effective system boundaries in terms of activities, time, and space is fundamental to the credibility of carbon footprint analysis for wood-based products and building systems (see Gustavsson and Sathre 2011; Sathre et al. 2012).

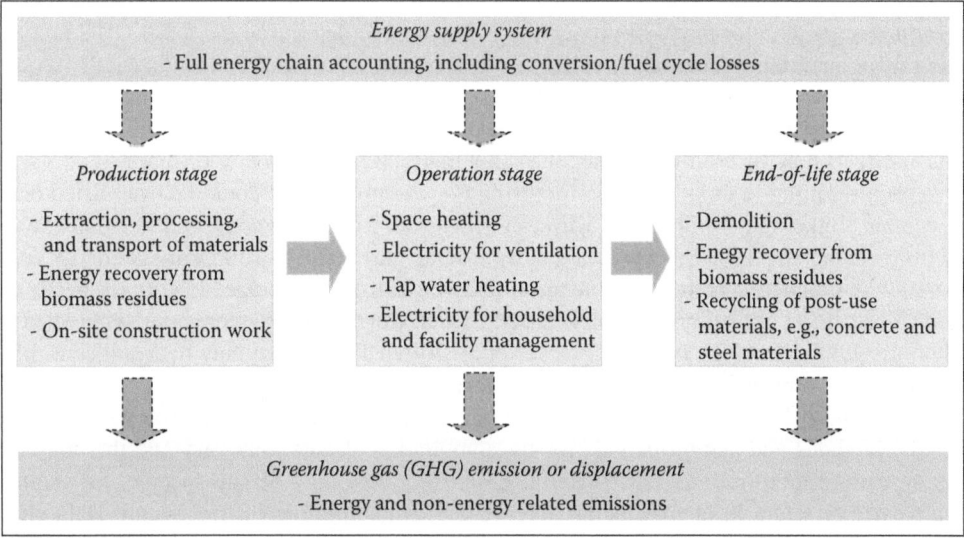

FIGURE 7.3 Key activities and GHG flows in life-cycle carbon footprint analysis of a residential building. (Reprinted from *Energy and Buildings*, 82, Dodoo A, Gustavsson L, Sathre R, Life-cycle carbon implications of conventional and low-energy multi-storey timber building systems, 194–210, Copyright 2014b, with permission from Elsevier.)

7.6.1 ACTIVITY-RELATED SYSTEM BOUNDARIES

Key activity-related life-cycle system boundary issues are connected to production, operation, and end-of-life stages of buildings (Figure 7.3).

7.6.1.1 Production Stage

Energy is expended during the production stage of buildings for a range of activities including acquisition of raw materials, transport and processing of raw materials into building materials, and fabrication and assembly of materials into a ready building. GHGs may be emitted from fossil-fuel combustion, land-use practices, and industrial process reactions when undertaking these activities. Biomass residues are produced during the production stage of wood-based product systems, for example, from forest thinning and harvesting, wood-processing industries, and construction sites. Accounting of carbon footprint of wood-based products needs to consider the flow of residues from the wood product chain (Schlamadinger et al. 1997). An appropriate system boundary for carbon footprint analysis for non-biological materials, for example, mineral ores, may be from the point of extraction. In the case of cultivated bio-based material such as sustainably produced wood, the system boundary needs to be defined to encompass the technological (i.e., human-directed) energy used for biomass production. This includes the GHG flows from fuels used for the management of forest land, the harvesting of timber, and the transport and processing of wood materials (Gustavsson and Sathre 2011).

CO$_2$ flows which occur during cement reaction need to be considered when modeling carbon footprint of cement-based products. Calcination emissions result from a chemical reaction that converts heated limestone (CaCO$_3$) to calcium oxide (CaO) and CO$_2$, during cement manufacture. With about 5% of the global anthropogenic CO$_2$ emission connected to cement production, of which about half come from the calcination process (Price et al. 2006), the climate implication of calcination is significant. Over the life cycle of cement-based products, parts of the CO$_2$ are re-absorbed into the concrete matrix by carbonation reaction. The net CO$_2$ emission from cement reaction includes the effects of the chemical reactions of calcination and carbonation. The net CO$_2$ emission from cement process reactions can be a significant part of the carbon footprint of cement-based products (Dodoo et al. 2009).

Specific energy data for material production is required to determine the fossil GHG emission resulting from material production. A variety of factors may affect the production energy requirement for building materials. These may include geographic-specific factors as well as factors linked to technological processes and primary fuels used to produce materials. Different technological options and fuels may be used to produce the same material in the same or different geographical regions, each resulting in unique carbon footprint. For example, wood could be oven-dried or air-dried, cement clinker may be produced with wet-kiln or dry-kiln technology, and steel may be produced from ore or scrap iron. Steel production from scrap may result in 70% primary energy saving compared with production from virgin material (Harvey 2010). On average, 35% more energy may be used when cement clinker production is based on wet-kiln technology compared with when it is based on dry-kiln technology (von Weizsäcker et al. 2009). Thus, there may be significant differences in carbon footprint for the same type of material.

ISO/TS 14067:2013 states that data "shall be representative of the processes for which they are collected." Methods and system boundaries for compilation of data may vary and this may have impact on carbon footprint analysis. Material production data may be site-specific, for example, from a particular sawmill, or may be an average across an entire industrial sector. Data choice should generally reflect the production structures, technologies, and fuels used to produce a material in a given geographical region. Data availability and quality are key challenges in carbon footprint analysis, and this is well noted in LCA literature (Brady 2005). For example, Tettey et al. (2014) found that the primary energy required for production of alternative insulation materials for elements of a building differs significantly when using different datasets (Figure 7.4). Björklund and

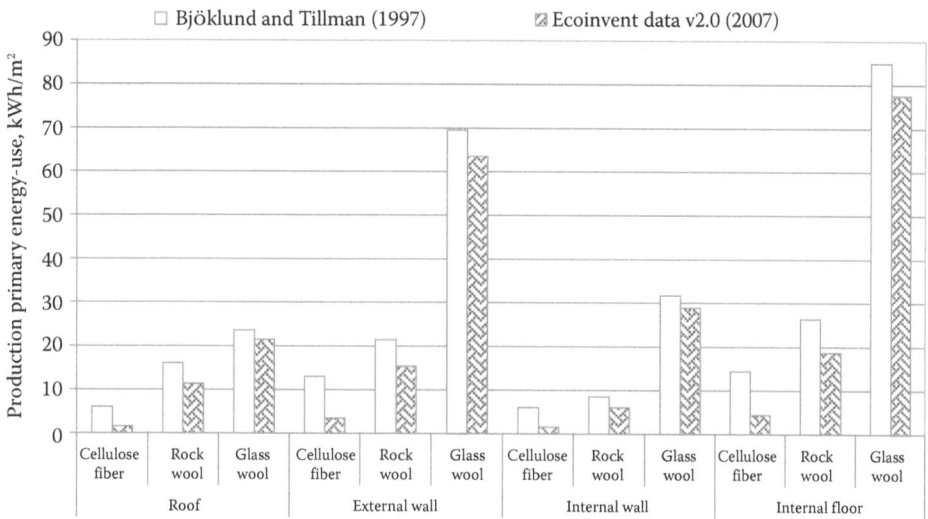

FIGURE 7.4 Comparison of primary energy required for production of alternative insulation materials for elements of a building when using different data sets. (Reprinted from *Energy and Buildings*, 82, Tettey, U.Y.A., Dodoo A, Gustavsson L, Effects of different insulation materials on primary energy and CO_2 emission of a multi-storey residential building, 369–377, Copyright 2014, with permission from Elsevier.)

Tillman's (1997) dataset is based on the Swedish building material industry and was compiled in the late 1990s. Ecoinvent's (2007) dataset is generally representative of the central European average situation and was compiled in the late 2000s. The efficiency of industrial technologies has generally improved over time, resulting in differences in energy requirements and emissions between materials processed by state-of-the-art technologies and those made in older factories. Variation is also seen geographically, as technological innovations diffuse across countries and regions. Data on industrial energy use can also vary depending on the methodology used to obtain the data. Nevertheless, both datasets in Figure 7.4 show consistency in the ranking of the insulation materials.

The impacts of off-site and on-site construction activities connected to buildings must also be included in carbon footprint accounting. The climate impacts from building construction may vary, depending on the method of construction, the type of building materials, and the parameters included, for example, fuel use to transport construction equipment, workers and off-site fabricated components, and so on. Adalberth (2000) estimated the energy for erection of a wood-frame building to be about half of that for a comparable concrete-frame alternative. To determine carbon footprint resulting from primary energy use for building construction activities, it is necessary to know the mix of on-site construction-related primary energy use, for example, its breakdown between end-use electricity and diesel fuel (Gustavsson et al. 2010).

7.6.1.2 Operation Stage and Service Life

Essentially, the carbon footprint from a building's operation stage depends on the building's final energy demand and supply systems. The operation stage is crucial as this typically dominates the life-cycle impacts of buildings (Sartori and Hestnes 2007). Activities in the operation stage, for example, of a residential building comprise space conditioning (heating and cooling), tap water heating, and electricity for ventilation, facility management, as well as for household appliances and lighting. The space conditioning and ventilation loads of a building are essentially linked to the building's construction system and energy efficiency level, in contrast to the other operation stage activities such as tap water heating and electricity for household purposes. A range of interrelated factors influence the space conditioning loads of buildings including those inherently linked to buildings' design, construction systems, HVAC systems, and operational schedules as well as those linked to climatic and geographic variables. To comprehensively account for the carbon footprint of operation stage of buildings, the dynamics of these factors need to be accurately characterized. Dynamic energy simulation models that can accurately characterize the complex interaction of factors which influence final energy demand of buildings are essential. Comprehensive simulation models can allow for one-, two-, and three-dimensional modeling of thermal transmissions of buildings' components and for detailed thermal bridge and thermal mass modeling. To ensure quality assurance, different standard tests and validation procedures have been developed for building simulation models. For example, BESTEST was developed by the International Energy Agency model evaluation and improvement expert group to evaluate numerical heat-transfer models in simulation software (Neymark et al. 2002). ANSI/ASHRAE Standard 140-2007 is another test developed by the American Society of Heating, Refrigerating, and Air-conditioning Engineers to evaluate building energy analysis computer models. A comprehensive directory of software commonly used for building energy simulation modeling can be found at the website of the US Department of Energy (2011). Some of the commonly used dynamic models are DEROB-LTH, DesignBuilder, Energyplus, ESP-r, IDA-ICE, TRNSYS, and VIP+.

The energy services required in a building can be provided by several kinds of technologies and supply systems which can result in significantly different carbon footprints. Activities and processes along the energy supply chain, from the extraction of raw material to refining, transport, conversion to heat and electricity, and distribution to the user can also be performed with different efficiencies and with varying GHG emissions (Gustavsson and Joelsson 2007). A comprehensive analysis of the carbon footprint of operation stage of a building needs to include the entire energy chain from natural resource extraction to final energy supply, taking into account the fuel inputs at each stage in the energy system chain and the energy efficiency of each process. Dodoo and Gustavsson (2013) showed that the carbon footprint over the life cycle of a building can vary by 17%–21% depending on whether the end-use heating is based on heat pump, district heat, or electric resistance heaters.

7.6.1.3 Maintenance and Renovation

During the service life of buildings, materials and components may be maintained, replaced, and renovated. The impacts of these activities might be significant (Cole and Kernan 1996) and are increasingly included in life-cycle studies. When analyzing carbon footprint over the life cycle of buildings, a key challenge often confronted is formulating credible maintenance and material replacement scenarios (IEA 2001).

7.6.1.4 End-of-Life Stage

GHG flows arising from the end-of-life stage of buildings can have significant impact from a life-cycle perspective (Dodoo et al. 2009; Salazar and Meil 2009; Wu et al. 2014). ISO/TS 14067:2013 indicates that "all the GHG emissions and removals arising from the end-of-life stage of a product shall be included in a [carbon footprint] study." For wood-based building systems, end-of-life management option is the single most significant variable in a complete carbon accounting (Gustavsson and Sathre 2006; Sathre and O'Connor 2010). End-of-life management activities of buildings include demolition or disassembly of materials and post-use management of material, for example, disposal, recycling, reuse, energy recovery, and so on. Different end-of-life management regimes for buildings may occur in the future, when today's new buildings reach their end-of-life, and this uncertainty poses a challenge in full carbon footprint analysis of buildings.

Few life-cycle studies, for example, Dodoo et al. (2009), have comprehensively analyzed the carbon implications of end-of-life management options for building materials. In studies where the end-of-life stage of a building is considered, plausible post-use material management scenarios reflecting the context of the studies are constructed and analyzed. Efficient management of post-use building materials is receiving increasing attention in many countries and regions, including in the European Union (EU) (European Commission 2011). Post-use concrete, steel, and wood materials have high recovery and recycling rates in many countries (European Commission 2001). Dodoo et al. (2009) assessed end-of-life scenarios where post-use wooden material is used as bioenergy, concrete is recycled as crushed aggregate, and steel is recycled as feedstock for production of new steel products. These scenarios reflect the typical end-of-life treatment for the materials in Sweden today. In an Australian study, Aye et al. (2012) explored a scenario involving reuse of concrete, steel, and wood materials at the end of buildings' life. In the EU, most post-use wooden materials are typically recovered of energy or used as raw material for further processing (Dodoo et al. 2014a). For example, 90% of recovered wood in Sweden is used as bioenergy, whereas 70% of recovered wood in France is used as raw material for further wood processing (Mantau et al. 2010). Typically, most post-use steel materials are recovered and used as scrap for production of new steel products. The World Steel Association (2013) reported that more than 70% of steel scraps generated domestically in North America are recycled. Krogh et al. (2001) noted that the production of new concrete reinforcement bars is based on scrap steel in Sweden. In Australia, it is reported that 90% of steel scraps from construction and demotion sites are recycled (Green Building Council of Australia 2010). Concrete from construction and demolition waste streams is increasingly recycled in many countries. Concrete recycling rates of 80 and 82% are reported for Japan and the United States, respectively (WBCSD 2009). Recycled concrete aggregates are typically used in below ground applications in Sweden (Engelsen et al. 2005). In principle, recycled concrete may be reused as aggregate in concrete. However, recycled aggregate concrete generally has high porosity and this increases water absorption and carbonation, compared with virgin aggregate concrete. This presents corrosion problems and prevents the use of recycled concrete aggregate in applications where a higher standard is required (Topcu and Sengel 2004). ISO/TS 14067:2013 suggests that assumptions for modeling the carbon footprint of end-of-life stage of products shall be "based on best available information, based on current technology, and documented in the [carbon footprint] study."

The quantification of GHG dynamics of end-of-life wood is not straightforward. Major methodological issues in the quantification of GHG implications of recovered wood include reference fuel or material replaced when post-use wood is recovered of energy or recycled, forest system boundaries, landfill dynamics, and uncertainties about how post-use wood products will be handled by future waste management systems (Gustavsson and Sathre 2011). Recovery of energy by burning the wood is a resource-efficient post-use option, where material reuse of recovered wood is not practical. The use of recovered post-use wood for bioenergy directly affects the carbon footprint of the material. The use of the bioenergy instead of fossil fuels, thus avoiding fossil carbon emissions, also affects the carbon footprint. In some areas, deposition in landfills is the most common fate for post-use wood material. For example, in North America, demolition waste including wood material is typically disposed in landfills (Salazar and Meil 2009), whereas landfilling of sorted post-use wooden building materials is not permitted in some states in the United States (DEFRA 2012). From the perspective of resource-use efficiency, landfilling of wood represents a waste of a potentially valuable energy source and leads to the use of other energy resources such as fossil fuels to provide the equivalent energy service.

Carbon dynamics in landfills are quite variable and uncertain and can have a significant impact on the carbon footprint of wood-based products (Micales and Skog 1997). A fraction of the carbon content in landfilled wood will likely remain in long-term storage, providing some climate benefits. Another fraction may decompose into methane, which has much higher GWP than CO_2. However, methane gas from landfills can be partially recovered and used to replace fossil fuels. The reported practical amounts of methane gas that can be recovered from landfills significantly differ in literature. The US Environmental Protection Agency reported that the current landfill gas capture technology can recover about 75% of the emitted landfill gas (USEPA 2006). However, an empirical study by Themelis and Ulloa (2007) suggested that maximum practical recovery of emitted landfill gas is about 35%. Various methodological treatments have been used in the analysis of GHG dynamics of wood in landfill. Franklin Associates (2004) compared various methodological treatments in life-cycle studies for wood in landfills and noted considerable differences in their outcomes and potential conclusions. The Intergovernmental Panel on Climate Change (IPCC 2000a,b,

2006) provides default methodology for landfill gas modeling based on a zeroth-order decay function. This is increasingly used to estimate GHG dynamics of wooden material in landfills (e.g., Gustavsson and Sathre 2004; Dodoo et al. 2014b).

Post-use management of concrete may entail crushing it into aggregate and stockpiling it for a period. This increases the surface area of the concrete exposed to air and thus facilitates the carbonation process. A complete analysis of the carbon footprint should consider such possible carbon dynamics of end-of-life management of concrete materials (Pommer and Pade 2005; Dodoo et al. 2009).

7.6.2 Time-Related System Boundaries

Temporally explicit GHG flows within the system should be taken into account in analysis of carbon footprint. Lippke et al. (2011) discussed the life-cycle impacts of wood utilization and indicated that consideration of time-related aspects is an important part of full accounting of the climate impact of wood-based products. Of particular interest are the dynamics of forest management, carbon sequestration in wood products, time dynamics of cement process reactions, and the service life of buildings as well as their components.

When modeling wood-based systems, a relatively long time frame is typically under consideration due to the nature of activities and processes associated with forestry and the wood supply chain, for example, seed germination, and forest growth, harvest, as well as regrowth, and so on. Forest management regimes can have significant impact on GHG balances of wood products (Gustavsson and Sathre 2011). Sustainable forestry, meaning that wood that is removed for product use does not exceed net forest growth, is fundamental for carbon benefit of wood-based products. Forests are managed sustainably in many countries, and forest covers are increasing over the years in some countries. For example, annual increment of Norwegian forests is about 25 million m^3, whereas total annual harvest is less than 50% of this growth (Nordic Forest Owners' Associations 2014). In Sweden, annual forest stemwood growth is about 120 million m^3, whereas the current total annual harvest ranges between 85 and 90 million m^3 (Nordic Forest Owners' Associations 2014). In the EU-27, net annual increment of forest was reported to be 630 million m^3, whereas annual felling was 469 million m^3, in 2010 (Forest Europe 2011). The corresponding values for the whole of Europe, including the Russian Federation, are 1552 and 683 million m^3, respectively (Forest Europe 2011).

Forests sequester carbon from the atmosphere during photosynthesis, and part of this carbon is stored in wood products. Depending on the service lifespan of wood products from sustainably managed forests and the forest rotation period, the harvested forest stand may regrow during the building life cycle, sequestering about the same amount of carbon as before the forest was harvested for wood products. At the end-of-life, carbon contained in the wood product may be released as CO_2 through burning for energy or natural decomposition.

Another temporal aspect, which is connected to buildings, relates to the cement process reaction of carbonation. Carbonation is a chemical process in which the CaO present in hardened cement products binds with CO_2 in the atmosphere to form calcium carbonate. The rate of concrete carbonation depends on a number of factors including the composition of the cement used to make the concrete, the temperature, and relative humidity of the environment, and the conditions and duration of concrete exposure to air (Gajda and Miller 2000). Figure 7.5 shows the carbon flows from the cement calcination reaction and the carbonation uptake for functionally equivalent concrete- and wood-frame buildings. The magnitude of the carbon flows from calcination and carbonation reactions in the buildings reflect the quantities of cement used in each building. The results suggest that carbonation uptake increases gradually over the service life of a building and increases considerably if post-use concrete material is crushed at the end of the service life and exposed to air. Nevertheless, carbonation uptake is always less than the initial calcination emission (Dodoo et al. 2009).

It is difficult to know how long a building will be in use as various factors will have a significant impact on the service life of a building, for example, quality of materials, construction, and

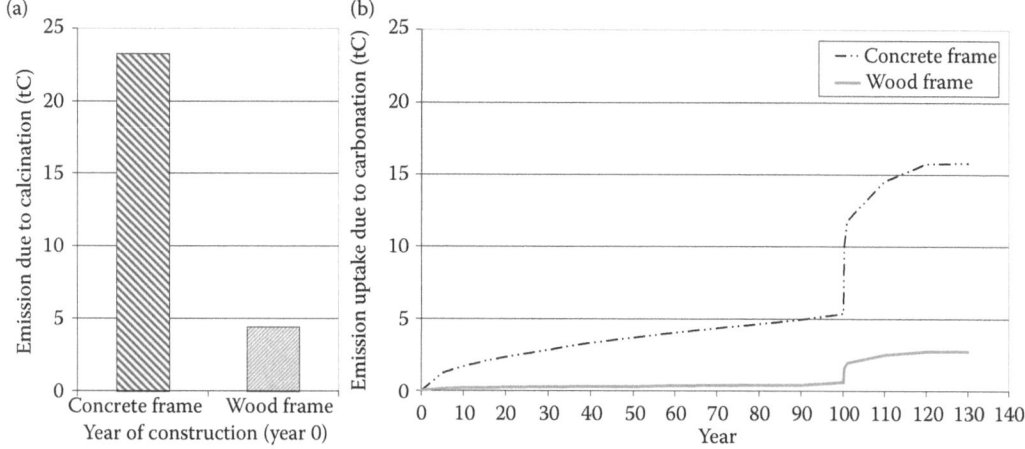

FIGURE 7.5 Carbon flows through cement calcination (a) and carbonation of concrete and cement mortar during the service life and after demolition (b) for the concrete- and wood-frame buildings. Concrete material is crushed at the end of service life of 100 years and exposed to air for 30 years. (Reprinted from *Resources, Conservation and Recycling*, 53(5), Dodoo A, Gustavsson L, Sathre R, Carbon implications of end-of-life management of building materials, 276–286, Copyright 2009, with permission from Elsevier.)

maintenance (O'Connor 2004). However, O'Connor (2004) noted that the main reason for building end-of-life is not related to materials. It is more related to economics and building function, and whether the building still plays a useful role in the built environment. The IEA (2011) reported typical lifespans of energy-related capital stock and noted the typical lifespan of buildings to range from 50 to 150 years, with 80 years as average. Assumed lifespans of 50, 75, and 100 years are commonly found in life-cycle studies. To assess the significance of uncertainties associated with a building's lifespan in carbon footprint analysis, sensitivity analysis can be conducted. For example, in a life-cycle study by Dodoo et al. (2014c), a 50-year lifespan was assumed in the base calculation, and this was varied to 80, 100, and 150 years in a sensitivity analysis.

7.6.3 SPACE-RELATED SYSTEM BOUNDARIES

Spatial system boundaries are relevant when assessing carbon footprint of wood-based products and buildings. The dynamics of forest carbon flow will differ depending on whether an analysis is done from a stand-level or landscape-level forest perspective. At a landscape level, the stand-level carbon flows associated with many different stands at different phases of their rotation cycles will be aggregated and produce overall trends. When a tree or stand is harvested, the carbon in living biomass is transferred into other carbon pools such as wood products and forest floor litter. At the landscape level, the dynamic patterns of the individual trees or stands are averaged over time as carbon flows into and out of various carbon pools associated with trees at differing stages of development. Thus, at the landscape level, the total carbon stock in living biomass can be fairly stable over time or even increasing, as the harvest of some trees during a given time period is compensated by other trees growing during the same period. For example, the standing stock of stem wood in Swedish forests has been increasing during the last 60 years and is expected to continue increasing during the coming 100 years due to improved forest management and the effects of climate change (Figure 7.6). This provides opportunities to increase use of renewable forest resources as part of a strategy to transition to a more sustainable, carbon-neutral society.

The choice between stand-level or landscape-level perspective may depend in part on the focus of a study. A landscape-level perspective is important to determine the overall effects of forest harvesting for wood-based products (Malmsheimer et al. 2011).

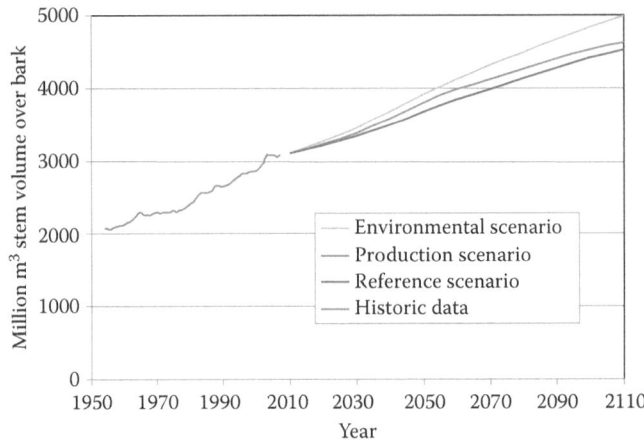

FIGURE 7.6 Historic and projected future standing stem volume on productive forest land in Sweden. (Adapted from Swedish Forest Agency. 2008. National Forest Inventory and Forest and Environment, Sweden, Jönköping.)

Wood-based buildings require greater amounts of biomass, and thus larger forest area, than nonwooden alternatives. Accurate and objective characterization of the implication of different land-use requirements when comparing wood versus non-wood product systems is a major challenge in carbon footprint analysis. Different approaches to deal with this challenge are presented by Gustavsson et al. (2006), who explored this issue in detail. One approach is to assume that the incremental wood material is produced through more intensive management of forest land, or from land that was not previously used for wood production. Another approach is to assume that an equal area of land is available to both the wooden and non-wooden systems and analyze the carbon balance impacts of biological carbon sequestration on the "surplus forest," or the part of the land not used.

Other relevant issues which fall within spatial boundaries include the technologies and infrastructure for production systems that may vary between different geographical regions.

7.7 ACCOUNTING FOR ELECTRICITY PRODUCTION

The choice of electricity production and supply systems are crucial in carbon footprint analysis as this steers the outcome of a study. Electricity production accounts for the single largest direct use of primary energy in OECD and non-OECD countries (Harvey 2010). Fossil fuels accounted for 67% of the share of fuels used for electricity production worldwide, in 2011 (IEA 2013a). In the EU, the electricity generation mix is comprised of 27% nuclear, 25% coal and lignite, 24% fossil gas, 21% renewables, and 3% oil (European Environment Agency 2013). Figure 7.7 shows current and projected trends in net electricity generation in the EU, up to 2050. Electricity production is typically dominated by standalone power plants, with large losses and excess waste heat. The average conversion efficiency for electricity production worldwide was 37% in 2005 (Harvey 2010).

Different electricity production systems and approaches for accounting for impacts from electricity are characterized by significant variation in GHG emissions and estimations. The average and marginal approaches are two distinct methods for accounting for GHG emission for electricity systems. The average accounting approach considers the emissions per unit of delivered electricity, based on the average power mix of a region or country. In contrast, the marginal accounting approach considers the marginal changes in electricity supply system in estimating the emission resulting from a unit change in electricity demand (Harmsen and Graus 2013; Hawkes 2014). A major discussion in energy and life-cycle analysis literature is the choice of accounting method to accurately capture the impacts of electricity demand and supply (see Sjödin and Grönqvist 2004;

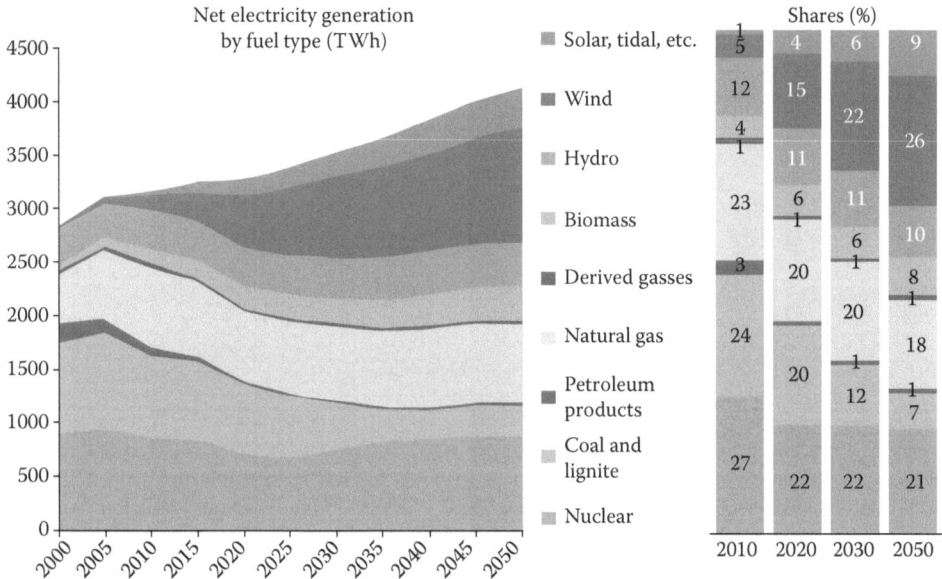

FIGURE 7.7 Current and projected net electricity generation trends in the EU by fuel type, up to 2050. The projection begins with current data gathered from EUROSTAT energy balances in 2013. (Adapted from European Commission. 2013. EU energy, transport and GHG emissions trends to 2050. Reference Scenario 2013 ©. http://ec.europa.eu/energy/observatory/trends_2030/doc/trends_to_2050_update_2013.pdf (December 2, 2014).)

Hawkes 2010, 2014; Lund et al. 2010; Harmsen and Graus 2013). The accounting method employed in a carbon footprint analysis must reflect the purpose and relevance of the study. Weidema et al. (1999) suggested the consideration of marginal technologies in prospective comparative life-cycle studies, as this gives the best reflection of the actual impact of a decision. The marginal accounting method captures the consequences of small changes due to variation in system parameters and reflects the technologies and inputs affected by a variation in a system. According to Weidema et al. (1999), data on marginal technologies are more precise and stable over time, in contrast to average technologies. The average accounting method is typically not suitable in a consequential-based carbon footprint analysis, because small changes do not readily reflect at the average level (Hawkes 2010, 2014). The determination of marginal technology for electricity production is influenced by the complex interplay of several factors (Lund et al. 2010). These include investments cost, GHG reduction policies, and strategic and security reasons (Gustavsson et al. 2006). In northern Europe, coal-fired condensing power plants are usually considered to be the marginal source of electricity at the moment (Sjödin and Grönqvist 2004; Gustavsson et al. 2006; Amiri and Moshfegh 2010).

7.8 TREATMENT OF ALLOCATION

In systems that coproduce more than one product, the issue of allocation between coproducts of burdens and benefits arises. The way allocation is treated can have a significant impact on the results of a carbon footprint analysis. However, allocation should be avoided as it can be complex and subjective (ISO 14044:2006). The ISO/TS 14067:2013 also states that "wherever possible, allocation should be avoided." Two procedures are suggested by the ISO/TS to avoid allocation: "dividing the unit process to be allocated into two or more sub-processes and collecting the input and output data related to these sub-processes" and "expanding the product system to include the additional functions related to the co-products."

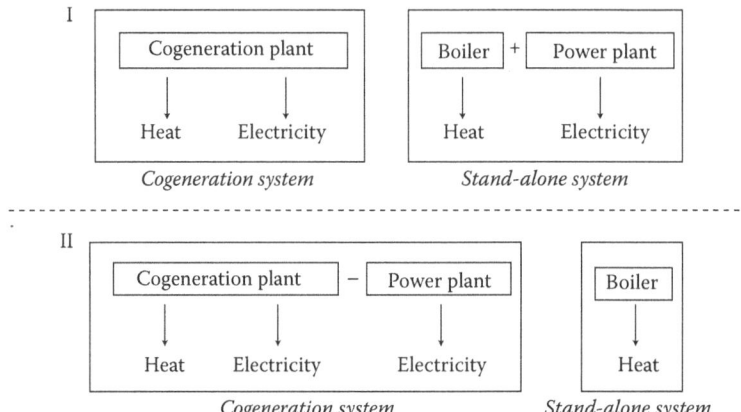

FIGURE 7.8 Different approaches to avoid allocation when analyzing systems involving a CHP plant. Approach I entails system expansion with multifunctional products in the functional unit, and Approach II entails system subtraction with cogenerated electricity subtracted so that the functional unit is heat. (With kind permission from Springer Science+Business Media: *Mitigation and Adaptation Strategies for Global Change*, CO_2 mitigation: On methods and parameters for comparison of fossil-fuel and biofuel systems, 11(5–6), 2006, 935–959, Gustavsson L, Karlsson Å; Reprinted from *Resources, Conservation and Recycling*, 24, Finnveden G, Ekvall T, Life-cycle assessment as a decision-support tool: The case of recycling versus incineration of paper, Copyright 235–256, 1998, with permission from Elsevier.)

When evaluating systems involving combined heat and power (CHP) plant, the issue of allocation is encountered. Figure 7.8 illustrates how allocation can be avoided in such systems, based on Finnveden and Ekvall (1998) and Gustavsson and Karlsson (2006). One way (I) is expanding the systems by adding an alternative means of producing heat or electricity to systems that produce only one of the energy carriers, thereby making the systems multifunctional. Another way (II) is subtracting either heat or electricity production from the CHP production. In this way, only electricity or heat will become the functional unit.

Allocation based on physical relationships (e.g., mass, volume, energy content) of coproducts is suggested where allocation cannot be avoided. Allocation based on economic value of coproducts might be used but this is suggested as the last option for dealing with allocation (ISO/TS 14067:2013), as economic values are market-driven and may not be stable over time. In this case, a sensitivity analysis should be conducted across a range of potential economic values of each coproduct.

7.9 KEY RECOMMENDATIONS AND OUTLOOKS

A system-wide life-cycle perspective is fundamental to analyze carbon footprint of wood-based products and buildings involving material and energy flows in different economic sectors, including forestry, manufacturing, construction, energy, and waste management. The closer integration of these flows is essential to reduce the overall climate impact of wood-based products and buildings. Partial carbon footprint analysis of wood-based products and buildings, based on appropriate system boundaries, may give misleading results. To understand the carbon implications of wood-based systems, full GHG flow accounting must be conducted, including flows from forest establishment and management to the production, use and post-use management of wood-based products and buildings. A robust and comprehensive analysis of the carbon footprint of wood-based products and buildings would be characterized by: definition of appropriate functional unit; selection of relevant characterization indicators; establishment of effective system boundaries in terms of activities, time, and place; careful consideration of impacts for energy supply systems affected by a decision; and transparent and justified treatment of allocation.

This chapter also emphasizes the need for a forward looking consequential-based approach that accounts for marginal changes when comparing different potential alternatives, for example, the use of wood building products and systems instead of non-wood building alternatives.

Reliable datasets are needed to accurately determine the carbon footprint of wood-based products and buildings. There are various databases that provide estimates of carbon footprint for materials and products. However, carbon footprint values for the same type of material can vary widely from one database to another, and this can have considerable impact on the conclusions of a study. Therefore, caution should be taken when selecting data for carbon footprint analysis. Key issues to be considered in the choice of data include the age of the data, the time over which the data were collected, the geographical scope of the data, the type of technical systems covered by the data, the precision of the data, the consistency of the data, the completeness and representativeness of the data, and the uncertainty of the quality of the data (ISO/TS 14067:2013). In a situation where data is not available in a study's context, data from closely related studies may be used. Sensitivity analyses may be conducted to determine the significance of the data uncertainty.

Current studies of carbon footprint of wood-based products and buildings are largely based on the GHG balance approach. In this approach, carbon footprint is calculated by summing all significant GHG emissions and removals that occur within a system, regardless of the time frame they occur. The main limitation of this approach is that it does not take into account the time dynamics of GHG flows. The time frame of GHG emissions and removal can affect atmospheric radiative forcing and hence climate impact. Further carbon footprint studies for wood-based product and buildings may take into consideration the effect of the timing of GHG flows, to give more accurate understanding of the climate impacts of wood substitution systems. CRF may be suitable evaluation indicator for such studies as it account for the time dynamics of the total amount of energy added to the earth system (Sathre and Gustavsson 2011; Sathre et al. 2013).

7.10 SUMMARY

This chapter has outlined the importance of wood-based products and buildings in creating a built environment with low-carbon footprint and has presented appropriate methods for comprehensive and robust analyses of carbon footprint of wood-based systems. Wood-based products and building materials from sustainably managed forests can play an important role in our transition to a low-carbon economy. If performed well, carbon footprint analysis can provide essential information for decision making from a climate mitigation perspective. A comprehensive understanding of the climate implications of wood-based products and buildings is needed for minimizing the carbon footprint of the built environment. This chapter emphasizes that a robust methodology incorporating life-cycle perspective is needed to analyze the carbon footprint of wood-product and building systems.

REFERENCES

Adalberth K. 2000. Energy use and environmental impact of new residential buildings. Ph.D. Dissertation. Sweden: Department of Building Physics, Lund University of Technology.

Amiri S. Moshfegh B. 2010. Possibilities and consequences of deregulation of the European electricity market for connection of heat sparse areas to district heating systems. *Applied Energy* 87(7):2401–2410.

ANSI/ASHRAE Standard 140-2007. Standard method of test for the evaluation of building energy analysis computer programs.

Aye L, Ngo T, Crawford RH, Gammampila R, Mendis P. 2012. Life cycle greenhouse gas emissions and energy analysis of prefabricated reusable building modules. *Energy and Buildings* 48:159–168.

Björklund T, Tillman A-M. 1997. LCA of building frame structures: Environmental impact over the life cycle of wooden and concrete frames. Technical Environmental Planning Report 1997:2. Chalmers University of Technology.

Brady J. (ed.). 2005. *Environmental Management in Organisations: The IEMA Handbook*. London and Sterling, VA, USA: Earthscan (Chapter 4.4, p. 230).

Buchanan AH, Honey BG. 1994. Energy and carbon dioxide implications of building construction. *Energy and Buildings* 20(3):205–217.

Cole RJ, Kernan PC. 1996. Life-cycle energy use in office buildings. *Building and Environment* 31(4):307–317.

DEFRA. 2012. Wood waste landfill restrictions in England call for evidence. http://www.defra.gov.uk/consult/files/consult-wood-waste-document-20120808.pdf (January 29, 2013).

Dodoo A, Gustavsson L. 2013. Life cycle primary energy use and carbon footprint of wood-frame conventional and passive houses with biomass-based energy supply. *Applied Energy* 112:834–842.

Dodoo A, Gustavsson L, Sathre R. 2009. Carbon implications of end-of-life management of building materials. *Resources, Conservation and Recycling* 53(5):276–286.

Dodoo A, Gustavsson L, Sathre R. 2014a. Lumber. Chapter 12. In: Worrell E, Reuter M (eds), *Handbook of Recycling: State-of-the-Art for Practitioners, Analysts, And Scientists*. Elsevier, Amsterdam. ISBN 978-0-12-396459-5.

Dodoo A, Gustavsson L, Sathre R. 2014b. Lifecycle carbon implications of conventional and low-energy multi-storey timber building systems. *Energy and Buildings* 82:194–210.

Dodoo A, Gustavsson L, Sathre R. 2014c. Lifecycle primary energy analysis of low-energy timber building systems for multi-story residential buildings. *Energy and Buildings* 81:84–97.

Ecoinvent data v2.0. 2007. In: Kellenberger D, Althaus H-J, Künniger T, Lehmann M, Jungbluth N, Thalmann P (eds), *Life Cycle Inventories of Building Products*, data v2.0. ecoinvent Report No. 7. Dübendorf: EMPA-Swiss Centre for Life Cycle Inventories.

Engelsen CJ, Mehus J, Pade C, Sæther DH. 2005. CO_2 uptake in demolished and crushed concrete. Norwegian Building Research Institute. www.byggforsk.no (January 28, 2008).

European Commission. 2001. Task Group 3: Construction and demolition waste. Final report of the construction and demolition waste working group, Brussels. http://ec.europa.eu (March 28, 2008).

European Commission. 2011. A resource-efficient Europe—Flagship initiative of the Europe 2020 Strategy. Brussels. http://ec.europa.eu (October 6, 2014).

European Commission. 2013. EU energy, transport and GHG emissions trends to 2050. Reference Scenario 2013. http://ec.europa.eu/energy/observatory/trends_2030/doc/trends_to_2050_update_2013.pdf (December 2, 2014).

European Environment Agency. 2013. Overview of the electricity production and use in Europe (ENER 038). http://www.eea.europa.eu/data-and-maps/indicators/overview-of-the-electricity-production/assessment (October 4, 2014).

Finnveden G, Ekvall T. 1998. Life-cycle assessment as a decision-support tool: The case of recycling versus incineration of paper. *Resources, Conservation and Recycling* 24:235–256.

Forest Europe. 2011. State of Europe's Forests, 2011—Status and trends in sustainable forest management in Europe. http://www.foresteurope.org/documentos/State_of_Europes_Forests_2011_Report_Revised_November_2011.pdf (December 2, 2014).

Franklin Associates. 2004. An analysis of the methods used to address the carbon cycle in wood and paper product LCA studies. Report No. 04-03. National Council for Air and Stream Improvement, 63 pp.

Fulton M, Mellquist N, Kitasei S, Bluestein J. 2011. *Comparing Life-Cycle Greenhouse Gas Emissions from Natural Gas and Coal*. Frankfurt: Deutsche Bank and Worldwatch Institute.

Gajda J, Miller FM. 2000. Concrete as a sink for atmospheric CO_2: A literature review and estimation of CO_2 absorption by Portland cement concrete. R&D Serial no. 2255. Skokie IL, USA: Portland Cement Association.

Green Building Council of Australia. 2010. Background and outcomes—Green star steel credit review. http://www.gbca.org.au/ (October 10, 2014).

GHG (Greenhouse Gas) Protocol. 2011. Product Life Cycle Accounting and Reporting Standard.

Gustavsson L, Joelsson A. 2007. Energy conservation and conversion of electrical heating systems in detached houses. *Energy and Buildings* 39(6):716–726.

Gustavsson L, Joelsson A, Sathre R. 2010. Life cycle primary energy use and carbon emission of an eight-story wood-framed apartment building. *Energy and Buildings* 42(2):230–242.

Gustavsson L, Karlsson Å. 2006. CO_2 mitigation: On methods and parameters for comparison of fossil-fuel and biofuel systems. *Mitigation and Adaptation Strategies for Global Change* 11(5–6):935–959.

Gustavsson L, Pingoud K, Sathre R. 2006. Carbon dioxide balance of wood substitution: Comparing concrete- and wood-framed buildings. *Mitigation and Adaptation Strategies for Global Change* 11(3):667–691.

Gustavsson L, Sathre R. 2004. Embodied energy and CO_2 emission of wood- and concrete-framed buildings in Sweden. *Second World Conference on Biomass for Energy, Industry and Climate Protection*, May 10–14, 2004, Rome, Italy.

Gustavsson L, Sathre R. 2006. Variability in energy and carbon dioxide balances of wood and concrete build-ing materials. *Building and Environment* 41(7):940–951.

Gustavsson L, Sathre R. 2011. Energy and CO_2 analysis of wood substitution in construction. *Climatic Change* 105(1–2):129–153.

Harmsen R, Graus W. 2013. How much CO_2 emissions do we reduce by saving electricity? A focus on meth-ods. *Energy Policy* 60(2013):803–812.

Harvey LDD. 2010. *Energy and the New Reality, Vol. 1: Energy Efficiency and the Demand for Energy Services*. Earthscan, London. ISBN 978-1849710725.

Hawkes AD. 2010. Estimating marginal CO_2 emissions rates for national electricity systems. *Energy Policy* 38(10):5977–5987.

Hawkes AD. 2014. Long-run marginal CO_2 emissions factors in national electricity systems. *Applied Energy* 125:197–205.

Hertwich EG, Peters GP. 2009. Carbon footprint of nations: A global, trade-linked analysis. *Environmental Science and Technology* 43:6414–6420.

IEA (International Energy Agency). 2001. LCA methods for buildings. Energy-related environmental impact of buildings. IEA Annex 31. www.ecbcs.org (August 6, 2011).

IEA (International Energy Agency). 2011. World Energy Outlook 2011. www.iea.org (September 10, 2012).

IEA (International Energy Agency). 2013a. Key world energy statistics. www.iea.org (October 3, 2014).

IEA (International Energy Agency). 2013b. *IEA statistics. CO_2 Emissions from Fuel Combustion: Highlights*, 2013 Edition. www.iea.org (September 28, 2014).

IPCC (Intergovernmental Panel on Climate Change). 2000a. *Special Report on Emissions Scenarios*. Cambridge, UK: Cambridge University Press.

IPPC (Intergovernmental Panel on Climate Change). 2000b. *Good Practice Guidance and Uncertainty Management in National Greenhouse Gas Inventories*. Cambridge: Cambridge University Press.

IPCC (Intergovernmental Panel on Climate Change). 2006. Guidelines for national greenhouse gas invento-ries. http://www.ipcc-nggip.iges.or.jp/public/2006gl/vol5.html (February 18, 2013).

IPCC (Intergovernmental Panel on Climate Change). 2007. Climate change 2007: Mitigation. Contribution of working Group III to the fourth assessment report. Cambridge, UK: Cambridge University Press.

IPCC (Intergovernmental Panel on Climate Change). 2013. Summary for policymakers. In: Climate Change 2013: The Physical Science Basis. Contribution of Working Group I to the Fifth Assessment Report of the Intergovernmental Panel on Climate Change. Cambridge, United Kingdom and New York, NY, USA: Cambridge University Press.

IPCC (Intergovernmental Panel on Climate Change). 2014. Summary for policymakers. In: Climate Change 2014, Mitigation of Climate Change. Contribution of Working Group III to the Fifth Assessment Report of the Intergovernmental Panel on Climate Change. Cambridge, United Kingdom and New York, NY, USA: Cambridge University Press.

ISO 14040. 2006a. Environmental management—Life cycle assessment—Principles and framework. Geneva: International Organization for Standardization.

ISO 14044. 2006b. Environmental management—Life cycle assessment—Requirements and guidelines. Geneva: International Organization for Standardization.

ISO/TS 14067:2013. Greenhouse gases—Carbon footprint of products—Requirements and guidelines for quantification and communication. Geneva: International Organization for Standardization.

Koch P. 1992. Wood versus non-wood materials in U.S. residential construction: Some energy-related global implications. *Forest Products Journal* 42(5):31–42.

Krogh H, Myhre L, Häkkinen T, Tattari K, Jönsson Å, Björklund T. 2001. Environmental data for production of reinforcement bars from scrap iron and for production of steel products from iron ore in the Nordic countries. *Building and Environment* 36(1):109–119.

Lenzen M, Treloar G. 2002. Embodied energy in buildings: Wood versus concrete—Reply to Börjesson and Gustavsson. *Energy Policy* 30(3):249–255.

Lippke B, Oneil E, Harrison R, Skog K, Gustavsson L, Sathre R. 2011. Life cycle impacts of forest management and wood utilization on carbon mitigation: Knowns and unknowns. *Carbon Management* 2(3):303–333.

Lund H, Mathiesen BV, Christensen P, Schmidt JH. 2010. Energy system analysis of marginal electricity sup-ply in consequential LCA. *International Journal of Life Cycle Assessment* 15(3):260–271.

Malmsheimer RW, Bowyer JL, Fried JS, Gee E, Izlar RL, Miner RA et al. 2011. Managing forest because carbon matters: Integrating energy, products, and land management policy. *Journal of Forestry* 109(7S):S7–S50.

Mantau U et al. 2010. EU wood—Real potential for changes in growth and use of EU forests. Methodology report. Hamburg/Germany, June 2010, 165 pp.

Matthews SH, Hendrickson C, Weber CL. 2008. The importance of carbon footprint estimation boundaries. *Environmental Science Technology* 42(16):5839–5842.

Micales JA, Skog KE. 1997. The decomposition of forest products in landfills. *International Biodeterioration and Biodegradation* 39(2–3):145–158.

Neymark J, Judkoff R, Knabec G, Lec H-T, Dürig M, Glass A, Zweifel G. 2002. Applying the building energy simulation test (BESTEST) diagnostic method to verification of space conditioning equipment models used in whole-building energy simulation programs. *Energy and Buildings* 34(9):917–931.

Nordic Forest Owners' Associations. 2014. Forests in Norway. http://www.nordicforestry.org/facts/Norway.asp (April 29, 2014).

O'Connor J. 2004. Survey on actual service lives for North American buildings. *Proceedings of Woodframe Housing Durability and Disaster Issues Conference*, Las Vegas.

PAS 2050:2011. Specification for the assessment of the life cycle greenhouse gas emissions of goods and services. London: British Standards Institution.

Pommer K, Pade C. 2005. Guidelines—Uptake of carbon dioxide in the life cycle inventory of concrete. Danish Technological Institute. http://www.dti.dk (April 16, 2008).

Plevin RJ, Delucchi MA, Creutzig F. 2014. Using attributional life cycle assessment to estimate climate-change mitigation benefits misleads policy makers. *Journal of Industrial Ecology* 18(1):73–83.

Price L, De la Rue du Can S, Sinton J, Worrell E. 2006. Sectoral trends in global energy use and GHG emissions. California, USA: Lawrence Berkeley National Laboratory.

Salazar J, Meil J. 2009. Prospects for carbon-neutral housing: the influence of greater wood use on the carbon footprint of a single-family residence. *Journal of Cleaner Production* 17(17):1563–1571.

Sartori I, Hestnes AG. 2007. Energy use in the life cycle of conventional and low-energy buildings: A review article. *Energy and Buildings* 39(3):249–257.

Sathre R. 2007. Life-cycle energy and carbon implications of wood-based products and construction. Ph.D. Dissertation, Department of Engineering, Physics and Mathematics. Östersund, Sweden: Mid Sweden University.

Sathre R, Dodoo A, Gustavsson L, Lippke B, Marland G, Masanet E, Solberg B, Werner F. 2012. Comment on "Material nature versus structural nurture: The embodied carbon of fundamental structural elements". *Environmental Science and Technology* 46(6):3595–3596.

Sathre R, Gustavsson L. 2011. Time-dependent climate benefits of using forest residues to substitute fossil fuels. *Biomass and Bioenergy* 35(7):2506–2516.

Sathre R, Gustavsson L, Haus S. 2013. Time dynamics and radiative forcing of forest bioenergy systems. Chapter 11. In: Kellomäki S, Alam A, Kilpeläinen A (eds), *Forest Bioenergy Production: Management, Carbon Sequestration and Adaptation*. New York, USA: Springer. ISBN 978-1-4614-8390-8.

Sathre R, O'Connor J. 2010. Meta-analysis of greenhouse gas displacement factors of wood product substitution. *Environmental Science and Policy* 13(2):104–114.

Schlamadinger B, Apps M, Bohlin F, Gustavsson L, Jungmeier G, Marland G, Pingoud K, Savolainen I. 1997. Towards a standard methodology for greenhouse gas balances of bioenergy systems in comparison with fossil energy systems. *Biomass and Bioenergy* 13(6):359–375.

Schlamadinger B, Marland G. 1996. The role of forest and bioenergy strategies in the global carbon cycle. *Biomass and Bioenergy* 10(5–6):275–300.

Sjödin J, Grönkvist S. 2004. Emissions accounting for use and supply of electricity in the Nordic market. *Energy Policy* 32(13):1555–1564.

Swedish Forest Agency. 2008. National Forest Inventory and Forest and Environment, Sweden, Jönköping.

Tettey, U.Y.A., Dodoo A, Gustavsson L. 2014. Effects of different insulation materials on primary energy and CO_2 emission of a multi-storey residential building. *Energy and Buildings* 82:369–377.

Themelis N, Ulloa P. 2007. Methane generation in landfills. *Renewable Energy* 32:1243–1257.

Topcu IB, Sengel S. 2004. Properties of concretes produced with waste concrete aggregate. *Cement Concrete Research* 34(8):1307–1312.

UNEP (United Nations Environment Programme). 2007. Buildings and climate change status, challenges and opportunities. www.unep.org (September 20, 2011).

Upton B, Miner R, Spinney M, Heath LS. 2008. The greenhouse gas and energy impacts of using wood instead of alternatives in residential construction in the United States. *Biomass and Bioenergy* 32(1):1–10.

US Department of Energy. 2011. Building energy software tools directory—Whole building analysis: Energy simulation. http://apps1.eere.energy.gov/ (October 1, 2014).

USEPA. 2006. Solid waste management and greenhouse gases: A life-cycle assessment of emissions and sinks. http://www.epa.gov (February 18, 2013).

von Weizsäcker E, Hargroves K, Smith M, Desha C, Stasinopoulos P. 2009. *Factor 5: Transforming the Global Economy through 80% Increase in Resource Productivity.* UK and Droemer, Germany: Earthscan, ISBN 978-1-84407-591-1.

WBCSD (World Business Council for Sustainable Development). 2009. The cement sustainability initiative: Recycling concrete. http://www.wbcsdcement.org/pdf/CSI-RecyclingConcrete-FullReport.pdf (October 10, 2014).

Weidema BP, Frees N, Nielsen A-M. 1999. Marginal production technologies for life cycle inventories. *International Journal of Life Cycle Assessment* 4:448–456.

Weidema BP, Thrane M, Christensen P, Schmidt J, Lokke S. 2008. Carbon footprint: A catalyst for life cycle assessment. *Journal of Industrial Ecology* 12(1):3–6.

Wiedmann T, Minx J. 2008. A definition of 'carbon footprint'. In: Pertsova CC (ed), *Ecological Economics Research Trends,* Chapter 1. Hauppauge NY, USA: Nova Science Publishers, pp. 1–11.

World Steel Association. 2013. Steel scrap: A world-traded commodity. http://www.worldsteel.org/media-centre/Steel-news/Steel-scrap—A-world-traded-commodity.html (October 10, 2014).

Wright L, Kemp S, Williams I. 2011. Carbon footprinting: Towards a universally accepted definition. *Carbon Management* 2(1):61–72.

Wu W. 2011. Carbon footprint—A case study on the municipality of Haninge. Master of Science thesis. KTH School of Technology and Health. http://www.diva-p (October 4, 2014).

Wu P, Xia B, Pienaar J, Zhao X. 2014. The past, present and future of carbon labelling for construction materials—A review. *Building and Environment* 77:160–168.

8 Applications of Carbon Footprint in Urban Planning and Geography

Taehyun Kim

CONTENTS

8.1 Introduction 163
8.2 Background of CF Applications in Urban Planning and Geography 164
8.3 Case Studies of CF Applications 168
    8.3.1 Household CF 168
    8.3.2 Individual CF 171
    8.3.3 Agricultural CF 177
8.4 Conclusions 179
References 180

8.1 INTRODUCTION

Human life, activity, and culture depend on their wider environment. While cities provide economic opportunities for goods and services, they cannot exist without food, energy, and water supply from their surroundings. Technical innovation of energy supply and transportation accelerates extension of urban areas and their physical separation from agricultural land. In consequence, the divide of urban-agricultural areas causes more energy demand for transportation of foods and goods between these areas (Owens 1986). As the energy resources are drawing from all over the world, environmental impact beyond city boundaries are growing, too. Although advances in energy and other technology may reduce the environmental impacts of consumption, there is still a gap between energy demand and supply with current technology, even in the most technologically advanced countries (Chambers et al. 2000). Therefore, reducing energy demand is more realistic than depending only on technological advances for sustainable development.

Within planning research, it is commonly assumed that design and land use are closely linked to energy efficiency levels for urban areas (Duvarci and Kutluca 2008) and make it possible to achieve a more sustainable consumption of energy for housing and transport (Breheny 1996). In 2002, a set of principles for sustainable cities was developed at an international charrette in Melbourne and endorsed by local governments at the Johannesburg Earth Summit. Of the 10 principles, the fourth is "enable communities to minimize their ecological footprint" (UNEP-DTIE-IETC 2002). From a perspective of climate change, carbon footprint (CF) can be derived from ecological footprint (EF) measurement due to the consumption categories of EF calculation, which includes components such as housing and travel energy uses that are required to calculate CF. The EF of a city measures wider impact of urban area on global environment and the scale of a city's metabolism (Newman and Jennings 2008). As a tool for urban designers and planners to evaluate proposals, EF has been applied extensively at international, national, and regional scales (Moos et al. 2006), as well as at a town scale as a measure of the environmental sustainability level of rural region planning (Liu et al. 2009).

The purpose of this chapter is to introduce CF applications in urban planning and geography. In these fields, CF has been assessed at different geographical scales, such as nation, city, region, household, and individual. The assessment for a nation and a city is available by using national- or city-level statistics of energy consumption categories. By means of CF calculation, it is possible to compare the ecological capacity and deficit among nations and cities. CF especially offers great insights into the geographical distribution of carbon intensity at smaller spatial levels.

The background of CF applications in urban planning and geography are followed by case studies showing how CF was applied to figure out sustainable land-use measures in urban planning and geography.

8.2 BACKGROUND OF CF APPLICATIONS IN URBAN PLANNING AND GEOGRAPHY

Many studies of CF concepts and calculation methods in urban planning and geography commonly assume that spatial factors are important to energy consumptions and carbon dioxide (CO_2) emissions. As a proxy measure of CF, the concept and calculation of EF have also been adopted by researches in these fields.

In urban studies, there is sufficient theoretical and empirical evidence supporting the idea that planning is associated with the level of energy consumption in urban areas.

Kim (1987) identified the relationship between energy consumption and spatial structure by estimating and comparing energy requirements for the journey to work, resulting from alternative population and employment growth patterns in a rural–urban regional settlement of Korea. The result demonstrated that the physical separation of activities affects energy consumption as major spatial determinants.

Alberti (1999) reviewed the empirical evidence on the relationships between urban patterns and various dimensions of environmental quality and performance. The author examined approaches to measuring urban environmental performance, drawing on the concepts of carrying capacity, EF, environmental space, and appropriated ecosystem area. Since cities affect and are affected by ecological systems far beyond their physical boundaries, interactions at the local, regional, and global scales in the definition of environmental performance were proposed. By reviewing the current literature on the relationship between four structural variables typically used to describe urban patterns—form, density, grain, and connectivity—and four dimensions of environmental performance—sources, sinks, ecological support systems, and human well-being, the author concluded that the scale of analysis affects the direction of observable urban impacts.

Recently, the EF is suggested to offer a concept and method that "can generate one of the most objective, nonbiased, aggregate, single dimension indicators for evaluating sustainability" (van den Bergh and Verbruggen 1999). EF analysis has in common that they can be applied at various spatial scales, ranging from the individual or household scale, through the village, town, or city up to provincial, national, continental, and global scale (Chambers et al. 2000; Hoekstra and Chapagain 2008).

Ryu (2005) conducted individual EF quiz surveys to measure individual EF in Dallas County. Adapting the multiple regression method, the study addressed the impact of several factors, including demography, environmental values, spatial attributes, and land-use patterns surrounding an individual, on per capita EF. The results demonstrated that a highly educated, nonmarried, older male living in a high-income household located in a low-population density area is more likely to have a larger personal composite footprint.

Holden (2004) examined the connection between physical urban planning and household consumption of two Norwegian cities by assuming that physical surroundings influence human behavior. The study focused on EF of housing because it has largely been designed and located via physical planning. Four housing-related consumption categories and planning factors were considered, respectively, in his study as follows.

Consumption categories: (1) energy consumption with regard to heating and operating housing, (2) a substantial amount of material housing consumption, (3) energy consumption relating to everyday transport, and (4) energy used for longer holiday and leisure trips.

Planning factors: (1) town size/national settlement pattern, (2) localization of houses within a town, municipality, or built-up area (the distance from the house to the center of town and relation to urban sprawl), (3) population and development density, and (4) type of housing.

The findings indicate that (1) town size is not important, (2) high-density, less urban sprawl, and less single-family housing are favorable for reducing a household's EF, and (3) nonphysical factors (number of people, car occupancy, and income) have a significant influence.

On the simulation approach, Moos et al. (2006) measured EF of residential site plan and consumption of residents as follows: (1) the aggregate EF based on the physical parameters including amount of built-up land (roads, buildings, parking, paths, private yards, parks) and consumption of building materials (building sizes, wood in the construction); (2) per-capita consumption including buildings, parking areas, and private yards divided by the number of residents; and (3) personal consumption such as food, automobile use, and utilities (home energy use and water consumption). The findings of their studies show that denser designs can reduce CFs of residents.

They emphasized the link between design and behavior rather than built form itself because consumption contributed most to the overall footprint. In that context, CF or EF could "play a useful role in conducting such assessments by documenting some of the behaviors that are most crucial to a person's total environmental impact and how they are related to design and built form" (Moos et al. 2006: 219).

However, although the impact of high-density development on reducing demand for car travel has been commonly accepted, the concept had been deconstructed and is in opposition to some research (Williams et al. 2000).

On the predictor side of view, some researchers expected that travel behavior may be more influenced by socioeconomic and attitudinal characteristics of people (Stead et al. 2000), travel preferences (Boarnet and Crane 2001), engine technology, taxes on gasoline and driving, and road pricing (Gordon and Richardson 1989; Boarnet and Crane 2001) rather than urban form (Holden and Norland 2005). Headicar (2000) claims that the location of development in relation to employment opportunities is more important than concentration itself. On the other side, Masnavi (2000) found that nonwork trip, as a dependent variable, is more influenced by compactness than commuting trip.

There are other doubts that proximity to everyday services and workplace might not reduce travel in highly mobile society (Owens 1992) and long leisure-time travel should be considered (Holden and Norland 2005). Simmonds and Coombe (2000) pointed out that proximity to a desired facility has only a weak effect on travel choices so that compact land-use pattern shows limited impact. Santamouris (2006: 9) argued that it is very difficult for land-use patterns to guide energy-efficient planning because of the inevitability of nonspatial variables: necessity of travel, choice of jobs and services, and the extent to which cars are used.

Regarding the determinants of home energy use, which is one of the major components of CF, it is much more difficult to demonstrate a clear empirical relationship between built form and heating requirements because of the number of variables involved. Studies based on a large variety of sources (including published statistics, records of public utilities, household surveys, and, in some cases, individual metering of dwellings) all show enormous variation in domestic energy requirements. Apart from the obvious influence of climate, other important determinants of household energy use are socioeconomic factors, especially income, as are behavioral variables, such as temperature preferences, daily life patterns, and even window opening habits (Owens 1986: 42). Clearly, space standards between different types of dwelling account for a large part of the variation. The findings from extant literatures on spatial-energy relationships are summarized in Table 8.1.

From an ecological perspective, sustainability can only be achieved when cities are approached as systems and components of nested systems in ecological balance with each other (Marcotullio et al. 2004). The sustainable development should be within ecological capacity since the ecology of

TABLE 8.1

Findings from Extant Literatures on Spatial-Energy Relationships

Variables	Findings	Literature
Physical separation: low physical separation of activities	Energy needs ↓	Owens (1990)
	Travel energy ↓	Kim (1987)
Density: higher density (densely populated design)	Public transport feasibility ↑	Owens (1990)
	Travel energy ↓	Ahn (1998)
	Energy efficiency ↑	Holden and Norland (2005)
	Composite, goods and services footprint ↓	Ryu (2005)
	Quality of life (QOL) ↓	Holden (2004) Moles et al. (2008)
Mixed land use (job/housing ratio)	District heating ↑	Owens (1986)
	Public transport feasibility ↑	
	Energy efficiency ↑	Duvarci and Kutluca (2008)
Age (built year): newer house	Energy efficiency ↑	Holden and Norland (2005)
Volume: a low surface area to volume ratio	Energy needs for heating ↓	Owens (1986) Owens (1990)
Orientation: building direction	Viable use of solar energy ↑	Owens (1986) Owens (1990)
Far from malls	Heating and cooling ↓	Owens (1986)
	Food, travel footprint ↓	Ryu (2005)
Income: low income	Composite, goods and services, travel footprint ↓	Ryu (2005)
Education: less educated	Composite, goods and services, housing footprint ↓	Ryu (2005)
Married	Composite, goods and services, housing footprint ↓	Ryu (2005)
Age: higher proportion of young people	Composite, goods and services, housing footprint ↓	Ryu (2005)
	Travel footprint ↑	Ryu (2005)
	Travel energy use ↑	Banister (1992)
Environmental awareness	Food footprint ↓	Ryu (2005)
Size: small housing size (heated ground floor area)	EF ↓	Moos et al. (2006)
Draught proofed, hot water tank, wall/loft insulation, double glazing, new boiler	Energy efficiency ↑	Jones et al. (2007)
Number of household member: more members	EF per household member ↓	Holden (2004)
Car: low car occupancy (ownership)	EF ↓	Holden (2004)
	Travel energy use ↓	Banister (1992)

a place refers to the pattern and balance of relationships between plants, animals, people, and the environment in that place. Thus, it is important to find an analysis method which is complemented with ecosystem planning approaches.

EF, which is "the land (and water) area that would be required to support a defined human population and material standard indefinitely" (Wackernagel and Rees 1996), provides a means of comparing overall and component footprints, including CF, for transport, energy, waste, water, and food flows by normalizing the area of productive land to the world average expressed in global hectares (Gha). EF measures how much of the biosphere's annual regenerative capacity is required to renew the natural resources used by a defined population in a given year (Venetoulis and Talberth 2008). Conceived in 1990 by Mathis Wackernagel and William Rees at the University of British Columbia, EF is now in wide use by various fields for advance sustainable development (GFN 2009).

Although CF and EF can be a useful tool in planning practice, there are some criticisms on calculating them. One inherent weakness of calculating footprint is that it is intended to measure impact rather than to get cause and effect (Moos et al. 2006: 218). It is acknowledged that footprinting provides a conservative estimate of overall environmental impact (Loh and Wackernagel 2004) and fails to account for social and economic dimensions of sustainable development (Van Vuuren and Smeets 2000).

Nevertheless, CF illustrates impact in a way that cannot be achieved by site-focused methods. CF could be used to make more visible decisions by estimating the environmental potential of communities. Assessing the effects of both built form and consumption encourages planners and designers to begin thinking about built form from an environmental consumption point of view. In addition, it may offer a tool for policy makers to approve proposed developments (Moos et al. 2006).

To minimize the CF of cities, changes to city form and life pattern of people are required at a local and bioregional scale. Because it is shaped by the populations of urban area, density of development is an important determinant in resource use, particularly transport energy (Newman and Jennings 2008).

The main source of data for estimating household energy use is a footprint quiz survey. To calculate an individual footprint, the EF quiz was first developed in 2002 with Earth Day Network by the nongovernmental organization, Redefining Progress. The quiz provides a simple way for people to measure their impact on the earth. The general methodology for the per capita figures is described in Venetoulis and Talberth (2008).

For example, Moles et al. (2008) employed EF as the most appropriate metric for comparing the sustainability of settlements. They calculated settlement footprints for each component (transport, waste, water, and food) and subsequently aggregated all components. Then, each per capita component footprint for each settlement was normalized to the score for the settlement with the lowest per capita footprint so that all flows were weighted equally within a dimensionless scale (Moles 2008).

The CF quiz usually estimates the amount of carbon dioxide required to support individual consumption of food, goods, services, housing, and energy and assimilate wastes, whereas the EF quiz estimates the area of land and ocean.

Now there are a number of online carbon/EF calculators in use provided by many organizations, including Center for Sustainable Economy (CSE) [www.myfootprint.org], Global Footprint Network (GFN) [www.footprintnetwork.org], World Wildlife Fund (WWF) [http://footprint.wwf.org.uk], Carbon Footprint Ltd. [http://www.carbonfootprint.com/calculator.aspx], and the Nature Conservancy [http://www.nature.org/greenliving/carboncalculator/]. Those individual/household footprint calculators commonly apply life-cycle data of individual consumption items and categorize the personal consumptions into food, housing (shelter), transportation (mobility), goods, and services. The personal footprint calculator consists of questions that increase or decrease different parts of those consumption categories relative to national footprint per person. More detailed explanations of personal footprint calculation method are available at Global Footprint Network homepage [http://www.footprintnetwork.org/en/index.php/GFN/page/methodology/].

Table 8.2 shows the EF, CF, and biocapacity of the South Korea, the United States, and the World, respectively, as of 2007.

The following section shows how these theoretical and methodological backgrounds of CF can be applied in urban planning and geography studies. As it was discussed above, considering geospatial units and boundaries are important for urban planners and geographers to analyze the effects on CF because it may vary a lot according to spatial units and/or boundaries especially at a subnational/municipal level, such as counties, household, and individuals. The following two case studies are examples of applying CF surveys to calculate household and individual CF, respectively, in urban planning. The third case study is an example of CF calculations, or calculating CO_2 emissions, from agricultural activities and a county-level spatiotemporal data analysis in geography.

TABLE 8.2
EF and Biocapacity in 2007

Population (million)[a]		South Korea	United States	World
		48	308.7	6671.6
EF (global hectares per capita)	EF of consumption	4.9	8.0	2.7
	Cropland footprint	0.75	1.08	0.59
	Grazing footprint	0.08	0.14	0.21
	Forest footprint[b]	0.26	1.03	0.29
	Fishing ground footprint	0.54	0.10	0.11
	CF[c]	3.17	5.57	1.44
	Built-up land[d]	0.07	0.07	0.06
Biocapacity (global hectares per capita)	Total biocapacity[e]	0.3	3.9	1.8
	Cropland	0.17	1.58	0.59
	Grazing land	0.00	0.26	0.23
	Forest	0.09	1.55	0.74
	Fishing ground	0.00	0.41	0.16
	Built land	0.07	0.07	0.06
	Ecological (deficit) or reserve	(4.5)	(4.1)	(0.9)

Source: All data from GFN, National Footprint Accounts, 2010 edition: October 13, 2010. The Ecological Footprint Atlas 2010, http://www.footprintnetwork.org/en/index.php/GFN/page/ecological_footprint_atlas_2010.
[a] Population data from the UN FAO.
[b] Forest footprint includes fuel wood.
[c] CF of a nations' consumption includes direct carbon dioxide emissions from fossil-fuel combustion, as well as indirect emissions for products manufactured abroad. It also includes carbon dioxide emissions associated with extraction of these fossil fuels, such as flaring of gas. Other consumption-related carbon dioxide emissions included in the accounts only of the global total are from cement production and tropical forest fires.
[d] Built-up land includes areas dammed for hydro power.
[e] Biocapacity includes built-up land (see row after EF).

8.3   CASE STUDIES OF CF APPLICATIONS

8.3.1   HOUSEHOLD CF (KIM ET AL. 2010)

The relationship between urban spatial structures and energy demands came to be considered as an important subject after the energy crisis of the 1970s. Rapid urbanization in metropolitan areas has led to local urban climate change (Coutts et al. 2008) and concentrations of energy demands for lighting, heating, cooling, and transportation of households (Santamouris 2006). Furthermore, increasing demand from India, China, and other developing countries that are embracing a Western-style car culture unbalances energy supply and demand (Ewing 2008).

Currently, the dominant energy resources for housing and travel in urban areas are not renewable and their uses contribute to the greenhouse effect by carbon dioxide (CO_2) emission. For this reason, the subject has acquired a new urgency in the context of climate change as a worldwide problem. The most commonly used approach in planning research is to enhance the energy efficiency of physical structures and energy supply systems in urban areas. The problem solving from the side of energy demand is important in urban areas because "it is people who use energy and who live in 'spatial structures'" (Owens 1986). As Owens (1992) argued, spatial factors such as density and the location and mixing of different land uses need to be taken into account at an early stage in the land-use planning process for a sustainable development. Knowledge about the spatial-energy relationship can be useful for planners to develop a climate change impact mitigation planning.

CF of residents has been closely linked to the development of local towns and the expansion of urban areas. Kim et al. (2010) investigated the relationship between the spatial features and household energy consumption patterns at the local and neighborhood level in a metropolitan area. The authors measured the CF of residents in eight cities on the outskirts of Seoul, Korea. They examined the path analysis model illustrating spatial–behavior relationships on the basis of a household survey and spatial data. The data from a survey estimating the CFs of 1382 households in Gyeonggi-do (the metropolitan area of Korea) were used to calculate the CF of each household. For collecting CF data for Gyeonggi-do, 2064 households were surveyed by the Korean Institute Center for Sustainable Development and Green Gyeonggi 21 in August 2008. Eighty-one percent (177) out of 218 neighborhoods in 8 cities were represented by at least one respondent.

Housing and travel energy uses are often selected as proxy measures of CF. Wiedmann and Minx (2007) proposed the definition of the term "carbon footprint" as "a measure of the exclusive total amount of carbon dioxide emissions that is directly and indirectly caused by an activity or is accumulated over the life stages of a product." Although CF should account for indirect carbon dioxide emission embedded in goods and services, it is hard to discriminate the influence of spatial features on consumption of goods and service from other factors.

The CF of this study includes only direct carbon dioxide emissions from the use of fossil fuel. The household EFs were calculated by using the weights of "Korean Institute Center for Sustainable Development (KICSD)" based on the national statistics of Korea in 2006. The carbon components of the household EF per year converted into tons of carbon dioxide (tCO_2) equivalent with conversion factor by multiplying carbon intensity of housing and travel footprint, respectively. The measures of variables are detailed in Table 8.3.

Housing CF (HCF) was calculated from the data of monthly energy use of electricity, natural gas (LPG, LNG), and oil on July and August in units of tCO_2. Travel CF (TCF) was calculated based on the number of kilometers for travel by private car, public transportation, and airplane in same unit (tCO_2). Total CF represents the sum of HCF and TCF.

TABLE 8.3
Measures of Variables

Variables	Measures and Units	Source
Total CF	Sum of housing and TCF, tCO_2	CF survey (KICSD)
HCF	Monthly energy use of electricity, natural gas (LPG, LNG), and oil on July and August, tCO_2	
TCF	The number of kilometers for travel by private car, public transportation, and airplane, tCO_2	
Income	Monthly income of household (7 levels)	
Apartment	1 = apartment, 0 = others (detached, raw house)	
Density	Populations in each village, people/km²	SGIS
Land-use mix in urban area (entropy)	Entropy index of each village, 0–1 $$\text{Entropy} = -\sum_{i=1}^{5}(P_i)\ln(P_i)/\ln(s)$$ P_i = ratio of land-use type i to urban area (P_1 = residential, P_2 = industrial, P_3 = commercial, P_4 = entertainment, P_5 = infrastructures) s = 5, the number of land-use types in urban area	EGIS
Distance	Distance to the center of Seoul (CBD) from the center of each village, meters	ArcGIS

Source: Adapted from Kim T-K, Kim H-K, Han S-K. 2010. *Journal of Asian Regional Association for Home Economics* 17(2):51–56.

The Ministry of the Environment of Korea provides electronic maps of land-cover classification based on land-use maps, field studies, and remote-sensed satellite images from IRS-1C and Landsat ETM+ through the Environmental Geographic Information System (EGIS). Electronic maps of land-cover and land-use allowed the measurement of the spatial features of each city, incorporating density, location, and land-use mix.

A path analysis, a structural equation modeling with observed variables, was undertaken to test many relationships simultaneously and to allow the indirect effects of spatial features on HCF and TCF to be calculated. As a set of regression analysis models, it enables the modeling of the relationship among several exogenous and endogenous variables. The four predictors (apartment, distance, density, and entropy) and one control variable (income) were incorporated in the path analysis model. In the CF quiz analysis, the measure of HCF was very sensitive to the relationship with household income level. While household income level showed a linear correlation with the HCF of households, it drew a parabola curve with HCF per capita, which divides HCF by the number of family members. This implies that the number of family members act as a parameter affecting the relationship between income levels and HCF.

The results of the path analysis provide statistical evidence that housing type is related to both housing and travel energy use. The interesting point is that those effects occur in opposite directions. Although respondents living in apartments use less housing energy (−0.195), their travel footprint is larger than those who live in other housing types (detached and row housing) (0.106). The sum of indirect effects confirms a slight decrease in total CF (−0.014). This result serves as evidence for the energy efficiency of apartment housing, but the specific factors affecting travel footprint, such as location or accessibility to public transportation.

The positive effect of the distance to the city center on TCF was identified (0.06), indicating that increases in distance from the city center are indeed associated with increases in travel energy use.

On the other hand, it was defined that high density reduces housing (−0.07) and travel (−0.05) energy use. The important point is that the population density has a negative correlation with mixed land use (−0.58). This implies that the impact of high density on reducing carbon emissions would be reduced if the area were occupied by one major land-use type, especially residential areas.

Land-use mix in an urban area (entropy) was found to have a significant negative effect on TCF (−0.070). It is likely that more mixed land-use tends to reduce the energy use for travel. This result is consistent with the assumption that mixed land use reduces the frequency of car travel for commuting and shopping. These findings support the common assumption that spatial features are linked to household energy consumption patterns. However, it is still unclear which travel pattern contributed to TCF more than others. It might be caused by short travel distance, or more use of public transportation rather than private car, or less travel time by airplane.

The path analysis model suggests the indirect and net impact of different spatial patterns and income level on household CF of housing and travel energy use in a metropolitan area.

The results of the analysis support the assumption that spatial features have a relationship with household energy use for housing and travel. Evidence for this conclusion was provided by the statistically significant relationship between household CF and several spatial features: housing type, distance to the city center, density, and land-use mix. For the methodological improvement of CF study, it is important to note that per capita HCF calculation may be insufficient to measure or infer individual HCF due to unanticipated changes in the relationship with income levels.

The results of the analysis provide empirical support for several planning strategies: (1) the location of development (close to the city center or infill redevelopment); (2) the energy efficiency of apartments themselves (as opposed to detached or row housing); (3) mixed land use (new urbanism and urban neighborhoods); and (4) high density (compact development).

In the context of climate change, these four planning strategies above would be more appropriate than existing policy for building new towns on undeveloped green or agricultural land in the suburbs of metropolitan areas.

The important implication is that the concept of a CF has a potential role for reducing intrinsic energy demands of urban residents and awakening them to recognize the impact of their fossil energy use on global warming and climate change.

The findings of this study are limited in seasonal-specific housing energy use and to several cities in the Seoul metropolitan area. Further work on this topic is needed to consider the significance of urban design to EF within additional geographical regions and at a more specific level, i.e. building or individual. Furthermore, qualitative methods should be developed to determine the "real reason" for the influence of spatial features on household consumption patterns.

This chapter has presented an advanced methodology for measuring the spatial-energy relationship by path analysis. It suggests that land-use policies have significant consequences for future energy consumption in a developing country.

8.3.2 Individual CF (Kim and Kim 2013)

This study analyzed the effects of the net and gross population densities on the CF of residents at an intra urban scale in Seoul, South Korea. Path analysis was conducted to verify the effects based on the results of individual surveys of 500 sample households.

A CF survey was conducted in this study for collecting data. The survey method was based mainly on the household EF quiz of the Korean Institute Center for Sustainable Development (KICSD), which has adopted the standardized footprint calculation method of the Global Footprint Network (GFN 2009). As stated by Vegara (2000), a study that measures the individual EF is needed because the EF per capita could underestimate the actual EF of a highly populated area. Therefore, some weights were modified by the authors to measure the individual EF. For example, the answers to the questions requiring the size of a house and the monthly household energy use were divided by the number of household members.

The capital city of Korea, Seoul, was selected as a study area because it represents typical consumption patterns of urban dwellers and individual EF survey has never been conducted in Seoul.

Seoul covers an area of 605 km^2 and consists of 25 Gu and 497 Dong which had a population of 10,288,000 in 2008 (The City of Seoul 2008). The population density of Seoul (17,000 per km^2) makes it one of the largest cities in the world.

About 60% of land is composed of urbanized built area, including residential, industrial, commercial, recreational, transportation infrastructure, and public establishment. Residential areas (49%) and transportation infrastructure (20%) comprise most land use. Commercial and public establishment uses cover 14% and 13%, respectively. Industrial and recreational uses cover only 3% of total land.

According to the energy statistics of Korea in 2008, residential and commercial sectors accounted for 55% of total primary energy requirements of Seoul in 2007, largely for space and water heating (see Table 8.4). The energy consumption for residential and commercial sectors accounted for more than a half (55%) of total final energy consumption of Seoul in 2007. Adding transportation (30%), those comprised most of the energy consumption (KEMCO 2008).

The housing and travel footprints were calculated by converting the carbon intensities of home and travel fossil-fuel use of respondents to the unit of metric ton of carbon dioxide (tCO$_2$) by multiplying the conversion factors equivalent to the carbon intensities of each energy source, such as electricity, natural gas, and gasoline (Ewing et al. 2008). Before conducting the survey, some ambiguous expressions, which can be misinterpreted among respondents, were modified based on the results of a pilot test for the survey questionnaire. For example, many participants appealed the difficulties of responding amount of monthly home energy use in each unit of energy resource, such as "kW h (kilowatts)" for electricity, "m^3 (cubic meter)" for LNG, and "kg (kilogram)" for LPG. Thus, those questions were presented after an amount of money or price for charging each kind of energy use.

The questions for "individual" travel behavior were employed instead of using "per capita" calculation method, which divides household footprint by the number of household members, because it is hard to reflect personal travel patterns.

TABLE 8.4
Final Energy Consumption in Demand Sectors of Seoul, 2007

	Residential/ Commercial	Transportation	Industry	Public and Others	Total (%)
Coal	144	—	—	—	144 (1%)
Petroleum	635	4590	1271	151	6647 (42%)
City Gas (LNG)	4579	157	60	156	4951 (31%)
Electricity	3035	124	209	328	3696 (23%)
Heat energy (CHP)	432	—	—	7	439 (3%)
Renewable and other	4	—	11	117	132 (1%)
Total (%)	8829 (55%)	4870 (30%)	1551 (10%)	758 (5%)	16,008 (100%)

Source: Adapted from KEMCO. 2008. Yearbook of Energy Statistics. The Korea Energy Management Cooperation.
Note: Unit, 1000 ton; CHP, combined heat and power.

The results of housing and travel footprints of the respondents show that most CO_2 emissions come from the use of electricity in the household both in summer (79%) and in winter (59%). As the proportion of CF from oil and LPG uses in summer and winter are almost same (6% and 7%, respectively), it can be said that the CO_2 emissions from LNG used for heating take the place of those of electricity by doubling in winter, when compared with summer (34 vs. 15% of the total) because the use of electricity for cooling decreases and the demand for LNG for heating increases in winter. The CO_2 emissions of the respondents from electricity and LNG make up almost all the housing energy uses (93%–94%). The average CO_2 emission from the respondents' housing energy use was 1.648 and 1.987 tCO_2 in summer and winter, respectively (Table 8.5).

The CO_2 emissions of travel by private-car account for much more than half (67%) of the total travel footprint of the respondents. The average CO_2 emission from the travel of the respondents was 1.66 tCO_2 (Table 8.6).

The major purpose of travel by public transportation was for commuting, whereas it was leisure for travel by private car. Respondents tend to travel by public transportation to commute (49%) rather than for business (21%), leisure (18%), and shopping (12%). On the other hand, they are likely to travel by private car for leisure (42%) rather than for commuting (21%), shopping (19%), and business (18%).

TABLE 8.5
Components of Housing Footprint in Summer and Winter

	Summer		Winter	
Housing Footprint (tCO₂/year)	Mean (%)	S.D.	Mean (%)	S.D.
Electricity	0.922 (79%)	0.494	0.919 (59%)	0.502
LNG	0.170 (15%)	0.146	0.528 (34%)	0.426
Oil	0.021 (2%)	0.078	0.028 (2%)	0.086
LPG	0.053 (4%)	0.136	0.077 (5%)	0.226
Total	1.648	0.585	1.987	0.678

Note: LNG, liquefied natural gas; Oil, kerosene; LPG, liquefied petroleum gas.

TABLE 8.6

Components of Travel Footprint

Travel Footprint (tCO$_2$/year)	Mean (%)	S.D.
Public transportation	0.651 (33%)	0.553
Private car	1.314 (67%)	1.847
Total	1.66	1.348

The spatial patterns of housing and travel energy use of residents as the components of CF are illustrated on the map (see Figures 8.1 through 8.3).

The general G-statistics[*] and z-scores of footprint variables are given in Table 8.7. The result indicated that almost all footprint values are randomly distributed in study area except for the carbon dioxide emission from the travel by public transportation and overall CF. The spatial statistics

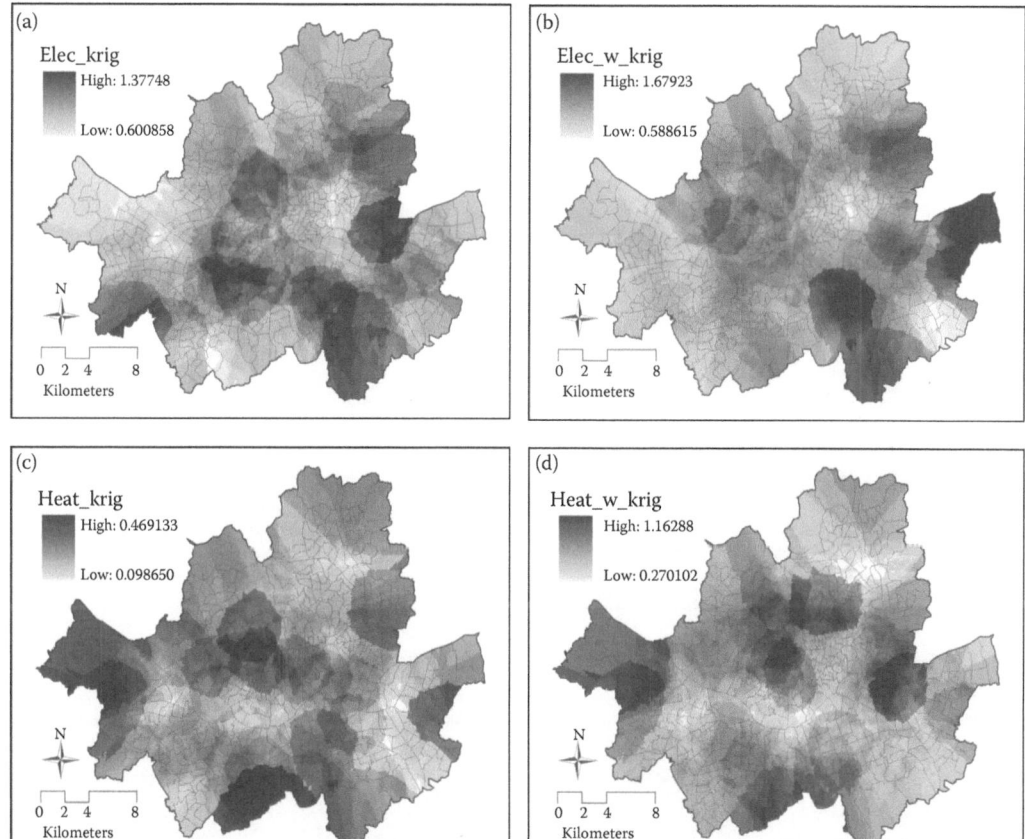

FIGURE 8.1 Spatial patterns of housing footprint components.

[*] The General G-statistic, developed by regional scientist Art Getis and statistician Keith Ord in the early 1990s, provides single statistic measuring concentrations of high or low values over the entire study area. Unlike Geary's c and Moran's I, which summarize the similarity of spatial pattern, the General G-statistic indicates whether high or low values are clustered (Mitchell, 2005: 108).

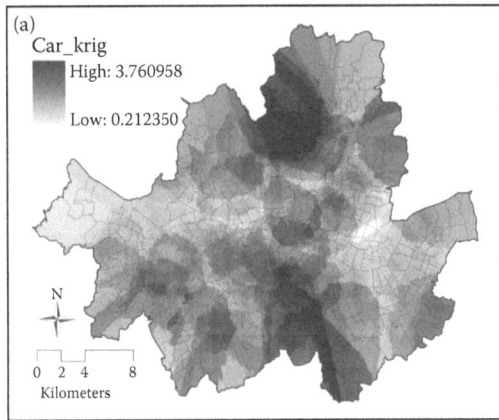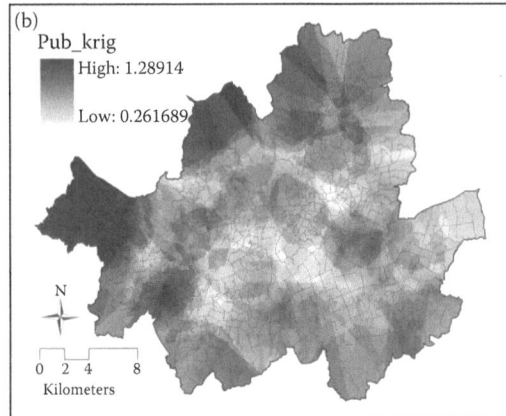

FIGURE 8.2 Spatial patterns of travel footprint components.

of public transportation footprint (Pub_tCO$_2$) and combined CF (HT_tCO$_2$) showed that low values are clustered within the study area at the significant level of 0.01 and 0.05, respectively.

Travel footprint (T1_tCO$_2$) and carbon dioxide emissions from the use of LNG, oil, and LPG for heating in summer (Heat_tCO$_2$) showed some clustering with low values, but the pattern may be due to random chance because those z-scores were not statistically significant at 0.05 level.

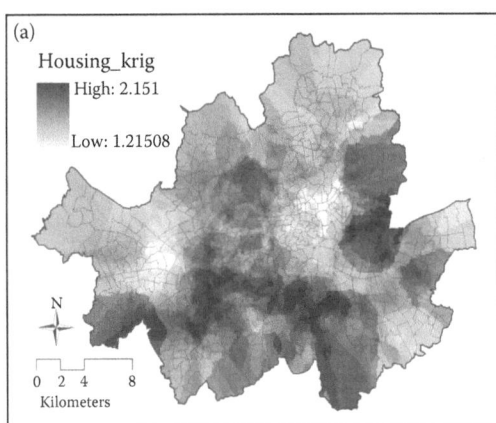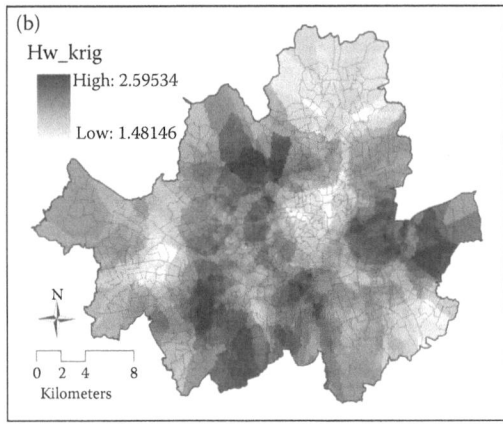

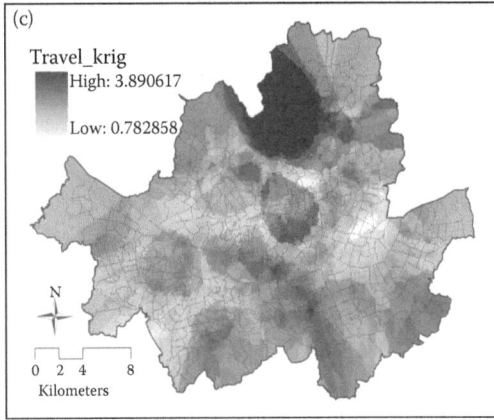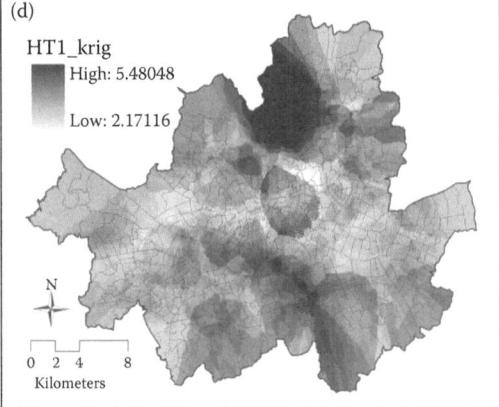

FIGURE 8.3 Spatial patterns of housing, travel, and combined footprint.

TABLE 8.7

The General G Statistics and *z*-Scores of Footprint Variables

	General G	*z*-Score	Type of Pattern
H_tCO$_2$	0.002011	−0.825314	Random
Hw_tCO$_2$	0.002012	−0.616312	Random
Elect_tCO$_2$	0.002018	0.280182	Random
Elect_W_tCO$_2$	0.002024	1.021448	Random
Heat_tCO$_2$	0.001990	−1.601222	Some low cluster
Heat_W_tCO$_2$	0.002014	−0.149206	Random
T1_tCO$_2$	0.001997	−1.626745	Some low cluster
Car_tCO$_2$	0.001999	−0.661760	Random
Pub_tCO$_2$	0.001977	−3.094411	Low cluster (99%)
HT_tCO$_2$	0.002000	−2.307725	Low cluster (95%)

The spatial patterns of housing footprint components illustrated in Figure 8.1 include carbon dioxide emissions from electricity, LNG, oil, and LPG uses in summer and winter. The carbon dioxide emissions from LNG, oil, and LPG uses were grouped as a factor of space heating at home.

The values of housing footprint components showed a tendency to increase in winter. Figure 8.1(a) and (b) shows that the areas with high values of electricity use have different patterns in summer and winter. Although electricity use was relatively high at the southwest side in summer (a), it was not in winter (b).

As shown in Figure 8.1(c) and (d), the north area illustrates relatively low heating energy use in winter and summer, where gross and net densities were relatively high in Figure 8.1(a) and (b). It may be explained by the possible connections between density and housing energy use for heating.

The spatial patterns of travel footprint components are illustrated in Figure 8.2, including carbon dioxide emissions from travel by private car and public transportation. As shown in Figure 8.2(a) and (b), the spatial patterns of private car and public transportation footprints were significantly different. The carbon dioxide emissions from travel by private car were relatively high at north and south, whereas those by public transportation were relatively high at west and northwest side.

The high value of public transportation at the west side can be associated with the proximity to the domestic airport of Gimpo. The area illustrated with dark shadow in Figure 8.2(a) could be an outlier as the spatial statistics showed the pattern is randomly distributed (see Table 8.7). If not, it may be related to the proximity or accessibility to public transportation facilities from the respondents who live in that area.

The spatial patterns of combined housing, travel, and CF are illustrated in Figure 8.3. The combined CF was illustrated by adding travel footprint to housing footprint in summer.

The spatial pattern of combined housing footprint is very similar with that of carbon dioxide emissions from the use of electricity in summer (a), because it is the major component of housing footprint. The housing footprint in winter is hard to explain by clustering values as a specific spatial pattern.

The combined CF in summer is reflecting the similar spatial pattern of travel footprint rather than housing footprint as the sum of tCO$_2$ from travel contributes more than that from housing.

The complete path diagram is illustrated in Figure 8.4. Two predictors (net and gross density), two mediators (job opportunity and service accessibility), and two response variables (housing and travel footprint) were incorporated into a path analysis. The labels, "S" and "W," of each path indicate "Summer" and "Winter," respectively, which were used for examining a moderating effect of season on those paths. For example, S_{NW} indicates a parameter of the path from net density to housing footprint.

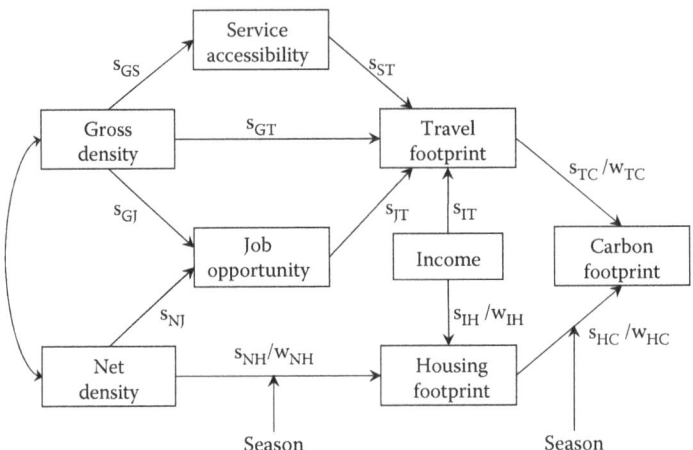

FIGURE 8.4 Path diagram.

The results of path analysis indicate that the net and gross population densities decreased the CF of the residents, whereas they have opposite impacts on job opportunities.

In summer, the results of the path analysis provide statistical evidence that the job opportunity is related to both net and gross density. The interesting point is that those effects have opposite directions. While more job opportunity was indicated where the people live in residential areas ($S_{NJ} = 0.220$, $p < 0.001$), the urban areas with high population showed less job opportunity ($S_{GJ} = -0.621$, $p < 0.001$).

The influence of gross density on accessibility to service facilities was not detected in this model. However, service accessibility affected the travel footprint although the size of the effect was not large ($S_{ST} = -0.093$, $p < 0.05$). As a result, the mediating effects of gross density on travel footprint were not confirmed.

The effect of the net density on reducing the housing footprint was not identified in summer (-0.038, $p = 0.394$), indicating that increases in population in residential areas are not associated with decreases in home energy use in summer. On the other hand, there was possibility that high gross density reduces travel energy use (-0.088, $p = 0.082$).

Individual incomes positively influenced the CF, whereas net and gross density, job opportunity, and service accessibility negatively influenced the CF.

These findings support the common assumption that spatial features are linked to household energy consumption patterns.

However, the mediating effects of service and job opportunity on the travel footprint were not identified in the model. Although it was anticipated that high gross density might increase service accessibility and job opportunity, neither of them related positively. Job opportunity increased with an increase in net density (0.220, $p < 0.001$) rather than gross density. Gross density was negatively linked to job opportunity (-0.621, $p < 0.001$). As service accessibility increased, travel footprint decreased (-0.093, $p = 0.032$), whereas job opportunity did not have significant influence on travel footprint (-0.054, $p = 0.285$).

Nevertheless, the link between gross density and travel footprint provides empirical evidence for spatial effect on travel behavior. The relationship indicates an increase in gross density might be associated with decreases in travel footprint (-0.088, $p = 0.082$).

Population density in urban area reduces travel footprint directly and highly correlated with land-use mix (job opportunity), but the mix of land use did not show the indirect effects reducing travel footprint. On the other hand, the reducing effect of accessibility to everyday service on travel footprint was verified while the connection with gross density was not significant.

Net population density decreased housing energy use only in winter, and housing energy use contributed to a larger CF for residents in winter than in summer.

The results of this study provide evidence for the energy efficiency of dense development in urban areas, but the specific factors affecting travel energy use, such as accessibility of public transportation, need to be identified in future related studies.

8.3.3 Agricultural CF (Kim and Dall'erba 2014)

Agriculture has an important role in the mitigation of greenhouse gases (GHGs) as a carbon sink, but it is also a source of GHGs. Indeed, significant amounts of the major GHGs—nitrous oxide (N_2O), methane (CH_4), and carbon dioxide (CO_2)—are released to and/or removed from the atmosphere because of agricultural activity (Paustian et al. 2004). According to Smith et al. (2008), agriculture accounts for 52% and 84% of global anthropogenic CH_4 and N_2O emissions, respectively. Although the net emission of CO_2 is small, it is mostly due to on-farm energy use and to the manufacture and transport of agricultural products (Snyder et al. 2009). Therefore, efforts to mitigate the role of agriculture in GHG emissions have a significant importance in the climate change policies.

While a significant amount of atmospheric carbon is fixed by the soil and crops in croplands (Hicke et al. 2004), it is respired back to the atmosphere as the harvested biomass is consumed by humans and domesticated animals (Ciais et al. 2007; West et al. 2009). Additionally, CO_2 emissions are generated in the production or manipulation of the inputs needed in the agricultural production process. Those inputs are fossil-fuel use for the operation of machinery, heating, crop drying, cultivation, planting, harvesting, irrigation, the manufacturing of fertilizers and pesticides as well as the transport and electricity used to carry and produce them (Paustian et al. 1998; West and Marland 2002). While terrestrial carbon stocks are both a sink and a source of carbon, fossil fuel used in the production, operation, and hauling of the above inputs is only a source of atmospheric CO_2 emission.

West et al. (2010) indicate that fossil fuel used in agricultural production released more CO_2 to the atmosphere than what was sequestered in soil in the United States for the year 2004. In addition, net CO_2 emissions, the difference between the source and the sinks, vary spatially. Indeed, net CO_2 emissions are governed by factors such as climate and soil characteristics (Polsky and Easterling 2001; Thornton et al. 2009), crop production practices, and land-use change (Paustian et al. 1998; Robertson et al. 2000; Smith et al. 2008; Nelson et al. 2009; West et al. 2010; Nalley et al. 2011). Geographical differences in the distribution of these variables influence on-going efforts to reduce emissions, enhance removals and avoid (or displace) emissions (Smith et al. 2008). As a result, even if it is well known that CO_2 emissions have a global climatic impact, understanding its local distribution and highlighting the nature and extent of its spatial dependence are still necessary steps to accurately estimate GHG mitigation potential and avoid misleading inference.

This article offers a spatiotemporal analysis of the distribution of CO_2 emissions, the main cause for GHGs, due to agricultural activities across U.S. counties. Based on a novel database, this study investigated how crop production output (measured in carbon) relates to CO_2 emitted in the production and transportation process of the inputs needed for crop production. The datasets of this study comes from the Carbon Dioxide Information Analysis Center (CDIAC). They offer county-level estimates of (1) total crop production (http://cdiac.ornl.gov/carbonmanagement/cropcarbon/) and (2) the associated amount of fossil-fuel CO_2 emissions across U.S. counties (http://cdiac.ornl.gov/carbonmanagement/cropfossilemissions/).

According to the author of the first dataset (West 2011), total crop production per county (in units of mega grams of carbon per year) has been calculated by integrating the amount of carbon in each type of crop within each county, following the statistical method described in Hicke et al. (2004), Hicke and Lobell (2004), and Prince et al. (2001). As Hicke et al. (2004) claim, this approach presents the advantage of allowing us to compare and aggregate production across different crop types under a common unit—carbon. It is important to note that this unit of measurement does not indicate pollution due to the production of crop. Pollution is measured in terms of carbon for CO_2

estimates in the denominator that captures the use, the fabrication, and transportation of inputs needed in the production of crops. The CO_2 emissions are allocated where the inputs (e.g., fertilizer), even when produced in a different county, are used.

The complete description of the CO_2 emissions measurement system for the second dataset is offered in Nelson et al. (2009). The authors include on-farm emissions (e.g., use of machinery during field operations) and off-farm emissions (e.g., use of electricity, energy, and emissions associated with the production and hauling of fertilizer and pesticide). The ratio of output to input that can be built as a result of the datasets above measures total crop production per CO_2 level of input. The greater it is for a county, the more biomass produced per fossil-fuel input.

Various spatial statistics were used to highlight the clusters of counties with similarities in the levels and growth of output per area, input per area, and productivity. At the same time, significant levels of heterogeneity are highlighted for all variables.

The authors performed exploratory spatial data analysis of the distribution of carbon embedded in produced crop (output), geogenic CO_2 emissions due to the use of fossil fuel for crop production (input), and the crop productivity measured as the ratio of the previous two elements (output/input). They also proposed a decomposition technique to identify the source of the regional differences in output–input levels and growth. The authors investigated temporal changes in the distribution of crop productivity by calculating the average over 1990–1992 (the initial period) and 2005–2007 (the final period).

All results for each of the two time periods and for the change between them are positive and significant at a 5% pseudo-significance level based on 999 permutations. Moran's I and APLE results indicate a significant presence of positive spatial autocorrelation: on average, neighboring counties tend to display similar levels of crop production, CO_2 input requirements, and ratios. General G indicates that counties with high-values cluster spatially. The results are consistent with the findings of Ciais et al. (2007) that the agricultural CO_2 flux distribution shows spatially coherent sinks of atmospheric CO_2 over croplands for cereals in the Midwest (or corn belt) where estimates of net primary production and harvested biomass are high relative to other crops (West et al. 2010).

The authors also note that the counties have become more similar to their neighbors in terms of output per area (increasing Moran's I and General G), whereas they have become less similar in terms of input per area and productivity level (decreasing Moran's I and General G). All statistics also report significant autocorrelation of the change over the overall period, but to an extent that is smaller than for the initial or final period.

The results of a decomposition method analysis uncover that the origin of interregional differences in productivity differs across the clusters of counties. The results for the entire sample indicate that, for each period, most of the variance in crop productivity comes from differences in output per area. Not only does its variance play a greater role than input per area, but so does its spatial lag. Among the covariances, the elements with the largest role are the covariance of output/area and input/area within county as well as the covariance of output/area and its spatial lag. As a result, increases in crop productivity are more sensitive to increases in output per area rather than in reduction in input per area.

Given the increasing interest in understanding the role of agriculture in GHGs, this study investigates the spatial distribution of carbon units per fossil-fuel CO_2 emission due to crop production across U.S. counties over 1990–2007. The dataset of this study provides us with details on the CO_2 emissions generated by the crop production process as well as the emissions due to the manufacturing and transportation of agricultural inputs, so that the authors can study the productivity levels (CO_2 due to outputs over CO_2 due to inputs) as well. The authors go beyond previous studies by relying on a set of spatial statistics to detect the presence of global and local spatial autocorrelations in the sample and for all time periods. While significant global spatial autocorrelation is found for all variables, the authors focus their measurement of local spatial autocorrelation on productivity and discover that it mostly takes the form of a large cluster of low productivity values (LL counties) in

Texas, along the Gulf of Mexico and the East coast (excluding Virginia and North Carolina). It could be due to soils and crops requiring more fertilizer and irrigation, hence leading to more fossil-fuel energy used, than in other states (Liska et al. 2009). They also find that the propensity of a county belonging to a cluster of low productivity to stay in its category is rather large (92%), which raises concerns since those are the counties wherein crop production requires more intensive use of fossil fuel than elsewhere.

A decomposition method was used to understand, for the sample as a whole and for each cluster (HH and LL), whether the spatial differences in crop productivity levels and changes are mostly due to differences in output per area, input per area, their respective spatial lag, or their covariance. It turns out that gains in crop productivity levels in the overall sample and, more especially, among the least productive regions are mostly attributed to gains in output per area as input per area increased less. In the absence of longitudinal data about crop types and management practices across counties, it was hypothesized that these gains are due to several factors such as "more effective fertilization and pest management, higher yielding cultivars, more favorable climate, shift to productive crop types" (Hicke et al. 2004: 18), moving set-aside lands to cropland production, changes in commodity prices, and government subsidies (Nelson et al. 2009) as well as agronomic innovation and improvements (Evans 1997).

Emissions due to the manufacturing and transportation of agricultural inputs generally follow the emissions due to on-site farming activities. However, their increase per acre is lower. In addition to the shift to less input-intensive crop types mentioned above, it was assumed that part of the difference is attributable to less polluting transportation means (reduction in the amount of pollution per miles driven), an increase in locally produced inputs (thus reducing the overall pollution; see, among others, Jones 2002; Milà i Canals et al. 2007) as well as production processes of inputs that have become less polluting.

The authors claimed that the spatial autocorrelation in the distribution and decomposition of GHG emissions highlighted in this paper should be considered when evaluating terrestrial ecosystem models (Hicke et al. 2004) and designing future agricultural and climate change policies. The latter could be delimited by the spatial boundaries of the clusters identified as they indicate the counties that belong to them present similar emissions characteristics. Clearly, administrative boundaries have no role on the processes at stake here. However, in order to clarify further the reasons and location for such clusters, future research should focus on the spatial and temporal differences in production practices (use of fertilizers, irrigation, machinery, pest management, etc.) and crop types because the ratio of fossil fuel emitted to biomass harvested varies with the latter two elements. For instance, cotton produces little biomass compared with corn, yet it is only 62% less energy-intensive (West and Marland 2002). This chapter did not explore these issues because of a problem of availability of continuous data over the period of interest.

Results of this study indicate that future mitigation policies should not fail to recognize inter-regional differences in the location, spatial extent, and origin of carbon emissions due to the crop production process.

8.4 CONCLUSIONS

The concepts and its applications of EF and CF were introduced in this chapter.

First of all, a survey of CF quiz results can be adopted to measure household and individual CF. For example, Kim et al. (2010) collected CF data from the survey measuring home energy use and travel behavior of 2064 households in 8 cities in Gyeonggi-do, Korea. Kim and Kim (2013) analyzed the effects of the net and gross population densities on CF of residents at an intraurban scale in the capital city of Seoul, Korea. In this study, individual CF of residents was calculated by converting the carbon intensities of home and travel fossil-fuel use of respondents to the unit of metric ton of carbon dioxide (tCO_2) by multiplying the conversion factors equivalent to the carbon intensities of each energy source, such as electricity, natural gas, and gasoline.

Second, Kim and Dall'erba (2014) analyzed the spatiotemporal distribution of CO_2 emissions from agricultural activities on cropland across U.S. counties. Based on a novel database, the authors investigate how crop production output (measured in carbon) relates to CO_2 emitted in the production and transportation process of the inputs needed for crop production by using various spatial statistics.

CF is an important notion for sustainable development as well as for climate change mitigation. As shown in the case studies, CF could be measured and applicable at variety of spatial units, including but not limited to, counties and neighborhoods.

Although only three case studies were introduced in this chapter, these examples may provide new perspectives on CF applications in planning and geography. Furthermore, additional concerns about consumptions of food, goods, and services (e.g., Kim et al. 2011) could be included in CF calculation in fields of urban planning and geography.

REFERENCES

Ahn KH. 1998. A study on the correlation between variables of urban form and energy consumption. *The Korea Spatial Planning Review* 27:19–30.

Alberti M. 1999. Urban patterns and environmental performance: What do we know? *Journal of Planning Education and Research* 19(2):151–163.

Banister D. 1992. Energy use, transport and settlement patterns. *Sustainable Development and Urban Form*. London: Pion, pp. 160–181.

Boarnet MG, Crane R. 2001. *Travel by design the influence of urban form on travel*. Cary: Oxford University Express.

Breheny MC. 1996. Centrists, decentrists and compromisers: Views on the future of urban form. In: Elizabeth Burton MJ, Williams K (eds), *The Compact City: A Sustainable Urban Form*. London: E&FN Spon.

Chambers N, Simmons C, Wackernagel M. 2000. *Sharing Nature's Interest: Ecological Footprints as an Indicator of Sustainability*. London: Earthscan/James & James.

Ciais P, Bousquet P, Freibauer A, Naegler T. 2007. Horizontal displacement of carbon associated with agriculture and its impacts on atmospheric CO_2. *Global Biogeochemical Cycles* 21 (GB2014).

Coutts AM, Beringer J, Tapper NJ. 2008. Investigating the climatic impact of urban planning strategies through the use of regional climate modelling: A case study for Melbourne, Australia. *International Journal of Climatology* 28(14):1943–1957.

Duvarci Y, Kutluca AK. 2008. Appraisal of energy efficiency of urban development plans: The fidelity concept on Izmir-Balcova case. *Paper Read at Proceedings of World Academy of Science: Engineering & Technology* 48:1155.

Evans LT. 1997. Adapting and improving crops: the endless task. *Philosophical Transactions of the Royal Society of London. Series B: Biological Sciences* 352(1356):901–906.

Ewing R. 2008. The impact of urban form on US residential energy use. *Housing Policy Debate* 19(1):1.

Ewing B, Reed A, Rizk SM, Galli A, Wackernagel M, Kitzes J. 2008. *Calculation Methodology for the National Footprint Accounts*, 2008 Edition. Oakland: Global Footprint Network.

GFN. 2009. *Ecological Footprint Standards 2009*. Oakland: Global Footprint Network.

Gordon P, Richardson HW. 1989. Gasoline consumption and cities: A reply. *Journal of the American Planning Association* 55(3):342–346.

Headicar P. 2000. The exploding city region: Should it, can it, be reversed? In: Williams K, Burton E, Jenks M (eds), *Achieving Sustainable Urban Form*. London: E&FN Spon.

Hicke JA, Lobell DB. 2004. Spatiotemporal patterns of cropland area and net primary production in the central United States estimated from USDA agricultural information. *Geophysical Research Letters* 31(20):L20502.

Hicke JA, Lobell DB, Asner GP. 2004. Cropland area and net primary production computed from 30 years of USDA agricultural harvest data. *Earth Interactions* 8(10):10.

Hoekstra AY, Chapagain AK. 2008. *Globalization of Water: Sharing the Planet's Freshwater Resources*. Malden: Blackwell.

Holden E. 2004. Ecological footprints and sustainable urban form. *Journal of Housing and the Built Environment* 19:91–109.

Holden E, Norland I. 2005. Three challenges for the compact city as a sustainable urban form: Household consumption of energy and transport in eight residential areas in the greater Oslo Region. *Urban Studies* 42:2145–2166.

Jones A. 2002. An environmental assessment of food supply chains: A case study on dessert apples. *Environmental Management* 30(4):560–576.

Jones P, Patterson J, Lannon S. 2007. Modelling the built environment at an urban scale—Energy and health impacts in relation to housing. *Landscape and Urban Planning* 83(1):39–49.

KEMCO. 2008. *Yearbook of Energy Statistics*. The Korea Energy Management Cooperation, Yongin City.

Kim KG. 1987. *Study of Energy and Spatial Structure—The case of Rural–Urban Regional Settlement*, Seoul, Seoul National University.

Kim T-K, Dall'erba S. 2014. Spatio-temporal association of fossil fuel CO_2 emissions from crop production across US counties. *Agriculture, Ecosystems & Environment* 183:69–77.

Kim T-K, Kim H-K. 2013. Analysis of the effects of intra-urban spatial structures on carbon footprint of residents in Seoul, Korea. *Habitat International* 38:192–198.

Kim T-K, Kim H-K, Han S-K. 2010. The relationship between spatial features and carbon footprints of households in Gyeonggi-Do, Korea. *Journal of Asian Regional Association for Home Economics* 17(2):51–56.

Kim T-K, Kim H-K, Han S-K. 2011. An analysis of the effects of spatial features of neighborhood on households' energy use for food, housing and travel—Focused on the ecological footprint survey in Gyeonggi-do. *Journal of the Korea Planners Association* 46(1):117–127.

Liska AJ, Yang HS, Bremer VR, Klopfenstein TJ, Walters DT, Erickson GE, Cassman KG. 2009. Improvements in life cycle energy efficiency and greenhouse gas emissions of corn-ethanol. *Journal of Industrial Ecology* 13(1):58–74.

Liu S, Dong Y, Wen M, Chen B. 2009. Quantify the landscape effect and environmental sustainability of rural region planning at town scale near metropolis. *Frontiers of Earth Science in China* 3(1):112–117.

Loh J, Wackernagel M. 2004. *Living Planet Report 2004*. Gland, Switzerland: WWF.

Marcotullio PJ, Boyle G, Ishii S, Karn SK, Suzuki K, Yusuf MA, Zandaryaa S. 2004. *Defining an ecosystem approach to urban management and policy development*. Edited by U. N. U. U. I. o. A. Studies (IAS).

Masnavi MR. 2000. The new millennium and the new urban paradigm: The compact city in practice. In: Williams K, Burton E, Jenks M (eds), *Achieving Sustainable Urban Form*. London and New York: E&FN Spon.

Milà i Canals L, Cowell SJ, Sim S, Basson L. 2007. Comparing domestic versus imported apples: A focus on energy use. *Environmental Science and Pollution Research* 14(5):338–344.

Mitchell A. 2005. *The ESRI Guide to GIS Analysis: Vol. 2. Spatial Measurement and Statistics*. Redlands, CA: ESRI Press.

Moles R, Foley W, Morrissey J, O'Regan B. 2008. Practical appraisal of sustainable development—Methodologies for sustainability measurement at settlement level. *Environmental Impact Assessment Review* 28(2–3):144–165.

Moos M, Whitfield J, Johnson LC, Andrey J. 2006. Does design matter? The ecological footprint as a planning tool at the local level. *Journal of Urban Design* 11(2):195–224.

Nalley L, Popp M, Fortin C. 2011. The impact of reducing green house gas emissions in crop agriculture: A spatial and production level analysis. *Agricultural and Resource Economics Review* 40:63–80.

Nelson RG, Hellwinckel CM, Brandt CC, West TO, De la Torre Ugarte DG, Marland G. 2009. Energy use and carbon dioxide emissions from cropland production in the United States, 1990–2004. *Journal of Environmental Quality* 38(2):418–425.

Newman P, Jennings I. 2008. *Cities as Sustainable Ecosystems: Principles and Practices*. Washington DC, Island Pr.

Owens SE. 1986. *Energy, Planning and Urban Form*. London: Pion.

Owens SE. 1990. Land use planning for energy efficiency. In: *Energy, Land & Public Policy—Energy & Environmental Policy*. New Brunswick, Transaction Publishers.

Owens SE. 1992. Energy, environmental sustainability and land-use planning. In: *Sustainable Development and Urban Form*. London, Pion.

Paustian K, Babcock B, Hatfield J, Lal R, McCarl BA, McLaughlin S, Mosier A, Rice C, Robertson GP, Rosenberg NJ. 2004. Agricultural mitigation of greenhouse gases: Science and policy options. CAST (Council on Agricultural Science and Technology) Report.

Paustian K, Cole CV, Sauerbeck D, Sampson N. 1998. CO_2 mitigation by agriculture: An overview. *Climatic Change* 40(1):135–162.

Polsky C, Easterling WE. 2001. Adaptation to climate variability and change in the US Great Plains: A multi-scale analysis of Ricardian climate sensitivities. *Agriculture, Ecosystems & Environment* 85(1–3):133–144.

Prince SD, Haskett J, Steininger M, Strand H, Wright R. 2001. Net primary production of US Midwest croplands from agricultural harvest yield data. *Ecological Applications* 11(4):1194–1205.

Robertson GP, Paul EA, Harwood RR. 2000. Greenhouse gases in intensive agriculture: Contributions of individual gases to the radiative forcing of the atmosphere. *Science* 289(5486):1922.

Ryu HC. 2005. Modeling the per capita ecological footprint for Dallas County, Texas: Examining demographic environmental value, land-use, and spatial influences. English, Texas A & M University.

Santamouris M. 2006. *Environmental Design of Urban Buildings: An Integrated Approach*. London, Earthscan.

Simmonds D, Coombe D. 2000. The transport implications of alternative urban forms. In: Williams K, Burton E, Jenks M (eds), *Achieving Sustainable Urban Form*. London and New York: E&FN Spon.

Smith P, Martino D, Cai Z, Gwary D, Janzen H, Kumar P, McCarl B et al. 2008. Greenhouse gas mitigation in agriculture. *Philosophical Transactions of the Royal Society B: Biological Sciences* 363(1492):789–813.

Snyder CS, Bruulsema TW, Jensen TL, Fixen PE. 2009. Review of greenhouse gas emissions from crop production systems and fertilizer management effects. *Agriculture, Ecosystems & Environment* 133(3):247–266.

Stead D, Williams J, Titheridge H, Williams K, Burton E, Jenks M. 2000. Land use, transport and people: Identifying the connections. *Achieving Sustainable Urban Form* 174–186. In: Williams K, Burton E, Jenks M (eds), Achieving Sustainable Urban Form. London and New York: E&FN Spon.

The City of Seoul. 2008. *Yearbook of Seoul*. The City of Seoul.

Thornton PK, Jones PG, Alagarswamy G, Andresen J. 2009. Spatial variation of crop yield response to climate change in East Africa. *Global Environmental Change* 19(1):54–65.

United Nations Environment Programme (UNEP), Division of Technology, Industry and Economics (DTIE), International Environmental Technology Centre (IETC). 2002. *Melbourne Principles for Sustainable Cities*. http://www.unep.or.jp/ietc/focus/melbourneprinciples/english.pdf.

van den Bergh Jeroen CJM, Verbruggen H. 1999. Spatial sustainability, trade and indicators: An evaluation of the ecological footprint. *Ecological Economics* 29(1):61–72.

Van Vuuren DP, Smeets EMW. 2000. Ecological footprints of Benin, Bhutan, Costa Rica and the Netherlands. *Ecological Economics* 34(1):115–130.

Vegara JM. 2000. *Footprint Computation: Three Common Errors*. Barcelona: Institut dEstudis Metropolitns de Barcelona.

Venetoulis J, Talberth J. 2008. Refining the ecological footprint. *Environment, Development and Sustainability* 10(4):441–469.

Wackernagel M, Rees WE. 1996. *Our Ecological Footprint: Reducing Human Impact on the Earth*. Gabriola Island, New Society Publishers.

West TO. 2011. *County-level estimates for carbon distribution in U.S. croplands, 1990–2005*. Oak Ridge National Laboratory, November 14, 2011. http://cdiac.ornl.gov/carbonmanagement/cropcarbon/.

West TO, Brandt CC, Baskaran LM, Hellwinckel CM, Mueller R, Bernacchi CJ, Bandaru V, Yang B, Wilson BS, Marland G. 2010. Cropland carbon fluxes in the United States: Increasing geospatial resolution of inventory-based carbon accounting. *Ecological Applications* 20(4):1074–1086.

West TO, Marland G. 2002. A synthesis of carbon sequestration, carbon emissions, and net carbon flux in agriculture: Comparing tillage practices in the United States. *Agriculture, Ecosystems & Environment* 91(1–3):217–232.

West TO, Marland G, Singh N, Bhaduri BL, Roddy AB. 2009. The human carbon budget: An estimate of the spatial distribution of metabolic carbon consumption and release in the United States. *Biogeochemistry* 94(1):29–41.

Wiedmann T, Minx J. 2007. A definition of carbon footprint. ISA Research & Consulting. ISA Research Report 07-01.

Williams Katie, Burton E, Jenks M. 2000. *Achieving Sustainable Urban Form*. London: E&FN Spon.

Section II

Modeling Aspects of Carbon Footprint

9 Quantifying Spatial–Temporal Variability of Carbon Stocks and Fluxes in Urban Soils
From Local Monitoring to Regional Modeling

V.I. Vasenev, J.J. Stoorvogel, N.D. Ananyeva, K.V. Ivashchenko, D.A. Sarzhanov, A.S. Epikhina, I.I. Vasenev, and R. Valentini

CONTENTS

9.1 Introduction 186
9.2 Urban Soils: A Potential Carbon Stock or Source? 187
    9.2.1 Soil as a Key Player of Carbon Cycle 187
    9.2.2 Urban Soils: Formation, Functioning, Specifics, and Impacts on Carbon Cycle 188
    9.2.3 Carbon Stocks in Urban Soils... 189
    9.2.4 Carbon Fluxes from Urban Soils 190
9.3 Methodological Aspects to Quantify Spatial–Temporal Variability of Carbon Stocks
    and Fluxes in Urban Soils 191
    9.3.1 Carbon Stocks... 191
        9.3.1.1 Field Sampling of Urban Soils: Challenges and Constraints.................... 191
        9.3.1.2 Laboratory Analysis of Urban Soil Carbon 193
        9.3.1.3 Quantifying, Modeling, and Mapping Carbon Stocks in Urban Areas..... 195
    9.3.2 Carbon Fluxes... 197
        9.3.2.1 *In Situ* Measurement of Soil Carbon Efflux 197
        9.3.2.2 Laboratory Measurements of Carbon Fluxes 198
        9.3.2.3 Spatial Analysis and Modeling of Soil Carbon Fluxes............................. 198
9.4 Case Studies.. 199
    9.4.1 Introduction 199
    9.4.2 Spatial Variability of Urban Soils' Microbial Carbon and Respiration in
    Contrast Bioclimatic and Functional Zones of Moscow Region 200
        9.4.2.1 Introduction..200
        9.4.2.2 Materials and Methods 201
        9.4.2.3 Results... 201
        9.4.2.4 Discussions...203
    9.4.3 Spatial–Temporal Variability of *In Situ* Respiration in Urban Soils of
    South-Taiga and Forest-Steppe Vegetation Zones 204
        9.4.3.1 Introduction..204
        9.4.3.2 Materials and Methods 206
        9.4.3.3 Results...206
        9.4.3.4 Discussion 207

9.4.4   Spatial Modeling and Mapping Carbon Stocks and Microbial Respiration in
Highly Urbanized Moscow Region 209
      9.4.4.1   Introduction...209
      9.4.4.2   Materials and Methods 209
      9.4.4.3   Results..210
      9.4.4.4   Discussions..211
9.5   Conclusions...213
References...213

9.1   INTRODUCTION

Carbon turnover is one of the major biogeochemical cycles on our planet. The carbon cycle determines the performance of many principal environmental functions and ecosystem services, including soil fertility, biodiversity, and climate change (Kudeyarov et al., 2007; FAO, 2013; IPCC, 2013). The largest pool of actively cycling carbon in terrestrial ecosystems is found in the soil that contains about 1500–2000 Pg C in various forms (Swift, 2001; Janzen, 2004). Carbon sequestration is therefore a widely accepted soil function (Blum, 2005; MEA, 2005). Soils are major carbon sinks, but they also have the potential to emit carbon to the atmosphere through carbon dioxide (CO_2) and methane (CH_4) effluxes (Houghton, 2003; Chapin et al., 2009; Levy et al., 2012). The storage of soil carbon as well as effluxes are influenced by a wide range of different bioclimatic conditions and land use.

In comparison with quite well-known effects of deforestation, reforestation, and land abandonment on the carbon cycle, the influence of urbanization on soil carbon is still poorly understood. Global and regional studies often neglected the role of urban soils in the carbon cycle (Kaye et al., 2005; Lorenz and Lal, 2009; Vasenev et al., 2013a,b,c, 2014). Specific features and heterogeneity of urban soils make it difficult to study the urban soils using conventional methods of analysis. Given that globally urban areas cover 2–3% of the land and that urbanization continues to take place at a rapid pace (Saier, 2007; Picket et al., 2011), a better understanding of carbon stocks and fluxes in urban soils is needed.

Urban soils experience a range of anthropogenic effects. Soil sealing, contamination, and translocation are very direct impacts on the soil system (Lal and Lorenz, 2009; Picket at al., 2011), whereas changes in the soil-forming factors are indirect effects on the soil in the urbanized environment (Gerasimova et al., 2003). As a result of anthropogenic changes, urban soils have characteristics (such as a high spatial heterogeneity, time dynamics, and profile distributions) that often differ from those in natural and agricultural ecosystems (Vasenev et al., 2013a,b,c). Urban ecosystems are very complex and heterogeneous (Vrscaj et al., 2008) and the soils present a high spatial variability in soil organic carbon (SOC; Pouyat et al., 2006; Lorenz and Lal, 2009; Vasenev et al., 2014). Studies into the spatial heterogeneity of urban SOC are further complicated by the specific profile distribution, including a considerable carbon storage in the subsoil. The profile distributions can include buried horizons, preurban zonal soils, or the so-called cultural layers formed by long-term residential activity (Beyer et al., 2001; Lorenz and Kandler, 2005; Dolgikh and Alexandrovskiy, 2010). The variability of carbon fluxes from urban soils is even higher, considering temporal dynamics caused mainly by seasonal and diurnal trends of soil temperature and water regime. Some local studies report surpassing urban soil respiration compared with adjacent natural or agricultural areas (Kaye et al., 2005); however, the mechanisms behind this tendency are still poorly understood.

High heterogeneity of urban areas and lack of legacy data on carbon stocks and fluxes from urban soils limit our knowledge of carbon balance in urban ecosystems. The considerable carbon stocks in urban soils together with reports of high soil respiration leave the question whether the urban soil is a net source or sink of carbon unanswered (Grimmond et al., 2002; Pataki et al., 2006), resulting in an increasing uncertainty in regional and global carbon assessments. Therefore the analysis of the carbon balance in the urban environment requires specific attention. This chapter focuses on the key theoretical aspects of urban soil formation and functioning (Section 9.2), approaches for the analysis of the carbon balance in the urban environment (Section 9.3), and several case studies in which these approaches have been implemented (Section 9.4). The methodological part of the chapter includes a consistent set of techniques to sample, analyze, quantify, and model carbon stocks and fluxes in urban soils from local to regional scale. Both conventional and advanced approaches to study soil carbon stocks and fluxes in the urban environment are observed. Implementation of the proposed methodology is illustrated in the case studies. The case studies mainly focus on the highly urbanized Moscow region, where urbanization is still taking place at a rapid pace. The first case study, which addresses the influence of anthropogenic disturbance on carbon stocks and fluxes, is a comparative analysis of soil microbial activity parameters in various ecosystems. The second case study shows the contribution of bioclimatic conditions and land use to the spatial–temporal variability of carbon fluxes. A spatial analysis including mapping of carbon stocks and basal respiration (BR) is conducted in the third case study. The literature review, methodology synthesis, and case studies aim to provide a full picture of the important role urban soils play in the carbon cycle and to improve our understanding of carbon stocks and fluxes in urban soil and their temporal variability from the level of local monitoring to regional modeling.

9.2   URBAN SOILS: A POTENTIAL CARBON STOCK OR SOURCE?

9.2.1   Soil as a Key Player of Carbon Cycle

Soil plays a key role in the carbon balance and makes a major contribution to carbon stocks and fluxes. Carbon accumulates in soils in different forms with different chemical features and mobility. SOC is the major stock in terrestrial ecosystems, surpassing estimates of atmospheric and biotic pools by a factor 2 or 3 (Schlesinger, 1995; Lal et al., 1998; Swift, 2001). Soil inorganic carbon (SIC) is mainly stored as carbonates and contributes considerably to soil carbon stocks in arid climates (Eswaran et al., 1993; Orlov et al., 1998; Dobrovolsky and Urussevskaya, 2004). In addition to more conventional SOC and SIC, soil black carbon (BC), resulting from pyrogenic processes and anthropogenic effects, receives increasing attention (Schmidt and Noack, 2000). Compared with other forms of soil carbon, microbial carbon (C_{mic}) contributes minimally to the total stocks with 1%–5% of total soil carbon (Sparling, 1992; Dalal, 1998). However, C_{mic} is the most labile form and influences key processes behind carbon sequestration and emission (Ananyeva, 2003; Ananyeva et al., 2008; Kurganova, 2010). The accumulation of soil carbon depends on a range of soil chemical, physical, and biological processes that are usually related to the soil type and bioclimatic conditions. Anthropogenic influences can contribute to the buildup of carbon stocks, but can also be a potential risk of carbon emission from soils (Smagin, 2005). The CO_2 emission from soils to the atmosphere through biogeochemical and physical processes is traditionally referred to as soil respiration (Rs) (Kudeyarov, 1995; Smagin, 2005). Soil respiration is one of the predominant terrestrial CO_2 fluxes and responsible for about 80 Pg/year of carbon emission back to the atmosphere (Raich et al., 2002; Schulze, 2006). Soil respiration includes two major flux components: autotrophic respiration (AR) by root systems and root-associated organisms, and heterotrophic respiration (HR) by free-leaving microorganisms in the soil (Chapin et al., 2006; Gomes-Casanovas et al., 2012). Although, the relative contribution by HR and AR varies between ecosystems (Hanson et al., 2000), HR usually dominates and contributes to 60%–75% of total respiration. Both components of soil respiration are very variable in time and space (Trumbore, 2006). Soil respiration has been correlated to a range of abiotic (soil temperature and moisture) and biotic factors (vegetation type, plant productivity, microbiological community, and soil features) (Davidson and Janssens, 2006; Bahn et al., 2010; Kuzyakov and Gavrichkova, 2010; Gomes-Casanovas et al., 2012). Temporal (diurnal and seasonal) dynamics in soil respiration usually correspond with changes in climatic conditions (Rey et al., 2002; Kurganova et al., 2003;

Kudeyarov et al., 2007), whereas the spatial variability mainly comes from different biomes and land use (Hamilton et al., 2002; Cruvinel et al., 2011; Fromin et al., 2012).

The second largest soil carbon efflux comes from CH_4 emission, which results from a combination of the two contrary processes: anaerobic decomposition of soil organic matter by methanogenic microorganisms (Denman et al., 2007; Levy et al., 2012), and methane oxidation by methanotrophic bacteria (Dalal and Allen, 2008; Lai, 2009). Net CH_4 efflux depends on bioclimatic and management conditions, and soil can act both as methane source and sink. Globally, wetlands and agricultural lands contribute mostly to the CH_4 emissions (Le Mer and Roger, 2001; Smith et al., 2008), although a high spatial heterogeneity occurs.

Land use is widely assumed as the most important factor affecting the soil carbon balance and is incorporated in most of global and regional models of carbon stocks and fluxes (Post and Known, 2000; Komarov et al., 2003; Sitch, 2003). So far, research on the soil carbon balance focuses on natural and agricultural ecosystems (Guo and Gifford, 2002; IPCC, 2001, 2007, 2013), on the consequences of land-use change (e.g., reforestation, deforestation, agricultural intensification, and land abandonment), and on carbon stocks and fluxes (Vuichard et al., 2008; Galford et al., 2010; Kurganova et al., 2014). However, little is known about the effect of urbanization on carbon stocks and fluxes.

9.2.2 Urban Soils: Formation, Functioning, Specifics, and Impacts on Carbon Cycle

Urbanization alters stocks and fluxes of energy and matter and results in irreversible changes in vegetation and soil cover (Milesi et al., 2005; Scalenghe and Marsan, 2009; Vasenev, 2011). Urban soils remained under-investigated for a while. Conventional soil surveys often focused on land-use planning in rural areas. In urban planning, deeper layers are more likely considered while dealing with the foundation of construction, whereas topsoil urban soils lack attention. The initial lack of interest in the urban soils can be illustrated by the fact that for a long time they were almost ignored by soil survey methodologies and soil classifications. The soil survey manual (Soil Survey Division Staff, 1993) classifies the urban soils under the undifferentiated areas. For a long time, soil classification systems such as the soil taxonomy (Soil Survey Staff, 1999), World Reference Base (WRB) for Soil Resources (FAO, 1988, 1998), and the Russian Soil Classification System (Shishov et al., 2004) did not pay specific attention to urban soils. Only recently, the increasing importance of urban areas and growing interest in their soil resources resulted in enhancements of WRB allowing consideration and diagnostic of urban and industrial soils, given as a new reference group of Technosols (FAO, 2007; Rossiter, 2007). The new edition of Russian soil classification is also expected to include taxons referring to urban soil types and subtypes (Prokofyeva et al., 2011).

The conditions in which urban soils are formed and function are very specific. Anthropogenic soil-forming factors dominate in an urban environment, and urban soils experience anthropogenic pressure through, for example, the translocation of soil material for construction, greenery work, soil compaction, sealing, and pollution. Indirectly, urban soils are also affected by humans by alteration of traditional soil-forming factors: climate, relief, parent material, vegetation, and time (Stroganova et al., 1997; Dobrovolsky and Urussevskaya, 2004). Mean air temperature in densely urbanized areas is likely to be higher compared with natural conditions. This effect is often described in literature as the "urban heat island effect" (Oke, 1973) that is caused by the produced heat (transport and heating) and the higher albedo of sealed surfaces (increasing absorption of solar radiation) (Arnfield, 2003; Weng et al., 2004). The effect also results in an increase in the temperature regime of the urban soils (Smagin, 2005; Smith et al., 2012), which in turn can result in an increase in microbiological activity and an acceleration of soil organic matter mineralization. Moisture regimes of urban soils are also often influenced by soil sealing, ground water control, mixing and translocation of parent material, and the use of artificial materials while building construction. The changes in runoff and infiltration may lead to temporal saturation of the urban soils (Kurbatova et al., 2004; Picket at al., 2011). The saturation may lead to an increased emission of methane (Kaye et al., 2004).

The vegetative cover is severely altered or converted into artificial cover. A limited number of tree species sustainable to anthropogenic pressure is used in urban greenery. A considerable part of the open areas is occupied by green lawns (Milesi et al., 2005). Ornamental trees and shrubs, lawns, and flower-beds require specific soil conditions and therefore receive specific soil management such as fertilization and irrigation.

Direct anthropogenic impact on urban soils can lead to even more serious consequences on carbon stocks and fluxes. Sealed soils are likely to be excluded from the carbon cycle owing to excavation of the most fertile topsoil and blockage of major soil fluxes. However, there is evidence that soil sealing results in an increase in CO_2 emission from neighboring open spaces as a result of lateral transfer (Smagin, 2005; Scalendhe et al., 2009). Soil compaction is another usual outcome of soil sealing owing to recreation (Craul, 1992; Jim, 1998; Gregory et al., 2006). Soil respiration in compacted urban soils is considerably lower than in undisturbed ones mainly because of unfavorable conditions for roots and soil microorganisms. Contamination is probably the most studied anthropogenic influence on urban soils. However, the influence of soil contamination on carbon cycle remains uncertain. For instance, it can result in both an increase in soil respiration (Khan and Scullion, 2002; Blagodatskaya et al., 2006; Vasenev et al., 2013a,b,c) and a reduction (Castaldi et al., 2004; Barajas-Aceves, 2005; Nwashukwu and Pulford, 2011) depending on the contaminant type and its concentration.

9.2.3 Carbon Stocks in Urban Soils

Compared with natural and agricultural soils, carbon stocks of urban soils remain very poorly studied. Most of the global and regional carbon assessments ignored the contribution from urban soils (Jobbagy and Jackson, 2000; Lal, 2004; Stockmann et al., 2013). In contrast, some local studies in the United States, China, Germany, and Russia found similar or even higher SOC concentrations and stocks of urban soils compared with adjacent natural and agricultural soils (Kaye et al., 2005; Pouyat et al., 2006; Zircle et al., 2011; Vasenev et al., 2013a,b,c). High carbon stocks in urban soils are given two alternative interpretations: (1) providing favorable conditions for carbon sequestration and (2) adding artificial substrates with a high amount of carbon (mainly, turf, peat, and organic composts). The first concept is indirectly stimulated by the high net primary productivity of urban trees, shrubs, and grasses attributable to the dominating juvenile growing phase, high soil fertility, and prolonged vegetation period caused by the heat island effect (Jo and McPherson, 1995; Greg, 2003; Nowak, 2006). For instance, field observations of chronosequences and modeling using CENTURY showed a potential carbon sequestration of up to 20–30 t C/ha on golf courses in Denver (Qian et al., 2002; Bandaranayke et al., 2003). Contrarily, it can be argued that the urban SOC is mainly imported. The latter is supported by a chemical analysis of urban soil carbon showing significantly higher concentrations of aromatic hydrocarbons and BC and by artificial inclusions of plastic particles in urban soils (Beyer et al., 2001; Lorenz et al., 2006). Urban SOC, whether sequestered locally or imported, varies depending on anthropogenic pressure and functional use (Mesheryakov et al., 2005).

The spatial variability in soil carbon of urban areas is driven by a combination of bioclimatic and urban-specific factors (Vasenev et al., 2014). Urban soils exist and function in zonal climatic conditions; they keep some zonal features and even contain buried horizons of initial natural or agricultural soils. As a result, carbon stocks in urban soils differ between various bioclimatic regions (Stroganova et al., 1997; Prokofieva et al., 2004; Vasenev et al., 2013a,b,c). At the same time, urban cities are also very heterogeneous referring to different functional use, history of management, and level of anthropogenic disturbance. Combinations of different functional zones (industrial, residential, and recreational) and areas of historical development (historical core, central part, and suburbs) contribute to high heterogeneity of urban soil carbon. Different levels of soil sealing, pollution, and compaction accomplish a "patchy" spatial distribution. Given the variability of soil-forming factors, urban soil carbon stocks are likely to vary also at short distances with a similar pattern.

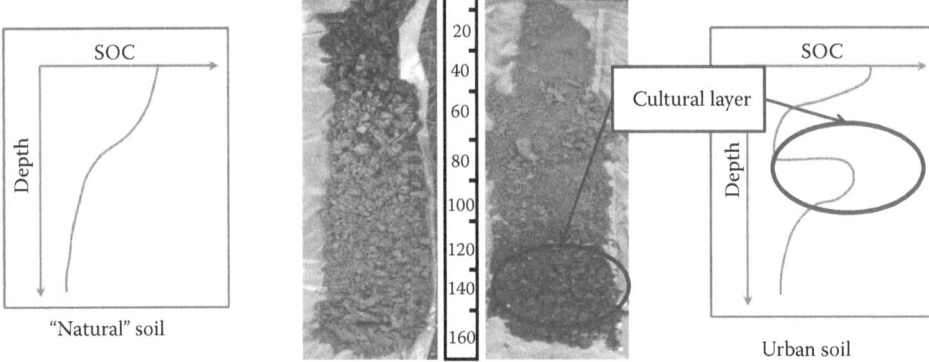

FIGURE 9.1 Sketch of SOC distribution in urban and natural soils. (Reprinted from *Catena*, 107, Vasenev, V.I., J.J. Stoorvogel, and I.I. Vasenev, Urban soil organic carbon and its spatial heterogeneity in comparison with natural and agricultural areas in the Moscow region, 96–102, Copyright 2013c, with permission from Elsevier.)

The analysis of spatial variability of urban soil carbon is further complicated by the very specific profile distribution (Wu et al., 1999; Beyer et al., 2001; Lorenz and Kandler, 2005). In contrast with natural soils in which SOC often decreases with depth, profiles of urban soils may display several SOC maximums (Gerasimova et al., 2003). The first one is connected with humus-accumulative horizon and the others correspond to the so-called cultural layer. The concept of this cultural layer originates in archaeological research, in which it was used to define the age of artifacts and describe the settlement history (Avdusin, 1980; Alexandrovskiy et al., 1998). A number of specific soil features are reported for cultural layers: high level of heavy metals' accumulation (Evdokimova, 1986) and soil microbiological communities, nontypical for topsoil (Marfenina et al., 2008). From the perspective of carbon stocks, cultural layers include wooden remains, coal, and buried nonurban horizons (Prokofieva and Stroganova, 2004) (Figure 9.1). As a result, SOC contents in the cultural layer may be as high as 3%–5% or even more (He et al., 2009; Dolgikh and Alexandrovskiy, 2010). The depth of the cultural layer depends on the age of the settlement and varies from 10 cm to several meters (Alexandrovskaya and Alexandrovskiy, 2000). Ignoring this cultural layer will lead to a considerable underestimation of the SOC stocks.

The spatial and profile variability of carbon stocks in urban areas requires a better analysis and needs to be incorporated in environmental assessment practices. Some studies report a considerable potential of urban soil to store carbon (Golubiewski, 2006; Poyat et al., 2006), which brings into question the widely assumed concept of urbanization as a threat to climate change (Kalnay and Cai, 2003; Grimm et al., 2008).

9.2.4 CARBON FLUXES FROM URBAN SOILS

Just as for stocks, carbon fluxes from urban soil are often ignored in regional and global assessments (Raich and Tufekcioglu, 2000; Hamilton et al., 2002; Houghton, 2003; Kudeyarov et al., 2007). A general view on urban soils as being artificial, polluted, and sealed is the main reason why carbon fluxes from urban soils are often neglected regionally and globally (Scalenghe and Marsan, 2009). This standpoint, however, is just a partial view of the overall picture. An actual percentage of the impervious areas varies in different parts of the city and functional zones and may differ from 7 to 90% with an average of 50% (Kurbanova et al., 2004; Vasenev et al., 2013a). Sport and ornamental green lawns occupy up to 40% of the nonsealed territory (Milesi et al., 2005). Moreover, for partly impervious areas with less than 90% of an area sealed, a decrease in carbon fluxes from impervious surfaces corresponds to a considerable rise in carbon emission from adjacent open spaces as a result of lateral migration of carbon fluxes beneath sealed surfaces (Smagin, 2005; Scalendhe et al., 2009).

Local studies conducted in Arizona and Baltimore, USA (Koerner and Klopatek, 2002; Kaye et al., 2005) and in Shanghai and Nanjing, China (Sun et al., 2009; Zhang el al., 2010) provide evidence for considerable urban soils' respiration, which was similar to or surpassing than that in natural areas. The drivers behind the urban soils' respiration are still poorly understood. Some authors explain this by higher SOC contents reported for urban soils (Pouyat et al., 2002; Zircle et al., 2011), including artificial organic substrates (compost and sewage materials) added during soil reclaiming and greenery work (Lorenz and Lal, 2009; Beesley, 2012). Others link higher urban soil respiration rates to urban heat island effect (Davidson, 1998) and alteration of soil physical properties (Chen et al., 2013).

Owing to the variation in disturbances, often alternative views can be found. Soil compaction likely causes a decrease in soil respiration owing to unfavorable conditions for microbiota and plant roots (Prokofieva et al., 2004). The effect of soil contamination remains uncertain. It can result in an increase but also a reduction in soil respiration. Soil pollution with gasoline and oil products likely causes rapid increases in respiration (Nwashukwu and Pulford, 2011; Vasenev et al., 2013a,b,c). The effect of soil contamination with heavy metals on soil respiration depends on the contaminant type and its concentration and vary from a slight increase to a marked decrease in soil respiration (Khan and Scullion, 2002; Castaldi et al., 2004; Barajas-Aceves, 2005; Blagodatskaya et al., 2006).

Soil carbon fluxes are very variable in time and space and largely depend on a range of soil abiotic and biotic parameters (Carlyle and Than, 1988; Gomes-Casanovas et al., 2012). The heterogeneous urban environment obviously causes a high spatial–temporal variability of urban soils' respiration. The spatial variability in the respiration likely depends on land-use structure (soil sealing, functional zoning, recreational load), soil chemical features (pH, SOM, and nutrient content), and management practices (irrigation, fertilization, green-keeping). Temporal dynamics in soil respiration strongly correlates with diurnal and seasonal changes in soil temperature and moisture regimes (Larionova et al., 2001; Rey et al., 2002; Kurganova et al., 2003).

9.3 METHODOLOGICAL ASPECTS TO QUANTIFY SPATIAL–TEMPORAL VARIABILITY OF CARBON STOCKS AND FLUXES IN URBAN SOILS

9.3.1 CARBON STOCKS

9.3.1.1 Field Sampling of Urban Soils: Challenges and Constraints

SOC is an important carbon stock. Therefore, the correct design of soil sampling campaigns is important. Urban conditions provide a number of specific features (Section 9.2) and contribute to a high spatial heterogeneity of urban soil. These specific features need to be considered when developing sampling protocols. Administrative and organizational reasons hamper the accessibility of urban areas for soil sampling. Soil sealing is one of the key urban factors influencing soil carbon stocks and fluxes and contributing to their spatial variability within a city. Sealed soils are referred to as Ekranozems or Ekranic Technosols (Rossiter, 2007; FAO, 2014). The contribution of Ekranic Technosols to carbon stocks and fluxes remains almost unknown. Sampling sealed soils is very labor consuming and it is difficult to get approval by local authorities. Very few studies on Ekranic Technosols report a high variability of carbon stocks from almost zero to considerable amounts depending on the surface construction type (buildings, roads, walking pathways, etc.) (Vasenev et al., 2013a). Carbon storage in Ekranic Technosols may become an important issue in investigation of urban soils in the coming future, although so far the absolute majority of research excludes sealed soils. Therefore, in the following text, we will refer to the unsealed soils while talking about the urban soils.

Functional zoning and history of the settlement are other important urban-specific factors that influence spatial variability of urban soils. Functional zoning is the stratification of the city territory into zones with contrasting, allowed, and regulated land management. The zones also differ in the level of anthropogenic pressure. Urban areas can typically be subdivided into industrial, residential, and recreational zones. Industrial areas refer to open spaces around factories and along highways. Residential zones include private and public courtyards. Recreational zones are parks and green zones within an urban infrastructure. Areas can also be subdivided into historical zones, which refer to different periods of city development and expansion. They include the historical core and other zones reflecting settlement boundaries in different historical periods. For example, five historical zones dating from the eleventh century up to now are distinguished in Moscow city (Kurbatova et al., 2004). Functional zoning puts more influence on the topsoil carbon through anthropogenic pressure and management practices, whereas historical zoning likely corresponds to differences in subsoil carbon owing to thickness and chemical features of the cultural layers. Both functional and historical zoning should be considered in sampling design, and relevant sampling scheme should include at least one plot from each of the functional zones within each historical zone (Vasenev et al., 2013c). Since internal spatial variability within functional zones can be high, a minimum of three to five sampling points are required at each location (Vasenev et al., 2013a). This stratified randomized sampling design is assumed as the most relevant to study urban soil carbon (Vasenev et al., 2014).

Some other limitations occur during the sampling itself. The traditional approach starts with a soil pit followed by a description based on soil horizons and collecting soil samples from the horizons. It is usually of a very limited use in cities. Digging a soil pit is traditionally accompanied with considerable disturbances of surface vegetation (including green lawns and ornamental species), which is hardly appropriate or allowed in a majority of urban areas. By locating the profile pits on the few allowed points, one deals with a convenience sample that lacks the basic advantages of the probability sample. Urban soils' profile usually contains a lot of artificial inclusions: wooden remains, bricks, and construction wastes, but also cables and pipelines. Digging soil pits (but also augering) can be limited for safety reasons. Finally, soil preparation of a soil pit is a very time- and labor-consuming approach to collecting soil samples and thus is not relevant for the studies requiring a considerable number of samples. For instance, at least 100–250 points, which can hardly be collected in an urban area for a reasonable time period by making soil pits, are needed in digital soil mapping (DSM) technique.

All these limitations make soil augering a more appropriate technique for sampling urban soils. Obviously, soil augering also has some constraints, such as mixed boundaries between horizons and slight compaction. These constraints are negligible for a spatial analysis of carbon stocks although they can be critical for some kinds of pedological analyses. Thus, the final decision on the sampling approach to be followed depends on the purpose and scale of the research.

Another issue to be decided is the sampling depth. Many global and regional analyses of soil carbon focus on the topsoil (Bui et al., 2009; Mishra et al., 2010; Martin et al., 2011), which is of limited relevance for urban soils, where a considerable amount of carbon is stored in the subsoil. Exact depth can depend on the settlement history. Older settlements likely possess thicker cultural layers and thus deeper sampling is needed (Vasenev et al., 2013a,c). In any case, at least 100–150 cm is a recommended depth to study carbon stocks in urban soils.

Field analysis of urban soils includes an *in situ* description of the soil profile and soil sampling. When describing soil profile, it is necessary to focus on the features required to classify urban soil and those used for further estimation of soil carbon stocks: depth of soil horizons, color, texture, structure, gravel content, soil inclusions (both natural and artificial ones), and different evidences of urbopedogenesis (Prokofyeva et al., 2011). Each sampling point is georeferenced for further spatial analysis and modeling. To deal with an extremely short distance, variability composite sampling is preferred in which various samples are taken from a, for example, 2 m², square plot (corners and center) and pooled into a single composite sample. Composite samples can be taken from genetic soil horizons or based on the depth layers. The first option is preferred to describe profile distribution of soil carbon; the second is more often used for a comparative analysis of carbon stocks. Soil microbiological features are very sensitive to depth factor; thus, samples for measuring soil microbial carbon are taken only on the layer basis. Collected soil samples are sent to the laboratory for further analysis.

9.3.1.2 Laboratory Analysis of Urban Soil Carbon

Preparation of the samples collected in field for laboratory analysis of soil chemical features usually includes air-drying, sieving, and pulverizing using a mortar (Vorobyova, 1998). However, these stages are excluded while preparing soil samples for microbiological analysis so as to avoid disturbance of soil microbiota (Creamer et al., 2014). A possible set of laboratory techniques to analyze urban soil carbon includes general approaches to quantifying total carbon content as well as more detailed measurements of the different fractions: SOC, water-soluble organic C (WSOC), hot-water-soluble organic C (HWSOC), readily oxidizable C (ROC), and microbial biomass C (MBC).

9.3.1.2.1 Chemical Analysis

Total soil carbon can be measured by direct and indirect approaches. Direct approaches are based on measuring CO_2 from soil combustion (dry combustion) or oxidation by chromic and sulfur acids (wet combustion) (Vorobyova, 1998; Fedorez and Medvedeva, 2009). Direct measurements of total soil carbon are usually carried out using CN Elementor analyzer (dry combustion) or by the Knop approach (wet combustion) (Vorobyova, 1998). Indirect approaches are based on quantification of oxygen, which is required for oxidation of soil organic matter. The examples of these indirect techniques include Turin's approach based on oxidation of soil organic matter with dichromate solution, and Turin's approach with Nikitin's modification based on dichromate oxidation with following spectrophotometric endpoint detection (Vorobyova, 1998).

The WSOC is measured by consistently adding distilled water, shaking in centrifuge to extract water-soluble phase, and further measuring of organic C concentrations in the extracts using an automated dry combustion (Liang et al, 1998). A fresh soil sample is placed into a 50-mL centrifuge tube, to which 40 mL of distilled water is added, and the sample is shaken at 120 rpm for 30 min at 25°C. After being centrifuged for 20 min at 4000g, soil samples are filtered through a 0.45-µm membrane filter. The organic C concentrations in the extracts are analyzed using an automated CN analyzer. The HWSOC is determined with a modified method of Sparling et al. (1998), including stage of water bath (80°C) for 16 h before shaking in centrifuge. Soil readily oxidizable C (ROC) is determined using the method of Graeme et al. (1995). A 2-g air-dried soil sample is placed into a 50-mL centrifuge tube, and 25 mL of 333 mmol L^{-1} $KMnO_4$ solution is added to the centrifuge tube. The centrifuge tube is tightly sealed, shaken for 1 h at 120 rpm, and centrifuged for 5 min at 5000g. The supernatant is diluted 250 times with deionized water. The absorbance of the diluted sample is read at 565 nm on a UV spectrophotometer. The $KMnO_4$ concentration in the diluted sample is determined from a standard curve. The difference between the amount of $KMnO_4$ added and remaining is used to calculate the amount of easily oxidized (labile) C concentration in the soil sample, assuming that 1 mmol $KMnO_4$ is consumed in the oxidation of 9 mg of C (Graeme et al., 1995). BC content is quantified using the chemical oxidation (CO) method applied by Lim and Cachier. In brief, the method consists of six steps: (1) weighing out 3.0 g of dried soil sample, sieved by 100-mesh sieve, (2) removal of carbonates (acid treatment with 3 mol/L HCl for 24 h), (3) removal of silicates (acid treatment with 15 mL of 10 mol/L HF: 1 mol/L HCl for 24 h), (4) removal of CaF_2 (acid treatment with 15 mL of 10 mol/L for 24 h), (5) removal of organic carbon (CO with 15 mL of 0.1 mol/L $K_2Cr_2O_7$: 2 mol/L H_2SO_4, 55°C for 60 h), and (6) defining the residual carbon as BC, quantified by dry combustion on CN analyzer.

9.3.1.2.2 Microbiological Analysis

Soil microbial biomass carbon is considered as the most dynamic and labile component of SOC. The pool of soil microbial carbon, its activity, and composition are the key parameters in soil processes, studied intensively in various ecological scenarios (Wardle 1992; Zvyaginzev 1994; Anderson and Domsch 2010). The methods to determine MBC are subdivided into two main groups: direct microscopic and indirect (biochemical, physiological) methods. Direct methods are used to determine quantity and contents of microorganisms, whereas indirect methods are implemented to estimate their activity, for example, respiratory activity after fumigation or adding substrate. The method of fluorescence microscopy makes it possible to determine the length of mycelium of fungi and actinomycetes, to quantify bacterial and fungal spores and to measure the diameter of mycelium and fungal spores in soil suspension on a glass mount stained with calcofluor white (for eukaryotes) or acridine orange (for prokaryotes) (Kozhevin 1989; Polyanskaya et al. 1995). However, implementing direct microscopy method to estimate soil microbial biomass is complicated by high subjectivity of the observer (Domsch et al. 1979; Stahl et al. 1995), and, therefore, the relevance and repeatability of the obtained results is questionable. Moreover, conversion of cells' biovolumes to their biomass units includes a few assumptions concerning the cells' density and water and carbon content (Ananyeva et al., 2008). In contrast, the application of the biochemical methods allows eliminating the influence of adsorption of cells by soil particles and their death due to dispersing of the soil. Moreover, the results of such determination of microbial biomass are less subjective than those obtained by direct microscopy methods. The limitations of biochemical methods for the determination of soil microbial biomass are related to the measurement of one of its compounds (ATP, DNA, muramic or diaminopimelic acid), as well as to the amount of easily mineralized (fumigation method) or functionally active (normal method) cellular components. Finally, the biochemical methods allow determination of different components of microbial cells and metabolic products and thus expand opportunities for a more accurate investigation of the soil microbial biomass and its activity.

Substrate-induced respiration (SIR) invented at the end of the 1970s by Anderson and Domsch (1978) is among the most widely used biochemical methods to quantify soil microbial biomass. It is based on the rate of initial maximal respiration of microorganisms after soil enrichment with an easily oxidized and universally available substrate—glucose. It is important that the incubation time of soil enrichment by glucose should not be more than six hours to exclude the reproduction of soil microorganisms (Mirchink and Panikov 1985). The heterotrophic microorganisms oxidize and co-oxidize the glucose for 3–5 h after its addition to the soil (Anderson and Domsch 1978; Ananyeva et al., 2008, 2011). The respiratory response for this time period (the initial maximal respiration is determined experimentally for each soil type) allows estimating a unit of soil microbial biomass (from μL CO_2 to μg C) by coefficient. This estimation is based on one of the main principles of microbiology, claiming that the respiratory response of microorganisms (the production of CO_2) is directly proportional to their biomass (Pirt, 1975). The SIR method supposes to follow the conditions first related to the amount of glucose applied to the soil, the temperature of the preincubation (before the application of glucose), and the incubation (after the application of glucose) of the soil samples. The glucose content providing the most active initial response of the soil microbial community in the form of CO_2 emission is determined experimentally for each soil type, which is from 1 to 20 mg glucose g^{-1} soil (Sparling and Williams, 1986; Priha and Smolander, 1994; Lin and Brookes, 1999a). The CO_2 production in the soils mainly resulting from the activity of heterotrophic microorganisms is known to be related to the hydrothermal conditions. As the carbon content in the soil microbial biomass is determined using the SIR method, the moisture of the soils should not be less than 60% of the total water holding capacity (optimal for microorganisms), and the incubation temperature should be 22°C (temperature of the upper soil horizons in summer in most of the European countries) (Anderson and Domsch, 1978).

The soil microbial biomass carbon can be analyzed by SIR measurements. Prior to the estimation of SIR, all soil samples (0.3–0.5 kg) are moistened up to 50%–60% water-holding capacity and preincubated in aerated bags at 22°C for 7 days to avoid an excess CO_2 production after mixing, sieving, and moistening of soil sample (Ciardi and Nannipieri, 1990; Ananyeva et al., 2008; Creamer et al., 2014). The SIR measurements are taken at least in triplicate and results are expressed per dry weight of soil (105°C, 8 h). To measure SIR, soil samples (2 g) are placed in a vial (15 mL volume) and a glucose solution is added dropwise (10 mg glucose g^{-1} soil, volume was 0.1 mL g-1 soil). The vial is kept tightly closed and the time course is recorded. The vial is incubated (3–5 h, 22°C) and an air sample is taken and injected into a gas chromatograph for measuring CO_2 production. SIR (µL CO_2 g^{-1} h^{-1}) is calculated according to

$$SIR = (CO_{2,soil} - CO_{2,standard}) \times V_{fl} \times 60 \times 1000/m \times \Delta T \times V_{as} \times 1000$$

where $CO_{2,soil}$ is the CO_2 content in the gaseous medium in the flask with the soil (vol%), $CO_{2,standard}$ the CO_2 content in the gas in the empty flask (without soil) (vol%), V_{fl} the volume of air in the flask with the soil (ml), m the mass of the dry soil (g), ΔT the time from the moment of closing the flask to take an air sample (min). and V_{as} the volume of the air sample introduced into the chromatograph (mL).

The microbial biomass carbon (C_{mic}) is estimated based on the SIR results, using the following formula:

$$C_{mic} = SIR \times 40.04 + 0.37 \text{ (Anderson and Domsch, 1978)}$$

There are different alternative views on the value of the coefficient to estimate microbial carbon from the respiration value. Possible values given in literature vary from 30 (Kaiser et al., 1992) to 50 (Sparling et al., 1990). However, the coefficient of 40.04 recommended by the original authors and confirmed by further research (West, 1986; Martens, 1987; Wardle and Parkinson, 1991) remains the most commonly used so far. The SIR method has the following advantages: sustainable quantitative determination of soil microbial biomass carbon, rapidity, high reproducibility, sensitivity, and lower labor costs compared with other methods (e.g., direct microscopy of soil suspension and fumigation-extraction method), which allows for a large number of measurements (Anderson and Domsch, 1978; Ananyeva et al., 2008). The method provides results with a relative error of less than 5% and takes into account only the active microbial biomass (Beare et al., 1990; Hassink, 1993; Wardle and Ghani, 1995). However, the SIR method also has some limitations. It is not appropriate for strong acidic (pH $\leq$ 2.5) and alkaline ($\geq$8.0) soils (Beck et al., 1997). Unlike practices in Russian studies determining the microbial biomass by direct microscopic counts of microbial cells (number of bacteria, spores of fungi, lengths of fungal and actinomycetal mycelia) with subsequent consideration of their biovolumes and density, the SIR method allows for determination of various additional physiological and metabolic parameters of the soil microbial community (Anderson, 1994).

9.3.1.3   Quantifying, Modeling, and Mapping Carbon Stocks in Urban Areas

Most of the carbon surveys focus on two kinds of carbon stocks: total amount of carbon stored in the area of interest and specific amount of carbon per unit of area. The specific carbon stocks in urban soils are estimated using the following formula:

$$SOC_{sp} = (SOC \times d \times K \times BD)/10$$

where SOC_{sp} is the specific SOC stock per area unit (kg/m^2), SOC the soil organic carbon content (%), d the depth of the layer (cm), BD the soil bulk density (g/cm^3), and K the correction coefficient for, for example, rock material and artificial inclusions.

Anthropogenic inclusions (bricks, concrete flags, and service tubes) are very likely in the urban environment. Depending on the sampling technique, they can just influence the bulk density of the soil sample (in the case of a soil pit) or even limit the sampling of some subsoil layers (when augering). The K coefficient allows correction for such inclusions and cut-off profiles and varies from 0 to 1 depending on the amount of artificial inclusion (Vasenev et al., 2014). Total SOC is calculated based on extrapolation of SOC_{sp} for a single unit of the study area (polygon or grid depending on the data type). In order to exclude sealed soils, a correction coefficient for the percentage of open (nonimpervious) area is introduced:

$$SOC_t = SOC_{sp} \times S \times P_{oa} \times 10^3$$

where SOC_t is the total soil organic carbon stock (Tg), P_{oa} the percentage of the open (nonimpervious) area, and S the area extent (km²).

High heterogeneity of urban soil results in considerable spatial variability of their carbon stocks. Capturing this variability requires mapping and spatial modeling approaches considering the specific condition of urban environment dealing with functional zoning, and the age and size of the settlements. Conventional soil surveys (Turin, 1959; Gavriluk, 1963; Fridland, 1972; Soil Survey Staff, 1999), presenting soil properties as qualitative data in the form of discrete maps, are not convenient for an urban environment characterized by short distance variability with many abrupt changes. In addition, traditional soil-surveying techniques are expensive as they require a large number of observations. In contrast, recently developed DSM (McBratney et al., 2000, 2003) aims to create continuous surfaces of soil properties based on a limited number of observations and quantitative, statistical relationships with auxiliary information. Following the classic Dokuchaev's concept of genetic soil science and soil forming factors, soil carbon stocks are described as a function of relief, parent material, climate, organisms (vegetation and land-use), and time (Kovda and Rozanov, 1988; Dobrovolsky and Urussevskaya, 2004). There are still some limitations for the assessment of soil carbon stocks by DSM in urban areas. Very high short-distance variability within the urban areas and long distances between settlements limit the use of traditional interpolation techniques. DSM typically uses a range of explanatory variables such as climate, topography, and land use (Dobrovolsky and Urussevskaya, 2004; Minasny et al., 2013) in a multiple regression approach (Kern, 1994; Ni, 2001; Wang et al., 2004). However, in the urban environment, more attention is needed for anthropogenic influence. Thus, additional anthropogenic factors—including soil sealing, functional zoning, size, and age of the settlements—are taken into account (Stroganova et al., 1997; Prokofieva and Stroganova., 2004; Lorenz and Lal, 2009).

Quite a few sources of auxiliary data are available. Climate conditions can be described by the 30 arc-second WorldClim database (Hijmans et al., 2005). Relief characteristics (elevation, slope) can be derived from the 90 m SRTM digital elevation model (Jarvis et al., 2008). The Land Cover Map for Northern Eurasia for 2000 (Bartalev et al., 2003) provides differences in land cover. The normalized difference vegetation index (NDVI) derived from a Landsat image for 2000 (30 m resolution) describes differences in vegetation. Soil and parent material data if possible can be obtained from existing exploratory regional soil and parent materials maps. Information on the urban-specific factors is more challenging to get. The size and age of the settlements can be derived from official statistical sources (for instance, RFSSS, 2012 for Russia). As a rule, spatially explicit information on functional zones is limited. As an option, approximate boundaries of the functional zones can be derived based on the NDVI, assuming that recreational, residential, and industrial functional zones differ in the vegetation and percentage of sealed areas (Vasenev et al., 2014).

DSM is based on the modeling of soil carbon stocks in new locations based on obtained statistical relations between existing data on SOC stocks for a limited number of locations and auxiliary data available for the whole area of interest. Normality of the distribution of SOC values and homogeneity of variances must be checked before starting the analysis. The correlation between explanatory variables should also be analyzed in advance in order to prevent multicollinearity. A statistical, multiple regression model, correlating SOC stocks to the explanatory variables, is used to predict carbon stocks in new locations. Auxiliary information includes both continuous and categorical data. Relationships between SOC stocks and secondary data can be analyzed using a general linear model (GLM). GLM is quite often used to predict SOC (Ingleby and Crowet, 2000; Meersmans et al., 2008). Some research compared GLM with generic geo-statistical models such as ordinary kriging and reported superior GLM performance (Kumar, 2013). Categorical variables are introduced as dummy variables (equal to 1 if a location is within a particular unit and 0 otherwise). Alternative GLMs are obtained by a backward, stepwise linear regression, keeping the statistically significant variables ($p < 0.05$). The predictive power of each statistical model is characterized by the coefficients of determination R^2 and R^2_{adj}. Statistical analysis is conducted using relevant software. DSM based on the GLM provides maps of soil carbon stocks with the level of details, depending on the resolution of the auxiliary data sources, as well as on the scale and specific goals of the research.

9.3.2 Carbon Fluxes

Urban soils can be a considerable source of carbon through CO_2 and CH_4 emissions (Kaye et al., 2004). Spatial variability and temporal dynamics of carbon fluxes from urban soils require approaches that consider the spatial–temporal distribution. The most common approach to determine soil carbon fluxes is based on direct field methods, where the effluxes from the soil surface are measured *in situ*, and indirect methods, where carbon fluxes are predicted based on auxiliary information or where R is measured under standardized conditions.

9.3.2.1 *In Situ* Measurement of Soil Carbon Efflux

Direct methods include conventional alkali absorption techniques (Buyanovsky et al., 1986) and a variety of chamber approaches (open-path, closed-path, and dynamic close chambers) (Nakadai et al., 1993; Bekku et al., 1997; Savage and Davidson, 2002). Currently used sampling and measuring procedures differ for CH_4 and CO_2 fluxes.

The CH_4 flux is typically measured by a chamber approach. At least 4 h before starting the measurements (to avoid disturbance), collars are mounted on the soil surface. Cylindrical chambers are placed on the top of collars and fixed by clips. The chamber is produced from chemically neutral material. A pressurized valve at the top of the chamber is used for gas sampling by a laboratory syringe. The chamber contains a temperature probe to control inside climatic conditions and a fan to mix the air inside. The carbon flux is derived from increasing or decreasing CH_4 concentrations inside the chamber. Gas samples are taken at least three times at, for example, 0, 30, and 60 min after starting the experiment start and placed into 10–15 mL gas vials with airproof lids. CH_4 concentrations in the vials are measured using gas chromatography. The trends in CH_4 concentrations are approximated by linear regression and the flux is estimated based on the ideal gas law, since the volume of the chamber and the extent of the analyzed soils surface are given (Smagin, 2005; Molchanov, 2007).

In situ measurement of soil respiration (CO_2 efflux) can be conducted similarly to CH_4 although the methodology needs to be adapted given the higher effluxes of CO_2 from the soil compared with the CH_4 effluxes. Currently more advanced approaches using an infra-red gas analyzers (IRGA) are used to measure soil respiration *in situ*. An IRGA provides measurements of CO_2 concentrations with a frequency up to 1 Hz or even higher and thus gives a much more accurate picture of the CO_2 flux.

In this approach, separate measurements were illustrated of CH_4 and CO_2. Alternatively, methods are available to distinguish between SOM-derived and root-derived respiration, both in laboratory and field conditions. The most frequently used are based on isotopic approaches (Pataki et al., 2003; Taneva and Gonzalez-Meller, 2011), trenching (Bond-Lamberty et al., 2011) and field segregation (Leake et al., 2004; Gavrichkova et al., 2010). Implementation of these approaches as well as comparison of the results for forest and crop ecosystems was reviewed by Hanson et al. (2000). Each of the techniques has its advantages and disadvantages. Isotopic approaches provide the most precise results. However, the measurements are expensive and the number of observations is typically limited. Field segregation provides a more general approach to quantify soil respiration *in situ*. It is based on the partitioning of soil respiration components by measuring two parallel plots, one of which is preliminary cleaned from existing roots and artificially isolated from intergrowth of outside roots (Leake et al., 2004; Gavrichkova et al., 2010). This approach is relatively cheap and simple and allows for a larger number of sampling points to consider spatial variability. However, soil disturbance during the experimental setup requires a long period before starting measuring to re-establish an equilibrium state (Hanson et al., 2000; Moyano et al., 2008). For all the approaches in which soil carbon fluxes are measured, the measurements usually include parallel monitoring of soil temperature and soil moisture as the key abiotic drivers of soil respiration.

9.3.2.2   Laboratory Measurements of Carbon Fluxes

Indirect methods estimate soil carbon fluxes as a function of proxy variables obtained from, for example, remote sensing (Guo et al., 2011; Huang et al., 2013) or measurements of soil microbiological activity under standardized conditions. Deriving carbon emission from remote sensing considerably expands the study area. It is therefore widely used for global analysis. The limitation of the methodology is that the remote-sensing image lacks a direct linkage to soil processes. Alternatively, spatial–temporal variability of soil respiration can be analyzed through a relatively easily measured proxy variable, referring to soil microbiological activity measured in standardized conditions. The rate of microbial (basal) respiration is one of the relevant proxies. The BR is the respiration of soil without plant roots and is therefore associated to the soil microbial activity. The basal soil respiration is often referred to as "microbial respiration" of the soil, as opposed to the "root respiration." The preincubation procedure for soil BR measurement is similar to one described earlier for SIR. BR is determined by the CO_2 evolution rate from the soil incubated for 24 h at 22°C as described for SIR. However, water (0.1 mL g^{-1} soil) is added to the soil samples instead of the glucose solution during SIR measurements. The BR is expressed in μg CO_2-C g^{-1} soil h^{-1}. Together with soil microbe biomass, BR is a commonly accepted indicator to quantify changes in the activity of the soil microbial community and soil quality (Winding et al., 2005; Bispo et al., 2009). Experimentally obtained values of microbial biomass carbon (C_{mic}) and its BR allow for the calculation of the "integrated" indicators of soil microbial community functioning:

1. The microbial metabolic quotient or qCO_2 (in μg CO_2-C mg^{-1} C_{mic} h^{-1}) is the ratio of BR/C_{mic} of a specific respiration activity of soil microbial biomass. The qCO_2 value characterizes the sustainability of soil microbial community to anthropogenic influences (Ananyeva, 2003);
2. The C_{mic}/C_{org} ratio (in %) is an additional parameter of soil quality (Anderson and Domsch, 1986).

The qCO_2 is an informative indicator of the ecophysiological status of soil microorganisms. A high qCO_2 value in disturbed soils indicates a high energy loss and carbon consumption in the turnover of nutrients (Insam and Haselwandter, 1989; Anderson and Domsch, 1990). The C_{mic}/C_{org} ratio is an indication of organic carbon availability (Anderson and Domsch, 1986; Insam and Domsch, 1988), and high values indicate the binding of organic matter in soil microbial biomass (Anderson and Domsch, 1989). Changes in C_{mic}/C_{org} reflect inputs of organic matter, conversion to microbial carbon, decomposition of carbon, and formation of organic carbon and mineral complexes in the soils (Wang et al., 2011). Therefore, these integral parameters allow getting additional information about the functioning of the soil microbial community under anthropogenic pressure.

9.3.2.3   Spatial Analysis and Modeling of Soil Carbon Fluxes

Both direct and indirect methods can be used to assess the spatial variability of soil carbon fluxes. However, the implementation of the research differs. Direct methods are relatively expensive and time and labor intensive. They are, therefore, only applied on a limited number of plots. To apply this approach for larger regions, the study area is stratified (e.g., based on soil or land-use type) with chambers installed at a limited number of representative sites (Nilsson et al., 2000; Kurganova, 2010). By relying on these representative sites, the spatial variation within each stratum is not considered. Whether direct measurements give satisfactory results in large and heterogeneous areas with a large number of different natural, rural, and urban ecosystems is questionable. Alternatively, indirect analysis of spatial variability of carbon fluxes through a relatively easily measured proxy variable allows for a larger number of observation points and thus expands possible research area. For instance, using BR, measured in the standardized experimental conditions, as a proxy eliminates the initial effect of field temperature and moisture regimes (Bloem et al., 2006). As a result, it allows for the comparison of different samples (e.g., taken at different locations or moments in time). Monitoring over a long periods is less important in contrast to *in situ* measurements, and many more samples can be taken throughout a region of interest within all the different strata. Considerable numbers of samples given by the BR approach allows for implementation of the most advanced approaches of spatial analysis, mapping, and modeling, including DSM technique discussed earlier.

Implementing DSM to analyze and map spatial variability of soil respiration is necessary at the regional level, especially for a highly urbanized region. In order to consider both natural and urban-specific factors in such region and to provide necessary data for DSM, a stratified sampling design can be implemented. It should represent the variability in bioclimatic conditions and consider short-distance variability within the settlements. A DSM approach can be implemented to map BR as a function of traditional (relief, climate, land-use, vegetation and soil type, and complexity) and urban-specific (functional zoning, size, and age of the city) factors. Since a strong correlation between SOC and BR is widely assumed and was reported for different ecosystems (Anderson and Domsch, 1986; Wardle, 1992; Ananyeva et al., 2008), the SOC content should be also added as an explanatory variable (Vasenev et al., 2014).

Land-use type, soil type, mean annual temperature, average annual precipitation, slope, normalized difference vegetation index (NDVI), and soil carbon content can be used as explanatory variables. To implement DSM in the urban areas, urban-specific factors are added, including functional zoning (derived from NDVI), age, and size of the settlements (derived from statistical open sources). Since the regression kriging is not available owing to the stratified sampling design statistical GLM, correlating BR to explanatory variables can be implemented to predict spatial patterns of BR in the region. On the basis of the GLM, separate maps for BR of different soil layers can be developed. Details on the implemented mapping and GLM approaches were published in Vasenev et al. (2014).

9.4   CASE STUDIES

9.4.1   INTRODUCTION

Urbanization influences soil carbon stocks and fluxes in different aspects and at multiple scales. Alteration of the conditions and functioning of the soil microbiological community results in changes in the soil microbial carbon stocks and microbial respiration. Changes in temperature and moisture regime as well as physical disturbances affect temporal dynamics of soil respiration. Soil sealing and functioning zoning contribute to the spatial patterns of soil carbon stocks. Variability in space and time caused by different bioclimatic conditions, management practices, and seasonal trends increases complexity of carbon stocks and fluxes in urban soils. The methodologies described in Section 9.3 are implemented in a number of case studies in the Moscow region to illustrate how they capture this spatial–temporal variability. The choice of the relevant technique depends on the scale of analysis, research area, and specific goals.

The first case study focuses on the microbial component of the carbon balance and analyzes the influence of land use and bioclimatic conditions on the functioning of soil microbial community. Microbial biomass, microbial (basal) respiration, and structure of microbial community, represented through contribution of fungi and bacteria to total SIR, were measured in natural, arable, and urban ecosystems within four administrative districts of the Moscow region, differing in soil and bioclimatic conditions. An analysis of these soil microbiological activity parameters and the integral indexes qCO_2 and C_{mic}/C_{org} illustrate the specifics of urban soils' capability to store and emit microbial carbon in a range of functional and bioclimatic zones. Considering that the major soil CO_2 efflux is coming from soil microbes, the case study contributes to understanding and quantification of the potential carbon emission from urban soils.

In contrast to the potential CO_2 emissions estimated based on the laboratory measurements in standard conditions, actual soil respiration measured *in situ* strongly depends on temperature and moisture parameters and thus varies between ecosystem, bioclimatic conditions, and over the season. The second case study compares soil respiration from industrial, residential, and recreational zones and from a reference urban forest in two settlements with different bioclimatic conditions: Moscow city located in south-taiga zone and Kursk city representing forest-steppe zone. Parallel observations of *in situ* soil respiration taken in June–November 2013, followed by measurement of air temperature, soil temperature, and moisture, provide an insight into the patterns of seasonal dynamics of soil CO_2 emissions and factors influencing it. Similar trends described for urban areas in different bioclimatic zones represent predominant influence of anthropogenic factor on soil respiration in urban environment over the bioclimatic conditions.

The combined influence of bioclimatic and urban-specific factors on the spatial variability of carbon stocks and fluxes is most evident when modeling and mapping them at the regional scale. The third case study demonstrates spatial patterns in SOC and BR distribution over the Moscow region, mapped using the DSM technique. Estimation of coefficient of variance per different combinations of soil types and land-use categories enables better understanding of the main factors behind heterogeneity in SOC and BR and represents urban areas as an important source of spatial variability in carbon stocks and fluxes at the regional scale.

9.4.2 SPATIAL VARIABILITY OF URBAN SOILS' MICROBIAL CARBON AND RESPIRATION IN CONTRAST BIOCLIMATIC AND FUNCTIONAL ZONES OF MOSCOW REGION

9.4.2.1 Introduction

The soil microbial community is the key driver behind soil functions (Martens, 1995; Anderson and Domsch, 2010) and plays a major role in the soil-atmosphere gas (CO_2 production) exchange (Cornard, 1996; Svirejeva-Hopkins et al., 2004; Zavarzin and Kudeyarov, 2006) and nutrient cycles. It is also very sensitive to various anthropogenic impacts (Ananyeva, 2003; Colloff et al., 2008; Macdonald et al., 2009). The soil microbial component is very variable in space owing to differences in the contents of organic matter and nutrients, pH, vegetation, and land use (Yan et al., 2003; Grinand et al., 2008; Gavrilenko et al., 2011). So far, major comparative studies of soil microbiological parameters are carried out at the local scale of coupled ecosystems, landscapes, and catenas ("horizontal" gradients). Research of microbiological properties of a specific area (e.g., at the regional or district level) in the context of the "spatial-horizontal" gradient of ecosystems, including urban ecosystems, is still very rare.

The contribution of the soil microbial component in the formation of carbon fluxes and stocks in urban ecosystems at the regional and global level is poorly known (Lorenz and Lal, 2009; Bond-Lamberty and Thomson, 2010). The urban environment influences the functioning of the soils' microbial community, which in turn affects the respiration activity, structure, and diversity of the community (Kaye et al., 2005; Macdonald et al., 2009; Lysak, 2013). This case study focused on the soil microbial component and its respiration activity (CO_2 formation) of urban soils for four administrative districts of Moscow region in comparison with natural (forest, meadow) and agroecosystems. The study addressed two main research questions: (1) Is the functioning of the soil microbial community deteriorated under the urban transformation? (2) Is an urban soil a potential source of atmospheric carbon dioxide?

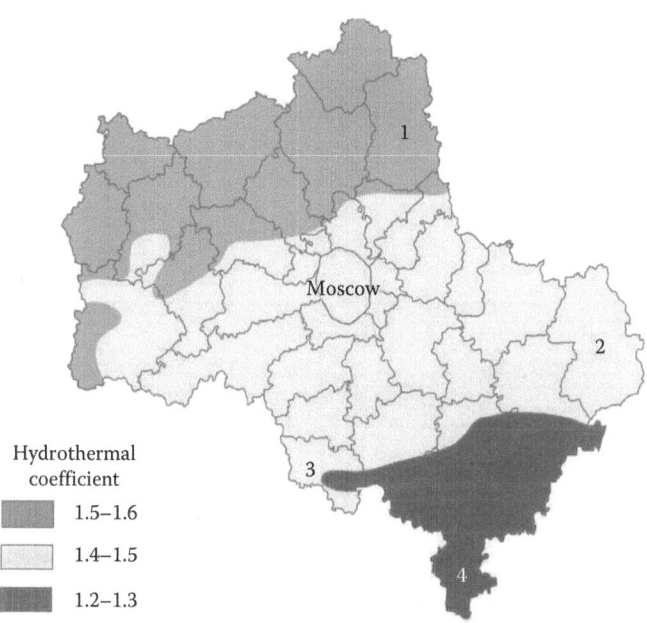

FIGURE 9.2 Bioclimatic zones of Moscow region. Districts: (1) Sergiev Posad; (2) Shatura; (3) Serpukhov; (4) Serebryanye Prudy. (From Ivashchenko K.V. et al. 2014. Biomass and respiration activity of soil microorganisms in anthropogenically transformed ecosystems (Moscow region). Pleiades Publishing, Ltd., 892–903. With permission.)

9.4.2.2 Materials and Methods

Soils of natural (forest, meadow) and anthropogenic-transformed (arable, city) ecosystems of the northeast and southwest of Moscow region (Sergiev Posad, Shatura, Serpukhov, and Serebryanye Prudy districts: 54°37′32.4″–56°26′57.5″N, 37°30′25″–39°39′3.9″E) located in different bioclimatic zones of the region were studied (Figure 9.2). Annual average air temperature for the region is around 4°C and annual precipitation amounts up to 570–700 mm. Soddy-podzolic soils prevail in the Sergiev Posad district; bog-podzolic, mucky-podzolic, bog-alluvial, and peat soils are predominant in the Shatura district; gray forest soil dominates in the Serpukhov district; and leached chernozem prevails in the Serebryanye Prudy district. Natural (forest, meadow) and anthropogenic-transformed (arable land, urbo-) ecosystems were selected in each district. Recreational, residential, and industrial zones in cities were differentiated. Three to five sampling points were selected randomly in each observed ecosystem and functional zone of district. In each of the 104 observation points, a composite topsoil (0–10 cm) sample was taken from a 2×2 m plot (corners and center). Soil samples were stored under field-moisture conditions with air exchange under cooled conditions (8–10°C) for no more than 4 weeks before the analysis.

Soil microbial biomass carbon (C_{mic}) and microbial respiration (MR) were analyzed by SIR and BR approaches, respectively (Anderson and Domsch, 1978; Ananyeva et al., 2011). Microbial metabolic quotient, qCO_2, and C_{mic}/C_{org} ratio were estimated (see Sections 2.1.2.2 and 2.2.2 for the detailed methodology). The contribution of fungi and bacteria to the total SIR of the soil was determined by the selective inhibition (SI) technique with antibiotics, and their ratio was calculated (Semenov et al., 2013).

9.4.2.3 Results

High spatial variability of C_{mic} was found in the study areas, and C_{mic} variance in urban ecosystems was generally higher than in arable and natural ecosystems (Figure 9.3). Average C_{mic} in forest and meadow soils (408–729 µg C g^{-1}) was 1.4–3.2 times higher than in arable ecosystems. C_{mic}

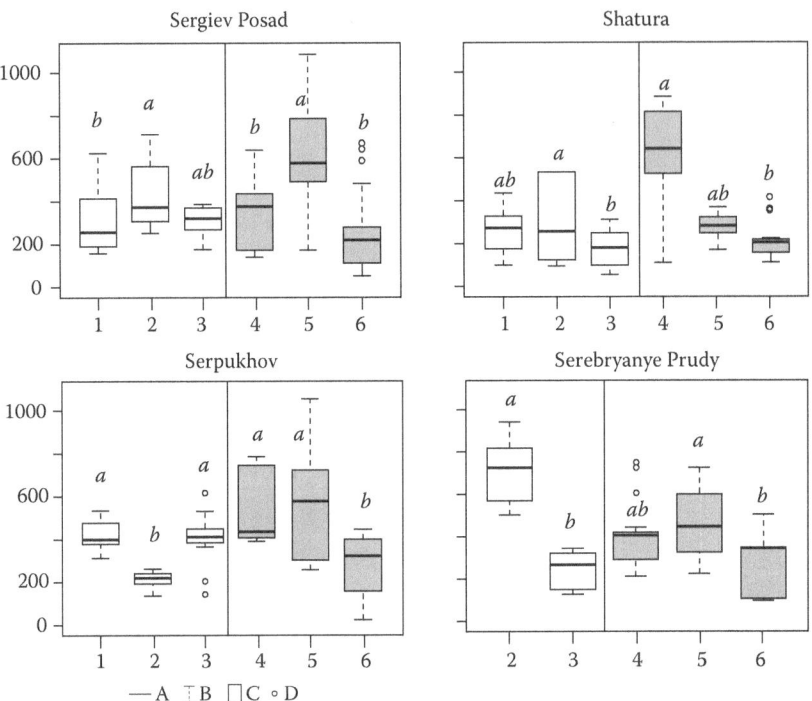

FIGURE 9.3 Distribution of microbial biomass carbon in the soils (0–10 cm layer, µg C g⁻¹) in different eco-systems of Moscow region. Here and below, the designations are as follows: (1) forest; (2) meadow; (3) arable; (4) recreational zone; (5) residential zone; and (6) industrial zone. The statistical characterization of the values is as follows: (A) median; (B) min–max; (C) 25–75%; (D) outliers. The values with different letters differ significantly ($p \leq 0.05$) within each district and its functional zones separately. (From Ivashchenko K.V. et al. 2014. Biomass and respiration activity of soil microorganisms in anthropogenically transformed ecosystems (Moscow region). Pleiades Publishing, Ltd., 892–903. With permission.)

values within the urban areas were also heterogeneous with the lowest values obtained for industrial zone, which were 1.7–2.7 times less compared to recreational and residential zones. A similar trend was shown for basal (microbial) respiration of the studied soils with the lowest in urban industrial areas 0.13 and the highest in forest ecosystems (0.13 and 3.08 µg CO_2-C g⁻¹ h⁻¹ correspondingly) (Table 9.1). Topsoil (0–10 cm) C_{mic} and MR in observed ecosystems were, respectively, 69%–83% and 75–85 from the total profile values (Table 9.2). The spatial variability of the qCO_2 value was also higher in the urban soils compared with the corresponding natural references, except Shatura district, where high qCO_2 variability in meadow and forest ecosystems was likely related to the patchiness of zonal soil cover (Figure 9.4). The C_{mic}/C_{org} was also sensitive to the land use with the lowest values in urban industrial zones (Figure 9.5). It should be noted that the urban soils were capable of higher microbial respiration than, for example, the arable ones. Therefore, the potential gas-production activity (CO_2 emission) by the urban areas is likely comparable to that of the soils in the nonurban ecosystems.

The contributions of fungi and bacteria to the total microbial biomass in the soddy-podzolic soil of the forest (C_{org} 4.92%; pH 3.84) and the urban soil of the recreational (C_{org} 3.50; pH 6.16) and industrial zones (C_{org} 3.62%; pH 7.05) in Sergiev Posad district were determined. It was shown that fungi prevailed in the soil microbial biomass, giving 63%–80% of its structure. The fungi-to-bacteria ratio of 3.5 for the forest zone, 3.3 for the recreational zone, and 1.4 for the industrial zone, respectively, indicated a significant decrease in the fungal component (by almost twice) under the increasing anthropogenic impact.

TABLE 9.1

BR (μg CO_2–C g^{-1} h^{-1}) in the Soils (0–10 cm) of Different Ecosystems and Functional Zones in Moscow Region

Ecosystem/Zone	District (Number of Sites)			
	Sergiev Posad (28)	Shatura (30)	Serpukhov (24)	Serebryanye Prudy (22)
Forest	0.85–3.08/1.61 *a*	0.63–1.44/0.95 *a*	0.76–1.42/1.10 *a*	N/A
Meadow	0.57–1.97/1.17 *a*	0.56–5.41/1.92 *a*	0.30–0.45/0.37 *b*	0.40–1.25/0.91 *a*
Arable	0.40–0.58/0.50 *b*	0.29–0.89/0.59 *b*	0.31–0.92/0.59 *b*	0.18–0.80/0.55 *b*
Recreational	0.45–1.13/0.79 *a*	0.23–1.37/0.99 *a*	0.48–0.87/0.63 *b*	0.33–0.92/0.63 *a*
Residential	0.61–1.06/0.84 *a*	0.22–0.45/0.33 *b*	0.90–2.04/1.27 *a*	0.40–1.36/0.81 *a*
Industrial	0.13–0.88/0.67 *a*	0.35–1.06/0.59 *a*	0.33–1.58/0.62 *b*	0.28–0.78/0.44 *b*

Source: From Ivashchenko K.V. et al. 2014. Biomass and respiration activity of soil microorganisms in anthropogenically transformed ecosystems (Moscow region). Pleiades Publishing, Ltd., 892–903. With permission.

Note: The ranges and the mean are given; the values marked with different letters significantly differ ($p \leq 0.05$) within the each district for ecosystems and functional zones separately.

TABLE 9.2

Microbial Respiration (g C m^{-2} $year^{-1}$) of Soils (0–10 cm Layer, Litter Excluded) in Different Ecosystems of Moscow Region

District (Number of Sites)	Forest	Meadow	Arable	City
Sergiev Posad (28)	1410	1025	438	674
Shatura (30)	832	1682	517	561
Serpukhov (24)	964	324	517	736
Serebryanye Prudy (22)	N/A	797	482	552

Source: From Ivashchenko K.V. et al. 2014. Biomass and respiration activity of soil microorganisms in anthropogenically transformed ecosystems (Moscow region). Pleiades Publishing, Ltd., 892–903. With permission.

9.4.2.4 Discussions

Natural (nondisturbed), arable (slightly disturbed), and anthropogenically transformed (urban) soils were studied under different bioclimatic conditions of the Moscow region. Such a spatial-horizontal gradient allowed tracing the changes in the microbial parameters of the soils under various anthropogenic impacts, considering their spatial variability. Microbial activity obtained for the anthropogenically transformed soils was generally lower compared with the natural analogues, which is in good coherence with the other researchers showing a significant decrease of C_{mic}, BR values and fungi-to-bacteria ratio in soils with increasing anthropogenic impact (Beyer et al., 1995; Lorenz and Kandeler, 2006; Gavrilenko et al., 2011).

The major part of the soil CO_2 emission results from the activity of heterotrophic soil microorganisms, giving, for example, almost 78% of its total emission for the arable land and 48% for the different-aged fallows (Kurganova et al., 2007). The microbial (basal) respiration of soil determined under laboratory conditions (optimum moisture content, temperature, disturbed structure) is a promising proxy for the potential carbon outflow from soil. Therefore, there is a basis to assess the microbial production of CO_2 upper most biologically active mineral soil. The rate of the microbial CO_2 evolution by 1 g of soil was recalculated per unit of area (1 m^2) of the 0.1 m thick soil layer. The density of the soil in the calculation was taken equal to 1 g cm^{-3} (Oreshkina, 1988). The found

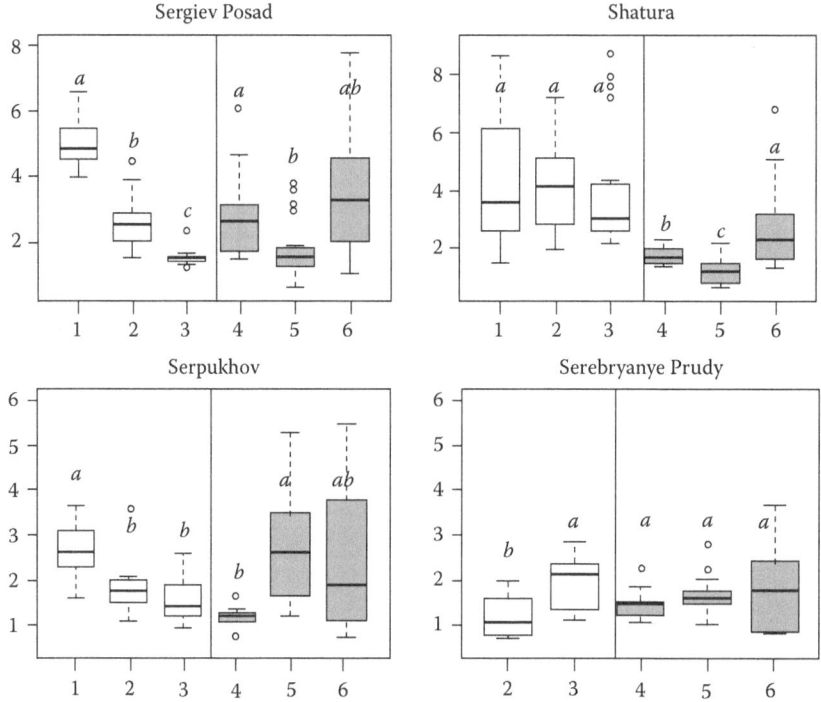

FIGURE 9.4 Distribution of the microbial metabolic quotient of the soils (0–10-cm layer, μg CO_2–C mg^{-1} C_{mic} h^{-1}) in different ecosystems and functional zones of Moscow region [the values with different letters differ significantly ($p \leq 0.05$) within each district and its functional zones separately]. (From Ivashchenko K.V. et al. 2014. Biomass and respiration activity of soil microorganisms in anthropogenically transformed ecosystems (Moscow region). Pleiades Publishing, Ltd., 892–903. With permission.)

values were the maximum in the soils under woody and herbaceous plants of natural ecosystems and lower by almost twice in the anthropogenically transformed soils (arable, city).

According to our calculations, the soils of arable lands and cities of Moscow region can potentially provide 438 to 736 g C m^{-2} $year^{-1}$ to the atmosphere (Table 9.2). The rate of potential microbial CO_2 emission from the urban soils is comparative to that from the arable ones. Our results are close to those obtained in long-term and annual field measurements of the CO_2 emission by the soils in the southern-taiga in Moscow region, giving 347–613 g C m^{-2} $year^{-1}$ for the arable land and the young fallow and 845 g C m^{-2} $year^{-1}$ for the sown meadow (Kurganova et al., 2007). When recalculated to the district areas, the highest emission of CO_2 was obtained in Sergiev Posad and Serebryanye Prudy districts (317.8 and 288.8 thousand tons C $year^{-1}$), which was twice as high as in Shatura and Serpukhov districts (140.0 and 178.5 thousand tons C $year^{-1}$) (Table 9.3). On average, the CO_2 emission from urban soils was 26% of the emission from the arable ones with 59% in the Sergiev Posad district. Considering that so far urban areas give less 10% of the region but the extent of urban areas is continuously growing, the importance of urban areas as a potential source of CO_2 emission will increase.

9.4.3 SPATIAL–TEMPORAL VARIABILITY OF *IN SITU* RESPIRATION IN URBAN SOILS OF SOUTH-TAIGA AND FOREST-STEPPE VEGETATION ZONES

9.4.3.1 Introduction

Urban soils can store a considerable amount of carbon resulting from adding of composts, wastes, intensive growth of root biomass, and "cultural layers" formed in result of historical residential

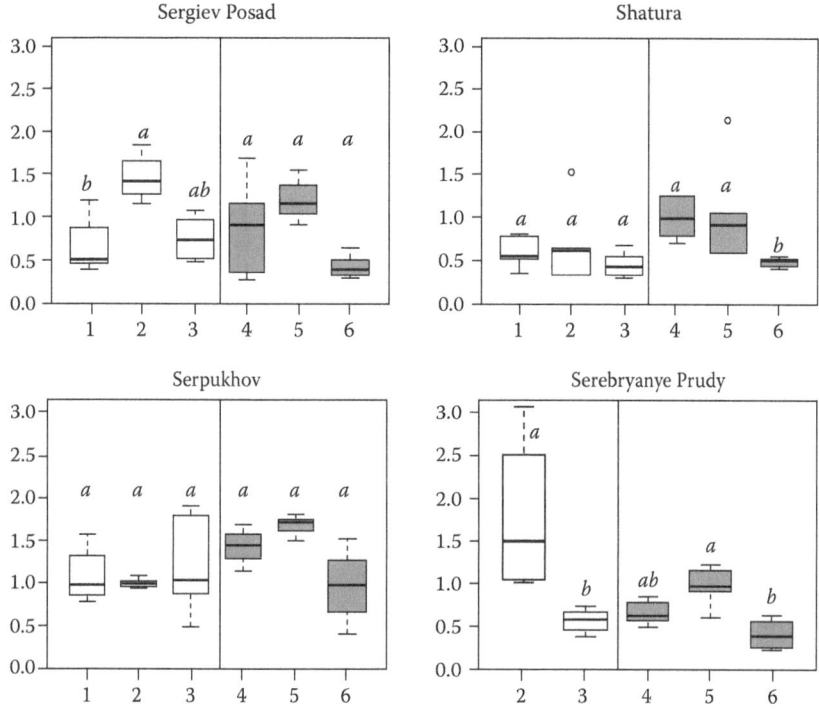

FIGURE 9.5 Distribution of the C_{mic}/C_{org} ratio (%) in the soils (0–10 cm layer) of different ecosystems and functional zones of Moscow region [the values with different letters differ significantly ($p \leq 0.05$) within each district and its functional zones separately]. (From Ivashchenko K.V. et al. 2014. Biomass and respiration activity of soil microorganisms in anthropogenically transformed ecosystems (Moscow region). Pleiades Publishing, Ltd., 892–903. With permission.)

activity. At the same time, urban topsoil has high potential to emit CO_2 through intensive mineralization of turf-based soil mixtures widely used in greenery, but rather nonstable in urban conditions. Biotic and abiotic factors in urban environment, including frequent temperature passage above and below zero, usual waterlogging, implementation of organic substrates and mineral fertilizers in greenery, create favorable conditions for mineralization of soil carbon (Kurbatova et al., 2004; Groffman, 2004). Rate of urban soil's respiration is determined by its main components:

TABLE 9.3
Area of Arable Lands and Settlements in the Studied Districts of Moscow Region and Their Microbiological Emission Activity

| District | Arable Lands | | Settlements | |
	Area (ha)	CO_2 Emission, Thousand Tons (C year⁻¹)	Area (ha)	CO_2 Emission, Thousand Tons (C year⁻¹)
Sergiev Posad	45,717	200	17,444	118 (59)
Shatura	21,331	110	5300	30 (27)
Serpukhov	29,637	153	3437	25 (16)
Serebryanye Prudy	58,104	280	1577	9 (3)

Source: From Ivashchenko K.V. et al. 2014. Biomass and respiration activity of soil microorganisms in anthropogenically transformed ecosystems (Moscow region). Pleiades Publishing, Ltd., 892–903. With permission.

HR of microorganisms and autotrophic root respiration (Kuzyakov and Larionova, 2005; Chapin et al., 2006; Gomes-Casanovas et al., 2012). Considerable amount of introduced vegetation species including green lawns contributes to root respiration (Milesi, 2005; Svirejeva-Hopkins et al., 2006), whereas considerable difference in biomass quantity and structure between urban and reference zonal soils determines corresponding differences in HR (Lysak, 2010; Vasenev et al., 2012).

Total soil respiration and its components are highly variable in space (Bahn et al., 2006; Kuzyakov and Gavrichkova, 2010). High heterogeneity reported for urban areas (Prokofieva and Stroganova, 2004; Vasenev et al., 2013a,b,c) contributes to this variability. Spatial variability of carbon fluxes in urban areas can be influenced by both conventional- (climatic, vegetation) and urban-specific (soil sealing functional zoning, soil pollution, and physical disturbance). Contrast anthropogenic pressure represented by different functional zones is a key urban-specific driver of spatial heterogeneity. Industrial, residential, and recreational zones are very different in percentage of the sealed areas, the level of anthropogenic disturbance, and chemical and biological soil features, which determine the difference in soil respiration between functional zones. At the same time, urban soils function in specific bioclimatic conditions and thus bioclimatic driver also contributes to soil respiration and its variability.

It is critical to understand driving factors influencing carbon fluxes in urban soils under various bioclimatic and management conditions to be able to model urban carbon and to analyze contribution of urban soils to regional and global carbon balance. In order to analyze contribution of "bioclimatic" and "urban-specific" factors to urban soil's respiration, we made a parallel comparative analysis of CO_2 efflux from urban soils of three contrast functional zones located in cities from different bioclimatic zones.

9.4.3.2 Materials and Methods

Respiration was measured in urban soils from the Moscow and Kursk cities. The Moscow city (N55°50′; E37°33′) is located in the south-taiga zone. Climatic conditions in the Moscow city are humid-continental with average July temperature of 19.1°C and average January temperature of −14.0°C. In winter, temperature normally drops to approximately −10.0°C, though there can be periods of warmth with temperature rising above 0.0°C. The average number of days with temperature below zero varies from 151 to 197 with the clear tendency to decrease during the last decades. The average annual precipitation is close to 700 mm (Naumov et al., 2009). Relief is represented mainly by moraine hilly plain, and moraine loam is the main parent material. Zonal soddy-podzolic soils (Eutric Podzoluvisols) are the most diffused zonal soil type in the area (FAO, 1988; Shishov and Voinovich, 2002). The Kursk city (N51°54′; E36°10′) is located 500 km to the SSW from the Moscow city. Climate of the area is temperate-continental (average July temperature 20.9°C and average January temperature of −6.4°C) with drier conditions than in the Moscow city (average annual precipitation less than 600 mm). Natural vegetation is represented by forest-steppe species. Grey forest and chernozemic soils dominate in the region (Dobrovolsky and Urussevskaya, 2004).

Sampling plots were chosen in the industrial, recreational, and residential functional zones of each city. Industrial areas were represented by green lawns near gas station and in the sanitary-hygienic zone of the rubber factory. Residential zone referred to university campuses, whereas recreational zones were represented by parks. Two urban forests (one in each city) were chosen as a semi-natural references. Soil respiration was measured by Li-820 IRGA (LI-COR Biosciences, USA) in 5–10 points from each experimental plot. Measurements were taken for the period of June–November 2013 to capture spatial variability. Measuring soil respiration was followed by observations of soil temperature and moisture.

9.4.3.3 Results

High temporal and spatial variability in soil respiration was reported for both of the settlements. Temporal dynamics were positively correlated with soil temperature, although correlation with soil moisture was not significant. Both in Kursk and in Moscow, the highest respiration was obtained at the end of July and at the beginning of August, and the lowest at the end of October. Average

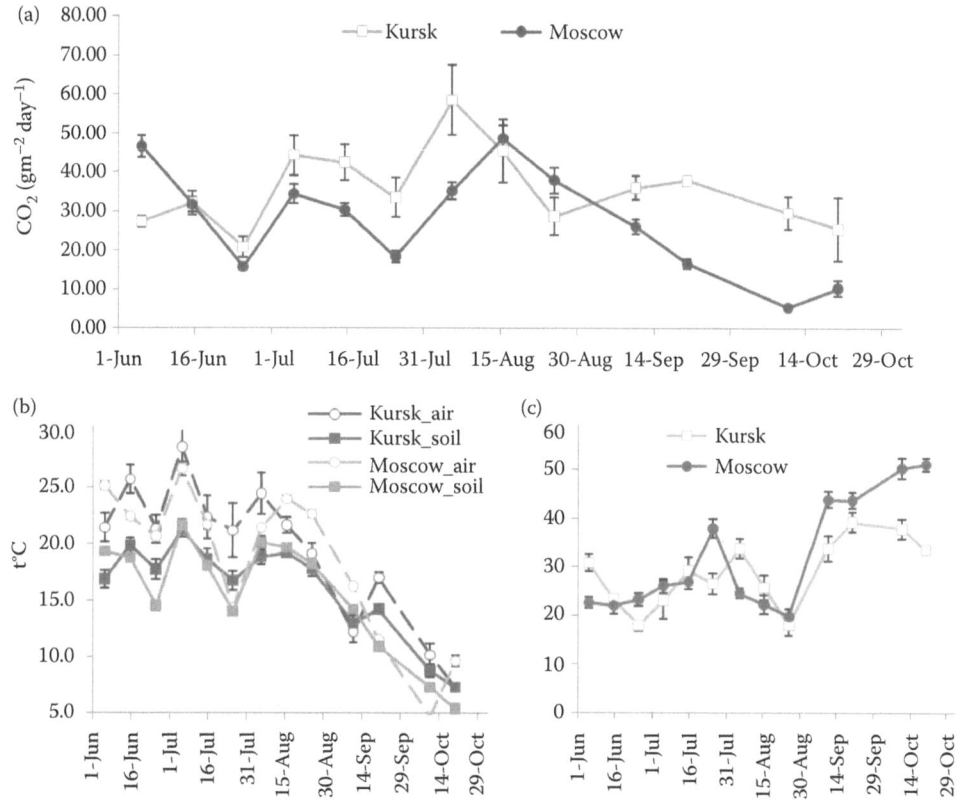

FIGURE 9.6   Temporal dynamic of *in situ* CO_2 emission (a), soil and air temperature (b), and soil moisture (c) averaged for analyzed soils in the Kursk and Moscow cities.

respiration in the Kursk city was 10%–15% higher than that in the Moscow city for a major part of the summer period, although the surpass reached 50% at the end of September–October, when the lowest temperatures and the highest moisture over the whole observation period were reported (Figure 9.6). A significant difference was found between averaged soil respiration obtained for different functional zones within the cities. For both settlement averages, soil respiration from more disturbed industrial and residential areas was 20%–50% higher than from recreational zones and urban forest where anthropogenic disturbance was less (Figure 9.7).

In spite of the reported high spatial–temporal variability, a significant difference in average soil respiration was found between urban functional zones and urban forests, and respiration rates obtained for urban areas were almost two times higher. In contrast, the difference between average soil respiration in the Moscow and Kursk cities was not statistically significant (Figure 9.8). These outcomes allow proposing intrazonal features in urban soils' respiration. Factorial ANOVA results confirm this hypothesis. "Urbanization" factor distinguished 27% of total soil respiration variance, whereas "bioclimatic" factor was not significant and contributed to less than 1% of total variance.

9.4.3.4   Discussion

In our study we observed higher respiration in urban soils in comparison with urban forest, which we used as a semi-natural reference. This outcome is in accordance with the few available comparable studies in Arizona and Baltimore (USA) (Koerner and Klopatek, 2002; Kaye et al., 2005), as well as Shanghai and Nanjing (China) (Sun et al., 2009; Zhang el al., 2010), which also report higher soil respiration in urban soils in comparison with agricultural and natural ones. Different

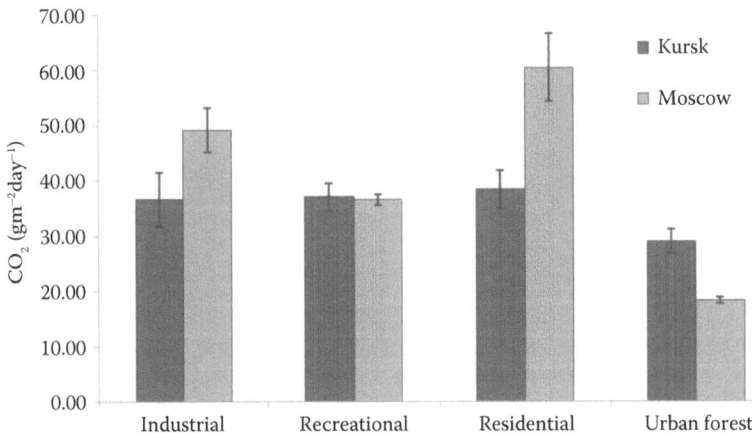

FIGURE 9.7 Averaged CO_2 emission from soils located in different urban functional zones and in urban forests of the Moscow and Kursk cities.

factors explaining this tendency are reported in literature. Some authors explain this by higher SOC contents reported for urban soils (Pouyat et al., 2002; Zircle et al., 2011), including artificial organic substrates (compost and sewage materials) added during soil reclaiming and greenery work (Lorenz and Lal, 2009; Beesley, 2012) and easily mineralizable by microorganisms. Another explanation links higher urban soil respiration rates to increases in average soil and air temperatures due to urban "heat island effect" (Davidson, 1998) and alteration of soil physical properties (Chen et al., 2013).

In both cities average air and soil temperature was cooler for urban forests compared to lawns of urban functional zones (industrial, residential, and recreational). This may have affected soil respiration. However, the difference in temperature between the forest and functional zones was rather small (1–1.5 C), although the difference in soil respiration between the plots was considerable (Lysak et al., 2010). Thus we assume that temperature was not the only factor influencing soil respiration for the case study, where lower values of microbial respiration were obtained in the urban soils compared with natural ones.

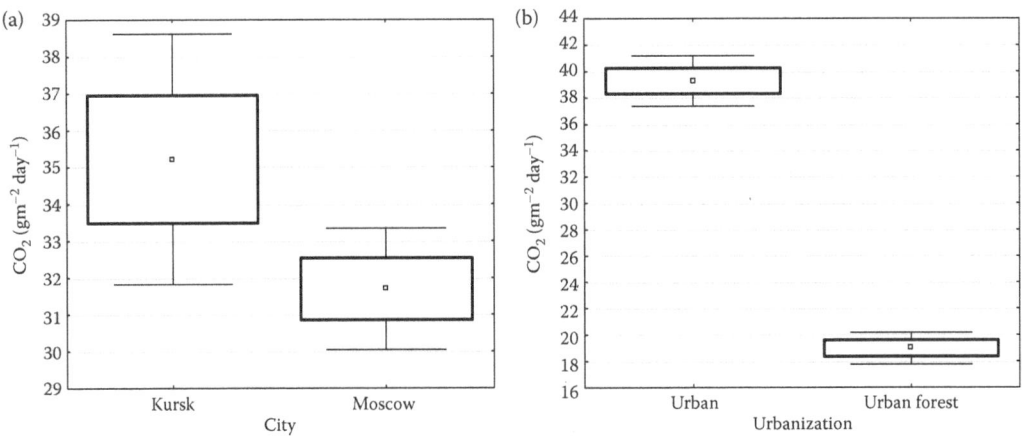

FIGURE 9.8 Mean, standard deviation, and 95% confidence interval of CO_2 emissions averaged for Moscow and Kursk cities (a) and for urban functional zones and urban forest plots (b).

9.4.4   SPATIAL MODELING AND MAPPING CARBON STOCKS AND MICROBIAL RESPIRATION IN HIGHLY URBANIZED MOSCOW REGION

9.4.4.1   Introduction

Spatial heterogeneity is probably the most evident feature of urban environment, contributing to spatial–temporal variability of carbon stocks and fluxes from urban soils. At the local level, diversity of specific features and processes—such as zoning, combustion, fertilization, sewage release (Lehman and Stahr, 2007), soil pollution, and compaction (Gerasimova et al., 2003; Prokofieva and Stroganova, 2004)—results in high short-distance heterogeneity and patchiness. At the regional level, urban-specific factors are further complicated by the influence of bioclimatic conditions referring to climate, parent materials, zonal vegetation, and soil types. If not properly considered, very variable urban soils can become an important source of uncertainty in regional models of carbon balance. There is a necessity in analyzing, modeling, and mapping spatial variability of urban soils' stocks and fluxes.

Conventional soil survey techniques are expensive as they require a large number of observations. In addition, the approach is not very suitable for an urban environment that is characterized by short distance variability with a lot of abrupt changes. DSM can give satisfactory results based on a limited number of field observations, but it still has some limitations for the assessment in urban areas. The limitations include: (1) high short-distance variability within the urban areas; (2) domination of the anthropogenic soil-forming factor, and (3) necessity to consider subsoil carbon in contrast to many regional and global analyses, in which the estimation was focused on the top layer (Bui et al., 2009; Mishra et al., 2010; Martin et al., 2011). Implementing DSM for soil respiration caused by its temporal dynamics will impose even more constraints. Traditional *in situ* measurements of soil respiration assume continuous monitoring of the same points during the season and are therefore only applied on a limited number of plots, which hardly give satisfactory results in large and heterogeneous areas with a large number of different land-use types. Alternatively, the spatial variability of Rs can be analyzed indirectly through a relatively easily measured proxy variable, such as microbial (basal) respiration (BR), measured in standardized conditions. Since the experimental conditions are standardized, the initial effect of field temperature and moisture regimes is eliminated (Bloem et al., 2006). Monitoring over a long period is less important and many more samples can be taken throughout a region of interest with all the different strata.

The study aimed to analyze and map soil carbon stocks and microbial respiration in a large, diverse, and urbanized Moscow region. So far, spatial patterns in soil carbon stocks and microbial respiration in this region remain poorly understood if one compares them with the EU and United States, where carbon stocks and fluxes are continuously measured through FLUXNET (Wilson and Baldocchi, 2000; Baldocchi et al, 2001).

9.4.4.2   Materials and Methods

Moscow region measures 46,700 km^2. Climate of the region has been described earlier (see Section 9.3.1.1). Parent material includes moraine loam and clay in the northern and central parts, fluvioglacial sands in the East, and cover loam in the South. Soils include orthic Podzols in the North, eutric Podzoluvisols in the center, orthic Luvisols and luvic Chernozems in the South, intrazonal dystric Histosols in the eastern part of the region, and eutric Fluvisols in the flood-plains of the Moskva and Oka rivers (Egorov et al., 1977; FAO, 1988; Shishov et al., 2004). Vegetation varies with climate and is represented by three bioclimatic zones: south-taiga, deciduous forests, and steppe-forest (Shishov and Voinovich, 2002). Anthropogenic-changed landscapes (agricultural, fallow, and urban lands) occupy nearly 60% of the territory. The urban area is rapidly expanding and currently occupies almost 10%, including 68 cities and towns with 18.8 million people (including Moscow city). Moscow is the largest city in Europe with a population of over 11.2 million people.

Following the DSM concept (McBratney et al., 2000), we assumed soil carbon stocks and microbial respiration to be a function of the soil-forming factors. The final list of available data contained (1) traditional soil-forming factors (elevation, climate, land-use, vegetation, and soil) and (2) urban-specific soil-forming factors (functional zoning, size, and history of the settlements, including age and population density (see Section 2.1.3 and Vasenev et al., 2014 for more details). Soil samples were taken in Moscow region in 2010–2011. Stratified sampling design was implemented considering specifics of urban soil survey (see Section 2.1.1). Sampling points were chosen in Moscow city and six settlements in the region in such a way that the different combinations of traditional and urban-specific factors distinguished earlier were represented (see Vasenev et al., 2013a for details on the sampling design). Samples were taken inside the settlements and from surrounding areas. Seven settlements with three surrounding land uses (arable land, pasture, and forest) and three functional zones (residential, recreational, and industrial) were sampled, which gave us a total of 244 locations (69 in Moscow city and 175 in its surroundings; 95 from urban areas and 80 from nonurban sires). After necessary preparation and preincubation, SOC and BR were measured following the methodology (see Sections 2.1.2 and 2.2.2).

Normality of the distribution of SOC and BR values was checked by Shapiro–Wilk's W test and homogeneity of variances was checked by Levene's test. One-way ANOVA was used to check significance of difference in SOC between different groups (soil types, land-use types, cities of different age, size, etc.). A statistical, multiple regression model correlated SOC and BR to the explanatory variables to predict the topsoil and subsoil stocks and fluxes in the area. Auxiliary information included both continuous and categorical data. Relationships between SOC and BR and secondary data were analyzed using GLMs. Categorical variables were introduced as dummy variables (equal to 1 if a location was within a particular unit and 0 otherwise). The correlation between explanatory variables was analyzed in advance to prevent multicollinearity. Estimation of carbon stocks based on SOC contents was conducted following the methodology discussed in Section 2.1.3. For many applications, the average carbon stocks and fluxes and its variability in a certain strata are more essential than the exact value at a specific location. Therefore, the primary results of the SOC stocks and BR were also aggregated to the land-use classes and soil types.

9.4.4.3 Results

SOC and BR maps were obtained separately for topsoil and subsoil layers with 771 m resolution. Highest spatial variability was captured by SOC maps. The urban areas and nearest suburbs were the "hotspots" of the largest topsoil and subsoil carbon stocks (Figure 9.9). The spatial patterns

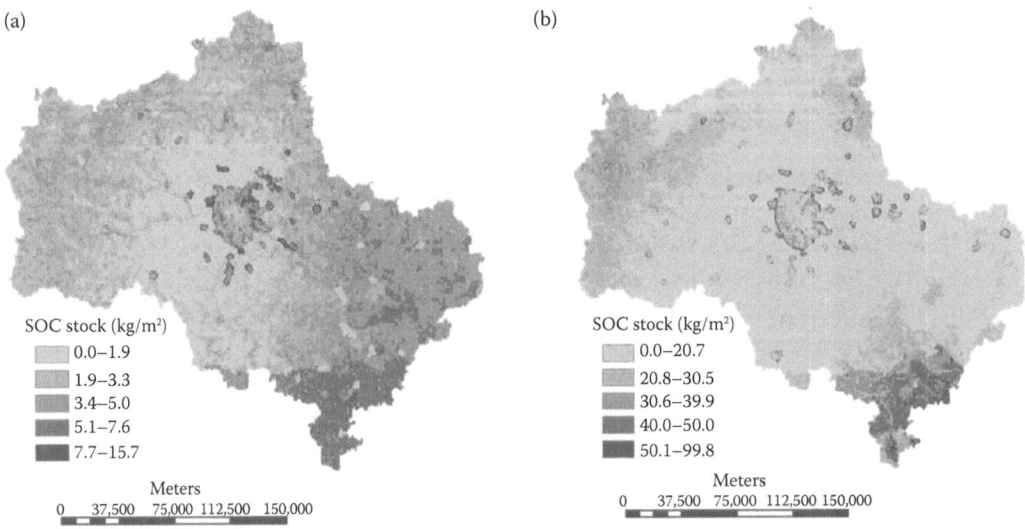

FIGURE 9.9 Modeled topsoil (a) and subsoil (b) SOC stocks in Moscow region. (Reprinted from *Geoderma*, 226–227, Vasenev, V.I. et al. How to map soil organic stocks in highly urbanized region? 103–115, Copyright 2014, with permission from Elsevier.)

(a)   (b)

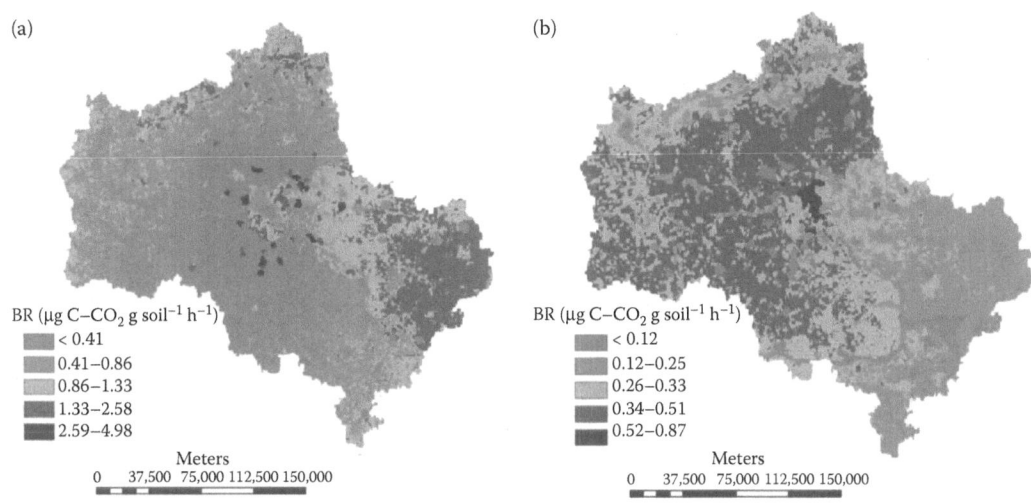

BR (μg C–CO$_2$ g soil^{-1} h^{-1})   BR (μg C–CO$_2$ g soil^{-1} h^{-1})
■ < 0.41   ■ < 0.12
■ 0.41–0.86   ■ 0.12–0.25
■ 0.86–1.33   ■ 0.26–0.33
■ 1.33–2.58   ■ 0.34–0.51
■ 2.59–4.98   ■ 0.52–0.87
            Meters   Meters
0   37,500 75,000 112,500 150,000   0   37,500 75,000 112,500 150,000

FIGURE 9.10   Modeled topsoil (a) and subsoil (b) microbial respiration in Moscow region.

of BR differed between the topsoil and subsoil maps, but patterns in BR corresponded to the patterns in soil types and land use for both. Topsoil BR was the highest in the East of the region with large areas occupied by bogs and dystric Histosols. High topsoil BR was also found for the orthic Podzols in the North and luvic Chernozems in the South. Urban areas and especially the Moscow city showed high variation in topsoil BR with higher values in the green spaces and lower in the central built-up parts (Figure 9.10a). Subsoil BR followed the same trends. In general, subsoil BR was less variable than topsoil BR with the highest values found in the West with eutric Podzoluvisols (Figure 9.10b). Analysis of BR averaged per land use and soil type provides information on the factors influencing its variability, but it does not give a clear picture of the spatial distribution. It is more valuable to analyze spatial variability per different strata representing interaction of various environmental and management conditions. In order to obtain this information, we aggregated the BR maps based on the combinations of traditional and urban-specific factors distinguished for the modeling and estimated CV values per each stratum. The highest variability of topsoil and subsoil BR was reported for the urban areas, which was clearly represented by hotspots on the maps, coinciding with the borders of settlements (Figure 9.11c and d). Areas with the highest CV in the topsoil carbon stocks also corresponded well with the settlement boundaries. However, for the subsoil, the opposite pattern was found—urban areas were more homogeneous, whereas the highest CV was obtained for the natural sites under orthic Podzols and eutric Podzoluvisols (Figure 9.11a and b).

9.4.4.4   Discussions

Depending on the research goal, information required from the regional carbon assessment may differ. When analyzing urban areas with extremely high heterogeneity, which is almost impossible to describe by regional analysis, understanding of carbon stocks and fluxes spatial variability becomes much more valuable than traditional data on mean carbon stocks. The maps of carbon stocks' spatial variability based on CV values, aggregated for different soil types, land-use classes, as well as for the urban areas of different size, age, and functional zoning, provided essential information on the patterns behind spatial distribution of carbon stocks in the Moscow region. In our study, urban areas were shown as the main source of spatial variance with the highest CV found in urban areas both for SOC and BR.

Obtained absolute values of carbon stocks and BR were rather uncertain, which was clearly shown by high CV values up to 70% and rather low R^2 of the GLM, referring to their quite poor performance. Such a high level of uncertainty could have been expected and rooted in high heterogeneity

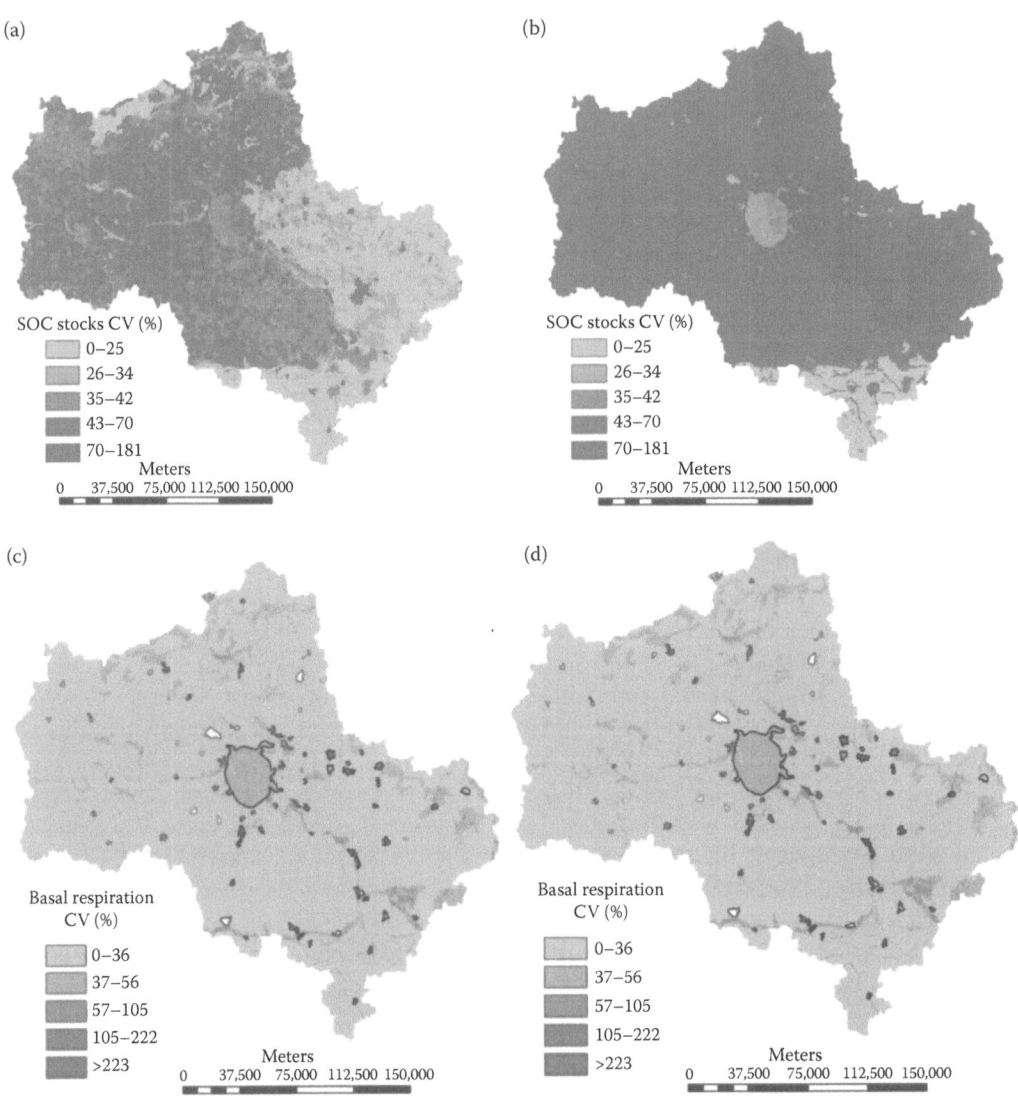

FIGURE 9.11 Spatial variability represented by CV (%) values for topsoil (a) and subsoil (b) carbon stocks (Reprinted from *Geoderma*, 226–227, Vasenev, V.I. et al. How to map soil organic stocks in highly urbanized region? 103–115, Copyright 2014, with permission from Elsevier.) and for topsoil (c) and subsoil (d) microbial respiration.

of the research area and quite a few assumptions were made at different steps of modeling and estimation processes: (1) the majority of sampling points were concentrated in the urban areas, covering approximately 10% of total research area; (2) residential and industrial functional areas were taken as a single unit in modeling; (3) contribution of the sealed soils was eliminated by reduction coefficient (see Vasenev et al., 2014). Comparison between topsoil and subsoil values also was not straightforward because of differences in sampling approaches and aggregating 10–150 cm subsoil in a single soil sample, which may provide very rough results. However, it gave us an opportunity to guess on the contribution of the subsoil to the variability of carbon stocks and microbial respiration, which is often left out of regional analysis.

Mapping carbon stocks and fluxes in a highly urbanized region is challenging owing to the specific characteristics of the urban environment. As a result, each step of modeling and mapping should change. In contrast to the majority of global and regional assessments (i.e., Nilsson et al., 2000; Schulp and Verburg, 2009), we demonstrated that urban carbon stocks and microbial respiration can be comparable to or even higher than those in natural areas and croplands. We also showed the considerable contribution of urban areas in spatial variability of carbon stocks and fluxes at the regional scale.

9.5   CONCLUSIONS

Urbanization is among the most important of the current land-use trends and its role will likely grow in the future (Pickett et al., 2011). Urbanization alters stocks and fluxes of energy and matter and results in irreversible changes in vegetation and soil cover (Milesi et al., 2005; Scalenghe and Marsan, 2009; Vasenev et al., 2013b). However, carbon stocks and fluxes of urban soils lack attention, and our understanding of their spatial–temporal variability and contribution to the regional carbon budget remains very limited and uncertain. In this chapter, we discussed the main current achievements and gaps in studying carbon stocks and fluxes in urban soils. Special attention was given to the methodological aspects of studying spatial–temporal variability of urban soil carbon, illustrated by the multiple scale case studies.

In our research, we demonstrated that an extremely variable urban environment, in terms of management practices and anthropogenic disturbance, leads to a very high spatial heterogeneity of carbon stocks and fluxes. This outcome was demonstrated for all investigated parameters, including microbial and SOC, basal, and *in situ* soil respiration and at the multiple scale. Urban soils became the major source of spatial variation at the level of the experimental plots, districts, or for the Moscow region, in general. This outcome highlights variability in carbon stocks and fluxes as an important characteristic of urban soils.

The significant difference of carbon stocks and fluxes in urban soils and agricultural or natural ones is another key outcome of comparative analysis between ecosystems and bioclimatic zones. Considering high internal variability discussed earlier, this significant difference of carbon stocks and fluxes in urban and nonurban soils may represent intrazonal features in urban carbon balance rather than zonal ones. However, we did not obtain similar spatial trends for different parameters of carbon cycle within urban areas. Soil microbial biomass and microbial respiration decreased when anthropogenic influence increased with the lowest values reported for the most disturbed industrial zones. In contrast, *in situ* soil respiration in industrial zones was higher than in residential and recreational ones. Apparently, microbiological activity parameters are sensitive to stressful conditions of urban environment and low values of microbial carbon and respiration indicate unfavorable conditions for microbial community. Higher *in situ* soil respiration is a factor of physical disturbance and dissipative metabolism versus immobilization of carbon into microbial biomass. In other words, urban environment creates conditions for high momentary carbon efflux, which can significantly decrease in future if new artificial adding of carbon to urban soils does not occur.

Although our results represent an overview of a limited number of individual studies and may remain rather uncertain, they contribute to our understanding of carbon stocks and fluxes in urban soils and their spatial–temporal variability. Our outcomes provide evidence that the contribution of urban soils to regional carbon balance will be progressively more important in the future when urbanization and pollution will be among the most important factors affecting soil quality and health.

REFERENCES

Alexandrovskaya, E.I. and A.L. Alexandrovskiy. 2000. History of the cultural layer in Moscow and accumulation of anthropogenic substances in it. *Catena* 41: 249–259.

Alexandrovskiy, A.L., O. Chichagova, J. Van Der Plicht et al. 1998. The early history of Moscow: ^{14}C dates from Red Square. *Radiocarbon* 40: 583–589.

Ananyeva, N.D. 2003. *Microbiological Aspects of Self-purification and Resilience of Soils*. Moscow: Nauka (in Russian).

Ananyeva, N.D., E.A. Susyan, and E.G. Gavrilenko. 2011. Determination of the soil microbial biomass carbon using the method of substrate-induced respiration. *Eurasian Soil Sci+* 44: 1215–1221.

Ananyeva, N.D., E.A. Susyan, O.V. Chernova, I.Y. Chernov, and O.L. Makarova. 2006. The ratio of fungi and bacteria in the biomass of different types of soil determined by selective inhibition. *Microbiology* 75: 702–707.

Ananyeva, N.D., L.M. Polyanskaya, E.A. Susyan, I.V. Vasenkina, S. Wirth, and D.G. Zvyagintsev. 2008. Comparative assessment of soil microbial biomass determined by the methods of direct microscopy and substrate-induced respiration. *Microbiology* 77: 356–364.

Anderson, J.P.E. and K.H. Domsch. 1973. Quantification of bacterial and fungal contribution to soil respiration. *Arch Microbiol* 93: 113–127.

Anderson, J.P.E. and K.H. Domsch. 1975. Measurement of bacterial and fungal contribution to respiration of selected agricultural soils. *Can J Microbiol* 21: 314–322.

Anderson, J.P.E. and K.H. Domsch. 1978. A physiological method for the quantitative measurement of microbial biomass in soils. *Soil Biol Biochem* 10: 215–221.

Anderson, T.-H. 1994. Physiological analysis of microbial communities in soil: Applications and limitations. In *Beyond the Biomass*, eds K. Ritz, J. Dighton, and K. Giller, pp. 67–76. West Sussex: Wiley.

Anderson, T.H. and K.H. Domsch. 1986. Carbon links between microbial biomass and soil organic matter. In *Perspectives in Microbial Ecology*, eds F. Megusar and M. Gantar, pp. 467–471. Ljubljana: Slovene Society for Microbiology.

Anderson, T.H. and K.H. Domsch. 1989. Rations of microbial biomass to total organic carbon in arable soils. *Soil Biol Biochem* 21: 471–479.

Anderson, T.H. and K.H. Domsch. 1990. Application of eco-physiological quotients qCO_2 and qD on microbial biomass from soils of different cropping histories. *Soil Biol Biochem* 22: 251–255.

Anderson, T.-H. and K.H. Domsch. 2010. Soil microbial biomass: The eco-physiological approach. *Soil Biol Biochem* 42: 2039–2043.

Avdusin, D.A. 1980. *Field Archaeology in the Soviet Union*. Moscow: Vyssh. Shk.

Bahn, M., I.A. Janssens, M. Reichstein, P. Smith, and S.E. Trumbore. 2010. Soil respiration across scales: Towards an integration of patterns and processes. *New Phytol* 186(2): 292–296.

Bailey, V.L., J.L. Smith, and H. Bolton. 2002. Fungal-to-bacterial ratios in soils investigated for enhanced C sequestration. *Soil Biol Biochem* 34: 997–1007.

Baldocchi, D., E. Falge, L. Gu et al. 2001. FLUXNET: A new tool to study the temporal and spatial variability of ecosystem-scale carbon dioxide, water vapor, and energy flux densities. *Bull Am Meteorol Soc* 82: 2415–2434.

Bandaranayake, W., Y.L. Qian, W.J. Parton, D.S. Ojima, and R.F. Follett. 2003. Estimation of soil organic carbon changes in turfgrass systems using the CENTURY model. *Agron J* 95: 558–563.

Barajas-Aceves, M. 2005. Comparison of different microbial biomass and activity measurement methods in metal contaminated soils. *Bioresour Technol* 96: 1405–1414.

Bardgett, R.D. and E. McAlister. 1999. The measurement of soil fungal: Bacterial biomass ratios as an indicator of ecosystem self-regulation in temperate meadow grasslands. *Biol Fertil Soils* 29: 282–290.

Bardgett, R.D., D.A. Wardle, and G.W. Yeates. 1998. Linking aboveground and below-ground interactions: How plant responses to foliar herbivory influence soil organisms. *Soil Biol Biochem* 30: 1867–1878.

Bardgett, R.D., P.J. Hobbs, and A. Frostegard. 1996. Changes in soil fungal: Bacterial biomass ratios following reductions in the intensity of management of an upland grassland. *Biol Fertil Soils* 22: 261–264.

Bartalev, S.A., A.S. Belward, D.V. Erchov, and A.S. Isaev. 2003. A new SPOT4-VEGETATION derived land cover map of Northern Eurasia. *Int J Remote Sens* 24(9): 1977–1982.

Beare, M.H., C.L. Neely, D. Coleman, and W.L. Hargrove. 1990. A substrate-induced respiration (SIR) method for measurement of fungal and bacterial biomass on plant residues. *Soil Biol Biochem* 22: 585–594.

Beare, M.H., R.W. Parmelee, P.F. Hendrix, W. Cheng, D.C. Coleman, and D.A. Crossley. 1992. Microbial and faunal interactions and effects on litter nitrogen and decomposition in agroecosystems. *Ecol Monogr* 62: 569–591.

Beck, T., R.G. Joergensen, E. Kandeler et al. 1997. An inter-laboratory comparison of ten different ways of measuring soil microbial biomass. *Soil Biol Biochem* 29: 1023–1032.

Beesley, L. 2012. Carbon storage and fluxes in existing and newly created urban soils. *J Environ Manage* 104: 158–165.

Bekku, Y., H. Koizumi, T. Oikawa, and H. Iwaki. 1997. Examination of four methods for measuring soil respiration. *Appl Soil Ecol* 5: 247–254.

Beyer, L., P. Kahle, H. Kretschemer, and Q. Wu. 2001. Soil organic matter composition of man-impacted sites in North Germany. *J Plant Nutr Soil Sci* 164: 359–364.

Bispo, A., D. Cluzeau, R. Creamer et al. 2009. Indicators for monitoring soil biodiversity. *Integr Environ Assess Manage* 5: 717–719.

Blagodatskaya, E.V., T.V. Pampura, E.G. Dem'yanova, and T.N. Myakshina. 2006. Effect of lead on growth characteristics of microorganisms in soil and rhizosphere of *Dactylis glomerata. Eurasian Soil Sci+* 39(6): 635–660.

Bloem, J., A.J. Schouten, S.J. Sørensen, M. Rutgers, A. Van der Werf, and A.M. Breure. 2006. Monitoring and evaluating soil quality. In *Microbiological Methods for Assessing Soil Quality*, eds J. Bloem, A. Benedetti, and D.W. Hopkins. Wallingford, Oxfordshire, UK: CAB International, 23–49.

Bloem, J., P.R. Bolhuis, M.R. Veninga, and J. Wieringa. 1995. Microscopic methods for counting bacteria and fungi in soil. In *Methods in Applied Soil Microbiology and Biochemistry*, eds K. Alef and P. Nannipieri, pp. 162–191. London: Academic Press.

Blum, W.E.H. 2005. Functions of soil for society and environment. *Rev Environ Sci Bio/Technol* 4: 75–79.

Bond-Lamberty, A. and A. Thomson. 2010. A global database of soil respiration data. *Biogeosciences* 7: 1915–1926.

Borovikov, V. 2003. *Art of Computer Data Analysis.* Saint-Petersburg, Russia, Piter.

Bui, E., B. Henderson, and K. Viergever. 2009. Using knowledge discovery with data mining from the Australian Soil Resource Information System database to inform soil carbon mapping in Australia. *Glob Biogeochem Cycl* 23. GB4033.

Burba, G. 2013. *Eddy Covariance Method for Scientific, Industrial, Agricultural and Regulatory Applications.* Li-Cor Biosciences, Linkoln, Nebraska.

Buyanovsky, G.A., G.H. Wagner, and C.J. Gantzer. 1986. Soil respiration in a winter wheat ecosystem. *Soil Sci Soc Am J* 50: 338–344.

Carlyle, J.C. and U.B. Than. 1988. Abiotic controls of soil respiration beneath an eighteen-year old *Pinus radiata* stand in south-eastern Australia. *J Ecol* 76: 654–662.

Castaldi, S., F.A. Rutigliano, and A. Virzo de Santo. 2004. Suitability of soil microbial parameters as indicators of heavy metal pollution. *Water Air Soil Pollut* 158: 21–35.

Chapin, F.S., III, G.M. Woodwell, J.T. Randerson et al. 2006. Reconciling carbon-cycle concepts, terminology, and methods. *Ecosystems* 9: 1041–1050.

Chen, W., X. Jia, T. Zha et al. 2013. Soil respiration in a mixed urban forest in China in relation to soil temperature and water content. *Eur J Soil Biol* 54: 63–68.

Ciardi, C. and P. Nannipieri. 1990. A comparison of methods for measuring ATP in soil. *Soil Biol Biochem* 22: 725–727.

Collof, M.J., S. Wakelin, D. Gomez, and S.L. Rogers. 2008. Detection of nitrogen cycle genes in soils for measuring the effects of changes in land use and management. *Soil Biol Biochem* 40: 1637–1645.

Conrad, R. 1996. Soil microorganisms as controllers of atmospheric trace gases (H_2, CO, CH_4, OCS, N_2O and NO). *Microbiol Rev* 60(4): 609–640.

Craul, P.J. 1992. *Urban Soil in Landscape Design.* New York: John Wiley.

Creamer, R.E., R.P.O. Schulte, D. Stone et al. 2014. Measuring basal soil respiration across Europe: Do incubation temperature and incubation period matter? *Ecol Indicat* 36: 409–418.

Cruvinel, E.F., M.M.C. Bustamante, A. Kozovits, and R.G. Zepp. 2011. Soil emission of NO, N_2O and CO_2 from croplands in the savanna region of central Brazil. *Agric Ecosyst Environ* 144: 29–40.

Dalal, R.C. and D.E. Allen. 2008. Greenhouse gas fluxes from natural ecosystems. *Aust J Bot* 56(5): 369–407.

Davidson, E.A. and I.A. Janssens. 2006. Temperature sensitivity of soil carbon decomposition and feedbacks to climate change. *Nature* 440 (7081): 165–173.

Dobrovolsky, G.V. and I.S. Urussevskaya. 2004. *Soil Geography.* Moscow: MSU.

Dolgikh, A.V. and A.L. Aleksandrovskii. 2010. Soils and cultural layers in velikii Novgorod. *Eurasian Soil Sci+* 43: 477–448.

Domsch, K.H., Th. Beck, J.P.E. Anderson, B. Soderstom, D. Parkinson, and G. Trolldenier. 1979. A comparison of methods for soil microbial population and biomass studies. *Z Pflanzenernaehr Bodenkd* 142: 520–533.

Egorov, V.V., V.M. Fridland, E.N. Ivanova, and N.I. Rosov (eds). 1977. *Classification and Diagnostics of Soils in USSR.* Moscow: Kolos.

Eswaran, H., E. Van den Berg, and P. Reich. 1993. Organic carbon in soils of the world. *Soil Sci Soc Am J* 57: 192–194.

Evdokimova, A.K. 1986. Heavy metals in the cultural layer of medieval Novgorod. *Vestn Mosk Un-ta Ser 5: Geogr* 3: 86–91.

FAO. 1988. *Soils Map of the World: Revised Legend.* Rome: Food and Agriculture Organization of the United Nations.

FAO. 2013. *Climate-Smart Agriculture. Sourcebook.* ISBN 978-92-5-107721-4.

FAO/ISRIC/ISSS. 1998. *World reference base for soil resources.* World Soil Resources Report, No. 84. FAO, Rome.

Fierer, N., J.A. Jackson, R. Vilgalys, and R.B. Jackson. 2005. Assessment of soil microbial community structure by use of taxon-specific quantitative PCR assays. *Appl Environ Microbiol* 71: 4117–4120.

Frey, S.D., E.T. Elliot, and K. Paustian. 1999. Bacterial and fungal abundance and biomass in conventional and no-tillage agroecosystems along two climatic gradients. *Soil Biol Biochem* 31: 573–585.

Frey, S.D., M. Knorr, J. Parrent, and R.T. Simpson. 2004. Chronic nitrogen enrichment affects the structure and function of the soil microbial community in a forest ecosystem. *Forest Ecol Manage* 196: 159–171.

Fromin, N., B. Porte, R. Lensi et al. 2012. Spatial variability of the functional stability of microbial respiration process: A microcosm study using tropic forest soil. *J Soil Sediment* 12: 1030–1039.

Frostegard, A. and E. Baath. 1996. The use of phospholipid fatty acid analysis to estimate bacterial and fungal biomass in soil. *Biol Fertil Soils* 22: 59–65.

Frostegard, A., A. Tunlid, and E. Baath. 2011. Use and misuse of PLFA measurements in soils. *Soil Biol Biochem* 43: 1621–1625.

Galford, G.L., J. Melillo, J.F. Mustard, C.E.P. Cerri, and C.C. Cerri. 2010. The Amazon frontier of land-use change: Croplands and consequences for greenhouse gas emissions. *Earth Interact* 14: 1–20.

Gavrichkova, O. 2010. *Drivers of soil respiration of root and microbial origin in grasslands.* Doctorate Thesis, Tuscia University, Viterbo.

Gavrilenko, E.G., E.A. Susyan, N.D. Anan'eva, and O.A. Makarov. 2011. Spatial variability in the carbon of microbial biomass and microbial respiration in soils of the south of Moscow oblast. *Eurasian Soil Sci+* 44: 1125–1138.

Gavriluk, F.Y. 1963. *Soil Survey and Mapping.* Moscow: V. Shkola.

Gerasimova, M.I., M.N. Stroganova, N.V. Mozharova, and T.V. Prokofieva. 2003. *Urban Soils.* Smolensk, Russia, Oykumena.

Golubiewski, N.E. 2006. Urbanization increases grassland carbon pools: Effects of landscaping in Colorado's front range. *Ecol Appl* 16: 555–571.

Gomes-Casanovas, N., R. Matamala, D.R. Cook, and M.A. Gonzalez-Meler. 2012. Net ecosystem exchange modifies the relationship between the autotrophic and heterotrophic components of soil respiration with abiotic factors in prairie grasslands. *Glob Change Biol* 18: 2532–2545.

Gregg, J.W., C.G. Jones, and T.E. Dawson. 2003. Urbanization effects on tree growth in the vicinity of New York City. *Lett Nat* 10: 183–187.

Grimm, N.B., H.F. Stanley, N.E. Golubiewski et al. 2008. Global change and the ecology of cities. *Science* 319: 756–760.

Grimmond, C.S.B., T.S. King, F.D. Cropley, D.J. Nowak, and C. Souch. 2002. Local-scale fluxes of carbon dioxide in urban environments: Methodological challenges and results from Chicago. *Environ Pollut* 116: 243–254.

Grinand, C., D. Arrouays, B. Laroche, and M.P. Martin. 2008. Extrapolating regional soil landscapes from an existing soil map: Sampling intensity, validation procedures, and integration of spatial context. *Geoderma* 143: 180–190.

Groffman, P.M., N.L. Law, K.T. Belt, L.E. Band, and G.T. Fisher. 2004. Nitrogen fluxes and retention in urban watershed ecosystems. *Ecosystems* 7: 393–403.

Guo, L.B. and R.M. Gifford. Soil carbon stocks and land use change: A meta-analysis. *Glob Change Biol* 8: 345–360.

Guo, X.L., S.C. Black, and Y.H. He. 2011. Estimation of leaf CO_2 exchange rates using a SPOT image. *Int J Remote Sens* 32: 353–366.

Hamilton, J.G., E.H. DeLucia, K. George, S.L. Naidu, A.C. Finzi, and W.H. Schlesinger. 2002. Forest carbon balance under elevated CO_2. *Oecologia* 131: 250–260.

Hanson, P.J., N.T. Edwards, C.T. Garten, and J.A. Andrews. 2000. Separating root and soil microbial contributions to soil respiration: A review of methods and observations. *Biogeochemistry* 48: 115–146.

Hassink, J. 1993. Relationship between the amount and the activity of the microbial biomass in Dutch grassland soils: Comparison of the fumigation-incubation method and the substrate-induced method. *Soil Biol Biochem* 25: 533–538.

Hijmans, R.J., S.E. Cameron, J.L. Parra, P.G. Jones, and A. Jarvis. 2005. Very high resolution interpolated climate surfaces for global land areas. *Int J Climatol* 25: 1965–1978.

Houghton, R.A. 2003. Why are estimates of the terrestrial carbon balance so different? *Glob Change Biol* 9: 500–509.

Huang, N., J.S. He, and Z. Niu. 2013. Estimating the spatial pattern of soil respiration in Tibetan alpine grasslands using Landsat TM images and MODIS data. *Ecol Indicat* 26: 117–125.

Insam, H. and K.H. Domsch. 1988. Relation between soil organic carbon and microbial biomass on chronosequences of reclamation sites. *Microb Ecol* 15: 177–188.

Insam, H. and K. Haselwandter. 1989. Metabolic quotient of the soil microflora in relation to plant succession. *Oecologia* 79: 174–178.

Intergovernmental Panel on Climate Change. 2001. Climate change 2001: The scientific basis. In *Contribution of Working Group I to the Third Assessment Report of the Intergovernmental Panel on Climate Change*, eds J.T. Houghton, Y. Ding, and D.J. Griggs, 881 pp. Cambridge, UK: Cambridge University Press.

IPCC. 2013. *Climate Change 2013: The Physical Science Basis. Contribution of Working Group I to the Fifth Assessment Report of the Intergovernmental Panel on Climate Change*, eds T.F. Stocker, D. Qin, G.-K. Plattner, M. Tignor, S.K. Allen, J. Boschung, A. Nauels, Y. Xia, V. Bex, and P.M. Midgley. Cambridge, UK: Cambridge University Press.

IPCC (Intergovernmental Panel on Climate Change). 2005. *IPCC Special Report on Carbon Dioxide Capture and Storage, Prepared by Working Group III of the Intergovernmental Panel on Climate Change*, eds B. Metz, O. Davidson, H.C. de Coninck, M. Loos, and L.A. Meyer, 442 pp. Cambridge, UK: Cambridge University Press.

Ivashchenko K.V., N.D. Ananyeva, V.I. Vasenev, V.N. Kudeyarov, and R. Valentini. 2014. Biomass and respiration activity of soil microorganisms in anthropogenically transformed ecosystems (Moscow region). Pleiades Publishing, Ltd., 892–903.

IUSS Working Group WRB. 2007. *World Reference Base for Soil Resources 2006; First Update 2007*. Rome: FAO.

Jacobi, G.A. and L. Gorin. 1967. The effect of streptomycin and other aminoglycoside antibiotics on protein biosynthesis. In *Antibiotics, Volume I. Mechanism of Action*, eds D. Gottlieb and P.D. Show, pp. 726–747. Berlin: Springer.

Janzen, H.H. 2004. Carbon cycling in earth systems—A soil science perspective. *Agric Ecosyst Environ* 104: 399–417.

Jarvis, A., H.I. Reuter, A. Nelson, and E. Guevara. 2008. Hole-filled SRTM for the globe. Version 4, CGIAR-CSI SRTM 90m Database. http://srtm.csi.cgiar.org.

Jim, C.Y. 1998. Soil compaction at tree-planting sites in urban Hong Kong. In *The Landscape Below Ground II: Proceedings of a Second International Workshop on Tree Root Development in Urban Soils*, eds D. Neely and G.W. Watson, pp. 166–178. Champaign, IL: International Society of Arboriculture.

Jo, H.K. and E.G. McPherson. 1995. Carbon storage and flux in urban residential greenspace. *J Environ Manage* 45: 109–133.

Jobbagy, E.G. and R.B. Jackson. 2000. The vertical distribution of soil organic carbon and its relation to climate and vegetation. *Ecol Appl* 10: 423–436.

Kaiser, E.A., T. Muller, R.G. Joergensen, H. Insam, and O. Heinemeyer. 1992. Evaluation of methods to estimate the soil microbial biomass and the relationship with soil texture and organic matter. *Soil Biol Biochem* 24: 675–683.

Kalnay, E. and M. Cai. 2003. Impact of urbanization and land-use change on climate. *Nature* 423: 528–531.

Kaye, J.P., I.C. Burke, A.R. Mosier, and J.P. Guerschman. 2004. Methane and nitrous oxide fluxes from urban soils to the atmosphere. *Ecol Appl* 14: 975–981.

Kaye, J.P., R.L. McCulley, and I.C. Burkez. 2005. Carbon fluxes, nitrogen cycling, and soil microbial communities in adjacent urban, native and agricultural ecosystems. *Glob Change Biol* 11: 575–587.

Kern, J.S. 1994. Spatial patterns of soil organic carbon in the contiguous United States. *Soil Sci Soc Am J* 58: 439–455.

Khan, M. and J. Scullion. 2002. Effects of metal (Cd, Cu, Ni, Pb or Zn) enrichment of sewage-sludge on soil microorganisms and their activities. *Appl Soil Ecol* 20(2): 145–155.

Kiryushin, V.I. and A.L. Ivanov (eds). 2005. *Land Agroecological Assessment, Projecting of Landscape-adapted Farming Systems and Agrotechnologies*. Moscow: Russian Ministry of Agriculture.

Koerner, B. and J. Klopatek. 2002. Anthropogenic and natural CO_2 emission sources in an arid urban environment. *Environ Pollut* 116: 45–51.

Komarov, A.S., O.G. Chertov, S.L. Zuidin et al. 2003. EFIMOD 2—A model of growth and element cycling of boreal forest ecosystems. *Ecol Model* 170: 373–392.

Kovda, V.A. and B.G. Rozanov (eds). 1988. *Soil Science*. Moscow: V. Shkola.

Kozhevin, P.A. 1989. *Microbial Populations in Nature*. Moscow: Moscow State University Press (in Russian).

Kudeyarov, V.N., F.I. Khakimov, N.F. Deyeva, A.A. Il'ina, T.V. Kuznetzova, and A.V. Timchenko. 1996. Evaluation of respiration in Russian soils. *Eurasian Soil Sci+* 28: 20–34.

Kudeyarov, V.N., G.A. Zavarzin, S.A. Blagodatsky et al. 2007. *Carbon Pools and Fluxes in Terrestrial Ecosystems of Russia.* Moscow: Nauka.

Kumar, S., R. Lal, and D. Liu. 2012. A geographically weighted regression kriging approach for mapping soil organic carbon stock. *Geoderma* 189: 627–634.

Kurbatova, A.S., V.N. Bashkin, Yu. Barannikova et al. 2004. *Ecological Functions of Urban Soils.* Smolensk, Russia, Manjeta.

Kurganova, I.N. 2010. *Emission and balance of carbon dioxide in terrestrial ecosystems of Russia.* Doctoral Thesis, Moscow State University, Moscow.

Kurganova, I.N., A.M. Yermolaev, V.O. Lopes de Gerenyu. 2007. Carbon balance in the soils of abandoned lands in Moscow region. *Eurasian Soil Sci+* 40: 51–58.

Kurganova, I.N., V.L. De Gerenuy, L. Rozanova, D. Sapronov, T. Myakshina, and V. Kudeyarov. 2003. Annual and seasonal CO_2 fluxes from Russian southern taiga soils. *Tellus, Ser B Chem Phys Meteorol* 55(2): 338–344.

Kurganova, I.N., V. Lopes de Gerenyu, J. Six, and Y. Kuzyakov. 2014. Carbon costs of collective farming collapse in Russia. *Glob Change Biol* 20(3): 938–947.

Kuzyakov, Y. and O. Gavrichkova. 2010. Review: Time lag between photosynthesis and carbon dioxide efflux from soil: A review of mechanisms and controls. *Glob Change Biol* 16: 3386–3406.

Kuzyakov, Y. and A.A. Larionova. 2005. Root and rhizomicrobial respiration: A review of approaches to estimate respiration by autotrophic and heterotrophic organisms in soil. *J Plant Nutr Soil Sci* 168(4): 503–520.

Lal, R. 2004. Agricultural activities and the global carbon cycle. *Nutr Cycl Agroecosyst* 70: 103–116.

Lal, R. and J.P. Bruce. 1999. The potential of world cropland soils to sequester C and mitigate greenhouse effect. *Environ Sci Policy* 50(3–4): 251–258.

Larionova, A.A., I.N. Kurganova, V.O.L. de Gerenyou, B.N. Zolotareva, I.V. Yevdokimov, and V.N. Kudeyarov. 2010. Carbon dioxide emissions from agrogray soils under climate changes. *Eurasian Soil Sci+* 43: 168–176.

Leake, J.R., D. Johnson, D.P. Donnelly, G.E. Muckle, L. Boddy, and D.J. Read. 2004. Networks of power and influence: The role of mycorrhizal mycelium in controlling plant communities and agroecosystem functioning. *Can J Bot* 82: 1016–1045.

Le Mer, J. and P. Roger. 2001. Production, oxidation and consumption of methane by soils: A review. *Eur J Soil Biol* 37: 25–50.

Lehmann, A. and K. Stahr. 2007. Nature and significance of anthropogenic urban soils. *J Soils Sediment* 7: 247–296.

Levy, P.E., A. Burdent, M.D.A. Cooper et al. 2012. Methane emissions from soils: Synthesis and analysis of a large UK data set. *Glob Change Biol* 18(5): 1657–1669.

Lin, Q. and P.C. Brookes. 1996. Comparison of methods to measure microbial biomass in unamended, ryegrass-amended and fumigated soils. *Soil Biol Biochem* 28: 933–939.

Lin, Q. and P.C. Brookes. 1999a. An evaluation of substrate-induced respiration method. *Soil Biol Biochem* 31: 1969–1983.

Lin, Q. and P.C. Brookes. 1999b. Comparison of substrate-induced respiration, selective inhibition and biovolume measurements of microbial biomass and its community structure in unamended, ryegrass-amended, fumigated and pesticide-treated soils. *Soil Biol Biochem* 31: 1999–2014.

Lorenz, K. and E. Kandeler. 2006. Microbial biomass and activities in urban soils in two consecutive years. *J Plant Nutr Soil Sci* 169: 799–808.

Lorenz, K. and R. Lal. 2009. Biogeochemical C and N cycles in urban soils. *Environ Int* 35: 1–8.

Lorenz, K., C.M. Preston, and E. Kandeler. 2006. Soil organic matter in urban soils: Estimation of elemental carbon by thermal oxidation and characterization of organic matter by solid-state C nuclear magnetic resonance (NMR) spectroscopy. *Geoderma* 130: 312–323.

Lysak, L.V., E.V. Lapygina, and I.A. Konova. 2013. Characteristics of bacterial communities of urban polluted soils in Moscow region. *Dokl. Ekologich. Pochvoved* 1: 202–213.

Lysak, L.V., E.V. Lapygina, I.A. Konova, and D.G. Zvyagintsev. 2010. Population density and taxonomic composition of bacterial nanoforms in soils of Russia. *Eurasian Soil Sci+* 43(7): 765–770.

MacDonald, C.A., N. Thomas, L. Robinson et al. 2009. Physiological, biochemical and molecular responses of the soil microbial community after afforestation of pasture with *Pinus radiata. Soil Biol Biochem* 41: 1642–1651.

Marfenina, O.E., A.E. Ivanova, E.E. Kislova, E.P. Zazovskaya, and I.Yu. Chernov. 2008. Fungal communities in the soils of early medieval settlements in the taiga zone. *Eurasian Soil Sci+* 41(7): 749–758.

Martens, R. 1987. Estimation of microbial biomass in soil by the respiration method: Importance of soil pH and flushing methods for the measurement of respired CO_2. *Soil Biol Biochem* 19: 77–81.

Martens, R. 1995. Current methods for measuring microbial biomass C in soil: Potentials and limitations. *Biol Fertil Soils* 19(2–3): 87–99.

Martin, M.P., M. Wattenbach, P. Smith et al. 2011. Spatial distribution of soil organic carbon stocks in France. *Biogeosciences* 8: 1053–1065.

McBratney, A.B., M.L. Mendonca Santos, and B. Minasny. 2003. On digital soil mapping. *Geoderma* 117: 3–52.

McBratney, A.B., I.O.A. Odeh, T.F.A. Bishop, M.S. Dunbar, and T.M. Shatar. 2000. An overview of pedometric techniques for use in soil survey. *Geoderma* 97: 293–327.

MEA—Millennium Ecosystem Assessment. 2005. *Ecosystems and Human Well-being: Synthesis.* Washington, D.C.: Island Press.

Meersmans, J., F. De Ridder, F. Canters, S. De Baets, and M. Van Molle. 2008. A multiple regression approach to assess the spatial distribution of soil organic carbon (SOC) at the regional scale (Flanders, Belgium). *Geoderma* 143: 1–13.

Meshcheryakov, P.V., E.V. Prokopovich, and I.N. Korkina. 2005. Transformation of ecological conditions of soil and humus substance formation in the urban environment. *Russ J Ecol* 36(1): 8–15.

Milesi, C. and S.W. Running. 2005. Mapping and modeling the biogeochemical cycling of turf grasses in the United States. *Environ Manage* 36: 426–438.

Minasny, B., A.B. McBratney, B.P. Malone, and I. Wheeler. 2013. Digital mapping of soil carbon. *Adv Agron* 118: 1–47.

Mirchink, T.G. and N.S. Panikov. 1985. Modern ways to estimate biomass and productivity of fungi and bacteria in soils. *Uspekhi Microbiol* 20: 198–226 (in Russian).

Mishra, U., R. Lal, D.S. Liu, and M. Van Meirvenne. 2010. Predicting the spatial variation of the soil organic carbon pool at a regional scale. *Soil Sci Soc Am J* 74: 906–914.

Moyano, F., W. Kutsch, and C. Rebmann. 2008. Soil respiration fluxes in relation to photosynthetic activity in broad-leaf and needle-leaf forest stands. *Agric Forest Manage* 48: 135–143.

Nakadai, T., H. Koizumi, Y. Usami, M. Satoh, and T. Oikawa. 1993. Examination of the methods for measuring soil respiration in cultivated land: Effect of carbon dioxide concentration on soil respiration. *Ecol Res* 8: 65–71.

Nakamoto, T. and S. Wakahara. 2004. Development of substrate induced respiration (SIR) method combined with selective inhibition for estimating fungal and bacterial biomass in humic andosols. *Plant Prod Sci* 7: 70–76.

Naumov, V.D. (ed.). 2009. 145 years to forest experimental station of RSAU-MTAA. RSAU-MTAA.

Ni, J. 2001. Carbon storage in terrestrial ecosystems of China: Estimates at different spatial resolutions and their responses to climate change. *Climatic Change* 49: 339–358.

Nilsson, S., A. Shvidenko, V. Stolbovoi, M. Gluck, M. Jonas, and M. Obersteiner. 2000. *Full Carbon Account for Russia.* Interim report. IR-00-021. Laxenburg, Austria.

Nwachukwu, O.I. and I.D. Pulford. 2011. Microbial respiration as an indication of metal toxicity in contaminated organic materials and soil. *J Hazard Mater* 185: 1140–1147.

Oke, T.R. 1973. City size and the urban heat island effect. *Atmos Environ* 7: 769–779.

Oreshkina, N.S. 1988. *Statistical Estimation of Spatial Variability in Soil Properties.* Moscow: Moscow State University Press.

Orlov, D.S. 1998. Organic substances of Russian soils. *Eurasian Soil Sci+* 31: 946–953.

Pataki, D.E., R.J. Alig, and A.S. Fung. 2006. Urban ecosystems and the North American carbon cycle. *Glob Change Biol* 12: 2092–2102.

Pickett, S.T.A., M.L. Cadenasso, J.M. Grove et al. 2011. Urban ecological systems: Scientific foundations and a decade of progress. *J Environ Manage* 92: 331–362.

Pirt, S.J. 1975. *Principles of Microbe and Cell Cultivation.* New York: John Wiley and Sons.

Polyanskaya, L.M., A.V. Golovchenko, and D.G. Zvyagintzev. 1995. Microbial biomass pool of the soil. *Dokl Akad Nauk* 344: 846–848 (in Russian).

Polyanskaya, L.M., V.V. Geiderbrekht, and D.G. Zvyagintsev. 1996. Fungal biomass in different soil types. *Eurasian Soil Sci+* 28: 43–52.

Post, W.M. and K.C. Kwon. 2000. Soil carbon sequestration and land-use changes: Processes and potential. *Glob Change Biol* 6: 317–327.

Pouyat, R.V., I.D. Yesilonis, and D.J. Nowak. 2006. Carbon storage by urban soils in the United States. *J Environ Qual* 35: 1566–1575.

Priha, O. and A. Smolander. 1994. Fumigation-extraction and substrate-induced respiration derived microbial biomass C, and respiration rate in limed soil of Scots pine sampling stands. *Biol Fertil Soils* 17: 301–308.

Prokofieva, T.V., I.A. Martynenko, and F.A. Ivannikov. 2011. Classification of Moscow soils and parent materials and its possible inclusions in the classification system of Russian soils. *Eurasian Soil Sci+* 44: 561–571.

Prokofieva, T.V. and M.N. Stroganova. 2004. *Soils of Moscow City (Soils in Urban Environment, Their Specifics and Environmental Significance)*. Moscow Biological. Moscow: GEOS.

Qian, Y., R.F. Follet, and J.M. Kimble. 2010. Soil organic carbon input from urban turfgrasses. *Soil Sci Soc Am J* 74(2): 366–371.

Raich, J.W. and A. Tufekcioglu. 2000. Vegetation and soil respiration: Correlations and controls. *Biogeochemistry* 48: 71–90.

Raich, J.W., C.S. Potter, and D. Bhagawati. 2002. Interannual variability in global respiration, 1980–94. *Glob Change Biol* 8: 800–812.

Rey, A., E. Pegorado, V. Tedeschi, I. De Parri, P.G. Jarvis, and R. Valentini. 2002. Annual variation in soil respiration and its components in a coppice oak forest in Central Italy. *Glob Change Biol* 8: 851–866.

Rossiter, D.G. 2007. Classification of urban and industrial soils in the world reference base for soil resources. *J Soils Sediment* 7: 96–100.

Russian Federation State Statistic Service (RFSSS). 2012. *Population of Russian Federation*. Moscow.

Saier, M.H. 2007. Are megacities sustainable? *Water Air Soil Pollut* 178: 1–3.

Savage, K.E. and E.A. Davidson. 2003. A comparison of manual and automated systems for soil CO_2 flux measurements: Trade-offs between spatial and temporal resolution. *J Exp Bot* 54: 891–899.

Scalenghe, R. and F.A. Marsan. 2009. The anthropogenic sealing of soil in urban areas. *Landsc Urban Plan* 90: 1–10.

Schleisinger, W.H. 1995. An overview of the carbon cycle. In *Soils and Global Change*, eds R. Lal, J. Kimble, E. Levine, and B.A. Stewart. Boca Raton: Lewis Publisher, 9–25.

Schmidt, M.W.I. and A.G. Noack. 2000. Black carbon in soils and sediments: Analysis, distribution, implications and current changes. *Glob Biogeochem Cycl* 14(3): 777–793.

Schulze, E.D. 2006. Biological control of the terrestrial carbon sink. *Biogeosciences* 2: 147–166.

Semenov, M.V., E.V. Stolnikova, N.D. Ananyeva, and K.V. Ivashchenko. 2013. Structure of the microbial community in soil catena of the right bank of the Oka river. *Biol Bull* 40: 266–274.

Shishov, L.L. and Voinovich (eds). 2002. *Soils of Moscow Region and Their Use*. Moscow: Dokuchaev Soil Science Institute.

Shishov, L.L., V.D. Tonkonogov, I.I. Lebedeva, and M.I. Gerasimova. 2004. *Classification and Diagnostics of Russian Soils*. Moscow: Dokuchaev Soil Science Institute.

Sitch, S., B. Smith, I.C. Prentice et al. 2003. Evaluation of ecosystem dynamics, plant geography and terrestrial carbon cycling in the LPJ dynamic global vegetation model. *Glob Change Biol* 9:161–185.

Smagin, A.V. 2005. *Soil Gas Phase*. Moscow: MSU.

Smith, L.C., C.A.M. de Klein, and W.D. Catto. 2008. Effect of dicyandiamide applied in a granular form on nitrous oxide emissions from a grazed dairy pasture in Southland, New Zealand. *New Zeal J Agric Res* 51: 387–396.

Soil Survey Division Staff. 1993. *Soil Survey Manual. Soil Conservation Service*. U.S. Department of Agriculture Handbook 18.

Soil Survey Staff. 1999. *Soil Taxonomy. A Basic System of Soil Classification for Making and Interpreting Soil Surveys*, 2nd edition. Agricultural Handbook 436, Natural Resources Conservation Service, USDA, Washington, D.C., USA.

Sparling, G.P. 1992. Ratio of microbial biomass carbon to soil organic carbon as a sensitive indicator of changes in soil organic matter. *Aust J Soil Res* 30: 195–207.

Sparling, G.P. and B.L. Williams. 1986. Microbial biomass in organic soils: Estimation of biomass C, and effect of glucose or cellulose amendments on the amounts of N and P released by fumigation. *Soil Biol Biochem* 18: 507–513.

Sparling, G.P., C.W. Feltham, A.W. West, and P. Singleton. 1990. Estimation of soil microbial C by a fumigation-extraction method: Use on soils of organic matter content, and a reassessment of the kEC-factor. *Soil Biol Biochem* 22: 301–307.

Stahl, P.D., T.B. Parkin, and N.S. Eash. 1995. Sources of error in direct microscopic methods for estimation of fungal biomass in soil. *Soil Biol Biochem* 27: 1091–1097.

Stockmann, U., M.A. Adams, J.W. Crawford et al. 2013. The knowns, known unknowns and unknowns of sequestration of soil organic carbon. *Agric Ecosyst Environ* 164: 80–99.

Strickland, M.S. and J. Rousk. 2010. Considering fungal: Bacterial dominance in soils—Methods, controls, and ecosystem implications. *Soil Biol Biochem* 42: 1385–1395.

Stroganova, M.N., A.D. Myagkova, and T.V. Prokofieva. 1997. The role of soils in urban ecosystems. *Eurasian Soil Sci+* 30: 82–86.

Sun, Q., H.-L. Fan, J. Liang et al. 2009. Soil respiration characteristics of typical urban lawns in Shanghai. *Chin J Ecol* 28(8): 1572–1578.

Susyan, E.A., N.D. Ananyeva, and E.V. Blagodatskaya. 2005. The antibiotic-aided distinguishing of fungal and bacterial substrate-induced respiration in various soil ecosystems. *Microbiology* 74: 336–342.

Svirejeva-Hopkins, A., H.J. Schellnhuber, and V.L. Pomaz. 2004. Urbanized territories as a specific component of the global carbon cycle. *Ecol Model* 173: 295–312.

Swift, S. 2011. Sequestration of carbon by soil. *Soil Sci* 166(11): 858–871.

Taneva, L. and M. Gonzalez-Meler. 2011. Distinct patterns in the diurnal and seasonal variability in four components of soil respiration in a temperate forest under free-air CO_2 enrichment. *Biogeosciences* 8(10): 3077–3092.

Trumbore, S. 2006. Carbon respired by terrestrial ecosystems—Recent progress and challenges. *Glob Change Biol* 12(2): 141–153.

Turin, I.V. 1959. *Soil Survey: Manual on Soil Field Investigation and Soil Mapping*. Moscow: Akad. Nauk USSR, Dokuchaev Soil Science Institute.

Vasenev, V.I., J.J. Stoorvogel, and I.I. Vasenev. 2013c. Urban soil organic carbon and its spatial heterogeneity in comparison with natural and agricultural areas in the Moscow region. *Catena* 107: 96–102.

Vasenev, V.I., J.J. Stoorvogel, I.I. Vasenev, and R. Valentini. 2014. How to map soil organic stocks in highly urbanized region? *Geoderma* 226–227: 103–115.

Vasenev, V.I., N.D. Ananyeva, and K.V. Ivachenko. 2013a. Influence of contaminants (oil products and heavy metals) on microbiological activity of urban constructed soils. *Russ J Ecol* 6: 475–483.

Vasenev, V.I., N.D. Ananyeva, and O.A. Makarov. 2012. Specific features of the ecological functioning of urban soils in Moscow and Moscow oblast. *Eurasian Soil Sci+* 45: 194–205.

Vasenev, V.I., T.V. Prokofieva, and O.A. Makarov. 2013b. Development of the approach to assess soil organic carbon stocks in megapolis and small settlement. *Eurasian Soil Sci+* 6: 1–12.

Velvis, H. 1997. Evaluation of the selective respiratory inhibition methods for measuring the ratio of fungal: bacterial activity in acid agricultural soils. *Biol Fertil Soils* 25: 354–360.

Vorobyova, L.A. 1998. *Soil Chemical Analysis*. Moscow: MSU.

Vrscaj, B., L. Poggio, and F. Marsan. 2008. A method for soil environmental quality evaluation for management and planning in urban areas. *Landsc Urban Plan* 88: 81–94.

Vuichard, N., Ciais, P., Belelli, L., Smith, P., Valentini, R. 2008. Carbon sequestration due to abandonment of agriculture in former USSR since 1990. *Global Biogeochem Cycles* 22: GB4018.

Wang, M., B. Markert, W. Shen, W. Chen, C. Peng, and Z. Ouyang. 2011. Microbial biomass carbon and enzyme activities of urban soils in Beijing. *Environ Sci Pollut Res* 18: 958–967.

Wang, S., M. Huang, R.A. Mickler, K. Li, and J. Ji. 2004. Vertical distribution of soil organic carbon in China. *Environ Manage* 33: 200–209.

Wardle, D.A. 1992. A comparative assessment of factors which influence microbial biomass carbon and nitrogen levels in soil. *Biol Rev* 67: 321–358.

Wardle, D.A. and A. Ghani. 1995. Why is the strength of relationship between pairs of methods for estimating soil microbial biomass often so variable? *Soil Biol Biochem* 27: 821–828.

Wardle, D.A. and D. Parkinson. 1991. Relative importance of the effect of 2,4-D, glyphosate, and environmental variables on the soil microbial biomass. *Plant Soil* 34: 209–219.

West, A.W. 1986. Improvement of the selective inhibition technique to measure eukaryote–prokaryote ratios in soils. *Microbiol Methods* 5: 125–138.

Wilson, K.B. and D.D. Baldocchi. 2000. Seasonal and interannual variability of energy fluxes over a broad-leaved temperate deciduous forest in North America. *Agric Forest Meteorol* 100: 1–18.

Winding, A., K. Hund-Rink, and M. Rutgers. 2005. The use of microorganisms in ecological soil classification and assessment concepts. *Ecotoxicol Environ Safe* 62: 230–248.

Yan, T., L. Yang, and C.D. Campbell. 2003. Microbial biomass and metabolic quotient of soils under different land use in Three Gorges Reservoir area. *Geoderma* 115: 129–138.

Zavarzin, G.A. and V.N. Kudeyarov. 2006. Soil as the key source of carbonic acid and reservoir of organic carbon on the territory of Russia. *Herald Russ Acad Sci* 76: 12–26

Zhang, G.-X., J. Xu, G.-B. Wang, S.-S. Wu, and H.-H. Ruan. 2010. Soil respiration under different vegetation types in Nanjing urban green space. *Chin J Ecol* 29(2): 274–280.

Zircle, G., R. Lal, and B. Augustin. 2011. Modeling carbon sequestration in home lawns. *HortScience* 46: 808–814.

Zvyaginzev, D.G. 1994. Vertical distribution of microbial communities in soils. In *Beyond the Biomass*, eds K. Ritz, J. Dighton, and K. Giller, pp. 29–37. West Sussex: Wiley.

10 Urban Carbon Footprint Evaluation of a Central Chinese City
The Case of Zhengzhou City

Lipeng Hou and Rongqin Zhao

CONTENTS

10.1 Introduction 224
10.2 Literature Review 224
    10.2.1 Researches on Carbon Emissions..224
    10.2.2 Researches on Carbon Footprint 224
    10.2.3 The Purpose of This Study 225
10.3 Data and Methods...225
    10.3.1 Data Sources 225
    10.3.2 Accounting Categories...226
    10.3.3 Evaluation Methods 226
        10.3.3.1 Carbon Footprint of Energy Activities 226
        10.3.3.2 Carbon Footprint of Industrial Production 227
        10.3.3.3 Carbon Footprint of Agricultural Activities 227
        10.3.3.4 Carbon Footprint of Land Use..227
        10.3.3.5 Carbon Footprint of Wastes 228
10.4 Results..229
    10.4.1 The Results of Urban Carbon Footprint 229
        10.4.1.1 Carbon Footprint of Energy Activities 229
        10.4.1.2 Carbon Footprint of Industrial Production 230
        10.4.1.3 Carbon Footprint of Agricultural Activities 231
        10.4.1.4 Carbon Footprint of Wastes 231
        10.4.1.5 Carbon Sinks of Land Use...232
        10.4.1.6 The Carbon Budget of Zhengzhou...233
    10.4.2 Urban Carbon Footprint Pressure Analysis..233
        10.4.2.1 Carbon Footprint Intensity and Carbon Footprint Productivity...............233
        10.4.2.2 Decoupling Analysis of Economic Growth and Carbon Footprint 234
    10.4.3 The Prediction of Urban Carbon Footprint 235
10.5 Conclusions and Policy Implications...235
    10.5.1 Conclusions..235
    10.5.2 Policy Implications 236
Acknowledgments..236
References..236

10.1 INTRODUCTION

Since 1990s, issues of carbon footprint and global climate change have increasingly become major technological as well as important societal and political challenges, which are closely related to energy generation and use (Pierucci, 2009). Anthropocentric carbon footprint from traditional fossil-fuel energy consumption is one of the main causes of global warming. The strategies for greenhouse effect mitigation and carbon footprint reduction are very important for each country to combat climate change. To explore the impact of human activities on global carbon cycles, carbon footprint caused by economic development and energy consumption has become one of the major concerns in academic circles (Liu et al., 2002; Qi et al., 2004; Zhang, 2006; Zhu et al., 2009). Because the production value of heavy industries accounts for a large proportion of the total GDP, there is huge energy consumption in China during industrial activities. With rapid economic development, CO_2 emission in China increased more than 73% from 1990 to 2003 to 17 million tons, and China has become the world's second largest carbon emitter (International Energy Agency, 1996; Zou et al., 2009). Therefore, in the efforts to decrease CO_2 emission, carbon emissions in China and their changes have become the focus of all countries across the world.

10.2 LITERATURE REVIEW

10.2.1 RESEARCHES ON CARBON EMISSIONS

In essence, the impact of human economic and energy activities on regional carbon cycles is largely achieved by changing the industrial space pattern. The alteration of industrial space structure and the regional differences will change the pattern and intensity of human energy consumption and further affect the rate of the regional carbon cycle. Therefore, carbon emissions from industrial activities have also become the concern of scholars at home and abroad. For example, Schipper et al. (2001) analyzed the carbon emission intensity of nine manufacturing sectors of 13 International Energy Agency countries, using the factor decomposition method; they explained the main reasons for growth in carbon emissions since 1990 and made evaluations combined with the targets of the *Kyoto Protocol*. Chang et al. (1998) studied the industrial carbon emission of Taiwan based on the input–output approach and decomposition model. Casler et al. (1998) used the model method to analyze the structure of U.S. carbon emissions and showed that the use of alternative energy was the major factor causing carbon emission decline. Yu et al. (2009) and Wei et al. (2009) used the input–output analysis to compare the carbon leakage and transfer of different industries in the study of carbon emissions embodied in international trade. In addition, some scholars carried out research on the relationship between different industries and carbon emissions (Zhang, 2005; Tan et al., 2008).

10.2.2 RESEARCHES ON CARBON FOOTPRINT

Carbon footprint was put forward based on the concept of ecological footprint. It is the measure of the amount of direct or indirect CO_2 emissions caused by an activity (or accumulation of a product in life cycle) (Wiedman et al., 2007). There are two views on the comprehension of carbon footprint. One view defines the carbon footprint as the carbon emission of human activities (Wiedman et al., 2007; Lee, 2011), that is, to measure it by emission amounts. In this view, Christopher et al. (2008) calculated the household carbon footprint in the United States by using the input–output model. Gary et al. (2008) found that the carbon footprint produced on Christmas day accounts for 5.5% of the whole year in England. Qi et al. (2010) estimated the carbon footprint in China from 1992 to 2007 and found a nearly twofold increase over that time period. The other viewpoint regards the carbon footprint as part of the ecological footprint, that is, the ecological carrying capacity required for absorbing CO_2 emissions from fossil-fuel combustion (Global Footprint Network, 2007; Wiedman et al., 2007), which measures in area. In this view, Kenny et al. (2009) compared and analyzed the performance of six carbon footprint models for use in Ireland. On the basis of global average net ecosystem production (NEP) of forest and grass, Xie et al. (2008) made an analysis of ecological footprint (carbon footprint) brought by fossil energy and electricity consumption in China. As the measurement of impact and pressure of human activities on the environment, carbon footprint has become the new focus in the field of ecology in recent years. For example, in the "Living Planet Report" (World Wildlife Fund, 2008) for calculating ecological footprint, carbon footprint as a separate category includes not only the direct carbon emissions caused by fossil-fuel combustion but also indirect carbon emissions brought by foreign imports. The results showed that the global ecological footprint per capita was 2.7 hm^2, in which carbon footprint was 1.41 hm^2, thus demonstrating that carbon footprint was an important factor causing human ecological impact. Sovacool et al. (2010) carried out an assessment and analysis on 12 metropolitan carbon footprints and put forward policy proposals to reduce carbon footprint. Kenny et al. (2009) compared and analyzed the performance of six carbon footprint models for use in Ireland. Schulz (2010) took Singapore as the case, estimated the direct and indirect greenhouse gas emission footprint of a small and open economic system, and suggested that indirect pressures of urban systems should be included in discussions of effective and fair adaptation and mitigation strategies. Some Chinese scholars carried out beneficial exploration on carbon footprint studies from the angle of carbon footprint accounting (Huang et al., 2009), carbon footprint per capita and carbon footprint products (Guo, 2009), and the infection and inductivity of carbon footprint (Lai et al., 2006). Overall, carbon footprint research is still in its early stages and further development is needed, especially in the field of regional differences in carbon footprint of various human energy activities. The studies of carbon sink were usually carried out on a small scale and the research results were quite different from each other. In China, Zhao et al. (2014a,b) studied the carbon footprint and carbon cycle pressure of Nanjing city. Generally, some research was conducted on the national or regional scale, but it adopted the global average carbon sink value (Xie et al., 2008), which cannot precisely represent the actual situation of different regions in China. Therefore, how to precisely evaluate the ecological carrying capacity of absorbing CO_2 emission measured in area still needs further research.

10.2.3 The Purpose of This Study

Zhengzhou city is the capital of Henan Province, which is located in the center of China. It is also the center of economic and transportation activity of China and has a very important strategic position in the country's regional development. Therefore, carbon footprint evaluation of economic and social life of Zhengzhou city has great significance in analyzing the carbon reduction potential.

Accounting the carbon footprint scientifically and reasonably is of great importance to determine the low-carbon development strategies and roads. There are many researches of carbon emissions on national or provincial level, whereas the studies on city-level are relatively few. The objectives of this study are: (1) to calculate and compare carbon footprint of different processes during human activities in Zhengzhou city; (2) to estimate carbon sinks of terrestrial ecosystems; (3) to calculate and compare the carbon footprint based on carbon budget; (4) to explore the strategies for reducing the carbon footprint of Zhengzhou city.

10.3 DATA AND METHODS

10.3.1 Data Sources

Presently, the main sources of energy are fossil energy, electricity, biomass, solar, hydraulic, wind and nuclear energy, and traditional energy, in which fossil energy is the representative and is the main cause of carbon emissions. Thus this paper will calculate the energy-related carbon footprint of energy activities, industrial production, agriculture and wasters, and so on. The data of energy consumption, land-use data, crop yields, forest, and animal husbandry in Henan provinces from

2000 to 2013 are adopted, which are taken from *China Energy Statistical Yearbook* (2000–2013) and *Zhengzhou Statistical Yearbook* (2000–2013).

10.3.2 ACCOUNTING CATEGORIES

According to the greenhouse gas inventory methods, which are provided by the *IPCC Guidelines for National Greenhouse Gas Inventories*, there are five main accounting categories of urban carbon footprint, as shown in Table 10.1.

10.3.3 EVALUATION METHODS

In this part, the carbon footprint of energy activities (fossil energy, and coal and ore mining) was estimated. Then, through the matching relationship between carbon emission items and relevant data, the carbon emissions of other processes were analyzed.

The following industries were considered to evaluate the carbon footprint of fossil energy (stationary source): coal mining and processing, metal mining industry, metal smelting, manufacturing industry (food and metal), petroleum industry, printing industry, and so on. The carbon footprint generated by railroad, highway, public transit system, private car, and aviation was estimated to explicate the carbon emission of mobile sources. The time span is 2000–2012.

10.3.3.1 Carbon Footprint of Energy Activities

Through the matching relationship between carbon footprint items and energy consumption, the carbon footprint of stationary sources and mobile sources was estimated by (Zhao, 2012)

$$CE_{energy\text{-}i} = Q_{energy\text{-}i} \times H_{energy\text{-}i} \times (C_{energy\text{-}i} + M_{energy\text{-}i}) \tag{10.1}$$

where $CE_{energy\text{-}i}$ is the carbon emission from energy consumption type i, $Q_{energy\text{-}i}$ is the consumption of energy type i, $H_{energy\text{-}i}$ is the net calorific value of energy type i, $C_{energy\text{-}i}$ is CO_2 emission factor of energy type i, $M_{energy\text{-}i}$ is CH_4 emission factor of energy type i, $H_{energy\text{-}i}$ is derived from the China Energy Statistical Yearbook, and carbon emission factors ($C_{energy\text{-}i}$ and $M_{energy\text{-}i}$) of each energy type were derived from IPCC (2006) (Zhao et al., 2014a,b) (Table 10.2).

TABLE 10.1
The Main Accounting Categories of Urban Carbon Footprint

Types	Contents	Calculations Items
Energy activities	Stationary sources	Fossil-fuel combustion
		Coal and ore mining
	Mobile sources	Highways, railways, public transport, private cars, aviation
Industrial production	Cement	Clinker production
	Steel	Fuel combustion
Agriculture	Animals	Intestinal fermentation and manure management
	Paddy	Rice cultivation
Land use	Cropland	Carbon sequestration of crops
	Forest and grass space	Photosynthesis of forest and grassland
	Waters	Carbon sequestration
Waste	Life	Solid waste, wastewater
	Industrial	Solid waste, wastewater

TABLE 10.2
Carbon Content and Carbon Emission Parameters of Each Energy Type

Energy Types	$H_{energy-i}$ kJ/kg (kJ/m³)	$C_{energy-i}$ (kg C/GJ)	$M_{energy-i}$ (kg CH₄/TJ)
Raw coal	20,908.00	25.80	1.00
Cleaned coal	26,344.00	26.21	1.00
Other washed coal	9408.50	26.95	1.00
Coal products	15,909.80	26.60	1.00
Briquette coal	15,909.80	26.60	1.00
Coal water slurry	9408.50	26.95	1.00
Pulverized coal	9408.50	26.95	1.00
Coke	28,435.00	29.20	1.00
Other coke chemicals	34,332.00	26.60	3.00
Coke oven gas	17,353.50	12.10	1.00
Blast furnace gas	2985.19	70.80	1.00
Other coal gas	16,970.33	60.20	1.00
Natural gas	38,931.00	15.30	1.00
Crude oil	41,816.00	20.00	3.00
Gasoline	43,070.00	18.90	3.00
Kerosene	43,070.00	19.60	3.00
Diesel	42,652.00	20.20	3.00
Fuel oil	41,816.00	21.10	3.00
Liquefied petroleum gas	50,179.00	17.20	1.00
Refinery gas	46,055.00	15.70	1.00
Coal tar	33,453.00	20.00	3.00
Other oil products	37,681.20	20.00	3.00
Heating power (kJ/kJ)	1	26.95	1.00
Electricity (kJ/kW h)	3596	26.95	1.00

Source: Adapted from Zhao RQ, Huang XJ, Liu Y. 2014a. *Journal of Geographical Sciences* 24: 159–176 and Zhao RQ et al. 2014b. *Renewable and Sustainable Energy Reviews* 33: 589–601.

10.3.3.2 Carbon Footprint of Industrial Production

The carbon footprint will be emitted from the production processes (except for fossil-fuel combustion) of several industrial products. For example, during the production of cement, CO_2 will be emitted from the lime combustion process. In this paper, carbon emission factors of 0.136 t CO_2/t (Fang et al., 1996) and 1.06 t CO_2/t (Cai et al., 2009) were used to estimate the carbon emissions from the production of cement and steel, respectively.

10.3.3.3 Carbon Footprint of Agricultural Activities

The carbon footprint of agricultural activities includes CH_4 emission from paddies and animals. The carbon footprint of animals was estimated by animal numbers as well as the carbon emission factors of intestinal fermentation and manure management. The emission factors are shown in Table 10.3.

Methane emission from rice was estimated by paddy area, rice production, and carbon footprint factors. Here, the parameters of carbon emission factor of paddies in Henan Province were adopted, which were derived from Tang et al. (2009). CH_4 emission from ruminant animals was estimated by CH_4 emission factors of each animal.

10.3.3.4 Carbon Footprint of Land Use

The carbon footprint of cropland, forest and grassland, and urban waters was considered in the evaluation of carbon footprint of land use.

TABLE 10.3

The Carbon Emission Factors of Animals (kg CH$_4$/year)

Categories	Cattle	Cow	Buffalo	House	Donkey	Mule	Camel	Pig	Goat	Mutton	Poultry
Intestinal fermentation	47	61	55	18	10	10	46	1	5	5	
Manure management	1.149	9.14	1.843	1.09	0.6	0.6	1.28	2.011	0.163	0.162	0.012

Source: Adapted from Zhao RQ, Huang XJ, Liu Y. 2014a. *Journal of Geographical Sciences* 24: 159–176.

Note: (1) Intestinal fermentation parameters were adopted, and the "other cattle factor" was used to replace the carbon emission factor of cattle. (2) Animal manure emission factors of Henan Province were selected to estimate the carbon footprint of agricultural activities.

Here, the carbon absorption of the cropland was calculated by (Zhao, 2012)

$$CI_{crop} = \sum_i CI_{crop\text{-}i} = \sum_i C_{crop\text{-}i} \times Y_{bio\text{-}i} \times (1 - P_{water\text{-}i})$$

$$= \sum_i C_{crop\text{-}i} \times (1 - P_{water\text{-}i}) \times \frac{Y_{eco\text{-}i}}{H_{crop\text{-}i}} \tag{10.2}$$

where CI_{crop} is the total carbon absorption of cropland, $CI_{crop\text{-}i}$ the carbon absorption of cropland type i, $C_{crop\text{-}i}$ the carbon absorption rate to synthesize organic unit (dry weight), $Y_{bio\text{-}i}$ the biomass production of crop type i, $Y_{eco\text{-}i}$ the economic emission of crop type i, $H_{crop\text{-}i}$ the economic factor of crop type i, and $P_{water\text{-}i}$ the water content of crop type i.

Carbon absorption of vegetation photosynthetic is calculated by (Zhao, 2012)

$$CI_{veg} = \sum C_{veg\text{-}i} \times Area_{veg\text{-}i} \tag{10.3}$$

where CI_{veg} is the carbon absorption of vegetation photosynthetic activity (except crops), $C_{veg\text{-}i}$ the carbon sink of unit area per year of vegetation type i, i (=1, 2) stands for forests and urban green spaces, respectively. Carbon productivity here refers to a variety of vegetation types' unit total photosynthetic rate per area and $Area_{veg\text{-}i}$ is the area of vegetation of type i.

Carbon absorption of waters is calculated by (Zhao, 2012)

$$CI_{water} = C_{unit\text{-}water} \times A_{water} + C_{unit\text{-}mud} \times A_{mud} + C_{sub} \times A \tag{10.4}$$

where CI_{water} is the total carbon absorption of waters, $C_{unit\text{-}water}$ and $C_{unit\text{-}mud}$ are the carbon sequestration rates of lakes and beaches, A_{water} and A_{mud} are the area of lakes and beaches, respectively, C_{sub} is the unit land area dry and wet deposition of carbon input to bring, and A is the area of urban waters. Carbon absorption and tidal waters here include the carbon absorption of aquatic organisms' photosynthesis.

10.3.3.5   Carbon Footprint of Wastes

The carbon footprint of wastes mainly comes from two aspects: the life rubbish and the industrial wastes. Therefore, the carbon footprint accounting list of wastes are further divided into two parts, that is, the forms of the wastes of life and industry are solid and wastewater.

10.3.3.5.1 Carbon Footprint of Solid Wastes

Two main final disposal ways of solid wastes are burning and burying. Carbon emission from the two processes can be estimated by the following equations (Zhao, 2012):

$$CE_{\text{waste-burn}} = Q_{\text{waste-burn}} \times C_{\text{waste}} \times P_{\text{waste}} \times EF_{\text{waste}} \qquad (10.5)$$

$$CE_{\text{waste-fill}} = Q_{\text{waste-fill}} \times 0.167 \times (1 - 71.5\%) \qquad (10.6)$$

where the $CE_{\text{waste-burn}}$ is the total carbon emission from burning process of solid wastes, $Q_{\text{waste-burn}}$ the amount of solid wastes that were burned, C_{waste} the carbon content rate of solid wastes (default value is 40%) (IPCC, 2006), P_{waste} the mineral carbon content rate of solid wastes (default value is 40%), EF_{waste} the burning efficiency (default value is 95%), and $CE_{\text{waste-fill}}$ the total CH_4 emission from buried solid wastes. $Q_{\text{waste-fill}}$ is the amount of solid wastes that were buried. The default CH_4 emission factor of buried solid wastes is 0.167 t/t, and the water content of solid wastes is 71.5% (Fang et al., 2007).

10.3.3.5.2 Carbon Footprint of Wastewater

Carbon emission from wastewater includes carbon emissions from life wastewater and industrial wastewater. Carbon emission from the two processes can be estimated by the following equation (Zhao, 2012):

$$CE_{\text{liv-water}} = Num_{\text{people}} \times BOD_{\text{capita}} \times SBF \times C_{\text{BOD}} \times 0.8 \times 365 \qquad (10.7)$$

$$CE_{\text{ind-water}} = Q_{\text{ind-water}} \times COD_{\text{ind-water}} \times C_{\text{COD}} \qquad (10.8)$$

where $CE_{\text{liv-water}}$ is the annual CH_4 emissions from life wastewater, Num_{people} the population number, BOD_{captia} the per capita content of biochemical oxygen demand (BOD) in life wastewater (60 g BOD/day), SBF the ratio of deposited BOD (50%), CBOD the CH_4 emission factor of BOD (0.6 g CH_4/g BOD), $CE_{\text{ind-water}}$ the CH_4 emissions from industrial wastewater, $Q_{\text{ind-water}}$ the yield of industrial wastewater, $COD_{\text{ind-water}}$ the content of chemical oxygen demand (COD) in industrial wastewater (kg COD/m³), which is taken from the *Zhengzhou Statistical Yearbook*. CCOD is the CH_4 emission factor of industrial wastewater (the default value is 0.25 kg CH_4/kg COD) (IPCC, 2006).

10.4 RESULTS

10.4.1 THE RESULTS OF URBAN CARBON FOOTPRINT

10.4.1.1 Carbon Footprint of Energy Activities

With the data of energy consumption provided by the Statistical Yearbook of Zhengzhou City and the above relevant parameters, the carbon footprint of energy activities can be obtained, and the results were shown in Table 10.4.

We can see that the carbon footprint of energy activities increased gradually since 2000. The carbon footprint of energy activities increased from 714.82×10^4 t in 2000 to 2455.99×10^4 t in 2012. As far as stationary sources were concerned, the carbon footprint was 618.29×10^4 t in 2000, and in the year of 2012 it was 1861.77×10^4 t, which represents a twofold increase. The carbon footprint of mobile sources increased from 96.52×10^4 t in 2000 to 594.22×10^4 t in 2012, which represents a fivefold increase. This was mainly attributed to the increasing consumption of energy of industries and the use of private cars.

TABLE 10.4
The Carbon Footprint of Energy Activities in Zhengzhou (10⁴ t)

Year	Stationary Sources	Mobile Sources	Energy Activities
2000	618.29	96.52	714.82
2001	678.18	103.54	781.72
2002	724.05	111.19	835.24
2003	805.47	116.70	922.17
2004	1003.56	130.23	1133.79
2005	1183.16	137.88	1321.04
2006	1278.72	157.66	1436.38
2007	1432.36	178.52	1610.88
2008	1394.08	278.74	1672.81
2009	1329.38	330.26	1659.63
2010	1477.11	413.47	1890.59
2011	1705.98	509.26	2215.24
2012	1861.77	594.22	2455.99

As to the composition of the carbon footprint of energy activities, the proportion of mobiles was less than the stationeries and the burning of fossil fuels of stationary source. The proportion of stationary sources was 86.5% in 2000 while it decreased to 75.8% in 2012, while the proportion of mobiles increased from 13.5% in 2000 to 24.2% in 2012. It was mainly caused by the proliferation of private cars.

10.4.1.2   Carbon Footprint of Industrial Production

With the data of industrial production provided by the Statistical Yearbook of Zhengzhou City and equations and the relevant parameters, the carbon footprint of industrial production can be obtained, and the results shown in Table 10.5.

The carbon footprint of industrial products increased drastically from 38.64×10^4 t in 2000 to 255.2×10^4 t in 2012, in which the carbon footprint of cement has increased by about 2.5 times (from 34.57×10^4 t in 2000 to 86.13×10^4 t in 2012), and the carbon footprint of steel increased from 4.07×10^4 t in 2000 to 139.08×10^4 t in 2012. It was mainly caused by the development of urban construction.

TABLE 10.5
The Carbon Footprint of Industrial Products in Zhengzhou (10⁴ t)

Year	Cement	Steel	Industrial Products
2000	34.57	4.07	38.64
2001	37.31	4.24	41.55
2002	38.20	5.09	43.30
2003	42.06	13.33	55.39
2004	47.40	30.01	77.41
2005	57.05	46.16	103.21
2006	59.05	62.39	121.44
2007	78.22	94.28	172.50
2008	70.40	103.70	174.10
2009	81.53	98.20	179.73
2010	77.74	108.99	186.74
2011	79.86	122.79	202.64
2012	86.13	139.08	225.20

Before 2006, the carbon footprint proportion of cement was higher than the steel industry, but the proportion attributed to the cement industry decreased after 2006. While the proportion of cement decreased from 89.46% in 2000 to 38.24% in 2012, the proportion of the steel industry gradually increased. In 2012 the carbon footprint of the two industries almost flat, which shows that the steel industry was in a prosperous stage, while the cement industry was declining. The proportion of steel rose from 10.54% in 2000 to 61.76% in 2012. The proportion of carbon footprint from cement production was declining, while the proportion from steel gradually increased, showing that the proportion of steel was higher than cement in 2012.

10.4.1.3 Carbon Footprint of Agricultural Activities

With the data of animals and paddies provided by the Statistical Yearbook of Zhengzhou City and the relevant parameters, the carbon footprint of agricultural activities can be obtained, and the results are shown in Table 10.6.

Generally, the carbon footprint of agricultural activities slightly increased from 31.39×10^4 t in 2000 to 36.84×10^4 t in 2012 with the increasing rate of about 17%. The carbon footprint of manure management was increasing while the intestinal fermentation and paddy cultivation were declining since 2000.

We can see that the carbon footprint of manure management accounted for the majority proportion in the total agricultural activities. The proportion of manure management increased from 93.9% in 2000 to 96.39% in 2012, while the proportion of intestinal fermentation and paddy cultivation were on decline. Therefore, the carbon footprint of manure management was the key component of carbon footprint of agricultural activities in Zhengzhou city.

10.4.1.4 Carbon Footprint of Wastes

With the data of wastes provided by the Statistical Yearbook of Zhengzhou City, the carbon footprint of wastes can be obtained (Table 10.7).

The carbon footprint of wastes was on increase. In 2000, the carbon footprint of wastes was 11.04×10^4 t and it was 31.87×10^4 t in 2012, an increase by about 1.8 times. The carbon footprint of solid wastes and wastewater in 2000 was 8.56×10^4 and 2.48×10^4 t, respectively. In 2012, the number was 28.56×10^4 and 3.22×10^4 t, respectively. The carbon footprint of solid wastes has increased 2.3 times and with the increasing rate of about 30%.

The proportion of solid wastes accounted for most of the carbon footprint, which was higher than the wastewater during 2000–2012. In 2000, the proportion of solid wastes was 77.5% and it

TABLE 10.6

The Carbon Footprint of Agricultural Activities in Zhengzhou (10^4 t)

Year	Intestinal Fermentation	Manure Management	Paddy Cultivation	Total Agricultural Activities
2000	1.63	29.47	0.28	31.39
2001	1.75	30.45	0.27	32.48
2002	1.79	31.67	0.25	33.70
2003	1.91	35.61	0.12	37.64
2004	1.99	37.70	0.11	39.80
2005	1.99	39.69	0.11	41.79
2006	2.01	39.28	0.08	41.37
2007	1.31	30.10	0.08	31.49
2008	1.43	33.90	0.05	35.38
2009	1.48	34.40	0.05	35.92
2010	1.43	33.80	0.03	35.26
2011	1.37	34.80	0.03	36.20
2012	1.31	35.51	0.02	36.84

TABLE 10.7
The Carbon Footprint of Wastes (10⁴ t)

Year	Solid Wastes	Wastewater	Wastes
2000	8.56	2.48	11.04
2001	10.62	2.54	13.15
2002	12.70	2.59	15.30
2003	14.83	2.63	17.46
2004	17.01	2.66	19.66
2005	19.26	2.68	21.94
2006	20.87	2.71	23.58
2007	25.39	2.72	28.11
2008	27.75	2.72	30.47
2009	24.21	2.73	26.94
2010	26.58	2.88	29.46
2011	27.62	3.04	30.66
2012	28.65	3.22	31.87

increased to 89.9% in 2012, whereas the proportion of wastewater dropped from 22.5% in 2000 to 10.1% in 2012.

10.4.1.5 Carbon Sinks of Land Use

In this section, the carbon sink generated by different plants was calculated and the results are shown in Table 10.8.

We can see that the total carbon sink of Zhengzhou city increased since 2000. It increased from 61.97×10^4 t in 2000 to 75.87×10^4 t in 2012 with an increasing rate of 22%, in which the carbon sink of cropland was highest (47.63×10^4 t in 2012). The carbon sink of forest was also on the rise, but the maximum value was 52.21×10^4 t, which appeared in 2009. Generally speaking, the proportion of forest and grassland showed a trend of decline. Estimating the carbon sink by farmland ecosystem in Zhengzhou city is very important, as it is the most important region of crops and food production in Henan province of China.

TABLE 10.8
The Carbon Sinks of Land Use in Zhengzhou (10⁴ t)

Year	Cropland	Forest Grassland	Waters	Total Carbon Sinks
2000	36.17	25.56	0.24	61.97
2001	38.14	16.45	0.24	54.83
2002	39.00	26.01	0.24	65.25
2003	39.15	26.94	0.24	66.33
2004	41.02	23.24	0.24	64.50
2005	42.68	39.61	0.24	82.53
2006	45.76	37.38	0.24	83.38
2007	42.70	41.51	0.23	84.45
2008	44.44	40.09	0.24	84.77
2009	44.65	52.21	0.25	97.11
2010	45.75	20.43	0.28	66.46
2011	46.67	25.99	0.31	72.98
2012	47.63	27.88	0.36	75.87

TABLE 10.9
The Carbon Budget of Zhengzhou (10^4 t)

Year	Energy Activities	Industrial Production	Agriculture Process	Wastes	Carbon Sink	Carbon Footprint
2000	714.82	38.64	31.39	11.04	61.97	795.89
2001	781.72	41.55	32.48	13.15	54.83	868.90
2002	835.24	43.30	33.70	15.30	65.25	927.54
2003	922.17	55.39	37.64	17.46	66.33	1032.66
2004	1133.79	77.41	39.80	19.66	64.50	1270.66
2005	1321.04	103.21	41.79	21.94	82.53	1487.98
2006	1436.38	121.44	41.37	23.58	83.38	1622.77
2007	1610.88	172.50	31.49	28.11	84.45	1842.98
2008	1672.81	174.10	35.38	30.47	84.77	1912.76
2009	1659.63	179.73	35.92	26.94	97.11	1902.22
2010	1890.59	186.74	35.26	29.46	66.46	2142.04
2011	2215.24	202.64	36.20	30.66	72.98	2484.74
2012	2455.99	225.20	36.84	31.87	75.87	2749.90

10.4.1.6 The Carbon Budget of Zhengzhou

By summarizing all the tables mentioned earlier, the urban carbon budget can be obtained and the results reported above. As shown in Table 10.9, the carbon budget of Zhengzhou increased from 733.92×10^4 t in 2000 to 2674×10^4 t in 2012, whereas the carbon sink increased from 61.97×10^4 t in 2000 to 75.87×10^4 t in 2012. Carbon sink was far less than carbon footprint, which suggested that the carbon sink was not enough to offset the carbon footprint.

The proportion of energy activities was the largest in the total. Generally speaking, the proportion of energy activities increased since 2000. And the proportion of industrial production and agriculture process was also increasing, whereas the proportion of wastes was decreasing in the past years. The development of economy was still relied on the large consumption of energy especially the traditional energy resources, which caused the rapid growth of carbon footprint.

10.4.2 URBAN CARBON FOOTPRINT PRESSURE ANALYSIS

Some indicators were chosen to evaluate the urban carbon footprint pressure of Zhengzhou: the carbon footprint intensity of urban systems and the carbon footprint production and intensity. The future decoupling status between economic growth and carbon footprint was also analyzed (Zhao et al., 2012).

10.4.2.1 Carbon Footprint Intensity and Carbon Footprint Productivity

With the carbon footprint and GDP data, the carbon footprint intensity and productivity of Zhengzhou can be obtained and the results are shown in Table 10.10.

In 2000, the carbon footprint intensity was 1.09×10^4 t/10^9 yuan and in 2012 it was 0.65×10^4 t/10^9 yuan. The carbon footprint productivity was 0.92×10^9 yuan/10^4 t in 2000 but the number increased to 1.53×10^9 yuan/10^4 t in 2012. The results suggested that the efficiency of carbon cycle was increasing gradually with the application of energy-saving and emission-reduction policies.

TABLE 10.10

The Carbon Footprint Intensity and Productivity of Zhengzhou during 2000–2012

Year	Carbon Footprint (10^4 t)	GDP (10^9 Yuan)	Carbon Footprint Intensity (10^4 t/10^9 Yuan)	Carbon Footprint Productivity (10^9 Yuan/10^4 t)
2000	795.89	728.38	1.09	0.92
2001	868.90	809.23	1.07	0.93
2002	927.54	897.44	1.03	0.97
2003	1032.66	996.16	1.04	0.96
2004	1270.66	1454.39	0.87	1.14
2005	1487.98	1682.73	0.88	1.13
2006	1622.77	1951.96	0.83	1.20
2007	1842.98	2264.28	0.81	1.23
2008	1912.76	2624.30	0.73	1.37
2009	1902.22	2944.46	0.65	1.55
2010	2142.04	3280.13	0.65	1.53
2011	2484.74	3706.55	0.67	1.49
2012	2749.90	4218.05	0.65	1.53

10.4.2.2 Decoupling Analysis of Economic Growth and Carbon Footprint

Here, the decoupling index method and elastic analysis were used to judge the degree of decoupling between economic growth and carbon footprint. The indicator can be calculated with the following equation:

$$DR_{t_0,t_1} = \frac{EP_{t_1}/EP_{t_0}}{DF_{t_1}/DF_{t_0}} \tag{10.9}$$

where the DR is the decoupling index, EP is the environmental pressure, which can be expressed with urban carbon footprint, DF is the quantity of GDP, t_0 and t_1 are the beginning and ending year, respectively. The results are shown in Table 10.11.

TABLE 10.11

Decoupling Analysis of Economic Growth and Carbon Footprint of Zhengzhou

Year	Carbon Footprint (10^4 t)	GDP (10^9 Yuan)	EP_{t1}/EP_{t0}	DF_{t1}/DF_{t0}	$DR_{t1,t0}$
2000	795.89	728.38			
2001	868.90	809.23	1.09	1.11	0.98
2002	927.54	897.44	1.07	1.11	0.96
2003	1032.66	996.16	1.11	1.11	1.00
2004	1270.66	1454.39	1.23	1.46	0.84
2005	1487.98	1682.73	1.17	1.16	1.01
2006	1622.77	1951.96	1.09	1.16	0.94
2007	1842.98	2264.28	1.14	1.16	0.98
2008	1912.76	2624.30	1.04	1.16	0.90
2009	1902.22	2944.46	0.99	1.12	0.89
2010	2142.04	3280.13	1.13	1.11	1.01
2011	2484.74	3706.55	1.16	1.13	1.03
2012	2749.90	4218.05	1.11	1.14	0.97

TABLE 10.12
The Predication of Total Carbon Footprint of Zhengzhou (10^4 t)

Year	Carbon Footprint
2016	5797.80
2017	6000.55
2018	6203.29
2019	6406.04
2020	6608.78
2021	6811.53
2022	7014.27
2023	7217.02
2024	7419.77
2025	7622.51
2026	7825.26
2027	8028.00
2028	8230.75
2029	8433.49
2030	8636.24

Generally speaking, during most of the years 2000–2012, the DR was on decoupling state except the year of 2009. The results suggested that the economic growth of Zhengzhou still relied on the large consumption of fuel, and it did not realize the goals of "low-carbon economy." Therefore, in order to promote the low-carbon development of urban areas, some measures should be taken: enhancing efficient energy use, reducing the use of conventional fuels (especially coals), promoting biological sequestration technologies, multicultivating plants and crops, reducing municipal waste incineration, and so on.

10.4.3 THE PREDICTION OF URBAN CARBON FOOTPRINT

Here, three indicators were used to predict the future carbon footprint: fuel consumption, population, and urban built-up area. The ternary linear regression model was chosen to predict, and the least squares method was used to solve the parameters. The equation used to solve is as follows:

$$Y = -902.72 + 3.56X_1 + 1.56X_2 + 3.9X_3 \qquad (10.10)$$

where the Y is the carbon footprint of future, X_1 is the consumption of fuel, X_2 is the population, and X_3 is the urban built-up area. The carbon footprint of the city in the future can be calculated with the model, and the results are shown in Table 10.12 above.

By prediction with ternary linear regression model, we see that the carbon footprint of Zhengzhou will be 5797.80×10^4 t in 2016 and 8636.24×10^4 t in 2030. It shows a growing trend in the next 20 years. But the growth rate will decline, which shows that the efficiency of carbon emission will increase in the future.

10.5 CONCLUSIONS AND POLICY IMPLICATIONS

10.5.1 CONCLUSIONS

1. The energy activities were the major sources of carbon footprint, accounting for more than 85%, followed by the carbon footprint of industrial production, which accounted for about

10%–15%. The proportion of the smallest carbon footprint was generated by the urban waste disposal. Therefore, optimizing energy structure, improving energy efficiency, and promoting the use of clean energy can prohibit the substantial increasing of urban carbon footprint.

2. CH_4 of the agriculture process generated by intestinal fermentation, manure management, and rice cultivation produced the largest proportion of the total CH_4 emissions, followed by waste disposal and wastewater. CH_4 emission of energy activities was very small; the amount of CH_4 produced by the industrial processes was very few, almost negligible. For the urban atmosphere, the warming potential of CH_4 was higher than CO_2; so the high-efficiency management of animal feeding and manure disposal can effectively curb the temperature growth too quickly.

3. The carbon footprint of Zhengzhou City increased year by year, but the growth rate decreased. The proportion of CO_2 produced by the energy activities decreased. The CH_4 generated by the agriculture has declined. The farming, woodland, urban green space, and urban waters were major carbon sinks in Zhengzhou City, but the carbon sinks were not enough to compensate for carbon footprint.

10.5.2 POLICY IMPLICATIONS

1. Optimizing the energy structure, improving energy efficiency, developing low-carbon and noncarbon energy sources, and promoting the diversification of energy supplies constitute the major steps to reduce urban carbon footprint.

2. To promote the low-carbon development of cities, it is important to strengthen scientific and technological innovation. Accelerating and promoting the existing low-carbon technologies also help us to promote the low-carbon development of urban areas.

3. Green plants are of great significance to human beings, and they are the generator of oxygen gas. So, cultivating more croplands and forest as well as protecting the waters is the most effective way to increase carbon sinks.

4. To reduce the carbon footprint, not only scientific and technological innovations are needed, but also legal measures. Establishing appropriate policies and a long-term mechanism of low-carbon is the city's duty. It calls for the efforts of government and every one all around.

ACKNOWLEDGMENTS

We acknowledge the National Natural Science Foundation of China (No. 41301633), Innovation Project for Undergraduate Students of North China University of Water Resources and Electric Power (201492) and the Human and Social Sciences project of the Education Department of Henan province (2015-GH-088) for funding this study.

REFERENCES

Cai BF, Liu CL, Chen CC. 2009. *City's Greenhouse Gas Emission Inventory Research*. Chemical Industry Press, Beijing, China.

Casler SD, Rose A. 1998. Carbon dioxide emissions in the U.S. economy: A structural decomposition analysis. *Environmental and Resource Economics* 11: 349–363.

Chang YF, Lin SJ. 1998. Structural decomposition of industrial CO_2 emission in Taiwan: An input–output approach. *Energy Policy* 26: 5–12.

Christopher L, Weber H, Scott M. 2008. Quantifying the global and distributional aspects of American household carbon footprint. *Ecological Economics* 66: 379–391.

Fang JY, Guo ZD, Piao SL et al. 2007. The estimation of carbon sink of terrestrial vegetation from 1981 to 2000 in China. *Science in China* (*D*) 37: 804–812.

Gary H, Anne O, Elena D et al. 2008. *The Carbon Cost of Christmas*. Stockholm Environment Institute, Stockholm.

Global Footprint Network. 2007. Ecological footprint glossary. http://www.footprintnetwork.org/gfn_sub.php?content=glossary.

Guo YG. 2009. *The Analysis on Calculation and Characteristics of Greenhouse Gas Emission in Mega-cities.* East China Normal University, Shanghai.

Huang XJ, Ge Y, Ye TL et al. 2009. *Circular Economics.* Southeast University Press, Nanjing.

International Energy Agency. 1996. *World Energy Outlook 1996.* Organization for Economic Cooperation and Development, Paris.

IPCC (Intergovernmental Panel on Climate Change). 2006. *IPCC Guidelines for National Greenhouse Gas Inventories.*

Kenny T, Gray NF. 2009. Comparative performance of six carbon footprint models for use in Ireland. *Environmental Impact Assessment Review* 29: 1–6.

Lai L, Huang XJ, Liu WL. 2006. Adjustment for regional ecological footprint based on input–output technique: A case study of Jiangsu Province in 2002. *Acta Ecologica Sinica* 26: 1285–1292.

Lee KH. 2011. Integrating carbon footprint into supply chain management: The case of Hyundai Motor Company (HMC) in the automobile industry. *Journal of Cleaner Production* 19: 1216–1223.

Liu H, Cheng SK, Zhang L. 2002. The international latest research of the impacts of human activities on carbon emissions. *Progress in Geography* 21: 420–429.

Pierucci S. 2009. PRES 2007: Carbon footprint and emission minimization, integration and management of energy sources, industrial application and case studies. *Energy* 33: 1477–1479.

Qi YC, Dong YS. 2004. Emission of greenhouse gases from energy field and mitigation countermeasures in China. *Scientia Geographica Sinica* 24: 528–534.

Qi Y, Xie GQ, Ge LQ et al. 2010. Estimation of China's carbon footprint based on apparent consumption. *Resource Science* 32: 2053–2058.

Schipper L, Murtishaw S, Khrushch M et al. 2001. Carbon emissions from manufacturing energy use in 13 IEA countries: Long-term trends through 1995. *Energy Policy* 29: 667–688.

Schulz NB. 2010. Delving into the carbon footprints of Singapore—Comparing direct and indirect greenhouse gas emissions of a small and open economic system. *Energy Policy* 38: 4848–4855.

Sovacool BK, Brown MA. 2010. Twelve metropolitan carbon footprints: A preliminary comparative global assessment. *Energy Policy* 38: 4856–4869.

Tan D, Huang XJ. 2008. Correlation analysis and comparison of the economic development and carbon emissions in the Eastern, Central and Western part of China. *China Population, Resource and Environment* 18: 54–57.

Tan D, Huang XJ, Hu CZ. 2008. Analysis of the relationship between industrial upgrading and carbon emission in China. *Sichuan Environment* 27: 74–78.

Wei BY, Fang XQ, Wang Y. 2009. Estimation of carbon emissions embodied in international trade for China: An input–output analysis. *Journal of Beijing Normal University (Natural Science)* 45: 413–419.

Wiedman T, Minx J. 2007. A definition of carbon footprint. http://www.censa.org.uk/docs/ISA-UK_Report_07-01_carbon_footprint.pdf.

World Wildlife Fund. 2008. Living planet report 2008. www.panda.org.

Xie HY, Chen XS, Lin KR. 2008. The ecological footprint analysis of fossil energy and electricity. *Acta Ecologica Sinica* 28: 1729–1735.

Yu HC, Wang LM. 2009. Research on the carbon emission transfer by Sino-US merchandise trade. *Journal of Natural Resources* 24: 1837–1846.

Zhang DY. 2005. *Progress in Estimation Method of Carbon Emission from Industrial Sector of China.* Beijing Forestry University, Beijing.

Zhang L. 2006. A changing pattern of regional CO_2 emissions in China. *Geographical Research* 25:1–9.

Zhao RQ. 2012. *Carbon Cycle of Urban System and Its Regulation Through Land Use Control.* Nanjing University Press, Nanjing.

Zhao RQ, Huang XJ, Liu Y. 2014a. Urban carbon footprint and carbon cycle pressure: The case study of Nanjing. *Journal of Geographical Sciences* 24: 159–176.

Zhao RQ, Huang XJ, Zhong TY et al. 2014b. Carbon flow of urban system and its policy implications: The case of Nanjing. *Renewable and Sustainable Energy Reviews* 33: 589–601.

Zhao RQ, Huang XJ, Zhong TY et al. 2011. Carbon footprint of different industrial spaces based on energy consumption in China. *Journal of Geographical Sciences* 21(2): 285–300.

Zhu YB, Wang Z, Pang L. 2009. Simulation on China's economy and prediction on energy consumption and carbon emission under optimal growth path. *Acta Geographica Sinica* 64:935–944.

Zou XP, Chen SF, Ning M et al. 2009. An empirical research on the influence factor of carbon emission in Chinese provincial regions. *Ecological Economics* 3:34–37.

11 Carbon Footprint Estimation from a Building Sector in India

Venu Shree, Varun Goel, and Himanshu Nautiyal

CONTENTS

11.1 Introduction 239
11.2 Carbon Footprint Methodologies 241
11.3 Past Studies Related to Carbon Footprint of Buildings 244
11.4 Case Studies in India 247
11.5 Energy Conservation in Buildings 252
11.6 Recommendations 254
11.7 Conclusion 255
References 255

11.1 INTRODUCTION

A significant variation in weather conditions, which is a part of climate change, has been observed on Earth. Although changes occur in climate due to natural phenomena (such as volcanoes, tectonic plates, variation in sun's energy, etc.), there are many domestic and commercial activities of human beings which are quite responsible for the increase in rate of climate change problems. The effect of climate change, which is referred to as global warming, can be observed as increase of earth's temperature year by year. Apart from increase in earth's temperature, natural disasters and extreme weather events are also being seen in many parts of the world. According to the Intergovernmental Panel on Climate Change (IPCC), greenhouse gas (GHG) emissions produced by human activities are quite responsible for global warming. The average surface temperature of earth increased from 0.6°C to 0.9°C per decade during 1906–2005. As a matter of fact, an increasd of 1°C temperature in earth is quite significant because rise in 1°C temperature has ample heat to warm the glaciers, oceans, lands, and atmosphere (www.earthobservatory.nasa.gov).

All climate change problems in the earth are due to increase in GHG concentration into the atmosphere. GHG emissions are also responsible for contaminating fresh air in the atmosphere due to which health of human beings is also affected. A study on the nature of air pollution and sources of emissions in India shows that industries, residential, transportation, power generation, and construction are the major sectors of emissions and are responsible for increase in pollution and health-related impacts in Indian cities (Guttikunda et al. 2014). The main constituents of GHG emissions are carbon dioxide (CO_2), methane (CH_4), nitrous oxide (N_2O), and fluorinated gases. Although CO_2 is a part of Earth's atmosphere, its proportion in earth is being increased by various human activities. CO_2 is the primary constituent of GHGs and mainly produced by the combustion of fossil fuel. In addition, other activities such as deforestation and so on also have a significant role in the increase of CO_2 gas in the atmosphere. Methane is another important GHG associated with waste management and use of agricultural by-products. Also, the excessive use of refrigerants of halocarbon series in heating, ventilation, and air-conditioning (HVAC) systems is responsible for emissions of fluorinated gases releasing into the atmosphere.

As the population of world is increasing drastically, an increase in energy demands has also been noticed. The growth in population, urbanization, and energy demands are the most significant factors to increase the level of CO_2 emissions in atmosphere (Levine et al. 1996). Conventional energy sources such as petroleum products and fossil fuels are extensively used to fulfill high energy demands of world population. The combustion of conventional energy sources produces high amount of emissions into the atmosphere. Although renewable energy-based power generation is being increased throughout the world, the dependency on conventional energy sources still exists. Figure 11.1 shows CO_2 emissions of the world in million tons from different sectors in past years. It can be seen from the graph that energy supply, transport, and residential sectors are significant contributors in total global emissions.

As discussed, the large amount of CO_2 in the Earth's atmosphere is responsible for climate change problems and other environmental issues. This is due to the interaction of heat flows in the environment (Raupach and Fraser 2011). Consequently, the problem of global warming is commonly observed in the earth. Therefore, it is quite necessary to minimize ecological threats by taking immediate and effective action at local and international levels. In 1997, the Kyoto Protocol came into existence with an objective to control the climate change problems. A mechanism popularly known as clean development mechanism was introduced under an article of the Kyoto Protocol, with an objective to control GHG emissions in developed countries (Annex I countries) and to promote sustainable development in developing countries (non-Annex I countries) (Nautiyal and Goel 2012). In addition, interest is being increased throughout the world to reduce carbon emissions associated with activities of human beings in various sectors, namely, power, building, transportation, and so on. Nowadays, the concept of low-carbon city is becoming quite popular, with a great interest to reduce GHG emissions (Zhang et al. 2014). A large portion of world's economic output is associated with cities as more than half of the world's population lives in cities and contributes about 70% of the total world's emissions. That is why policy makers are quite interested to reduce the emissions of cities, and consequently the concept of low-carbon city is coming up (Sullivan et al. 2012). To achieve a low-carbon city, improvement of energy efficiency of buildings within the boundary is essentially required (Chavez and Ramaswami 2014).

The consequences of large-scale construction, transportation, and so on have a big role in the increase of GHG emissions in Earth's atmosphere. Building sector is one of the most important sectors having a significant role in the generation of high amount of GHG emissions. As a matter of fact, a building contributes a large amount of GHG emissions in all the stages of its life cycle, that is, from its construction to demolition (EIA, Annual Energy Outlook 2008). All buildings, that is, residential and commercial buildings, account for high energy consumption and generation of GHG emissions, which are continuously increasing throughout the world (Dhaka et al. 2012). About

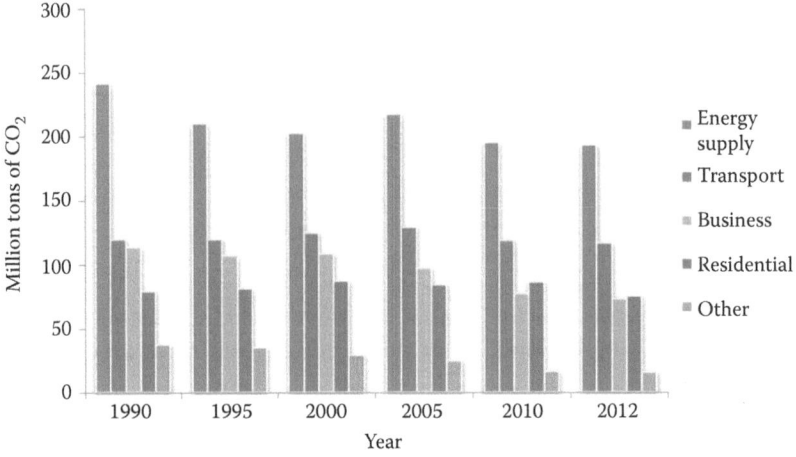

FIGURE 11.1 CO_2 emissions in world from various sources (www.gov.uk/decc).

30%–40% of the primary energy is consumed in construction, operation, and maintenance phase of buildings throughout the world (Bansal et al. 2014). The energy consumed in the construction, operation, and demolition of a building is referred to as direct energy. The energy consumption is also associated with the production of materials in the different tasks of construction and technical installations of buildings (Sartori and Hestnes 2007). Consequently, buildings are responsible for generating a high amount of GHG emissions into the atmosphere. Therefore, it is very important to explore and detect those sources related to buildings, which are responsible for generating GHG emissions. It is also important to analyze the performance of a building and its elements before construction for achieving sustainable development (Srinivasan et al. 2014). But above all, there is a great need to explore appropriate tools and techniques to calculate environmental impacts through GHG emissions, so that environmental sustainability of buildings can be improved. Also, it is important to design and develop buildings with high performance so that GHG emissions can be reduced significantly for achieving sustainable environment.

In order to control GHG emissions, the first important step is to estimate or quantify GHG emissions associated with a product and its process during entire product's life cycle. The concept of carbon footprints helps in estimating GHG emissions, so that the issues of different climate change problems can be solved. The term "carbon footprint" is becoming very popular throughout the world because the problem of climate change has become a serious issue of concern. The concept of carbon footprint originated basically from the term "ecological footprint," proposed by Wackernagel and Rees (1996). Carbon footprint is related to GHG emissions generated by a product and its process and activities over its entire life cycle. It may also be defined as the amount of CO_2 and its equivalent produced by different anthropogenic activities (Ramachandra and Shwetmala 2012). Several definitions of carbon footprint have been proposed by the researchers. According to Wiedmann and Minx (2008), "The carbon footprint is a measure of the exclusive total amount of carbon dioxide emissions that is directly and indirectly caused by an activity or is accumulated over the life stages of a product." Also, Wiedmann and Minx (2007) have discussed uniformity in the definitions of carbon footprints. According to the Parliamentary Office of Science and Technology (POST 2006), "A carbon footprint is the amount of CO_2 and other greenhouse gases emitted over full life cycle of a process or product." It is expressed as grams of CO_{2eq} per kilowatt hour of generation (gCO_{2eq}/kWh), which accounts for different global warming effects of other GHGs (Wiedmann and Minx 2008).

From various definitions of carbon footprint, it can be found that it is the estimation of GHG emissions associated with a product. As discussed, there are many gases, other than CO_2, that are constituents of GHG emissions; therefore, estimation of GHG emissions means estimation of all GHGs. As many gases are not based on carbon as well as due to unavailability of data, they are difficult to be quantified. In addition, CO_2 is an important component of earth's natural cycle and also comes from natural occurrences. Therefore, it is very difficult to estimate the total amount of carbon footprints due to the unavailability of many data.

In this chapter, discussions begin with different methods to estimate carbon footprints of buildings. The past studies pertaining to estimation of emissions for residential and commercial buildings are discussed. Some case studies on the estimation of emissions associated with buildings in India are also discussed. In addition, a brief discussion on the techniques to reduce the carbon footprint of buildings is also introduced.

11.2 CARBON FOOTPRINT METHODOLOGIES

As discussed, carbon footprint refers to GHG emissions associated with a product during its complete life cycle or from its cradle to grave. Life-cycle analysis (LCA) is an effective and popular technique to estimate GHG emissions associated with a product and its process. It provides complete information of all inflows and outflows related to a product with respect to release of GHG emissions. It can be described as a tool for comparative analysis and is used to estimate environmental hazards and resource consumption associated with a process and its product and activity over complete life cycle (SETAC 1991; Tshudy 1994; Basket et al. 1995). Input–output and process-based analyses are the most common methodologies used to carry out LCA. LCA helps in determining carbon footprints by estimating GHG emissions generated by a product during its different stages of life cycle. Therefore, LCA has been a widely accepted technique to calculate environmental impacts of different products and processes (Sharma et al. 2011; Proietti et al. 2013). The technique of LCA was introduced from the comprehensive environmental assessment of different products and is first used in the United States and Europe (Boustead 1996). LCA studies are competent to address different environmental aspects of products systems, from acquisition of raw material to its final disposal (ISO 14040 2006). In addition, it is quite helpful to provide large information related to environment by data collection and analysis in the form of several indicators such as carbon footprint, energy consumption, toxicity, and so on. Therefore, these studies are important to explore several critical facts and issues associated with environmental and climate change problems. However, the details and depth of LCA studies depend on their definitions of goal and scope. Nowadays, advent of LCA software such as Gabi, SimaPro, Building for Environmental and Economic Sustainability (BEES), and Building Life-Cycle Cost (BLCC) is also becoming popular and helps in estimating and studying environmental impacts (www.buildingecology.com).

The main phases included in the LCA are the goal scope and definition, inventory analysis, life-cycle impact assessment (LCIA), and result interpretation, as shown in Figure 11.2. In addition, there are many parameters that come under goal and scope definition, that is, functional unit, system boundaries, and so on. Many times recycling of products has been performed, and the same material is successively used in some other products. In such situations, a problem known as allocation problem arises. This allocation problem is described in most LCA guidelines, and an allocation procedure is also recommended (ISO 14041 1998). There are many ISO standards for the estimation of carbon footprints related to the product.

The first phase of an LCA process is the definition of goal and scope, as shown in Figure 11.2. In this phase, objective, scope, and system of the study are defined. As discussed earlier, details

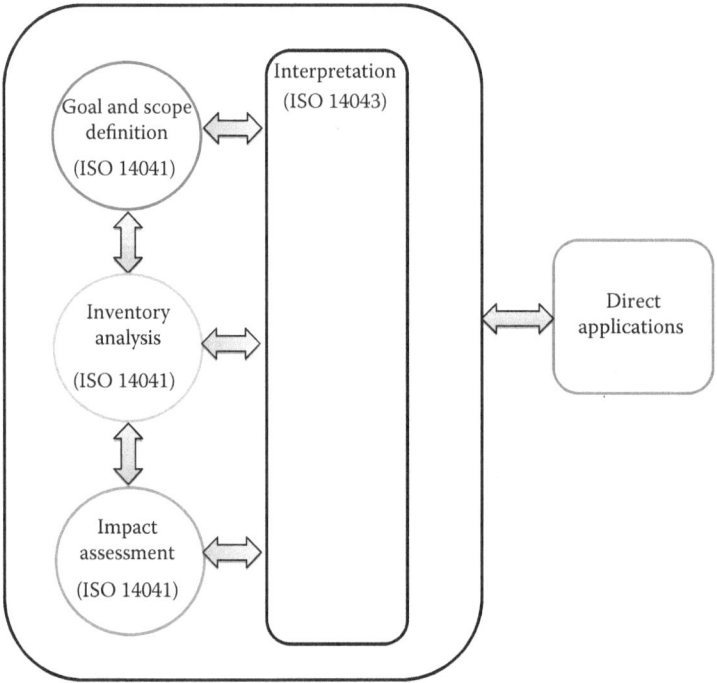

FIGURE 11.2 LCA framework. (Adapted from Varun, Bhat, I.K., and R. Prakash. 2009. *Renewable and Sustainable Energy Reviews* 13:1067–73.)

and depth of the LCA study depend on its goal and scope. In the next step, which is inventory analysis, consumption of different resources and associated GHG emissions are estimated for all stages of a product's life cycle (from extraction of raw material to its disposal). Data are collected and analyzed to estimate inputs and outputs of a process. In the impact assessment phase, the study of environmental consequences associated with a product during its life cycle is carried out. In this phase, characterization models known as impact categories are selected to study the environmental issues. Various indicators such as carbon footprints, energy consumption, and toxicity are used to describe the results (Varun 2010). The main objective of the impact assessment phase is to study a product according to the environmental point of view, with the help of category indicators. This phase is also helpful in providing information related to the interpretation phase (ISO 14042 2000). The final phase of LCA is described as life-cycle interpretation. In this phase, results obtained from the inventory analysis and LCIA (if conducted) are compiled and discussed according to the conclusions and recommendation with respect to definition of goal and scope (ISO 14043 2000).

GHG emissions associated with buildings are mainly in their energy consumption, transportation to and from buildings, energy used in water supply, and construction phase, as shown in Figure 11.3. Carbon footprints of the building greatly depend on its energy consumption. Generally, the energy used in the building is electrical as it is the most convenient source of energy. Consequently, the amount of carbon footprints associated with energy consumption of buildings depends on the source of electricity generation. Electricity generated from fossil fuels will have more carbon footprints compared with that generated from renewable energy sources. It is important to promote the use of renewable energy-based electricity generation in buildings for appreciable reduction in carbon footprints. However, it is very difficult to differentiate electricity generated from fossil fuels and renewable-based sources. Emission associated with transportation to and from the buildings is also an important contributor of carbon footprints. While designing a building, some considerations must be taken into account to reduce the emissions from transport. A high amount of energy is also consumed by water supply in buildings. In addition, it is important to include the amount of recycled materials used in buildings while calculating carbon footprints (www.expresstowers.in).

Barnett et al. (2014) conducted a study to compare the methods (input–output and process models) of estimating carbon footprints of a product. The input–output model used in this study was developed by the Centre for Sustainable Accounting and used by PLC (UK). PAS2050 (GHG protocol) was the process model selected in this study. The analysis was carried out using data from 365 products provided by Carbon Trust, UK, and it was found that no significant correlation can be determined between the two carbon footprint methods. Apart from this, many countries have developed and implemented their own guidelines and set of rules for accounting GHG emissions such as the Environmental Protection Agency in the United States and Carbon Trust in the United Kingdom (Pandey et al. 2011).

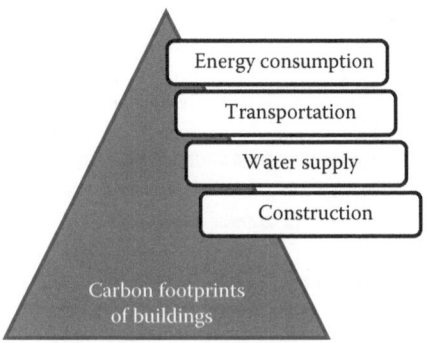

FIGURE 11.3 Factors associated with the estimation of carbon footprints of buildings (www.expresstowers.in).

11.3 PAST STUDIES RELATED TO CARBON FOOTPRINT OF BUILDINGS

Several studies have been conducted to calculate carbon footprints of buildings, and several methodologies have been proposed for this purpose. The evaluation of carbon footprint involves direct as well as indirect emissions generated from different life-cycle stages of a product (Wiedmann and Minx 2008). Therefore, the LCA tool is quite effective and useful to find out direct and indirect emission sources related to the building. There are a lot of carbon footprint studies related to buildings which have been carried out by different researchers and scientists, such as household buildings, hotels, corporate/commercial buildings, and so on. Table 11.1 shows the estimated GHG emissions of different buildings in some previous studies.

Almost all stages of building, that is, extraction of raw materials, construction, operation, maintenance, and demolition, consume large amounts of energy, and consequently a lot of GHG emissions are associated with buildings during their entire life cycle. In many cases, buildings are reconstructed during the operation phase, which again increases their energy consumption and emissions. The construction phase and the operation phase of the building account for a large amount of energy consumption. In the construction phase, the energy involved in the extraction of raw materials, their manufacturing and production, transportation, and so on is quite high (Reddy and Jagadish 2003). Further, HVAC systems, lighting systems, and other activities of occupants increase the energy consumption in the operation phase. A study was carried out to analyze GHG emissions produced by consumption of different goods and services for 73 nations and 14 regions of the world (Hertwich and Peters 2009). The study concluded that shelter with its construction is one of the most important consumption categories in producing GHG emissions. Also, the role of indirect impacts is greater than that of direct impacts in the household category. A lot of environmental issues are associated

TABLE 11.1
GHG Emissions from Buildings (Sharma et al. 2011)

S. No.	Building Specification	Place	Year	Type	Life (Years)	Floor Area (m²)	GHG Emissions ($CO_{2eq}/$ m² 50 year)
1	Malmo (Adalberth et al. 2001)	Sweden	1996	Residential	50	700	1.30 ton
2	Helsingborg (Adalberth et al. 2001)	Sweden	1996	Residential	50	1160	1.35 ton
3	Vaxjo (Adalberth et al. 2001)	Sweden	1996	Residential	50	1190	1.51 ton
4	Stockholm (Adalberth et al. 2001)	Sweden	1996	Residential	50	1520	1.40 ton
5	High-end (Junnila and Horvath 2003)	South Finland	2003	Commercial	50	15,600	48,000 ton
6	Sam Wyly Hall, University of Michigan (Scheuer et al. 2003)	Michigan, USA	2003	Commercial	75	7300	67,500 ton
7	Via Garrone Building (Blengini 2009)	Turin, Italy	2004	Residential	—	6110	3340 kg
8	Low-density building (Norman et al. 2006)	Toronto, Canada	2006	Residential	50	—	5365 kg
9	High-density building (Norman et al. 2006)	Toronto, Canada	2006	Residential	50	—	3885 kg
10	Office building (Kofoworola and Gheewala 2008)	Thailand	2008	Commercial	50	60,000	5,600,000 ton

Source: Adapted from Sharma, A. et al. 2011. *Renewable and Sustainable Energy Reviews* 15:871–5.

with the construction sector. This is due to the exploitation of use of land, conventional energy sources, and energy consumption in all stages of life cycle (Proietti et al. 2013).

The selection of appropriate building materials and components, low environmental costs, and achievement of best performance during construction and operation stages are the main objectives of a new sustainable design of buildings (Proietti et al. 2013). A study on an office building in Bangkok, Thailand shows that construction materials such as steel and concrete have significant impacts on the environment during its manufacturing phase (Kofoworola and Gheewala 2008). This study also concludes that the operation phase of the building accounts for about 52% of the total global warming potential. A study performed on four multifamily buildings in Sweden found that about 70%–90% of the environmental impact is due to the occupation phase of the buildings (Adalberth et al. 2001). The environmental impacts were studied using an LCA technique evolved at the Danish Building Research Institute (Peterson 1997). The study suggests that by choosing proper options of construction and installation, reduced environmental impacts during occupation phase can be achieved.

The maximum GHG emissions and the energy consumption are associated with the operational phase (Adalberth et al. 2001). A study was conducted to determine environmental impacts of a new high-end office building in Southern Finland (Junnila and Horvath 2003). The result shows that electricity consumption and manufacturing of building materials (especially electricity used in lighting and HVAC systems, manufacturing and maintenance of steel, and office waste management) had a big role in environmental impacts. Also, another study was conducted in Northern Mendoza (Argentina) to analyze and compare various building technologies applied in a rural school building for thermal comfort with minimal use of energy from fossil fuels (Arena and Rosa 2008). One more comparative study was performed using LCA and economic input–output (EIO)-based LCA approach to analyze environmental effects of two different types of buildings, namely, steel- and concrete-framed buildings in the Midwestern USA (Guggemos and Horvath 2004). It was found that concrete-framed buildings account for more GHG emissions and energy consumption than steel-framed buildings. This is all due to longer installation process associated with concrete-framed buildings. Ibn-Mohammed et al. (2013) discussed the relationship between operational and embodied emissions of a building as they are the components of total emission associated with the building during its life cycle.

A study was carried out to estimate the carbon footprint of residential and commercial buildings in the United States as per carbon accounting standards and scopes of the World Resources Institute (WRI). The emissions for construction, operation, and disposal phases of the buildings were estimated using a comprehensive hybrid EIO-LCA. The study shows emission from electricity consumption (purchase electricity) as having highest carbon footprints in buildings of the United States. Also, the construction supply chain also contributes to carbon footprints of buildings in the United States. The GHG emissions in the operation phase were highest among all the life-cycle phases of the building (Onat et al. 2014).

Ozawa-Meida et al. (2013) conducted a study to estimate emissions and to find the hot spots in different departments in De Montfort University, UK. The carbon footprint study was predicated on the consumption-based carbon footprint methodology, which includes scope 1, 2, and 3 emissions. Direct and indirect emissions released by the consumption of fossil fuels and use of electricity in the buildings, stationary and mobile emission sources (scope 1 and 2 emissions), and the procurement and business travel emissions (scope 3 emissions as per the classification of Greenhouse Gas Protocol Corporate Standard by the WRI and World Business Council for Sustainable Development) were estimated. The study showed that scope 3 emissions contributed 79% of the total GHG emissions of the university. In addition, the consumption-based methodology is recommended to other universities for effective analysis to reduce GHG emissions.

Nowadays, awareness is being increased to involve parameters such as sustainable design and human well-being in the performance of buildings (Steemers and Manchanda 2010). A significant reduction in energy consumption and carbon footprints in buildings can be achieved by promoting use of renewable energy sources. The use of efficient electricity generation technologies (if used effectively) leads to big savings of fossil fuels and considerable reduction in GHG emissions (Halmann and Steinberg 1998). Airaksinen and Matilainen (2011) conducted a study on a nine-storey office building, developed by Skanska Commercial Development Finland, in Helsinki. The study concluded that the lowest amount of emissions is achieved if energy used in an office is based on renewable energy sources, nuclear, or biosources.

Good comfort conditions and reduction in energy consumption of HVAC systems were achieved with appropriate thermal insulation of buildings. Proietti et al. (2013) have presented a study according to LCA principles to analyze carbon footprint of reflective foil material as a thermal insulation of building envelope. A comparative analysis was also performed by considering application to wall with hypothetical cavity and exterior insulation and finishing system for buildings. Another study has been carried out to estimate primary energy consumption and carbon footprints for a wood frame apartment of a building (conventional and passive houses) in Sweden with different end-use heating systems using LCA. The heating systems were also compared with biomass-based energy supply. All life-cycle phases of buildings, namely, extraction and processing of raw materials and their fabrication and assembly, operation and demolition phases, as well as the post usage of building materials, were considered in this analysis. The study shows that the operation phase of the building is the highest contributor to the primary energy consumption during its life cycle. It was found that primary energy use and CO_2 emissions are reduced in the passive house for heating. In addition, the reduction in primary energy and emissions depends on the type of the heating system. It was also concluded that the use of biomass-based systems with cogeneration yields low primary energy consumption and CO_2 emissions for conventional houses (Dodoo and Gustavsson 2013).

A study was carried out in New York City using computer modeling, data sets, and help of experts to check whether climate change problems can be controlled by improving the energy efficiency in the building sector with presently available technologies. The study discussed ways, including change of fuel-based HVAC and hot-water systems by electrically powered equivalents, to reduce energy consumption of buildings in New York City. The study also concluded that key to a sustainable future is the potential of energy efficiency measures (Wright et al. 2014).

Proper variations in the design of a building are also significant to minimize energy usage and carbon footprint. A study was carried out to estimate energy consumption rate and carbon footprint by analyzing the effects of changes of multiglazed windows in a hut in Kuala Lampur, Malaysia, using BIM software. The number of glazed layers, filled gases, sizes, and orientation of windows were considered as parameters in the study. Important points were discussed regarding the design of building to reduce energy consumption and carbon footprint (Tahmasebi et al. 2011).

The urban density, living area, and number of residents in a building also have a significant effect on the generation of GHG emissions into atmosphere. A comparative study to analyze energy consumption and GHG emissions in high- and low-populated buildings was carried out in Toronto, Canada (Norman et al. 2006). The study shows that material production across supply chain for low-density buildings accounts for 1.5 times more GHG emissions and embodied energy than high-density buildings. Another study, carried out on a six-storey building at University of Michigan Campus, Ann Arbor, Michigan, USA, shows that only the operation phase of the building contributes more than 83% of the total environmental impact (Scheuer et al. 2003). High-energy consumption and carbon footprints are associated with the occupation phase of buildings; therefore, the optimization of the performance of the operation phase in building design should be focussed. A study related to use of water in multi-occupancy buildings in United States shows that the use of efficient fixtures and appliances, instead of conventional fixtures and appliances, is quite helpful to reduce energy consumption in buildings (Arpke and Hutzler 2005).

Hotels provide examples of multi-occupancy buildings and account for high-energy consumption and GHG emissions. A study of three representative hotels in Hong Kong was conducted to estimate emissions released, and carbon footprints of all three hotels were compared. Regression models were developed to estimate emissions with variation in outside temperatures. Electricity was found as the most dominant source of carbon emission, and the study concluded that efforts should be applied to minimize electricity consumption. This can be achieved by proper energy audits for hotels. The study recommends the use of methodology for developing benchmarks of carbon emissions, so that these can be useful to optimize and check carbon footprints of hotels (Lai 2013).

In addition, a study was conducted to estimate CO_2 emissions for two hotels in Poole, Dorset, UK, using life-cycle energy analysis. The results were compared with those of the other studies based on different methodologies by previous authors. The results show that energy usage and carbon production are found to be less than those of other buildings discussed in previous studies, which imply the progress in energy efficiency of hotels over time. The current practices on energy consumption of the hotels were studied, and suggestions were provided to improve the energy performance and to reduce the carbon footprint (Filimonau et al. 2011). A study was carried out using environmental extended input–output methodology to estimate carbon footprint of Norwegian University of Technology and Science (NTNU). The results show that joint services, renting of premises, construction work, and other miscellaneous installations of buildings have significant contribution in total carbon footprints of NTNU (Larsen et al. 2013).

Apart from construction and operation phases of buildings, studies have been carried out to determine environmental effects during the demolition phase of a building. A study was conducted to analyze the demolition phase and recycling potential of a building, which was demolished in 2004 in Turin, Italy (Blengini 2009). The study concluded that recycling of building waste is quite sustainable, feasible, and profitable from economic point of view. In the case of wooden-frame houses, the residues of the wood products chain are large after demolition and can be used as fuels (Dodoo and Gustavsson 2013).

In order to reduce the energy consumption in residential and commercial buildings, research has been carried out to promote new innovative technologies. A study was carried out on an energy harvesting technique viz. gravity-energized wastewater system in high-raised buildings in India. In this system, the energy of gray water falling from the floors of building is utilized to run a micro- or picoturbine at ground floor to produce electricity. The study shows that the proposed design is commercially feasible for most major Indian cities (Sarkar et al. 2014). This type of electricity generation can help reduce of fossil fuel-based electricity consumption in buildings.

Regulations of buildings also help in estimating and controlling GHG emissions. However, more strict energy-efficient building regulations do not guarantee noticeable emission reductions (Nordby 2011). A study was carried out by Nordby (2011) to analyze the effectiveness of regulatory measures to control environmental impacts of buildings. The study was performed using Norway's wood-fired mountain cabin (self-service) in three different scenarios to explore the importance of strictness regarding energy-efficient requirements in building regulations. The emissions associated with the use of extra material as per new regulations are estimated. In addition, the estimated emissions were compared with reduction in emissions due to the reduction in operational energy requirements over a 50-year life cycle. It was observed from the results that emissions associated with the use of extra material as well as their transport are found to be greater than the emissions savings due to reduction in energy requirements.

With respect to the concept of net zero energy and carbon neutral in residential building sectors, a study was carried out by Newton and Tucker (2011) on hybrid buildings. The concept of hybrid buildings, which is associated with net zero energy and zero-carbon status, was discussed. In the study, carbon footprints of a hybrid building with alternative configurations were estimated. The study explored and discussed the ways to achieve zero-carbon housing in existing five-star project built homes.

11.4 CASE STUDIES IN INDIA

The carbon emissions in India are mainly associated with power generation, transportation, domestic energy consumption, industries, and agriculture (Ramachandra and Shwetmala 2012). After industries, the electricity consumption in buildings is the highest, that is, 30% of the electricity consumption in India (Tulsyan et al. 2013). More attention in associations between pollutants and economic activities has been found in developing countries such as India due to increase in concern about the climate change and high carbon emissions into the atmosphere (Yang and Zhao 2014). The high energy consumption in India poses a major threat for sustainable development and CO_2 emissions (Parikh 2012). The consequence of rapid growth in population and urbanization is increase of buildings and other physical infrastructure requirements for residential, commercial, and industrial activities (Sabapathy and Maithel 2013). In addition, rise in income with urbanization is also responsible for increasing building energy service demands and energy consumption in India (Chaturvedi et al. 2014). It is quite important to use the energy costing principles in buildings, and there is a great need to explore the methods to reduce the energy consumption in buildings during their entire life cycle (Tiwari 2001).

A study was carried out by Ramachandra and Shwetmala (2012) to analyze CO_2, carbon monoxide (CO), and methane (CH_4) emissions using region-specific emission factors and carbon sequestration capacity in different states of India. It was estimated that CO_2, CO, and CH_4 emissions produced yearly in India are 965.9, 22.5, and 16.9 million metric tons, respectively. Also, it was found that the maximum amount of CO_2 emissions in India associated with electricity generation. Another study using input and output analysis was carried out by Das and Paul (2014) to determine direct and indirect CO_2 emissions associated with fuel consumption in households for years 1993–1994, 1998–1999, 2003–2004, and 2006–2007 in India. The results showed that the main reasons of high CO_2 emissions are the activity, structure, and population effects.

A study was carried out to analyze the importance of renewable energy sources to achieve targets of climate change mitigation in India by 2050 using the TIAM-UCL optimization model. The study shows that increasing the use of renewable energy sources is an effective solution to reducing CO_2 emissions, especially in power sectors (Anandarajah and Gambhir 2014).

As discussed in the previous sections, the materials used in the building construction make a significant contribution in GHG emissions. Bricks are one of the most important materials for building construction. A study was carried out in Western Maharashtra, India using LCA technique to estimate and analyze environmental impacts in terms of emissions, depletion of resources, and biodiversity loss associated with traditional bricks. The study used SimaPro 7.3.3 software to carry out LCA impact studies. All phases including the acquisition of raw materials and transportation of bricks were included in the scope of the methodology. It was found from the results that the CO_2 emissions had the highest contribution on all release to the atmosphere. The major emissions produced were associated with the combustion of coal in kiln and diesel in transportation. However, the quality of coal used in kiln is also responsible for other environmental impacts such as high value of acidification and so on. The study also recommends that there is a great need to incorporate more ecofriendly technology in brick production (Kumbhar et al. 2014).

A study was carried out by Varun et al. (2012) to estimate environmental effects of an educational building in Northern India. It was the building of Mechanical Engineering Department at National Institute of Technology, Hamirpur, Himachal Pradesh, India and has three stories with a floor area of 3960 m^2 and projected service life of 50 years. The building was commissioned in the year 1994, and its second floor was commissioned in the year 2007. The LCA technique was used to estimate GHG emissions associated with the building. Three main phases of life cycle, namely, construction, operation, and maintenance, were considered. The study was carried out with the assumption that no further work of extension and reconstruction will be done during the 50-year life cycle.

To estimate GHG emissions, first, inventory data of the material used (which are major contributors in the building emissions) in the building were collected. Second, energies (energy required for extraction, manufacturing, assembly, and transportation of a product) for all materials for all the three floors were estimated individually. Finally, the GHG emissions were estimated using Indian electricity scenarios. The total emissions associated with the buildings considered were found to be 1764 ton CO_{2eq}. It was also found that the largest portion of GHG emissions was associated with the operation phase. About 59% of the energy is used during the operation phase of the building due to the excess use of HVAC systems, computers, and machines used in laboratories. The emissions associated with different materials used in the construction phase for all three floors are shown in Figures 11.4 through 11.6. The construction phase of second floor has highest contribution in GHG emissions due to more use of steel and aluminum. This shows that selection of materials during construction of a building has a significant effect on GHG emissions. The study recommends exploring some alternative solutions, which help in selecting materials with less GHG emissions for the construction of buildings. In addition, consideration must be taken to reduce GHG emissions during the operation phase. This can be performed by promoting the use of renewable energy-based technologies during the operation phase.

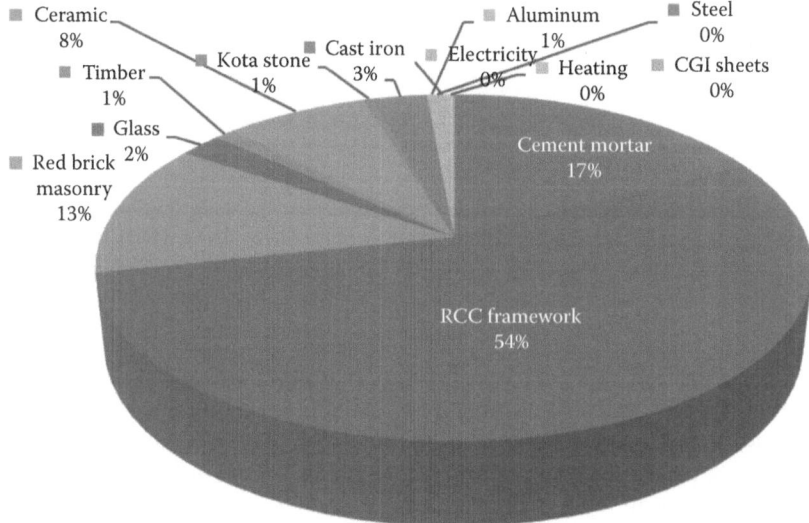

FIGURE 11.4 GHG emissions in the ground floor. (Adapted from Varun, V. et al. 2012. *Sustainable Cities and Society* 4:22–8.)

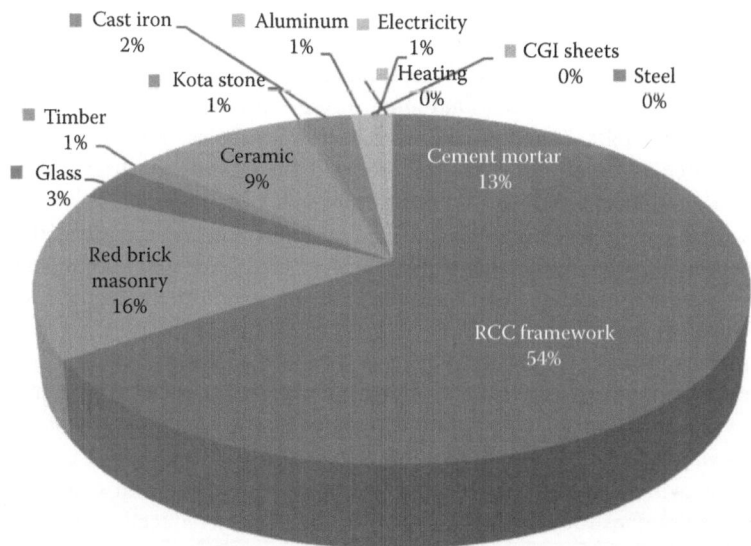

FIGURE 11.5 GHG emissions in the first floor. (Adapted from Varun, V. et al. 2012. *Sustainable Cities and Society* 4:22–8.)

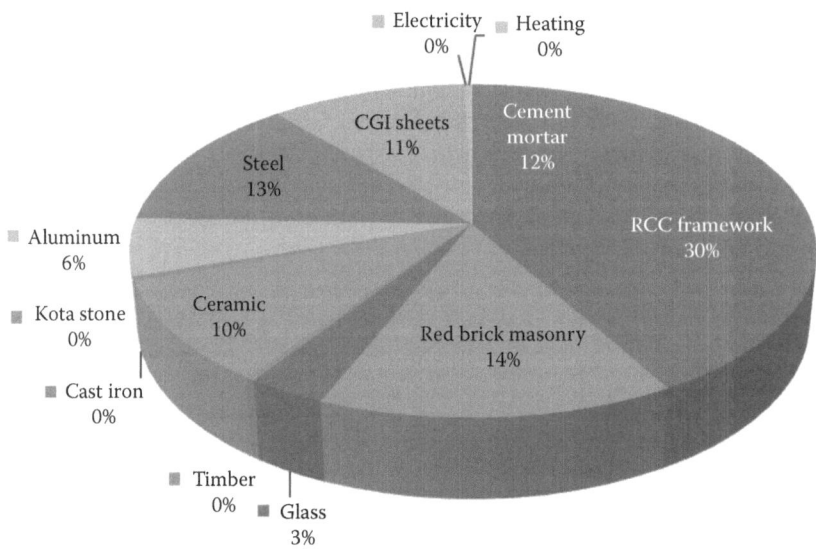

FIGURE 11.6 GHG emissions in the second floor. (Adapted from Varun, V. et al. 2012. *Sustainable Cities and Society* 4:22–8.)

The selection of appropriate materials in the construction of building is quite effective in controlling the carbon footprints of a building. Some studies on the use of low-cost and low-energy-intensive materials in the construction of buildings in India show significant energy savings and reduction in emissions by these alternative materials. A study was carried out by Shukla et al. (2009) to estimate the embodied energy of the adobe house located at Solar Energy Park, Indian Institute of Technology, Delhi, India. The adobe house was made up of different low-energy-intensive materials. The results show that reduction in CO_2 mitigation by 101 tons/year is achieved by using low-energy-intensive materials.

Tiwari (2001) conducted a study to introduce a comprehensive optimization model [energy-efficient housing options evaluation model (ENEHOPE)] for energy estimation in different techniques of building construction. Pucca techniques and low-cost construction techniques were studied and compared with common practices for a representative room of dimensions 3.5 m × 3.5 m × 3.14 m. Different types of activities, namely activities associated with primary resources (coal and electricity) needed to produce intermediate and final inputs used in the construction and activities associated with different technological options of construction of a building, were considered. The results were discussed in three different scenarios: base (for minimization of construction cost), coal (constraint of coal use), and electricity (constraint of electricity consumption). Figure 11.7 shows the comparison of carbon emissions associated with the requirement of resources in different construction techniques in order to meet the housing shortage in India. The study recommends structural changes in construction and energy policies for significant reduction in energy consumption and emissions associated with buildings.

Another study was conducted by Tiwari et al. (1996) to analyze various aspects of low-cost construction techniques in buildings. A representative room having dimensions of 3.5 m × 3.5 m × 3.14 m was constructed using low-cost techniques. The results showed that the construction cost of the room was reduced by 34% compared with the cost associated with conventional construction techniques without lack of any technical standard functional utility. In addition, the CO_2 emissions for the room constructed with low-cost techniques are reduced to 2.82 tons from 5.88 tons for conventional techniques. It was also found that the emissions can be further reduced by 23%, with increase in cost by 0.03%. The study concluded that the use of materials which are easily available and found in abundance effectively reduces the construction cost of the building as well

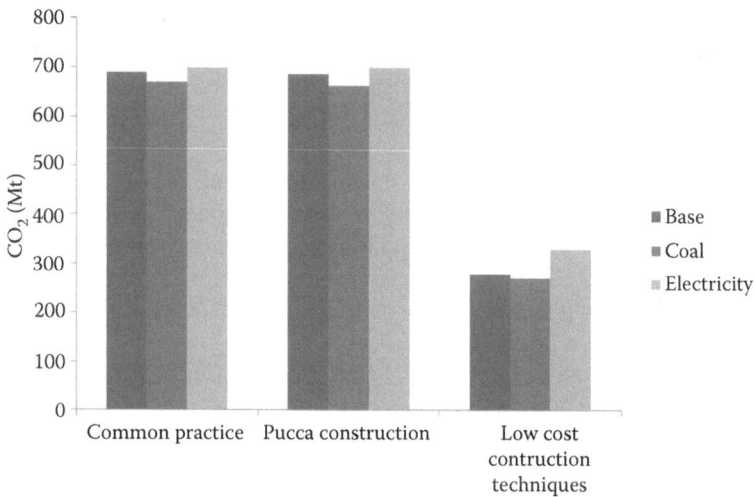

FIGURE 11.7 Comparison of CO_2 emissions associated with different construction techniques. (Adapted from Tiwari, P. 2001. *Building and Environment* 36:1127–35.)

as associated CO_2 emissions. A social benefit associated with the promotion of low-cost construction techniques is the increase of employment because of the high use of labor-intensive techniques associated with it. The study recommended the use of low-cost and easily available materials shown in Table 11.2.

A life-cycle energy analysis of a residential building, with a usable floor area of 85.5 m² at Hyderabad, Andhra Pradesh, India, was carried out under different climatic conditions (hot and dry, warm and humid, composite, cold, and moderate). The building was studied with walls made up of conventional fire clay materials and different alternative materials such as hollow concrete, soil cement, fly ash and aerated concrete with varying thicknesses, and insulation on wall and roof. The results showed that the use of alternative wall materials in buildings is able to reduce the life-cycle energy demand by 1.5%–5%. Also, the energy performance of aerated concrete was found to be better than other materials. In addition, the study concluded that the use of insulation on wall and roof gives significant energy savings, but it also depends on the climatic conditions (Ramesh et al. 2012a).

Another life-cycle energy analysis, using dynamic energy simulation tools such as Design Builder, e-Quest, and Energy Plus, was carried out in India on different types of residential buildings, namely, one-storey, two-storey, and duplex and multistorey. The designs of 10 houses were studied using energy-saving techniques such as thermal insulation on wall and roof and double pane glass in windows. The life-cycle energy of the buildings was found to be between 240 and 280 kWh/m² year, depending on the type of building and climatic conditions. The results show that the use of

TABLE 11.2
Use of Low-Cost Materials in Building Construction

Construction	Construction Material	Substitute
Masonry	Bricks	Mud blocks
		(lime, bitumen, molasses for stabilizer in the absence of good soil)
	Cement	Lime, surkhi, fly ash
Plastering	Cement	Lime, surkhi, fly ash
Roof construction	Concrete slabs	Filler slabs with Mangalore tiles

Source: Adapted from Tiwari, P., Parikh, J., and V. Sharma. 1996. *Building and Environment* 31(1):75–90.

energy-saving techniques (thermal insulation on wall and roof and double pane glass in windows) yields 5%–30% savings in life-cycle energy. Also, one of the studied buildings was analyzed with onsite power generation using renewable energy sources (solar photovoltaic and wind turbine). The results showed that the use of onsite power generation helps to achieve net zero operating energy building (Ramesh et al. 2012b).

The reduction in energy consumption during the operation phase can also be achieved by adopting proper energy management and efficient use of lighting systems. A study was conducted to determine energy conservation measures in a building of the technical institution in Tamil Nadu, India. The study shows that proper energy management and efficient lighting in the building, which leads to the reduction in gap between electricity generation and demand, are useful to reducing the electricity consumption in the building (Loganthurai et al. 2011).

Another study was carried out by Devi and Palaniappan (2014) on LCA of a multistorey residential building in Southern India. The building is located in Indian Institute of Technology Madras, Tamil Nadu and has 96 identical apartments. Four blocks were considered in the study, each having a ground floor above which there were six floors. Every floor, excluding ground floor (used for parking), had four identical apartments, and the usable floor area of each apartment was 112.15 m². In addition, each block had two lifts and two staircases. The service life of the building considered in the study was 50 years. The energy consumption associated with different phases of life cycle, that is, extraction of materials, building construction, operation, and demolitions, was determined. Energy consumption in material production and construction phase was determined using embodied energy calculations of drawing and specification of materials used in the building. The energy associated with the operation phase was assumed to be same through the life cycle. Also, the energy in the demolition phase was assumed to be 3% of the initial embodied energy. The study excluded the energy associated with maintenance and renovation of the building. The results show that material production and energy operation phase consumes high amount of energy during the entire life cycle of the building. Although carbon footprints were not calculated in the study, the high energy use associated with material production and operation phase of the building contributes high GHG emissions in the environment.

11.5   ENERGY CONSERVATION IN BUILDINGS

The case studies presented illustrate that the energy consumption of the building must be reduced as much as possible so that the associated emissions can be minimized significantly. A lot of researches have been carried out on energy conservation in buildings through appropriate construction, design, operation, and maintenance of buildings. In this regard, the promotion of green buildings is one of the most effective methods to control the impact of building on environment, society, and economy as well (Zuo and Zhao 2014). The concept of "green buildings" was introduced with the main objective of reducing the impact of buildings on natural environment. This can be performed by using efficient use of energy, water, and other resources in buildings. In addition, reduction of GHG emissions by promoting renewable energy sources, reducing waste, and environmental degradation is also a part of green buildings. The use of sustainable materials in the construction of buildings helps us to reduce carbon footprints in the construction phase of a building. Sustainable materials used in green buildings include recycled content or those fabricated with the usage of renewable energy sources. Also, healthy indoor conditions with reduction in emissions and reduction of water usage of feature landscaping are important in reducing carbon footprints by promoting green buildings (Sharma et al. 2010). A lot of studies are being carried out throughout the world to assess and mitigate the environmental impact of buildings. A review was carried out by Zuo and Zhao (2014) on green buildings, which shows that the present studies on green buildings significantly focus on their environmental aspects.

The use of solar energy in green buildings is used for passive heating, cooling, and day lighting. This is carried out by designing buildings in such a manner that it can receive ample amount of solar radiations. Direct, indirect, and isolated gains and heat storage techniques are used in passive heating of buildings using solar radiations (Chiras 2002). There are several environmental, economic, and social benefits associated with the development of green buildings. The reduction in carbon footprints, conservation of natural sources, and controlling climate change problems are achieved with green buildings. Also, reduction in operation costs is an important aspect of green buildings. Additionally, improvement of comfort and health of occupants is a big social factor related to the promotion of green buildings.

A green building certification system namely, Leadership in Energy and Environmental Design, was incorporated by the U.S. Green Building Council to give a framework for recognition and execution of measurable green buildings design including building design construction, interior design construction, operation, and maintenance (www.usgbc.org). In India, a national rating system for green buildings is Green Rating for Integrated Habitat Assessment (GRIHA) developed by the Ministry of New and Renewable Energy (MNRE) and The Energy and Resources Institute (TERI). Around 525 projects registered in India under GRIHA are having a big role in reducing GHG emissions (www.grihaindia.org). Also, the Indian Green Building Council formed in 2001 provides ratings to promote sustainable development in India (www.igbc.in).

In India, several buildings having energy conscious design are contributing in carbon footprints reduction with the help of passive solar heating and cooling techniques, sustainable materials, and other renewable energy technologies. An example of such type of building in India is that of Himachal Pradesh State Cooperative Bank. It is a three-storey building located at Shimla, Himachal Pradesh, India, having its longer axis facing an east-west direction. A sectional view of the building with different passive solar techniques is shown in Figure 11.8. The building has solar collectors in the roof to provide warm air through ducts, and the south-facing Trombe wall and sunspace are also used to warm the internal space of the building. The protection of the north face from winter winds is performed using the cavity wall. In addition, insulation and double glazing are provided to the western wall. Light shelves and skylight on the terrace provide daylighting. The use of air lock lobbies helps reduce the exchange of heat. Due to these characteristics, the building of Himachal Pradesh Cooperative Bank saves around 84,720 kWh energy per annum compared with a conventional building. Figure 11.9 shows component-wise energy saving per year in the Himachal Pradesh Cooperative Bank building. This noticeable amount of energy saving in the building significantly helps in reducing considerable carbon footprints (www.mnre.gov.in).

Other examples of these types of energy-saving buildings in India are Auroville Ecohouse, Auroville, Centre for Application of Science and Technology for Rural Areas, Bangalore, Solar Energy Centre, Gurgaon, LEDeG Trainees' Hostel, Leh, Solar Energy Centre, Gwal Pahari,

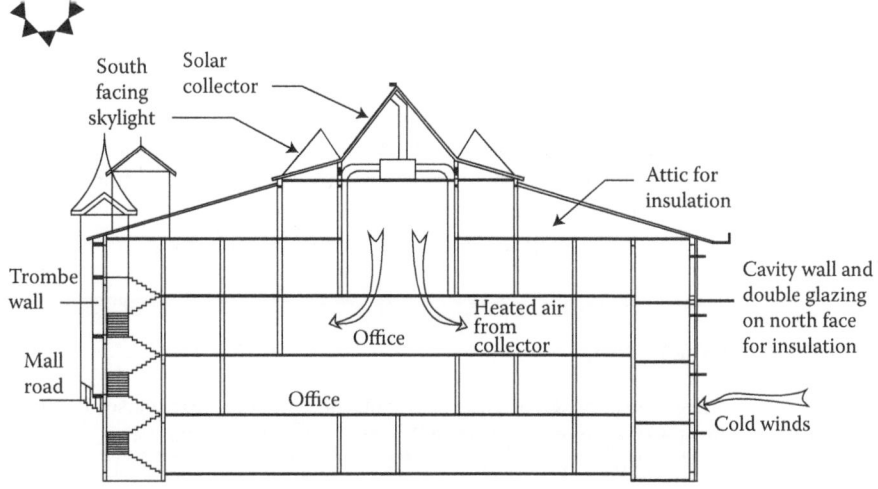

FIGURE 11.8 Section of the Himachal Pradesh Cooperative Bank, Shimla (MNRE www.mnre.gov.in).

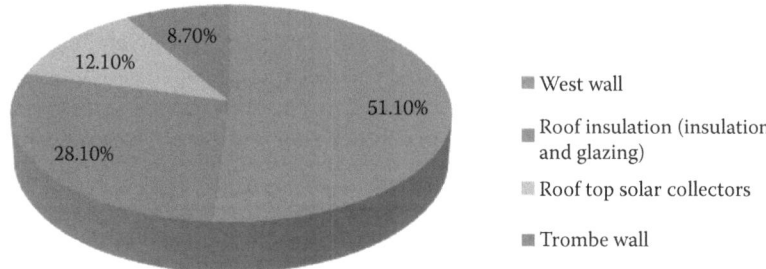

FIGURE 11.9 Componentwise energy saving per year, Himachal Pradesh Cooperative Bank, Shimla (MNRE, www.mnre.gov.in).

TABLE 11.3
Energy Consumption and GHG Emissions of Buildings in India

Building, Location/Year	Location	Floor Area (m²)	Life Span (Years)	Operational Energy (kWh/m²/year)	Life Cycle Energy (kWh/m²/year)	GHG Emissions
Solar Energy Park, IIT Delhi/2009 (Shukla et al. 2009)	New Delhi	100	40	24	61	Reduced by 101 tons per year
NIT, Hamirpur/2012 (Varun et al. 2012)	Hamirpur	3960	50	1180	1987	0.45 ton
Residential Building/2012 (Ramesh et al. 2012a)	Hyderabad	85.5	75	167–174	194–203	—
Residential Apartment, IIT Madras/2014 (Devi and Palaniappan 2014)	Chennai	10,766.4	50	116.42	185.72	—

Gurgaon, and so on (MNRE, www.mnre.gov.in, MNES and TERI 2001). These buildings play a significant role in energy conservation as well as in carbon footprints reduction.

Chel et al. (2008) carried out a study on a honey storage building with a Trombe wall for heating in the winter season. The building was located at Gwalior, Madhya Pradesh, India. TRANSYS simulation software was used to determine the heating potential of the Trombe wall used in the building. The analysis showed that about 3312 kWh of energy can be saved, and around 33 tons of CO_2 emissions can be reduced per year using the Trombe wall. In addition, economics of retrofitting for the building was also studied, and the simple payback period of retrofitting was found to be about 7 months, which made it affordable as well.

Table 11.3 shows the energy consumption and associated GHG emissions estimated in previous studies in India.

11.6 RECOMMENDATIONS

It can be seen from the discussion presented that construction and operation phases of a building are responsible for maximum energy consumption and emissions. The emissions are associated with the buildings even from the stage of procurement of raw materials for construction and transportation. It is quite necessary to consider the appropriate selection of materials for building construction, and it is also important to promote the use of raw materials which are easily available and demand less transportation. It can further be seen from the studies reported in this chapter that emissions can be significantly reduced by proper selection of raw materials as well as by appropriate construction methods. Also, there is a great need to change the design of the buildings and to include energy-efficient techniques to control energy consumption and emissions in the occupation phase. In addition, emphasis must be given to make these solutions more cost-effective and attractive, so that they can be accepted and implemented throughout the world to control the carbon footprints of buildings.

11.7 CONCLUSION

As problems of climate change are being observed in earth, it is important to find out and control those factors that are related to human activities and have significant impacts on the environment. High amounts of GHG emissions generated by various activities of human beings are responsible for increasing the global warming effect on earth. Consequently, a great need arises to explore methods and tools to estimate and compare the environmental impact caused by the activities of a product (goods and services) during its entire life cycle. The building sector is one of the most important sectors associated with high amount of energy consumption and GHG emissions. Buildings consume a lot of resources and release considerable amount of GHG emissions during its life cycle (from construction to demolition). In order to achieve sustainable development, it is important to make buildings more efficient and have less emission. Therefore, it is important to know the environmental impact of a building during its entire life cycle in order to make it environmentally conscious. Carbon footprint is an important tool and parameter to evaluate environmental impacts associated with a product. Carbon footprint is related to GHG emissions associated with a product during its entire life cycle (from extraction of raw materials to disposal of a product). The most common techniques used to estimate carbon footprints are input–output and process analyses. LCA is an effective technique to estimate carbon footprints of buildings. In addition, advent of computer software is becoming very useful to save time and effectiveness of LCA.

The materials used in the construction of a building contribute greatly to its carbon footprints. Selection of appropriate materials for building construction and introduction of efficient design consideration have become important to reducing carbon footprints of buildings. The operation phase of a building contributes a major portion of total GHG emissions to its entire life cycle. High-energy consumption in the operation phase due to excessive use of HVAC systems, lighting, and so on is responsible for a high amount of GHG emissions. This factor also forces us to think about the optimization of building performance during the operation phase. Many previous studies reveal that the role of indirect impacts is as significant as direct impacts, especially in the household category. Therefore, it is important to consider indirect emission associated with buildings in the carbon footprint analysis to increase depth of estimation of carbon footprints. Energy-efficient technologies for electricity generation are essential to reduce the consumption of fossil fuels and emissions. The promotion of renewable energy technologies helps in reducing energy consumption in buildings; consequently, carbon footprints can be reduced considerably. The concept of green buildings is also coming up to promote energy-efficient buildings. In many studies, it was found that technologies such as thermal insulation of buildings are quite effective to save energy consumed in HVAC systems. The recycling of building materials after its demolition is also feasible and economical in order to achieve low-carbon footprints. This is due to the fact that the material obtained after demolition of buildings can be used in other applications which save energy requirement and emissions associated with the material that is being saved.

REFERENCES

Adalberth, K., Almgren, A., and E.H. Peterson. 2001. Life-cycle assessment of four multifamily buildings. *International Journal of Low Energy and Sustainable Buildings* 2:1–21.
Airaksinen, M., and P. Matilainen. 2011. A carbon footprint of an office building. *Energies* 4:1197–210.

Anandarajah, G., and A. Gambhir. 2014. India's CO_2 emission pathways to 2050: What role can renewables play? *Applied Energy* 131:79–86.

Arena, A.P., and C.de. Rosa. 2008. Life cycle assessment of energy and environmental implications of the implementation of conservative technologies in school building in Mendoza-Argentina. *Building and Environment* 38:359–68.

Arpke, A., and N. Hutzler. 2005. Operational life-cycle assessment and life-cycle cost analysis for water use in multi-occupant buildings. *Journal of Architectural Engineering* 11(3):99–109.

Bansal, D., Singh, R., and R.L. Sawhney. 2014. Effect of construction materials on embodied energy and cost of buildings—A case study of residential houses in India up to 60 m² of plinth area. *Energy and Buildings* 69:260–6.

Barnett, A., Barraclough, R.W., Becerra, V., and S. Nasuto. 2014. A comparison of methods for calculating the carbon footprint of a product. http://www.reading.ac.uk (accessed September 12, 2014).

Basket, J.S., Lacke, C.J., Weizt, K.A., and J.L. Warren. 1995. *Guidelines for Assessing the Quality of Life Cycle Inventory Analysis*. Springfield: Research Triangle Institute, National Technical Information Service.

Blengini, G.A. 2009. Life cycle of buildings, demolition and recycling potential: A case study in Turin, Italy. *Building and Environment* 44:319–30.

Boustead, I. 1996. LCA—How it came about: The beginning in the UK. *International Journal of LCA (Springer)* 1(1):4–7.

Chaturvedi, V., Eom, J., Clarke, L.E., and P.R. Shukla. 2014. Long term building energy demand for India: Disaggregating end use energy services in an integrated assessment modeling framework. *Energy Policy* 64:226–42.

Chavez, A., and A. Ramaswami. 2014. Progress toward low carbon cities: Approaches for transboundary GHG emissions' footprinting. *Carbon Management* 2(4):471–82.

Chel, A., Nayak, J.K., and G. Kaushik. 2008. Energy conservation in honey storage building using Trombe wall. *Energy and Buildings* 40:1643–50.

Chiras, D. 2002. *The Solar House: Passive Heating and Cooling*. Chelsea Green Publishing Company, Canada.

Das, A., and S.K. Paul. 2014. CO_2 emissions from household consumption in India between 1993–94 and 2006–07: A decomposition analysis. *Energy Economics* 41:90–105.

Devi, L.P., and P. Palaniappan. 2014. A case study on life cycle energy use of residential building in Southern India. *Energy and Buildings* 80:247–59.

Dhaka, S., Mathur, J., and V. Garg. 2012. Combined effect of energy efficiency measures and thermal adaptation on air conditioned building in warm climatic conditions of India. *Energy and Buildings* 55:351–60.

Dodoo, A., and L. Gustavsson. 2013. Life cycle primary energy use and carbon footprint of wood-frame conventional and passive houses with biomass-based energy supply. *Applied Energy* 112:834–42.

EIA. 2008. Annual energy outlook. Energy Information Administration, U.S. Department of Energy.

Filimonau, V., Dickinson, J., Robbins, D., and M.A.J. Huijbregts. 2011. Reviewing the carbon footprint analysis of hotels: Life cycle energy analysis (LCEA) as a holistic method for carbon impact appraisal of tourist accommodation. *Journal of Cleaner Production* 19:1917–30.

Guggemos, A.A., and A. Horvath. 2004. Comparison of environmental effects of steel- and concrete-framed buildings. *Journal of Infrastructure Systems* 11(2):93–101.

Guttikunda, S.K., Goel, R., and P. Pant. 2014. Nature of air pollution, emission sources, and management in the Indian cities. *Atmospheric Environment* 95:501–10.

Halmann, M.M., and M. Steinberg. 1998. *Greenhouse Gas Carbon Dioxide Mitigation: Science and Technology*, CRC Press, US.

Hertwich, E.G., and G.P. Peters. 2009. Carbon footprint of nations: A global, trade-linked analysis. *Environmental Science and Technology* 43:6414–20.

http://www.buildingecology.com (accessed September 13, 2014).

http://www.expresstowers.in. Sustainable buildings and the carbon footprint—An overview (accessed September 12, 2014).

http://www.gov.uk/decc, 2013 UK Greenhouse Gas Emissions, Provisional Figures and 2012 UK Greenhouse Gas Emissions. Final figures by fuel type and end-user statistical. 2014, Department of Energy and Climate Change, London (accessed September 6, 2014).

http://www.grihaindia.org, Green Rating for Integrated Habitat Assessment, India (accessed September 12, 2014).

http://www.igbc.in, Indian Green Building Council (accessed September 12, 2014).

http://www.usgbc.org, US Green Building Council USA (accessed September 12, 2014).

Ibn-Mohammed, T., Greenough, R., Taylor, S., Ozawa-Meida, L., and A. Acquaye. 2013. Operational vs. embodied emissions in buildings—A review of current trends. *Energy and Buildings* 66:232–45.

ISO 14040. 2006. Environmental management—Life cycle assessment—Principles and framework.

ISO 14041. 1998. Environmental management—Life cycle assessment—Goal and scope definition and inventory analysis.

ISO 14042. 2000. Environmental management—Life cycle assessment—Life cycle impact assessment.

ISO 14043. 2000. Environmental management—Life cycle assessment—Life cycle interpretation.

Junnila, S., and A. Horvath. 2003. Life-cycle environmental effects of an office building. *Journal of Infrastructure Systems* 9(4):157–66.

Kofoworola, O.F., and S.H. Gheewala. 2008. Environmental life cycle assessment of a commercial office building in Thailand. *International Journal of Life Cycle Assessment* 13:498–511.

Kumbhar, S., Kulkarni, N., Rao, A.B., and B. Rao. 2014. Environmental life cycle assessment of traditional bricks in western Maharashtra, India. 4th International Conference on Advances in Energy Research 2013 (ICAER 2013). *Energy Procedia* 54:260–9.

Lai, J.H.K. 2013. Carbon footprints of hotels: Analysis of three archetypes in Hong Kong. *Sustainable Cities and Society* 14:334–41.

Larsen, H.N., Pettersen, J., Solli, C., and E.G. Hertwich. 2013. Investigating the carbon footprint of a university—The case of NTNU. *Journal of Cleaner Production* 48:39–47.

Levine, M.D., Price, L., and N. Martin. 1996. Mitigation options for carbon dioxide emissions from buildings. *Energy Policy* 24(10/11):937–49.

Loganthurai, P., Parthasarathy, S., Selvakumaran, S., and V. Rajasekaran. 2011. Energy conservation measures in a technical institutional building in Tamil Nadu in India. 2nd International Conference on Advances in Energy Engineering (ICAEE 2011). *Energy Procedia* 14:1181–6.

MNES (Ministry of Non-conventional Energy Sources) and TERI (Tata Energy Research Institute). 2001. Representative designs of energy-efficient buildings in India.

MNRE (Ministry of New and Renewable Energy), Government of India. http://www.mnre.gov.in (accessed September 13, 2014).

Nautiyal, H., and V. Goel. 2012. Progress in renewable energy under clean development mechanism in India. *Renewable and Sustainable Energy Reviews* 16:2913–9.

Newton, P.W., and S.N. Tucker. 2011. Hybrid buildings: A pathway to carbon neutral housing. *Architectural Science Review* 53(1):95–106.

Nordby, A.S. 2011. Carbon reductions and building regulations: The case of Norwegian mountain cabins. *Building Research and Information* 39(6):553–65.

Norman, J., MacLean, H.L., and C.A. Kennedy. 2006. Comparing high and low residential density: Life-cycle analysis of energy use and greenhouse gas emissions. *Journal of Urban Planning and Development* 132(1):10–21.

Onat, N.C., Kucukvar, M., and O. Tatari. 2014. Scope-based carbon footprint analysis of U.S. residential and commercial buildings: An input–output hybrid life cycle assessment approach. *Building and Environment* 72:53–62.

Ozawa-Meida, L., Brockway, P., Letten, K., Davies, J., and P. Fleming. 2013. Measuring carbon performance in a UK university through a consumption based carbon footprint: De Montfort University case study. *Journal of Cleaner Production* 56:185–98.

Pandey, D., Agrawal, M., and J.S. Pandey. 2011. Carbon footprint: Current methods of estimation. *Environmental Monitoring and Assessment* 178(1–4):135–60.

Parikh, K. 2012. Sustainable development and low carbon growth strategy for India. *Energy* 40:31–8.

POST, 2006. Carbon footprint of electricity generation. POST note 268, October 2006, Parliamentary Office of Science and Technology, London, UK.

Peterson, E.H. 1997. Database and inventory tool for building components and buildings environmental parameters. SBI Report 275, Danish Building Research Institute, Denmark.

Proietti, S., Desideri, U., Sdringola, P., and F. Zepparelli. 2013. Carbon footprint of a reflective foil and comparison with other solutions for thermal insulation in building envelope. *Applied Energy* 112: 843–55.

Ramachandra, T.V., and Shwetmala. 2012. Decentralised carbon footprint analysis for opting climate change mitigation strategies in India. *Renewable and Sustainable Energy Reviews* 16:5820–33.

Ramesh, T., Prakash, R., and K.K. Shukla. 2012a. Life cycle energy analysis of a residential building with different envelopes and climates in Indian context. *Applied Energy* 89:193–202.

Ramesh, T., Prakash, R., and K.K. Shukla. 2012b. Life cycle approach in evaluating energy performance of residential buildings in Indian context. *Energy and Buildings* 54:259–65.

Raupach, M., and P. Fraser. 2011. Climate and greenhouse gases. Climate change book. http://www.csiro.au (accessed September 13, 2014).

Reddy, B.V.V., and K.S. Jagadish. 2003. Embodied energy of common and alternative building materials and technologies. *Energy and Buildings* 35:129–37.

Sabapathy, A., and S. Maithel. 2013. A multi-criteria decision analysis based assessment of walling materials in India. *Building and Environment* 64:107–17.

Sarkar, P., Sharma, B., and U. Malik. 2014. Energy generation from grey water in high raised buildings: The case of India. *Renewable Energy* 69:284–9.

Sartori, I., and A.G. Hestnes. 2007. Energy use in the life cycle of conventional and low energy building: A review article. *Energy and Buildings* 39:249–57.

Scheuer, C., Keoleian, G.A., and P. Reppe. 2003. Life cycle energy and environmental performance of new university building: Modelling challenges and design implications. *Energy and Buildings* 35:1049–64.

SETAC (Society for Environmental Toxicology and Chemistry). 1991. A technical framework for life-cycle assessment. Workshop Rep., SETAC, Washington, D.C.

Sharma, A., Nautiyal, H., Varun, V., and V. Shree. 2010. Green building, a step in sustainable development. Proceedings of International Conference on Advances in Renewable Energy, MANIT, Bhopal, June 24–26, pp. 558–66.

Sharma, A., Saxena, A., Sethi, M., Shree, V., and V. Varun. 2011. Life cycle assessment of buildings: A review. *Renewable and Sustainable Energy Reviews* 15:871–5.

Shukla, A., Tiwari, G.N., and M.S. Sodha. 2009. Embodied energy analysis of adobe house. *Renewable Energy* 34:755–61.

Srinivasan, R.S., Ingwersen, W., Trucco, C., Ries, R., and D. Campbell. 2014. Comparison of energy-based indicators used in life cycle assessment tools for buildings. *Building and Environment* 79:138–51.

Steemers, K., and S. Manchanda. 2010. Energy efficient design and occupant well-being: Case studies in the UK and India. *Building and Environment* 45:270–8.

Sullivan, R., Gouldson, A., and P. Webber. 2012. Funding low carbon cities: Local perspectives on opportunities and risks. *Climate Policy* 13(4):514–29.

Tahmasebi, M.M., Banihashemi, S., and M.S. Hassanabadi. 2011. Assessment of the variation impacts of window on energy consumption and carbon footprint. 2011 International Conference on Green Buildings and Sustainable Cities. *Procedia Engineering* 21:820–8.

Tiwari, P. 2001. Energy efficiency and building construction in India. *Building and Environment* 36:1127–35.

Tiwari, P., Parikh, J., and V. Sharma. 1996. Performance evaluation of cost effective buildings—A cost, emissions and employment point of view. *Building and Environment* 31(1):75–90.

Tshudy, J.A. 1994. Environmental life cycle analysis—The foundation for understanding environmental issues and concerns. Proceedings, International Symposium on Resource, Conservation and Environmental Technologies in Metallurgical Industries, 20–25 August 1994, Lancaster USA, 229–37.

Tulsyan, A., Dhaka, S., Mathur, J., and J.V. Yadav. 2013. Potential of energy savings through implementation of energy conservation building code in Jaipur city, India. *Energy and Buildings* 58:123–30.

Varun, V. 2010. Life Cycle Assessment of Small Hydropower Systems in India. Ph.D. Thesis, National Institute of Technology, Hamirpur, India.

Varun, V., Sharma, A., Shree, V., and H. Nautiyal. 2012. Life cycle environmental assessment of an educational building in Northern India: A case study. *Sustainable Cities and Society* 4:22–8.

Varun, Bhat, I.K., and R. Prakash. 2009. LCA of renewable energy for electricity generation systems—a review. *Renewable and Sustainable Energy Reviews* 13:1067–73.

Wackernagel, M., and W.E. Rees. 1996. *Our Ecological Footprint: Reducing Human Impact on the Earth.* New Society Publishers, Gabriola Island.

Wiedmann, T., and J. Minx. 2007. A definition of carbon footprint. ISAUK Research Report 07-01, Durham, ISAUK Research and Consulting.

Wiedmann, T., and J. Minx. 2008. A definition of "carbon footprint." In: C.C. Pertsova, ed. *Ecological Economics Research Trends*, Chapter 1, pp. 1–11, Nova Science Publishers, Hauppauge, NY, USA.

Wright, D., Leigh, R., Kleinberg, J., Abbott, K., and J. Scheib. 2014. New York City can eliminate the carbon footprint of its buildings by 2050. *2014 ACEEE Summer Study on Energy Efficiency in Buildings* 10:365–77.

www.earthobservatory.nasa.gov, Earth Observatory, NASA, Global Temperature (accessed September 6, 2014).

Yang, Z., and Y. Zhao. 2014. Energy consumption, carbon emissions, and economic growth in India: Evidence from directed acyclic graphs. *Economic Modelling* 38:533–40.

Zhang, Y., Zheng, H., and B.D. Fath. 2014. Analysis of the energy metabolism of urban socioeconomic sectors and the associated carbon footprints: Model development and a case study for Beijing. *Energy Policy* 73:540–1.

Zuo, J., and Z.Y. Zhao. 2014. Green building research—Current status and future agenda: A review. *Renewable and Sustainable Energy Reviews* 30:271–81.

12 The Carbon Footprint of Dwelling Construction in Spain

Jaime Solís-Guzmán, Patricia González-Vallejo,
Alejandro Martínez-Rocamora, and Madelyn Marrero

CONTENTS

12.1 Introduction 259
12.2 Spanish Dwelling Construction...260
12.3 Quantifying Resource Consumption 261
12.4 Methodology...263
    12.4.1 Emission Factors...264
    12.4.2 CF of Electricity 265
    12.4.3 CF of Water Consumption 266
    12.4.4 CF of Manpower: Food Consumption 267
    12.4.5 CF of Manpower: Mobility...268
    12.4.6 CF of Manpower: MSW..269
    12.4.7 CF of Construction Materials 269
    12.4.8 CF of Machinery...271
    12.4.9 CF of Construction and Demolition Waste...271
12.5 Results...272
    12.5.1 Project Budget and Resources Database...272
    12.5.2 General Project Characteristics 273
    12.5.3 Overall Results..276
12.6 Conclusions..277
12.7 Limitations, Advantages, and Further Research 278
Acknowledgments...278
References..278

12.1 INTRODUCTION

During the past decade, Spain has experienced a huge construction bubble, which has caused the biggest economic crisis since democracy was introduced (COAAT Barcelona 2010). Not only does this high construction rate exert an impact in the economy, but it also causes an impact on the Spanish environment, which has yet to be evaluated. It is necessary to assess this exigency through indicators, so that the weight of environmental impacts can be qualified and quantified. The tools that analyze these impacts generally follow the methodology of life-cycle analysis (LCA) (Malmqvist and Glaumann 2009; Zabalza Bribián et al. 2011). However, other approaches exist such as the emergy analysis (Meillaud et al. 2005) and the material flow analysis (Sinivuori and Saari 2006). Currently, there is a tendency to use simpler methodologies that can be more easily understood by society, among which feature the ecological footprint (EF) (Wackernagel and Rees 1996) and the carbon footprint (CF) (Weidema et al. 2008). Both can be adapted to the unique characteristics of the construction sector (Solís-Guzmán et al. 2013, 2014), but the latter has been chosen for its comprehensibility, transparency, and adaptability (Cagiao et al. 2011).

The CF is largely used in the business environment for its utility in energy planning and as a marketing tool. Furthermore, its compatibility with the Kyoto Protocol has provided a major incentive for its application. This indicator measures the total amount of greenhouse gas (GHG) emissions caused directly and indirectly by an individual, event, organization, or product and is expressed in equivalent units of mass of CO_2 (Weidema et al. 2008). The Kyoto Protocol is considered equivalent to the category of global warming potential of LCA methodologies and is usually calculated according to the GHG Protocol and PAS 2050 methodologies (Pérez Leal 2012).

In this chapter, Spanish dwelling construction is studied through 92 different projects by applying the CF methodology to their construction process (Solís-Guzmán et al, 2014). This research, therefore, aims to bring all previous knowledge related to the CF indicator to analyze the construction phase of dwellings and to determine the advantages and disadvantages that this indicator may yield in the analysis of environmental impact on the building sector.

The following procedure for the assessment of the CF in the construction of residential buildings is based on the project bill of quantities and on the systematic classification of construction work (Marrero and Ramirez-de-Arellano 2010): materials, manpower, and machinery. An approach to the calculation of the CF, valid for any residential building construction, is described in detail therein.

For the selection of projects, the following steps were taken.

1. Identify the average dwelling types, which were built in Spain, by means of statistical data collection and information analysis (Spain MD 2011).
2. Search for current projects that represent the most common typologies.
3. Create a resources quantification database by means of systematic classification (ACCD 2010), which defines materials, machinery, and manpower used.

Once all project data were processed, the CF was calculated, based mainly on the authors' previously established methodology (Solís-Guzmán et al. 2014) and on new approaches to the quantification of resources, energy consumption of machinery, and water consumption during construction work.

12.2 SPANISH DWELLING CONSTRUCTION

During the construction bubble in Spain, an unprecedented surplus of activity took place, resulting in thousands of empty dwellings throughout the whole country (Spain MD 2013). This construction is analyzed in Table 12.1 from official statistical data generated by the Ministry of Development in Spain (2011). The typical construction characteristics can be obtained from the same report and are summarized in Table 12.2.

TABLE 12.1
Buildings Constructed in Spain from 2007 to 2010

	New Buildings				Old Buildings			
	Residential							
	Collective Residence							
Year	Dwellings	Permanent	Temporary	Total	Nonresidential	Total	Rehab.	Demolished
2006	208,016	306	309	208,631	21,413	230,044	35,856	2848
2007	165,833	197	292	166,322	20,825	187,147	33,359	26,141
2008	79,467	126	159	79,752	13,926	93,678	34,807	14,573
2009	39,349	102	113	39,564	12,180	51,744	33,267	7984
2010	34,317	183	610	35,110	9671	44,781	31,910	8084
Total	526,982	914	1483	529,379	78,015	607,394	169,199	59,630

TABLE 12.2

Most Common Construction Characteristics in Spain, Period 2007–2010

Building Characteristics								
Dwelling Type	Building Type	Floors above Ground	Floors below Ground	Floor Space (m²)	Bedrooms	Bathrooms	Garage	Entity
Multifamily	Two or more dwellings	Four or more	1	72	3	2	Yes	Private

Constructive Characteristics								
Vertical Structure	Horizontal Structure	Roof	Facade	Carpentry	Floor	Doors	False Ceiling	Exterior Blinds
Concrete	Unidirectional	Ceramic	Ceramic	Aluminum	Ceramic	Wood	Yes	Yes

The economic impact of this construction is well known and resulted in an economic crisis that started in 2008 and continues to the present day, but little has been done to measure its environmental impact. In this chapter, a general approach is presented to address this environmental impact by means of the CF indicator.

12.3 QUANTIFYING RESOURCE CONSUMPTION

The present CF assessment focuses on the implementation and construction phase of residential buildings: research into the other two phases of the life cycle of buildings, those of use and demolition, does not constitute part of the present analysis. A second assumption is also considered: the only activity that exerts an impact on the study area is that which corresponds to the construction of the residential buildings considered earlier.

Once the boundaries of the analysis are established, the project budget becomes the main source of information for the quantification of the resources consumed during the building construction. The construction industry normally organizes works clearly in the project budget because cost control is always a major issue. The budgets of 92 different dwelling projects, which met the construction characteristics evaluated in the previous section, are analyzed and classified; their budgets are reorganized in a construction-work breakdown system (CBS) that enables comparison. This system, for the organization of the work and its quantities, has been successfully applied to control the generation of construction waste (Solís-Guzmán et al. 2009). In Table 12.3, the measurement of the projects is presented along with the average quantities per square meter constructed. These are grouped per number of floors above ground. The sample sizes are: four projects with one floor, two with two floors, 22 with three floors, 22 with four floors, 20 with five floors, and 22 with 10 floors.

Once the quantities per project characteristic are defined, the next step is to translate them into specific consumed resources, which is possible by means of the CBS employed. Each Q_i listed in Table 12.3 is transformed into three types of resources: manpower, materials, and machinery. The systematic classification of the Andalusia Construction Cost Database (ACCD 2010) is used. The ACCD structure is hierarchical, with clearly defined levels, whereby each group is divided into subgroups of similar characteristics (Marrero and Ramirez-de-Arellano 2010). As an example, in Table 12.4, the quantity Q_i 05F "Concrete and ceramic slab" is shown, and its corresponding resources are listed and grouped into: materials, manpower, and machinery. Five different unitary costs, obtained from the ACCD, 05FUA000018, 05FUA00118, 05FUS00018, 05FUS00108, and 05FUS00118, are used to define the average quantity of each item which forms part of the unitary cost, corresponding to the last column in Table 12.4. The same procedure is carried out for each

TABLE 12.3
Resources, Q_i, per Square Meter Constructed

			Project Quantities, Q_i (U/m²)					
			Number of Floors above Ground					
Code	Measurement Unit and Short Description	Specific Resources	1	2	3	4	5	10
02	Earthwork							
02E	m³, excavations	Loader	0.59	0.32	0.54	0.45	0.44	0.30
02R	m³, pads	Mechanical means	0.19	0.09	0.09	0.08	0.06	0.03
02T	m³, transport	Mechanical means	0.52	0.33	0.68	0.57	0.55	0.38
03	Foundation							
03A	kg, reinforced steel		6.36	4.30	7.08	6.20	5.51	4.67
03P	m, piles		0.00	0.00	0.12	0.13	0.12	0.04
03E	m², wooden formwork	Wood	0.46	0.24	0.32	0.25	0.23	0.12
03HA	m³, reinforced concrete	Shed with crane	0.12	0.07	0.10	0.08	0.07	0.06
03HM	m³, concrete		0.24	0.11	0.05	0.05	0.05	0.05
03H	m³, foundation concrete beams		0.07	0.03	0.03	0.03	0.03	0.02
04	Water disposal							
04A	U, sewer/catch basins	*In situ*	0.05	0.03	0.01	0.01	0.01	0.01
04C	m, collectors	Concrete	0.13	0.09	0.06	0.05	0.04	0.02
04B	m, vertical pipelines	PVC	0.12	0.14	0.09	0.10	0.10	0.10
05	Structures							
05F	m², concrete slab	Concrete block	1.92	1.46	1.12	1.09	1.06	1.02
05HA	kg, steel reinforcement		7.56	11.75	12.13	11.67	11.23	12.50
05HE	m², formwork/concrete cast	Metallic	0.55	0.91	0.79	0.76	0.73	0.78
05HA	m³, reinforced concrete		0.07	0.10	0.10	0.09	0.08	0.10
06	Partitions							
06DC	m², wall chambers		0.77	1.00	0.60	0.65	0.67	0.73
06DT	m², wall partitions		0.90	0.88	0.64	0.70	0.72	0.80
06LE	m², exterior brick		0.88	1.08	0.85	0.86	0.86	0.90
06LI	m², interior brick		0.61	0.34	0.29	0.30	0.31	0.33
07	Roofs							
07H	m², horizontal roof	Passable	1.11	0.00	0.18	0.13	0.10	0.05
07I	m², inclined roof	Ceramic tile	0.00	0.55	0.18	0.15	0.12	0.07
08	Installations							
08EC	m, circuits		0.71	0.61	0.55	0.59	0.61	0.66
08ED	m, derivations	PVC	0.03	0.02	0.10	0.12	0.12	0.21
08EL	U, light points		0.13	0.12	0.11	0.11	0.11	0.12
08ET	U, sockets		0.23	0.22	0.18	0.20	0.20	0.22
08EP	m, ground connection		0.17	0.14	0.09	0.10	0.10	0.17
08FC	m, hot water pipe	Copper	0.26	0.23	0.15	0.17	0.17	0.19
08FD	U, drains		0.09	0.08	0.06	0.07	0.07	0.08
08FF	m, cold water pipe	Copper	0.39	0.34	0.30	0.33	0.34	0.45
08FG	U, tap		0.07	0.07	0.05	0.06	0.06	0.06
08FS	U, bathroom appliances	Porcelain	0.07	0.06	0.04	0.05	0.05	0.05
08FT	U, thermos/heaters	Electric	0.01	0.01	0.01	0.01	0.01	0.01
09	Insulation							
09T	m², thermal insulation	Polystyrene	2.39	1.36	0.61	0.65	0.66	0.73

(Continued)

TABLE 12.3 (*Continued*)
Resources, Q_i, per Square Meter Constructed

			Project Quantities, Q_i (U/m²)					
			Number of Floors above Ground					
Code	Measurement Unit and Short Description	Specific Resources	1	2	3	4	5	10
10	Finishes							
10AA	m², tiling	Cement mortar	0.51	0.44	0.34	0.37	0.39	0.43
10CE	m², plaster		1.70	1.81	1.45	1.50	1.57	1.66
10CG	m², whitewash	Plaster	3.35	2.93	2.15	2.37	2.43	2.70
10S	m², floors	Ceramic	0.79	0.75	0.77	0.79	0.79	0.83
10SS	m², pavers		0.12	0.00	0.16	0.13	0.12	0.06
10T	m², ceiling	Metallic frame	0.05	0.05	0.06	0.07	0.07	0.08
10R	m, finishing	Limestone	0.16	0.15	0.08	0.08	0.09	0.09
11	Carpentry							
11CL	m², aluminum carpentry		0.15	0.14	0.11	0.11	0.11	0.12
11M	m², wood carpentry		0.02	0.02	0.00	0.00	0.00	0.00
11MA	m², closets		0.02	0.00	0.00	0.00	0.00	0.00
11MP	m², wood doors		0.12	0.11	0.11	0.12	0.12	0.13
11B	m², bannister	Steel	0.00	0.05	0.05	0.06	0.05	0.06
11P	m², blinds		0.07	0.06	0.05	0.06	0.06	0.07
11R	m², safety bars		0.08	0.05	0.02	0.04	0.01	0.01
12	Glass and Polyes							
12A	m², glass		0.12	0.11	0.09	0.10	0.11	0.11
13	Coating							
13PE	m², exterior paint		1.67	1.59	1.17	1.21	1.23	1.35
13PI	m², interior paint		3.51	3.36	2.57	2.77	2.81	3.09

of the classification elements: 02T, 02R, 02P, 03A, 03P, and so on in Table 12.3. The final result is established as a resource quantification database (RQDB), which enables the evaluation of all typical Spanish dwelling constructions from the point of view of consumed resources (González-Vallejo et al. 2015).

12.4 METHODOLOGY

Manpower, construction materials, and machinery are not the only resources consumed during building construction. Waste generation, electricity, and water consumption also generate a CF. These are defined from the project's general characteristics such as the project total cost or the building typology (see the second level in Figure 12.1).

Manpower is the first resource defined from the project budget data. In building construction, the operators consume food and fuel in their transportation back and forth from the construction site, and they also generate municipal solid waste (MSW) during their working hours. These impacts are identified in the third level of Figure 12.1.

The second resource consumption obtained from the budget is building materials, which, through manufacturing processes, transportation, and installation (Figure 12.1), consume energy from different sources (necessary for the material extraction, manufacture, and commissioning). For the CF assessment of materials, after a quantitative study using the RQDB, their corresponding GHG emissions are calculated.

The third resource determined from the project budget is the machinery and tools used in the construction work. Machinery can be grouped into two families depending on its engine: combustion

TABLE 12.4

Quantities in Concrete Slabs per ACCD

05F		m^2, Concrete Slab					
Unit	Component	05FUA00018	05FUA00118	05FUS00018	05FUS00108	05FUS00118	Average
	Materials						
kg	Steel bars B500S	1.990	2.990	2.190	3.490	3.490	2.830
U	Concrete block	5.400	4.860	5.400	4.860	4.635	5.031
m^3	Concrete HA-25/P/20/Iia	0.105	0.115	0.110	0.120	0.120	0.114
m^3	Pine wood panel	0.001	0.001	0.001	0.001	0.001	0.001
m	Prestressed beam	1.397	2.338	1.397	2.338	2.338	1.962
U	Small material	1.000	1.000	1.000	1.000	1.000	1.000
	Manpower						
h	First-class official ironwork	0.026	0.046	0.030	0.054	0.054	0.042
h	First official	0.630	0.053	0.107	0.120	0.121	0.206
h	Specialized labor	0.306	0.356	0.356	0.406	0.406	0.366
	Machinery						
h	Vibrator	0.070	0.070	0.080	0.075	0.080	0.075

or electric. The former consumes gasoline or diesel in order to perform on the construction site. The liters of petrol or gasoil consumed are calculated from the RQDB.

In contrast, the electricity consumed by electric tools is considered a separate resource because it is not only consumed by electric machines but also by the construction site lighting, air conditioning, heating, and domestic hot water. All these aspects depend on the general characteristics of the project, as occurs with the water consumption on site. Finally, the building construction generates waste, usually referred to as construction and demolition waste (CDW), and is considered a separate impact defined from the general construction project characteristics.

Various coefficients transform the previously defined consumptions and the intermediate elements into partial footprints. In the case of the electricity, the country's electric mix is needed. For the food calculations, energy intensities are needed, and for the water, GHG emissions per volume are required. The calculations for construction materials are based on the emission factor for each material used, and those for the waste need the generation coefficients and conversion factors. The same happens with mobility, which requires a mobility conversion factor.

Partial and total footprints, generated in the construction phase of the dwelling sector, are defined by means of the intermediate elements and their corresponding coefficients. These are located in the bottom level of Figure 12.1.

12.4.1 EMISSION FACTORS

Given that a considerable amount of the total CF will be due to energy consumption in the form of electricity or fuel combustion, an emission factor for each type of consumption is needed.

The emission factor applied for the mobility of workers (E_f) comes expressed in kg CO_2 eq/l and is as specified in Table 12.5. The same is used for machinery, as Gasoil A or B is assumed to be its energy supply.

For the national energy mix (E_e), the emission factor is expressed in kilograms of CO_2 per gigajoule, and accounts for the CO_2 emissions from fossil fuels consumed for electricity generation,

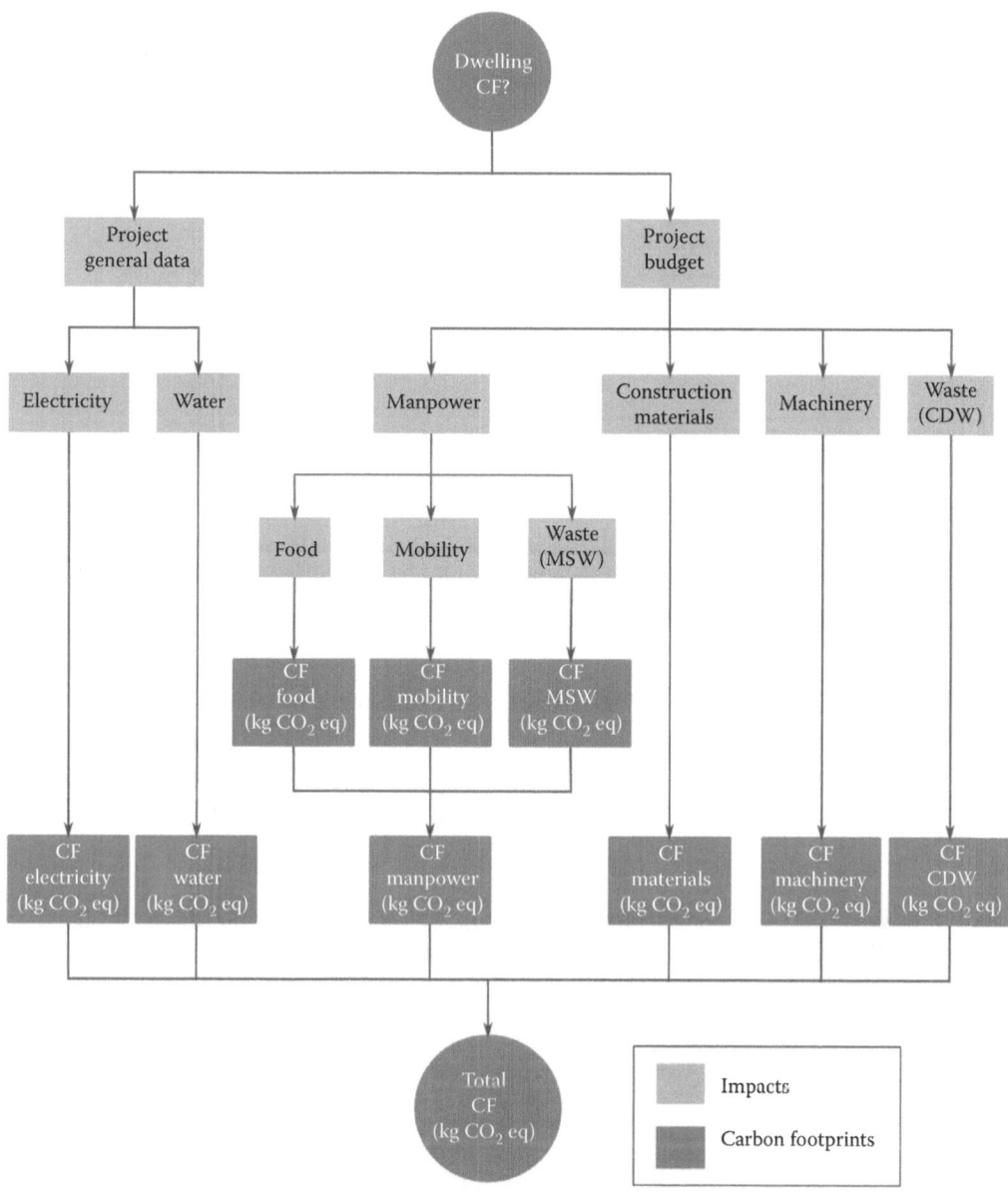

FIGURE 12.1 Methodology flowchart.

in both electricity-only and combined heat and power plants, divided by output of electricity generated from all fossil and nonfossil sources. For our 92 projects, this value is 80.83 kg CO_2/GJ.

To sum up, the emission factors used in this methodology are presented in Table 12.6, except those of construction materials, which are listed in Section 4.7.

12.4.2 CF OF ELECTRICITY

The current electricity consumption is not available for the 92 projects analyzed. Instead, the electricity consumed in the construction is determined by means of a polynomial formula (Spain MP 1970, 1981), which estimates the energy cost as a percentage of the total costs for different types

TABLE 12.5

Fuel Consumption and Emission Coefficients of Cars in Spain

Fuel	Consumption (l/100 km)	CO_2 Emissions (kg CO_2 eq/l)
Gasoline	6.80	2.35
Gasoil	5.42	2.64

Source: Adapted from Instituto para la Diversificación y Ahorro de la Energía (IDAE). 2011. *Guía Práctica de la Energía: Consumo Eficiente y Responsable* (*Practical Energy Guide: Efficient and Responsible Consumption*). Madrid, Spain: IDAE. http://www.idae.es/index.php/mod.pags/mem.detalle/relcategoria.1161/id.542/relmenu.64 (accessed November 1, 2013).

of construction work (roads, canals, railways, buildings, etc.). Thus, the polynomial formula says that 9% of the total production cost (TPC) is dedicated to energy consumption. The total energy consumption of the work execution is considered to be shared by electricity and fuel consumption. Therefore, once the total energy consumption cost is obtained from the formula and the fuel consumption cost is determined in the corresponding section, then the difference between these costs can be considered to be the electricity cost (Solís-Guzmán et al. 2013). The electricity cost in Spain was 0.14256 €/kWh (39.6 €/GJ) in 2012. Losses due to transmission and distribution of electricity are accounted for by assuming an efficiency factor (E_{ef}) of 3 GJ/GJ, in order to convert the electricity consumption to primary energy (IDAE 2011). The emission factor is that specified in Table 12.6.

The formula used is:

$$CF_e = \frac{TC_e}{C_e} \times E_{ef} \times E_e \tag{12.1}$$

where CF_e is the CF of energy consumption (kg CO_2 eq), TC_e the total cost of electricity consumption (€), C_e the cost of electricity (€/GJ), E_{ef} the efficiency factor for electricity production (3 GJ/GJ), and E_e the emission factor of energy mix (kg CO_2 eq/GJ).

12.4.3 CF OF WATER CONSUMPTION

The current data of water consumed by the 92 projects remain unavailable, but water consumption data from Seville's Water Supply Company for 90 other projects have been analyzed and the corresponding average water consumption per project type determined. The results in cubic meters per square meter constructed are summarized in Table 12.7 (Trigo-Ledesma 2013).

Energy consumption per volume of water consumed (E_w), which, according to EMASESA (2005) is 0.44 kWh/m³ or 0.001584 GJ/m³, is the key for this calculation. This is the energy employed to

TABLE 12.6

Emission Factors Used in the Present Study

Emission Factor	Value	Source
E_f (fuel combustion for the mobility of operators and the use of machinery)	2.35 kg CO_2/l (gasoline) 2.64 kg CO_2/l (gasoil)	(IDAE 2011)
E_e (national energy mix)	80.83 kg CO_2/GJ	(IEA 2012)

TABLE 12.7

Water Consumption per Dwelling Type

No. of Floors above Ground	Sample Size	Average Consumption (m³/year · m²)
1	22	0.29
2	31	0.24
3	21	0.13
4	15	0.15
5	3	0.15
10	2	0.07

conduct water to the dwellings, for drinking water, and treatment of wastewater. Therefore, the formula employed for the calculation of the CF of water consumption is:

$$\text{CF}_\text{w} = W \times E_\text{w} \times E_\text{ef} \times E_\text{e} \tag{12.2}$$

where CF_w is the CF of water consumption (kg CO_2 eq), W the water consumption (m³), E_w the energy consumption per volume of water consumed (GJ/m³), E_ef the efficiency factor for electricity production (3 GJ/GJ), and E_e the emission factor of energy mix (kg CO_2 eq/GJ).

12.4.4 CF OF MANPOWER: FOOD CONSUMPTION

The initial assumption of this section is that workers' food is attributed to the CF of the building construction as this activity takes place on the worksite in the same way as in the methodology developed by Domenech Quesada (2007), in which business meals are allocated to the corporate EF. To this end, the total number of manpower hours worked for the entire construction project must first be calculated and is obtained from the RQDB.

The CF is calculated using the expression:

$$\text{CF}_\text{fd} = \frac{N_\text{h}}{h_\text{me}} \times \text{CF}_\text{me} \tag{12.3}$$

where CF_fd is the CF of food consumption (kg CO_2 eq), N_h the total number of hours worked, and h_me is the 8 h/meal. One meal per working day is assumed as CF_me: CF per meal (kg CO_2 eq/meal).

Therefore, it is necessary to obtain the CF of various types of food that make up the daily meal of every worker. This CF is generated due to the energy required to process them, taking into account all the inputs, fertilizers, pesticides, and treatments. This energy is expressed in the following equations as energy intensity (EI_i) of each type of food. Eventually, it is transformed into CO_2 emissions with the formula:

$$\text{CF}_\text{me} = C \times \text{EI} \times E_\text{e} \tag{12.4}$$

where CF_me is the CF per meal (kg CO_2 eq/meal), C the food consumption (t/meal), EI the energy intensity (GJ/t), and E_e the emission factor of energy mix (kg CO_2 eq/GJ).

Taking an estimated cost per meal at 10 €, the percentage of this cost corresponding to each type of food ($\%F_\text{i}$), the conversion factor of tons for each 1000 € (C_i), and the energy intensities of the different kinds of food (EI_i) are represented in Table 12.8 (Domenech Quesada 2007):

$$C \times \text{EI} = \frac{C_\text{me}}{1000} \times \sum \frac{\%F_\text{i}}{100} \times C_\text{i} \times \text{EI}_\text{i} \tag{12.5}$$

TABLE 12.8
Parameters for the Calculation of the Food Footprint

Foods	F_i (%)	C_i (t/1000 €)	EI_i (GJ/t)
Meat	25	0.65	80
Fish	25	0.50	100
Cereals	12	4.69	15
Beverages	10	0.34	7
Vegetables	8	1.45	10
Sweets	6	0.70	15
Oil	5	0.71	40
Dairy	5	0.93	37
Coffee	4	0.54	75

Source: Adapted from Domenech Quesada, J.L. 2007. *Huella Ecológica y Desarrollo Sostenible (Ecological Footprint and Sustainable Development).* Madrid, Spain: AENOR.

where C_{mc} is the cost per meal (10 €/meal), $\%F_i$ the percentage of the meal cost that each type of food represents (%), C_i the consumption in tons per 1000 € (t/1000 €), and EI_i the energy intensities (GJ/t).

12.4.5 CF OF MANPOWER: MOBILITY

In order to determine the CF related to the mobility of workers, the following assumptions are made:

1. Private vehicles are established as the only means of transport, as it is assumed that the worksite is placed in a remote area away from the city center.
2. The total number of days worked is equivalent to the total number of workers participating in the construction process.
3. The average distance traveled by the vehicles is 15 km each way.
4. The average vehicle occupancy is 1.2 people per vehicle (IDAE 2011). In order to determine the number of workers, the total number of hours worked must be known (calculated in the previous section for food CF). Eight working hours per day helps define the number of trips to the worksite.
5. The average fuel consumption of vehicles is 5.42 l/100 km (IDAE 2011).

Taking into account the previous assumptions, the calculation of the CF of fuel consumption is made as follows:

$$\mathrm{CF_{mo}} = \left(\frac{N_h/H_d}{V_o} \right) \times D_t \times F_c \times E_f \tag{12.6}$$

where $\mathrm{CF_{mo}}$ is the CF of the mobility of workers (kg CO_2 eq), N_h the total number of hours worked (h), H_d the hours per worker each day (8 h/worker), V_o the average vehicle occupancy (1.2 workers/trip), D_t the distance per trip to the worksite, back and forth (30 km/trip), F_c the average fuel consumption of vehicles (5.42 l/100 km), and E_f the emission factor of fuel (kg CO_2 eq/l) from Table 12.6.

TABLE 12.9
Energy Intensities, Recycling, and Energy Saving Rates for MSW

	Organic	Paper	Plastic	Glass
EI_x (GJ/t)	20	30	43.75	20
$\%R_x$	13	50	40	40
$\%SE_x$	100	50	70	40

12.4.6 CF OF MANPOWER: MSW

The determination of the CF of MSW is based on statistical data (Spain ME 2001; Andalusia ME 2009; IDAE 2011) and on the energy intensity of materials (Solís-Guzmán et al. 2013). In order to simplify the calculations of the energy needed for waste treatment, a conversion rate, CR_x, is calculated as follows using the data summarized in Table 12.9:

$$CR_x = EI_x \times \left(1 - \frac{\%R_x}{100} \times \frac{\%SE_x}{100} \right) \tag{12.7}$$

where CR_x is the conversion rate of MSW (GJ/t) and EI_x is the energy intensity of the production of the material from which the waste is made. For these values, the energy intensities of the materials to be recycled must be known ($\%R_x$ is the recycling rate of waste x and $\%SE_x$ is the percentage of energy saved by recycling).

The quantities of each kind of MSW (organic, paper/cardboard, plastics, and glass) generated by workers are assumed to be half the total MSW per capita in 1 year. Thus, the total generation of MSW in tons of each waste stream "i," G_i is obtained by multiplying the aforementioned statistical data, G_x, by the number of workers on the construction site and dividing by 2. Both G_x and CR_x are shown in Table 12.10.

The CF of MSW is calculated using the following equation:

$$CF_{MSW} = \sum \frac{N_h}{H_d} \times \frac{G_i/365}{2} \times CR_i \times E_e \tag{12.8}$$

where CF_{MSW} is the CF of MSW (kg CO_2 eq), N_h the total number of hours worked (h), H_d the hours worked per day (8 h/day), G_i the quantity of waste i generated (t/cap·year), CR_i the conversion rate of waste i (GJ/t), and E_e the emission factor of electricity (kg CO_2 eq/GJ).

12.4.7 CF OF CONSTRUCTION MATERIALS

The footprint of construction materials is determined using the following expression:

$$CF_m = \sum Q_i \times C_i \times E_i \tag{12.9}$$

TABLE 12.10
Conversion Factors for MSW

	Organic	Paper	Plastic	Glass
G_x (t/cap·year)	0.264	0.126	0.066	0.042
CR_x (GJ/t)	17.4	22.5	31.5	16.8

where CF_m is the CF of construction materials (kg CO_2 eq), Q_i the quantity of material i (U), C_i the conversion coefficient of material i (kg/U), and E_i the emission factor of material i (kg CO_2 eq/kg).

The example shown in Table 12.11 corresponds to the most representative materials due to their high volumes. The consumption of materials (by weight) is determined through the RQDB. Column 2 of Table 12.11 shows the unit in which the resource is measured in construction budgets. In order to change the measurement units of the resources (m, m², m³, etc.) to weight (kg), the coefficients calculated by Mercader (2010) are used as listed in column 3.

The emission factor values (column 4) were obtained from various databases (Ecoinvent Centre 2013; ELCD 2013; ITeC 2013; PlasticsEurope 2013), by taking the most suitable values according to the origin of the data, its transparency, and comprehensiveness (Martínez-Rocamora 2012). The data for CO_2 emissions are calculated by applying the GHG Protocol methodology to the life-cycle inventory of each material.

TABLE 12.11
Examples of Emission Factors of the Most Representative Materials in the Study

Materials	U	C_i (kg/U)	E_m (kg CO_2 eq/kg)	Source Database	Source Study
			Concrete		
Cover with concrete fence	m²	100.91	0.098	Base Carbone	SNBPE (2012)
Concrete	m³	2500.00			
Concrete HM	m³	2500.00			
Concrete collectors	m	67.00			
			Steel		
Steel	kg	1.00	1.03	ELCD	WSA (2011)
Plastic steel rolling guide	m	0.10			
Steel	m²	12.26			
Tap	U	1.36			
Steel panel	U	4.63			
			Ceramic		
Brick	U	1853.00	0.219	Ecoinvent	Kellenberger et al. (2007)
White tiling 15 × 15 cm	U	0.30	0.57	BEDEC	—
Ceramic floor tile	U	0.74			
			Polystyrene		
Polystyrene	m³	12.00	3.39	PlasticsEurope	Boustead (2006)
Rigid polystyrene plates	m³	12.00			
Sockets II + T 10/16 A C/Plate	U	0.09			
Universal sockets	U	0.03			
Simple switch	U	0.07			
			PVC		
Rolling blind	m²	6.07	3.24	PlasticsEurope	Ostermayer and Giegrich (2006)
Siphon trap	U	0.20			
Vertical pipeline	m	1.39			
Pipe	m	2.13			
Drain	U	0.31			

By performing a similar analysis with all the materials measured in the design project, their consumption and emission factors are obtained, which lead to the CF of the construction materials.

12.4.8   CF OF MACHINERY

In order to determine the energy consumption due to the machinery involved in the work, the total hours of machinery used are calculated from the project bill of quantities and the RQDB, and then the economic cost of the machinery energy can be calculated, together with its total consumption. In the previous work (Solís-Guzmán 2011), the whole machine cost obtained from the polynomial formula mentioned in Section 4.2, which includes machine depreciation, maintenance, and fuel, was considered as fuel cost and was then transformed into energy consumed. In the present work, a revised methodology is used after performing a machinery cost analysis (Ramírez-de-Arellano-Agudo 2006), whereby it is established that only 20% of the total machinery cost can be considered energy cost, which includes fuel (5%) and maintenance cost (15%) but not equipment depreciation. The liters of fuel consumed are determined using the current cost of fuel in Spain, which is 1.365 €/l (CORES 2013).

The CF of machinery can be expressed as the sum of the CF of fuel consumption and the CF of the electricity consumed for machinery maintenance:

$$CF_{mac} = CF_{fm} + CF_{em}$$ (12.10)

which are calculated as follows:

$$CF_{fm} = \frac{TC_f}{C_f} \times E_f$$ (12.11)

where CF_f is the CF of fuel consumption by machinery (kg CO_2 eq), TC_f the total cost of fuel consumption (€), C_f the cost of fuel (€/l), and E_f the emission factor of fuel (kg CO_2 eq/l) and in the case of electricity for maintenance:

$$CF_{em} = \frac{TC_e}{C_e} \times E_{ef} \times E_e$$ (12.12)

where CF_{em} is the CF of electricity consumption for the maintenance of machinery (kg CO_2 eq), TC_e the total cost of electricity consumption for maintenance (€), C_e the cost of electricity (€/GJ), E_{ef} the efficiency factor for electricity production (3 GJ/GJ), and E_e the emission factor of the energy mix (kg CO_2 eq/GJ).

As mentioned in Section 4.2, these results must be extracted from the total energy cost obtained using the polynomial formula in order to determine the electricity cost to be used in that section.

12.4.9   CF OF CONSTRUCTION AND DEMOLITION WASTE

In this section, the environmental impact of CDW is analyzed. These calculations are based on Alcores software tool (Ramírez-de-Arellano-Agudo et al. 2008; Solís-Guzmán et al. 2009). Two types of waste are considered in accordance with the management models of the CDW treatment plants in Spain: excavated earth and mixed CDW. Mixed CDW groups the remains of materials generated during the execution of the work units and the packaging of construction materials. In the new construction work, excavated earth may represent over 80% of CDW, whereas the mixed CDW is distributed among the remains of materials and the packaging (Solís-Guzmán et al. 2009).

The determination of the CF of CDW is made using Equation 12.13 and is based on the following assumptions.

1. The recycling of construction materials (mixed CDW) has been already accounted for through their emission factors, as LCA methodologies include a percentage of recycled material in their inventories.
2. About 70% of the total excavated earth and none of mixed CDW are re-used on site with no extra emissions.
3. The remaining waste is transported to a landfill or treatment plant by trucks with a maximum charge of 24 tons.
4. The trucks used for transport consume 26 L of fuel each 100 km (IDAE 2006).
5. The average distance to the landfill or treatment plant is assumed to be 10 km, which makes 20 km per trip (back and forth):

$$\mathrm{CF_{CDW}} = \sum \frac{\left(1 - {}^{\%R_i}\!\big/\!_{100}\right) \times G_i}{T_w} \times T_d \times T_c \times E_f \tag{12.13}$$

where $\mathrm{CF_{CDW}}$ is the CF of CWD (kg CO_2 eq); $\%R_i$ the percentage of the CDW i that is re-used on site; G_i the total weight of CDW i generated (t); T_w the maximum weight charge of trucks (t/truck); T_d the trip distance to the landfill or treatment plant, back and forth (km); T_c the fuel consumption by trucks (l/100 km); and E_f the emission factor of gasoil as fuel, from Table 12.6 (kg CO_2 eq/l).

12.5 RESULTS

As shown in Figure 12.1, the data needed to calculate the CF of the construction process have two main sources: the project budget and the general characteristics of the project itself such as the TPC or the size of the building. From the former, the manpower, construction materials, machinery, and waste generated will be obtained, whereas from the latter we will extract the electricity and water consumption used for the construction process. In the following two sections, intermediate results for an example project will be presented. To this end, project number 25 out of the 92 analyzed has been chosen; it is a five-storey building with 5551.78 m² of constructed area and a TPC of 2,618,866.95 €. Finally, some overall and statistic results depending on the number of floors are shown in order to draw the final conclusions.

12.5.1 Project Budget and Resources Database

As mentioned before, the manpower on the construction site generates a significant CF due to food, mobility, and MSW generated. From the project budget and the RQDB, the total number of

---

TABLE 12.12
CF of Food Consumption by Workers

CF of Food Consumption	Values
C_{me}, cost per menu (€)	10
CF_{me}, carbon footprint per menu (kg CO_2 eq/menu)	32.92
N_h, total number of hours worked (h)	49,880
h_{me}, number of hours worked per menu eaten (h/menu)	8
Total number of menus eaten (menu)	6235
Total CF of food consumption (kg CO_2 eq)	205,271.56

TABLE 12.13
CF of Mobility of Workers

CF of Mobility	Values
N_h, total number of hours worked (h)	49,880
H_d, hours worked per day (h/day)	8
Total number of working days (or workers) (day or worker)	6235
V_o, average vehicle occupancy (worker/trip)	1.20
Total number of trips to the worksite, back and forth (trip)	5195.83
D_t, average distance per trip, back and forth (km/trip)	30
Total distance traveled (km)	155,875
F_c, average fuel consumption (l/100 km)	5.42
Total fuel consumption (l)	8448.43
E_f, emission factor of fuel (kg CO_2/l)	2.64
Total CF of mobility (kg CO_2 eq)	22,303.84

manpower hours for the entire work can be calculated. Thus, the CF from the different foods that make up the daily meals of the workers is obtained, as can be seen in Table 12.12. Following the corresponding guidelines raised in the methodology, the CF of mobility and MSW generated by workers are shown in Tables 12.13 and 12.14, respectively. It is noticeable that food is the main source of impact within the manpower section.

The CF of construction materials is shown in a summary table with a selection of materials (Table 12.15), which entail 93.20%, and the remaining 6.80% is produced by less representative materials such as sand, glass, gypsum, copper, stone, and so on. The intermediate calculations made to reach these results are not presented for reasons of size, but basically the first step was the quantification of each material through the RQDB and eventually their conversion to the CF by means of the corresponding emission factor, which was obtained from several LCA databases, as explained in Section 4.7. Also extracted from the project budget, the CF of machinery and CDW are calculated in Tables 12.16 and 12.17, respectively, resulting both less significant than manpower.

12.5.2 General Project Characteristics

The CF of electricity and water consumption are determined from the general project characteristics. As the energy consumption of machinery is already calculated, the difference between total energy consumption and fuel consumption is therefore the electricity consumption.

TABLE 12.14
CF of MSW

CF of Municipal Solid Waste	Values
N_h, total number of hours worked (h)	49,880.16
H_d, hours worked per day (h/day)	8.00
Total number of working days (day)	6235
Half of MSW generated in 1 day or year are produced at lunch	0.5
Days in a year (day/year)	365
Virtual years worked (year)	8.54
G_i, generation of waste i per year and person (t/year)	Table 12.10
CR_i, conversion rate for waste i (GJ/t)	Table 12.10
E_e, emission factor of energy mix (kg CO_2 eq/GJ)	80.83
Total CF of MSW (kg CO_2 eq)	7050.06

TABLE 12.15
CF of Construction Materials

Materials	Carbon Footprint (kg CO_2 eq)	Percentage (%)
Cement	355,074.15	19.54
Ceramics	419,928.74	23.11
Concrete	599,804.80	33.01
Polystyrene	70,558.18	3.88
Painting	156,056.75	8.59
Steel	92,161.24	5.07
Others	123,657.76	6.80
Total CF of construction materials	1,817,241.62	100.00

TABLE 12.16
CF of Machinery

CF of Machinery	Values
Total cost of machinery (€)	9761.37
TC_f, total cost of fuel consumption (€), 5%	488.07
TC_e, total cost of electricity for maintenance (€), 15%	1464.21
C_f, cost of fuel (€/l)	1.365
C_e, cost of electricity (€/GJ)	39.60
E_f, emission factor of fuel (kg CO_2 eq/l)	2.64
E_{ef}, energy efficiency factor (GJ/GJ)	3.00
Primary energy consumption (GJ)	110.925
E_e, emission factor of energy mix (kg CO_2 eq/GJ)	80.83
CF_{fm}, carbon footprint of fuel for machinery (kg CO_2 eq)	943.96
CF_{em}, carbon footprint of electricity for maintenance (kg CO_2 eq)	8966.07
Total CF of machinery (kg CO_2 eq)	9910.03

TABLE 12.17
CF of CWD

CF of CWD	Values
G_i, total excavated earth generated (t)	1832.09
G_i, total mixed CDW generated (t)	832.77
R_i, on-site reused percentage excavated earth (%)	70.00
R_i, on-site reused percentage of mixed CDW (%)	0.00
T_w, maximum weight charge of trucks per trip (t/trip)	24.00
Total number of trips full of excavated earth (trip)	23.00
Total number of trips full of mixed CDW (trip)	35.00
T_d, average trip distance to landfill or treatment plant, back and forth (km/trip)	20.00
T_c, fuel consumption by trucks (l/100 km)	26.00
E_f, emission factor of fuel (kg CO_2 eq/l)	2.64
CF of transport of excavated earth (kg CO_2 eq)	315.74
CF of transport of mixed CDW (kg CO_2 eq)	480.48
Total CF of CDW (kg CO_2 eq)	796.22

First, the electric energy CF is obtained on the basis of the polynomial formula (Solís-Guzmán et al. 2013), which determines the percentage of the TPC that corresponds to energy consumption (9%). This energy, as explained in Section 4.2, is shared between machinery and electricity. Therefore, the remaining cost is assumed to be the total electricity cost. This cost is transformed into GJ through the cost of electricity given by the power supplier. Eventually, the CF of electricity consumption is calculated following the guidelines in Section 4.2, as shown in Table 12.18, resulting in the second source of CF after construction materials.

In the case of water, the consumed volume has been recently estimated by Trigo-Ledesma (2013) for different sizes of buildings, and that consumption is converted to CF terms, as can be seen in Table 12.19. To sum up, general results for project number 25 are presented in Table 12.20, along with the percentage of the total CF that each category of impact source represents.

TABLE 12.18
CF of Electricity Consumption

CF of Electricity Consumption	Values
TC_e, total cost of electricity (€)	233,745.75
C_e, cost of electricity (€/GJ)	39.60
E_{ef}, energy efficiency factor (GJ/GJ)	3.00
Total primary energy consumed (GJ)	17,708.01
E_e, emission factor of energy mix (kg CO_2 eq/GJ)	80.83
Total CF of electricity consumption (kg CO_2 eq)	1,431,338.56

TABLE 12.19
CF of Water Consumption

CF of Water Consumption	Values
W, total water consumption (m³)	831.66
E_w, energy consumption per volume of water consumed (GJ/m³)	0.001584
E_{ef}, energy efficiency factor (GJ/GJ)	3.00
E_e, emission factor of energy mix (kg CO_2 eq/GJ)	80.83
Total CF of water consumption (kg CO_2 eq)	319.44

TABLE 12.20
Total CF of Project Number 25

Impact Source	Total CF (kg CO_2 eq)	Percentage (%)
Food	205,271.56	5.87
Mobility	22,303.84	0.64
MSW	7050.06	0.20
Materials	1,817,241.62	52.01
Machinery	9910.03	0.28
CDW	796.22	0.02
Electricity	1,431,338.56	40.96
Water	319.44	0.01
Total	3,494,231.33	100.00

TABLE 12.21
CF of Residential Buildings by Number of Floors above Ground

	CF (kg CO_2 eq/m²) Number of Floors above Ground					
Impacts	1	2	3	4	5	10
Food	57.178	48.217	41.803	38.915	38.519	38.971
Mobility	6.213	5.239	4.542	4.228	4.185	4.234
MSW	1.075	1.656	1.336	1.337	1.323	1.305
Materials	712.763	497.345	393.210	370.970	356.139	340.933
Machinery	3.176	1.978	3.921	3.293	2.960	1.928
CDW	0.202	0.157	0.214	0.192	0.181	0.144
Electricity	410.164	337.091	243.133	255.857	259.459	253.143
Water	0.113	0.092	0.051	0.058	0.058	0.026
Total CF (kg CO_2 eq/m²)	1190.88	891.78	688.21	674.85	662.82	640.68
Dwelling area (m²)	181	143	72	70	72	72
Bedrooms	5	5	3	3	3	3
People per dwelling	6	6	4	4	4	4
CF per person (kg CO_2 eq/person)	35,924.98	21,253.99	12,387.80	11,809.89	11,930.84	11,532.30

12.5.3 OVERALL RESULTS

When taking into account the 92 projects under study, a tendency to decrease is observed regarding the CF of construction materials and electricity consumption (per square meter) when increasing the number of floors over ground, whereas data show some stabilization starting at three-storey buildings (Table 12.21). This tendency does also show up when calculating CF per capita, based on the number of people per dwelling (Spain MH 2006) and the number of bedrooms per dwelling as obtained from the Ministry of Development in Spain (2011). Percentages that each category represents of the total CF for the 92 projects are presented in Figure 12.2, which are majorly controlled by materials (56%) and electricity (37%), only followed by the food branch of manpower.

Regarding the CF of manpower, the third impact source of the total CF (Table 12.22), it is clearly dominated by food, representing 87% of the manpower CF, whereas mobility takes another 9.5% and MSW remains less significant.

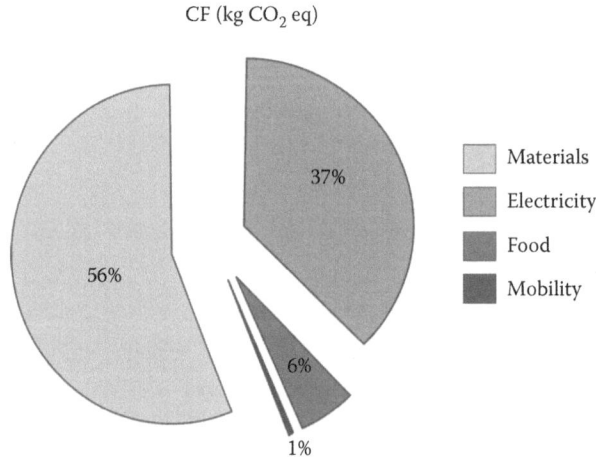

CF (kg CO_2 eq)

□ Materials
■ Electricity
■ Food
■ Mobility

FIGURE 12.2 Main factors generating a CF in the 92 projects analyzed.

TABLE 12.22

CF of Manpower in the Construction of Residential Buildings

	Manpower CF (kg CO$_2$ eq/m^2) Number of Floors above Ground					
Impacts	1	2	3	4	5	10
Food	57.18	48.22	41.80	38.92	38.52	38.97
Mobility	6.21	5.24	4.54	4.23	4.19	4.23
MSW	1.08	1.66	1.34	1.34	1.32	1.31
Manpower	64.47	55.11	47.68	44.48	44.03	44.51
Total CF (%)	5.41	6.18	6.93	6.59	6.64	6.95

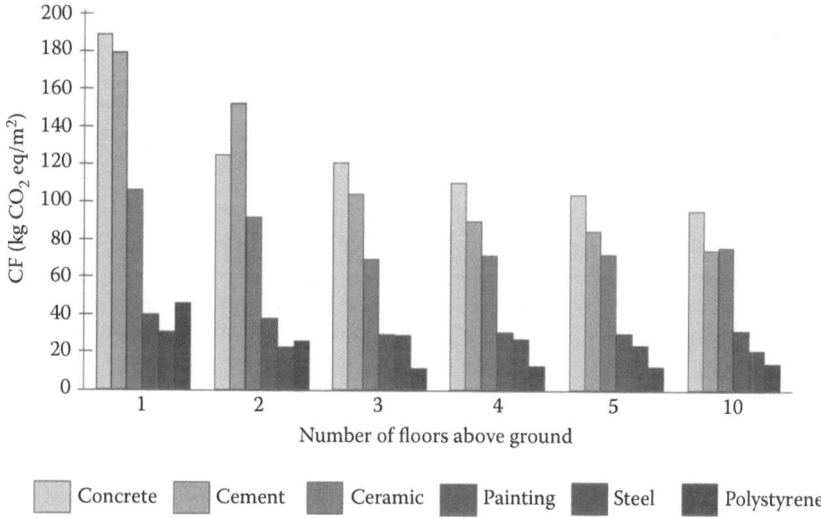

FIGURE 12.3 CF of the most representative materials.

Furthermore, in Figure 12.3, the percentages for the most representative materials are shown according to the number of floors. Concrete, cement, ceramic, painting, steel, and polystyrene represent 90% of the total CF of materials, independent of the building type. This implies that around 50% of the total CF is controlled by these six construction materials.

The fourth controlling factor of the total CF is the fuel consumption by machinery. In this case, the impact of combustion engine machinery on the construction site is very low, as this type of machinery is mainly used in the first stages of the work (earthworks and foundation), whereas in the following construction stages, electric machinery is used. Again, water and CDW represent less than 1% of the total CF, thus being considered negligible.

12.6 CONCLUSIONS

The CF of dwelling construction projects in Spain has been evaluated successfully. The following can be concluded: the CF is sensitive to changes in building characteristics, such as whether the dwelling is detached, is of one or two floors, and is a multifamily dwelling, and these changes significantly affect the results. For example, detached houses generate more CF than multifamily buildings.

The CF evaluation in the dwelling construction sector has made it necessary to define a database of resources per construction system. This has been carried out for the most common building characteristics in Spain, where the Andalusian Construction Cost Database, classification system and quantification model have been used as reference. Resources have been easily divided into three categories: materials, manpower, and machinery.

The evaluation of manpower and construction materials wields significant impact on the total CF, as observed when various scenarios are analyzed, such as those involving detached houses and multifamily buildings.

The results show that the most important factors in the CF calculation are: construction materials (56%), on-site electricity consumption (37%), and manpower (7%), whereas the least significant are machinery, CDW, and water.

Six construction materials control more than a half of the total CF, these being: concrete, cement, ceramics, painting, steel, and polystyrene. Strategies in order to reduce dwelling construction CF can focus on using recycled concrete and steel, reused ceramics, and insulation materials with low embodied energy.

Another factor which can be significantly reduced is the electricity consumption on the construction site by using presence sensors to regulate lighting.

12.7 LIMITATIONS, ADVANTAGES, AND FURTHER RESEARCH

Several project characteristics, such as renewable energy, sewage treatment installations, and domestic solid waste installations, remain excluded from the analysis and will become part of future research in order to render the present model more robust.

A methodology and a structure have been defined which can be adapted to other countries through local construction breakdown systems following the proposed steps: define the budget quantities into construction systems; then transform these into impacts such as materials, manpower, and machinery; and finally, determine the resources consumed and their corresponding CF.

In future work, the model sensitivity can be improved by identifying a greater variety of construction systems for foundation, structure, building envelope, or roof. Another important step involves relating energy consumption (electricity and fuel) to machinery classification of the resource database created. In this way, quantification of energy consumption can be directly generated on the basis of machine power, efficiency, and working hours.

ACKNOWLEDGMENTS

The authors thank the Ministry of Innovation and Science, by concession of the R + D + I: Evaluation of the ecological footprint of the building in the residential sector in Spain. (EVALHED) 2012–2014 (Ministerio de Innovación y Ciencia, por la concesión del Proyecto I + D + i: Evaluación de la huella ecológica de la edificación en el sector residencial en España (EVALHED). 2012–2014).

REFERENCES

Andalusia ME (Ministry of Environment). 2009. Informe de Medio Ambiente 2008. http://www.juntadeanda-lucia.es/medioambiente/site/web/menuitem.318ffa00719ddb10e89d04650525ea0/?vgnextoid=3b32db0 dee134210VgnVCM1000001325e50aRCRD (accessed August 15, 2011).

Andalusian Construction Cost Database (ACCD). 2010. (Base de Costes de la Construcción de Andalucía, 2008. Consejería de Obras Pública y Vivienda de la Junta de Andalucía). http://www.juntadeandalucia.es/ obraspublicasyvivienda/portalweb/web/areas/vivienda/texto/bcfbb3af-ee3a-11df-b3d3-21796ae5a548 (accessed November 1, 2013).

Boustead I. 2006. Eco-profiles of the European plastic industry. Polystyrene (expandable) (EPS). http:// www.plasticseurope.org/plasticssustainability/eco-profiles/browseby-flowchart.aspx?LCAID=r300 (accessed July 10, 2013).

Cagiao, J., B. Gómez, J.L. Domenech, S. Gutiérrez Mainar, and H. Gutiérrez Lanza. 2011. Calculation of the corporate carbon footprint of the cement industry by the application of MC3 methodology. *Ecological Indicators* 11:1526–40.

Colegio de Arquitectos Técnicos de Barcelona (COAAT Barcelona). 2010. Congreso Internacional. Rehabilitación y Sostenibilidad (International Conference. Rehabilitation and Sustainability). Barcelona, Spain. http://www.rsf2010.org (accessed November 1, 2013).

Corporación de Reservas Estratégicas de Productos Petrolíferos (CORES). 2013. Informe estadístico anual. Año 2012 (Annual Statistical Report 2012). http://www.cores.es/pdf/Informe_Estadistico_Anual_2012 .pdf (accessed December 1, 2013).

Domenech Quesada, J.L. 2007. *Huella Ecológica y Desarrollo Sostenible (Ecological Footprint and Sustainable Development)*. Madrid, Spain: AENOR.

Ecoinvent Centre. 2013. Ecoinvent database website. http://www.ecoinvent.org/database (accessed January 3, 2013).

ELCD Core Database II Website. 2013. ELCD. http://lca.jrc.ec.europa.eu/lcainfohub/datasetArea.vm (accessed January 9, 2013).

EMASESA. 2005. Sostenibilidad y gestión. Así éramos, así somos. 1975–2005. (Sustainability and management. How we were, how we are. 1975–2005). Seville, Spain.

González-Vallejo, P., M. Marrero, and J. Solís-Guzmán. 2015. The ecological footprint of dwelling construction in Spain. *Ecological Indicators* 52:75–84.

Instituto para la Diversificación y Ahorro de la Energía (IDAE). 2006. Guía de Gestión combustible. http://www.idae.es/uploads/documentos/documentos_10232_Guia_gestion_combustible_flotas_ carretera_06_32bad0b7.pdf (accessed September10, 2014).

Instituto para la Diversificación y Ahorro de la Energía (IDAE). 2011. *Guía Práctica de la Energía: Consumo Eficiente y Responsable (Practical Energy Guide: Efficient and Responsible Consumption)*. Madrid, Spain: IDAE. http://www.idae.es/index.php/mod.pags/mem.detalle/relcategoria.1161/id.542/relmenu.64 (accessed November 1, 2013).

Instituto Tecnológico de la Construcción de Cataluña (ITeC). 2013. BEDEC website. http://www.itec.es/ nouBedec.e/bedec.aspx (accessed January 3, 2013).

International Energy Agency (IEA). 2012. CO_2 emissions from fuel combustion. 2012 Edition.

Kellenberger, D., H.-J. Althaus, N. Jungbluth, T. Künniger, M. Lehmann, and P. Thalmann. 2007. Life cycle inventories of building products. Final report Ecoinvent data v2.0, no. 7. EMPA Dübendorf, Swiss Centre for Life Cycle Inventories, Dübendorf, CH http://www.ecoinvent.org (accessed January 14, 2013).

Malmqvist, T., and M. Glaumann. 2009. Environmental efficiency in residential buildings—A simplified communication approach. *Building and Environment* 44:937–47.

Marrero, M., and A. Ramírez-de-Arellano. 2010. The building cost system in Andalusia: Application to construction and demolition waste management. *Construction Management and Economics* 28:495–507.

Martínez-Rocamora, A. 2012. *Influencia de las bases de datos de ACV en el cálculo de la huella ecológica en edificación. (Influence of lca databases in the calculation of the ecological footprint in building)*. Master thesis, University of Seville.

Meillaud, F., J. Gay, and M.T. Brown. 2005. Evaluation of a building using the emergy method. *Solar Energy* 79(2):204–12.

Mercader, P. 2010. Cuantificación de los recursos consumidos y emisiones de CO_2 producidas en las construcciones de Andalucía y sus implicaciones en el Protocolo de Kyoto. (Quantification of the resources consumed and of CO_2 emissions on the construction sites of Andalusia and its implications for the Kyoto Protocol). PhD thesis, Universidad de Sevilla, Seville, Spain. http://fondosdigitales. us.es/tesis/tesis/1256/cuantificacion-de-los-recursos-consumidos-y-emisiones-de-co2-producidas- en-las-construcciones-de-andalucia-y-sus-implicaciones-en-el-protocolo-de-kioto/ (accessed January 14, 2013).

Ostermayer, A., and J. Giegrich. 2006. Eco-profiles of the European Plastics Industry—Polyvinylchloride (PVC) (Suspension polymerisation). http://www.plasticseurope.org/plasticssustainability/eco-profiles/ browse-by-flowchart.aspx?LCAID=r43 (accessed July 10, 2013).

Pérez Leal, M.M. 2012. *Huella de carbono. Herramienta de gestión ambiental, empresarial y social (Carbon footprint. Environmental, business and social management tool)*. Master thesis, University of Seville, Spain.

PlasticsEurope. 2013. Eco-Profiles Website. PlasticsEurope. http://www.plasticseurope.org/plastics-sustain- ability/eco-profiles.aspx (accessed January 10, 2013).

Ramírez-de-Arellano-Agudo, A. 2006. *Presupuestación de Obras. 3a Edición (Giving a Construction Estimate)*, 3rd ed. Seville, Spain: Secretariado de Publicaciones de la Universidad de Sevilla.

Ramírez-de-Arellano-Agudo, A., J. Solís-Guzmán, and J. Pérez Monge. 2008. *Generación de RCD versión 2.0 (Software de Evaluación de RCD para Tramitación de Licencias Municipales)* (*CDW Generation 2.0: CDW Evaluation Software for Processing of Municipal Licences*). Universidad de Sevilla, Spain.

Sinivuori, P., and A. Saari. 2006. MIPS analysis of natural resource consumption in two university buildings. *Building and Environment* 41(5):657–68.

Solís-Guzmán, J. 2011. *Evaluación de la huella ecológica del sector edificación (uso residencial) en la comunidad andaluza.* (*Assessing the ecological footprint of the building sector (residential use) in Andalusia*). PhD thesis, Universidad de Sevilla, Seville, Spain. http://fondosdigitales.us.es/tesis/tesis/2051/evaluacion-de-la-huella-ecologica-del-sector-edificacion-uso-residencial-en-la-comunidad-andaluza / (accessed December 15, 2013).

Solís-Guzmán, J., M. Marrero, M.V. Montes-Delgado, and A. Ramírez-de-Arellano. 2009. A Spanish model for quantification and management of construction waste. *Waste Management* 29(9):2542–8.

Solís-Guzmán, J., M. Marrero, and A. Ramírez-de-Arellano. 2013. Methodology for determining the ecological footprint of the construction of residential buildings in Andalusia (Spain). *Ecological Indicators* 25:239–49.

Solís-Guzmán, J., A. Martínez-Rocamora, and M. Marrero. 2014. Methodology for determining the carbon footprint of the construction of residential buildings. In: *Assessment of Carbon Footprint in Different Industrial Sectors, Volume 1*, ed. S.S. Muthu, pp. 49–83. Singapore: Springer Science + Business Media.

Spain MD (Ministry of Development). 2011. Construcción de edificios 2006–2010 (Construction of buildings 2006–2010). Madrid, Spain. http://www.fomento.gob.es/MFOM/LANG_CASTELLANO/ESTADISTICAS_Y_PUBLICACIONES/ (accessed December 1, 2013).

Spain MD (Ministry of Development). 2013. Viviendas por comunidades autónomas y provincias (Housing by regions and provinces). http://www.fomento.gob.es/MFOM/LANG_CASTELLANO/ESTADISTICAS_Y_PUBLICACIONES/ (accessed December 1, 2013).

Spain ME (Ministry of Environment). 2001. Plan Nacional de Residuos de Construcción y Demolición 2001–2006 (National C&D Waste Plan 2001–2006). Ministry of the Environment, Madrid, Spain.

Spain MH (Ministry of Housing). 2006. Código Técnico de la Edificación (Building Technical Code). Madrid, Spain. http://www.codigotecnico.org/web/ (accessed December 1, 2013).

Spain MP (Ministry of the Presidency). 1970. Real Decreto 3650/1970, de 19 de Diciembre, por el que se aprueba el cuadro de fórmulas-tipo generales de revisión de precios de los contratos de obras del Estado y Organismos autónomos para el año 1971 (Royal Decree 3650/1970, of December 19, through which is approved the set of general formulae of the revision of costs for works contracts of the State and autonomous bodies for the year 1971). Madrid, Spain.

Spain MP (Ministry of the Presidency). 1981. Real Decreto 2167/1981, de 20 de agosto, por el que se complementa el Decreto 3650/1970, de 19 de diciembre, sobre fórmulas-tipo generales de revisión de precios de los contratos de obras del Estado y Organismos autónomos (Royal Decree 2167/1981 of August 20, through which Decree 3650/1970 of December 19 is supplemented on general formulae for the revision of costs for works contracts of the State and autonomous bodies). Madrid, Spain.

Syndicat National du Bèton Prêt à l'Emploi (SNBPE). 2012. Declarations environnementales et sanitaires de produits du bèton conforme a la norme NF P 01-010 (Environmental and sanitaire declarations of concrete products according to NF P 01-010). http://www.inies.fr (accessed January 10, 2013).

Trigo-Ledesma, R. 2013. *Evaluación del consumo de agua en obras de edificios residenciales y su aplicación a la huella ecológica* (*Assessment of water consumption in residential buildings works and its application to the ecological footprint*). Master thesis, University of Seville, Seville, Spain.

Wackernagel, M., and W. Rees. 1996. *Our Ecological Footprint: Reducing Human Impact on the Earth.* Gabriola Island, British Columbia: New Society.

Weidema, B.P., M. Thrane, P. Christensen, J. Schmidt, and S. Løkke. 2008. Carbon footprint. *Journal of Industrial Ecology* 12:3–6.

World Steel Association (WSA). 2011. Life cycle inventories of steel products. http://www.worldsteel.org (accessed February 20, 2013).

Zabalza Bribián, I., A. Valero Capilla, and A. Aranda Usón. 2011. Life cycle assessment of building materials: Comparative analysis of energy and environmental impacts and evaluation of the eco-efficiency improvement potential. *Building and Environment* 46:1133–40.

13 Carbon Footprint
Calculations and Sensitivity Analysis for Cow Milk Produced in Flanders, a Belgian Region

Ray Jacobsen, Valerie Vandermeulen,
Guido Vanhuylenbroeck, and Xavier Gellynck

CONTENTS

13.1 Introduction 282
13.2 Background..283
13.3 Methodology...284
    13.3.1 Standard and Method Used 284
    13.3.2 Scope and System Boundaries..284
    13.3.3 Functional Unit 285
    13.3.4 Allocation Method...285
    13.3.5 Land Use and Land-Use Change 285
13.4 Data Sources 286
    13.4.1 Raw Materials and Farm Level 286
    13.4.2 Dairy Processing...288
    13.4.3 Dairy Processing: Data Description 288
13.5 Data Analysis..289
    13.5.1 Emissions from Fodder Production 289
        13.5.1.1 Purchased Fodder..290
        13.5.1.2 Home-Grown Roughage 291
    13.5.2 Emissions from Cattle Breeding...292
        13.5.2.1 Energy Consumption Farm 292
        13.5.2.2 Animal Emissions: Rumen Fermentation..........................292
    13.5.3 Emissions—Manure Storage and Usage 292
        13.5.3.1 Methane 292
        13.5.3.2 Laughing Gas...293
        13.5.3.3 Manure Usage for Crop Production 294
    13.5.4 Emissions from Transport 294
    13.5.5 Emissions from Milk Processing...294
13.6 Results..295
    13.6.1 The CF of Milk..295
    13.6.2 Milk Processing...296
    13.6.3 Total Result..296
    13.6.4 CF Sensitivity 296
        13.6.4.1 Impact of Estimating Indirect Laughing Gas Emissions...........297
        13.6.4.2 Feed and Herd Characteristics..297
        13.6.4.3 Manure Storage/Disposal 297
        13.6.4.4 Influence of Allocation Method between Milk and Meat 298

      13.6.4.5 Energy in the Dairy Processing Plant and the Use of Heat
           Recuperation 298
      13.6.4.6 Influence of the Allocation Method between Several Dairy Products 298
      13.6.4.7 CF Range 298
   13.6.5 Mitigation Measures 299
13.7 Conclusion 299
Acknowledgments 300
References 300

13.1 INTRODUCTION

There are many carbon footprinting initiatives currently in development, providing a clear methodology. Since 2008, a number of standards for calculating the carbon footprint (CF) of products have been developed. The present study describes the most common internationally (worldwide and European) available standards.

- PAS2050, October 2008, British Standard Institute, UK—http://www.bsigroup.com/upload/Standards%20&%20Publications/Energy/PAS2050
- BPX30-323, February 2010, AFNOR[2]-ADEME[3], France—http://affichage-environnemental.afnor.org/
- ISO 14067, March 2010 (draft), ISO, International
- GHG protocol initiative; Product Accounting and Reporting Standard, November 2010 (draft), WRI and WBCSD, International

The standards mentioned earlier rely on the international LCA standards ISO 14040 and ISO 14044.

The standards also refer to the ILCD (International Reference Life Cycle Data System) handbook developed by the Joint Research Center of the European Community. Moreover, the work of Intergovernmental Panel on Climate Change (IPCC) is taken into account.

The International Dairy Federation (IDF) has proposed a common approach for the calculation of the CF for dairy products. The IDF guidelines for the calculation of CFs have been integrated in the comparative study: the IDF guide to standard lifecycle assessment methodology for the dairy sector, November 2010, IDF, http://www.idf-lca-guide.org/Files/media/Documents/445-2010-A-common-carbonfootprint-approach-for-dairy.pdf

Achieving sustainable development can be established by limiting agricultural GHG emissions in order to reach a stabilization of GHG emissions (Dalgaard et al. 2011).

Achieving sustainable production hence proves the need for evaluating the current situation and assesses where the production system needs improvement (Eriksson et al. 2005). If one wants to identify where along the production chain improvements can be made, it is necessary to quantify all emissions during the life cycle. Carbon footprinting is one of the methods able to calculate the impact of livestock products on climate change (Espinoza-Orias et al. 2011). A CF quantifies the climate change impact of an activity, product, or service. Within the CF, all important agricultural GHG emissions [carbon dioxide (CO_2), methane (CH_4), and nitrous oxide (N_2O)] are combined. It is a measure of the total amount of greenhouse gas (GHG) emissions of a system or activity, considering all relevant sources, sinks, and storage within the spatial and temporal boundaries of the population, system, or activity of interest. A CF is calculated as the carbon dioxide equivalent using the relevant 100-year global warming potential (GWP100) (Wright et al. 2011).

Given the importance of milk in terms of world consumption and livestock production, a study was ordered by the Flemish Government to calculate the CF of Flemish milk production in order to benchmark with other countries. Moreover, milk is an interesting case to examine because

TABLE 13.1
Overview Methodology of the Studies Discussed Regarding the CF of Milk

Reference	Protocol CF	System Boundaries	Functional Unit	Allocation Method	LUC Included?
Williams et al. (2006)	ISO Protocol	Cradle to farm gate	1 kg milk	Economic	Yes
Van der Werf et al. (2009)	ISO Protocol	Cradle to shelf	1 kg milk	Economic	Yes
Blonk et al. (2007)	PAS2050	Cradle to farm gate	1 kg FPCM	Economic	Yes
De Vries and de Boer (2010)	ISO 14040	Cradle to shelf	1 kg FPCM	Economic	Yes
Muller-Lindenlauf et al. (2010)	Own method	Cradle to farm gate	1 kg milk	Not mentioned	Yes
Thoma et al. (2010)	ISO Protocol	Cradle to farm gate	1 kg FPCM	Biological (dairy products)	No
Leip et al. (2010)	Own method	Cradle to farm gate	1 kg milk (4% fat)	Physical (feed)	Yes

Note: FPCM, fat and protein-corrected milk (IDF 2010).

that estimations for the CF of milk do abound (Blonk et al. 2008b; Muller-Lindenlauf et al. 2010; Sonesson et al. 2009; Thoma et al. 2010; Van der Werf et al. 2009). The Flemish region of Belgium is hence interested to establish a benchmark for the climate change impact of cow's milk production with its neighboring countries.

Most studies on CF are not clear in terms of methodology or standard, nor the chosen system boundaries or system definition. Stakeholders with different backgrounds and interests might draw incorrect conclusions. Indeed, different approaches in methodology prevent fair comparisons of CFs between products and sectors, because different calculations are used and hence one compares apples and oranges. A CF is calculated by means of a life-cycle assessment (LCA) (Finkbeiner 2009). The fact that each LCA has to deal with many different issues (such as allocation method, scope, system boundaries, data, inclusion of land use (change), to name a few) (Finkbeiner 2009) makes it necessary to properly describe each of these aspects. In our own study (CF methodology applied to livestock produce in Flanders 2011), a literature study was conducted on the state of the art in terms of existing LCA or CF studies on milk production and found that in several cases important information was missing [e.g. Dalgaard et al. (2007) and Leip et al. (2010) did not indicate the used allocation method]. This problem was also mentioned by de Vries and de Boer (2010), who had to exclude sources from their meta-analysis due to lack of data. Table 13.1 indicates the information filtered from the literature. Yet again, we had to reveal and identify the information ourselves. It was not clear in the body of the text which methodology as such was used.

International literatures on CF calculations for livestock products resulted in 30 studies for milk, described in 25 reports or articles. It should be mentioned that this list is not exhaustive. Table 13.1 summarizes the methodological choices of a number of recent studies.

13.2 BACKGROUND

The results of this chapter were obtained through a study conducted for the Flemish Government, Department of Agriculture and Fisheries. The purpose of this chapter is to estimate the CF and furthermore to identify hot spots in the life cycle of beef, swine, and milk production in Flanders although the primary focus is on milk production. In Flanders, the environmental pressure from livestock production abounds, with a major impact on climate change from large emissions of GHGs. In Flanders, there are 3835 dairy cattle farms, which is 13% of all farms (Boerenbond 2011–2012).

13.3 METHODOLOGY

13.3.1 STANDARD AND METHOD USED

A CF quantifies the total amount of GHG emissions for which a product, organization, or process is responsible. It is a measure of the contribution of persons, products, and organizations on the GHG effect. Figure 13.1 presents the different steps that are run through when calculating a CF.

An LCA is a method to determine the total environmental impact of a product during the whole chain or life cycle of the product. Carbon footprinting differs from LCA in one aspect: it focuses solely on quantifying GHG emissions causing climate change. Determining the CF is hence a choice to focus on one environmental indicator (FAO, 2008).

A product CF comprises all emissions related to each phase of the product's life cycle, from cradle to grave. In practice, the boundaries of CF calculations are often shortened. The choice of the system boundaries depends on the goal and application.

For the study, we made use of the IPCC (2006a) guidelines in line with the National Inventory Report of Belgium (VMM et al. 2010). Albeit the IPCC (2006b) directive gives a description of the calculation of the total amount of GHG emissions, it does not include the allocation of GHG emissions to a particular product. In order to tackle this, a specific methodology such as PAS2050 is needed (Espinoza-Orias et al. 2011). Currently, the Publicly Available Specification (PAS2050, BSI 2011) is one of the most profound methods (among others, e.g. ISO 14067), and, therefore, this was chosen.

Based on the IPCC 2007 (IPCC, AR4 2007), the global warming potentials for methane and nitrogen gas emissions are defined as follows: 1 kg of methane (CH_4) equals 25 kg of CO_2 and 1 kg of nitrogen gas (N_2O) equals 298 kg CO_2.

13.3.2 SCOPE AND SYSTEM BOUNDARIES

PAS2050 states that emission factors contributing less than 1% of the total CF are negligible (BSI 2011). The lion's share of GHG emissions occurs at the farm level. Therefore, the ultimate steps in the dairy chain (Blonk et al. 2008b; Campens et al. 2010) are not included in the calculations of the CF. Table 13.1 gives an overview of the included emission sources throughout the chain.

The study included the GHG emissions, as shown in Figure 13.2. Production of materials, energy, and transport steps are included. The system boundary excludes production of capital goods, similar to most international studies (Jacobsen et al. 2013).

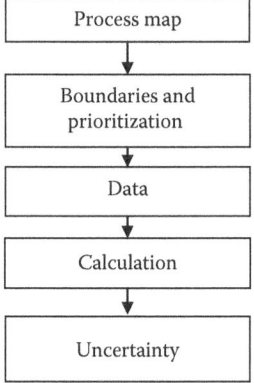

FIGURE 13.1 Five necessary steps for calculating a CF. (Adapted from British Standards Institute (BSI). 2011. PAS2050 specification for the assessment of the life cycle greenhouse gas emissions of goods and services. England.)

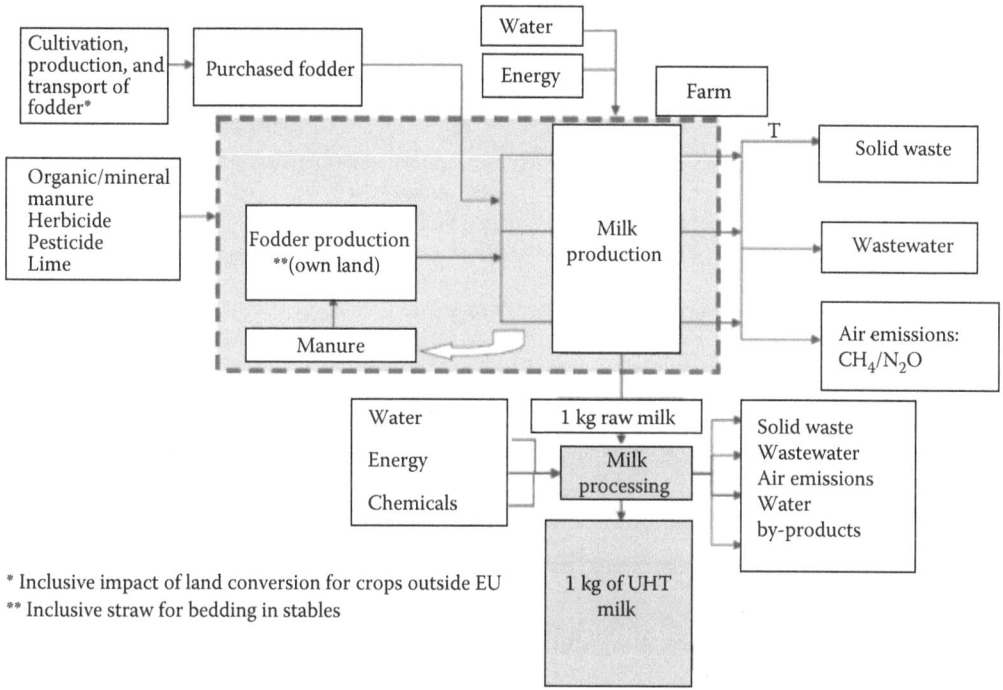

FIGURE 13.2 System boundaries. The boxes with dashed lines represent a process and those with full lines a product flow. The shaded boxes are the foreground system.

13.3.3 FUNCTIONAL UNIT

Several functional units were defined upon agreement with the guiding committee of the project. This allowed better identification of the hot spots along the milk production chain. For dairy products, the following functional units were used:

- 1 kg raw milk with 4% fat and 3.3% protein
- 1 kg ultrahigh temperature (UHT) milk with 1.5% fat

13.3.4 ALLOCATION METHOD

Another very important assumption describes the allocation of GHG emissions between various by-products emerging from a process. There are, in fact, two major ways to define allocation: physical and economic. With regard to the project, a combination of both methods was applied, depending on the chain level. Manure contributes in the production of crops, and therefore, physical allocation was used to allocate the GHGs from manure among crops and animal production (Table 13.2). Overall, a combination of physical and economic allocation based on several other references was used (Blonk 2008a). In order to determine the allocation method, international standards and best practices have been followed (Table 13.3).

13.3.5 LAND USE AND LAND-USE CHANGE

According to the PAS2050 methodology, land-use change (LUC) should be considered if land conversion took place in the last 20 years. This is not the case for agriculture in the European

TABLE 13.2
Overview of Emission Sources within the Covered System Boundaries

Name	GHG	Description
Feed mixtures (purchased)	CO_2 and N_2O	Farming, transport, processing, and land conversion included
Animal	CH_4	IPCC method (Tier 2)
Manure storage and disposal	CH_4 and N_2O	IPCC method (Tier 2)
Manure application (not used for own feed mixtures)	CH_4 and N_2O	Allocation between animal (40%) and vegetable production system (60%) based on nitrogen uptake by plants
Energy and water consumption	CO_2, CH_4, and N_2O	Energy consumption (electricity (red); diesel; gas). Water consumption (tap and ground water)
Transport of goods	CO_2, CH_4, and N_2O	Assumptions made for the goods entering and leaving the farm
Processing materials	CO_2 and refrigerant	Cleansing products and refrigerants

Source: Adapted from Jacobsen, R. et al. 2013. *International Journal of Agricultural Sustainability* 1:1–18. DOI:10.10 80/14735903.2013.798896.

TABLE 13.3
Overview of the Applied Allocation Methods

Process	Products	Allocation Method
Farming of crops	Products for human consumption (such as flour); products for animal consumption (such as wheat starch); and other products (such as straw)	Economic allocation
Dairy sector	Milk and meat	Physical relation registered by the IDF
Dairy products	Low-fat (skimmed), medium-skimmed, and whole milk, cream, milk powder, yoghurt, butter, etc.	Physical relation registered by the IDF
Manure production	Stock farming products and crop farming	Physical relation

Union, and thus LUC is zero. However, part of the feed is imported overseas, and for those, FAO (2010) statistics are used to define the LUC in the past 20 years. The total emissions from this LUC are calculated, and 1/20th is attributed to each forthcoming year [(Blonk 2008a; Ecoinvent 2011) and LCA Food (Nielsen et al. 2010)].

Land use as such (or carbon sequestration in the soil) is not considered. In Flanders, considerable uncertainty remains with regard to the net effect (absorption or emissions) from land use; therefore, the net effect was not included. International standards and guidelines for carbon footprinting also exclude it from the necessary calculations (ERM 2010).

13.4 DATA SOURCES

PAS2050 has specific rules for using primary over secondary data (BSI 2011). Primary data were extended with secondary data from reports. Data were collected for the year 2009. However, certain data required use of more recent values, for example, for the feed compound composition which changes daily.

13.4.1 RAW MATERIALS AND FARM LEVEL

The data on the representative conventional farm were collected through the Farmers Union dataset (Boerenbond 2011–2012). This dataset, used to select these farms, specialized in milk production.

The database allowed us to identify the average of the data, in order to obtain a representative and existing farm. Outliers in the data were not used, and hence the average represents a real farm with the following characteristics:

- 63 milk cows
- 55 female young cattle, of which 26 are between 1 and 2 years of age, and 29 calves less than 1 year
- 29 male young cattle, kept only 15 days at the farm

The milk production on the farm amounts to 492.764 l/year. Each milk cow, hence, has a production of 7.822 l/year. The biggest share of this production is delivered to dairy processors (about 97%). The remaining amount is being used as milk for the calves (2%) or used in the household or sold on the farm (1%).

Farms are confronted with loss of animals. The mortality rate amounts 1.25% for mature animals and 10% for calves. The replacement rate amounts 25%, whereby the reform cows are sold. During the last years, a decrease in this rate has taken place; farms are growing and more and more pregnant heifers are available.

Data on manure production were collected from reports of the Flemish Centre for Manure Processing and were linked to the number of animals on the farm (VLM 2011).

The fodder applied partly originates from own production and is partly purchased. Table 13.4 presents the overview per farm. The average composition of the feed concentrate was given by BEMEFA, the Union of Belgian Feed Compound Processors (personal communication, April–October 2011). The composition of the feed compound was given for October 4, 2010—randomly chosen during the course of the project. It varies daily according to the availability of components on the market. One can be sure that the feed used has an appropriate composition for the animals. Per milk cow it means: 195 kg of grains, 13.365 kg of corn, 10.436 kg of grass, and green fertilizer per year, of which the latter two are completely ensilaged. Besides the aforementioned amounts, the following produce is being purchased: 1.301 kg of wet pulp from sugar beets and 385 kg of wet by-products, 14 kg of dry sugar beet pulp, 1.302 kg of feed concentrate, 85 kg of soymeal, 64 kg of proteins and vitamins, and 24 kg of minerals.

TABLE 13.4
Yearly Consumption of Feed Compound per Dairy Farm

Resources	Kilogram Product	Yield[a] (kg Product/ha)
Soy meal[b]	5376	
Sugar beet pulp (dry)	886	
Sugar beet pulp (wet)	72.800	
Wet by-products	106.640	
Composite milk fodder	85.736	
Protein core, vitamins, and minerals	5.566	
Milk (own production)	10.355	
Grass silage and fresh grass (home-grown)	660.095	23.600[c]
Wheat (home-grown)	12.309	8.985
Maize (home-grown)	845.398	48.670

[a] Home-grown.

[b] Origin of imported soy: 53% Brazil, 11% Argentina, 21% United States, and 16% Canada. Calculated per milk cow, the calves consume 164 L of raw milk.

[c] Weighted average meadow/temporary grassland.

Emission factors were derived from BlonkMilieuAdvies (Blonk et al. 2008a), LCA Food (Nielsen et al. 2010), and Ecoinvent (2011).

The animals are kept in stables on a bed of home-grown straw. Suckler cows and female young cattle (1–2 years) stay outside 24 h a day, during a period of 6 months per year. Female calves remain outside 24 h a day, during a period of about 4 months. The male young cattle and the male calves stay inside. The animals produce 623 kg of manure per day. About 32% of this manure ends on the grassland during grazing, 60% is preserved as stable, and 8% as mixed manure (IPCC, 2006c).

The farmer possesses 19.7 ha of grass and cropland: 17.4 ha of maize, 1.5 ha of ryegrass, and 6.8 ha of green manure. In total, the farm uses 650 kg of fertilizer per hectare of grassland (containing 170 U of N) and 100 kg of starter fertilizer per hectare of maize per year (20 U of P and 20 U of N). Moreover, the farm consumes 31 kg of herbicides per year and about 1000 kg of lime.

The farm yearly consumes 24.889 kWh of electricity and 6.250 l of oil fuel. In terms of water consumption, the farm consumes 2.579 m³ of ground water.

13.4.2 Dairy Processing

Given that the number of dairy processing companies in Flanders is rather limited, we did not work with a single company, but rather with a weighted average of a great number of large companies.

The Belgian Dairy Confederation was willing to cooperate and requested information from its members. Four different companies delivered the requested information. After analysis, it became clear that one company differed significantly from the other three, in that it produces a smaller amount of UHT milk (<1% of the total amount). Therefore, we only took into account the three major dairy processors because they produce about an equal share of UHT semi-skimmed milk. Each of the companies was requested to provide the following information:

* Total annual amount of processed raw milk
* Fat percentage of the raw milk
* Amount of applied industrial milk
* Yearly production of semi-skimmed UHT milk
* Yearly production of all other dairy produces
* Distance covered by truck during collection
* Amount of collected milk per truck
* Energy consumption of the milk-processing company (water and electricity)
* Consumption of cleansing products

After the data collection, a meeting was held with the respective organizations in order to assure representativeness and correctness of the data. Attention was given to the following:

* Allocation among the dairy products (application of the IDF standard)
* Product properties (density and fat content)
* Description of the UHT semi-skimmed milk production process

Such consultations with experts or cooperation are necessary to assure a true picture of the CF.

13.4.3 Dairy Processing: Data Description

In total, the three companies process 786 million liters of raw milk, with a fat percentage of 4.1%. Besides this amount of milk, an amount of 88 million liters of industrial milk is used. This leads to a production of the following:

- 45 million liters of skimmed UHT milk
- 237 million liters of semi-skimmed UHT milk (the final product of which the CF is determined)
- 93 million liters of full UHT milk
- 53 million liters of semi-skimmed pasteurized milk
- 10 million liters of fully pasteurized milk

The companies produce other products besides regular milk: chocolate milk, coffee milk, milk powder, buttermilk, desserts, and so on. In total, these three companies produce a yearly total of 25 million liters of cream with a fat percentage of about 40%.

Milk is transported to the processing plant by trucks that collect milk at several farms. The first round occurs with an empty truck. Once the truck is fully loaded, it returns to the processing plant. The distances covered depend on the size of the vehicle and the considered company. On average, the total distance for one round amounts 100 km, where 22 tons of milk are being collected.

The three companies yearly consume 91 million kWh of electricity, which is almost completely purchased from the net. Besides electricity, the companies consume a yearly amount of 2.498 million liters of water, of which 73% is ground water, 19% tap water, and 8% other sources (wastewater, gray water, or recuperation water).

Per 1000 l of processed milk (industrial and fresh), companies use 5 kg of cleansing materials, of which 90% is caustic soda.

A closer look was taken at the production process in order to determine how much raw milk is needed to produce 1 kg of semi-skimmed UHT milk. Taking that into account, we obtain the following:

- 40 g of fat per liter of raw milk, 16 g of fat per liter of semi-skimmed milk, and 400 g/l of cream.
- The density of raw milk amounts to 1.03 kg/l, of semi-skimmed milk 1.0325 kg/l, and of cream 0.991 kg/l.
- There is a loss of 1.5% of raw milk during the production.

The following can be stated:

- There are 38.8 g/kg of raw milk, 15.5 g of fat per kilogram of semi-skimmed milk, and 403.6 g of fat per kilogram of cream.
- In order to produce 1 kg of semi-skimmed UHT milk with a 15.5 g fat content, one needs 1.061 kg of raw milk with a fat content of 41.2%.
- Besides UHT milk, the company also produces 0.064 kg of cream with a fat content of 25.7 g.
- Taking into account the loss in the production system, there is a need for 1.077 kg of raw milk per kilogram of semi-skimmed UHT milk.

Additionally, according to the IDF guidelines, one can make an allocation of the upstream emissions among semi-skimmed milk (82.9%) and cream (17.1%). In the section on the sensitivity analysis, the impact of the allocation method on the overall result is examined.

13.5   DATA ANALYSIS

13.5.1   EMISSIONS FROM FODDER PRODUCTION

A distinction is made between home-grown and purchased fodder. The production of fodder also comprises the production and transportation of resources to sow, grow, and harvest the crops (seeds, fertilizers, pesticides, and diesel). The accompanying emissions are allocated to the crops.

Land use during cultivation of the crops results in extra GHG emissions. Laughing gas is the most important GHG for land use.

For purchased fodder, crops are transported to a processing plant. Emissions accompanying transport and processing (crushing, mixing, etc.) are included. Table 13.3 gives an overview of the yearly consumption of fodder on a dairy farm. The considered dairy farm consists of 63 dairy cows, 25 young cattle (between 1 and 2 years of age), and 30 female calves less than 1 year. There are also 30 male calves that remain for a short period on the farm (10 days). These male calves drink 5 l of milk per day during their stay. The dairy farm was described more in detail above.

13.5.1.1 Purchased Fodder

Purchased fodder is not grown on the farm itself. The resources of feed concentrate are processed to concentrate, consisting of a mix of components. The resources of composed concentrate are processed to fodder consisting of different components. The Belgian Feed Compound Union (BEMEFA) was contacted to identify the composition. Databases were checked on August 4, 2011, and furthermore, feed specialists were consulted. Table 13.4 indicates the representative composition of about 80% of feed concentrate for a dairy farm (discussed earlier).

Emissions accompanying the production of feed compound are calculated on the basis of emission factors of the resources. The used emission factors are literature-based values, in most cases from reports and studies from BlonkMilieuAdvies and Wageningen University. The data were extended with values from the following databases:

- Ecoinvent database 2011
- LCA Food Database Denmark 2010
- Carbon Trust Information U.K. 2011

Table 13.5 provides information on the most important emission factors. LUC is a complex issue in LCAs. As recommended by the IDF guidelines, the PAS2050 methodology was used. This method in turn relies on the IPCC methodology, whereby the land conversion is depreciated on a 20-year period. Several CF studies mention LUC separately. The uncertainty remains too high. A recent Swedish study by Sonesson et al. (2009) states that it is not reasonable for soy products to take Latin-American deforestation into account without changing the current calculation methods. The uncertainty lies in the fact that the initial carbon content of the rain forest is inadequately known. Moreover, the deforested area is not being monitored for the full 20 years. According to Sonesson, it gives rise to new, secondary forests that are able to capture carbon (Table 13.6).

TABLE 13.5

Emissions Accompanying the Purchased Fodder per Kilogram of Product (See also Table 13.4)

Resources	kg CO_2eq/kg Product	% LUC
Soy meal	3.06	71
Sugar beet pulp (dry)	0.11	
Sugar beet pulp (wet)	0.03	
Wet by-products	0.03	
Composed feed compound dairy cattle	1.24	51
Minerals, protein core, and vitamins	0.57	

TABLE 13.6

Emissions Accompanying the Cultivation of Own Crops

Resources	Area (ha)	kg CO_2eq/year
Wheat (home-grown) for straw and grains	1.4	2.817
Maize (home-grown)	17.4	33.648
Milk (own production)	—	—
Grass silage (home-grown)	28	55.271

13.5.1.2 Home-Grown Roughage

Farm land is applicable as grassland and moreover for the cultivation of fodder crops. The yield of the own crops is used as roughage. Table 13.7 presents the calculated emissions.

Energy and fuels used for machinery and transportation are included in the total energy consumption of the farm. They are not mentioned in Table 13.5. GHG emissions accompanying production and transportation of fertilizers, herbicides, insecticides, and fungicides are also included. Emission factors were calculated from the Ecoinvent database. Emissions due to the application of these substances are mentioned in Table 13.7.

13.5.1.2.1 Laughing Gas Emissions

N_2O emissions due to crop cultivation are calculated according to the IPCC 2006 method. Nitrogen sources applied to land are in this case fertilizers, natural fertilizers (from own cattle and purchased pig manure), and crop residue.

"Direct" laughing gas emissions are the result of denitrification. It is assumed that 1% of all nitrogen applied to land converts to laughing gas (uncertainty interval 0.3%–3%). The IPCC value (1.25%) is still applied in the national inventories until 2014. Current research in Flanders points out that 3.16% of all nitrogen converts to N_2O. "Indirect" N_2O emissions due to nitrogen evaporation as ammonia (NH_3) and NO_x are calculated with the same data of the NIR Belgium (2009). The amount of nitrogen evaporated as NH_3 or NO_x depends on the nitrogen source.

1. *Fertilizers* in Flanders: average NH_3 evaporation amounts to 3.3% and the NO_x evaporation amounts to 1.5% (NIR 2010).
2. *Organic fertilizers* in Flanders: average nitrogen evaporation as NH_3 or NO_x amounts to 20% (NIR 2010).

According to the IPCC calculation method, 1% (0.2%–5%) of the evaporated nitrogen (as NH_3 or NO_x) is converted to N_2O.

13.5.1.2.2 Lime Application

Lime is applied on land to increase the soil pH. This causes CO_2 emissions (bicarbonate dissolute into water and CO_2). For calculating these emissions, the IPCC Tier 1 method was used.

TABLE 13.7

Emission Factors: Production, Transportation of Fertilizers, Herbicides, and Lime

Name	Value	Unit
Fertilizer (calcium ammonium nitrate)	8.81	kg CO_2eq/kg N
Herbicide	10.73	kg CO_2eq/kg
Lime (calcium carbonate)	0.02	kg CO_2eq/kg

TABLE 13.8

Emission Factors: Electricity and Gasoline Oil

Name	Value	Unit	Source
Electricity	0.40	kg CO_2eq/kWh	Energy covenant
Gasoline oil	2.66	kg CO_2eq/kg	Energy covenant[a]

[a] After the energy crisis in the 1970s, there was an agreement among several industrialized countries to operate more energy efficient, implemented into an energy-covenant document.

The dairy farm uses 1.000 kg of lime yearly. The used emission factor is 0.48 kg CO_{2eq}/kg lime (e.g. dolomite).

13.5.2 Emissions from Cattle Breeding

13.5.2.1 Energy Consumption Farm

The energy consumption is included as a whole and not allocated. In Table 13.8, emission factors are presented. Each mature cow unit yearly consumes about 1.625 kWh of energy (26% electricity and 74% gasoline oil).

13.5.2.2 Animal Emissions: Rumen Fermentation

Emissions due to rumen fermentation are calculated on the basis of the IPCC guidelines (Tier 2 method). For calculating the necessary gross energy (GE) uptake per animal, the daily need, growth, and gestation are included. Distinction was made between lactating and nonlactating animals. The digestible energy (DE) is expressed as % GE. An adapted value is calculated on the basis of the fodder and the number of grazing days. It is calculated that per lactating cow about 316 MJ of GE is needed. The DE is calculated on the basis of the fodder and amounts on average 75% GE. Table 13.9 represents the DE per feed component. The time spent on the grassland is included in order to determine an adapted DE content of animals' diet. According to the IPCC calculation method, 6.5% of the GE is converted to methane gas. The uptake of GE is calculated on the basis of the fodder composition (roughage and feed concentrate in the stable and fresh grass on land) and the DE content of each feed component.

13.5.3 Emissions—Manure Storage and Usage

Manure production takes place on the meadow and in the barn. Dairy cows stay about 76 days per year on the meadow, young cattle (between 1 and 2 years) about 152 days per year, and calves (<1 year) about 30 days. Manure produced in the stable is stored temporarily. It is assumed that 80% of the manure production in the barn is being stored as stable manure. Manure disposal on grassland and manure storage are accompanied with methane and N_2O emissions. The calculations are explained subsequently, based on the IPCC 2006 guidelines (Tier 2).

13.5.3.1 Methane

Methane emissions related to manure production depend on the excreted volatile solids, the maximum methane production capacity of the manure, and the storage. The excreted volatile solids are calculated by using the IPCC (2006a) formula. Moreover, the IPCC 2006 reference values for the urine fraction (4%) and dry matter content (8%) were used (IPCC 2006b). Allocation of manure production between meadow and stable is presented in Table 13.10.

TABLE 13.9
DE Values for Different Types of Fodder

Name	DE	Unit	Source
Wheat/barley	86	% GE	FAO
Maize-roughage	72	% GE	NIR Belgium
Soy meal	80	% GE	FAO
Beet pulp/citrus pulp	81	% GE	FAO
Wet by-products	78	% GE	FAO
Composite fodder	80	% GE	NIR Belgium
Protein, vitamins	80	% GE	Proxy: composite fodder
Feed concentrate	80	% GE	Proxy: composite fodder
Composite young feed	80	% GE	Proxy: composite fodder
Fodder for young cattle	80	% GE	Proxy: composite fodder
Full milk	90	% GE	NIR Belgium
Grass silage	72	% GE	NIR Belgium
Fresh grass (grazing)	79	% GE	NIR Belgium

Source: Adapted from FAO. 2010. Greenhouse gas emissions from the dairy sector, a life cycle assessment. Food and Agriculture Organization of the United Nations—Animal Production and Health Division, Rome, Italy, 1998; NIR Belgium. NIR Belgium (2009). National institute for the Accounts in Belgium. NIR (2011). National Inventory Report for greenhouse gases (NIR) of Belgium (2009). Farmer's Union: Boerenbond (2011). Bedrijfseconomische boekhouding (CD-ROM). Leuven (2009).

13.5.3.2 Laughing Gas

Through a combination of nitrification and denitrification, N_2O was released from stored manure or was disposed on land. The amount of produced laughing gas emissions depends on the nitrogen excretion of the animals (N_{ex}). The excreted nitrogen per type of animal is taken from the NIR report of Belgium (Table 13.11).

The amount of N_2O from the total amount of nitrogen depends on the manure storage. It is assumed that 0.5% of the total nitrogen is converted to N_2O during manure storage. For mixed manure stored underneath the slatted floor, it is assumed that 0.1% of the total nitrogen is converted to N_2O during storage. For manure disposed on the meadow by animals, it is assumed 2% conversion to N_2O ("direct emissions").

TABLE 13.10
Methane Conversion Factors and Manure Storage Systems

Name	Stable Manure (%)	Mixed Manure (%)	Manure Disposal on Grassland (%)
Methane conversion factors	2	19	1
% manure dairy cows (lactating)	8	71	21
% manure young cattle (1–2 years)	6	52	42
% manure calves (<1 year)	55	37	8

Source: IPCC (Intergovernmental Panel on Climate Change), NIR Belgium (2009). National institute for the Accounts in Belgium. NIR (2011). National Inventory Report for greenhouse gases (NIR) of Belgium (2009). Farmer's Union: Boerenbond (2011). Bedrijfseconomische boekhouding (CD-ROM). Leuven (2009).

TABLE 13.11

N_{ex} Per Type of Animal

Animal Category	N_{ex} (kg/head year)
Calves (<1 year)	33
Young cattle (1–2 years)	58
Dairy cattle	110

Source: NIR Belgium (2009). National institute for the Accounts in Belgium. NIR (2011). National Inventory Report for greenhouse gases (NIR) of Belgium (2009). Farmer's Union: Boerenbond (2011). Bedrijfseconomische boekhouding (CD-ROM). Leuven (2009). NIR Belgium/Manure Database.

Indirectly, there are N_2O emissions formed through volatilized NH_3 and NO_x. The amount of NH_3 and NO_x formed from the manure depends on storage. Table 13.12 presents how much of the total nitrogen converts to NH_3 and NO_x. It is assumed that 1% of indirect nitrogen losses convert to laughing gas ("indirect laughing gas emissions").

13.5.3.3 Manure Usage for Crop Production

When manure is used on agricultural land for growing crops, emissions are allocated among crops and livestock. All produced manure is disposed of on own land. Accompanying emissions are included in Section 5.1.

13.5.4 EMISSIONS FROM TRANSPORT

Feed components are transported to the processing plant. Distances are limited within Europe (<1000 km). The soy component is transported overseas. Emissions related to both types of components are included in the applied emission factors and covered by the production of purchased fodder. For home-grown roughage, the necessary amount of fuel in the total energy consumption (Section 5.2.1) is included.

Secondly, feed is transported from the fodder-processing plant to the farm (average distance of 30 km). The related emissions are covered within the farm data. Thirdly, milk is transported from the farm to the processing plant (average distance amounts 25 km).

13.5.5 EMISSIONS FROM MILK PROCESSING

Emissions originating from processing relate to electricity consumption, fuel, cleansing products, water usage, and waste processing. Transport at other levels in the chain is also included (raw milk

TABLE 13.12

Nitrogen Losses from Manure as NH_3 or NO_x as a Function of Manure Storage Systems (IPCC 2006)

Name	N Volatilization (%) (NH_3/NO_x)
Dairy cattle—stable manure (fixed manure)	30
Dairy cattle—mixed manure storage	28
Dairy cattle—manure disposal on grassland	20

TABLE 13.13
Allocation of Emissions for Semi-Skimmed UHT Milk

Emission Source	Impact Assigned to Semi-Skimmed UHT Milk (%)
Energy	30–47
Water use	1–7
Chemicals/process materials	25–35

from farms, industrial milk, and other process materials). Emission factors are derived from Table 13.8 and from the Ecoinvent database (2011).

For the allocation of different emission sources among different products, the IDF guidelines were used (IDF 2010). The allocation depends on the total gamma of products. Table 13.13 provides an indication of the impact borne by the semi-skimmed UHT milk. The share of semi-skimmed UHT milk varies between 27% and 56% of the produced final volumes.

13.6 RESULTS

13.6.1 THE CF OF MILK

Results are presented in Figure 13.3 and summarized as follows: a kilogram of raw milk creates a "CF of 1.02 kg CO_{2eq}," including land conversion. Rumen fermentation (40.9%) has the lion's share in the overall CF. This is a large share determined by the supposed feed uptake and digestibility. The feed consumption is based on primary data; however, reference values for digestibility are accompanied with a higher uncertainty (secondary data). As of the summer of 2015, there were no recent literature values regarding the digestibility of the fodder.

Fodder production is responsible for 38.7% of the CF. Albeit that only 12% of the fodder is purchased, the impact is equally compared with home-grown crop cultivation. The following needs to be stated:

- The lion's share of the purchased fodder has a higher dry matter and energy content, compared with the home-grown roughage.
- The contribution of land conversion is 50% of the total impact of the purchased fodder. The method is not yet internationally accepted.

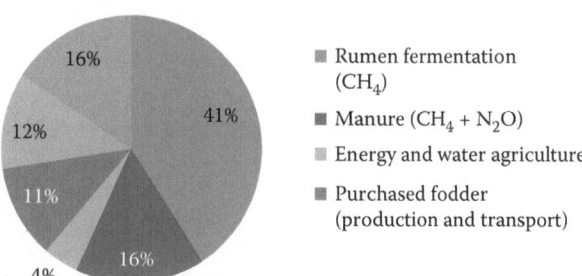

Milk—contribution of 1 kg raw milk (including LUC)

- ■ Rumen fermentation (CH₄)
- ■ Manure (CH₄ + N₂O)
- ▨ Energy and water agriculture
- ■ Purchased fodder (production and transport)

FIGURE 13.3 CF of 1 kg of raw milk in Flanders.

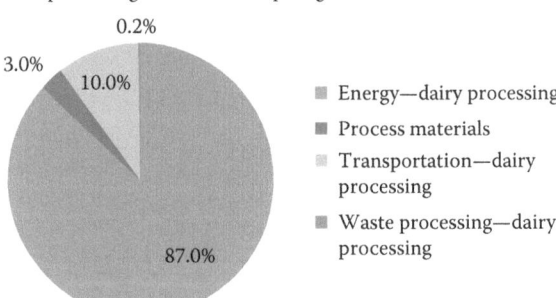

Milk processing—contribution per kg semi-skimmed UHT milk

- Energy—dairy processing
- Process materials
- Transportation—dairy processing
- Waste processing—dairy processing

FIGURE 13.4   CF of 1 kg semi-skimmed UHT milk.

Manure storage and application on grassland account for 16% of the emissions (69% due to methane and 31% due to N_2O emissions).

Energy and water represent 5.7% (electricity consumption represents 37.3% of this impact, gasoline oil 62.4%, and water 0.3%).

13.6.2   MILK PROCESSING

The contribution of milk processing per kilogram of semi-skimmed UHT milk amounts to 0.13 kg CO_{2eq}. The biggest contribution originates from energy consumption (87%). About 30%–40% is due to electricity consumption. The remaining parts are due to the combustion of fossil fuels (gas and heavy and light oil fuels). Transportation of resources to the processing plant accounts for 10%. Transport of raw and industrial milk represents the largest share (95%–98%). Distribution of the final product among supermarkets is not included. The production of process materials (3%) and waste processing (0.2%) are almost negligible.

13.6.3   TOTAL RESULT

Results are given per kilogram of raw milk and per kilogram of semi-skimmed UHT milk. For each kilogram of semi-skimmed UHT milk, 1.077 kg of raw milk is needed. About 82.9% of the upstream emissions of raw milk are assigned to semi-skimmed UHT milk (Figure 13.4). The remaining emissions are assigned to cream. The results (including land conversion) are summarized as follows:

- 1.02 kg CO_{2eq}/kg raw milk
- 1.04 kg CO_{2eq}/kg semi-skimmed UHT milk

Dairy processing (converting raw milk to semi-skimmed UHT milk) contributes for about 13% of the total CF of semi-skimmed UHT milk (including LUC). Table 13.14 provides a detailed breakdown of the CF.

13.6.4   CF SENSITIVITY

The single outcome of CF calculations should be used with caution. Giving a range of figures in which the CF is expected to be gives a more realistic insight (Flysjo et al. 2011b). Therefore, a sensitivity analysis* is conducted to define fluctuations.

---

* A statistical sensitivity analysis was not carried out because not enough information was available to calculate the standard deviation on the secondary data used or on the final result.

TABLE 13.14

Details of the Results per Kilogram of Raw Milk and Semi-Skimmed UHT Milk

Emission Sources	Kilogram Raw Milk (kg CO₂eq/FU)	Kilogram of Semi-Skimmed UHT Milk (kg CO₂eq/FU)
Purchased fodder—production and transport	0.11	0.10
Purchased fodder—LUC	0.12	0.11
Home-grown roughage for cattle—production	0.16	0.14
Rumen fermentation (CH_4)	0.42	0.37
Manure ($CH_4 + N_2O$)	0.16	0.15
Energy and water—agriculture	0.05	0.04
Dairy processing	—	0.13
Total	1.02	1.04

Note: FU, functional unit.

13.6.4.1 Impact of Estimating Indirect Laughing Gas Emissions

If the 2006 IPCC reference value of 1% nitrogen losses for direct emissions is being replaced by the value of 3.16%, it gives rise to an increase in the CF of raw milk of about 21%.

13.6.4.2 Feed and Herd Characteristics

Table 13.15 presents trends in feed and herd characteristics and their possible impact on the CF. The result is only marginally influenced by the mortality rate as the number of adult cows is not being influenced, but only the number of calves and young cattle between 1 and 2 years. As the latter do not produce milk yet, their impact on the final result is very limited.

13.6.4.3 Manure Storage/Disposal

In the original model, an allocation was defined between manure disposal on grassland (24%) and in the barn. In the barn, part of the manure is kept as stable manure (19%) and the remaining part as mixed manure (57%). Three scenarios are considered to test the sensitivity of these parameters. It is supposed that 100% of the manure production is disposed on grassland (scenario 1), stored as stable manure (scenario 2), and kept as mixed manure (scenario 3).

The results (Table 13.16) are little influenced by these parameters. It is striking that storage as stable manure provides the least emissions. With extended time on grassland, animals consume more energy. The calculated GE is higher and hence the rumen fermentation.

TABLE 13.15

Applied Sensitivity Analysis: Impact of Changes in Herd and Feed Concentrate Parameters on the CF

Parameter	Initial Value	Min	Shift in CF	Max	Shift in CF
Mortality rate animals <1 year	10%	5%	+0.01	15%	−0.01
Mortality rate animals >1 year	1.25%	0.5%	—	3%	—
Replacement rate	25%	20%	+0.03	30%	−0.03
Milk production (liter per cow per year)	7822	6500	+0.13	8500	−0.12
DE content of fodder (% GE)	75%	68%	+0.11	83%	−0.09

TABLE 13.16
Variation of Parameters Regarding Manure Storage

Parameter	Initial Value	Scenario 1	Scenario 2	Scenario 3
Disposal grassland (%)	24	100	0	0
Stable manure (%)	19	0	100	0
Mixed manure (%)	57	0	0	100
Result (relative toward initial)	1	0.96	0.95	1.02

TABLE 13.17
Overview Variation of Allocation Method between Milk and Meat

Allocation Method	Milk (%)	Meat (%)
Mass allocation	98	2
IDF (empirical formula)	88	12
FAO (protein)	85	15
Biological allocation	78	22

13.6.4.4 Influence of Allocation Method between Milk and Meat

The allocation among meat and milk is done by means of the IDF guidelines. About 12% of the emissions is assigned to meat and 88% to milk. From a French case study (Dollé and Bertrand 2010), it appears that this allocation can vary between 78% and 98% for milk and between 2% and 22% for meat (calf and cow). Table 13.17 presents the variation in the allocation method.

It is striking that the CF per output milk is very robust and only influenced to a small extent by the choice of the allocation method. The CF of meat products per kilogram of live weight is much more sensitive because specialized dairy farms produce only a limited amount of meat.

13.6.4.5 Energy in the Dairy Processing Plant and the Use of Heat Recuperation

Heat cogeneration is the combined generation of heat and power, mostly in the form of steam and electricity. Mostly, this is more energy-efficient than the conventional separate generation. The use of heat cogeneration is hence seen as an energy-saving option. Electricity and heat yields in the installation determine the energy efficiency of the whole company. In order to conduct the simulation of a heat cogeneration installation in a dairy company, it is assumed that the energetic efficiency of fossil fuels increases by 50%. The CF of dairy processing is therefore lowered by 27.8%. Hence, the total footprint of semi-skimmed UHT milk (including LUC) decreases by about 2.5%.

13.6.4.6 Influence of the Allocation Method between Several Dairy Products

At the level of dairy processing, more products besides semi-skimmed UHT milk are being produced. As described in the previous sections, during the milk production process, cream is produced as well. The initial calculations, based on the IDF guidelines, allocate GHG emissions among milk and cream. Table 13.18 presents the variations.

13.6.4.7 CF Range

Based on the sensitivity analysis, the overall estimated CF of milk production in Flanders falls between 0.90 and 1.23 kg CO_{2eq}/kg of raw milk. For semi-skimmed UHT milk, the CF ranges between 1.03 and 1.36 kg CO_{2eq}/kg.

TABLE 13.18
Overview of the Variation of Allocation Method between Milk and Cream

Allocation Method	Semi-Skimmed milk (%)	Cream (%)
Mass allocation	95	5
IDF (empirical formula)	82.9	17.1
Dry matter content	79	21

13.6.5 Mitigation Measures

Three huge hot spots in the production of milk were revealed: rumen fermentation, fodder production, and manure production and the usage of it. Some opportunities to reduce the CF of milk were defined.

In particular, the composition of feed has a very big impact on the overall CF. Within Europe, the use of soy bean in feed concentrates has increased rapidly. However, the use of soy has a negative impact on the CF [negative LUC impact and transportation of feed components over long distances (Hortenhuber et al. 2011)]. Therefore, replacement with regional products can reduce the CF. When overseas products are to some extent indispensable, priority should be given to products produced in a sustainable way with a restricted impact on LUC. However, the composition of the feed depends more on availability, price, and the characteristics of the components. Price and availability are major economic factors influencing the final price of the feed and the possible usage by farmers. Hortenhuber et al. (2011) clearly indicate that regional and local products are not always at people's disposal. Shifting production in Europe toward these alternatives might lead to LUC effects in Europe (Steinfeld et al. 2006). However, one cannot remove all carbon-negative components, as this limits the economic sustainability of farming practices.

Therefore, parameters such as price, availability, and feed component characteristics need to be taken into account alongside the CF to ensure that milk production is not compromised in an effort to reduce the GHG emissions (Espinoza-Orias et al. 2011). This economic aspect is often neglected in other literatures, as described by Verspecht et al. (2012). Manure production, storage, management, and usage are the second largest contributors to the overall CF. In Flanders, one type of manure management, which is the most popular method, involves the separation of liquid and solid components. The solid part gives rise to a quantity similar to nitrate emissions as the storage and use of untreated animal manure would do.

Both things exemplify the possible tradeoffs between dealing with GHG emissions and other aspects of sustainability, put in a larger perspective. Sustainability consists of three pillars: environmental protection, economic growth, and social equity. A mitigation measure only has a positive effect when all aspects lead to better or higher sustainability. Moreover, it is important to stress that the CF is a good indicator for GHG emissions as one environmental indicator, but it is not an indicator for environmental impact in general.

13.7 CONCLUSION

The CF of beef estimated in our study using the PAS2050 methodology (BSI 2011) ranges from 0.90 to 1.23 kg CO_{2eq}/kg of raw milk. For semi-skimmed UHT milk, the CF ranges between 1.03 and 1.36 kg CO_{2eq}/kg. The main hot spots were found in rumen fermentation and fodder production, accounting for the greatest proportion of the total CF. Furthermore, manure management is another important hot spot in the production chain. These hot spots reveal where measures can be taken in order to decrease GHG emissions along the chain. Our study helps to fill the void in the CF literature, which existed around livestock produce. Moreover, the chapter reports on the methodology and assumptions that have been used, the chosen system boundaries, and the system definition. This makes it possible to follow a similar method and to estimate the CF of milk in other regions, allowing better and fairer comparisons (Flysjo et al. 2011a) and hence assisting the definition of a benchmark for the CF. This in turn will stimulate the search for opportunities to reduce the CF within the framework of international targets such as the 2011 Durban Accord (Dalgaard et al. 2011).

Flanders is required to implement European policy measures with regard to agriculture. From this perspective, our study will assist Flemish policy makers in achieving their aims for the period 2012–2020. During this period, GHG emissions for EU sectors that do not fall under the transferable emission system have to decrease emissions by 15%. Therefore, this study helps to reveal hot spots in the chain and potential strategies to decrease their impact in terms of GHG emissions. However, it should be noted that an integrated sustainability approach is necessary, whereas this study focuses solely on the environmental impact of one indicator: namely, climate change.

ACKNOWLEDGMENTS

This study was funded by the Flemish Administration—Department of Agriculture and Fisheries. The authors gratefully acknowledge this funding.

REFERENCES

Blonk, T.J., Kool, A. and Vlaar, L.M.C. 2007. Landbouw en klimaat in Brabant. Culemborg: CLM Onderzoek en advies.

Blonk, H., Kool, A., and B. Luske. 2008a. Environmental effects of Dutch consumption of protein-rich products, consequences of animal protein replacement anno 2008 (in Dutch). Blonk Milieuadvies, Gouda.

Blonk, H., Luske, B., and C. Dutilh. 2008b. Greenhouse gas emissions of meat—Methodological issues and establishment of an information infrastructure. Blonk Milieuadvies, Gouda.

Boerenbond. 2011–2012. Corporate economic accounting (in Dutch [CD-ROM], 2009), Leuven.

British Standards Institute (BSI). 2011. PAS2050 specification for the assessment of the life cycle greenhouse gas emissions of goods and services. England.

Campens, V., Van Gijseghem, D., Bas, L., and I. Van Vynckt. 2010. Climate and livestock (in Dutch), Department of Agriculture and Fisheries, Division Monitoring and Study, Brussels.

Dalgaard, R., Halberg, N., and J. Hermansen. 2007. Danish pork production: An environmental assessment. DJF Animal Science, 82.

Dalgaard, T., Olesen, J.E., Petersen, S.O., Petersen, B.M., Jorgensen, U., Kristensen, T., Hutchings, N.J., Gyldenkaerne, S., and J.E. Hermansen. 2011. Developments in greenhouse gas emissions and net energy use in Danish agriculture—How to achieve Substantial CO(2) reductions? *Environmental Pollution* 159(11):3193–203.

de Vries, M. and de Boer, I.J.M. 2010. Comparing environmental impacts for livestock products: A review of life cycle assessments. *Lifestock Science* 12:1–11.

Dollé, B., and S. Bertrand. 2010. Greenhouse gas emissions from the dairy sector. A life cycle assessment, FAO, Rome.

Ecoinvent. 2011. Life Cycle Inventory (LCI) database [CD-ROM] (1998–2013). Swiss Centre for Life Cycle Inventories, St.-Gallen, Switzerland.

Eriksson, I.S., Elmquist, H., Stern, S., and T. Nybrant. 2005. Environmental systems analysis of pig production—The impact of feed choice. *International Journal of Life Cycle Assessment* 10(2):143–54.

ERM. 2010. *Carbon Footprint Methodology Applied to the Flemish Livestock Sector.* Brussels: ERM Issue, 184 p.

Espinoza-Orias, N., Stichnothe, H., and A. Azapagic. 2011. The carbon footprint of bread. *International Journal of Life Cycle Assessment* 16(4):351–65.

FAO. 2008. World meat markets at a glance. *Food Outlook*, Global Market Analysis. http://www.fao.org/docrep/011/ai482e/ai482e01.htm#39.

FAO. 2010. Greenhouse gas emissions from the dairy sector, a life cycle assessment. Food and Agriculture Organization of the United Nations—Animal Production and Health Division, Rome, Italy, 1998.

Finkbeiner, M. 2009. Carbon footprinting—Opportunities and threats. *International Journal of Life Cycle Assessment* 14(2):91–4.

Flysjo, A., Cederberg, C., Henriksson, M., and S. Ledgard. 2011a. How does co-product handling affect the carbon footprint of milk? Case study of milk production in New Zealand and Sweden. *International Journal of Life Cycle Assessment* 16(5):420–30.

Flysjo, A., Henriksson, M., Cederberg, C., Ledgard, S., and J.E. Englund. 2011b. The impact of various parameters on the carbon footprint of milk production in New Zealand and Sweden. *Agricultural Systems* 104(6):459–69.

Hortenhuber, S.J., Lindenthal, T., and W. Zollitsch. 2011. Reduction of greenhouse gas emissions from feed supply chains by utilizing regionally produced protein sources: The case of Austrian dairy production. *Journal of the Science of Food and Agriculture* 91(6):1118–27.

IDF. 2010. A common carbon footprint approach for dairy—the IDF guide to standard life cycle assessment methodology for the dairy sector, Brussels: Bulletin of the International Dairy Federation.

IPCC. 2006a. Intergovernmental Panel on Climate Change [online]. Established by United Nations Environmental Programme and World Meteorological Organization, Geneva, Switzerland. http://www.ipcc.ch/index.htm#.UOVJmrXLe_0 (accessed January–August 2011).

IPCC. 2006b. Chapter 10: Emissions from livestock and manure management. 2006 IPCC Guidelines for National Greenhouse Gas [online], Geneva, Switzerland. http://www.ipcc-nggip.iges.or.jp/public/2006gl/pdf/4_Volume4/V4_10_Ch10_Livestock.pdf (accessed January–August 2011).

IPCC. 2006c. Chapter 11: Emissions from managed soils, and CO_2 emissions from lime and urea application. 2006 IPCC Guidelines for National Greenhouse Gas [online], Geneva, Switzerland. http://www.ipcc-nggip.iges.or.jp/public/2006gl/pdf/4_Volume4/V4_11_Ch11_N2O&CO2.pdf (accessed January–August 2011).

IPCC. 2007. Intergovernmental Panel on Climate Change [online]. Established by United Nations Environmental Programme and World Meteorological Organization, Geneva, Switzerland. http://www.ipcc.ch/index.htm#.UOVJmrXLe_0 (accessed July–December 2011).

Jacobsen, R., Vandermeulen, V., Vanhuylenbroeck, G., and X. Gellynck. 2013. The carbon footprint of pigmeat in Flanders. *International Journal of Agricultural Sustainability* 1:1–18. DOI:10.1080/14735903.2013.798896.

Leip, A., Weiss, F., Wassenaar, T., Perez, I., Fellmann, T., Loudjani, P., Tubiello, F., Grandgirard, D., Monni, S., and K. Biala. 2010. Evaluation of the livestock sector's contribution to the EU greenhouse gas emissions (GGELS)—Final report. European Commission, Joint Research Centre.

Muller-Lindenlauf, M., Deittert, C., and U. Kopke. 2010. Assessment of environmental effects, animal welfare and milk quality among organic dairy farms. *Livestock Science* 128(1–3):140–8.

Nielsen, P., Nielsen, A.M., Weidema, B.P., Frederiksen, R.H., Dalgaard, R., and N. Halberg. 2010. LCA Food Database [online]. Faculty of Agricultural Sciences, Danish Institute for Fisheries Research, Denmark. http://www.lcafood.dk/ (accessed January–August 2012).

NIR Belgium 2009. National institute for the Accounts in Belgium.

NIR 2011. National Inventory Report for greenhouse gases (NIR) of Belgium (2009).

Sonesson, U., Cederberg, C., and M. Berglund. 2009. Greenhouse gas emissions in milk production: Decision support for climate certification. Klimatmarkning for mat, Sweden: SIK-Institutet för Livsmedel och Bioteknik.

Statistics Belgium. 2010. *Farm Counting (Landbouwtelling)*, Brussels.

Steinfeld, H., Gerber, P., Wassenaar, T., Castel, V., Rosales, M., and C. De Haan. 2006. Livestock's long shadow: Environmental issues and options. FAO, Rome.

Thoma, G., Popp, J., Nutter, D., Shonnard, D., Ulrich, R., Matlock, M., Kim, D.S. et al. 2010. Regional analysis of greenhouse gas emissions from milk production practices in the United States. Seventh International Conference on Life Cycle Assessment in the Agri-Food Sector, Bari, Italy.

Van Der Werf, H.M.G., Kanyarushoki, C., and M.S. Corson. 2009. An operational method for the evaluation of resource use and environmental impacts of dairy farms by life cycle assessment. *Journal of Environmental Management* 90(11):3643–52.

Verspecht, A., Vandermeulen, V., Ter Avest, E., and G. Van Huylenbroeck. 2012. Review of trade-offs and co-benefits from greenhouse gas mitigation measures in agricultural production. *Journal of Integrative Environmental Sciences* 9(1):147–57.

VLM. 2011. Manure bank dataset (in Dutch) [CD-ROM] (2009). Brussels.

VMM, VITO, AWAC, IBGE-BIM, Federal Public Service of Health, Food Chain Safety and Environment, IRCEL-CELINE, ECONOTEC. 2011. Belgium's greenhouse gas inventory (1990–2008). National Inventory Report submitted under the United Nations Framework Convention on Climate Change and the Kyoto Protocol, April 2010, Belgium.

Williams, A.G., Audsley, E., and D.L. Sandars. 2006. Determining the environmental burdens and resource use in the production of agricultural and horticultural commodities. Main Report, Defra Research Project IS0205. Bedford: Cranfield University and Defra. www.silsoe.cranfield.ac.uk; www.defra.gov.uk.

Wright, L., Kemp, S., and I. Williams. 2011. "Carbon footprinting": Towards a universally accepted definition. *Carbon Management* 2(1):61–72.

14 Digitizing the Assessment of Embodied Energy and Carbon Footprint of Buildings Using Emerging Building Information Modeling

F.H. Abanda, A.H. Oti, and J.H.M. Tah

CONTENTS

14.1 Introduction 303
14.2 Challenges Hindering the Uptake of BIM in Construction...304
14.3 Sustainability Appraisal Using BIM Software: An Overview of Current Practices..........306
14.4 Embodied Energy and CO_2 Analysis: An Overview...308
14.5 Embodied Energy and CO_2 Analysis: Computational Steps 309
14.6 Implementation in Revit 310
    14.6.1 Description of the Model 311
    14.6.2 Quantity Takeoff..311
14.7 Exploiting Model in Revit for Decision-Making Purposes 315
    14.7.1 Scenario 1: Model Is Complete and Perfect...316
    14.7.2 Scenario 2: Model Is Complete and Minor Changes Are to Be Made 317
    14.7.3 Scenario 3: Model Is Complete and Component Information Needs
            to Be Changed...318
            14.7.3.1 Carrying Out Changes through the Properties Palette 318
            14.7.3.2 Carrying Out Changes through the Edit Type Function....................318
14.8 Implementation in MS Excel 319
14.9 Results and Analysis...322
14.10 Conclusion 326
References...327

14.1 INTRODUCTION

Sustainability has become a hot topic in the AEC (Architecture, Engineering, and Construction) industry. Recent studies reveal the importance of incorporating sustainability in construction projects (Cheung et al. 2012; Shrivastava and Chini 2012; Motawa and Carter 2013; Wong and Fan 2013; Abanda et al. 2014; Alwan and Jones 2014; Bragança et al. 2014; Wong and Kuan 2014). However, construction projects comprising so many materials and components are often too complex, making it a challenge to assess their sustainable performance. As a result of the complexity and variety of different specializations in construction, many building information modeling (BIM) software packages have been developed for different applications. Based on current practice, modeling software (e.g. Revit, ArchiCAD, Edificius, PriMus, and PriMus-To), energy analysis software [e.g. Green Building

Studio, Ecotect, and Integrated Environmental Solutions (IES)], cost estimating tools (e.g. Vico and Cato), and planning (e.g. Synchro, Navisworks, Visual Simulation, and ProjectWise Navigator) exist as separate entities. However, these software packages depend on interoperable languages or plug-ins for exchanging construction project information. In practice, a building may have been modeled in Revit and then exported via gbXML to Green Building Studio before performing an energy analysis on the model. In most cases, once a building model has been exported, no modification in BIM energy analysis software can be done. This means that, although sustainability analysis software systems such as Green Building Studio are great for what they are developed, they do not provide the means to alter design parameters for making alternative choices. If changes are required, then the changes will have to be made on the original model and then re-exported to the sustainability analysis software. Also, most often, the exported data need to be modified to conform to the input data of the sustainability analysis software (Shrivastava and Chini 2012; Wong and Fan 2013). Furthermore, the challenges with interoperability between software systems have been noted in the literature (Bynum et al. 2013). Data loss is a very common example. To avoid time-consuming operations, going back and forth between sustainability analysis and BIM design software to explore suitable alternatives, the building should be properly and accurately modelled in the BIM authoring software before being exported into any other application software. An option will be to explore various alternatives in the early design stages in the BIM design software. Furthermore, consideration of sustainability and other construction performance factors at the early design stage has proven to be beneficial in the overall performance of the construction project (Cheung et al. 2012; Bragança et al. 2014).

The aim of this study is to explore the capability of BIM software in conducting sustainability assessments in the early design stage to support decision making. To facilitate understanding, the sustainability assessment capability of BIM software will be compared with conducting the same sustainability assessment in Microsoft Excel, a very popular software package often used in the AEC industry. This popularity is attributed to the fact most professionals are already using MS Excel and will easily be able to perform "like-for-like" comparison. The focus will be on embodied energy and embodied CO_2.

14.2 CHALLENGES HINDERING THE UPTAKE OF BIM IN CONSTRUCTION

The construction sector across the globe looks forward to when BIM becomes fully matured and accepted as a medium for presentation of all construction information and transactions. In the United Kingdom, a BIM working group was commissioned by the government to examine the construction and post occupancy benefits of BIM for the building and infrastructure market. It was recommended that there should be a structured government/construction sector strategy to increase the uptake of BIM over a 5-year horizon (BIM-IWG 2011). This is geared toward the plan to improve government estates in terms of cost, value, and performance. To ensure a clear articulation of the levels of competence expected in the BIM adoption process with supporting standards and guidance, a BIM maturity model (Figure 14.1) has been devised. It was initially developed by Bew and Richards in 2008, but is now receiving support from the government client group on its applicability to projects and the contract industry (BIM-IWG 2011). The model defines levels from 0 to 3 in order to categorize types of technical and collaborative working for a concise description and understanding of processes, tools, and techniques to be used. This model creates a clear and transparent view of BIM with respect to the building supply chain for the client's understanding and progress made to-date in construction IT applications. Globally, most construction firms are believed to be currently operating at maturity level 2, in which a managed 3D environment is held in separate BIM disciplines and with possibility of holding and managing 4D program data as well as 5D cost elements. The challenges characterizing this maturity level include lack of high-level integration and collaboration, including real-time synchronous processes. Challenges also exist with the integration of BIM into existing contractual arrangements. This arouses legal implications relating to data translation and interoperability, design liability issues, delegation and ownership, risk allocation, and so on

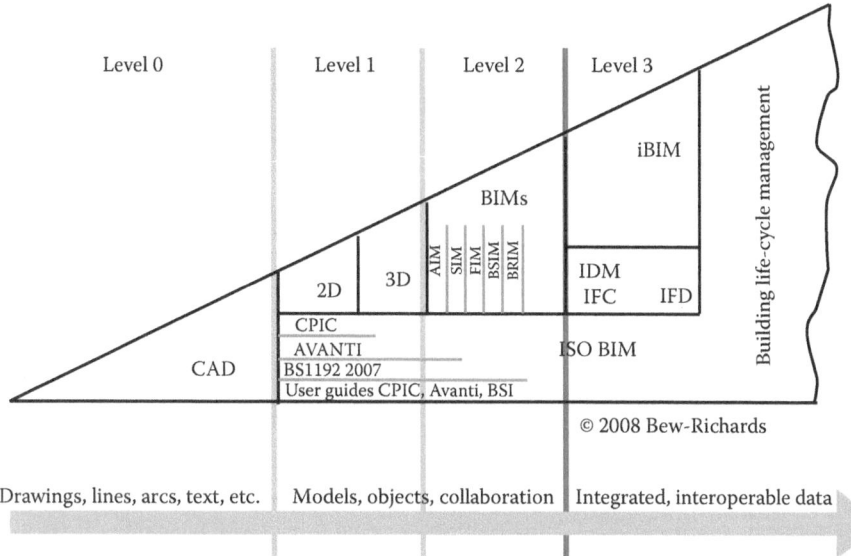

FIGURE 14.1   BIM maturity levels (BIM-IWG 2011).

(Ashcraft 2008; McAdam 2010). Thus, these issues themselves cannot be explicitly divulged from technology or how technology is being used (Ashcraft 2008). The Joints Contracts Tribunal Public Sector Supplement acknowledges the promises of BIM and has made provision for how BIM can be incorporated into its existing contract arrangements. As a step forward and in response to the U.K. Government BIM Strategy, the Construction Industry Council (CIC) has developed a BIM protocol to enable the production of building information models at defined stages of a project (CIC 2013). This protocol, targeted at level 2 BIM maturity, is new and therefore requires convincing demonstration of its successful use in projects for best practice lessons to be learnt.

To reduce project cost and delivery time, the construction industry is looking forward to advancing beyond traditional collaborative practices that are often based on nonintelligent/noninteroperable building data and 2D drawings. Contemporary collaborative working approaches place emphasis on data interoperability. However, interoperability may not be a hindrance to collaboration using BIM tools from a single product family, but remains a key to collaboration across different product families. In either case, affordability in terms of costs associated with acquiring software packages and requisite training needs remains a common hindrance factor. Although the benefits of BIM have been well envisaged (Eastman et al. 2008; Azhar 2011; Abanda et al. 2013), there exists some degree of skepticism that it should not be another replacement of old problems with new ones. The deployment of IT within businesses has often been faced with such skepticisms and fear of not realizing expected benefits (Willcocks 1992; Love and Irani 2004). The industry is optimistic that emerging best practice experiences and innovations will mitigate these challenges to the barest minimum.

As such, firms that utilize best practice lessons and are innovative are more likely to successfully deploy BIM and realize its value as an IT facility (Lee and Runge 2001; Dibrell et al. 2008). A recent survey on BIM implementation in the U.K. construction project life cycle acknowledges that BIM requires significant cost and training investments (Eadie et al. 2013). This can be a huge hindrance to the uptake of BIM, especially for small businesses. For instance, research stipulates that small and medium enterprises (SMEs) take-up of IT innovations is highly influenced by the large amount of investment required (Acar et al. 2005; Benjaoran 2009; Gledson et al. 2012; Abanda and Tah 2014). However, the findings of the survey by Eadie et al. suggest that the resulting additional costs are still within the reach of SMEs. This is obviously dependent on the strength of investment garnered by such firms over time. With respect to construction SMEs, the literatures centered on their vulnerability in the emerging market scenarios of innovative IT applications and the need to develop a defined collaborative working approach and requisite training to overcome inherent challenges. In a study on developing a roadmap for BIM implementation in the United Kingdom, Khosrowshahi and Arayici (2012) identified the following key barriers and challenges militating against BIM implementation.

- Firms are not familiar enough with BIM use.
- Firms are reluctant to initiate new workflows or to train staff.
- Benefits from BIM implementation do not outweigh the costs to implement it.
- Benefits are not tangible enough to warrant its use.
- BIM does not offer enough financial gain to warrant its use.
- There is insufficient capital to invest in training having started with hardware and software.
- BIM is too risky from liability standpoint to warrant its use.
- There is resistance to culture change.
- There is no demand for BIM use.

The first two barriers are directly related to the lack of technical know-how required for operation of BIM applications. A closer examination of the remaining seven barriers still reveals some connections with technical know-how or knowledge about the operation of BIM technology. The ensuing argument is that if extensive knowledge (gained from requisite training) about the operation BIM technology is put to work, the resulting output will outweigh associated costs. This will yield tangible benefits to the establishment in question, induce some cultural changes, and also create demand for its use. This line of argument is portrayed in the 2012 National BIM Survey Report, which stipulates that respondents currently using BIM rose from 13% in 2010 to 31% in 2011 (NBS 2012). A similar survey suggests technical skills and financial capacity as two key limiting factors to BIM implementation (NFB 2012). It is the intention of this chapter to enhance the aspect of technical skill to encourage sustainability implementation in BIM. The authors are of the opinion that bringing the manipulations in BIM closer to packages, such as spreadsheets, widely and commonly used in the industry will be beneficial for BIM uptake. This will help stakeholders realize that the operation of BIM is not as complex or complicated as may have been imagined.

14.3 SUSTAINABILITY APPRAISAL USING BIM SOFTWARE: AN OVERVIEW OF CURRENT PRACTICES

Views on what actually constitutes sustainable development are varied (Kua and Lee 2002; Fiksel 2003); however, many of such views are built on the three cores of economic, environmental, and social foundations of human growth. These three aspects summarize the intrinsic relationship sustainable development has with the construction industry and more specifically the building. Environmental sustainability may be achieved through the protection of resources and the ecosystem. Long-term resource productivity and low use cost can contribute to the economic aspects, although the building also serves as source of contribution to health, comfort, social, and cultural values of the society. Figure 14.2, which is a best practice illustration of the concept of sustainability assessment, further reflects the various important elements in building sustainability. It combines clients' requirements, regulatory requirements, functional requirements, and technical requirements with those of the environment, economic, and social elements for the building. Integrated building performance encompasses environmental, social, and economic performance as well as the technical and functional performance, which are intrinsically related to each other (BS EN 15643-1 2010). The building sustainability arm of European Committee for Standardization (CEN/TC 350) is working on ways to standardize aspects related to assessment procedures and communication of results from defined indicators. However, the integration of standardized building sustainability assessment procedures into contemporary IT tools, such as BIM, for building design and modeling is still in its infancy.

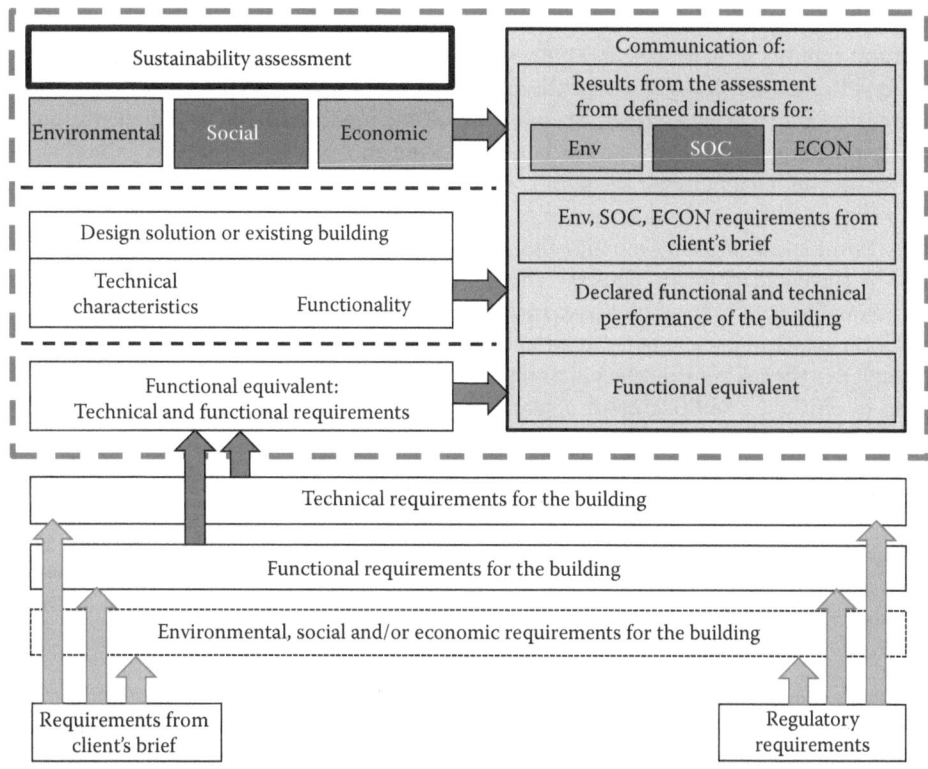

FIGURE 14.2 The concept of sustainability assessment of buildings (BS EN 15643-1 2010).

In the planning and design stage, the benefits of the early incorporation of sustainability principles in guiding project decisions and design iterations have been well emphasized (Kohler and Moffatt 2003). One area of challenges has been the development of sustainability appraisal tools to guide professionals in making conceptual design decisions among alternative solutions. Although a number of sustainability assessment tools exist, it has been difficult for engineers to apply them to design iterations via the emerging BIM process. The Building Research Establishment Environmental Assessment Method, used in the United Kingdom, is yet to be incorporated into BIM. It is currently being used to guide project development and to rank already completed buildings. In the United States, research efforts to incorporate Leadership in Energy and Environmental Design (LEED) criteria into BIM tools have been ongoing. Nguyen et al. (2010) have attempted to use BIM to evaluate sustainability of architectural designs by storing the LEED criterion indicators as project parameters in Revit Architecture software. These parameters are extracted when applied to a project to compute the maximum possible LEED ratings. Although this work targets architectural designs, it is limited to the LEED sustainability parameters and will not be of direct benefit to the structural or services engineer. Similar to LEED, Wong and Kuan (2014) demonstrated that BIM can also be integrated with the Hong Kong's "BEAM Plus" sustainable building rating. This study explored the possibility of using BIM (Revit) to achieve 26 out of 80 credit points in BEAM Plus sustainable rating of a residential building. On buildings' carbon emissions, the integration of the estimation process into BIM tools to establish impacts early in the design stage is recommended (Iddon and Firth 2013; Oti and Tizani 2015). In developing a BIM tool that simultaneously estimates embodied and operational carbon over 60 years life span for a typical four-bedroom detached house, Iddon and Firth (2013) stressed the need for a more robust and universal methodology for measuring embodied carbon in sustainability assessment. This opinion is shared by Oti and Tizani (2015), who incorporated life-cycle costing, ecological footprint, and carbon footprint into Revit to estimate sustainability credentials of alternative structural design solutions. The research concentrated on the structural framing of buildings and suggests that all building professionals ought to be responsible for the information on the sustainability performance of their design specifications as they do for the integrity of their designs. Also, Abanda et al. (2014) employed Revit to validate the quantities of building materials used in the assessment of embodied energy and CO_2 by incorporating associated information into the building modeling process.

Thus, IT is the kernel for modeling, storing, and exchanging information/data within and across platforms. Information probably remains the most invaluable construction "material," which must be shared by stakeholders in the industry (Tolman 1999). Dawood and Sikka (2006) assert that the construction industry is information-based by nature. In overcoming the associated shortcomings with AEC information management, researchers have had the vision of capturing all the information embedded in the building product in a single information model. This has developed gradually, improving in efficiency and degree of application overtime, and has given rise to contemporary BIM. Literatures show that there is increasing research on BIM-related issues including sustainability. The consensus on this emerging technology is that BIM-based software provides a medium to capture complete design and project information and to make the best use of associated project data. Thus, BIM has the tendencies for continuous expansion to closely mimic, as much as possible, the vast amount of information embedded in a typical building project.

14.4 EMBODIED ENERGY AND CO_2 ANALYSIS: AN OVERVIEW

Although embodied energy has recently gained ground, the concept has been around for at least three decades, perhaps used for different purposes. Input–output techniques developed by Leontief (1986) for analyzing industrial interdependencies in a national or regional economic system today constitute solid theoretical underpinnings for embodied energy assessment (e.g. Treloar 1997; Treolor et al. 2003). Since then, it has been common to also find embodied energy concepts being used in industrial and chemical processes. Van Gool (1980) evaluated the product ("process" plus "embodied") energy required for chemical process equipment, often termed "unit operations."

Before the recent applications of input–output techniques (e.g. Treloar 1997; Treolor et al. 2003) in the embodied energy analysis, the technique had already featured in environmental applications (Leontief 1970) in the 1970s. Specifically, Hannon (1973) adopted the input–output technique (Leontief 1986) in the embodied energy analysis to describe ecosystem energy flows. Similar to Leontief's input–output theory, energy inputs to a system are aggregated from all subsidiary pathways to yield the total embodied energy or gross energy requirement (Cabeza et al. 2013).

There are two forms of embodied energy in buildings (Harries 2007; Shrivastava and Chini 2012). The initial embodied energy consists of nonrenewable energy required to extract and process its raw materials (indirect energy) and the energy used to transport the finished product to the job site and install it (direct energy) (Harries 2007; Shrivastava and Chini 2012). The recurring embodied energy of a building component can be defined as the nonrenewable energy consumed to maintain and replace it, as well as recycle or disposing it at the end of its useful life (Harries 2007; Shrivastava and Chini 2012). Operational energy of buildings is the energy required for maintaining comfort conditions and day-to-day maintenance of the buildings by operating processes such as heating and cooling, lighting and appliances, and air conditioning (Ramesh et al. 2010). For many years, the embodied energy content of a building was assumed to be small compared with the operational energy (Pacheco-Torgal et al. 2013; Cabeza et al. 2014). Pacheco-Torgal et al. (2013) reported that embodied energy represents between 10% and 15% of the operational energy. This is close to Cabeza et al.'s (2014) report that embodied energy constituted 10%–20% of the life-cycle energy of a building. Some studies have reported even very low figures, as low as 2%. For example, Sartori and Hestnes (2007) reported that embodied energy could account for 2%–38% of the total life-cycle energy of a conventional building and 9%–46% for a low-energy building. Consequently, most energy-related research efforts have been directed toward reducing operational energy, largely by improving energy efficiency of the building envelope. In addition to embodied energy, the production of building materials (e.g. extraction, transportation, and manufacturing processes) releases CO_2 mainly due to the use of fuel or electricity. This is often called embodied CO_2. Thormark (2006) reported that embodied energy in traditional buildings can be reduced by approximately 10–15% through proper selection of building materials with low environmental impacts. González and García Navarro (2006) estimated that the selection of building materials with low impacts can reduce CO_2 emissions by up to 30%. In the United Kingdom, Sturgis and Roberts (2010) predict the proportion of embodied carbon[*] to increase from 30% to 95%, whereas operational energy will reduce to 5% from 70% for a domestic dwelling over the coming 7–10 years with improved legislation. The effective implementation of policies such as the Energy Performance Building Directive could see a significant reduction in operational energy, whereas embodied energy could increase to almost 400% of the operational energy in the near future (Cabeza et al. 2013). Thus, embodied energy and CO_2 are quite important in the environmental building assessment. Cabeza et al. (2014) provide an extensive review of studies about the selection of environmentally preferred materials for buildings.

14.5   EMBODIED ENERGY AND CO_2 ANALYSIS: COMPUTATIONAL STEPS

In order to facilitate understanding the steps required for the computation of embodied energy and CO_2, it is important to revisit the underlying mathematical model often used:

$$EE_k = \sum_{k=1}^{n} (1 + \xi_k) \cdot Q_k \cdot I_k \tag{14.1}$$

$$EC_k = \sum_{k=1}^{n} (1 + \xi_k) \cdot Q_k \cdot I_k \tag{14.2}$$

where
   EE_k and EC_k are the embodied energy and embodied CO_2 of material type k with units MJ and
      $kgCO_2$, respectively.
   ξ_k is the waste factor (dimensionless) of material type k.
   Q_k is the total functional quantity of material.
   I_k is the embodied energy factor or embodied CO_2 factor with units MJ/functional unit and
      $kgCO_2$/functional unit of the material, respectively.

In the computation of environmental emissions, the British Standard recommends that the data sources and key assumptions in the use of data be stated explicitly. This is to facilitate the verification of the quantification of environmental emissions at the end of the assessment.

There are eight steps involved in the computation of embodied energy and CO_2. First, a clear idea and knowledge of the object of assessment needs to be fully understood. In the case of this study, the building is the object of assessment. Secondly, a system boundary needs to be established. According to ISO 14040, the system boundary is a set of criteria specifying which unit processes are part of a product system (EN ISO 14040 2006). It also describes the limits of what is included or not included in the assessment of the whole life cycle for a new building or any remaining cycle stages for the existing building. Although there is a lack of consensus as to the different types of phases in a building life cycle, generally, the product phase (raw materials supply, transport, and manufacturing), construction phase (transport and construction–installation onsite processes), use phase (maintenance,

---

[*] Embodied carbon is often confused with embodied CO_2. In this study, we strictly stick to embodied CO_2, and embodied carbon can be computed from embodied CO_2 using molar mass relationships of the constituent elements.

repair and replacement, refurbishment, and operational energy use: heating, cooling, ventilation, hot water, and lighting and operational water use), and end-of-life phase (deconstruction, transport, recycling/re-use, and disposal) (Blengini and Di Carlo 2010) are quite common and encompass most other classifications. In the case of this study, the building construction phase will be considered. In practice, two major categories of activities that have impacts on the environment occur in this phase. The first category consists of activities that are aimed at erecting the building. The activities are the site installation, the transportation of plants/equipment, plant/equipment use, and the use of temporary materials. The second category is induced activities that occur during the erection of the building. Depending on whether the building is prefabricated or not, onsite construction waste may be assumed to be negligible or considered. Another constraint is the fact that construction waste data are scarce and often assumed, based on the experience of professionals, to be a percentage of the total material consumed in the building. Based on this challenge, construction waste is not considered in this study. Hence, the waste factors of Equations 14.1 and 14.2 are assumed to be zero. Also, by considering the onsite construction phase, one assumes that the product phase is included. The product phase includes the extraction of the material from its original sources: the transportation to the production unit in which manufacturing into different products is undertaken. The manufactured components are then transported to the construction site where the construction phase begins. Thus, the physical boundary considered is from cradle to gate. This choice has a major advantage, in that it provides the possibility to directly use the inventory of carbon and energy for construction materials in the United Kingdom (Hammond and Jones 2008). The latest version is ICE V2.0 of January 2011. With respect to the boundary on processes or activities, material extraction, material transportation, and material transformation and transportation of components from the gate to the construction site have been considered in this project. The third step is to identify the components or parts of the building to be assessed. For example, windows and doors could be the components of the object to be assessed. Fourth, the data to be used in the computation need to be identified. This requires an understanding of the variables in Equations 14.1 and 14.2. The first data refer to the dimensions (e.g. volumes). This will be taken from architects' specifications or generated from BIM software. In this study, the latter was chosen to benefit from the automatic generation of volumes. The second data types are densities of materials, which can be obtained from some established sources. In this study, and for consistency purposes, the density of materials will be obtained from the Bath ICE V2.0. The fifth step consists of determining the quantities of construction materials upon which emission calculations are based. With regard to the mass or quantity of the material, it is advisable to revisit classical physics, which clearly defines the relationship between mass and quantity and volume and density. The quantities are multiplied by the densities of the respective materials to generate their masses or quantities. Sixth, given that the object of assessment is in the United Kingdom, the Bath ICE V2.0 was chosen as the appropriate emission factor inventory. Seventh, based on the Bath ICE V2.0, the relevant emission factors are extracted. Eighth, different data values are inputted into Equations 14.1 and 14.2 to generate emissions for each component and a sum of the different components undertaken to determine the total emission of the building. To facilitate understanding, the computation steps are summarized in Figure 14.3.

14.6 IMPLEMENTATION IN REVIT

Revit stands for REVise InsTantly (Stine 2012). It is a BIM software package for architects, structural engineers, MEP engineers, designers, and contractors. It is used to design a building and structure in three dimensions and allows users to enrich the 3D model with rich data that can be used for other applications. Examples are data such as the unit cost material and embodied energy intensity for computing model cost and embodied energy, respectively.

The current version of Revit comes as a suite in three components. The three components are structural, architectural, and systems (MEP) and are installed once with the possibility of switching between the three. The latest version is Revit 2016.

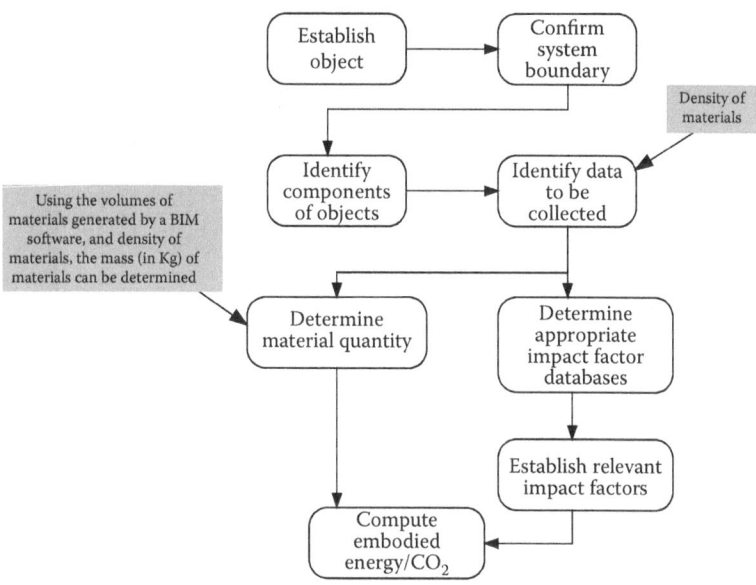

FIGURE 14.3   Steps for the computation of embodied energy and CO_2.

14.6.1 Description of the Model

The model is a hypothetical building assumed to be in the United Kingdom. This choice has also been made because it is a model that has been used for teaching undergraduate and postgraduate students in the Department of Real Estate and Construction, Oxford Brookes University. Recently, the model was used in teaching professionals in the construction industry as part of the FutureFit Built Assets project (http://www.secbe.org.uk/futurefit). As a result, the authors are conversant with the model. The model is developed using the steps described in Krygiel and Vandezande (2014). The model in Revit 2016 is presented in Figures 14.4 and 14.5.

14.6.2 Quantity Takeoff

Takeoff is a part of the cost-estimating process. The process involves "taking off" or extracting real-world quantities from project design data in order to prepare a list of resources (materials, equipment, labor, etc.) that are needed to build the project. Examples of takeoffs include determining the number of light fixtures needed in a building plan or the amount of concrete that is needed for the piling. The quantity takeoff (QTO) steps are as follows (Figure 14.6).

- QTO or material takeoff
- Go to View > Schedules > Material takeoff
- QTO or material takeoff
- Go to View > Schedules > Material takeoff

As indicated in Figure 14.7, the term "multi" implies that QTO is being taken on the whole building. If QTO is to be taken on any component (e.g. door), then it should be selected under the category field. By clicking on OK on Figure 14.7, the software provides the opportunity to select the parameters to be included in the QTO output. The output of the operation in Figure 14.8 is presented in Figure 14.9.

The computation of embodied energy and CO_2 depends on the mass of the material. Mass is defined as a new parameter in the program. However, in reality, it is difficult to quantify the actual

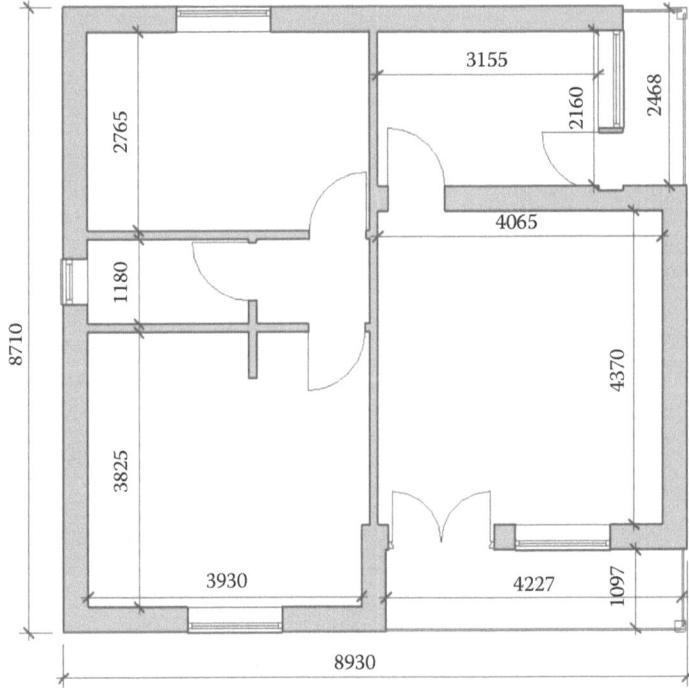

FIGURE 14.4   Floor plan of the hypothetical case study model.

weight of each element and to input the data into the model. Instead of direct input of the mass of elements, it is recommended to input the densities of the element materials and to calculate the mass using the formula: mass = density × volume.

Based on Figures 14.8 and 14.9, it can be noted that mass (kg), embodied energy, and CO_2 intensity parameters required for the computation of embodied energy and CO_2 are not available as one of the fields. Before discussing how these parameters will be entered into the Revit system, it is important to define the different types of parameters allowable in Revit. They are shared, system, project, and family parameters.

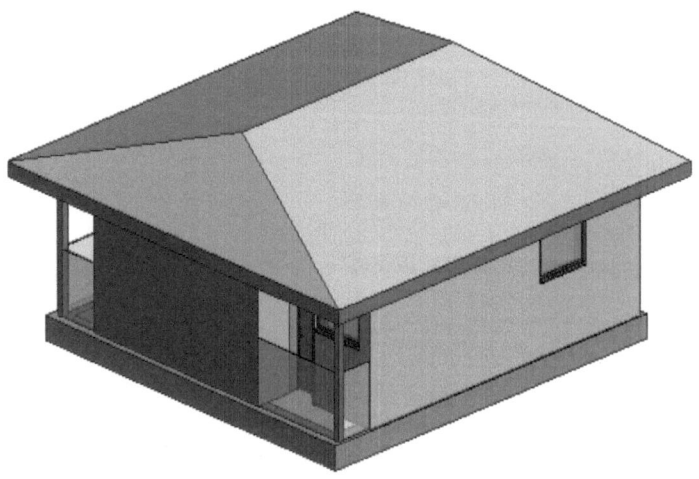

FIGURE 14.5   3D view of the case study model.

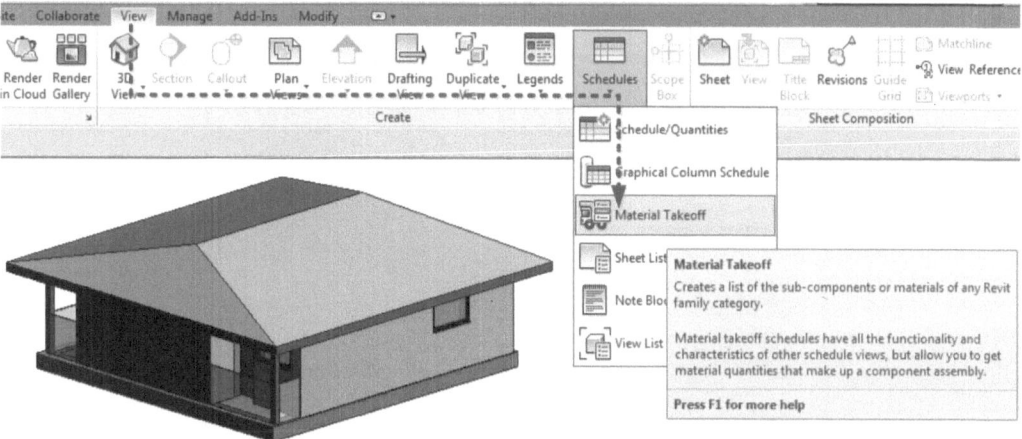

FIGURE 14.6   Preliminary QTO process.

Shared parameters are parameter definitions that can be used in multiple families or projects. Consequently, they are stored in a file independent of any family file or Revit project; this allows you to access the file from different families or projects. System parameters are preexisting parameters defined inside of every family. Most system parameters can appear in Schedules and Tags. New parameters such as sustainability indicators can be created and stored as project parameters that can be used to determine the level of sustainability of a building project (Wong and Kuan 2014). However, this is limited to a specific project and cannot be reused in other projects. Family parameters are parameters created in the Family Editor mode, which cannot appear in Schedules or Tags. Family parameters are applicable to the family for which they were created. For the purposes of this, we do not plan to use parameters created for the case study model on other models; hence, only project parameters will be created.

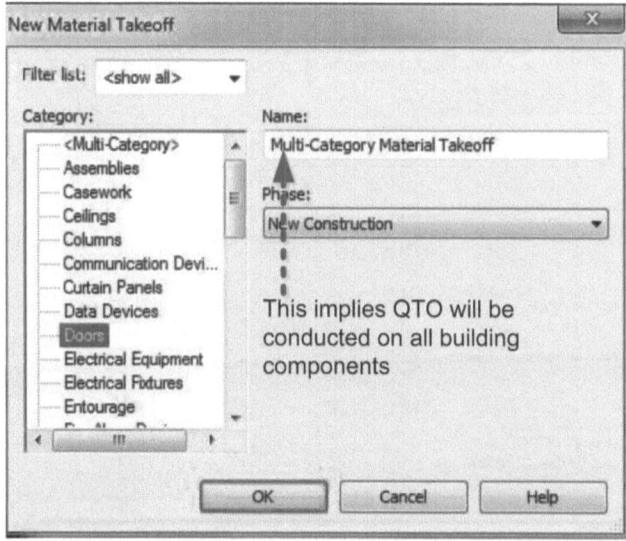

FIGURE 14.7   Building component(s) selection for QTO.

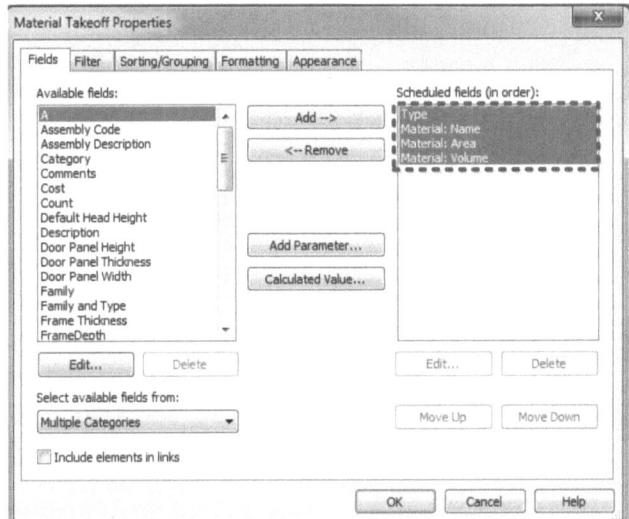

FIGURE 14.8 Selection of fields for QTO.

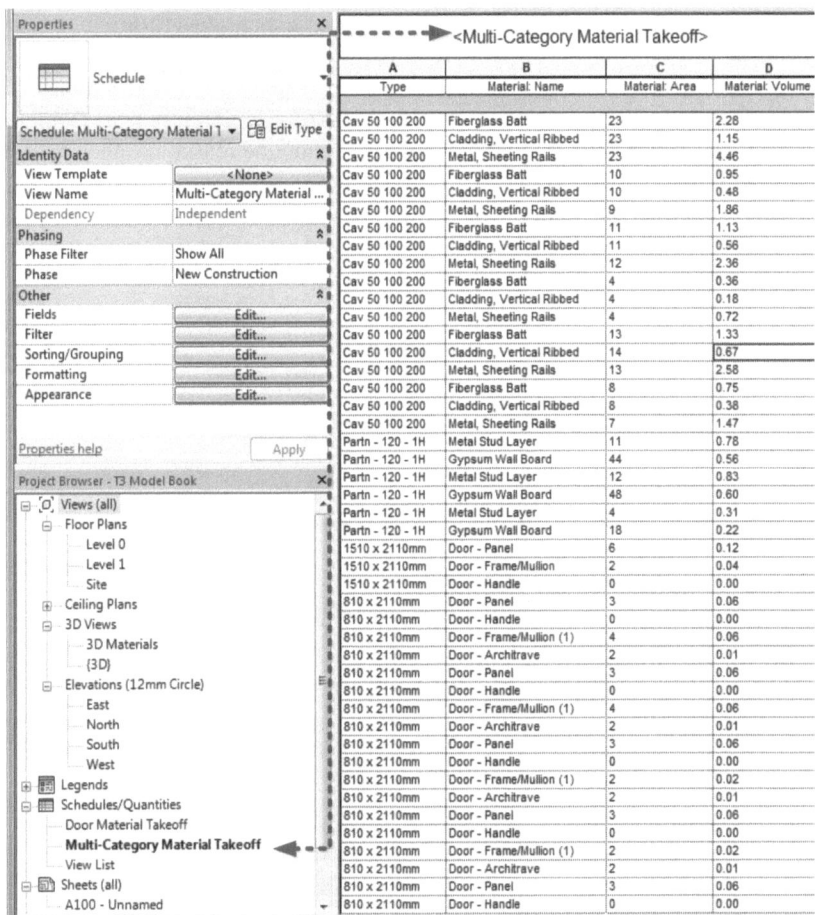

FIGURE 14.9 QTO output.

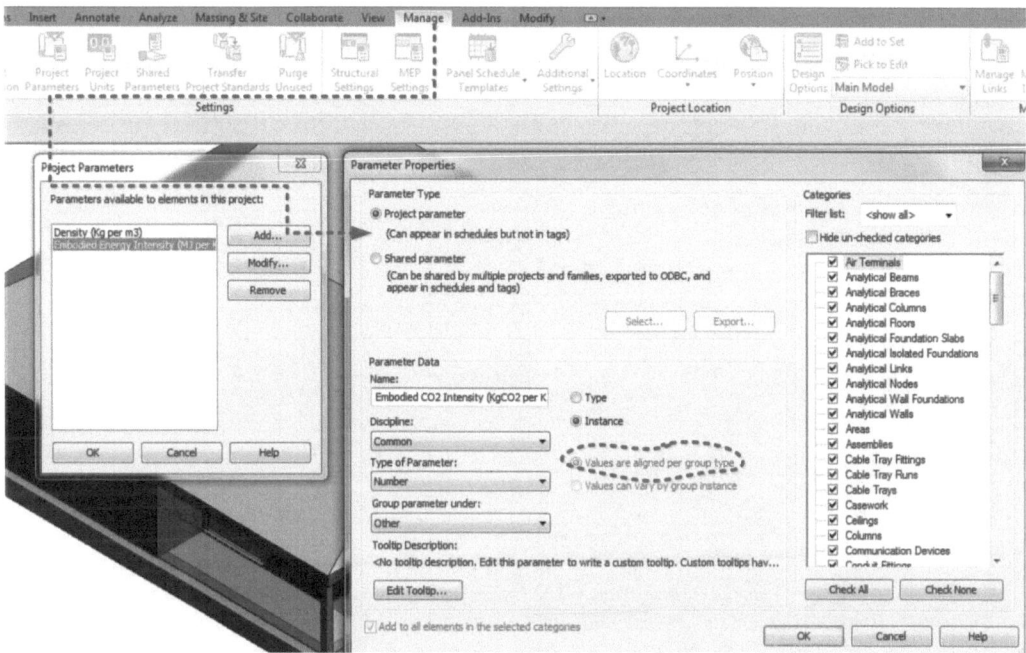

FIGURE 14.10 Creating project parameters.

The following project parameters were added to the building materials:

- Density of materials
- Mass of materials
- Embodied energy intensity of materials from Bath ICE (MJ/kg)
- Embodied CO_2 intensity of materials from Bath ICE (kgCO$_2$)
- Construction waste coefficient (%)

The process of creating the project parameters is indicated in the screenshot in Figure 14.10. After creating the project parameters, we will now focus on creating a formula that calculates the mass of each material in kilograms. It is important to note that although mass (kg), embodied energy (MJ), and embodied CO_2 (kgCO$_2$) can be created as project parameters, it is preferable to create them as dependent variables. This implies creating them as "calculated values."

Figure 14.11 depicts the modeling of the formula for computing the mass of any given material component of the building model. Similarly, the formulae for embodied energy and CO_2 are created using the calculated value function in Revit, as shown in Figure 14.11.

Densities, embodied energy intensities, and embodied CO_2 intensities of various materials are taken from established repositories and edited in the Schedule interface of Figure 14.12. In this case, the Bath ICE was used. As an example, the density, embodied energy, and embodied CO_2 values of 10 kg/m^3, 28 MJ/kg, and 1.54 kgCO$_2$/kg, respectively, were taken from the Bath ICE V2.0 and edited in the Material schedule, as depicted in Figure 14.13.

14.7 EXPLOITING MODEL IN REVIT FOR DECISION-MAKING PURPOSES

As mentioned earlier, specialized BIM software for applications can be efficient for specific tasks (e.g. cost energy analysis) for which they were developed. However, their limitation is that they depend on the complete and perfect design of model in other BIM software programs (e.g. Revit).

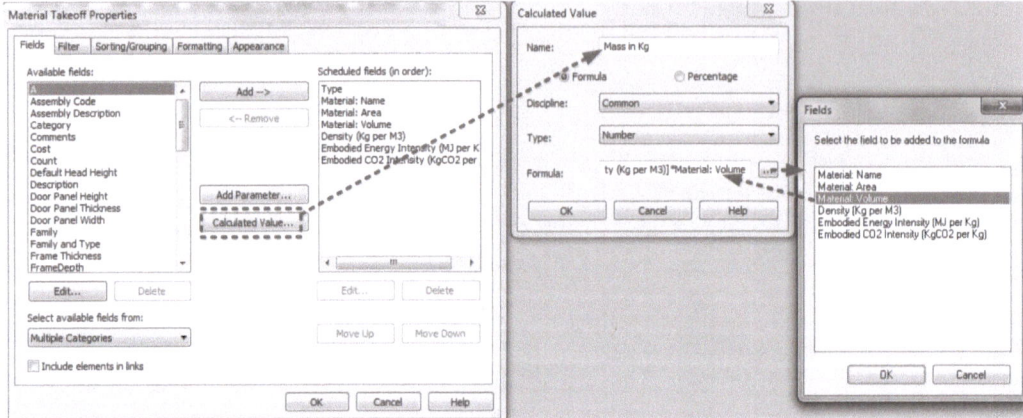

FIGURE 14.11 Modeling formulae in Revit.

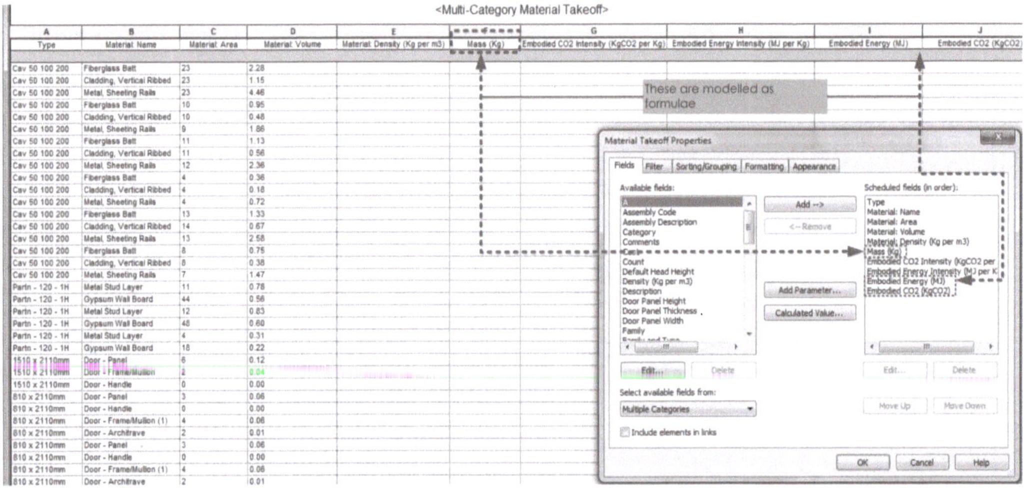

FIGURE 14.12 Output of project parameters.

If changes are to be made, then they have to be made in the original software environment in which the design has been carried out before being re-exported. Even in circumstances in which the designed model is perfect, in order to make informed decisions about the use of alternative materials, different material systems cannot be iterated in real time in the analysis BIM software. The key question therefore is: "How can a model be altered in a BIM software environment to meet certain sustainability requirements?" To explore this question further, this section illustrates some typical scenarios.

14.7.1 Scenario 1: Model Is Complete and Perfect

In this scenario, it is assumed that the building model is complete and the interest is only in the investigation of the impact of alternative building materials. This can be performed by replacing the selected component in the model by the desired alternative in the family. If it does not already exist, it can be created. Once this is done, then the embodied energy and the CO_2 analysis can be conducted. The steps in this scenario are depicted in Figure 14.13.

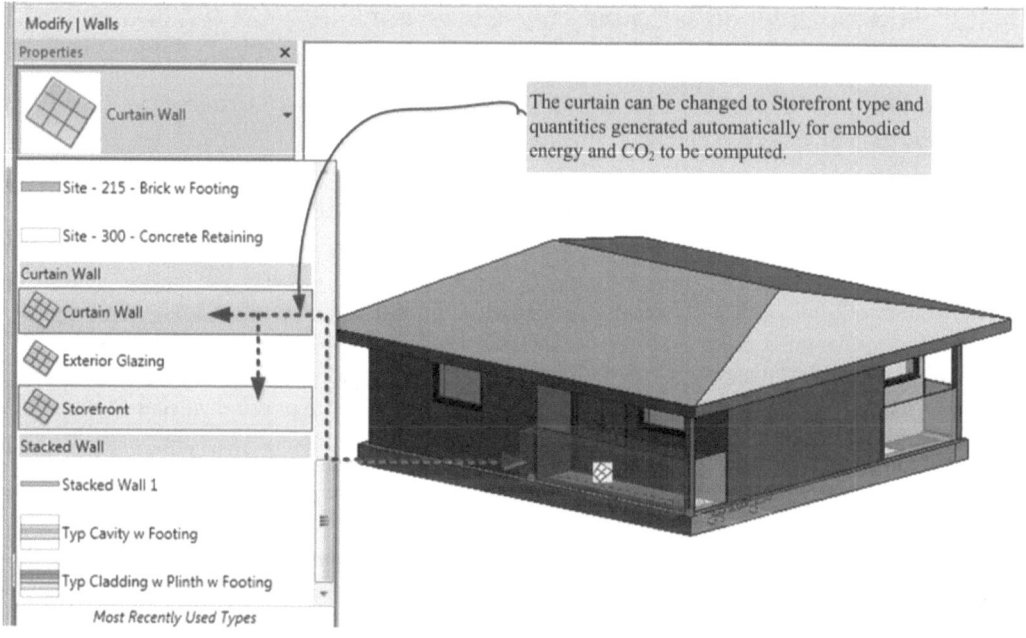

FIGURE 14.13   Changing components by replacing with preferred available types.

14.7.2   Scenario 2: Model Is Complete and Minor Changes Are to Be Made

In this scenario, it is assumed that the model is complete, and embodied energy and CO_2 analyses have been performed but there is a need to delete certain building components. This can be done in the spreadsheet (e.g. Figure 14.12) that contains the text data. Whatever is deleted in this interface is deleted in the geometric model. The steps in this scenario are presented in Figure 14.14.

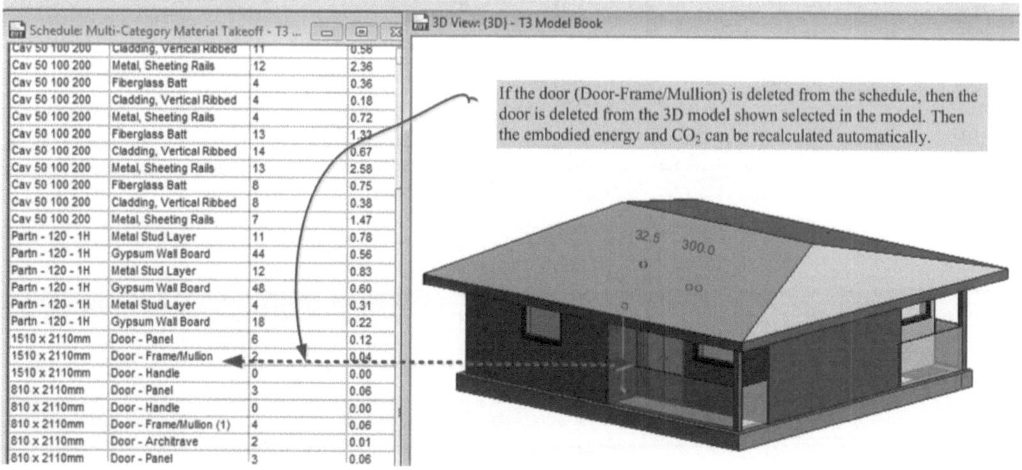

FIGURE 14.14   Changing quantities or replacing components through the material schedule.

14.7.3   SCENARIO 3: MODEL IS COMPLETE AND COMPONENT INFORMATION NEEDS TO BE CHANGED

There are two ways to change component information in this scenario. Changes can be effected through the properties palette and the Edit Type function.

14.7.3.1   Carrying Out Changes through the Properties Palette

Where component information such as density, unit embodied energy, or CO_2 emission values needs to be changed, the properties palette provides the required access to carry out desired alterations. As illustrated in Figure 14.15, the various properties' information becomes active on the properties palette when an element of interest is selected. Designers can then edit the particular embodied CO_2 information specified. On completion of all desired alterations, the overall building embodied CO_2 value can then be recomputed. Note that the edited information of a particular component applies to the component only and not to the family type.

14.7.3.2   Carrying Out Changes through the Edit Type Function

The Edit Type function also provides access to selected component properties and to their family type (Figure 14.16). When a component is selected, the Edit Type function (if clicked) brings up

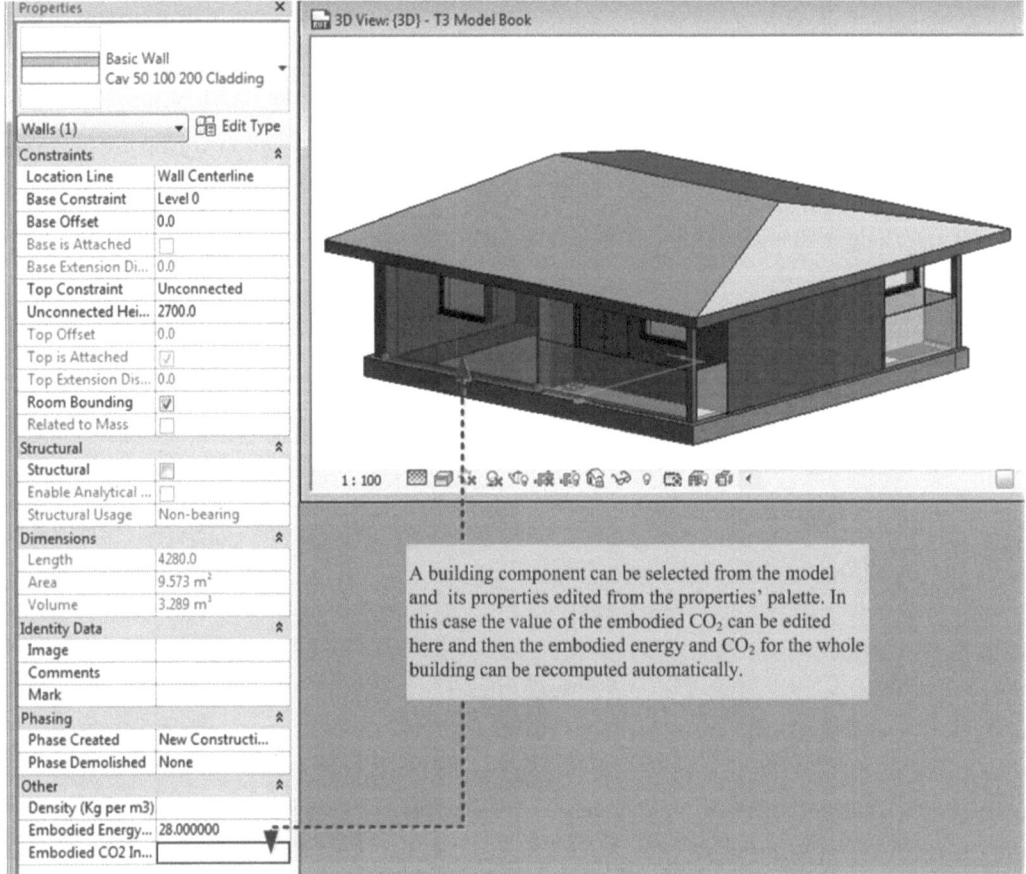

FIGURE 14.15   Changing material properties through the Properties Palette.

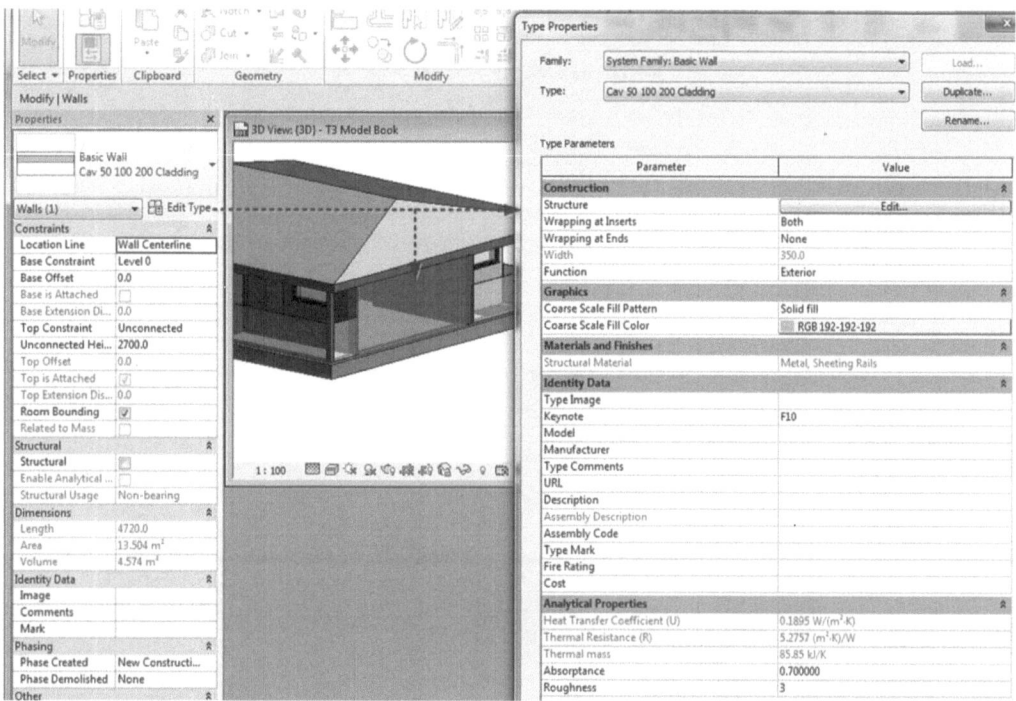

FIGURE 14.16 Changing material properties through the Edit Type function.

Type Properties information where Type parameters can be edited. Such edited information will be applicable to all components within the family type.

14.8 IMPLEMENTATION IN MS EXCEL

MS Excel is one of the oldest and popular software programs used in the construction industry. In order to perform embodied energy and CO_2 in MS Excel, based on BIM, the building must be modeled first in BIM software, in this case Revit. The QTO operations must be performed as depicted in Figure 14.9. Based on Figure 14.9, the QTO schedule is exported into MS Excel. The steps are described subsequently.

Step 1: With the QTO sheet still displayed go to

- Application Menu > Export > Reports > Schedule

The screenshot of these steps is presented in Figure 14.17.

Step 2: Saving the QTO schedule

- Based on the last operation of the preceding step, save the file with any name of your choice. In this case, we maintain the name QTO sheet generated in Revit, that is, "Multi-Category Material Takeoff."
- Save it as a Delimited text (.txt) anywhere in the computer system.
- Click on Save as indicated in Figure 14.18, and the resulting window will be as shown in Figure 14.19.
- Click on OK as indicated in Figure 14.19.

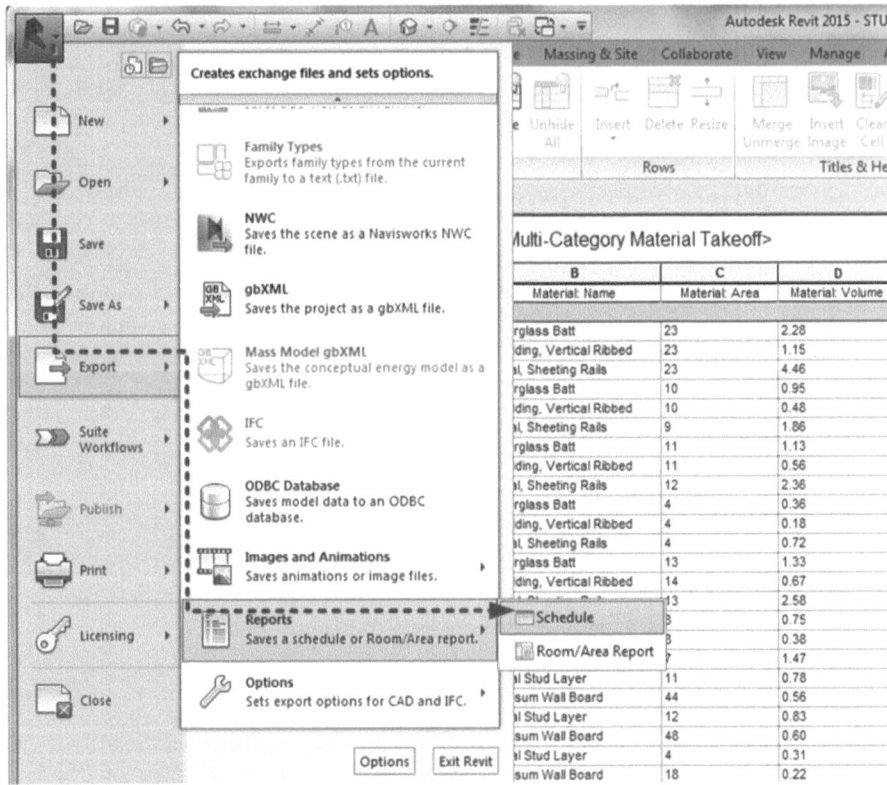

FIGURE 14.17   Preliminary export process to MS Excel.

Step 3: Opening the saved file in MS Excel

- Go to Start on your computer > Go to All Programmes > Microsoft Office > Click on Microsoft Excel 2010.
- With an empty MS Excel sheet opened, Click Open > Change file type to "Text" and then select the file "Multi-Category Material Takeoff," as shown in Figure 14.20.

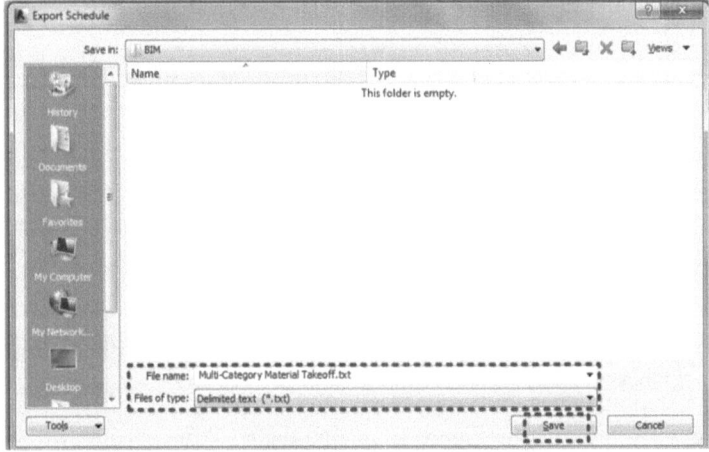

FIGURE 14.18   Saving QTO as a delimited text file.

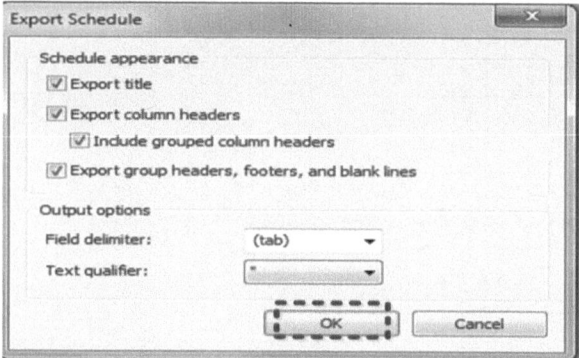

FIGURE 14.19   Choosing output options.

Step 4: Opening the delimited file

- Click on Open on the right of "Tools," as shown in Figure 14.20.
- Click on "Next" > "Next" > Next" > "Finish."
- The intermediary steps are as shown on Figure 14.21.
- The result will be as shown in Tables 14.1 and 14.2.

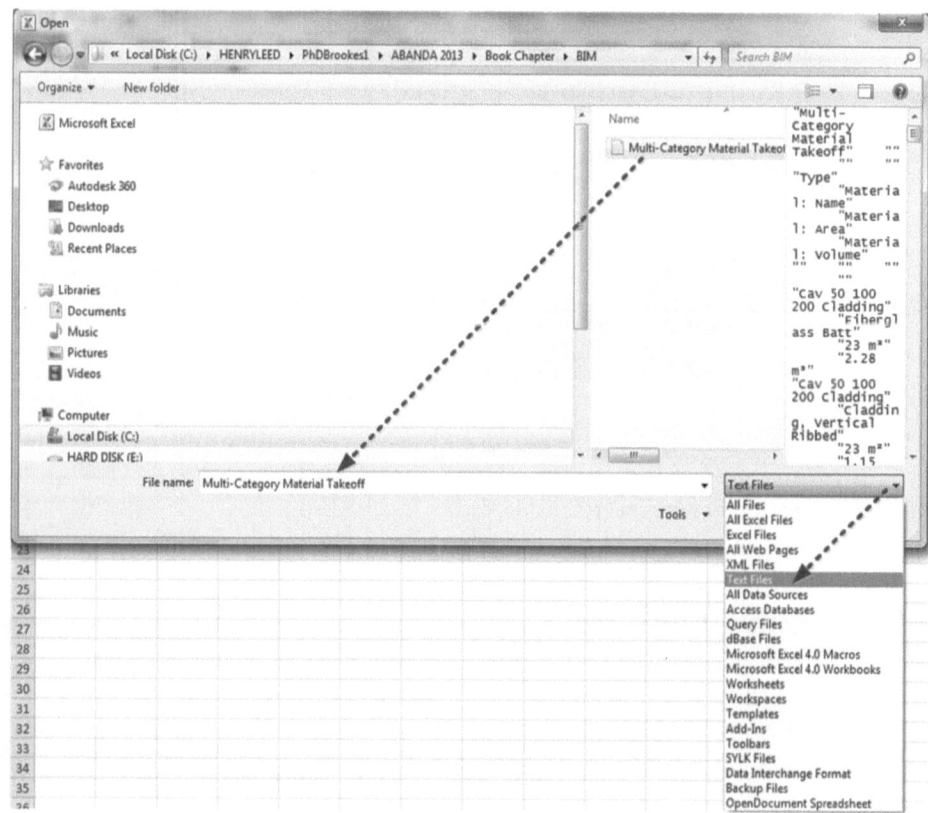

FIGURE 14.20   Specification of file type.

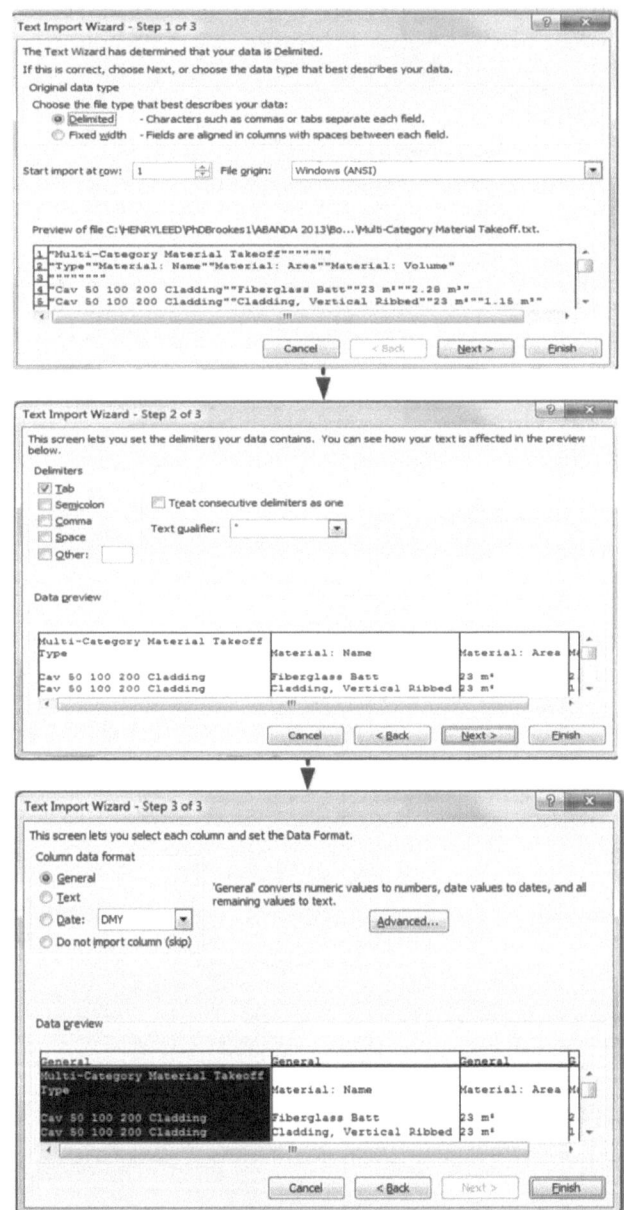

FIGURE 14.21   Final export stages of QTO to MS Excel.

14.9   RESULTS AND ANALYSIS

It is well documented that BIM software packages can be used in visualization, coordination, and storage of geometric and nongeometric data of construction projects (Eastman et al. 2008). Most of the commercially available BIM software can allow these manipulations to varying degrees of ease, detail, and completeness of information. This may also depend on the proficiency level of the person carrying out various operations. In some cases, external programs may be installed as plug-ins to carry out certain operations. For example, Revit provides an application programming interface that facilitates the bolting-on of add-in programs by independent software developers and researchers.

TABLE 14.1
Output from MS Excel

Multi-Category Material Takeoff

Type	Material: Name	Material: Area	Material: Volume
Cav 50 100 200 Cladding	Fiberglass batt	23	2.23
Caw 50 100 200 Cladding	Cladding, vertical ribbed	23	1.15
Cav 50 100 200 Cladding	Metal, sheeting rails	23	4.46
Cav 50 100 200 Cladding	Fiberglass batt	10	0.95
Cav 50 100 200 Cladding	Cladding, vertical ribbed	10	0.48
Cav 50 100 200 Cladding	Metal, sheeting rails	3	1.S6
Cav 50 100 200 Cladding	Fiberglass batt	11	1.13
Cav 50 100 200 Cladding	Cladding, vertical ribbed	11	0.56
Cav 50 100 200 Cladding	Metal, sheeting rails	12	2.36
Cav SO 100 200 Cladding	Fiberglass batt	4	0.36
Cav 50 100 200 Cladding	Cladding, vertical ribbed	4	0.13
Cav 50 100 200 Cladding	Metal, sheeting rails	4	0.72
Cav 50 100 200 Cladding	Fiberglass batt	13	1.33
Cav 50 100 200 Cladding	Cladding, vertical ribbed	14	0.67
Cav 50 100 200 Cladding	Metal, sheeting rails	12	2.53
Cav 50 100 200 Cladding	Fiberglass batt	B	0.75
Cav 50 100 200 Cladding	Cladding, vertical ribbed	S	0.38
Cav 53 100 200 Cladding	Metal, sheeting rails	7	1.47
Partn—120—l h	Metal stud layer	11	0.73
Partn—120—l h	Gypsum wall board	44	0.56
Partn—120—l h	Metal stud layer	12	0.83
Partn—120—l h	Gypsum wall board	48	0.6
Partn—120—l h	Metal stud layer	4	0.31
Partn—120—l h	Gypsum wall board	18	0.22
1510 × 2110 mm	Door—panel	e	0.12
1510 × 2110 mm	Door—frame/mullion	2	0.04
1510 × 2110 mm	Door—handle	0	0
810 × 2110 mm	Door—panel	3	0.06
810 × 2110 mm	Door—handle	0	0
810 × 2110 mm	Door—frame/mullion (1)	4	0.06
810 × 2110 mm	Door—architrave	2	0.01
810 × 2110 mm	Door—panel	3	0.06
810 × 2110 mm	Door—handle	0	0
810 × 2110 mm	Door—frame/mullion (1)	4	0.06
810 × 2110 mm	Door—architrave	2	0.01
810 × 2110 mm	Door—panel	3	0.06
810 × 2110 mm	Door—handle	0	0
810 × 2110 mm	Door—frame/mullion (1)	2	0.02
810 × 2110 mm	Door—architrave	2	0.01

For sustainability analysis, this chapter explored the option of creating additional parameters within a BIM software package and utilizing embedded BIM capabilities to estimate embodied energy and CO_2 measures. The study used Revit, a BIM software package in decision making to determine the sustainability performance of alternative building material systems within the modeling environment and with the aid of spreadsheets.

TABLE 14.2
Modified Output in MS Excel

Multi-Category Material Takeoff

Type	Material: Name	Material: Area	Material: Volume	Density (kg/m³)	Mass (kg)	Embodied Energy (EE) Intensity (MJ/kg)	Embodied CO$_2$ (EC) Intensity (kgCO$_2$/kg)	EE (MJ)	EC Emissions (kgCO$_2$)
Cav 50 100 200 Cladding	Fiberglass batt	23	2.28						
Cav 50 100 200 Cladding	Cladding, vertical ribbed	23	1.15						
Cav 50 100 200 Cladding	Metal, sheeting rails	23	4.46						
Cav 50 100 200 Cladding	Fiberglass batt	10	0.95						
Cav 50 100 200 Cladding	Cladding, vertical ribbed	10	0.48						
Cav 50 100 200 Cladding	Metal, sheeting rails	9	1.86						
Cav 50 100 200 Cladding	Fiberglass batt	11	1.13						
Cav 50 100 200 Cladding	Cladding, vertical ribbed	11	0.56						
Cav 50 100 200 Cladding	Metal, sheeting rails	12	2.36						
Cav 50 100 200 Cladding	Fiberglass batt	4	0.36						
Cav 50 100 200 Cladding	Cladding, vertical ribbed	4	0.18						
Cav 50 100 200 Cladding	Metal, sheeting rails	4	0.72						
Cav 50 100 200 Cladding	Fiberglass batt	13	1.33						
Cav 50 100 200 Cladding	Cladding, vertical ribbed	14	0.67						
Cav 50 100 200 Cladding	Metal, sheeting rails	13	2.58						
Cav 50 100 200 Cladding	Fiberglass batt	8	0.75						
Cav 50 100 200 Cladding	Cladding, vertical ribbed	8	0.38						
Cav 50 100 200 Cladding	Metal, sheeting rails	7	1.47						
Partn—120—1 h	Metal stud layer	11	0.78						
Partn—120—1 h	Gypsum wall board	44	0.56						
Partn—120—1 h	Metal stud layer	12	0.83						
Partn—120—1 h	Gypsum wall board	48	0.6						
Partn—120—1 h	Metal stud layer	4	0.31						
Partn—120—1 h	Gypsum wall board	18	0.22						
1510 × 2110 mm	Door—panel	6	0.12						

(Continued)

TABLE 14.2 (Continued)
Modified Output in MS Excel

		Multi-Category Material Takeoff							
Type	Material: Name	Material: Area	Material: Volume	Density (kg/m³)	Mass (kg)	Embodied Energy (EE) Intensity (MJ/kg)	Embodied CO$_2$ (EC) Intensity (kgCO$_2$/kg)	EE (MJ)	EC Emissions (kgCO$_2$)
1510 × 2110 mm	Door—frame/mullion	2	0.04						
1510 × 2110 mm	Door—handle	0	0						
810 × 2110 mm	Door—panel	3	0.06						
810 × 2110 mm	Door—handle	0	0.01						
810 × 2110 mm	Door—frame/mullion (1)	4	0.06						
810 × 2110 mm	Door—architrave	2	0.01						
810 × 2110 mm	Door—panel	3	0.06						
810 × 2110 mm	Door—handle	0	0						
810 × 2110 mm	Door—frame/mullion (1)	4	0.06						
810 × 2110 mm	Door—architrave	2	0.01						
810 × 2110 mm	Door—panel	3	0.06						
810 × 2110 mm	Door—handle	0	0						
810 × 2110 mm	Door—frame/mullion (1)	2	0.02						
810 × 2110 mm	Door —architrave	2	0.01						
810 × 2110 mm	Door—panel	3	0.06						

When manipulating a model within the BIM software environment for the purposes of calculating embodied energy and CO_2, one needs to be conversant with the various steps of defining and adding the new parameters into the model in Revit. The first decision to be made is whether to add the new parameters (embodied energy and CO_2) as shared parameters for all projects or as project parameters applicable to specific projects (Figure 14.10). Another hurdle in the process is in the definition of the formula to automatically compute respective component values into schedule tables. There appears to be some limitation to the level of sophistication a formula can be defined and the units available in the Type for the Calculated Value (Figure 14.11). For example, the closest unit to embodied CO_2 unit ($kgCO_2$) for the Calculated Value is "Mass Density." The easiest way to get around this issue is to specify the Type of the Calculated Value as Number. The actual unit can then be specified as part of the Field, as seen in Figure 14.12. As one of the valuable BIM features, manipulation of models in Revit allows real-time computation of the total embodied value when desired changes are made to explore varying scenarios of combining alternative materials. Even in situations in which components are to be deleted or added, the total embodied energy or CO_2 of the model can automatically be recomputed. However, the additional challenge is in the rigor of adding new parameters for each building component and/or new building project. It is therefore advisable to specify such parameters as shared parameters to avoid repeating the process for other projects, although not required for this study.

In the case in which the implementation of embodied energy and CO_2 is performed using a spreadsheet such as Excel, the limitations relate mainly to the fact that the BIM cannot be updated automatically with changes made on the spreadsheet. This will have to be manually effected in the model as there is yet to be a direct two-way communication between Excel and BIM environments. This may turn out to be cumbersome if the parameters of each building component will have to be altered in the model. In contrast, spreadsheets are endowed with the provision of extensive mathematical manipulations. As such, they can be useful for producing quick results of embodied energy and CO_2 measures once the required information has been exported.

Thus, this study demonstrated that changes could be effected through the geometrical model and/or material schedule view and the quantities automatically adjusted with respect to the changes. In contrast, in exporting the model to MS Excel, it was possible to perform computations. However, a key limitation here is that if changes are to be made, this has to be done in a BIM environment (i.e., Revit in this case) before exporting to MS Excel. As such, either approaches or a combination of both could be employed by designers depending on their intentions and inherent limitations discussed.

14.10   CONCLUSION

The expectation in the AEC industry is that BIM applications should help foster collaboration among project partners, reduce project delivery time and costs, and improve overall project efficiencies including design facilitation and reduction of errors. However, there are numerous factors that come into play in the BIM implementation process. A combination of economic, social, and technology/technical know-how issues governs these factors. For instance, BIM software packages are expensive to acquire; staff need to be well trained and prepared to use such packages; and the technology is still expanding in scope through research. Sustainability assessment of buildings has been one of the active areas of such research. Thus, the growing environmental concern in the AEC industry has placed demand on the incorporation of sustainable building design principles into BIM applications. One of the key aspects being examined is on making informed decisions about the use of different materials in building projects. The extent to which BIM software can be used to accurately compare different building materials for the purpose of sustainability analysis has not been fully explored. As such, this chapter detailed two approaches to using BIM to perform sustainability analysis based on embodied energy and CO_2 measures of building components. One approach explored manipulating a model within the Revit BIM software environment by defining and adding new component parameters into the model in Revit. The other approach entailed exporting pre-organized information in a BIM software environment to a spreadsheet program (Excel) for the completion of the computation process.

The study demonstrates that either approaches or a combination of both can be used to achieve desired intentions but not without limitations. The option of carrying out all the analyses in a BIM software package has the advantage of automatically synchronizing all alterations and modifications of the model in real time. However, limitations exist with the level of formulae sophistication that can be embedded into the model. When exported to Excel to complete the sustainability analysis process, these limitations are outrightly reversed. Thus, the experience from this study corroborates the conclusion that BIM technology provides the construction industry with the opportunity to make the best use of building data to capture complete design and project information.

REFERENCES

Abanda, F., Zhou, W., Tah, J., and Cheung, F. 2013. Exploring the relationships between linked open data and building information modelling. Proceedings of the Sustainable Building Conference, Coventry University, UK, July.

Abanda, F. H., Elambo, N. G., Tah, J. H. M., Ohandjia, E. N. F., and Manjia, M. M. 2014. Embodied energy and CO_2 analyses of mud–brick and cement–block houses. *AIMS Energy* 2(1): 18–40.

Abanda, F. H. and Tah, J. H. M. 2014. Free and open source building information modelling for developing countries. ICT for Africa 2014 Conference, Yaoundé, Cameroon, October 1–4, 2014.

Acar, E., Kocak, I., Sey, Y., and Arditi, D. 2005. Use of information and communication technologies by small and medium-sized enterprises (SMEs) in building construction. *Construction Management and Economics* 23(7): 713–722.

Alwan, Z. and Jones, P. 2014. The importance of embodied energy in carbon footprint assessment. *Structural Survey* 32(1): 49–60.

Ashcraft, H. W. 2008. Building information modeling: A framework for collaboration. *Construction Law* 28: 5.

Azhar, S. 2011. Building information modeling (BIM): Trends, benefits, risks, and challenges for the AEC industry. *Leadership and Management in Engineering* 11(3): 241–252.

Benjaoran, V. 2009. A cost control system development: A collaborative approach for small and medium-sized contractors. *International Journal of Project Management* 27(3): 270–277.

BIM-IWG. 2011. BIM management for value, cost and carbon improvement. A report for the Government Construction Client Group, BIM Working Party Strategy Paper, BIM Industry Working Group (BIM-IWG).

Blengini, G. A. and Di Carlo, T. 2010. The changing role of life cycle phases, subsystems and materials in the LCA of low energy buildings. *Energy and Buildings* 42(6): 869–880.

Bragança, L., Vieira, S. M. and Andrade, J. B. 2014. Early stage design decisions: The way to achieve sustainable buildings at lower costs. *The Scientific World Journal* 2014.

BS EN 15643-1. 2010. Sustainability of construction works—Sustainability assessment of buildings: General framework, British Standard. BS EN 15643-1:2010.

Bynum, P., Issa, R. R., and Olbina, S. 2013. Building information modeling in support of sustainable design and construction. *Journal of Construction Engineering and Management* 139(1): 24–34.

Cabeza, L. F., Barreneche, C., Miró, L. et al. 2013. Low carbon and low embodied energy materials in buildings: A review. *Renewable and Sustainable Energy Reviews* 23: 536–542.

Cabeza, L. F., Rincón, L., Vilariño, V., Pérez, G., and Castell, A. 2014. Life cycle assessment (LCA) and life cycle energy analysis (LCEA) of buildings and the building sector: A review. *Renewable and Sustainable Energy Reviews* 29: 394–416.

Cheung, F. K., Rihan, J., Tah, J., Duce, D., and Kurul, E. 2012. Early stage multi-level cost estimation for schematic BIM models. *Automation in Construction* 27: 67–77.

CIC. 2013. Building information model (BIM) protocol. Construction Industry Council (CIC), UK.

Dawood, S. and Sikka, S. 2006. The value of visual 4D planning in the UK construction industry. *Intelligent Computing in Engineering and Architecture Lecture Notes in Computer Science* 4200: 127–135.

Dibrell, C., Davis, P. S., and Craig, J. 2008. Fueling innovation through information technology in SMEs. *Journal of Small Business Management* 46(2): 203–218.

Eadie, R., Browne, M., Odeyinka, H., McKeown, C., and McNiff, S. 2013. BIM implementation throughout the UK construction project lifecycle: An analysis. *Automation in Construction* 36: 145–151.

Eastman, C., Teicholz, P., and Sacks, R. 2008. *BIM Handbook: A Guide to Building Information Modelling for Owners, Manager, Designers, Engineers, and Contractors.* USA: John Wiley & Sons, Inc.

EN ISO 14040. 2006. Environmental management—Life cycle assessment—Principles and framework. International Standardization Organization, EN ISO 4040:2006.

Fiksel, J. 2003. Designing resilient, sustainable systems. *Environmental Science and Technology* 37(23): 5330–5339.

Gledson, B., Henry, D., and Bleanch, P. 2012. Does size matter? Experiences and perspectives of BIM implementation from large and SME construction contractors In: 1st UK Academic Conference on Building Information Management (BIM) 2012, 5–7 September 2012, Northumbria University, Newcastle upon Tyne, UK.

González, M. J. and García Navarro, J. 2006. Assessment of the decrease of CO_2 emissions in the construction field through the selection of materials: Practical case study of three houses of low environmental impact. *Building and Environment* 41(7): 902–909.

Hammond, G. P. and Jones, C. I. 2008. Embodied energy and carbon in construction materials. Proceedings of the Institution of Civil Engineers, ICE 161(2): 87–98.

Hannon, B. 1973. The structure of ecosystems. *Journal of Theoretical Biology* 41(3): 535–546.

Harries, H. 2007. Embodied energy: Masonry walling in South Africa—Technical Note No. 7. http://www.claybrick.org/content/07-embodied-energy-masonry-walling-south-africa (accessed May 10, 2010).

Iddon, C. R. and Firth, S. K. 2013. Embodied and operational energy for new-build housing: A case study of construction methods in the UK. *Energy and Buildings* 67: 479–488.

Khosrowshahi, F. and Arayici, Y. 2012. Roadmap for implementation of BIM in the UK construction industry. *Engineering, Construction and Architectural Management* 19(6): 610–635.

Kohler, N. and Moffatt, S. 2003. Life cycle analysis of the built environment. *Sustainable Building and Construction*, UNEP Industry and Environment: 17–21.

Krygiel, E. and Vandezande, J. 2014. *Mastering Autodesk Revit Architecture 2015: Autodesk Official Press*, John Wiley & Sons, Canada.

Kua, H. and Lee, S. 2002. Demonstration intelligent building—A methodology for the promotion of total sustainability in the built environment. *Building and Environment* 37(3): 231–240.

Lee, J. and Runge, J. 2001. Adoption of information technology in small business: Testing drivers of adoption for enterpreneurs. *Journal of Computer Information Systems* 42(1): 44–57.

Leontief, W. 1970. Environmental repercussions and the economic structure: An input–output approach. *The Review of Economics and Statistics* 25(3): 262–271.

Leontief, W. 1986. *Input–Output Economics.* Oxford University Press, UK, London.

Love, P. E. and Irani, Z. 2004. An exploratory study of information technology evaluation and benefits management practices of SMEs in the construction industry. *Information and Management* 42(1): 227–242.

McAdam, B. 2010. Building information modelling: The UK legal context. *International Journal of Law in the Built Environment* 2(3): 246–259.

Motawa, I. and Carter, K. 2013. Sustainable BIM-based evaluation of buildings. *Procedia—Social and Behavioral Sciences* 74: 419–428.

NBS. 2012. National BIM Report 2012. National Building Specification (NBS).

NFB. 2012. BIM: Ready or not? The NBF BIM—Readiness Survey 2012. The National Federation of Builders (NBF) and Room 4 Consulting Ltd.

Nguyen, T., Shehab, T., and Gao, Z. 2010. Evaluating sustainability of architectural designs using building information modeling. *Open Construction and Building Technology Journal* 4: 1–8.

Oti, A. H. and Tizani, W. 2015. BIM extension for the sustainability appraisal of conceptual steel design. *Advanced Engineering Informatics* 29(1): 28–46.

Pacheco-Torgal, F., Faria, J., and Jalali, S. 2013. Embodied energy versus operational energy. Showing the shortcomings of the Energy Performance Building Directive (EPBD). Materials Science Forum, Trans Tech Publications.

Ramesh, T., Prakash, R., and Shukla, K. 2010. Life cycle energy analysis of buildings: An overview. *Energy and Buildings* 42(10): 1592–1600.

Sartori, I. and Hestnes, A. G. 2007. Energy use in the life cycle of conventional and low-energy buildings: A review article. *Energy and Buildings* 39(3): 249–257.

Shrivastava, S. and Chini, A. 2012. Using building information modeling to assess the initial embodied energy of a building. *International Journal of Construction Management* 12(1): 51–63.

Stine, D. J. 2012. *Design Integration Using Autodesk Revit 2013.* Schroff Development Corporation, US.

Sturgis, S. and Roberts, G. 2010. *Redefining Zero: Carbon Profiling as a Solution to Whole Life Carbon Emission Measurement.* London: The Royal Institution of Chartered Surveyors.

Thormark, C. 2006. The effect of material choice on the total energy need and recycling potential of a building. *Building and Environment* 41(8): 1019–1026.

Tolman, F. P. 1999. Product modeling standards for the building and construction industry: Past, present and future. *Automation in Construction* 8(3): 227–235.

Treloar, G. J. 1997. Extracting embodied energy paths from input-output tables: Towards an input-output based hybrid energy analysis method. *Economic Systems Research* 9(4): 375–391.

Treloar, G. J., Gupta, H., Love, P. E. D., and Nguyen, B. 2003. An analysis of factors influencing waste minimisation and use of recycled materials for construction of residential buildings. *Management of Environmental Quality: An International Journal* 14(1): 134–145.

Van Gool, W. 1980. Thermodynamic aspects of energy conservation. *Energy* 5(8): 783–792.

Willcocks, L. 1992. Evaluating information technology investments: Research findings and reappraisal. *Information Systems Journal* 2(4): 243–268.

Wong, K.-D. and Fan, Q. 2013. Building information modelling (BIM) for sustainable building design. *Facilities* 31(3/4): 138–157.

Wong, J. K.-W. and Kuan, K.-L. 2014. Implementing "BEAM Plus" for BIM-based sustainability analysis. *Automation in Construction* 44: 163–175.

Section III

Carbon Footprint Assessment—
Case Studies

15 Product Carbon Footprint
Case Study of a Critical Electronic Part (Subassembly of a Product)

Winco K.C. Yung and Subramanian Senthilkannan Muthu

CONTENTS

15.1 Introduction 334
15.2 Methodology 334
    15.2.1 Scope 334
        15.2.1.1 Product System and Its Function(s) 334
        15.2.1.2 Functional Unit 334
        15.2.1.3 Data and Data Quality 334
        15.2.1.4 Cutoff Criteria and Cutoff 335
        15.2.1.5 Allocation Procedures 335
        15.2.1.6 Geographical and Time Boundary of Data 335
        15.2.1.7 System Boundary 335
        15.2.1.8 Relevant Assumptions in this Study 336
        15.2.1.9 Treatment of Electricity 337
15.3 Carbon Footprint Analysis 337
    15.3.1 Carbon Footprint Calculation 337
15.4 LCI for PCF 339
    15.4.1 LCI of "Raw Material" Stage 339
        15.4.1.1 Sources of Data 339
        15.4.1.2 Database Selection 339
    15.4.2 LCI of "Manufacturing" Stage 339
    15.4.3 LCI of "Transportation" Stage 341
15.5 PCF Analysis 341
    15.5.1 Carbon Footprint Results in Component/Activity 341
15.6 Life-Cycle Interpretation 346
    15.6.1 Results of Life-Cycle Interpretation 346
    15.6.2 Specific GHG Emissions and Removals 346
    15.6.3 Limitations 346
        15.6.3.1 Focus on a Single Environmental Issue 346
        15.6.3.2 Limitations Related to Assumptions 347
        15.6.3.3 Limitations Related to Methodology 347
    15.6.4 Disclosure and Justification of Value Choices 347
    15.6.5 Sensitivity Analysis 348
        15.6.5.1 Sensitivity Analysis of Significant Input Data in Material Stage 348
        15.6.5.2 Sensitivity Analysis of Significant Input Data in Manufacturing Stage 348
    15.6.6 Uncertainty Analysis 349
15.7 Conclusion 350
Acknowledgments 350
References 350

15.1   INTRODUCTION

Handling climate changes and cutting down carbon emissions is definitely the need of the hour. If manufacturers and industries consider this need as an opportunity to do green business rather than feeling pressured, then accepting this challenge will definitely yield them better profits and improve the global economy.

When industries understand the carbon footprint created by their products, to which they must take steps to reduce emissions, then they can both create an ecofriendly product and acquire a competitive edge over others in the market. Product carbon footprint (PCF) analysis of a liquid crystal display (LCD) panel, a semifinished product used as one of the components in a household thermometer, performed and reported here is the first of its kind in this area of research. The results will definitely help manufacturers to reduce the carbon emissions of their products.

The goal of this PCF study is to calculate the potential contribution of the chosen product, LCD module (hereafter named as "the LCD module"), to global warming expressed as grams or kilograms of CO_2 eq., by quantifying all significant greenhouse gas (GHG) emissions and removals over the product's life cycle (only "Cradle to Gate" stages are considered in this study). This report follows the standard of ISO/TS 14067: 2013 "Greenhouse gases—Carbon footprint of products—Requirements and guidelines for quantification and communication" (ISO 14067, 2013). The results, data, methods, assumptions, and life-cycle interpretations are presented in sufficient detail to allow the reader to comprehend the complexities and tradeoffs inherent in the PCF study. The PCF value is measured by quantifying the GHG emissions for the entire product life cycle expressed in carbon dioxide equivalent value (kg CO_2 eq.), as per ISO/TS 14067.

$$CF_{total} = CF_{RM} + CF_M + CF_D + CF_U + CF_{EOL}$$

where CF_{RM} is the carbon footprint of the raw material stage (kg CO_2 eq.); CF_M the carbon footprint of the manufacturing stage (kg CO_2 eq.); CF_D the carbon footprint of the distribution stage (kg CO_2 eq.); CF_U the carbon footprint of the use stage (kg CO_2 eq.); and CF_{EOL} the carbon footprint of the end-of-life stage (kg CO_2 eq.).

15.2   METHODOLOGY

15.2.1   SCOPE

15.2.1.1   Product System and Its Function(s)

The product selected for this study is an LCD panel, a semifinished product that is used as one of the components in a household thermometer. Figure 15.1 shows the outlook of the LCD model.

15.2.1.2   Functional Unit

As there is no specific product category rule (PCR) for an LCD panel, the functional unit (f.u.) of the LCD panel is referenced to the PCRs of similar electronic components in Korea, Japan, and Taiwan. The functional unit is defined as one piece of LCD module (net weight: 5.5 g) with primary (polystyrene styrofoam board) and secondary packaging (PET (Polyethylene Terephthalate) tray and carton box). The reference flow is defined as one piece of LCD module.

15.2.1.3   Data and Data Quality

15.2.1.3.1   Description of Data

The report covers "Cradle to Gate" stages of product life cycle of the LCD module. Data for raw material, manufacturing, and transportation were primarily collected by Clover Display Limited, which had financial and operational control of the unit processes. The emission factors used were as

FIGURE 15.1 LCD module.

localized as possible with geographical, technological, and time-related considerations. Table 15.1 shows the details of individual data.

15.2.1.3.2 Characterization of Data Quality

The best quality data were used to reduce bias and uncertainty as far as practically possible. Primary and secondary data were selected to meet the goal and scope of the PCF study.

15.2.1.4 Cutoff Criteria and Cutoff

No materials or processes have been omitted in this study; thus, no cutoff criteria were established.

15.2.1.5 Allocation Procedures

Allocation of emissions is based on the ratio of the manufactured quantities of the studied product and the total production in the factory in Fo Shan City from January 2013 to December 2013. The allocation was applied to the electricity, water, and consumables used in the manufacturing stage (Section 4.2).

15.2.1.6 Geographical and Time Boundary of Data

The LCD modules were manufactured in Fo Shan City from January 2013 to December 2013 and are sold to other manufacturers in China.

15.2.1.7 System Boundary

As a manufacturer, Clover Display Limited provides products to other manufacturers (B2B business model).

The system boundary for this study covers "Cradle to Gate" of the products' life-cycle stages of the LCD module, from raw material extraction, manufacturing, to product distribution stage. However, in this study, the calculation of emission in the distribution stage is not included in the final PCF results and is reported for reference only. Regarding the waste generated from the material

TABLE 15.1

Description of Individual Data

Life-Cycle Stage	Type of Data	Data
Raw material stage	Primary data (site-specific)	Amount of materials and components used
	Secondary data	Emission factors of materials and components
Manufacturing stage	Primary data (site-specific)	Electricity consumption of manufacturing process
	Secondary data	Emission factors of electricity
Transportation stage	Primary data (site-specific)	Transportation tools used
	Secondary data	Emission factors of transportation route
		Distance of transportation routes

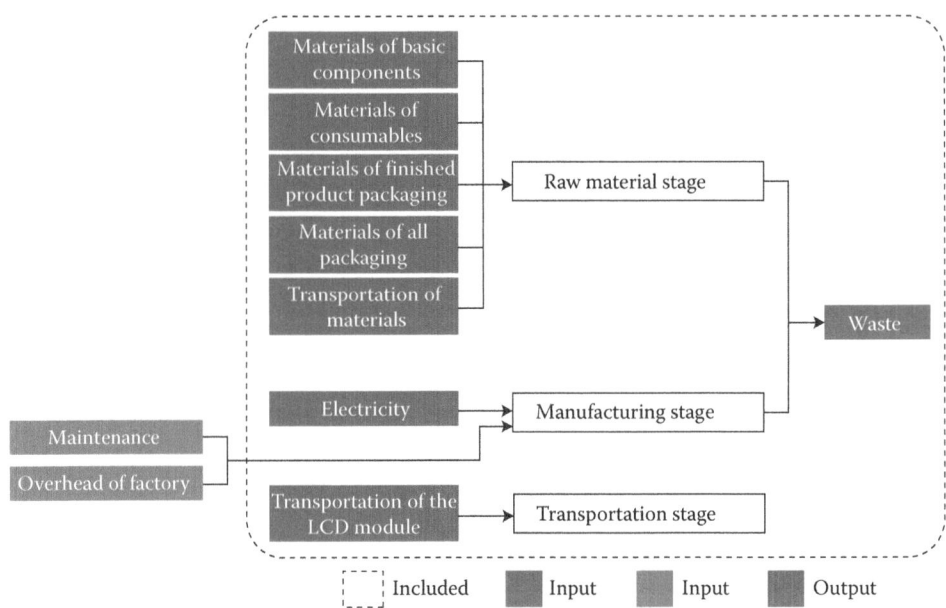

FIGURE 15.2   System boundary of carbon footprint analysis of "the LCD module."

and manufacturing stages, the transportation of the waste to the recycling site and landfill and the emission from disposal activities were included in this report, but the environmental burden from the recycling processes was not included in the calculation. Figure 15.2 shows the system boundary of this study.

All unit processes included in this study directly contributed to the end result (GHG emissions) from "Cradle to Gate" life-cycle stages. The overheads of the factory and equipment maintenance in all life-cycle stages were excluded in this study, as they were not significant and not directly related to the GHG emissions and removal of the LCD module.

15.2.1.8   Relevant Assumptions in this Study

15.2.1.8.1   Assumptions in "Raw Material" Stage

- The density of the deionized (DI) water was assumed to be 1 kg/l.
- There are differences among the weighing data for some of the components and consumables; the variances may be due to different suppliers of the components and consumables.
- Mean weight of materials for four samples is taken as the actual weight used for the production.
- The emission factor of the unknown organic chemicals was assumed on the basis of database from Swiss Ecoinvent data version 2.2—chemicals organic, at plant (GLO).
- The emission factor of the unknown inorganic chemicals was assumed on the basis of database from Swiss Ecoinvent data version 2.2—chemicals inorganic, at plant (GLO).
- The closest and most similar emission factor of materials has been chosen for the emission factor selection, for example, emission factors for butyl carbitol, divinylbenzene, and so on.
- Silver paste was assumed to have 60% of silver and 40% of organic solvent content (Electron Microscopy Sciences 2014).
- The transportation from supplier factory before loading to aircraft is left out as no specific addresses are provided and its contribution to the final carbon footprint result is rather small.
- Breach glue was assumed to have been made using 33.3% acrylic acid, 33.3% bisphenol A, and 33.3% epoxy resin.

- The distance of the high-contamination water treatment plant from the manufacturing plant was estimated by the engineer of the manufacturing plant.
- Data estimation and calculation were made by the engineer of the company.
- Supplementary data for materials and transportation activities were provided by the factory.
- All distances in each wastewater treatment transportation route taken were obtained from Google Maps.

15.2.1.8.2 Assumptions in "Manufacturing" Stage

- The energy consumption of each LCD module manufactured in the factory was assumed to be the same.
- The consumables' consumption of each LCD module manufactured in the factory was assumed to be the same.
- The annual energy consumption is 12 months' data as the LCD module was only manufactured from January 2013 to December 2013. No seasonal factor of production was included in this study.
- Electricity consumption of lift usage was included in the production.
- Electricity consumption of the canteen, residential hall for workers, and office was excluded from the electricity consumption of the LCD module production.
- All the solid wastes from manufacturing are transported to landfill disposal. All wastewater is discharged to sewage. All highly contaminated wastewater is treated in the factory wastewater treatment plant.
- Wastes from packaging materials are either taken to the local recycling plant (20 km away from factory site) or transported to landfill (120 km away from factory site). The emission from transportation of waste and disposal processes was included, but the environmental burden from the recycling processes was excluded from the calculation.
- Data estimation and calculation were made by the engineer of the company.

15.2.1.8.3 Assumptions in "Transportation" Stage

- The distribution of products in this study was assumed to be in Singapore (address: 70 Alps Avenue), as the majority of production orders comes from this customer (70.92%).
- Distances of transportation activities were obtained from Google Maps.
- Data estimation and calculation were made by the engineer of the company.

15.2.1.9 Treatment of Electricity

The emission factor of electricity used for manufacturing is from Chinese Government National Development and Reform Commission (NDRC). The emission factor for southern China in 2013 was 0.9223 kg CO_2 eq./kWh (National Development and Reform Commission 2012).

15.3 CARBON FOOTPRINT ANALYSIS

15.3.1 Carbon Footprint Calculation

According to the data provided by Clover Display Limited, calculation of the "Cradle to Gate" assessment of PCF is based on the life-cycle assessment (LCA) approach, which includes the raw material stage, the manufacturing stage, and the distribution stage. The analysis is based on the concept of LCA and ISO/TS 14067: 2013 (ISO 14067, 2013) and allows computation of the total amount of carbon emission within the selected system boundary. This PCF study, according to ISO/TS 14067: 2013, includes four phases of LCA, that is, goal and scope definition, life-cycle inventory (LCI), life-cycle impact assessment (LCIA), and life-cycle interpretation (LCI). Figure 15.3 shows the framework of LCA from ISO 14040 (ISO 14040, 2006).

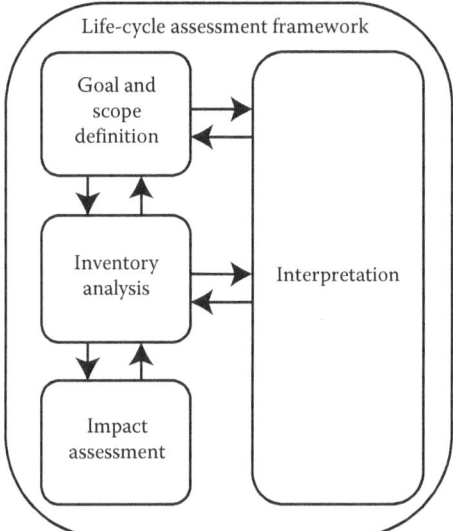

FIGURE 15.3 Framework of LCA. (Adapted from International Organization for Standardization (ISO). 2006. ISO 14040: 2006 Environmental management—Life-cycle assessment—Principles and framework.)

The unit processes comprising the product system are grouped into selected life-cycle stages in this study, that is, raw material acquisition, manufacturing, and transportation. GHG emissions and removals from the product's life cycle are assigned to the life-cycle stage in which the GHG emissions and removals occur. Partial PCFs may be added together to quantify the PCF, provided that the same methodology is used and there are no gaps and overlaps. The PCF is calculated and presented as a carbon dioxide equivalent (kg CO_2 eq.) using the relevant 100-year global warming potential (GWP 100). Findings in the impact assessment and the sensitivity and contribution analyses, performed as part of the interpretation, help identify the most relevant and contributing ("key") processes and basic flows of the system:

$$\text{Carbon Emission} = \text{Activity Data} \times \text{Emission Factor} \times \text{Global Warming Potential}$$
$$(\text{European Commission 2012})$$

The GWP should be excluded, where the unit of the "emission factor" was "kg CO_2 eq./unit." The activity data collected from each unit process were given as an input into an online PCF analyzer (G-BOM Analyzer provided by The Hong Kong Polytechnic University) and the product carbon footprint assessment was done. The working principle, the methodology behind the software, and access can be found in the following link: http://www.pctech.ise.polyu.edu.hk/ecodesign/gbom_analyzer.html.

The Green Bill of Materials (G-BOM) analyzer is online software based on the structure of the bill of materials. The G-BOM analyzer allows manufacturers to input their product information/data according to the product life cycle, which includes raw materials, manufacturing, distribution, usage, and end-of-life stages, and simply calculates the PCF using the GHG emissions on the basis of the database provided. Users can define the material names, emission factors, and quantities according to their LCI. The GHG emission database is pregrouped into five life-cycle stages. The emission factor can be selected manually in life-cycle groups or by a keyword search engine.

During the quantification of this study, values were rounded off to four decimal places in the LCI data input. The results of the PCF in the impact assessment were rounded off to two significant figures (BSI 2011).

15.4 LCI FOR PCF

15.4.1 LCI OF "RAW MATERIAL" STAGE

15.4.1.1 Sources of Data

The sources of data include the following: NDRC—Chinese Government National Development and Reform Commission and Ecoinvent—Swiss Ecoinvent data version 2.2.

15.4.1.2 Database Selection

This was based on credibility and geographical, technological, and time-related considerations. Emission factor for electricity consumption is obtained from NDRC 2013, and the data used for components, consumables, packaging materials, and transportation are obtained from Ecoinvent version 2.2.

The materials used (components and product packaging materials) and their weight are listed in Table 15.2.

15.4.2 LCI OF "MANUFACTURING" STAGE

- The process flow in the manufacturing stage is shown in Figure 15.4.
- All activity data of resources and energy consumption in the "manufacturing" stage were provided by Clover Display Limited.

TABLE 15.2
Weight Details of Components and Product Packaging Materials

No.	Name of Component (ISO 2013)	Weight (g)
	Components	
1	ITO glass	3.90114
2	PI (polyimide) solution	0.00586
3	Pad agents (glue)	0.00003
4	Epoxy resin	0.00459
5	Glass dust	0.00003
6	Silver paste	0.00276
7	Polymer powder (mixed)	0.00001
8	Liquid crystal	0.00887
9	Breach glue	0.00740
10	Polarizer A	0.21913
11	Polarizer B	0.25211
12	Carbon paste	0.00271
13	Metal pin	1.26244
14	Mixing glue	0.05247
15	Mixing glue	0.01749
Total weight of LCD module		5.73704
	Product Packaging Materials	
16	Carton box	0.00144
17	Foam (EPS)	0.00016
18	PET tray	0.00630
Total weight of packaging		2.59300

Note: The name of the component and the materials used are provided by Clover Display Limited.

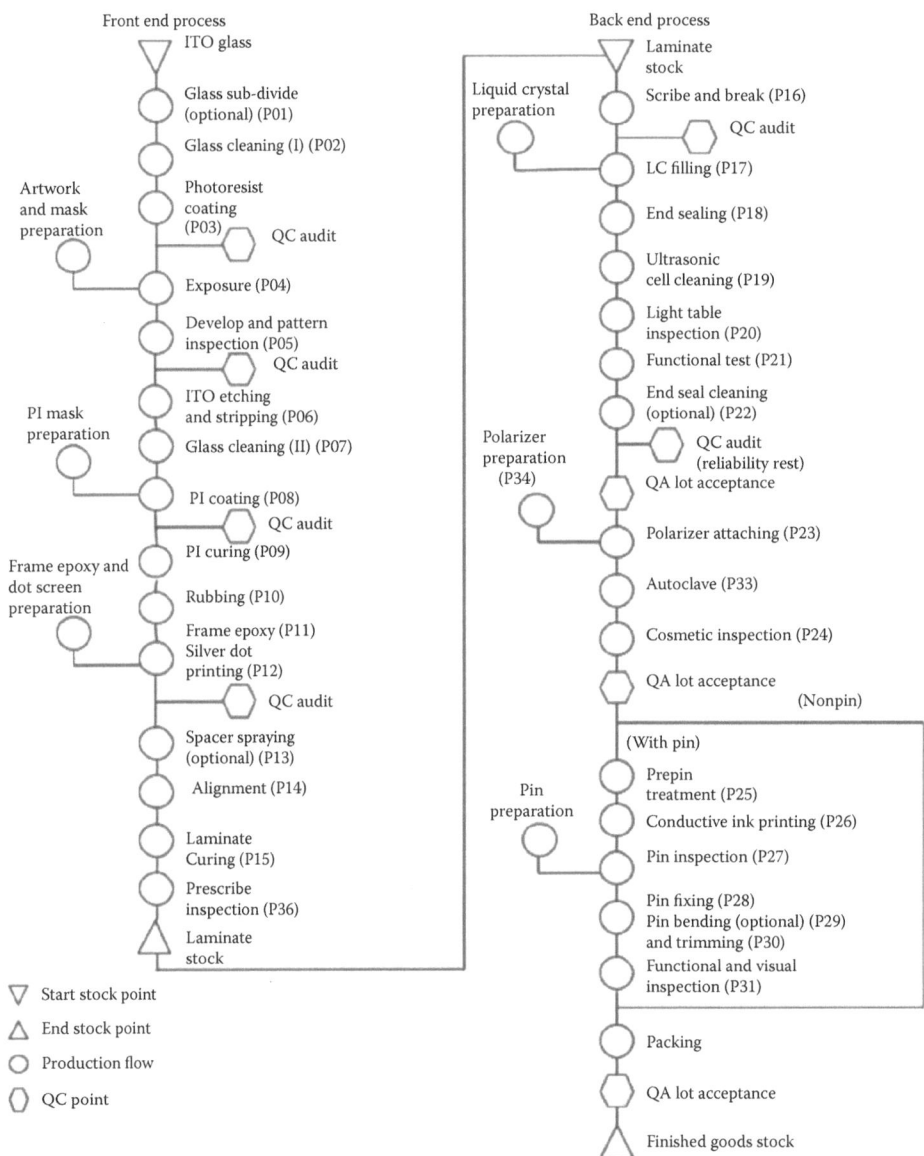

FIGURE 15.4 Process flow in the manufacturing stage.

- The emission factor of electricity used is from Chinese Government National Development and Reform Commission (NDRC). The emission factor for southern China in 2013 was 0.9223 kg CO_2 eq./kWh.
- The electricity consumption in the manufacturing stage was provided by Clover Display Limited. It was based on the allocation of total electricity consumption according to the production volume of the selected product and total production from January 2013 to December 2013. Nonproduction usage of electricity is excluded.
- The energy consumption for manufacturing each LCD module was 0.1048 kWh.
- Some of the packaging used in materials and consumables are sent to a recycling factory 20 km away from the manufacturing plant. The remaining packaging materials are sent to a landfill site that is 150 km away from the manufacturing plant. The emission from transportation of waste and disposal processes was included, but the environmental burden

TABLE 15.3

Transportation and Recycling Activities of Packaging Material Waste

	Item	Subtotal (kg)	Total Amount (kg)	Transportation Tool	Distance (km)
Recycling	Packaging of basic material and component and packaging materials	0.00090	0.00123	8 tons lorry	20
	Consumables packaging	0.00033			
Disposal	Packaging of basic material and component	0.00001	0.00007	8 tons lorry	150
	Consumables packaging	0.00006			

TABLE 15.4

Disposal Activities of Packaging Material Waste

	Item	Subtotal	Total Amount (kg)
Disposal solid waste	Packaging of basic material and component	0.00001	0.00007

TABLE 15.5

Traveling Distance and Transportation Tools of Finished Product to Singapore

	Routes	Transportation Tool	Distance (km)	Amount (tkm)
1	China Factory in Fo Shan Shi to Hong Kong Warehouse in Kwun Tong	8 tons lorry	140	0.00117
	Hong Kong Ware House in Kwun Tong to HK Airport	Van	37.3333	0.00031
	HK International Airport to Singapore Airport	Plane	2500	0.02083
	Singapore Airport to 70 Alps Avenue	Van	10	0.00008

from the recycling processes was excluded from the calculation. Tables 15.3 and 15.4 show the transportation (for both recycling and disposal) and disposal activities of the packaging material waste, respectively.

15.4.3 LCI OF "TRANSPORTATION" STAGE

- Transportation routes and tools for distributing finished products provided by Clover Display Limited are as shown in Table 15.5. Mass of the final product is 5.74 g and packaging is 2.59 g; therefore, total mass = 5.74 + 2.59 = 8.33 g will be taken for calculating the carbon footprint at the transportation stage.
- Distances of routes are obtained from Google Maps.

15.5 PCF ANALYSIS

15.5.1 CARBON FOOTPRINT RESULTS IN COMPONENT/ACTIVITY

Calculation of the carbon footprint of the LCD panel was based on a comprehensive coverage of raw material, manufacturing, and transportation stages. By assigning quantities of the input and output in each process, the carbon footprint of each input, output, and process was obtained. The results are tabulated in Table 15.6.

TABLE 15.6

Carbon Footprint Results in Different Life-Cycle Stages

No.	Item		Carbon Footprint (kg CO_2 eq./f.u.)	Percentage of Total PCF
		Raw Material Stage		
		Components		
1	ITO glass		0.01701	11.23
2	PI solution		0.00002	0.01
3	Pad agents (glue)		0.00000	0.00
4	Epoxy resin		0.00003	0.02
5	Glass dust		0.00000	0.00
6	Silver paste		0.00043	0.17
7	Polymer powder (mixed)		0.00000	0.00
8	Liquid crystal		0.00002	0.01
9	Breach glue		0.00003	0.02
10	Polarizer A		0.00057	0.38
11	Polarizer B		0.00066	0.44
12	Carbon paste		0.00001	0.01
13	Metal pin		0.00743	4.90
14	Mixing glue A		0.00024	0.16
15	Mixing glue B		0.00008	0.05
	Subtotal		0.02653	17.40
		Product Packaging		
16	Carton box		0.00199	1.31
17	Foam (EPS)		0.00014	0.09
18	PET tray		0.00630	4.16
	Subtotal		0.00842	5.56
		Packaging of Components		
1	ITO glass	Wood box	0.00024	0.16
		Carton paper	0.00001	0.01
		Carton paper	0.00003	0.02
		PP film	0.00001	0.01
		Foam (EPS)	0.00001	0.01
2	PI solution	Plastic bottle	0.00000	0.00
3	Pad agents (glue)	Plastic bottle lid	0.00000	0.00
		Glass bottle body	0.00000	0.00
4	Epoxy resin	Plastic bottle	0.00003	0.02
5	Glass dust	Plastic bottle lid	0.00000	0.00
		Glass bottle body	0.00000	0.00
6	Silver paste	Plastic bottle	0.00000	0.00
7	Polymer powder (mixed)	Plastic bottle lid	0.00000	0.00
		Glass bottle body	0.00000	0.00
8	Liquid crystal	Plastic bottle lid	0.00000	0.00
		Aluminum bottle body	0.00004	0.03
9	Breach glue	Plastic bottle	0.00000	0.00
10	Polarizer A	Carton box	0.00011	0.08
		Aluminum foil bag	0.00008	0.05
		Carton paper	0.00003	0.02
11	Polarizer B	Carton box	0.00013	0.09
		Aluminum foil bag	0.00009	0.06
		Carton paper	0.00003	0.02
12	Carbon paste	Plastic bottle	0.00000	0.00

(Continued)

TABLE 15.6 (*Continued*)
Carbon Footprint Results in Different Life-Cycle Stages

No.	Item		Carbon Footprint (kg CO_2 eq./f.u.)	Percentage of Total PCF
13	Metal pin	Roll paper	0.00013	0.09
14	Mixing glue A	Iron bottle	0.00002	0.02
15	Mixing glue B	Plastic bottle	0.00000	0.00
	Subtotal		0.00101	0.67
Packaging of Packaging Materials				
16	Carton box	Carton box	0.00052	0.34
17	Foam (EPS)	Plastic bag	0.00000	0.00
18	PET tray	Plastic bag	0.00011	0.07
	Subtotal		0.00063	0.42
Transportation of Basic Materials, Components, and Packaging Materials				
1	ITO glass (road)		0.00026	0.17
2	PI solution (road)		0.00000	0.00
3	Pad agents (glue) (road)		0.00000	0.00
4	Epoxy resin (road)		0.00000	0.00
5	Glass dust (road)		0.00000	0.00
6	Silver paste (road)		0.00000	0.00
7	Polymer powder (mixed) (road)		0.00000	0.00
8	Liquid crystal (road)		0.00002	0.01
9	Breach glue (road)		0.00000	0.00
10	Polarizer A (road)		0.00001	0.01
11	Polarizer B (road)		0.00001	0.01
12	Carbon paste (road)		0.00000	0.00
13	Metal pin (road)		0.00008	0.05
14	Mixing glue A (road)		0.00000	0.00
15	Mixing glue B (road)		0.00000	0.00
16	Carton box (road)		0.00000	0.00
17	Foam (EPS) (road)		0.00000	0.00
18	PET tray (road)		0.00006	0.04
	Subtotal		0.00045	0.30
	Subtotal of raw material stage		0.03688	24.34
Manufacturing Stage				
Energy Consumption				
1	Electricity consumption for production of one unit studied product		0.0966	63.78
Consumables				
1	Detergent MG		0.00033	0.22
2	Photoresist		0.00014	0.09
3	Acetone		0.00006	0.04
4	ITO mask		0.00001	0.01
5	Plastic frame film		0.00001	0.01
6	Silver spot film		0.00001	0.01
7	Sodium hydroxide (NaOH)		0.00033	0.22
8	Hydrochloric acid (HCl)		0.00034	0.23
9	Nitric acid		0.00000	0.00
10	Photographic film		0.00001	0.01
11	Screen (for screen printing)		0.00001	0.01

(*Continued*)

TABLE 15.6 (*Continued*)
Carbon Footprint Results in Different Life-Cycle Stages

No.	Item		Carbon Footprint (kg CO_2 eq./f.u.)	Percentage of Total PCF
12	Yellow glue		0.00001	0.00
13	BC (butyl carbitol) solution		0.00000	0.00
14	NMP solution		0.00003	0.02
15	Velvet cloth		0.00003	0.02
16	Alcohol		0.00022	0.14
17	Detergent 500		0.00061	0.40
18	Transparent plastic tape		0.00152	1.00
19	Carbon slurry thinners		0.00000	0.00
20	320 sand paper		0.00015	0.10
21	1.8 cm tape		0.00072	0.48
22	Clean white cloth (front)		0.00079	0.52
23	Clean white cloth (back end)		0.00079	0.52
24	DI water		0.00553	3.65
	Subtotal		0.01166	7.70
		Packaging of Consumables		
1	Detergent	Plastic bottle	0.00004	0.03
2	Photoresist	Plastic bottle	0.00003	0.02
3	Acetone	Plastic bottle lid	0.00000	0.00
		Glass bottle body	0.00005	0.03
4	ITO mask	Carton box	0.00000	0.00
5	Plastic frame film	Carton box	0.00002	0.01
		Plastic bag	0.00000	0.00
6	Silver spot film	Carton box	0.00002	0.01
		Plastic bag	0.00000	0.00
		Plastic bag	0.00000	0.00
7	Sodium hydroxide	Carton cardboard	0.00002	0.01
	(NaOH)	Aluminum bottle	0.00004	0.03
8	Hydrochloric acid (HCl)	Plastic bottle	0.00013	0.08
9	Nitric acid	Plastic bottle lid	0.00000	0.00
		Glass bottle body	0.00000	0.00
10	Photographic film	Plastic bottle	0.00000	0.00
11	Screen (for screen	Carton tube	0.00000	0.00
	printing)	Plastic bottle	0.00000	0.00
12	Yellow glue	Iron pail	0.00000	0.00
13	BC solution	Plastic bottle	0.00000	0.00
14	NMP solution	Plastic bottle	0.00000	0.00
15	Velvet cloth	Carton box	0.00000	0.00
16	Alcohol	Plastic bottle lid	0.00001	0.00
		Glass bottle body	0.00033	0.22
17	Detergent 500	Plastic bottle	0.00007	0.04
18	Transparent plastic tape	Carton box	0.00006	0.04
19	Carbon slurry thinners	Iron bottle	0.00000	0.00
20	320 sand paper	Plastic bag	0.00010	0.07
21	1.8 cm tape	Carton box	0.00001	0.01
22	Clean white cloth (front)	Plastic bag	0.00000	0.00

(*Continued*)

TABLE 15.6 (*Continued*)
Carbon Footprint Results in Different Life-Cycle Stages

No.	Item		Carbon Footprint (kg CO_2 eq./f.u.)	Percentage of Total PCF
23	Clean white cloth (back end)	Plastic bag	0.00000	0.00
24	DI water	Nil		
	Subtotal		0.00094	0.62
Transportation of Consumables				
1	Detergent (road)		0.00001	0.01
2	Photoresist (road)		0.00000	0.00
3	Acetone (road)		0.00000	0.00
4	ITO mask (road)		0.00000	0.00
5	Plastic frame film (road)		0.00000	0.00
6	Silver spot film (road)		0.00000	0.00
7	Sodium hydroxide (NaOH) (road)		0.00001	0.01
8	Hydrochloric acid (HCl) (road)		0.00002	0.02
9	Nitric acid (road)		0.00000	0.00
10	Photographic film (road)		0.00000	0.00
11	Screen (for screen printing) (road)		0.00000	0.00
12	Yellow glue (road)		0.00000	0.00
13	BC (butyl carbitol) solution (road)		0.00000	0.00
14	NMP solution (road)		0.00000	0.00
15	Velvet cloth (road)		0.00000	0.00
16	Alcohol (road)		0.00001	0.00
17	Detergent 500 (road)		0.00002	0.01
18	Transparent plastic tape (road)		0.00003	0.02
19	Carbon slurry thinners (road)		0.00000	0.00
20	320 sand paper (road)		0.00000	0.00
21	1.8 cm tape (road)		0.00000	0.00
22	Clean white cloth (front end) (road)		0.00000	0.00
23	Clean white cloth (back end) (road)		0.00000	0.00
24	DI water		0.00000	0.00
	Subtotal		0.00011	0.07
Waste Management				
1	Wastewater from production		0.00342	2.26
2	Wastewater (highly contaminated) from production		0.00006	0.04
3	Solid waste from production		0.00176	1.16
4	Disposal activities of materials packaging		0.00004	0.03
5	Transportation activities of recycling materials		0.00001	0.01
6	Transportation activities of disposal materials		0.00000	0.00
	Subtotal		0.00529	3.49
	Subtotal of manufacturing stage		0.10432	75.66
Transportation Stage (for Reference Only)				
1	Transportation of final product		0.02377	NA
	Total product carbon footprint (excluding distribution)		0.1515	100

15.6   LIFE-CYCLE INTERPRETATION

15.6.1   Results of Life-Cycle Interpretation

- The carbon footprint per functional unit (one unit of LCD module with packaging) is 0.2132 kg CO_2 eq. The summary of carbon footprint in each life-cycle stage is presented in Table 15.7 and Figure 15.5.
- The "manufacturing" stage is the most significant stage as it contributes a significant amount of carbon emission (75.66% of total PCF) in the whole life cycle of a product. In the "manufacturing" stage, the electricity consumption has a significant contribution, as shown in Table 15.7 (63.78% of the total PCF).
- The "raw material" stage contributes 24.34% of the total product carbon emissions in the whole life cycle.

15.6.2   Specific GHG Emissions and Removals

Table 15.8 separately shows the specific GHG emissions and removals that arise in this study.

15.6.3   Limitations

15.6.3.1   Focus on a Single Environmental Issue

The PCF only reflects the sum of GHG emissions and removals of a product system, expressed as CO_2 eq., associated with raw material acquisition, manufacturing, and transportation of a product. Although the PCF can be an important environmental aspect of the life cycle of a product affecting "climate," a product's life cycle may have other environmental impacts that are also of concern (e.g. resource depletion, air, water, soil, and ecosystems).

TABLE 15.7
Contribution to Total Carbon Footprint at Each Stage

Life-Cycle Stages		Carbon Footprint (kg CO_2 eq./f.u.)		Contribution (%)	
Raw material stage	Components	0.02637	0.037	17.40	24.34
	Product packaging	0.00842		5.56	
	Packaging of components	0.00101		0.67	
	Packaging of product packaging	0.00063		0.42	
	Transportation of components and product packaging	0.00045		0.30	
Manufacturing stage	Electricity consumption of production	0.09662	0.11	63.78	75.66
	Consumables	0.01166		7.70	
	Packaging of consumables	0.00094		0.62	
	Transportation of consumables	0.00011		0.07	
	Waste management	0.00529		3.49	
Transportation stage (for reference)		0.02377		NA	
Total product carbon footprint (excluding transportation)		0.1515		100	

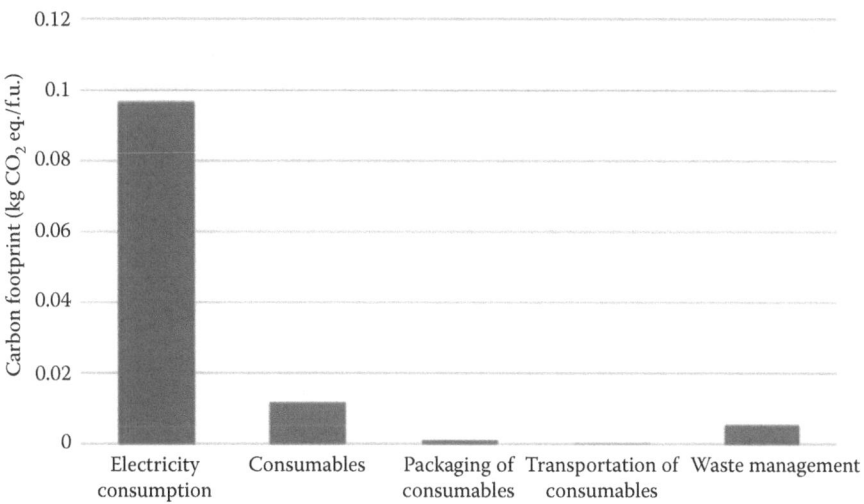

FIGURE 15.5   Contribution to carbon footprint from each process in the manufacturing stage.

15.6.3.2   Limitations Related to Assumptions

Assumptions were made in each stage of this study, which may affect the absolute accuracy of the final result.

15.6.3.3   Limitations Related to Methodology

The PCF was calculated on the basis of the LCA methodology. ISO 14040 and ISO 14044 address its inherent limitations and tradeoffs of LCA methodology. These include the establishment of a functional unit and the system boundary, the availability and selection of appropriate data sources, allocation rules, and assumptions regarding the transportation. Some of the chosen data may be limited to a specific geographical area (e.g. national electricity grid) and/or may vary in time (e.g. seasonal variations). The choice of the functional unit and allocation rules may also be important in more accurately modeling the life cycle.

15.6.4   Disclosure and Justification of Value Choices

One of the important value choices is the database selection. The priority for the database selection was based on credibility, geographical, technological, and time-related considerations. In this report, the emission factor for electricity consumption is obtained from NDRC 2013; the emission factor for sewage treatment is obtained from Xie and Wang (2012) of School of Environment of Tsinghua University; and the data used for components, consumables, packaging materials, and transportation are obtained from Ecoinvent version 2.2.

TABLE 15.8
Specific GHG Emissions and Removals That Arise in This Study

Source	Emission and Removals
Fossil carbon sources and sinks	NA
Biogenic carbon sources and sinks	NA
Direct land use change	NA
Aircraft transportation	NA

TABLE 15.9
Result of Changing Emission Factor of ITO Glass Used in the Material Stage

Material	Weight	Database	Name in Database	Emission Factor (kg CO$_2$ eq./kg)	Carbon Emission (kg CO$_2$ eq.)/f.u.	Final Carbon Footprint Result (kg CO$_2$ eq./f.u.)
ITO glass	3.90114 g	Ecoinvent 2.2	LCD glass, at plant (GLO)	4.36	0.01701	0.1515
		Korean LCI 2012	Plate glass	0.789	0.00308	0.13757

15.6.5 SENSITIVITY ANALYSIS

This section describes the sensitivity analysis regarding the influence of choice of emission factor database on the elements with most significant contributions to the GHG emissions, and electricity, on the final result.

15.6.5.1 Sensitivity Analysis of Significant Input Data in Material Stage

As shown in Tables 15.6 and 15.7, ITO (Indium Tin Oxide) glass used in the material stage contributes the largest amount of carbon emissions in this stage, and its emission factor is secondary data, which are not localized. Sensitivity analysis was conducted for this inventory. Different emission factors were used to analyze the sensitivity. Table 15.9 and Figure 15.6 show the difference in carbon footprint results on changing the emission factor of ITO glass from Ecoinvent to Korea LCI.

15.6.5.2 Sensitivity Analysis of Significant Input Data in Manufacturing Stage

As shown in Tables 15.6 and 15.7, electricity used in the manufacturing stage contributes the large amount of carbon emissions, and its emission factor is secondary data, which are not localized. Sensitivity analysis was conducted for this inventory. Different emission factors were used to analyze the sensitivity. Table 15.10 and Figure 15.7 show the difference in carbon footprint results on changing the emission factor of electricity provided by NDRC, Ecoinvent 2.2, and Hou et al. (2012).

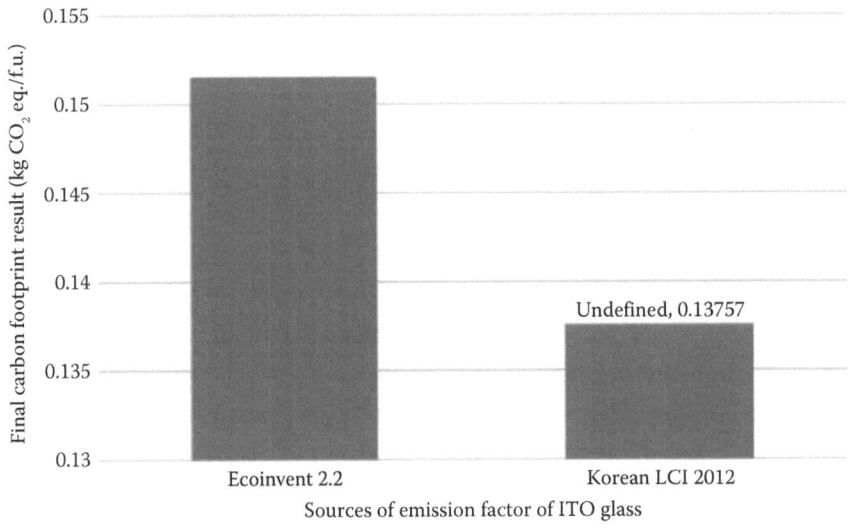

FIGURE 15.6   Result of the changing emission factor of electricity used in the material stage.

TABLE 15.10

Result of Changing Emission Factor of Electricity Used in the Manufacturing Process

Electricity Consumption	Database	Emission Factor (kg CO$_2$ eq./kWh)	Carbon Emission (kg CO$_2$ eq./f.u.)	Final Carbon Footprint Result (kg CO$_2$ eq./f.u.)
0.09359	NDRC	0.9223	0.08632	0.1515
	Ecoinvent 2.2	1.51	0.14132	0.2065
	Hou et al. (2012)	0.853	0.07983	0.1450

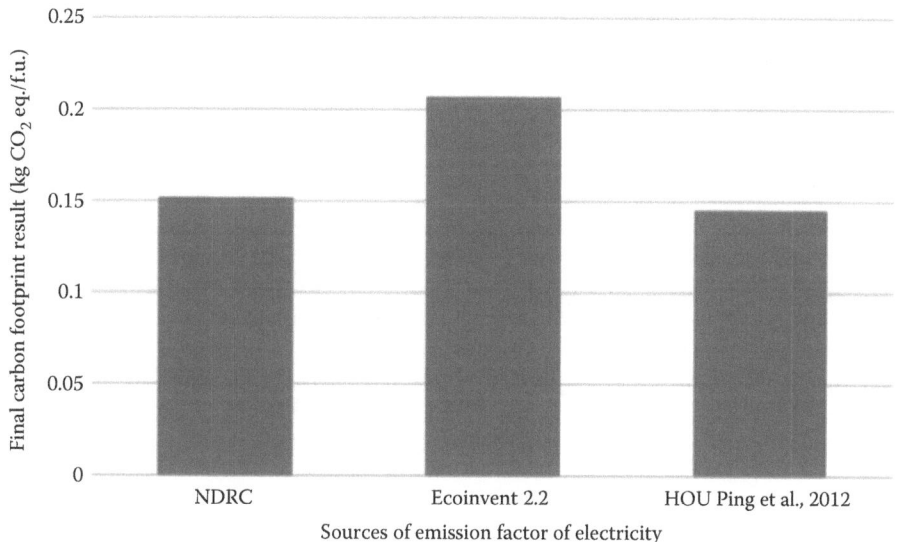

FIGURE 15.7 Result of the changing emission factor of electricity used in the manufacturing stage.

The emission factor that is obtained from a Chinese research journal (Hou et al. 2012) was used in the manufacturing stage.

15.6.6 UNCERTAINTY ANALYSIS

In this study, primary and secondary data are included. Some uncertainties may exist. Table 15.11 summarizes the type of data and the possible errors that may contribute to the uncertainty of the results.

TABLE 15.11

Summary of Data Types and Possible Error That May Contribute to Result Uncertainty

Type of Data	Type of Error Contributing Uncertainty to Results
Material data	Variation of the packaging materials from different suppliers
Process data	Allocation of electricity for manufacturing one unit of LCD module
Process data	Seasonal factor of production
Emission factor	Localized emission factor is the first choice and other emission factors should be adopted only if localized emission factor is not applicable or available

15.7   CONCLUSION

According to the results, electricity consumption (63.78%) makes a large contribution to the total GHG emission from "Cradle to Gate" life-cycle stages. Alternative electricity production methods may be employed as substitutes for reducing the consumption of electricity in order to reduce GHG emissions. The carbon footprint result reflects that any improvement in the manufacturing stage (75.66%) may effectively reduce the carbon footprint of the product. Thus, a better design and development of manufacturing processes is critical in order to reduce the PCF. Low-carbon economy is the future. Therefore, manufacturers who wish to get an edge over competitors in the market have to work toward delivering cleaner products with lesser carbon footprints and reduced overall GHG emissions. This will not only improve their business but also secure the environment. This kind of study is just a first step to provide directions to researchers in this field for further work and to encourage more manufacturers and industries to cut down carbon emissions in their manufacturing process.

ACKNOWLEDGMENTS

We would like to express our thanks to SGS Hong Kong Limited for the support provided to the Project "Development of an embedded Greenhouse Gas (GHG) emissions database with a G-BOM analyzer and a SME advisory kit for electrical and electronic industries to respond to the implementation and compliance of ISO 14067 (carbon footprint of products)." This case study report of the product carbon footprint of an LCD module (Brand: Clover Display Limited; Product model: C51486A) was made possible by the technical and professional contribution of our collaborating organization: SGS Hong Kong Limited. We would like to thank the above organization for giving valuable technical support and professional advice and for participating in site assessment.

REFERENCES

British Standards Institution (BSI). 2011. PAS 2050:2011 Specification for the assessment of the life cycle greenhouse gas emission of goods and services.

Electron Microscopy Sciences. 2014. Scanning electron microscope supplies. https://www.emsdiasum.com/microscopy/products/sem/colloidal.aspx/.

European Commission. 2012. Directive 2012/19/EU of the European Parliament and of the Council on Waste Electrical and Electronic Equipment (WEEE) recast.

Hou, P., Wang, H.T., Zhang, H., Fan, C.D., and Huang, N. 2012. Greenhouse gas emission factors of Chinese power grids for organization and product carbon footprint. *China Environmental Science* 32: 961–967.

International Organization for Standardization (ISO). 2006. ISO 14040: 2006 Environmental management—Life cycle assessment—Principles and framework.

International Organization for Standardization (ISO). 2013. ISO/TS 14067:2013 Greenhouse gases—Carbon footprint of products—Requirements and guidelines for quantification and communication.

National Development and Reform Commission. 2012. 2012 Baseline emission factors for regional power grids in China. cdm.ccchina.gov.cn/WebSite/CDM/UpFile/.

Xie, T. and Wang, C.W. 2012. Wastewater treatment plant greenhouse gas emission assessment. *Journal of Tsinghua University (Science and Technology)* 52(4): 473–477.

16 GHG Emissions from Municipal Wastewater Treatment in Latin America

Leonor Patricia Güereca-Hernandez, María Guadalupe Paredes-Figueroa, and Adalberto Noyola-Robles

CONTENTS

16.1 Introduction 351
16.2 Wastewater Treatment and GHG Emissions 352
    16.2.1 Carbon Dioxide Emissions 352
    16.2.2 Methane Emissions 353
    16.2.3 Nitrous Oxide Emissions 354
16.3 Wastewater Management in LAC 354
    16.3.1 Wastewater Treatment Technologies 354
    16.3.2 Opportunities for Improving Sanitation in the Region 355
16.4 CF of the Most Representative Wastewater Technologies in Latin America 356
    16.4.1 GHG Emission Calculation for LAC Representative Technologies (Baseline) 358
    16.4.2 CF for Cubic Meter of Wastewater Treated in LAC 359
16.5 Technological Improvements for CF Reduction 360
    16.5.1 CF for the Wastewater Treatment Scenarios with Technological Improvements 361
16.6 GHG Emissions by the Municipal Wastewater Treatment Sector in Mexico 362
    16.6.1 Current Municipal Wastewater Treatment in Mexico 363
    16.6.2 GHG Emissions Generated by the Management and Treatment of Wastewater
        in Mexico in 2010 and Its Projection for 2030 363
16.7 Economic Aspects 364
16.8 Final Remarks 366
References 367

16.1 INTRODUCTION

In 1950, over 150 million people in Latin America and the Caribbean (LAC) lived in cities. Due to a significant rural migration, the number grew to more than 360 million at the end of the twentieth century. This fact has generated a deep concern about water and soil resources. Facing this challenge, governments have made great efforts to improve an urban developing plan, focusing on water supply and sewer systems but leaving wastewater treatment behind.

About 13% of Latin American people living in urban areas and 37% in rural areas have inappropriate wastewater systems, resulting in health risks to around 120 million people (WHO/JMP/UNICEF 2013).

According to the Water and Sanitation Program in 2007 in LAC countries, only 15% of wastewater was treated (WSP 2007). This proportion may be estimated to be between 20% and 25% at present.

It is evident that economic requirements for increasing the water supply coverage and sanitation in LAC countries are enormous. In addition, about 74% of the population live in urban centers, which implies the necessary centralization of sanitation.

The challenge of improving the lives and health of all people in developing countries requires innovative solutions based on a clear understanding of the specific context of the region.

It is imperative to develop and implement new solutions to the lack of infrastructure in order to achieve an appropriate wastewater management in LAC countries. The new systems have to consider the limitations and capabilities of each region, with a high dose of innovation and adaptation.

This is a difficult challenge to face, as the decision of which wastewater treatment technology to install is not simple; in many cases, the plants stop operating just a few months after their start-up, which is due to many reasons, including the installation of a technology that is not appropriate for the type of water to be treated or environmental conditions not favorable to their optimal operation. However, in many cases, the reason for abandoning or neglecting a wastewater system is based on high operating costs that cannot be covered by the municipalities.

Moreover, treatment plants are considered "environmentally friendly" installations, as they reduce the impact associated with eutrophication and the risks to human health and aquatic toxicity. However, they generate other environmental impacts such as direct and indirect greenhouse gas (GHG) emissions, which together with relevant technical, economic, and social issues should be carefully analyzed in order to make the best decisions regarding treatment technologies (Noyola et al. 2013a).

Based on this, it is necessary to promote the effective treatment options in technical terms, feasible from the economic point of view and capable of reducing GHG emissions while generating social benefits for the community.

In this chapter, an analysis of GHG emissions from municipal wastewater treatment in Latin America is presented. This document integrates the original results of several research projects, which have been conducted during the last 4 years at the Engineering Institute at the Universidad Nacional Autónoma de México.

This chapter identifies the sources of GHG emissions in the wastewater treatment systems and then presents the wastewater treatment situation in the LAC region, showing the coverage and the technologies used, as well as some opportunities for improving the sanitation in the region. Then, the carbon footprint (CF) of 1 m^3 of wastewater treated is estimated, considering the nine representative technological scenarios of the LAC region. Based on the previous results, technological improvements are identified in order to reduce the CF, and potential reductions of GHG emissions are obtained. Finally, the GHG emissions corresponding to the Mexican wastewater policy of 100% treatment coverage at year 2030 are projected, and the economic costs of GHG emission mitigation technologies within the wastewater treatment systems are estimated, considering the Mexican economic data.

16.2 WASTEWATER TREATMENT AND GHG EMISSIONS

Municipal wastewater treatments are used to diminish the pollutant load to the environment and to protect public health. This is accomplished through physical, chemical, and biological processes. However, because carbon dioxide (CO_2), methane (CH_4), and nitrous oxide (N_2O) emissions are produced, those systems cause global warming. This section describes the sources of these gases, as well as their importance as GHGs.

16.2.1 CARBON DIOXIDE EMISSIONS

CO_2 emissions are produced by municipal wastewater treatments both directly and indirectly. Carbon dioxide direct emissions take place on the aerobic treatment process, particularly in the aeration tank (Rittman and McCarty 2001), in which reactions cause the biological oxidation of the organic fraction present in wastewater. Another direct CO_2 source can be identified on those systems in which the sludge is treated anaerobically and biogas produced, which, according to best practices, is burnt to generate CO_2 (Hospido et al. 2005). In these two cases, as CO_2 comes from a biogenic origin, it is not responsible for global warming (IPCC 2006). This is the reason that it is not quantified within the GHG inventories (Rodriguez-García et al. 2012). However, it is important to point out that according to Griffith et al. (2009), up to 20% of the wastewater carbon present in municipal wastewater treatments may have a fossil origin, mostly due to detergents, so this should be reviewed and considered in the GHG inventories.

Indirect carbon dioxide emissions come from the electricity production based on fossil fuels (electricity mix), and this is used in different electromechanical elements of the treatment system. Usually, this is an important item, being the fifth from the total consumed energy registered from a typical U.S. country. This power demand will increase as the water consumption rises and the regulations become stronger (Means 2004).

The indirect CO_2 emissions associated with electricity production are highly relevant in the GHG inventories (Mo and Zhang 2012).

16.2.2 Methane Emissions

In wastewater treatment plants (WWTPs), methane comes from the anaerobic breakdown of the organic fraction in the sewage, and preliminary treatment mainly takes place at the anaerobic treatment process. It is important to point out that an aerobic treatment plant, whether designed or operated wrongly, can produce methane by creating anaerobic zones into the aeration tank or in the sludge zone in settling tanks (Noyola et al. 2013a).

Usually, in national GHG emission inventories, there are not enough data relating to methane production coming from wastewater treatments or sludge digesters. According to Abdulla and Al-Ghazzawi (2000), WWTPs may produce total methane emissions between 8% and 11%. However, there is a need to have quantitative data which can describe methane emissions coming from these specific processes in order to propose effective GHG mitigation strategies.

Methane emissions from WWTPs can be estimated both theoretically and experimentally, starting from the organic matter in wastewater that goes through anaerobic degradation. The theoretical approach considers that the total organic fraction removed from the wastewater is converted under anaerobic conditions to methane. However, this approach shows an overestimation because it does not take into account key factors such as hydrolysis rates and extent, nutrient limitation, biological inhibition, and physicochemical interactions, among others (El-Fadel et al. 1996).

The experimental method is based on the data from actual methane emissions produced in pilot digesters of laboratories in which a homogeny mixture and a controlled environment for methanogenesis are present (El-Fadel and Massoud 2001). This method may also overestimate methane emissions, which is applied to real wastewater treatment systems because potential leaks and real operating conditions are not taken into account.

The IPCC method is also another commonly used technique because it is applied to quantify national GHG inventories. It consists of three steps. First, it determines the total organic fraction present in wastewater entering into each treatment system. Secondly, it proposes an emission factor expressed as kg CH_4/kg of degradable organic fraction removed; finally, it multiplies the emission value by a correction factor for each treatment system (default values) (IPCC 2006). Nevertheless, this method underestimates the methane emission value, which can be explained because it is based on biochemical oxygen demand (BOD), whereas other methods use chemical oxygen demand (COD) (El-Fadel and Massoud 2001).

It is important to point out two issues: first, methane emissions coming from municipal wastewater treatments vary among countries because of the use of different technologies, in addition to the fact that in tropical and subtropical regions, the temperature is close to 35°C. This is the optimal value for methanogenesis and as a consequence, high methane emissions are produced and compared with mild latitudes (IPCC 1996; Halsnaes et al. 1998). Secondly, the resultant sludge is also a significant methane source, which produces up to 50% of emissions depending on the wastewater treatment technology used (Arvizu 2008).

In fact, sludge management in tanks and storage ponds keeps an important biological activity, so methane may be produced, depending on the organic fraction and environmental conditions (usually anaerobic conditions).

A common sludge treatment system in big treatment facilities is anaerobic digestion, in which methane is produced and burned under controlled conditions. This burning reduces GHG emissions as CH_4 is converted to CO_2, and at the same time, if a cogeneration unit is part of the facility, indirect GHG emissions associated with electricity production from fossil fuels are reduced.

In developed countries, anaerobic digestion is a cost-effective sludge treatment process, whereas in developing countries, anaerobic process treatments are applied in treatment pond systems, mainly due to land availability, favorable climate conditions, simpler operation, and very low-energy requirements, resulting in affordable treatment cost for municipalities (Noyola et al. 2012).

16.2.3 NITROUS OXIDE EMISSIONS

Nitrous oxide can be generated through aerobic treatment processes, particularly when reactions involving the nitrogen cycle are carried out (nitrification and denitrification). Its production in wastewater plants is significantly lower compared with indirect CO_2 emissions and direct methane emissions when an anaerobic process is considered (Noyola et al. 2013a). However, this gas has a very relevant role in global warming due to its high calorific power (310 times greater than that of CO_2) (Préndez and Lara-González 2008).

Regularly, nitrous oxide is not monitored in wastewater treatment systems, and it is unknown whether there are systems that produce large amounts of N_2O (Greenfield and Batstone 2005).

Sources of nitrous oxide emissions and their magnitude on wastewater treatments are still unknown, and it is a particular discussion issue in scientific literatures due to its high inconstant flow, which depends on several operational parameters and environmental conditions (Greenfield and Batstone 2005).

IPCC guides (2006) are commonly used to quantify N_2O. They recognize that direct N_2O emissions from wastewater treatment systems are lower than those produced by nitrification and denitrification on receiving water bodies (estuaries and rivers). The N_2O direct emission factor is 0.5% of the nitrogen content in WWTP effluents (Kampschreur et al. 2009).

16.3 WASTEWATER MANAGEMENT IN LAC

16.3.1 WASTEWATER TREATMENT TECHNOLOGIES

In order to describe the coverage and representative wastewater treatment systems in LAC countries, Noyola et al. (2012) conducted a study in which the current state of water treatment throughout the region was assessed. They considered a 2734 WWTP sample in six countries, according to the following: 702 facilities in Brazil (total estimated facilities: 2985), 177 in Chile (total estimated facilities: 263), 139 in Colombia (unknown total estimated facilities), 32 in Guatemala (estimated full facilities: 87), 1653 in Mexico (estimated full facilities: 1833), and 31 in Dominican Republic (total estimated facilities: 56).

The information obtained from the WWTP in the six countries allowed us to identify four flow ranges (Q1, Q2, Q3, and Q4), according to the size of the population served: small (0–25 L/s), medium (25.1–250 L/s), big (250.1–2500 L/s), and huge (>2500 L/s), which are the most representative wastewater technologies in the region.

Table 16.1 shows that the representative technologies used in the LAC region are, by number, stabilization ponds (38%), activated sludge (26%), and up-flow anaerobic sludge blanket (UASB) reactor (17%). These technologies account for 80% of the sample. According to the treated flow,

TABLE 16.1

Number of WWTP and Accumulated Treated Flow by Wastewater Treatment Technologies in LAC Region

Technologies	No. of Installed Facilities	Accumulated Treated Flow by Technology (m³/s)
Stabilization pond	1106	27.1
Activated sludge	760	104.1
UASB	493	14.2
Aerated pond	140	10.3
Wetland	137	0.4
Trickling filter	126	6.4
Imhoff tank	84	0.9
Anaerobic filter	54	0.3
Enhanced primary treatment	18	16.1
Aerobic submerged filter	10	0.7
Rotating biological contactor	5	0.4

activated sludge and stabilization ponds turn out to be the most significant (Table 16.1), with a clear contribution of activated sludge (58%) followed by stabilization ponds (17%) (Noyola et al. 2012).

According to the results, the construction and operation of WWTPs of small scale (0–25 L/s) are very common practices in LAC countries, even in big cities, representing 67% of the facilities in the sample.

If the GHG emissions of the existing wastewater systems in LAC countries are analyzed, they may offer information that could be useful in developing and implementing comprehensive GHG mitigation strategies, allowing us to identify opportunities for the selection of technologies with better environmental performances.

16.3.2 Opportunities for Improving Sanitation in the Region

One of the requirements for this sector is the need for new treatment infrastructures in order to reduce the existent lag. This is a big challenge because of the huge investments that surpass the financing capabilities of most economies in Latin America, as well as in other developing regions.

In 2012, the World Health Organization (WHO 2012) found that the water and sanitation sector in LAC would require 68 billion dollars to construct and maintain the infrastructure necessary to have 100% of water and sanitation service coverage in the region.

This is a clear indication that there is an enormous municipal water treatment investment need. The problem is exacerbated if we consider that the increase in population will demand more water for urban use, thus requiring more treatment infrastructures in the coming years.

The abandonment of existing plants, their inefficient operation, and the use of inappropriate technologies are other relevant aspects in the wastewater treatment in LAC.

Improving the performance of these systems requires applying a set of criteria to consider the suitability of the technology and its proper operation in order to get the expected performance. This appears to be a difficult challenge, as the decision of which technology to install is not simple; in many cases, the plants stop operating just a few months after their start-up.

Some of the criteria that support the selection of appropriate technologies, thereby improving water treatment in the region (Noyola et al. 2013a), are described subsequently.

1. *Wastewater typology*: Municipal wastewaters have a homogeneous composition; however, when they come from combined sewer systems, wastewater can contain water from industries, businesses, or rainwater, which end up mixing with them. In this case, it is essential to

work on proper characterization of the water together with the regulatory requirements and the type of reuse or disposal, in order to identify the constituents that should be removed for the best water quality to be reached.

2. *Use or disposal of treated water.* In a treatment plant, the main product is the treated water, which must meet the characterization requirements established by the disposal or reuse rules. Therefore, it is necessary to identify clearly the quality of treated water as it determines the treatment train configuration.

3. *Technology offers diversity.* Although apparently there are many treatment technologies, proposals show very slight variations within the same technology. They only comply with one marketing strategy. In this regard, it is noteworthy that the treatment technologies and their integration into different trains are reduced to about only eight options so that there is the need to identify the efficiencies offered by each of them for treatment, their energy requirements, and their inputs in general.

4. *Investment, operation, and maintenance costs.* The initial investment cost is one of the most important criteria in selecting a treatment technology, but it must be closely related to operation and maintenance costs on a long-term horizon corresponding to the useful plant life.

5. *Sludge generation and treatment.* The generation of sludge in wastewater treatments is inevitable, and this is an important factor for operating costs involved in its management, which can reach up to 50% of the total investment. Depending on the quantity and quality of the sludge generated, it can, in turn, generate subproducts such as compost and biogas (if treated by anaerobic digestion) or become sanitary landfill once it has been stabilized.

6. *Odors.* Odor emissions in a WWTP cause a significant impact because it creates dissatisfaction in the community and social pressure very often forces the plant to stop operation. In this sense, we must identify the proximity of communities and consider filter installation that is capable of reducing odor.

7. *Environmental conditions.* Temperature is an important parameter in the reaction speed that takes place in the treatment processes because the bacterial cells' metabolism depends heavily on this criterion. For each process, temperature ranges are to be observed for efficient operation, and this should be taken into account when selecting a technology, which must properly operate within the range of ambient temperatures in the locality.

8. *Available area.* There are processes that are more compact and therefore require less area. This aspect becomes important in areas where land is expensive, scarce, and/or topographically rugged. Generally, a wastewater treatment system with low operating and maintenance requirements (extensive systems) requires more area than the one that demands more resources for operation and maintenance (compact process).

9. *Personnel requirements.* An important aspect to consider is the specialization of the required personnel to operate the treatment plant, especially in regions in which it is difficult to have qualified technicians.

16.4 CF OF THE MOST REPRESENTATIVE WASTEWATER TECHNOLOGIES IN LATIN AMERICA

As mentioned, wastewater treatment systems are a GHG emission source due to the indirect CO_2 produced at the electricity generator facility and also due to methane and nitrous oxide direct emissions from biological treatment. However, it is important to note that the wastewater treatment reduces the GHG emissions that would occur if wastewater were not treated, due to uncontrolled methane emissions in sewage pipes, channels, watercourses, and receiving water bodies (mainly those with low aeration conditions).

Taking into account the current infrastructure delay in LAC and the projected population growth, the present rate in building municipal wastewater treatment systems will be maintained or even increased in the coming years. As a result, among many benefits for public health and water protection issues, a decrease in GHG emissions should be accomplished, considering that no uncontrolled methane emissions will be produced from receiving water bodies.

However, if GHG global emissions are analyzed, the WWTPs may produce between 8% and 11% of the total methane emissions (Abdulla and Al-Ghazzawi 2000). Therefore, the selection of a wastewater treatment system must take into account not only the technical and economic aspects related to the construction, operation, and maintenance of facilities, but also the emissions of GHG generated by the adopted technologies.

Thus, quantification of GHG emissions and the CF associated with the existing wastewater treatment technologies in the LAC region are a relevant research concern. Within this context, the Institute of Engineering of the National Autonomous University of Mexico, with support from the International Development Research Center of Canada, developed the research project: "Water and sanitation: LAC cities adapting to climate change by making better use of bioenergy resources available" (Noyola et al. 2013b), which included, among others, objectives to quantify the GHG emissions generated from the municipal wastewater treatment sector in LAC and to identify technological mitigation proposals.

In order to achieve the above objectives, the most representative municipal wastewater treatment in LAC scenarios was identified (Noyola et al. 2012), as presented in Table 16.2.

A conceptual engineering analysis was carried out on the nine representative wastewater treatment scenarios. This was achieved from mass and energy balances considering the water quality of the influent and effluent, the organic matter removal efficiency, regulatory requirements, and the characteristics of sludge production and air emissions.

This step was carried out with a spreadsheet (simulator), developed specifically for this project, which includes, in addition to the data collected, a database based on literature references and on empirical data obtained from the expert judgment and the experience of the research group.

It is noteworthy that the study was performed considering three different representative flows of wastewater treatment: 13, 70, and 620 L/s, corresponding to a population of 7000, 38,000, and 335,000 inhabitants. The characterization of the influent municipal wastewater was defined according to the statistical analysis of data collected in the region. The effluent characterization was defined on the basis of the result of an analysis of the discharge standards in each of the selected countries.

TABLE 16.2
Representative Scenarios of Municipal Wastewater Treatment for LAC

Scenario	Description	Inflow (L/s)
1. EA	Extended aeration coupled with drying beds, small flow	13
2. SP	Stabilization ponds, small flow	13
3. UASB+F	UASB with trickling filter coupled with drying beds, flame burning biogas, small flow	13
4. EA	Extended aeration coupled with drying beds, medium flow	70
5. SP	Stabilization ponds, medium flow	70
6. UASB+SP	UASB with stabilization ponds, flame burning biogas, medium flow	70
7. AS	Conventional activated sludge process coupled with anaerobic sludge digestion and centrifugation, flame burning biogas, large flow	620
8. SP	Stabilization ponds, large flow	620
9. UASB+AS	UASB with activated sludge, coupled with sludge centrifuge, flame burning biogas, large flow	620

The quality of the resulting sludge was defined in accordance with the provisions for application to agricultural land (Class B) in EPA regulations (USEPA 1992, 1999).

The proposed scenarios are determined by the flow treated and by the type of technology, considering sludge management. Stabilization ponds in this work are understood as a set of three ponds in series: anaerobic, facultative, and polishing units.

16.4.1 GHG Emission Calculation for LAC Representative Technologies (Baseline)

For each wastewater treatment technological scenario, GHG emissions were quantified on the basis of the methodologies taken from clean development mechanism (CDM), which was proposed by the United Nations Framework Convention on Climate Change (UNFCCC 2011). In order to quantify emissions, CDM methodologies were chosen due to several reasons. First, a methodology capable of quantifying baseline emissions (such as wastewater treatment systems currently operating) and emissions with technological improvement proposals (projected scenario) was required. Secondly, this mechanism was a trigger for the implementation of technological improvements leading to GHG mitigation in all sectors, despite the fact that there are some flaws; thirdly, it was a valid and adopted strategy during the development period of the research project.

The emission quantification obtained supports the decision-making process aimed at GHG mitigation and could be the basis for wastewater treatment systems to participate in other carbon market mechanisms, considering that methane is a powerful short-lived climate forcer (Ramanathan and Carmichael 2008).

Within the UNFCCC methodologies, those corresponding to Sector 13, which corresponds to management and disposal of waste, were used as well as those from Sector 01 related to generation of renewable/nonrenewable energy on a small scale. Specifically, AMS.III.H "Methane recovery in wastewater treatment" (UNFCCC 2011) and AMS.ID "Network connected to the generation of renewable energy" (UNFCCC 2011) were considered.

The characteristics of influent wastewater and treated effluent are shown in Table 16.3. The environmental parameters influencing the design are: average ambient temperature 22°C (summer) and 16°C (winter), average wastewater temperature 20°C, and height above sea level (average) 902 m.

These parameters were estimated from the information collected at wastewater treatment facilities during the years 2010 and 2011 in the countries of the sample in LAC (Noyola et al. 2012). Additionally, a process simulator was developed in order to facilitate the calculation of mass balances, equipment and power requirements, and sludge generation, among others. Based on this information, the electricity consumed per cubic meter of treated wastewater was identified for each scenario (Table 16.4).

The quantification of GHG considers the wastewater treatment scenarios defined previously (Table 16.2) and assumes the following considerations.

- The global warming potential of 25 is adopted for methane.
- For electricity use, the emission factor of Dominican Republic, corresponding to 0.00063 tCO_2eq/kWh, is considered because it is the country of the sample with the

TABLE 16.3
Influent Features

Parameter	Inflow (mg/L)	Outflow (mg/L)
BOD	240	30
COD	520	70
TSS	220	30
VSS	165	20

TABLE 16.4
Electricity Consumption per Cubic Meter
of Wastewater Treated for Each Scenario

Scenario	kWh/m³
1. EA	0.80
2. SP	0.0
3. UASB+F	0.43
4. EA	0.69
5. SP	0.04
6. UASB+SP	0.09
7. AS	0.44
8. SP	0.02
9. UASB+AS	0.31

highest CO_2e emissions (according to the data published by the Latin American Energy Organization in 2010).

- The emission calculations for sludge treatment are determined from dry matter, as recommended by the UNFCCC (2011).
- It is considered that the treated wastewater is discharged into a water body for all scenarios.
- The final disposal of the dried sludge is assumed in a landfill without biogas recovery.
- About 10% of the biogas fugitive emission is considered (UNFCCC 2011).
- Biogas burners have an efficiency of 95%.
- In order to adjust methane emissions according to the treatment system, the methane correction factors proposed by the IPCC (2006) and accepted by UNFCCC (2011) are used.
- An uncertainty factor of 0.89 is assumed for baseline and project emissions.

The calculation of the baseline is defined by the following equation:

$$BE_y = \{BE_{power,y} + BE_{ww,treatment,y} + BE_{s,treatment,y} + BE_{ww,discharge,y} + BE_{s,final,y}\}$$

where

BE_y is the baseline emission during the year y (tCO_2e)

$BE_{power,y}$ is the baseline emission from electricity consumption or fuel consumption during the year y (tCO_2e)

$BE_{ww,treatment,y}$ is the baseline emission from the wastewater treatment systems in year y (tCO_2e)

$BE_{s,treatment,y}$ is the baseline emission of the sludge treatment systems during the year y (tCO_2e)

$BE_{ww,discharge,y}$ is the baseline methane emission from degradable organic carbon in the treated wastewater discharged into water bodies during the year y (tCO_2e)

$BE_{s,final,y}$ is the baseline methane emission produced by sludge anaerobic decomposition during the year y (tCO_2e)

16.4.2 CF FOR CUBIC METER OF WASTEWATER TREATED IN LAC

Results indicate that the CF per cubic meter of treated wastewater is within the range of 0.64–1.30 $kgCO_2e$ (Table 16.5). These results do not show a significant difference in emissions related to the size of the plants.

The stabilization pond wastewater treatment systems (2. SP, 5. SP, and 8. SP) generate the highest CF, which is due to methane produced in anaerobic lagoons and emitted directly to the atmosphere.

TABLE 16.5

CF for Cubic Meter of Wastewater Treated by the Representative Treatment Scenarios of LAC

Scenario	GHG Emissions (kg CO$_2$e/m^3 Wastewater Treated)
1. EA	0.918
2. SP	1.281
3. UASB+F	0.648
4. EA	0.851
5. SP	1.305
6. UASB+SP	0.762
7. AS	0.942
8. SP	1.284
9. UASB+AS	0.816

UASB scenarios (3. UASB+F, 6. UASB+SP, and 9. UASB+AS) present the lowest CF because in this technology the methane emissions produced anaerobically are recovered and burned in a flare, emitting CO$_2$ instead of CH$_4$.

The extended aeration scenarios (1. EA and 4. ES) and the activated sludge (7. AS) present CF values higher than those of UASB and lower than those of stabilization ponds. This system represents greater electricity consumption per cubic meter of wastewater treated (Table 16.4) due to the use of electricity in the aeration tank, which is necessary for aerobic degradation of the organic fraction in wastewater. In this technology, the use of electricity per cubic meter treated is less as the flow rate increases according to the economy of scale concept discussed by Noyola et al. (2012).

16.5 TECHNOLOGICAL IMPROVEMENTS FOR CF REDUCTION

The nine scenarios representing the usual practice for municipal wastewater treatments in LAC countries were analyzed in order to identify potential technological improvements to achieve CF reduction.

From this analysis, it was identified that methane emissions are negligible for the extended aeration scenarios (1. EA and 4. EA). However, they require large amounts of oxygen supplied with electromechanical equipment, which consumes high amount of electricity. In these cases, the proposal could be to reduce electricity consumption by adopting energy-efficient measures. Other possibilities are to use cleaner or renewable energy sources such as wind, solar, or microhydropower, among others, depending on the specific location. Another option is to replace and add an anaerobic pretreatment step with biogas recovery followed by the aerobic system, resulting in lower aeration needs.

According to the above, any technological change in these aerobic systems should be associated with the source of electricity, but it is not possible to propose improvements focused on the treatment process itself (except improving the efficiency of the aeration equipment). For this reason, technological improvements to reduce CF are not considered for these wastewater treatment scenarios.

In order to identify technological improvements in other scenarios (2. SP, 3. UASB+F, 5. SP, 6. UASP+SP, 7. AS, 8. SP, and 9. UASB+AS), the UNFCCC methodologies related to wastewater treatments were selected:

- AMS-III.H. Methane recovery in wastewater treatment
- AMS-I.D. Network connected to the renewable energy generation

The wastewater treatment recovery methane methodology (AMS-III.H) is applied for the biogas recovery coming from the anaerobic decomposition of the organic fraction.

This is achieved by introducing an anaerobic treatment system for wastewater or for sludge digestion, through which biogas generated is used as an energy source of low GHG emissions, replacing a high GHG emission energy source.

The methodology of the network connected to the generation of renewable energy (AMS-ID) is based on the operation of a power plant that uses renewable energy displacing electricity, which could be provided by a high GHG emission source. In these cases, the electricity production is calculated considering a methane capture efficiency of 90% and a factor of net electricity generation of 3156 kWh/tCH$_4$.

For the stabilization pond scenarios (2. SP, 5. SP, and 8. SP), it is considered that the ponds will be covered to capture the produced biogas, with a high-density polyethylene (HDPE) geomembrane of 1.5 mm thickness.

After capturing the biogas, electricity was generated, taking into account that biogas has a methane content of 65% (Walsh et al. 1988), a calorific value of 9464 kcal/m^3, and an efficiency of 30% on the generating equipment. A production of 2.15 kWh/m^3 is estimated. The methane emission factor used was 0.6 kg CH$_4$/kg BOD removed (IPCC 2006).

For scenarios considering a UASB reactor followed by three different post-treatment processes (3. UASB+F, 6. UASB+SP, 7. AS, and 9. UASB+AS), methane is recovered and used to generate electricity, either for internal consumption in the WWTP or to incorporate it in the grid.

16.5.1 CF for the Wastewater Treatment Scenarios with Technological Improvements

Figure 16.1 shows the CF per cubic meter of wastewater treated for the representative systems and for the technologically improved scenarios (mitigation scenarios). The results show that the mitigation scenarios achieve GHG emission reductions in all improvement options. As mentioned, scenarios 1. EA and 4. EA were not eligible for improvement; therefore, no change in the value of their emissions is observed.

Improved stabilization ponds (2.SP, 5.SP, and 8.SP) were the scenarios that showed greater GHG emission reductions after the proposed improvements, as, in the baseline scenario, methane

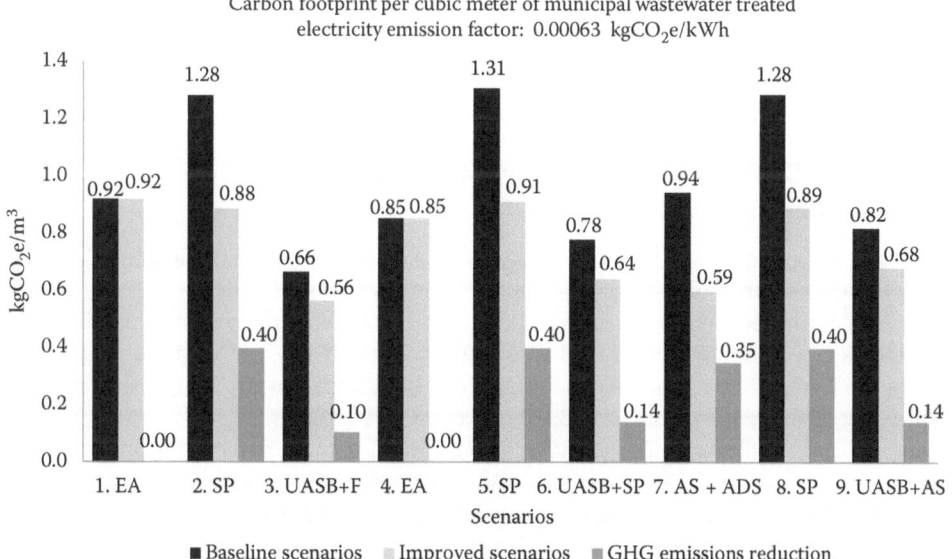

FIGURE 16.1 CF per cubic meter of wastewater treated for baseline scenarios (representative wastewater treatment systems), improved scenarios (mitigation scenarios), and estimated CF reductions.

is released directly into the atmosphere. In these scenarios, electricity production increases the benefits of the mitigation strategy.

The 2. SP improved scenario presents less electricity production because the flow treated is small and therefore the production of biogas is also less. This scenario is designed without electromechanical equipment, so it does not consume electricity. With the capture of biogas, 19,691 kWh/year could be produced and delivered to the grid. The technological improvement proposed for this system could reduce 0.40 $kgCO_2e/m^3$ of treated water.

For the mitigation scenario 5. SP, it is estimated that the captured biogas can generate 106,029 kWh/year. The technological improvement proposed would reduce 0.40 $kgCO_2e/m^3$ of treated water.

The 8. SP improved scenario generates approximately 2.4 times the energy it consumes. It has the capacity to produce 1,939,117 kWh/year, and it consumes 391,280 kWh annually. The technological improvement proposed for this system would reduce the CF by 0.40 $kgCO_2e/m^3$ of treated water.

The improved UASB scenarios (3. UASB+F, 6. UASB+SP, and 9. UASB+AS) decreased, to a lesser extent, their emissions when compared with stabilization ponds. This is due to the fact that in these systems, the methane was already captured and burned in a flare, diminishing the total GHG emissions. As in the case of stabilization ponds, the electricity generated from biogas, reducing the consumption from the grid, is a benefit of the mitigation option proposed.

Scenario 3. UASB+F requires 177,755 kWh/year and could generate 66,162 kWh annually because biogas could produce 37% of the electrical energy consumed by the system in a year. The technological improvement proposed for this system will reduce by 0.10 $kgCO_2e/m^3$ of treated water.

In scenario 6. UASB+SP, the electricity production is 75% more than that required due to the treatment train configuration, as it is a technology that combines UASB and a covered lagoon system, which has very low electricity requirements. The technological improvement proposed for this system reduces the CF by 0.14 $kgCO_2e/m^3$ of treated water.

Scenario 9 requires 6,002,425 kWh/year for operation, but the technological improvements can produce 3,972,480 kWh/year, which means savings in GHG emissions of 0.14 $kgCO_2e/m^3$ of treated water.

Finally, scenario 7.AS already includes the biogas capture from its original design, which is burned in order to maintain the optimal temperature in the biodigester (35°C). The mitigation scenario considers electricity cogeneration, so that the heat energy produced in the electric generator is used to maintain the digester temperature. The baseline scenario requires 8,652,325 kWh annually, and the improved system produces 6,939,123 kWh/year, which results in a CF reduction of 0.35 $kgCO_2e/m^3$ of treated water.

16.6   GHG EMISSIONS BY THE MUNICIPAL WASTEWATER TREATMENT SECTOR IN MEXICO

In Mexico, in recent years, the water issue has been a priority topic at the national level. It has become a key element in public, environmental, economic, and social policies.

Water availability per capita in Mexico fell, between 2000 and 2005, from 4841 to 4573 m^3/year, and the studies carried out by the National Water Commission (CNA), as well as population projections from the National Population Council (CONAPO) indicate that, by 2030, the average water availability per capita will be reduced to 3430 m^3/year (CNA 2013).

It is clear that the abatement of the existing backlog in infrastructure for water supply, the establishment of drainage and sewerage services, and wastewater treatment are among the important challenges that Mexico faces. Thus, it is essential to invest in technologies that allow us to make better use of this resource.

There are great challenges established in the country for the coming years in the field. The intention is to achieve 100% treatment coverage by 2030 and to increase water reuse in various sectors such as agriculture and industry (CNA 2011a). Therefore, existing treatment plants and those to be built in the future must operate efficiently to ensure that effluent meets the environmental standards established in the legal framework.

From a sustainable human development approach, in 2013, Mexico presented the National Development Plan (Plan Nacional de Desarrollo, 2013–2018), which includes the environmental sustainability axis, including action lines on water and, for the very first time, policies related to climate change mitigation and adaptation. The main objective is to reduce GHG emissions in strategic sectors.

International concerted action is essential to face a problem that no country will be able to solve alone. In this context, in 2011, Mexico generated 1.4% GHG emissions (0.432 Gg CO_2e), a figure that ranks Mexico's emissions as the 10th second largest worldwide (SEMARNAT 2013).

The National Emissions Inventory of Greenhouse Gases (INEGEI) in 2006 indicated that emissions in units of carbon dioxide equivalent (CO_2e) for Mexico were 709.005 Gg (INE 2009). The contribution by category is as follows: waste 14.1% (99,627.5 Gg); land use, land use change, and forestry 9.9% (70,202.8 Gg); industrial processes 9% (63,526 Gg); agriculture 6.4% (45,552.1 Gg); and energy 60.6% (430.097 Gg). GHG gases, measured in CO_2e units, were: CO_2, 492,862.2 Gg (69.5%); CH_4, 185,390.9 Gg (26.1%); N_2O, 20,511.7 Gg (2.9%); and the remaining 1.4% is made up of 9586.4 Gg of HFCs (hydrofluorocarbons) and 654.1 Gg of SF_6 (sulfur hexafluoride).

The wastewater treatment and management subsector directly contributes to GHG generation through carbon dioxide (CO_2), nitrous oxide (N_2O), and methane (CH_4) production and also indirectly through CO_2 emissions produced from the energy required for the treatment. In 2006, this subsector contributed with a total of 48,228 CO_2 eq, being CH_4 the gas with the highest percentage emission (96%) and N_2O the lowest (4%) (INE 2009).

16.6.1 Current Municipal Wastewater Treatment in Mexico

Table 16.6 shows the distribution of different technologies used in Mexico for municipal wastewater treatment. The main processes used correspond to stabilization ponds (34.7%), activated sludge (29.6%), septic tank (7.7%), and UASB (6.9%) nationally. These four technologies represent 78.9% (1724 WWTPs) of the total number of municipal WWTPs existing all around the country (CNA 2011b).

The percentage of different technologies according to their design flow places activated sludge process as the most important (52%), followed by stabilization ponds (14.5%), together representing 66.5% of the total treatment capacity, as shown in Figure 16.2.

16.6.2 GHG Emissions Generated by the Management and Treatment of Wastewater in Mexico in 2010 and Its Projection for 2030

The methodology established in the IPCC guidelines (IPCC 2006), Volume 5, Waste, Chapter 6: Wastewater treatment and disposal, is applied to calculate GHG emissions. This methodology is used to establish the reference for the GHG emission generation of national inventories.

In Table 16.7, the trend parameters used for the calculation of CO_2e emissions and their projection for 2030 (Noyola et al. 2014) are shown.

In a business scenario as usual, keeping the current population and treatment infrastructure growth rates, with the same mix of treatment technologies reported in 2010, the GHG emissions from municipal WWTPs in Mexico, in units of CO_2e, would go from 13,334 Gg of CO_2 in 2010 to 11,867 Gg of CO_2 in 2030 (a reduction of 1467 Gg CO_2e). These data are presented in Figure 16.3. According to this scenario, the treatment coverage of collected sewage would rise from 45% (2010) to 87% (2030) (Noyola et al. 2014).

The GHG emission reduction which took place during the 5-year period from 2010 to 2015 is attributed to the commissioning of Atotonilco WWTP in 2015, which will treat about 60% of the wastewater generated in the Valley of Mexico (Mexico's Metropolitan Area) and produce a treated effluent that will be used for crop irrigation. This huge facility has a design flow capacity of 23 m³/s (with peaks up to 35 m³/s). The treatment process is activated sludge with anaerobic digestion of

TABLE 16.6
Technologies Used for Municipal Wastewater Treatment Facilities in Mexico in 2010

Type of Treatment	Number of Facilities	%
Dual[a]	17	0.8
Trickling filter	52	2.4
Septic tank	168	7.7
Septic tank + trickling filter	11	0.5
Septic tank + wetland	73	3.3
Aerated pond	33	1.5
Stabilization pond	759	34.7
Activated sludge	647	29.6
Enhanced primary treatment	16	0.7
Primary treatment or sedimentation	21	1.0
UASB	150	6.9
UASB + trickling filter	29	1.3
UASB+ wetland	9	0.4
Primary sedimentation + wetland	18	0.8
Imhoff tank	67	3.1
Imhoff tank + trickling filter	13	0.6
Wetland	75	3.4
Oxidation ditch	28	1.3
Total	2186	100

Source: Adapted from CONAGUA (Comisión Nacional del Agua). 2011a. Agenda del Agua 2030, Secretaría de Medio Ambiente y Recursos Naturales, México; CONAGUA (Comisión Nacional del Agua). 2011b. Inventario nacional de plantas municipales de potabilización y de tratamiento de aguas residuales en operación. México.

[a] Combined treatment system or double stage, usually a trickling filter followed by an activated sludge.

excess sludge coupled with a cogeneration facility. At peak flow (rainy season), a parallel physico-chemical process (12 m³/s) will provide additional treatment capacity (De la Peña et al. 2013).

In addition to this trend scenario analysis, five technological scenarios were analyzed for a 100% treatment coverage at year 2030, considering full aerobic and a combination of anaerobic treatments followed by aerobic post-treatment (Noyola et al. 2014). Results show a GHG emission reduction as high as 34% compared with the baseline scenario if the combined anaerobic–aerobic processes with 95% methane burning efficiency and electricity cogeneration in facilities with treatment capacity above 500 L/s are considered.

16.7 ECONOMIC ASPECTS

The technology implementation which mitigates GHG emissions within the wastewater treatment systems has an associated economic cost.

The economic analysis specifically refers to Mexico because of the ease in obtaining information. It definitely gives an idea for the entire LAC region.

Table 16.8 shows the required investments and the operating annual expenses for the operation to take place as well as for the biogas use for the most representative technologies.

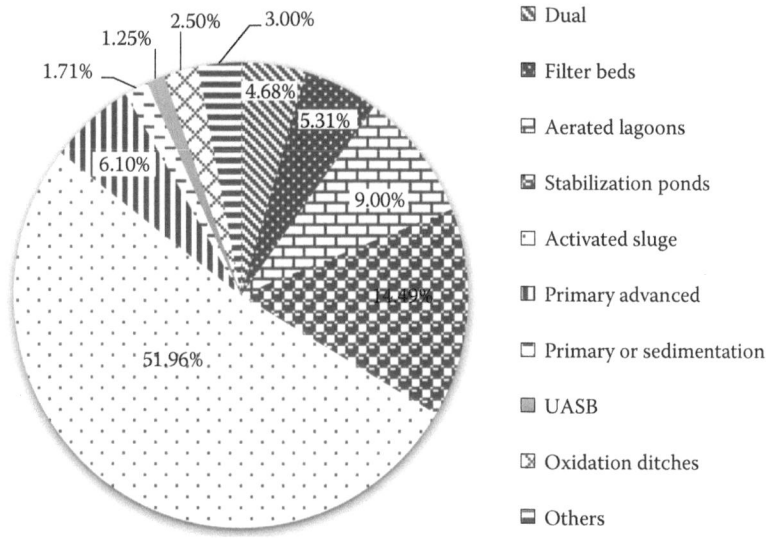

FIGURE 16.2 Treatment capacity by technology in Mexico in 2010.

TABLE 16.7
Parameters Used for the Calculation of CO_2e Emissions by Sector Wastewater Treatment in Mexico

Parameter	Units	2010	2015	2020	2025	2030
Population[a]	People	112,336,538	115,762,289	118,692,987	120,928,075	122,348,728
Wastewater generated	m³/s	234	241	247	251	254
Collected wastewater[b]	m³/s	208	214	220	224	227
Treated wastewater	m³/s	94	142	161	179	198
% of treated flow[c]	%	45	66	73	80	87
Generated BOD[d]	kg BOD/year	2,460,170,182	2,535,194,129	2,599,376,415	2,648,324,843	2,679,437,143
BOD that enters the WWTP[e]	kg BOD/year	709,998,981	1,077,259,668	1,217,612,529	1,357,965,389	1,498,318,249
BOD removal by the WWTP[f]	kg BOD/year	606,001,912	932,351,856	1,053,825,121	1,175,298,386	1,296,771,652
BOD untreated	kg BOD/year	1,854,168,270	1,602,842,273	1,545,551,294	1,473,026,457	1,382,665,491

[a] CONAPO (2007). http://www.conapo.gob.mx/.
[b] Per capita generation wastewater 180 L/hab/day. CONAGUA (2010).
[c] Percentage of wastewater collected in municipal sewage systems being treated.
[d] Per capita generation DBO de 60 g/hab/day. CONAGUA (2010).
[e] [DBOe] 242 mg/L obtained from a representative sample of 124 MWWTPs.
[f] [DBOs] 150 mg/L discharge to water body and 30 mg/L for irrigation, depending on the type of discharge of each WWTP. NOM-001-SEMARNAT-1996.

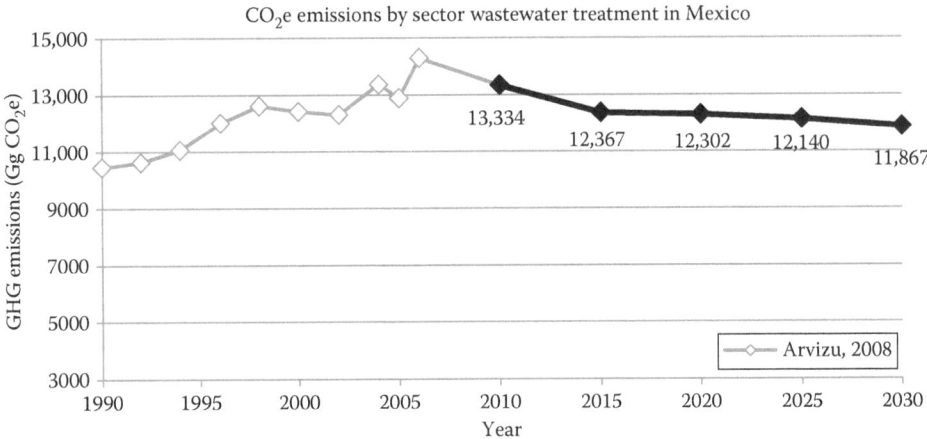

FIGURE 16.3 CO_2e emissions generated by sector wastewater treatment in Mexico in 2010 and projection for 2030.

TABLE 16.8
Investment and Operating Expenses for New WWTP US Dollars

Scenarios	2. SP	3. UASB+F	5. SP	6. UASB+SP	7. AS	8. SP	9. UASB+AS
Inflow (L/s)	13	13	70	70	620	620	620
Infrastructure investment (USD)	3,278,990	1,370,240	4,825,082	6,573,792	16,908,269	49,869,757	20,032,101
Cogeneration unit (USD)	85,479	56,395	224,513	165,247	434,025	7818	434,025
Geomembrane cover (USD)	57,826	—	176,109	—	—	1,048,455	—
WWTP total investment (USD)	3,422,295	1,426,635	5,225,704	6,739,039	17,342,294	51,700,012	20,466,126
Annual operative expenses WWTP (USD)	8998	44,621	58,993	73,074	1,721,219	104,274	1,441,970

It is important to note that investment amounts and operating costs vary between scenarios because they are specifically linked to the flow of each scenario. First, the investment increases as the flow does the same. Secondly, operating expenses are not directly related to the scale of the project.

As a GHG mitigation strategy, WWTPs have potential GHG emission abatement. On the one hand, they can reduce power consumption; on the other, they also prevent the methane release into the atmosphere.

According to the results, improving the operational performance of the WWTP (energy efficiency), cogenerating electricity from biogas in order to reduce its consumption coming from fossil sources, as well as the sale of treated wastewater would sharply increase the economic feasibility of WWTPs. Therefore, it is necessary that the guidelines and criteria for decision making about investing in sanitation projects prioritize WWTP capabilities to generate electricity for their own consumption and to produce treated water for sale.

16.8   FINAL REMARKS

The emission factor for electricity production plays an important role in assessing technology and its feasibility in order to implement GHG mitigation strategies. This means that improvements should be identified within the context in which the decision maker and wastewater treatment facilities are located.

The stabilization pond scenarios show greater mitigation potentials for GHG emissions, but UASB systems are better environmental options even without electricity generation from biogas.

When applying the default values defined by the IPCC, the uncertainty level associated with the default values is accepted. But, it is important that future measurements will be performed in order to calculate the specific emission factors for each country with the intention of minimizing the uncertainty and obtaining emission data closest to the existing conditions.

The wastewater treatment subsector may be considered as a field of opportunity for adopting efficient GHG mitigation strategies, according to the national GHG emissions inventory calculated for the wastewater sector in Mexico, under different technological scenarios.

REFERENCES

Abdulla, F. A. and Al-Ghazzawi, Z. D. 2000. Methane emissions from domestic waste management facilities in Jordan applicability of IPCC methodology. *Journal of the Air and Waste Management Association* 50(2): 234–9.

Arvizu, J. L. 2008. Actualización del Inventario Nacional de Gases de Efecto Invernadero 1990–2006 en la categoría de desechos. Instituto Nacional de Ecología (INE), México.

CONAGUA (Comisión Nacional del Agua). 2010. Estadísticas del Agua en México, Secretaria de Medio Ambiente y Recursos Naturales, México.

CONAGUA (Comisión Nacional del Agua). 2011a. Agenda del Agua 2030, Secretaría de Medio Ambiente y Recursos Naturales, México.

CONAGUA (Comisión Nacional del Agua). 2011b. Inventario nacional de plantas municipales de potabilización y de tratamiento de aguas residuales en operación. México.

CONAGUA (Comisión Nacional del Agua). 2013. Estadísticas del Agua en México, Secretaría de Medio Ambiente y Recursos Naturales, México.

CONAPO (Consejo Nacional de Población). 2007. Proyecciones de la Población 2010–2030. Series de estimaciones demográficas y proyecciones de población. Secretaria de Gobernación. México. http://www.conapo.gob.mx/es/CONAPO/Proyecciones_Datos

De la Peña, M. E., Ducci, J., and Zamora, V. 2013. Tratamiento de aguas residuales en México, Banco Interamericano de Desarrollo.

El-Fadel, M., Findikakis, A., and Leckie, J. 1996. Estimating and enhancing methane yield from municipal solid waste management. *Hazardous Waste and Hazardous Materials* 13: 309–31.

El-Fadel, M. and Massoud, M. 2001. Methane emissions from wastewater management. *Environmental Pollution* 114: 177–85.

Greenfield, P. F. and Batstone, D. J. 2005. Anaerobic digestion: Impact of future greenhouse gases mitigation policies on methane generation and usage. *Water Science and Technology* 52(1–2): 39–47.

Griffith, D. R., Barnes, R. T., and Raymond, P. A. 2009. Inputs of fossil carbon from wastewater treatment plants to U.S. rivers and oceans. *Environmental Science and Technology* 43: 5647–51.

Halsnaes, K., Callawy, M., and Meyer, J. J. 1998. Economics of greenhouse gas limitations. UNEP Collaborating Center on Energy and Environment, Denmark.

Hospido, A., Moreira, T., Martín, M., Rigola, M., and Feijoo, G. 2005. Environmental evaluation of different treatment processes for sludge from urban wastewater treatments: Anaerobic digestion versus thermal processes. *International Journal of Life Cycle Assessment* 10: 336–45.

Informes del Inventario Nacional de Emisiones de Gases de Efecto Invernadero 1990–2006. 2009. Instituto Nacional de Ecología. México. Disponible en. http://www.ine.gob.mx/cpcc-lineas/929-inem-1990–2006 (accessed June 5, 2015).

IPCC. 1996. The revised guidelines for national Greenhouse Gas Inventories. Vol. II. Workbook: Waste. University Press. Cambridge, United Kingdom.

IPCC. 2006. Guidelines for National Greenhouse Gas Inventories. Vol. 5, Waste. IPCC National Greenhouse Gas Inventories Programme, Institute for Global Environmental Strategies, Hayama, Kanagawa, Japan.

Kampschreur, M. J., Temmink, H., Kleerebezem, R., Jetten, M. S. M., and van Loosdrecht, M. C. M. 2009. Nitrous oxide emission during wastewater treatment (Review). *Water Research* 43: 4093–103.

Means, E. 2004. Water and wastewater industry energy efficiency: A research roadmap. IWA Research Foundation.

Mo, W. and Zhang, Q. 2012. Can municipal wastewater treatment systems be carbon neutral? *Journal of Environmental Management* 112: 360–67.

Noyola, A., Padilla-Rivera, A., Morgan-Sagastume, J. M., Güereca, P., and Hernández-Padilla, F. 2012. Typology of wastewater treatment technologies in Latin America. *Clean-Soil, Air, Water* 40: 926–32.

Noyola, A., Morgan-Sagastume, J. M., and Güereca, L. P. 2013a. Selección de tecnologías para el tratamiento de aguas residuales municipales. Guía de apoyo para ciudades pequeñas y medianas. Universidad Nacional Autónoma de México, Instituto de Ingeniería, 126 pp.

Noyola, A., Güereca, L. P., Morgan-Sagastume, J. M., Hernández-Padilla, F., Padilla, A., Carius, C., Cisneros, M., and Villalba, E. 2013b. Water and sanitation: LAC cities adapting to climate change by making better use of bioenergy resources available. Final Report, International Development Research Center (IDRC). http://hdl.handle.net/10625/52761 (accessed October 16, 2014).

Noyola, A., Paredes, M. G., Morgan, J. M., and Güereca, L. P. 2014. Reduction of greenhouse gas emissions from municipal wastewater treatment by selection of more sustainable technologies in Mexico. *Clean-Soil, Air, Water* (submitted).

Plan Nacional de Desarrollo 2013–2018. VI. 4. México Próspero. Estrategia 4.4.3. Diario Oficial de la Federación. Presidencia de la Republica. México. 132–136 pp. http://pnd.gob.mx/wp-content/uploads/2013/05/PND.pdf

Préndez, M. and Lara-González, S. 2008. Application of strategies for sanitation management in wastewater treatment plants in order to control/reduce greenhouse gas emissions. *Journal of Environmental Management* 88:658–64.

Ramanathan, V. and Carmichael, G. 2008. Global and regional climate changers due to black carbon. *Nature Geoscience* 1: 221–7.

Rittman, B. E. and McCarty, P. E. 2001. *Environmental Biotechnology. International.* Singapore: McGraw Hill Higher Education.

Rodriguez-García, G., Hospido, A., Bagley, D. M., Moreira, M. T., and Feijoo, G. 2012. A methodology to estimate greenhouse gases emissions in life cycle inventories of wastewater treatment plants. *Environmental Impact Assessment Review* 37: 37–46.

SEMARNAT. 2013. National Climate Change Strategy, 10-20-40 Vision, Ministry of the Environment and Natural Resources (Secretaría de Medio Ambiente y Recursos Naturales), Mexico.

UNFCCC. 2011. CDM Methodology Booklet. Mitigation of greenhouse gases emission with treatment of wastewater in aerobic wastewater treatment plants. Approved baseline and monitoring methodology AM 0080, United Nations Framework Convention on Climate Change.

USEPA. 1992. US Sewage Sludge Regulations 40 CFR Rule 503. Environmental Protection Agency, Washington, D.C.

USEPA. 1999. Control of pathogens and vector attraction in sewage sludge. Environmental Protection Agency, Washington, D.C.

Walsh, J., Roos, C., Smith, M., Harper, S., and Wilkins, A. 1988. *Handbook on Biogas Utilization.* USA: Environment Health and Safety Division, Georgia Tech Research Institute.

WHO (World Health Organization). 2012. Global costs and benefits of drinking-water supply and sanitation interventions to reach the MDG target and universal coverage. http://www.un.org/spanish/waterfor-lifedecade/sanitation.shtml.

WHO/JMP/UNICEF. 2013. Progress on sanitation and drinking-water: 2013 Update. World Health Organization/Joint Monitoring Programme for Water Supply/United Nations Children's Fund. Paris, France. http://www.who.int/water_sanitation_health/publications/2013/jmp_report/en/.

WSP. 2007. Informe Latinosan 2007. Water and sanitation program. www.wsp.org/wsp/sites/wsp.org/files/11142007111614_Latinosan_Final.pdf.

17 Carbon Footprint of the Operation and Products of a Restaurant
A Study and Alternative Perspectives

Benjamin C. McLellan, Yoshiki Tanaka,
Thi Thu Huyen Dinh, Huong Long Dinh, Piradee Jusakulvijit,
Faizi Ashley Taro Freemantle, and Aibek Hakimov

CONTENTS

17.1 Introduction 369
    17.1.1 Background of the Study Target...370
    17.1.2 Notes on the Educational Process..370
    17.1.3 Literature Review 371
17.2 Methodology 372
    17.2.1 Scoping 372
        17.2.1.1 Functional Unit 373
        17.2.1.2 System Boundaries 373
    17.2.2 Data Collection 374
        17.2.2.1 Initial Data Collection Plan 374
    17.2.3 Inventory..376
17.3 Results...378
    17.3.1 Product-Based CF...378
    17.3.2 Business-Based CF 380
17.4 Discussion...382
    17.4.1 Equipment...382
    17.4.2 Sensitivity Analysis 383
    17.4.3 Comparing the Two Footprints..384
    17.4.4 Limitations of the Study 384
17.5 Conclusion 385
Acknowledgments..385
Appendix...385
References..387

17.1 INTRODUCTION

This chapter presents a carbon footprint (CF) study into a small café-restaurant in Kyoto City, Japan. Commercial businesses, and particularly in multi-product systems such as the retail food sector, are of significant interest from the perspective of CF studies because they offer multiple perspectives

370 The Carbon Footprint Handbook

on the CF process. Such businesses can be assessed from the viewpoint of the organization and its operations as a whole or on the functional unit of the product that is served. In many cases, this product has a variety of alternatives (one-option menus are scarce); hence, there is ample variability and complexity to be considered. Moreover, considering energy use, greenhouse gas emissions, and sustainability more broadly, small-to-medium enterprises are of interest as they contribute a large proportion of the number of businesses operating in most countries and can be significant contributors to national energy demand. Additionally, they often lack the capacity to undertake a carbon footprinting exercise on their own, owing to a lack of time, skills, and budget to internally undertake the study or because of a lack of financial ability to employ a consultant.

Taking all these factors into account, the current study target is a useful demonstration of the CF process. Furthermore, the context in which this study was undertaken was as a class project for Master's students at Kyoto University's Graduate School of Energy Science, studying carbon footprinting. As an educational project undertaken within a limited timeframe and with relatively inexperienced project members guided by the professor, a small business located near to the University with willing owner-managers was ideal. This work represents a confined study (with minimal internal complexity in the business itself) which also provides a good foundational example.

17.1.1 BACKGROUND OF THE STUDY TARGET

The study target in this case was a restaurant-cafe in close vicinity (within 1 km) of Kyoto University. This restaurant is somewhat typical of small businesses in the city, run by an elderly married couple in the middle of a mixed residential-commercial area near Kyoto University for approximately the past 40 years. For much of its time, the business was run as a cafe during the day and a bar at night. It has supported the local student community for many years as a venue for the student bands to play (particularly jazz) and for the students to meet for lunch away from crowded cafeterias. The bar is no longer operating, but the cafe still continues. Keeping the students in mind, the cafe is focused on providing reasonably priced food with a balance of nutrients. The result is what is locally known as a "one coin," "one plate" lunch—a "one coin" lunch in Japan generally means 500 yen, while it is "one plate" in that the main dish, salad, dessert, and pasta are all served on one plate (actually a second plate is used for rice or bread).

The business has only two staff—the owners—who cook, clean, serve, and run all other aspects of the business. The business is housed in a stand-alone, two-story building that is perhaps 50–60 years old. The building is fitted with appliances of various ages, but generally in excess of 10 years, and these appliances are typical of household or residential appliances rather than being commercial items.

17.1.2 NOTES ON THE EDUCATIONAL PROCESS

As already mentioned, this study was conducted by a group of Master's students as a part of the coursework component of their degree. The course itself was divided into three phases:

1. Theory of carbon footprinting
2. Practical project
3. Project report writing

The first of these phases introduced the students to the concepts of life-cycle assessment, the importance of the CF, and its uses in various spheres of society. A footprinting survey tool (developed by the lead author of this study for specific utilization in the context of students in Japan) was used to help the students understand their own individual footprint. The PAS 2050 standards [1] were then introduced as the basis for the project, with an initial overview and then detailed working through the theory of each stage. In this way, students were brought up to a similar level of theoretical background on which to base the project assessments.

The second phase was the most critical with regard to the study itself, as the students were engaged in the footprinting project, applying the theory to a real study. Although the study was quite contained in its scope, it offered most of the important elements that could be found in a more complex study. Without this practical element, the students find understanding the footprinting process more difficult, as it remains abstract. Within the project phase, the students were required to apply a variety of techniques for gathering and estimating data, utilizing a variety of software packages, and negotiating with the client. Most importantly perhaps, the students come to understand the challenges of real-life projects as opposed to theoretical studies, which remain highly within the "ideal" realm.

The third phase of the project was to introduce students to the report-writing process, aiming to instill an understanding of the logical framing of a consultant-style report on a CF project. This process also enabled the collation of results and the solidification of the linkage between the initial theory, the practical project, and the outcomes as reflecting the theory in practice.

Within the Graduate School of Energy Science at Kyoto University, the student base is highly interdisciplinary (social science through to engineering and natural sciences) and the students are from various national backgrounds. This causes a variety of challenges in forming an appropriate course—particularly in achieving an appropriate level of knowledge acquisition. In this context, the current project was highly useful as a practical (rather than theoretical) approach, enabling students of all backgrounds to understand through various styles of teaching and peer-learning. For the remainder of this chapter, the students will generally be referred to as "the project team."

17.1.3 Literature Review

This section presents a brief review of literature on carbon footprinting studies relevant to the study at hand. Potential topics and themes that were reviewed for this study included general studies on food LCA and CFs, hospitality-industry-related studies and conceptual or theoretical examinations of carbon footprinting techniques among other literature related to the broad topic at hand. All these cannot be presented here, but some of the key contextualizing literature is presented.

The CF is the total sum of GHG emissions caused directly or indirectly by an activity or those accumulated over the life stages of a product. It is often expressed (as in this case) in terms of carbon dioxide equivalents of all GHG emissions [2]. From CF data, consumers' awareness of the environmental impacts involved in the production of products will be enhanced, thereby encouraging them to find the potential ways to reduce their CF. Some other studies have examined food menus to identify low impact or efficient meals to assist consumers in their choices [3].

A variety of carbon footprinting methodologies and tools exist and have been widely reviewed by other authors [4] and by others focusing on the specific system boundaries [5]. One particular methodology, using a top-down, national input–output methodology, has been examined for food portions in Finland [6]. This study examined selected nutritionally balanced lunch meals, indicating a wide range of impacts—particularly for meals including meat—with a large share of impacts from domestic agriculture in this case. The most common approach to carbon footprinting of foods is to examine individual ingredients or components of a meal on a bottom-up basis, for example, in a study in India [7] or other lunch plate studies [8]. A particularly relevant study includes reviews of the LCA of bread, dairy and meat, rice, packaging systems, and food waste management for a variety of foods [9]. Other studies particularly focusing on rice were used in this study [10]. Other life-cycle examinations using products similar to the current study have also been undertaken—for example, tomato sauce and bread [11].

The global growth in food consumption continues to cause increasing pressure on the environment and has attracted political and social attention in recent years. The production of consumed food products causes emissions of greenhouse gases across the supply chain, with significant impacts from fertilizer production, utilization, and land clearing. On the global level, 72% of GHG emissions are related to household consumption and food accounts for 20% of GHG emissions [12]. A study on Dutch food consumption indicated that CO_2 contributes 75% of the total greenhouse emissions, while CH_4 and N_2O contribute 19.5 and 5.5%, respectively [13]. The main sources of CO_2 emissions are fuel combustion activities, whereas CH_4 emissions are mainly due to enteric fermentation in ruminant animals and N_2O emissions, which is mainly due to nitrogen fertilizer application [14]. The specific production routes and types of food typical of different countries can vary the locally specific high-impact foods [15]. The variability in both methods of carbon footprinting and the specific production methods has been widely examined (particularly to identify the difference between organic and inorganic produce) [16–18].

On the level of the business, a variety of studies have examined the hospitality industry, including tourism [19], identifying food as one of the key impact elements of the CF. Some studies have examined the possibility of developing sustainability standards for restaurants, again with food procurement identified as a hotspot [20]. Waste is another area of importance for reducing the CF— due to both the emissions associated with land fill or incineration of waste and the implication that wasted food has for over-production at the agricultural stage.

In summary, it was identified that food is an important aspect of the global and residential CF. A variety of approaches have been used for the examination of food products in different countries—in the current case, we will use a bottom-up estimation, but input–output studies at the economy level have also been used. Several studies have examined lunch or meal plates, as this study does, but none that we identified have covered both the business and meal footprints. Waste and agriculture have been widely identified as important hotspots, which we also examine in this work.

17.2 METHODOLOGY

As a project with both educational and scientific aims, the methodology applied closely followed the approach of the British Standards Institute's (BSI) Publically Available Specification (PAS 2050) [1]. This methodology is outlined in Figure 17.1. In this case, the project team undertook the activities in steps 1 and 4 as a combined group, utilizing workshop techniques to come to a consensus on the approach and outcomes. In steps 2 and 3, the team worked as three independent groups, collecting data and undertaking the footprint calculations within the agreed framework. The specific methods are described in the following sections.

17.2.1 Scoping

The scoping phase of the study was undertaken as a series of workshops, gradually working through the required elements of PAS 2050 in order to obtain a near-to-standard performance in assessing the CF. Project team members were tasked with researching specific elements deemed by the team or team leader to be required for clarifying the scope.

The scope of the study included the assessment of a restaurant as a business and the product (selected menu items). In the case of the business, Scopes 1 and 2 emissions were considered as the key operational contributors to CF. This can be effectively attributable to the utility usage of the restaurant. Scope 3 emissions were found to be associated with waste disposal, and the PCF of the materials used in the product (from the calculation of the second substudy) were also included, while other Scope 3 emissions were considered to be immaterial to the business PCF. In the case of the menu items, the PCF associated with the supply chain from cradle to point of consumption is considered. In both cases, the CF has been calculated for a specific geographical location—Kyoto, Japan—and all material supply chains have been calculated to the same location. The year of the study was 2013, and secondary data from 2011 were used as the preferred basis year in this case.

The basis of the emissions factors used in the CF was a global warming potential over a 100-year period. Carbon dioxide equivalent emissions are used throughout.

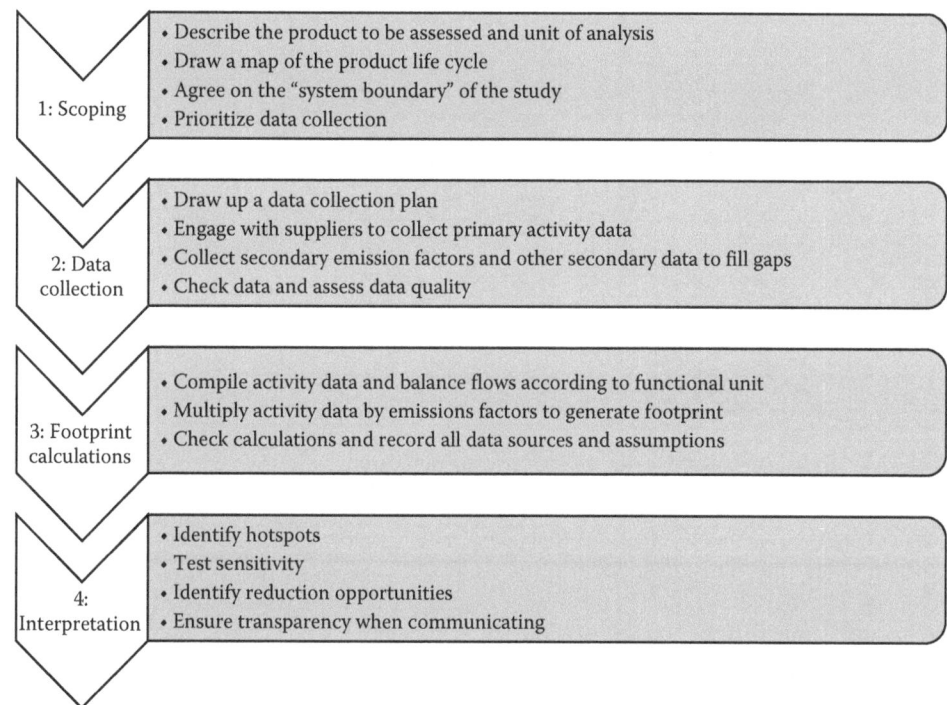

1: Scoping
- Describe the product to be assessed and unit of analysis
- Draw a map of the product life cycle
- Agree on the "system boundary" of the study
- Prioritize data collection

2: Data collection
- Draw up a data collection plan
- Engage with suppliers to collect primary activity data
- Collect secondary emission factors and other secondary data to fill gaps
- Check data and assess data quality

3: Footprint calculations
- Compile activity data and balance flows according to functional unit
- Multiply activity data by emissions factors to generate footprint
- Check calculations and record all data sources and assumptions

4: Interpretation
- Identify hotspots
- Test sensitivity
- Identify reduction opportunities
- Ensure transparency when communicating

FIGURE 17.1 Procedure followed (diagram after PAS 2050). (Adapted from BSI. 2011. Publicly Available Specification: PAS 2050: Specification for the Assessment of the Life Cycle Greenhouse Gas Emissions of Goods and Services. London: British Standards Institute, p. 38.)

17.2.1.1 Functional Unit

Two alternative functional units were used for the assessment in this study—(a) the restaurant as a business and (b) selected menu items as a product. This enabled the project team to examine two alternative results of footprinting within the same overall study subject.

In the first case, all functions of the business are considered. The current case study being a small operation, this is relatively straightforward. The ownership is clear and simple—no stakeholders are involved, and there is no allocation according to share or subfunction.

In the second case, there was potential for more complexity, as the full menu contained a wide variety of items, and there were also daily specials, which could be assessed. Being a cafe, there are also a wide variety of beverages offered. To maintain a relatively restricted scope, the team assessed the CF of two selected menu items that are available every day of operation and that are the most popular with clientele.

17.2.1.2 System Boundaries

As a business, the operational boundaries of the study were drawn around those operations that are under the main control of the cafe. The CF of all inputs and outputs of the business was calculated for the operating phase of the business lifetime, that is, production CF of consumable inputs, operational releases of greenhouse gases, and the disposal of wastes were considered. Most of the operational activities of the business are undertaken onsite at the cafe; however, some cooking is undertaken offsite, and this activity is included, while it is expected that the error involved may be somewhat higher than that in the cafe-based activities, which can be directly measured or monitored.

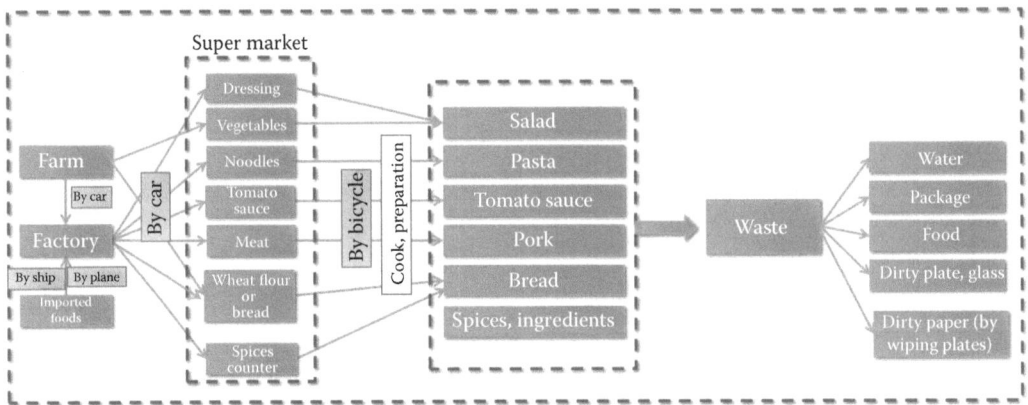

FIGURE 17.2 System boundaries—the product (menu item) case.

Considering the two menu selections that were being assessed, the functional unit was the meal, as provided at the cafe. The meals themselves are readily assessed, being relatively simple in their constituents. The ingredients for the meal were assessed for their CF back to their country of origin (Japan has only around 40% food self-sufficiency, so much of the produce is sourced from offshore). Waste estimates in the cooking phase and from uneaten food were made, and the end of life considered. In this case, most of the organic waste is sent to the clean center for incineration, whereas packaging materials (although minimal) are largely recycled—cans, plastic, and so on. In the case of the product or menu item, the system boundaries are displayed in Figure 17.2.

17.2.2 DATA COLLECTION

Data collection in this project consisted of two types: (1) operational data collection and (2) CF data and emissions factors for calculation. The data collection plan was required to take into account the specific conditions and context of Kyoto city and the operation itself. Culturally and commercially, consideration to the need for privacy in the collection of operational data was considered quite important, particularly given the family nature of the operation. However, the owner was open to data collection and no difficulties with disclosure were obtained. CF data were obtained through the use of Japan-specific databases for use in carbon footprinting [21], as well as being supplemented with data from previous carbon footprinting studies if required.

Both direct data collection and literature-based or theoretical calculations were used, although the preference was for primary data; much of the energy use in a typical small business is limited to electricity and city gas, both of which can be readily converted to a CF with direct emissions measurement. No other direct gaseous emissions are considered onsite.

17.2.2.1 Initial Data Collection Plan

The initial data collection plan was developed over two workshops, between which the project team members undertook independent research on the given project target, identified the required data, and developed questions that might be appropriate to ask so as to gather the data effectively or to understand the available sources of data. As the study target was an operating business, the team preparation was made as thorough as possible so as to not disturb business operations any more than necessary. Table 17.1 shows the questions that were devised for the initial data collection. The questions are split into the two alternative scopes being examined in this study—the business as a whole, and the product or individual meal. Some questions are applicable across both scopes, as indicated in the table. Blank cells in some categories indicate where the questions were concluded as not being relevant for a given scope.

TABLE 17.1

General Categories and Questions for Data Collection

Category	Business	Product
Operating patterns	What are the times of operation?	
	Are these constant or seasonal?	
	Is there a weekly fluctuation?	
	How many customers?	
Operational processes	What are the heating/cooling requirements?	
	What devices are used?	
	How often are they used?	
	What lighting is used?	
	How often is it used?	
	How is cooking undertaken?	
	(How often? What processes? When? Where?)	
	How much water is used by the shop?	How much water is used in making the product?
		a. How are plates washed? (By hand? By machine? When?)
		b. Is water used in cooking? (How? How much?)
Energy inputs	How much electricity is used?	Where is electricity used in the cooking process? (What devices? For what length of time? What setting? Power usage?)
	Can we get seasonal representation?	How long are feedstocks stored? How?
	How much gas is used?	How much gas is used in cooking?
	Is it piped in or delivered in tanks?	(What devices? What length of time?)
	What is it used for? (cooking/heating?)	
	How much oil is used for heating?	
	When is it used? What devices?	
Feedstock		What feedstocks are delivered?
		How often are deliveries made?
		What quantities are delivered?
		How is feedstock delivered?
		(Packaging? Transportation?)
		How much is consumed per serve?
		How much is wasted?
		How many serves per day?
		How is water cooled? How much?
		Where do the various feedstock materials come from?
		a. Wholesale/retail location
		b. Nation or region of origin
		c. Organic/conventional
		How is feedstock stored?
		How long is feedstock stored?
Waste	What type of waste is produced?	How much feedstock is wasted in preparation?
	How is waste removed?	How much is wasted by consumers?
	How much is paid for waste removal?	
Equipment	How much breakage of crockery or loss of utensils?	
	How often do they change large equipment items?	
Employee commuting	How many employees?	
	How do employees commute?	
	If motorized, how far?	
Advertising	How is the shop advertised?	
Accounting	Where/how is accounting undertaken?	

TABLE 17.2

Key Responses to the Initial Data Collection Interview

Category	Business
Operating patterns	General operation: 11:00–20:00
	Preparation included: 8:00–20:30
	Open Monday–Saturday except public holidays
	Lunch is the busiest time
	30–40 clients per day
	Little seasonal variation
Operational processes	Oil heater
	Oil heater used for about 5 months in Winter, morning to lunch or all day in mid-Winter
	Air con—end of June until September, on only when customers come. Air con old (about 14 years)
	4× 100 W lights in the main operating area, usually at half power
	On all through working hours
Waste	Mainly kitchen waste, 3 × 30 L bags per week. Clear bag 30 L once a week
Equipment	Minimal breakage of crockery
	Minimal refurbishment
Employee commuting	Two employees commuting by bicycle
Advertising	Advertising by word of mouth

With regard to the practical collection of the initial data, an interview was set up with the manager in conjunction with a site visit, in a typically slow period of the operational pattern—between the lunch and early evening rush. The project team included native Japanese and other fluent speakers so that language barriers were not involved in the interview process—although the manager is also quite fluent in English. Notes were taken by four members of the project team (two focused on English and two focused on Japanese responses), and the final notes were collated for comparison.

Some of the questions were further specifically tuned to the Kyoto city situations, while a number of the answers could be deduced owing to regulatory and infrastructure restrictions (e.g., the location of waste processing is the closest "Clean Center" or waste facility), thus reducing the time requirement for the interview.

The project team further undertook a guided inventory of the facilities, identifying all the main energy-utilizing items, particularly those relevant to the product—refrigerators and freezers for storage of ingredients, microwave, stove for heating of meals, and so on. Other energy-intensive items, mostly for space heating and space cooling, were identified as potential areas for efficiency gains—particularly as the equipment was typically old, and therefore less efficient than modern appliances. Lighting was also considered, as the current installed system is high wattage incandescent bulbs (100W) set in an adjustable lighting system, therefore typically operating at 50W, but still highly inefficient in comparison with modern alternatives. Water, gas, and electricity bills, and an estimate of fuel oil usage were obtained. Table 17.2 shows the main important responses.

The product-related data were mostly obtained through the initial interview and follow-up interviews by project team members during the second phase of data gathering. As many ingredients were involved, the data are indicated in the inventory section of this chapter. The Scope 3 emissions from the production of equipment, advertising materials, and the commuting of the employees were considered immaterial based on the responses in Table 17.2.

17.2.3 INVENTORY

The initial interview provided a starting point for the project team to refine the data requirements and seek corroborating evidence or secondary sources where necessary. For the menu items,

TABLE 17.3

Important Data Regarding the Product CF

Category		Product
Feedstock	What feedstocks are delivered?	Bread, rice
		(others are bought and transported by bicycle)
	How often are deliveries made?	Bread—once every 2 days
		Rice—once per week
	What quantities are delivered?	Bread—750 g × 2–4 /time
		Rice—10 kg/time
	How is feedstock delivered?	By truck
	How much is consumed per serve?	(see menu specifics)
	How much is wasted?	Almost none
	How many serves per day?	30–40
	Nation or region of origin	Beef—United States or Australia or Mexico
		Pork—Mexico or United States
		Chicken—Japan (Tottori prefecture)
		Pasta—Italy
		Rice—Japan
		Vegetables—Japan (sometimes China)
		Fruit (dessert)—Japan (seasonal or frozen onsite)
		Organic/conventional—not taken into
		consideration in purchasing
	How is feedstock stored?	Refrigerated onsite (apart from bread and rice)
	How long is feedstock stored?	One week maximum

the important primary data include the quantity of each item constituting a meal (Table 17.3). Such data were obtained by direct measurement. The owners indicated that the serves varied somewhat in size, but the tested range was relatively minimal, although the sample size (two samples of two menu options) was not large enough to provide any statistical accuracy. The inventory of the main menu items is shown in Table 17.4. The specific fruit dessert varies daily; so, in this case, a frozen apple dessert was used in the CF.

TABLE 17.4

Menu Combinations Considered—Everyday Menu

Items on the Menu	Mass Range of Tested Serves (g)
Standard side menu items (all meals):	
i. Salad (mixed lettuce, carrot, and spinach)	30–35 g
ii. Pasta (macaroni)	32–70 g (cooked) (16–35 g uncooked)
iii. Fruit dessert	41–43 g
Main dish (A or B)	
A. Chicken set	
Chicken (chicken filét with sauce and spices)	130–142 g (including 5 g of sauce)
B. Pork set	
Pork (Pork meat with tomato sauce)	129–153 g
Choice of bread or rice:	
C. Bread (dark bread): 2 pieces	100–130 g
D. Rice (white standard rice, boiled)	111–178 g (unboiled rice approx. 50% of mass)

TABLE 17.5
Emissions Factors for Utilities

Utility	Emission Factor	Source
Water (kg CO_2-eq./m³)	0.376	Japan Ministry of Environment http://www.env.go.jp/council/06earth/y060-69/mat09-5.pdf
Electricity (kg-CO_2-eq./kW h)	0.45	Kansai Electric Power Company (KEPCO) http://www.kepco.co.jp/pressre/2012/0724-2j.html
Gas (kg-CO_2-eq./m³)	2.08	J&T Systems http://www.jt-sys.co.jp/business/keisuu.html

CFs of most items were obtained from Japan-specific data bases [21], including the emissions for transportation, and for the production of the products themselves. Onsite cooking energy usage was estimated, and the refrigeration or freezing of some items was typically considered to be 1 day (shopping is undertaken each day for fresh foods). Pasta was imported from Italy, pork from Mexico or United States, and wheat flour (for making bread) was from Australia or the United States, but the remaining items were domestically grown (a full list of transportation routes identified in Appendix). All goods requiring international transportation were expected to arrive in a ship, while domestic transport of goods was undertaken by a truck. Most fresh ingredients were purchased and transported to the cooking site by bicycle, while bread, rice, pasta, and oil were delivered by small commercial vans.

Emissions factors were obtained from the Japanese domestic databases [21] or from other local CF studies as a preference. Distances for domestic transport were estimated with Google Maps™. Emission factors for energy were sourced from the electricity or gas providers as a preference, then from other institutional sources (Emissions factors are given in Table 17.5).

17.3 RESULTS

The results of the study are presented here in their compiled form, but discussion will mostly follow in the next section. Short descriptions are given for clarity reasons.

17.3.1 PRODUCT-BASED CF

On the basis of the product or menu set meal, the assessment of the CF yielded the following results. These are displayed here in several ways to enable analysis. The calculations based on secondary data for production and transportation emission factors, primary data for serve sizes, and distance calculations resulted in the effective emissions factors for each of the meal components were shown in Table 17.6. These were used for the remainder of the study calculations of the CF of each meal.

TABLE 17.6
Effective Calculated Emission Factors
(kg CO_2/kg) for Each Meal Component

Apple	0.47
Bread	1.04
Rice	1.90
Pasta	4.19
Pork	3.65
Chicken	1.26
Salad	1.90

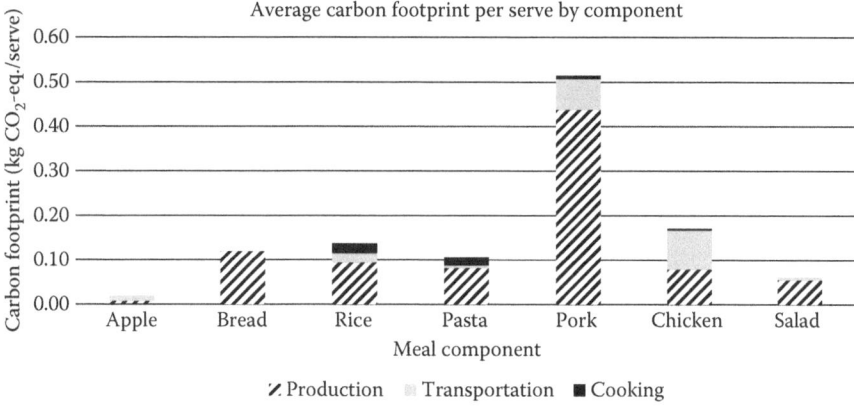

FIGURE 17.3 Total CF of each item on the alternative menus.

In Figure 17.3, the total CF for the serving size of the alternative components that can be combined to make the respective pork or chicken set meals is shown. It is apparent that the pork, pasta, and rice contribute significantly—in the case of pork and rice, this is associated with the large serving size, while pork is exacerbated owing to its larger PCF per unit. With regard to pasta, there is a large contribution from the transportation of pasta from Italy (typical of the actual product used).

In Figure 17.4, the four meal set alternatives are displayed—the two choices of pork or chicken and the alternatives of rice (set 1) and bread (set 2). It is apparent that the pork has a higher CF than the chicken sets in both cases, and that the rice sets have a higher impact than bread.

To illustrate some of the details of the ingredient item CFs, the breakdown of CF contributions from the various lifecycle components is shown in Figure 17.5. In this case, a small percentage of the rice and bread CF is attributable to disposal of wasted food and packaging, while the majority of the impact is in the raw materials. Owing to the alternative production processes, the location of impacts is different. The production phase of bread includes baking offsite locally in Kyoto, then in the use phase it is toasted onsite, while for rice the use phase involves cooking onsite, but distribution of the rice from other parts of Japan is a large contributor as well.

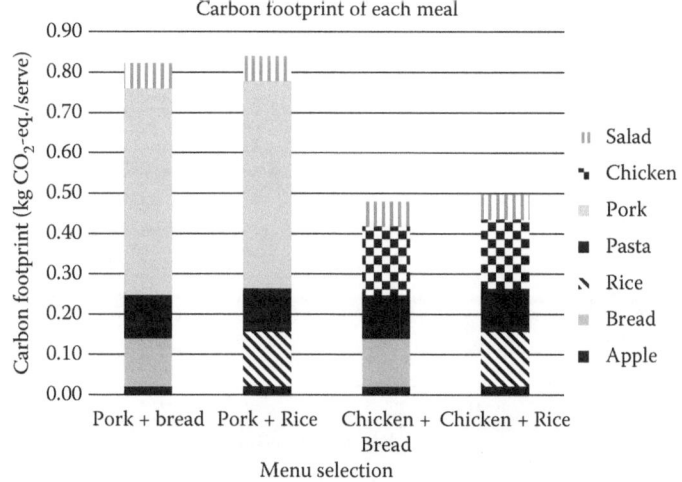

FIGURE 17.4 CF for the alternative menu items—set 1 with rice, set 2 with bread.

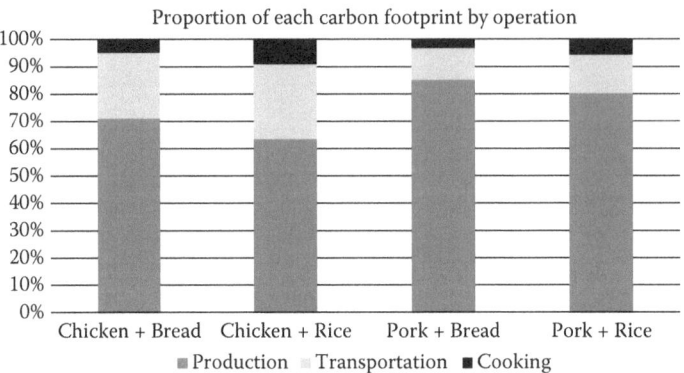

FIGURE 17.5 Contributions to the life-cycle CF of life-cycle stages.

Figure 17.5 indicates that the production (manufacture) of the ingredients is the major component of the CF, followed by transportation. The production impact is the most difficult to improve, as it is largely dependent on the agricultural practices and in this study we have not indicated a differentiation in these practices by nation. The transportation can be reduced by sourcing foods locally, but Japan as a whole has a low rate of food self-sufficiency, and some of these items can only be sourced from other countries. Particularly in the case study at hand, there is minimal cost flexibility to source (typically expensive) Japanese meat.

17.3.2 BUSINESS-BASED CF

The approach taken to the business CF does not include the "Scope 3" emissions from the production and disposal phases of the products used in the cafe. Instead, it is focused on the usage of major utilities, which, for the small business example where no other major emissions are created onsite, is the most effective and efficient method of estimation. Three alternative methods were used for obtaining the electricity, gas, and water-usage estimates—firstly, by interviewing; secondly, by asking for the official account statements; and thirdly, by monitoring (with permission) the gas and electricity meters on a daily basis. The other major elements of the business CF were the use of heating oil in the winter and the emissions associated with waste (although this is considered out of scope for the present study, we have estimated this below).

Figure 17.6 shows the emissions from the business due to electricity and gas usage. As mentioned already, the business uses gas mainly for hot water and cooking onsite, whereas lighting, cooling,

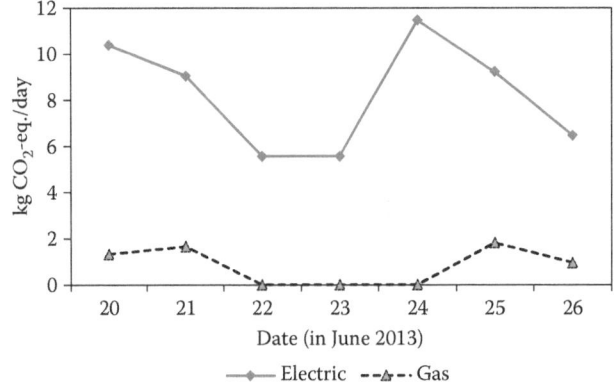

FIGURE 17.6 CO_2 emissions per day of electricity and gas.

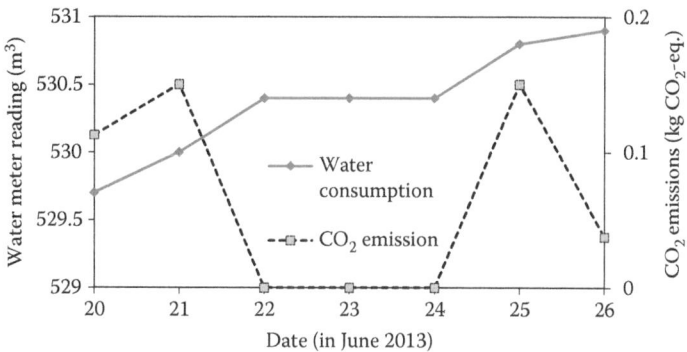

FIGURE 17.7   Water meter reading and equivalent CO_2 emissions.

refrigeration, music, a beverage vending machine, and the microwave oven are major electricity-drawing items. The 23rd of June was a Sunday, indicating that the draw of electricity and gas was much lower when compared with that on an operating day. Figure 17.7 shows the equivalent graph for the consumption of water and the equivalent CF associated with that consumption.

Table 17.7 indicates the climate data for the days within the period of monitoring, and the emissions for that period. It is important to note that, due to the limited period of monitoring, in a relatively mild climatic season, the energy use in this period cannot be used to represent the entire year's energy utilization. In lieu of this, and given the availability of only the most recent utilities bills, the annual CF of the business was forced to rely heavily on the initial interview estimates of electricity, gas, water, and oil monthly bills and their variability (Table 17.8).

TABLE 17.7

Climate and Emissions Data Recorded over a 1-Week Period

June (Date)	Temperature (°C)	Humidity (%)	CO_2 Emission (kgCO$_2$-eq.)		
			Water	Electricity	Gas
20	20.2	91	0.1128	10.395	1.32704
21	20	88	0.1504	9.045	1.65776
22	22	74	0	5.58	0.00208
23	23	74	0	5.58	0.00208
24	23.7	73	0	11.475	0
25	25	67	0.1504	9.225	1.8096
26	21.3	88	0.0376	6.48	0.9568

TABLE 17.8

Annual CF for the Business Based on the Utilities Costs

Utility	Usage	Annual Emissions (kg CO_2-eq.)	Contribution to CF (%)
Electricity	1000 kW h/month	5400	70
Gas	40 m³/month	970	13
Fuel oil	100 L/month (5 months)	1290	17
Total		7660	100

Japan's energy mix has been a topic of particular interest since the Fukushima accident of 2011. The failure to restart nuclear power plants to date has increased the carbon intensity of grid electricity significantly from the relatively low emissions when the full nuclear fleet was active. At the time of this study in 2013, the electricity mix of Japan consisted of only 2% produced from nuclear energy, LNG had increased up to 48%, and oil has increased up to 16% of the share from 8%. A variety of alternative future scenarios are possible, with nuclear power or without it, and this has a particularly strong influence on the CF of businesses such as the cafe in this study, as they are not in the financial situation to be able to consider renewable energy (at the current cost levels), and are therefore restricted to grid electricity. The only reasonable opportunity for reducing CF then is efficiency.

Scope 3 emissions from the product and from the treatment of waste from the operations were also calculated. When the onsite cooking-related emissions are removed (so as not to double-count), the emissions from the product are on average 7 t CO_2-eq./year with a range of uncertainty between 4 and 12 t CO_2-eq. per year (for 30–40 serves per day and 300 days operation).

17.4 DISCUSSION

The CF of the chicken meal was found to be 0.48 kg CO_2 equivalent for the chicken meal with bread, and 0.5 kg CO_2 equivalent if with a plate of rice instead of bread. The largest single contributor was the meat component. Meat is generally considered to be bad for the environment but this research shows that the CF associated with chicken in Japan is in fact relatively low compared with rice and pasta (especially pasta because it is imported). This is partly down to the fact the chicken is produced domestically, but in fact the largest proportion of CF for meat and agricultural products is in the production stages. Pork was found to have a large CF, contributing 0.47 CO_2-eq. per meal.

However, there are still areas where the CF could be reduced. The cooking stage was identified as a hotspot for pasta and meat; so the cafe could consider using a more efficient gas cooker. The CF of drinks was not considered here but it is likely that these would contribute significantly, especially from the refrigeration stage. The refrigerators used in the cafe are very old and therefore not energy efficient, and are of course left on 24 h to keep the drinks cool. The drinks, unlike most of the food, are not bought and consumed daily but delivered occasionally and left in the refrigerators. Along with drinks to order, a glass of water with ice is given with every meal. The energy and CF of this was not considered as part of the meal in this study, but the use of ice may be a significant contributor.

By way of comparison, the impact of a single lunch portion in the Finnish situation, calculated with a top-down approach, ranged from 0.65 to 3.80 kg CO_2-eq. [6]. This places the meals here in the lower end of the range—most likely due to the lower meat and dairy content.

17.4.1 EQUIPMENT

Mapping out the equipment and the energy consumption of each item give some idea of which are the most energy-intensive processes (therefore the ones with the highest CF). Some solutions to lower the CF would be to invest in more energy-efficient, and therefore eco-friendly equipment. Below are the results for each major equipment item, with our estimation of how many hours per day (on average) it is used as well as monthly and yearly consumption.

Of the items in Table 17.9, it is apparent that the largest consuming items are the large storage refrigerators, the air conditioner, lighting, and the vending machine. Of these items, the large (upstairs) storage refrigerator is the most intensive, and most of these equipment items are old and inefficient. From the financial perspective, it may not make sense to replace these items at present, owing to the low revenue and the expectation that the end of operations may not be too many years in the future. However, some immediate efficiency measures could be undertaken at relatively low cost for reasonable return on investment.

TABLE 17.9

Major Equipment Items and Utilization

Name of the Equipment (Operating Hours)	Power Consumption/ Month (kW h)	Power Consumption/ Year (kW h)
Big fridge (Fujitsu) (24 h/day)	46	552
Fridge upstairs (Aurora ER-177) (24 h/day)	110	1322
Air conditioner Corona—S220 (3 h)	61	734
Vending machine KUBOTA (24 h)	810	9720
Microwave TKS-8402A5 (1 h/day)	28	331
Lamps 4× 100 W (half-dimmed) 10 h/day	48	576

It would be recommended that the upstairs refrigerator be turned off if possible, and the storage reduced to a minimum. The turnover and the daily supply of most feedstocks would indicate that there is room for this type of action. The replacement of the lamps with LEDs at full voltage would still be expected to reduce the power consumption to perhaps 25% of the current pattern. The vending machine could also be considered, particularly if the revenue is not large. Heating and cooling loads could potentially be reduced by retrofitting appropriate insulation or improving the heating devices; again these would be costs that would be unlikely to have sufficient benefit to be worth the investment.

17.4.2 SENSITIVITY ANALYSIS

In the case of the business footprint, one of the most important factors is the emissions associated with electricity utilized. This is of particularly relevance in the post-Fukushima environment in Japan, in which the effect of the shut-down of nuclear power is acutely impacting the emissions factors of electricity. In the current study, the 2011 emissions factor for the monopoly local supplier was utilized however, since the completion of the study the emissions have risen owing to minimal nuclear power in the grid. A sensitivity analysis is used to identify the quantitative impacts of Japan's political decisions, which may lead to more or less nuclear power in the grid, as has been undertaken in previous footprinting studies [22]. Table 17.10 shows the effect that electricity emissions have on the overall PCF for the business or the product.

As electricity in this case provides 70% of the base-case CF for the business, the impact of grid mix is particularly significant. Importantly, under a near-zero-nuclear scenario, represented by the 2013 emissions, the PCF increases by around 11%, whereas a return to pre-Fukushima nuclear generation leads to a 22% reduction in the operational CF.

Examining the uncertainty brought into the CF due to the variability of serve sizes (across the range shown in Table 17.4), the contribution of each of the meal components is shown in Figure 17.8. The uncertainty is greater for those items demonstrated to have a relatively large variability in the serving size. This could in some cases change the priority of CF reduction, for example, the comparison between bread, rice, and pasta may change.

TABLE 17.10

Emission Factors and Sensitivity of the PCF (Study Basis Year Was 2011)

Year	Emissions Factor (kg CO_2-eq./kW h)	Fraction of Basis Year	Change in PCF	
			Business (Scope 1 + 2)	Product (Scope 3)
2013	0.522	1.16	+11%	
2011	0.450	1	0	0
2010	0.311	0.69	−22%	

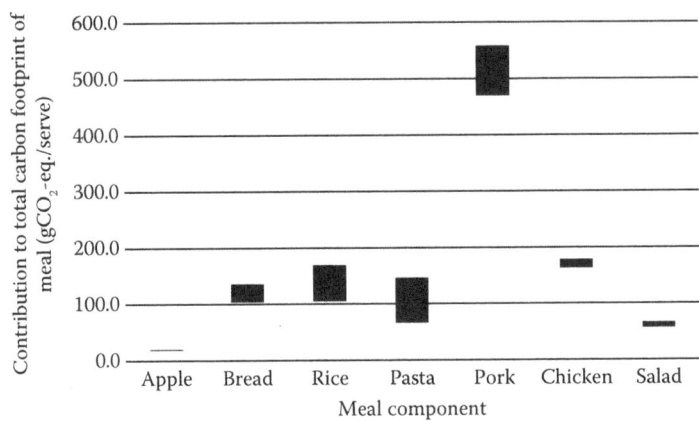

FIGURE 17.8 Uncertainty analysis based on food proportion variation (range of potential contributions).

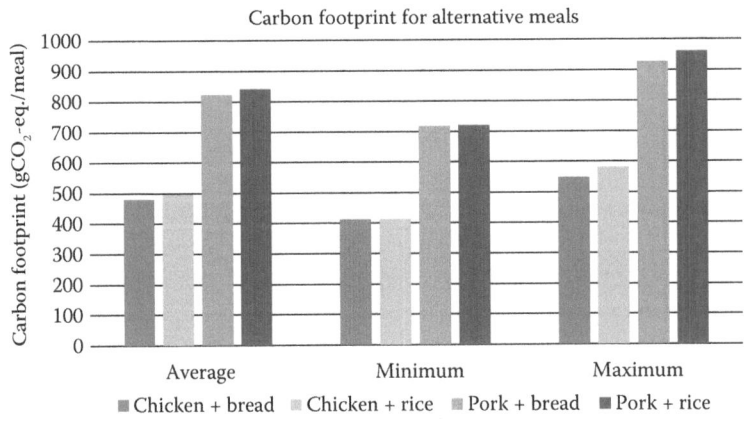

FIGURE 17.9 Variability of CF of meals under uncertainty analysis based on range of portion sizes.

Figure 17.9 shows the variability under the same uncertainty assessment using the range of serving sizes. The overall variability is from 13 to 17%, with the largest variation associated with the chicken and rice meal. Further uncertainty analysis could look at the effect of the different sources of feedstock on the CF, but that was not undertaken owing to the limitations of time.

17.4.3 COMPARING THE TWO FOOTPRINTS

Making a comparison of the two different CFs, the annual business footprint of around 7700 kg CO_2-eq. is not high. If the meals of the average 30–40 customers daily are taken into account, the equivalent footprints of the sold products are around 4000–12,000 kg CO_2-eq. annually. If the company were to consider including such emissions in the scope of its CF, then it would be another 1–2 times the direct emissions associated with operations.

17.4.4 LIMITATIONS OF THE STUDY

The study aimed to be comprehensive and detailed, but there were apparent limitations associated with the timing, available data, and methods. These limitations can largely be attributed to the nature of the business (a small business with aging infrastructure and minimal budgetary scope for improvement) and to the nature of the study as a project run over a set period of the year with importance placed on not disturbing the operations to an excessive extent. The primary data were only able to be directly measured over a single week of the study—in a particularly mild period of the year—meaning that the operational variations of hot summer and cold winter heating and cooling requirements could only be estimated from the cited energy expenditures.

17.5 CONCLUSION

This study has demonstrated the carbon footprinting process applied to a small restaurant cafe in Kyoto, Japan. The study is considered to be an effective demonstration of the alternative methods of footprinting a business—using both the operation and the product as examples.

The direct footprint of business operations was calculated using the monthly estimates of electricity, gas, and heating oil use, while the footprint of two menu items was calculated with direct measurements and secondary data on emissions factors. The direct impact of the business (Scopes 1 and 2) is estimated to be in the range of 0.5–1 times the Scope 3 emissions owing to the consumption of the products they produce.

The business footprint was limited to an estimation based on interview data and emissions factors; although some monitoring of the utilities consumption was undertaken, there was little long-term and seasonally representative data available.

The production stage of the food inputs (from agriculture through processing) was found to have the highest impact in the CF, followed by transportation. While the transportation element can be reduced somewhat by buying locally, the production component is unavoidable, and could only be varied with the ability to footprint individual farms or national agricultural sectors.

This study has shown some interesting challenges when dealing with a small enterprise that has limited means with which to promote environmental improvement or to manage its data for this purpose. This is reflected in the limited opportunities for improvement, which do not extend much beyond the purchase of some efficient light bulbs and perhaps turning off old appliances. The budget for refurbishment or retrofit is unlikely to be available or acceptable for spending in the lifetime of the cafe.

ACKNOWLEDGMENTS

The authors sincerely appreciate the cooperation of the "Master" and his wife, owners of the small cafe restaurant, Zaco, used as an example in this study. Without their assistance in providing data, and acceptance of frequent interruptions from the project team, the CF project would have lacked the applied aspect. Educationally and scientifically from the point of value of the study, this access to empirical operational data and the interaction with the proprietors was invaluable.

APPENDIX

CALCULATIONS

Transportation calculations using appropriate emissions factors were developed to give proportional representation of the routes for supplying the main materials. For example, transportation of apples was based on the mean distance to the local supermarket (840 km) and EF of a 1.5 t truck (0.351 kg-CO_2/(t km)). This mean distance was calculated as $\Sigma(D_i \times P_i)$, where D_i is the distance from each Japanese prefecture (i) that produces many apples, and P_i is the proportion of the shipment or production in prefecture (i) to the total (Table 17.11).

This approach of mean distance $\Sigma(D_i \times P_i)$ was also applied to calculate CFs of transportation of chicken, salad, and pork in the United States (from States i to Long Beach).

TABLE 17.11

Example Calculation—Distance of Apple Transportation for Dessert

	Prefecture (i)	Proportion of Shipment (Pi)	Distance (Di) (km)
1	Aomori	0.550	1027
2	Nagano	0.185	400
3	Iwate	0.069	963
4	Yamagata	0.062	721
5	Akita	0.044	834
6	Fukushima	0.044	708
7	Gunma	0.012	467
8	Hokkaido	0.011	1443

Source: Adapted from http://www.japan-now.com/article/187570142.html

TABLE 17.12

Transportation Emissions Factors

Mode	EF Value	Unit
Truck (1.5 t, 100% load)	3.51E–01	tkm
Truck (freezer, 1.5 t)	1.02E+00	tkm
Truck (freezer, 20 t, 100% load)	7.01.E–02	tkm
Ship (Container ship, >4000 TEU)	9.52E–03	tkm
Ship (freezer ship, >4000 TEU)	9.58E–03	tkm
Water (piped)	3.48E–01	m^3
Chicken (production)	1.23E+00	kg
Salad (production)	1.90E+00	kg
Apple (production)	4.65E–01	kg

Routes of Transportation were assumed as following:

- Apple, Chicken, Salad: Japanese Prefectures—supermarket
- Bread: Bakery (Kyoto)—Zaco
- Rice: Shiga—Kyoto, Yamaya (Kyoto)—Zaco

TABLE 17.13

CF of Components of the Set Menu (kg-CO_2/Serve)

Component	Production	Transportation	Cooking	Total
Apple	0.01	0.013	0	0.02
Bread	0.14	0.00010	0	0.14
Rice	0.23	0.054	0.053	0.34
Pasta	0.24	0.018	0.057	0.31
Pork	0.40	0.064	0.0061	0.47
Chicken	0.082	0.093	0.0038	0.18
Salad	0.058	0.0081	0	0.067

- Pasta: Tesco factory (London)–London–Genoa (Italy)–Osaka—supermarket
- Pork: United States–Long Beach (USA)–Tokyo–Osaka—supermarket

Transportation emissions factors utilized were as shown in Table 17.12, which were taken from the JEMAI Carbon Footprint Databases [21]. Table 17.13 gives the CF data on each meal component broken down by life-cycle stage.

REFERENCES

1. BSI. 2011. *Publicly Available Specification: PAS 2050: Specification for the Assessment of the Life Cycle Greenhouse Gas Emissions of Goods and Services.* London: British Standards Institute, p. 38.
2. Wiedmann, T. and J. Minx. 2008. A definition of "carbon footprint." *Ecological Economics Research Trends* 1: 1–11.
3. Carlsson-Kanyama, A., M.P. Ekström, and H. Shanahan. 2003. Food and life cycle energy inputs: Consequences of diet and ways to increase efficiency. *Ecological Economics* 44(2): 293–307.
4. Čuček, L., J.J. Klemeš, and Z. Kravanja. 2012. A review of footprint analysis tools for monitoring impacts on sustainability. *Journal of Cleaner Production* 34: 9–20.
5. Matthews, H.S., C.T. Hendrickson, and C.L. Weber. 2008. The importance of carbon footprint estimation boundaries. *Environmental Science and Technology* 42(16): 5839–5842.
6. Virtanen, Y. et al. 2011. Carbon footprint of food—Approaches from national input–output statistics and a LCA of a food portion. *Journal of Cleaner Production* 19(16): 1849–1856.
7. Pathak, H. et al. 2010. Carbon footprints of Indian food items. *Agriculture, Ecosystems and Environment* 139(1): 66–73.
8. Kurppa, S. et al. 2009. Environmental impacts of a lunch plate-challenges in interpreting the LCA results. Paper in proceedings of the conference "Future of the Consumer Society" 28–29 May, Tampere, Finland, 185–193.
9. Roy, P. et al. 2009. A review of life cycle assessment (LCA) on some food products. *Journal of Food Engineering* 90(1): 1–10.
10. Roy, P. et al. 2009. Life cycle inventory (LCI) of different forms of rice consumed in households in Japan. *Journal of Food Engineering* 91(1): 49–55.
11. Andersson, K., T. Ohlsson, and P. Olsson. 1994. Life cycle assessment (LCA) of food products and production systems. *Trends in Food Science and Technology* 5(5): 134–138.
12. Hertwich, E.G. and G.P. Peters. 2009. Carbon footprint of nations: A global, trade-linked analysis. *Environmental Science and Technology* 43(16): 6414–6420.
13. Kramer, K.J. et al. 1997. Greenhouse gas emissions related to Dutch food consumption. *Energy Policy* 27(4): 203–216.
14. Weber, C.L. and H.S. Matthews. 2008. Food-miles and the relative climate impacts of food choices in the United States. *Environmental Science and Technology* 42(10): 3508–3513.
15. Mungkung, R. et al. 2012. Product carbon footprinting in Thailand: A step towards sustainable consumption and production? *Environmental Development* 3: 100–108.
16. Lynch, D., R. MacRae, and R. Martin. 2011. The carbon and global warming potential impacts of organic farming: Does it have a significant role in an energy constrained world? *Sustainability* 3(2): 322–362.
17. Röös, E., C. Sundberg, and P.-A. Hansson. 2010. Uncertainties in the carbon footprint of food products: A case study on table potatoes. *The International Journal of Life Cycle Assessment* 15(5): 478–488.
18. van Rikxoort, H. et al. 2014. Carbon footprints and carbon stocks reveal climate-friendly coffee production. *Agronomy for Sustainable Development* 1–11.
19. Gössling, S. et al. 2011. Food management in tourism: Reducing tourism's carbon "footprint." *Tourism Management* 32(3): 534–543.
20. Baldwin, C., N. Wilberforce, and A. Kapur. 2011. Restaurant and food service life cycle assessment and development of a sustainability standard. *The International Journal of Life Cycle Assessment* 16(1): 40–49.
21. JEMAI. 2012. Carbon footprint of products. http://www.cms-cfp-japan.jp/calculate/verify/permission.php (accessed July 23, 2012).
22. Hassard, H.A. et al. 2014. Product carbon footprint and energy analysis of alternative coffee products in Japan. *Journal of Cleaner Production* 73(0): 310–321.

18 Cultivation of Microalgae

Implications for the Carbon Footprint of Aquaculture and Agriculture Industries

Kirsten Heimann, Samuel Cires, and Obulisamy P. Karthikeyan

CONTENTS

18.1 Introduction 389
18.2 Biology of Microalgae: Cell Physiology and Photosynthesis 392
    18.2.1 Hetero- and Mixotrophy 393
    18.2.2 Photosynthesis (Autotrophy) 393
    18.2.3 Accessory Pigments for Light Harvesting 395
    18.2.4 Fertilization 396
18.3 Critical Factors for Microalgal Growth 398
    18.3.1 Light and Temperature 398
    18.3.2 Fertilization 400
    18.3.3 Trace Metals 401
18.4 Mass Cultivation Systems 401
18.5 Bioproducts from Microalgae 410
18.6 Case Study: Aquaculture Industry 413
    18.6.1 Goal and Scope 414
    18.6.2 Methodology 415
18.7 Case Study: Agricultural Industry 417
18.8 Conclusions 419
References 420

18.1 INTRODUCTION

Dwindling oil and phosphorous reserves, climate change-induced unstable weather patterns, rising seawater levels, and sea surface temperatures are global challenges interlinked with and exacerbated by a rapidly growing human population with increased energy, fuel, and food demands. The world population is predicted to increase from 7 to 9 billion people by 2050—an increase of 47% from the year 2000 (UN 2004), requiring secure energy, fuel, and food resources, which are prerequisites for stable economic growth and social stability (world peace). Clean freshwater resources and arable land are already scarce in some developing nations and the impact of climate-induced changes in global rainfall patterns, temperatures, weather stability, and rising sea levels, which are thought to be tightly linked with rising greenhouse gas (GHG) emissions, are predicted to add enormous pressures on these scarce and life-enabling resources, particularly in countries with densely populated coastal areas and little inland arable land availability, more intense population growth, and island/delta communities (e.g., North Africa, the Middle East, South and Central Asia and several areas in North and South America) (AWC 2009). Africa is likely to be especially hard hit by climate change-induced freshwater availability, as its population is predicted to increase to 20% of the world population, up by 7% from previous estimates (UN 2004). In addition to negative impacts on socio-economic development, particularly in already arid or semi-arid regions, climate change is also expected to adversely affect agricultural productivity and therefore food security and health, the latter through increase in disease and decreasing freshwater quality (IPCC 2007; IDA 2007).

Algae, that is, microalgae, are often hailed as the "miracle cure" for the world's ailments, as they can be cultivated on nonarable land in water sources not suitable for human consumption or irrigation (e.g., sea-, brackish water, and polluted water, i.e., metals, nutrients) (ATSE 2008) and, like terrestrial and aquatic plants, convert carbon dioxide (CO_2) into organic biomass through photosynthesis (Andersen 2005). In addition to potential for GHG and wastewater remediation, microalgae could provide new protein-rich diets and invaluable resources for long chain polyunsaturated omega-3 fatty acids (LC-ω-3-PUFA) like docosahexaenoic acid (DHA) and eicosapentaenoic acid (EPA) and antioxidants (oxygen radical scavengers), which are essential for maintaining human health and nutrition of the growing global population (Raposo et al. 2013). DHA and EPA are essential fatty acids for human/animal development and cardiovascular health (Huerlimann and Heimann 2013). As synthesis of these fatty acids is extremely slow in animals and humans, diets need to be supplemented with sources enriched with these PUFAs, such as oily fish or extracts (Huerlimann and Heimann 2013). Often, nonprotein sulfonic acid (i.e., taurine) is overlooked and lumped with amino acid profiles. Taurine is essential for carnivorous fish and not available in any land plants. It has been reported in the macroalgae *Laminaria, Undaria, Porphyra,* the prasinophye *Tetraselmis,* the red alga *Pophyridium,* the heterotrophic dinoflagellate *Oxyrrhis,* and the diatom *Nitzschia* (Al-Amoudia and Flynn 1989; Flynn and Flynn 1992; Jackson et al. 1992; Dawczynski et al. 2007). Thus, supplementation of agricultural conventional foods with microalgae could ensure high-quality foods for the growing population in the face of scarce freshwater and arable land resources.

Feed formulation for aquaculture is being studied intensely, specifically focusing on substitution of fishmeal and microalgae with other protein, lipid and pigment sources due to dwindling supplies of marine oils and the costly production of photosynthetic microalgae, which can account for close to 50% of the overall production cost of the industry (e.g., Wassef et al. 2009; Babalola and Apata 2012; Bauer et al. 2012; Poot-Lopez et al. 2012). Several companies have shifted their focus from algal biodiesel production to high-value products such as ω-3 and protein-rich biomass as animal feed (e.g., Aurora Algae, MBD, Cellana) (Adarme-Vega et al. 2012) despite microalgal cultivation of a few select species for feed in aquaculture is well established (a few recent examples of microalgal diet development in bivalve hatcheries are listed in Table 18.1). It is impossible within the content of a book chapter to list all species and target aquaculture species, which, for many of the latter important species (i.e., crustaceans, fish larvae, etc.), includes the rearing of feed zooplankton on microalgal diets. A microalgae in aquaculture review lists the most commonly used species, such as the haptophytes *Isochrysis* sp. (T-ISO), *Prymnesium* spp. and *Pavlova lutheri,* the diatoms *Chaetoceros calcitrans, C. muelleri* (synonym *C. gracilis*), *Thalassiosira pseudonanna, Skeletonema* spp., *Navicula* spp., *Nitzschia* spp., *Cocconeis* spp., *Amphora* spp., the green microalgae *Tetraselmis suecica* and *Pyramimonas* spp., the cryptophyhte *Rhodomonas* spp. and the eustigmatophyte *Nannochloropsis* spp. for use as feed organisms for rearing of bivalve moluscs, crustacean larvae, juvenile abalone (benthic diatoms *Navicula* spp., *Nitzschia* spp., *Cocconeis* spp., and *Amphora* spp.), and the rearing of zooplankton, as feed for crustaceae and fish larvae (Brown 2002). These genera literally present only a handful of the several thousand existing genera that might be suitable as feed or food-supplement organisms.

Microalgae has been demonstrated to be beneficial as nutrient supplements, boosting the immune system, improving lipid metabolism, showing antiviral and antibacterial actions in fish, and providing stress resistance of the feed animals (Güroy et al. 2011; Nath et al. 2012; Sheikhzadeh et al. 2012). Nutritional values of the most commonly used microalgae in aquaculture were reviewed recently (Brown and Blackburn 2013) and the fatty acid profiles of 12 strains belonging to the cyanobacteria, red algae, cryptophytes, diatoms, green microalgae, xanthophyceae, and eustigmato-phytes have been investigated (Patil et al. 2007). However, as shown in Table 18.1, mixed diets may show better results in terms of survival, growth, or biochemical composition of the target aquaculture species. In addition, as briefly outlined later (Section 18.2), biochemical profiles, while species-specific with regard to dominant carbon storage form (carbohydrate vs. lipid) and fatty acids (e.g., contents of LC-ω3-PUFA), are greatly influenced by other environmental variables and also growth phase (Thompson et al. 1993; Renaud et al. 2002; Xu et al. 2008; Huerlimann et al. 2010; Khoeyi et al. 2012; Islam et al. 2013; Marchetti et al. 2013).

TABLE 18.1
Examples of Microalgal Diet Development for Bivalve Hatcheries

Target Aquaculture Species	Microalgal Diet	Characteristics Positively Influenced	Reference
European oyster, *Ostrea edulis* juveniles	*Nannochloropsis oculata, Pavlova lutheri*	Growth and PUFA profile (DHA, EPA)	Ronquillo et al. (2012)
Brown mussel, *Perna perna*	Mixed diet of *Isochrysis galbana* with diatoms: (a) *Chaetoceros calcitrans*; (b) *Phaeodactylum tricornutum*; (c) *Skeletonema costatum* vs. monospecific diet of *I. galbana*	Higher growth rates on mixed diets driven by EPA and DHA contents	Aarab et al. (2013)
Juvenile lions-paw scallop, *Nodipecten subnodosus*	Monospecific diet of *Isochrysis galbana* vs. 3 binary diets: (a) *I. galbana* + *Pavlova salina*, (b) *I. galbana* + *Chaetoceros muelleri*, (c) *P. salina* + *C. muelleri* and one ternary diet of *I. galbana* + *P. salina* + *C. muelleri*	Scallops grew significantly faster and larger when fed *P. salina* + *C. muelleri* at 28 and 24°C	Saucedo et al. (2013)
Postlarvae taquilla clam, *Mulinia edulis*	3 monospecific diets: (a) *Isochrysis* aff. *galbana* (clone T-ISO), (b) *Phaeodactylum tricornutum*, (c) *Tetraselmis suecica*; and an *I. galbana* (clone T-ISO) binary diet (+ *P. tricornotum*) and trinary diet (+ *P. tricornotum* + *T. suecica*)	Survival and growth: growth was best on monospecific T-ISO diet; diet had no effect on survival	Oliva et al. (2013)
Larvae flat oyster, *Ostrea angasi*	24 diets: monospecific diets of (a) *Isochrysis* sp. (T-ISO), (b) *Nannochloropsis oculata*, (c) *Tetraselmis chuii*, (d) *Pavlova lutheri*; binary diet of *T. chuii* + T-ISO; ternary diets of *T. chuii* + T-ISO + *P. lutheri* or *N. oculata*	Listed diets had the best effect on larval growth and metamorphosis	O'Connor et al. (2012)
Pacific oyster, *Crassostrea gigas* spat	*Pavlova pinguis, Rhodomonas salina, Tetraselmis* sp. CS-362 and *Nannochloropsis*-like sp. CS-246 + *Isochrysis galbana* (T-ISO)	Growth best on mixed diet, for monospecific diets *P. pinguis* was the best choice	Brown et al. (1998)
Oyster spat, *Crassostrea virginica*	5 strains of high lipid *Tetraselmis* spp. compared to *Chlamydomonas* and *Isochrysis* sp. (T-ISO)	Fastest growth on *Tetraselmis* spp., due to higher contents EPA and the sterols 24-methylcholesterol and/or 24-methylenecholesterol	Wikfors et al. (1996)

The requirement for renewable fuel and aquaculture crude protein needs for larval rearing was recently addressed for the rearing of juvenile red drum, *Sciaenops coellatus* (Patterson and Gatlin 2013). In this study, the diatom *Navicula* sp., the green microalga *Chlorella* sp., and the eustigmatophyte *Nannochloropsis salina* were lipid extracted, simulating lipid use for microalgal biodiesel production. The lipid-extracted biomass was then used in separate feeding trials replacing up to 25% of soy and fish meal crude protein, the traditionally arable land- or fisheries-derived protein source of control diets, which competes with arable land use for food production and fuel use, respectively. In essence, this study established that 10% microalgal meal replacement of crude protein did not significantly affect fish performance (weight gain, feed efficiency, survival and condition indices, such as hepatosomatic index, intraperioteneal fat ratio and muscle ratio, and whole body proximate condition) (Patterson and Gatlin 2013). While this study highlights that it is possible to use microalgal production for both biodiesel and protein-rich aquaculture diets, it also shows that whole algal diets performed better, whereas lipid-extracted protein replacement was limited as higher crude protein substitutions had negative effects on protein and energy retention, mainly by lowering the digestibility coefficient (Patterson and Gatlin 2013). While the microalgal industry for fuel production is in its infancy, commercial production of microalgae for specialty commodities was established decades ago. Commercial enterprises, bio-products, and target markets will be briefly described in sections on growth requirements (Section 18.3) and cultivation systems/products (Sections 18.4 and 18.5), as applicable.

Given the multiple potential benefits of microalgal cultivation at commercial scales, the remediation potential of GHGs and wastewater, in the context of global challenges faced by mankind in the near future, it is imperative to evaluate the carbon footprint and water use in relation to product development of the nascent microalgal biomass industry.

18.2 BIOLOGY OF MICROALGAE: CELL PHYSIOLOGY AND PHOTOSYNTHESIS

Microalgae are important primary producers and provide the nutritional basis directly or indirectly for all aquatic life. They occur in all types of habitat as long as sufficient water/moisture qualities meeting their specific requirements are available (Round 1981). Algae are classified based on their structure, chloroplast architecture (e.g., number of envelope membranes, thylakoid stacking), and accessory pigment composition, the latter is often unique to specific genera (see relevant later text) (Margulis et al. 1989). However, positive morphological identification cannot be based on one criterion alone and all characteristics must be taken into account. For example, it was shown that red algae do not possess a unique carotenoid profile, but xanthophylls are the main accessory pigments, followed by the carotenoid lutein in some species or zeaxanthin, antheraxanthin, or violaxanthin in others (Schubert et al. 2006). Similarly, the xanthophyll fucoxanthin is the dominant accessory pigment in the golden algae, diatoms, and the brown macroalgae, whereas β-carotene is dominant in green algae, similar to higher plants. Indeed, the green algae, now classified as Viridiplantae (Simon et al. 2006), are closely related to land plants (Adl et al. 2005) and hence their accessory pigment profiles are similar (Strain and Sherma 1972). Positive morphological identification of microalgae for species selection, especially for commercial use, remains important, despite recent advances using molecular techniques, as molecular identification relies on sequence similarities with depositions in molecular data bases. Deposited sequences of *Chlorella* spp., a genus of growing importance for commercial production and potential fuel generation (Benemann 2013; Islam et al. 2013), however, are frequently based on misidentification (Hoshina and Fujiwara 2013). Thus, such data bases may contain genetic sequences declared to belong to a specific genus, when, as a matter of fact, the sequence can belong to a different genus, which can have downstream consequences for microalgal producers and the marketability of whole-cell-based products.

Some algae have very narrow optima for temperature (e.g., artic algae), water quality (e.g., salinity, tolerance to pollutants, e.g., metals), light quantity and quality, and nutrients, while others have broad optima (Round 1981). For commercial applications, high biomass productivity and end-product content are crucial criteria that strongly influence profitability of product-orientated microalgal industries, as there is no profitable price yet on the use of algae for GHG emission and wastewater management. Irrespective of this, even if a remediation business could be established, efficient remediation capacity would still be tightly linked with biomass productivity and a use for the generated biomass would need to be found. For commercial production of high-volume, low-value products (e.g., fuel), it is therefore essential to select species with broad tolerance ranges. For example, salinity from 0 to 36 ppt had no effect on growth, nutrient utilization, biochemical profile (total lipid, protein, and carbohydrate contents and fatty acid profile) of the green euryhaline freshwater microalga *Picochlorum atomus*, whereas nutrient status significantly affected these criteria (see Section 18.3) (von Alvensleben et al. 2013). This broad salinity tolerance is remarkable and allows cultivation of this microalga, which has a suitable fatty acid profile for biodiesel production (von Alvensleben et al. 2013), in a wide range of locations, an important feature to be considered for large volume biomass production. Typically, in GHG or wastewater remediation application of microalgal biomass industries, the water source cannot be selected, particularly for high-volume end products (e.g., biofuels) and therefore, salinity effects for end-product optimization will not be further considered here (please refer to Campenni et al. 2013; Pal et al. 2013; Yoshimura et al. 2013 for some recent publications). In general, most marine algae will not tolerate salinities below a certain threshold (i.e., <18 ppt), while freshwater species will have an upper limit tolerance (<11 ppt) (Heimann unpublished).

18.2.1 HETERO- AND MIXOTROPHY

Although most microalgae are photosynthetic (capable of CO_2 fixation using sun energy), many can also be grown heterotrophically (in the absence of light and provision of organic carbon), mixotrophically (low light and supply of organic carbon), and autotrophically [light and inorganic carbon (CO_2)] (Perez-Garcia et al. 2011). Acetate, glucose, and glycerol are typically used for heterotrophic and mixotrophic microalgal cultivation, but acetate promotes strong growth only in freshwater species, while it inhibits biomass production in marine microalgae (Wood et al. 1999; Orosa et al. 2001; Xu et al. 2004; Garcia et al. 2005). Biomass productivities and product accumulation are typically significantly greater under heterotrophic and mixotrophic conditions, as light penetration into dense cultures limits photosynthetic carbon acquisition through self-shading (Wen and Chen 2000; Chen and Chen 2006; Heredia-Arroyo et al. 2011). Despite positive effects on biomass and product productivities, carbon use is an important consideration for high-volume products such as fuel, that is, the use of sugarcane glucose and corn starch for bioethanol production is already competing with human uses and glycerol is a key ingredient in the food industry (e.g., as a sweetener for low carbohydrate foods) (Stevens 2000), in medical and pharmaceutical industries (Spiegel and Noseworthy 1963), as well as in the automotive industry (e.g., anti-freeze) (Hudgens et al. 2007), and forms the base for conversion to other products, for example, hydrogen (Marshall and Haverkamp 2008), epoxy resins and acrolein (Ott et al. 2006; Watanabe et al. 2007; Dow 2007), and ethanol (Yazdani and Gonzalez).

18.2.2 PHOTOSYNTHESIS (AUTOTROPHY)

Compared with terrestrial crops, microalgae have higher areal productivities and are more efficient in photosynthetic carbon acquisition (Huerlimann et al. 2010). This is mainly due to simpler structure and living in an aquatic habitat, which avoids having to invest into structure-stabilizing compounds (i.e., lignin) and the need for transport of nutrients and water in specialized conduction systems. In addition, water and gases are also directly accessible (e.g., no stomata for gas exchange are required). Yet, oxygenic photosynthesis follows the ancient biochemical pathway that evolved in the cyanobacteria, which later formed the photosynthetic organelles in algae and plants through endosymbiosis (Huerlimann and Heimann 2013). In essence, photosynthetic carbon acquisition is divided into two reactions: the light reaction and the carbon reaction (formerly called dark reaction) (Falkowski and Raven 2007). In principle, photosynthesis in algae is no different to plants, and the light reaction uses the sun's energy to drive the formation of reducing equivalents (NADPH) and chemical energy in form of ATP through step-wise electron transport from photosystem II (PSII) to photosystem I (PSI) (also called the Z-scheme of photosynthesis or Hill reaction) (Raven and Girard-Bascou 2001). The key components of the photosynthetic redox chain residing in the thylakoid membranes of the chloroplast are PSII (where water splitting and oxygen evolution occurs), the cytochrome b_6f complex, plastoquinone, plastocyanine pool, and lastly PSI. Light energy excites an electron in the reaction center chlorophyll a of PSII and I, which flow through step-wise redox reactions to form NADPH at PSI through electron donation from reduced ferredoxin. ATP production is the result of the ATP synthase using the proton-motive force (PMF) generated by protons from the water splitting reaction and influx via plastoquinone into the thylakoid lumen (Raven and Girard-Bascou 2001). The net equation of the photosynthetic light reaction is shown in Equation 18.1:

$$2H_2O + 2NADP + 3ADP + 3P_i \rightarrow O_2 + 2NADPH + 3ATP \qquad (18.1)$$

CO_2 assimilation occurs in the stroma of chloroplasts via the Calvin cycle and is still linked to the light reactions of photosynthesis through the use of photosynthetically derived NADPH and ATP, which in turn regulates generation of these molecules by cyclic electron transport around PSI to avoid overproduction (Peltier et al. 2010). Cyclic electron transport then generates more ATP, as 1.5 ATP molecules per NADPH molecules are required to assimilate CO_2 (Falkowski and Raven 2007). In the Calvin cycle, CO_2 binds to the five carbon sugar ribulose-1,5-bisphosphate (RuBP) and this reaction is catalyzed by the enzyme ribulose-1,5-bisphosphate carboxylase/oxygenase (RubisCO), a slow process that requires many molecules of RubisCO to make photosynthetic carbon acquisition effective. For this reason, RubisCO is thought to be the most abundant protein on earth and can indeed account for 50% of total leaf protein in plants (Falkowski and Raven 2007). CO_2 fixation results in the formation of two 3-carbon phosphoglycerate (PGA) molecules. While the majority of PGA molecules are used to reform RuBP in the Calvin cycle, some are converted to glucose in a step-wise reaction via glyceraldehyde-3-phosphate (G-3-P) and fructose-1-phosphate, requiring ATP and NADPH (the products of the light reaction) (Falkowski and Raven 2007). If production is sufficient, glucose is stored as an insoluble and osmotically inactive polymer (starch) in the chloroplast (Falkowski and Raven 2007). The summary formula for CO_2 fixation is shown in the following equation:

$$3CO_2 + 3RuBP + 9ATP + 6NADPH \rightarrow G\text{-}3\text{-}P + 3RuBP + 8P_i + 9ATP + 6NADP^+ \quad (18.2)$$

Because light indirectly affects CO_2 fixation through activation of five enzymes of the Calvin cycle, including the key RubisCO, and allows influx of Mg (further activating enzymatic activity of RubisCO) and 3-carbon intermediates (required for the Calvin cycle) into chloroplasts (Falkowski and Raven 2007), the term dark reaction is misleading. The dependency of the carbon fixation machinery on light-mediated processes highlights light as a driving factor that needs to be well managed in commercial production and for optimizing GHG emission reductions using algae.

The summarizing equation of both photosynthetic processes, shown in Equation 18.3, requires six CO_2 molecules to yield one molecule of glucose. The process also yields six oxygen molecules, but consumes water through water splitting at PSII:

$$6CO_2 + 12H_2O + light \rightarrow C_6H_{12}O_6 + 6H_2O + 6O_2 \qquad (18.3)$$

A little more than eight photons of light are required to fix 1 mol of CO_2. The core photosynthetic pigment chlorophyll a absorbs light in the red regions (680 nm), which has a radiant energy level of 167 kJ mol^{-1}. As eight photons are required and the energy value of stored carbon being

479 kJ mol^{-1} CO_2, photosynthetic carbon fixation is extremely efficient (slightly less than 34% efficiency) (Nobel 1999). Given some loss of fixed carbon through night-time respiration and secretion of organic materials, net carbon fixation has been estimated to be 1.826 g CO_2 per g dry weight (DW) of photosynthetic biomass generated (Chiang et al. 2011). This factor can now be easily used in techno-economic models and life-cycle analyses, to estimate GHG emissions or abatement of various industries coupled to either aqua- or agriculture (see Sections 18.6 and 18.7, respectively).

18.2.3 ACCESSORY PIGMENTS FOR LIGHT HARVESTING

Photosynthetic activity is further influenced by pigments such as carotenoids and xanthophylls (also called accessory pigments), which, as lipid-soluble compounds, form part of the total lipid content. These pigments that form part of the antenna complexes of the light-harvesting centers (LHCs) of the photosystems (Grossman et al. 1995; Nobel 1999) are high-value products for the nascent algal industry. The LHC of PSI contains a chlorophyll *a* dimer, P_{700}, that is unique to PSI and allows absorption of light in the far red region (Nobel 1999). A special monomeric LHC [PSI-fucoxanthin-chlorophyll protein (FCP)] was found in the diatom *Phaeodactylum tricornutum* with a chlorophyll *a* to P_{700} ratio of ~200:1 (Veith and Buchel 2007). The FCP complex transfers light energy from the accessory chlorophyll c to the PSI core. Interestingly, the PSI-LHC shares greater similarities with those found in green algae, but shows lower similarities with those found in another diatom, *Cyclotella cryptica* (Veith and Buchel 2007). Lipid-soluble accessory pigments are carotenoids, which are divided into two groups: the carotenes (isoprene compounds containing no oxygen) and the xanthophylls (oxygenated isoprene compounds) (Nobel 1999), which are present in characteristic phyla- and/or class-specific signatures in the chloroplasts of eukaryotic algae and higher plants (Round 1981; Raven and Girard-Bascou 2001). Some accessory pigments are water-soluble phycobilins (phycoerythrin, allophycocyanin, and phycocyanin), present in cyanobacteria, the eukaryotic glaucocystophytes, and red algae (Gantt 1996). Owing to their water-soluble nature, they are assembled through protein linking into phycobilisomes that sit as structurally identifiable globules on top of the thylakoid membrane (Wildman and Bowen 1969; Morschel 1991; Bhattacharya and Schmidt 1997; Heimann et al. 1997). In addition to serving as accessory pigments, they also possess antioxidant properties to protect the photosynthetic machinery against reactive oxygen species (ROS) generated as a by-product of photosynthesis, particularly in response to high light (Vershinin 1999; van den Berg et al. 2000; Havaux et al. 2004; Gantar et al. 2012). Owing to the latter properties and their high-protein content and/or high PUFA content (Spolaore et al. 2006), whole-cell microalgal markets are dietary supplements.

As part of the antenna complexes of the LHCs, carotenoids absorb in the blue and green region (~400–540 nm) of the visible light spectrum (400–700 nm), thereby closing the absorption gap, enabling the use of the entire visible light spectrum for photosynthesis (Nobel 1999). This is particularly important for aquatic species, such as microalgae and cyanobacteria, as light penetration into dense cultures does not only inhibit light quantity *per se*, but more high-energy radiation (blue to green wavelengths) will penetrate further than low-energy orange-to-red wavelengths, thus altering the light spectrum quality a culture at various depths may receive (Falkowski and Raven 2007; Dai et al. 2013). Accessory pigment and chlorophyll content respond to changes in light quantity and quality, mistakenly termed chromatic adaptation for responses of cyanobacteria to nitrogen starvation (Magnus and Schindler 1912; Boresch 1913; Schindler 1913). However, light does influence pigment composition and quantity (Titlyanov and Lee 1978; Czeczuga 1986; Lopezfigueroa 1991; Kotzabasis and Dornemann 1998) and in case of cyanobacteria the composition of the pigment compounds in the phycobilisomes and regulation of phycobilisome size (Fujita 1996). Self-shading, for example, leads to higher chlorophyll and accessory pigment contents owing to increased antenna size and/or through increases in abundance of photosystems (Johnsen and Sakshaug 1996), in order to maximize light utilization. However, temperature effects can mimic light-induced changes in pigment composition (see Section 18.3). Since temperature and light are parameters difficult to control in large-volume microalgal production, yet their impacts are quick and often species-specific, it will be essential to understand responses to these parameters for the organism under cultivation and with regard to end-product requirements (e.g., biochemical profile guarantees for feed and food, antioxidant content, and so on).

18.2.4 FERTILIZATION

Like plants, algal cultivation requires provision of nutrients, such as macronutrients (carbon, nitrogen, phosphorous), but also micronutrients (minerals in particular Fe, Mn, Zn, Cu, etc.), as these act as co-factors in many enzymes or are part of other compounds. Many books have been written on algal cultivation (Andersen 2005; Borowitzka 2013b; Moheimani et al. 2013), focusing on isolation, methods for biomass measurements, cultivation, storage (e.g., freezing in liquid nitrogen, plating on agar and maintenance under low illumination), and cultivation recipes, as many are formulated to provide specialist diets favoring specific species or to ensure long-term culture maintenance to reduce manpower required for large microalgal collections (Andersen 2005). In essence, the first nutrient to become limiting will limit growth, leading to increased carbon storage, unless carbon is the limiting nutrient, but algal health can also be affected (e.g., Mg, iron deficiencies). The Redfield ratio describes the elemental contribution to all types of biomass (ranging from algae to mammals) as 106C:16N:1P (Redfield 1934), which was later extended to include Fe (0.1–0.001), with the variability of Fe content reflecting Fe availability (Broecker et al. 1982). Even though use of the Redfield ratio to derive fertilization needs of phytoplankton is not 100% accurate (von Alvensleben et al. 2013), as some compounds are secreted and hence lost from the biomass, it still reflects that in addition to carbon, more nitrogen will be required compared with phosphorous, which, for fertilizer development, can then be amended based on species-specific needs.

Inorganic carbon sources (e.g., HCO_3^- and CO_2) are often overlooked as being part of the required nutrient mix, yet inorganic carbon levels fluctuate in algal cultures in line with photosynthetic activity and respiration, leading to changes in medium pH on a diurnal basis, unless buffered or supplemented with CO_2 (Raven 1995) (Figure 18.1) and, on a large-volume industrial scale, it has proved difficult to enrich microalgal suspensions with sufficient CO_2 from GHG sources, such as coal-fired power stations, in an economically sustainable way (Heimann, personal communication). The three

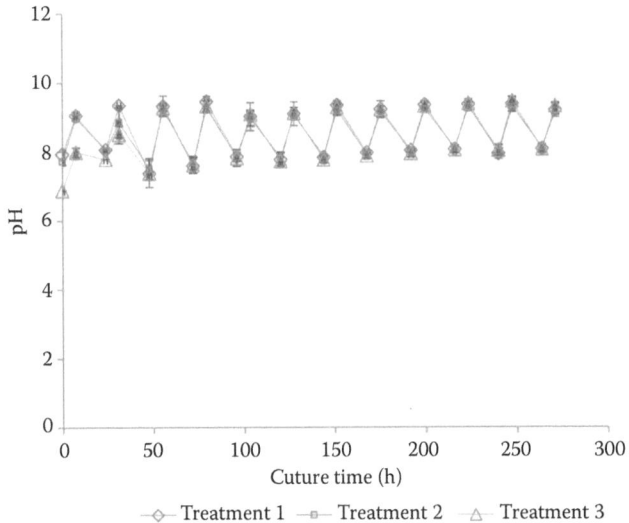

FIGURE 18.1 Diurnal pH fluctuations in aerated batch cultures of freshwater microchlorophytes and the effect of nitrogen status. Media NO_3^- levels: 53, 26.7, and 0 mg L^{-1} for treatments 1–3, respectively.

forms of dissolved carbon are shown in Equation 18.4 and the conversion from one form to another is typically slow and dominance profiles change with pH (Moroney et al. 2001). At a pH below 6.4, dissolved CO_2 dominates, while HCO_3^- dominates between pH 6.4 and 10.3, while carbonate (CO_3^{2-}) dominates at a pH above 10.3 (Moroney et al. 2001) and presents a nonusable form of carbon. It is thus imperative to control the culture pH (see Sections 18.3 and 18.4), as algal photosynthesis quickly increases pH through the use of dissolved bicarbonate often reaching pH of more than 10 in carbon-limited cultures (Heimann, personal communication).

$$CO_2 + H_2O \leftrightarrow HCO_3^- + H^+ \leftrightarrow CO_3^{2-} + 2H^+ \tag{18.4}$$

Algae possess carbonic anhydrases that assist in supplying bicarbonate for carboxylation reactions (photosynthesis, RubisCO reaction) and transport inorganic carbon away from actively respiring cells (decarboxylation reactions) (Moroney et al. 2001), but they also assist in carboxylation reactions in the cytosol, such as carboxylation of phosphoenol-pyruvate and carboxylation of acetyl-CoA carboxylation (Raven 1995) the starting point for fatty acid synthesis in the plastid (Huerlimann and Heimann 2013) and to form malonyl-CoA for fatty acid elongation in the smooth endoplasmic reticulum for the production of LC-PUFAs (Huerlimann and Heimann 2013). RubisCO is also an oxygenase enzyme that binds oxygen and CO_2 with roughly equal affinity (Andersson 2008) (see Section 18.3), and a high CO_2 concentration needs to be maintained for the enzyme to direct enzymatic activities in favor of carboxylation, which is achieved by using carbon-concentrating mechanisms (CCMs) (Coleman 1991; Badger and Price 1992). In cyanobacteria and microalgae, an inorganic carbon transporter exists and a CO_2 environment for RubisCO is maintained by packaging RubisCO with carbonic anhydrase in discrete cellular locations, carboxysomes in cyanobacteria, and pyrenoids (where present) in microalgae (Badger and Price 1992; Meyer and Griffiths 2013).

In phytoplankton, 91% of cellular nitrogen is present in proteins and 14%–21% in the nitrogenous bases of DNA and RNA (Vincent 1992), while phosphate is essential for the formation of energy stores (ATP), enzyme and cell signaling regulation through phosphorylation/dephosphorylation (and GTP), membrane integrity, and for cell division, as it forms the backbone of DNA and RNA (Raghothama 1999; Martinez et al. 1999; Powell et al. 2008). The cyanobacteria are a remarkable group of oxygenic photosynthetic organisms, as some species, termed diazotrophic cyanobacteria, can assimilate atmospheric nitrogen, making their large-scale cultivation potentially independent of other nitrogen sources (Chakdar et al. 2012; Huo et al. 2012; Muro-Pastor and Hess 2012; Tiwari and Pandey 2012). N_2 fixation is oxygen sensitive, as the enzyme nitrogenase contains the co-factors molybdenum and iron (Berman-Frank et al. 2003). To overcome this limitation in an oxygen-enriched atmosphere, cyanobacteria have developed two strategies a) structural and biochemical separation in specialist cells (heterocysts) (Muro-Pastor and Hess 2012) and b) temporal separation, where nitrogen fixation occurs at night when photosynthesis is inactive and photosynthetic oxgen evolution ceases (Shi et al. 2012). In heterocysteous cyanobacteria, atmospheric nitrogen fixation can occur during the day, as the cells are thick-walled and the oxygen-evolving PSII is lacking from the thylakoids. Instead, the thylakoids do contain PSI, which does contribute the energy through cyclic electron transport required to drive the energy expensive N_2-fixation process (16 mol of ATP are required for 1 mol N_2 fixation) (Bohme 1998; Bothe et al. 2010; Herrero et al. 2013). Therefore, use of nitrogen-fixing cyanobacteria could be of interest for low-value, high-volume products in areas with limited water and fertilizer resources.

Nitrogen assimilation in algae has been less studied than in plants, but it has been shown that algae can utilize a variety of nitrogen compounds such as ammonium, nitrate, nitrite (Heimann unpublished), and urea (Eustance et al. 2013), which are typically present in high concentrations in eutrophic wastewaters (e.g., secondary treated sewage, piggery waste, aquaculture and agriculture effluents, etc.), making algae ideal in the remediation of such wastewaters. However, high ammonium concentrations led to decreasing culture pH, and buffering at industrial-scale cultivation is not an option (Eustance et al. 2013). Therefore, concentrations of ammonium must be limited to those tolerated by the algae to be cultivated. In contrast, nitrate uptake is an active process (driven by the PMF of the $2H^+/NO_3^-$ symporter) and high- and low-affinity uptake systems have been described with their expression, depending on nitrate availability in the medium (Crawford et al. 2000). Once uptaken, nitrate is reduced to nitrite in the cytosol by nitrate reductase, then nitrite is transported to the chloroplast for reduction to ammonium, which is used for the synthesis of amino acids (Crawford et al. 2000). The reduction of nitrite by nitrite reductase requires the provision of reduced ferredoxin from PSI and as such, this process only operates effectively during day light hours and in photosynthetically active cultures (Crawford et al. 2000). Since nitrate uptake is not restricted to day light hours, oversupply of nitrate will typically result in the secretion of nitrite (as nitrite cannot be stored) by the algal cells, leading to build up in the medium (Malerba et al. 2012; von Alvensleben et al. 2013). Similarly, phosphate uptake is an active process and also occurs via a proton symporter (Raghothama 1999) and low-affinity and high-affinity systems are expressed depending on the concentrations present in the cultivation broth (Keenan and Auer 1974).

Trace element supplementation is equally important to ensure culture health, coloration and optimal photosynthetic activities, as well as biomass and biochemical productivities (Kropat et al. 2011; Hsieh et al. 2013; Malasarn et al. 2013), as these constitute essential co-factors for enzymatic reactions. In particular, redox reactions require Cu and Fe, whereas the water-splitting and oxygen evolution step in photosynthesis requires manganese, and Zn is essential for proper regulation of transcription. Hence any deficiency will result in reduced biological function, while trace metal excess can also limit or abolish culture performance (Hughes and Poole 1991; Neff 1997; Bell et al. 2002).

18.3 CRITICAL FACTORS FOR MICROALGAL GROWTH

It is well established that microalgal biomass productivity and biochemical profiles are species-specific and influenced by growth phase (Xu et al. 2008; Huerlimann et al. 2010; Islam et al. 2013), environmental factors such as temperature (Renaud et al. 2002), light (Thompson et al. 1993; Khoeyi et al. 2012; Marchetti et al. 2013), nutrient provision, and status (Huerlimann et al. 2010; Islam et al. 2013). These parameters will therefore affect selection of production sites and site-specific end product range, that is, the potential suitability of microalgal species for nutrition and/or fuel production, in addition to species-specific impacts.

18.3.1 LIGHT AND TEMPERATURE

Light and temperature are almost impossible to control in large-volume microalgal cultivation. For open systems, evaporative cooling will typically provide for ambient temperatures, while closed systems, unless cooled, can reach temperatures exceeding the general permissive biomass production temperatures of 28–30°C (Figure 18.2). In addition, responses to temperature can often simulate responses to light. For example, low temperature (5°C) mimicked responses to high light conditions, leading to increased chlorophyll a/b ratios, lower chlorophyll content (five times lower) and increased xanthophyll content, but higher light saturation levels and up to ~4-times higher oxygen evolution rates compared with cells grown at 27°C in *Chlorella vulgaris* (Maxwell et al. 1994). When 5°C cells were additionally exposed to low light, the pigment content and composition were similar to those in cells grown at 27°C and high light, suggesting that such responses could be due to excitation pressure on PSII in general with the redox state of Q_A acting as a signal for photosynthetic acclimation (Maxwell et al. 1994). This shows that the effect of light and/or temperature can be exploited for optimizing pigment content in commercial settings.

For irradiance, on the other hand, it was shown that higher light quantities significantly increased saturated fatty acid (SFA), while lowering monounsaturated (MUFA) and polyunsaturated fatty acid (PUFA) content in the green microalga *Chlorella vulgaris* (Khoeyi et al. 2012), while light quality

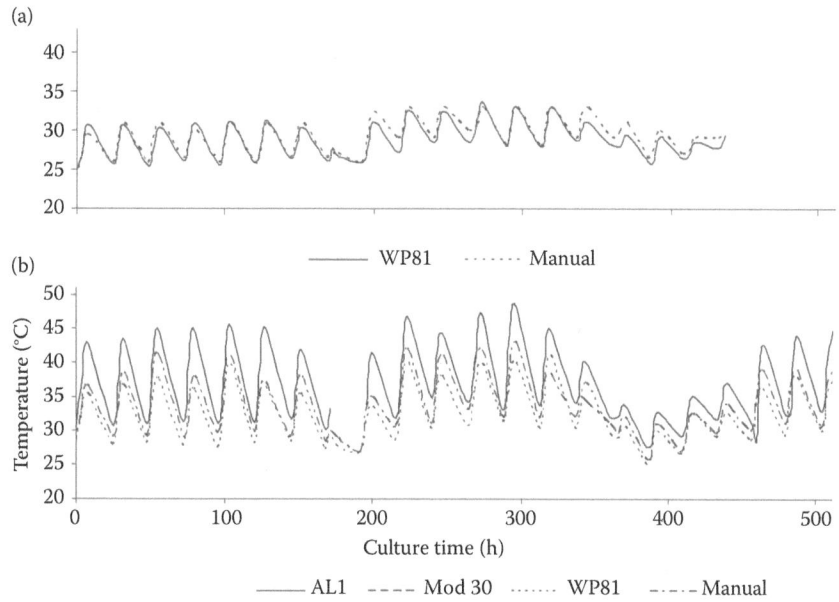

FIGURE 18.2 Typical temperature fluctuations in a 1.3 m³ open pond (a) and a 4 m³ closed tubular photobioreactor (b) during a 21-day growth period in the tropical wet season at James Cook University, Townsville, Australia. Measurements were recorded using the data-logging capabilities of the WP81 (a and b), in-built reactor probes (AL1, Mod 30 in b), and probe checks were conducted with a mercury hand-held thermometer (manual).

(blue light) changed the biochemical profile of *Isochrysis* (T-ISO) to being protein- and chlorophyll-rich while lowering carbohydrate content without affecting growth performance but increasing photosynthetic efficiency (Marchetti et al. 2013) and hence CO_2 remediation potential. Changes in biochemical profiles, however, do affect end-use potential, with conditions increasing SFA and short chain MUFAs making the biomass more suitable for renewable fuel production, while those conditions that increase protein and PUFA, especially LC-ω3-PUFA, contents render biomass more suitable for feeds in aquaculture and for human diet supplements to (a) ensure a healthy diet and (b) prevent illnesses arising from diet imbalance such as cardio-vascular impact, which are increasing worldwide with negative socio-economic flow on effects (e.g., increased medical costs) (Raposo et al. 2013).

High light can lead to photodamage of the photosystems, but the xanthophyll cycle acts as a photoprotectant (Schubert et al. 1994; Niyogi et al. 1997). *Chlorella pyrenoidosa* reacted to photodamaging light conditions by epoxidizing violaxanthin to anthreaxanthin and zeaxanthin (Schubert et al. 1994), thus care must be taken if the biomass generated for a pigment and/or whole cell product market, whereas such responses do not affect biofuel production potential. In contrast, UV irradiation did not activate the xanthophyll cycle in the green microalga *Micrasterias denticulate* (Lutz et al. 1997). Futhermore, in *Haematococcus lacustris*, extraplastidial astaxanthin, canthaxanthin, and echinenone function as a photo-anti-oxidant, as the pigments are accumulated in lipid vacuoles around the nucleus acting as a physico(shading function)-chemical (radical oxygen scavenger function) barrier protecting the genome against ROS damage (Hagen et al. 1993). In terms of CO_2 mitigation, photosynthetic activity is very important and directly influenced by light quality and quantity. The antenna size of the LHCs has been shown to be directly linked with photosynthetic efficiency, where high light-photo-damaged cells of the important β-carotene producer, *Dunaliella salina*, had reduced antenna sizes and chlorophyll content, but showed enhanced photosynthetic efficiency, which could be even further increased when high light cultures were enriched with CO_2 (Melis et al. 1998).

18.3.2 FERTILIZATION

Other than light, CO_2 levels and phosphate availability affect the enzymatic activity of RubisCO, which has been reviewed comprehensively by Spreitzer (2003); Evans (2013); Moroney et al. (2013); Tcherkez (2013); Spreitzer and Salvucci (2002), influencing culture pH, which, in turn, governs CO_2 availability (Section 18.2). To sustain biomass productivity and GHG remediation potential, control of CO_2 availability is crucial, as the efficiency of the carboxylation step can be reduced by 50% if the pool of RuBP is depleted through the oxygenase activity of RubisCO (Andersson 2008). Oxygenation of RuBP leads to the formation of 2-phosphoglycolate, a compound of little use in most organisms and only part of the carbon that can be salvaged through photorespiration, which is energetically expensive (Andersson 2008). However, under photoinhibitory light conditions, this same salvage pathway may be beneficial in dissipating excess reducing equivalents, in particular when CO_2 is limiting (Osmond 1981).

Other than the carbon status of cultures, nitrogen limitation also critically affects culture color (loss of pigmentation) and thus photosynthesis and CO_2 fixation, which has flow on the effects on biomass productivity. For example, the commercial production of phycobilins (phycocyanin and phycoerythrin) either as fluorescent tags in biomedical research or as antioxidant enrichment in dietary supplements relies on sufficient nitrogen levels, as these pigments contain nitrogen, just like chlorophylls. As already shown at the start of the nineteenth century, nitrogen starvation will lead to reutilization of nitrogen from phycobili pigments and also chlorophylls, leading to culture color changes (mistaken as a response to light) from a healthy blue-green to yellow or even brown, as other nonnitrogen containing pigments are unmasked (Magnus and Schindler 1912; Boresch 1913; Schindler 1913). Such material would not be usable for whole organism dietary supplements, the main market niche for the commercially produced cyanobacterium *Spirulina platensis*, and destruction of the biomass would release assimilated CO_2, if dietary supplements or phycobili pigments/chlorophylls were the only product niche of the business. Similarly, although the enzyme RubisCO is stored in excess, particularly in the pyrenoids of algae, if present, nitrogen deficiency can lead to RubisCO loss, through reutilization of the stored nitrogen, which can negatively affect CO_2 fixation potential (Ahlgren and Hyenstrand 2003). Nitrate limitation also reduced the synthesis of chloroplastic proteins, altering pigment profiles; reduced chlorophyll but increased carotenoid content (Falkowski et al. 1989; Geider et al. 1998).

It is well known that several algae of commercial importance respond to nitrogen limitation with increased lipid content, as photosynthetically acquired carbon will not be invested into growth (Roessler 1990; Zhila et al. 2005; Li et al. 2008; Huerlimann et al. 2010). This is often accompanied by changes in the fatty acid methyl ester profiles, showing decreased amounts of LC-PUFAs, the main constituents of membrane lipids and increased contents in short chain fatty acids, specifically saturated (SFAs) or monounsaturated (MUFAs), the typical components of storage fats (triacylglyceride (TAG)) (Huerlimann et al. 2010; Islam et al. 2013; von Alvensleben et al. 2013). Nitrogen-deprivation, introduced in the second cultivation stage when maximal biomass was reached under nonlimiting nutrient conditions in phase 1 (two-phase cultivation, see Section 18.4), was shown to be more effective than potassium-phosphate, and Fe-deprivation to elicit total lipid increase in *Chlorella* sp. strain BU11008 (Praveenkumar et al. 2012). Lipid productivity, another important performance characteristic for biodiesel production, reached ~54 and ~49 mg L^{-1} d^{-1} under nitrogen deprivation or under conditions limiting all three nutrient resources, respectively, making this organism and cultivation approach an ideal candidate for biodiesel production (Praveenkumar et al. 2012). Nitrogen limitation also enhanced astaxanthin content in *Haematococcus pluvialis* (Zhekisheva et al. 2002), which could be used to enhance the production of this important feed for salmonoids and as antioxidant in human nutrition and cosmetics (Cerón et al. 2007).

For optimal biomass productivity, on the other hand, it is essential to maintain adequate nitrogen supplies, as most microalgae of commercial interests, in contrast to some diatoms capable of nitrogen storage in vacuoles (Bode et al. 1997), do not possess large vacuoles and are therefore not able to store nitrogen (Dortch et al. 1984; Lomas and Glibert 2000). In contrast to nitrogen storage, microalgae are capable of phosphate storage and can thus sustain growth when temporarily deprived of phosphorous (Keenan and Auer 1974); however, phosphorous starvation does ultimately lead to reduction of biomass productivity through the depletion of energy stores and the incapacity to provide RNA, DNA, and membranous phospholipids for cell division (Raghothama 1999). Phosphate availability also regulates the activity of RubisCO and therefore affects the carbon sequestration efficiency through algal biomass production (Murali 1992).

18.3.3 TRACE METALS

Provision of nutrients and essential trace elements (fertilization) are critical considerations for the mass cultivation of microalgae, in particular if biomass productivity forms the basis for business case evaluation or carbon remediation potential. Microalgae have different nutritional requirements. For example, the addition of Germanium dioxide will inhibit growth of diatoms, but will not affect the cultivation of dinoflagellates (Shea and Chopin 2007). High steady-state cell densities were achieved in semi-continuous cultures of *Tetraselmis suecica*, an important feed organism in aquaculture, when the cultivation medium was enriched with Mg, Sr, and Si and fatty acid content, and antioxidant activity was enhanced with Sr addition (Ulloa et al. 2012). Trace metal deficiencies typically lead to culture underperformance. For example, zinc deficiency leads to copper deficiency and reduced carbonic anhydrase activity in *Chlamydomonas rheinhardtii*, impacting negatively on CO_2 assimilation rates (Malasarn et al. 2013). Similarly, manganese is a critical co-factor in the water-splitting and oxygen-producing step at PSII and deficiency will result in loss of photosynthetic function (Hsieh et al. 2013). Likewise, iron plays a critical part in photosynthetic and respiratory redox reactions and iron-deficiency will result in yellowish cultures (Heimann unpublished), as Fe deficiency results in loss of chlorophyll, which in addition to its light-harvesting function is also a critical component for the structural integrity of the LHCs (Yadavalli et al. 2012). In particular, in *Chlamydomonas rheinhardtii* and most likely all algae, reduced chlorophyll content restricts the formation of large PSI-LHC complexes required for photosynthetic performance (Yadavalli et al. 2012). Thus for optimal photosynthetic performance, inorganic carbon assimilation, and biomass productivities, it is essential to maintain adequate levels of these important trace elements, which could be easily achievable by use of or blending with trace metal-enriched wastewaters, depending on concentrations of desired (Cu, Zn, Mn, Fe) and undesirable metals (i.e., those with no biological function Cd, Hg, etc.) present.

18.4 MASS CULTIVATION SYSTEMS

Laboratory-scale cultivation of microalgae dates back ~140 years (Borowitzka 2013a), while mass cultivation technologies were being developed at the Massachusetts Institute of Technology ~60 years ago. This project investigated the potential of *Chlorella* for human food supplement production using closed cultivation systems (photobioreactors; PBRs) (Benemann 2013). The mass cultivation of microalgae under controlled conditions ("microalgae farming") is therefore a nascent industry compared with the thousands-of-years history of agricultural plant production.

Successful mass cultivation of microalgae requires adequate light, temperatures, macro- and micro-nutrients (Sections 18.2 and 18.3). There are, however, some advantages of mass cultivation of microalgae over higher plant production. In principle, compared with the agriculture of plants, microalgae show less seasonal variability, require less freshwater input, can be cultivated on nonarable land, and do not require the use of herbicides and pesticides. In this regard, mass cultivation of microalgae does not compete with human food production. Still, the nascent microalgal industry faces several economic (e.g., the cost of nitrate and phosphate for fertilization, except when grown in wastewater; CO_2 provision, unless production is co-located with a CO_2 emitter) and energy bottlenecks, in particular the cost and energy demands for dewatering, required for most end-product applications, compromising its definitive take-off. These problems can be partially overcome by the capacity of microalgae to efficiently use nutrients and CO_2 from a variety of urban and industrial wastewaters and flue gases (McGinn et al. 2011; Cai et al. 2013; Zhao and Su 2014). Furthermore, mass cultivation of N_2-fixing cyanobacteria and using biofilm attached bioreactors can overcome nitrogen and dewatering costs.

Although microalgae have the capacity to remediate wastewaters (removal of nutrients and metals) and gases (CO_2-sequestration), the main purpose for mass cultivation is focused on the production of bioproducts or biofuels. An overview of the mass cultivation process for microalgae, targeting the production of animal/aquaculture feeds, is shown in Figure 18.3.

The first step is the mass cultivation of microalgal biomass using one of the different cultivation technologies available (open, closed, or hybrid systems), the choice of which is dependent on quality requirements for the targeted market, and cultivation is either done in batch, continuous, or semi-continuous mode, depending on biomass productivity and harvest capacity. Ninety percent of the biomass weight is typically water, which needs to be removed through centrifugation, flocculation/coagulation, or filtration, which vary in efficiencies and energy requirements (Table 18.2).

The recovered biomass will then be subjected to further dewatering steps, which vary depending on the final water content required for processing of the target end product. Steps generally followed are: (1) primary concentration (concentration factor 10–20), (2) thickening (additional concentration factor of 10), (3) dewatering, generating a paste with 10%–25% solids, and (4) drying, generating dry solids (Pahl et al. 2013). Owing to this complexity, the harvesting/dewatering process can represent up to 70% of the total costs and also consumes the most energy (30% of the total energy input) negatively affecting the carbon footprint, if a dry product is required (Klein-Marcuschamer et al. 2013; Medeiros et al. 2013; Slade and Bauen 2013) (Table 18.2). Most target products can be produced from wet biomass (e.g., biofuels through hydrothermal liquefaction), but others such as aquaculture feeds require drying so as to reduce the cost of transport or to allow mixing the dry microalgal biomass with traditional feedstock or fish meal. Drying can be performed inexpensively albeit slowly using solar drying or by means of more energy demanding driers (e.g., spray driers; freeze driers, drum driers). Unless end-use industries are co-located with production facilities, the final biomass then needs to be distributed either to specialist facilities processing the biomass further to the desired end product (e.g., hydrothermal liquefaction plants for biofuel production) or to the direct user farms or aquaculture facilities, if no further processing is required. Recent business models show that a large part of the production cost, and hence economic feasibility, is adversely affected by transport distance to consumers (Slaby 2013), which advocates on-site production of microalgae in agri- and aquaculture to maintain positive economic, energy, and carbon footprint perspectives.

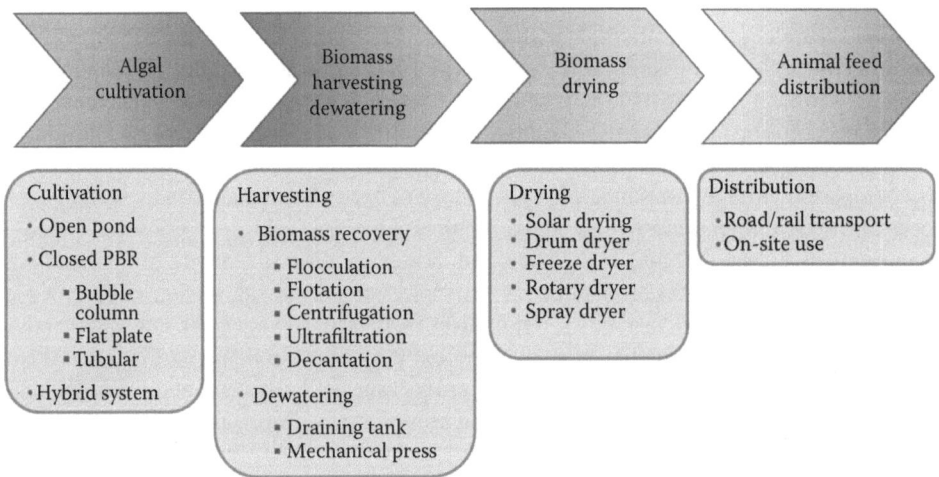

FIGURE 18.3 Flow diagram of aquaculture/animal feed production from microalgal biomass.

TABLE 18.2

Energy Inputs and Estimation of CO_2 Emissions for Different Microalgal Cultivation Systems Including Harvesting and Dewatering

Parameter	Unit	Cultivation Systems		
		Open Ponds/ Raceways	Suspension-Based PBRs	Biofilm PBR
Energy input[a]	kWh kg^{-1} microalgae	0.19–1.05	0.23–7.27	1.3
GHG emissions for cultivation[b]	kg CO_{2eq} kg^{-1} microalgae	0.085–0.47	0.102–3.24	0.6
GHG net balance	kg CO_2 kg^{-1} biomass[c]	−1.7 to −1.4	−1.7 to +1.4	−1.2
	kg CO_2 kg^{-1} protein[d]	7.03	3.83	—
	kg CO_2 MJ^{-1} biodiesel[e]	−16.1 to 534	−75.3 to 344.3	—
NER on biomass (direct harvest)[f]	—	0.3	0.5	0.2
NER[f] on biomass including dewatering[f]	—	0.1–2	1.1 – 12.5	0.2

[a] Ranges for cultivation of *Chlorella vulgaris*, *Nannochloropsis* sp. and *Dunaliella* sp. obtained from Lardon et al. (2009), Campbell et al. (2011), Clarens et al. (2010a), Stephenson et al. (2010), and (Medeiros et al. 2013).

[b] Calculated using a conversion factor of 0.4455 for kW h to CO_2-eq. (electricity derived).

[c] Estimated as CO_2 inputs minus CO_2 sequestration assuming carbon fixation of 1.826 kg CO_2 kg^{-1} biomass (Chiang et al. 2011), where negative values indicate carbon sequestration overcomes carbon emissions for cultivation.

[d] Protein production by *T. pseudonana* and *P. tricornutum* (Draaisma et al. 2013).

[e] Gao et al. (2013) and references therein.

[f] Net energy ratio defined as energy inputs divided by the energy content of the dry biomass (NER < 1 indicates that the process produces more energy than it consumes). (Data obtained from Jorquera et al. (2010), Slade and Bauen (2013) and Ozkan et al. (2012).)

Among the variety of extant microalgae (i.e., ~35,000 according to latest records (Borowitzka 2013b), the best candidates for large-scale cultivation should (a) contain sufficient quantities of the desired end product (see Section 18.5) and have (b) high biomass productivities, (c) high temperature tolerance (30–35°C), (d) high photosynthetic efficiencies, (e) resilience to high irradiation, (f) resistance to contamination, (g) high salinity and shear force resistance, (h) high-specific gravity, and (i) no production of auto-inhibitors. As no strain meets all criteria, strain selection criteria should be informed by the cultivation conditions and end-product requirements (cultivation system, geographic location, dewatering requirements of the end product). Pure strains cultivated in open systems are more prone to contamination (Table 18.3), but these system/strain constraints can be overcome by selecting species that favor growth in restrictive environments, for example, *Dunaliella salina* has a very high salinity tolerance or *Spirulina platensis* has a high tolerance to increased alkalinity. Generally, the most successful approach is to use strains native to the cultivation site. In recent years, a number of model strains of cyanobacteria have been genetically engineered to maximize biofuel production and excretion (Ducat et al. 2011), although their current use in outdoor cultivation is presently compromised by regulations on the release of genetically modified organisms (GMOs) into the environment. Other approaches gaining support are the use of N_2-fixers to reduce fertilization costs (Benemann 2013), biofilm-forming strains to reduce dewatering costs (Ozkan et al. 2012), and mixed consortia that are less prone to contamination and perform better than single strains in bioremediation applications (Renuka et al. 2013).

Once strains are selected, biomass cultivation occurs in open, closed or hybrid systems (Figure 18.4) with choice being primarily driven by end-product quality demands and value. Open systems

TABLE 18.3

Productivity of Outdoor Mass Cultivation Systems for Microalgae

Cultivation System	Species	Location	Productivity (g DW m⁻² d⁻¹)	Reference
			Open Systems	
Raceway	*Anabaena* sp.	Spain	9.4–23.5	Moreno et al. (2003)
Raceway	*Dunaliella salina*	Perth, Australia	20–37	Moheimani and Borowitzka (2006)
Raceway	*Pleurochrysis carterae*	Perth, Australia	16–33.5	Moheimani and Borowitzka (2006)
Raceway	*Spirulina platensis*	Israel	15–27	Richmond et al. (1980)
Raceway	*Spirulina platensis*	California, USA	8.2	Belay (1997)
Raceway	*Tetraselmis suecica*	Italy	5–26	Pedroni et al. (2004)
Inclined thin layer pond	*Chlorella* sp.	Czech Republic, Spain	10–30	Doucha and Livansky (2006)
Circular central pivotal pond	*Chlorella* sp.	Japan	1.6–14.7	Kanazawa et al. (1958)
			Closed Flat-Panel PBRs	
Vertical flat-plate	*Synechocystis aquatilis*	Japan	53	Zhang et al. (1999)
Flat panel (Green Wall Panel, GWP)	*Nannochloropsis* sp.	Italy	6.7–17.2	Bondioli et al. (2012)
Flat panel (Green Wall Panel, GWP)	*Tetraselmis suecia*	Italy	7.4–7.8	Bondioli et al. (2012)
			Closed Tubular PBRs	
Horizontal tubular	*Nannochloropsis* sp.	Spain	20	Zittelli et al. (2013)
Inclined tubular	*Synechocystis aquatilis*	Japan	57	Ugwu et al. (2005)
Inclined tubular (3D Matrix System, 3DMS)	—	USA (Arizona)	98–170	Pulz (2007)
Horizontal serpentine (Algae Link NV)	*Tetraselmis* sp.	Australia	82	Van den Dorpel (2010)
			Hybrid Systems	
Open ponds and horizontal bags (BAGS system, MBD Ltd.)	*Nannochloropsis oculata*	Australia	2	Stammbach et al. (2010)
Open ponds and tubular PBRs	*Haematococcus pluvialis*	—	10.4	Huntley and Redalje (2007)

Source: With kind permission from Springer Science+Business Media: Energy from microalgae: A short history, In: *Algae for Biofuels and Energy*, 2013a, pp. 1–15, Borowitzka MA.

Note: PBR, photobioreactor.

allow direct exchange with the environment (e.g., evaporative water loss is very high, temperature can be highly variable (dependent on location), weather dependence and space requirements are also high, the latter due to low biomass productivities, etc.), yet they are most likely the ideal system for large-volume, low-value commodities such as biofuel production due to low capital investment and the use of various nonpotable water options (marine, nutrient-rich or other industrial wastewaters), while closed photobioreactors have a high capital price tag in exchange for a more controlled environment (Table 18.4).

FIGURE 18.4 Mass production systems for microalgae. (a) Raceway flue gas carbon dioxide remediation with *Nannochloropsis oculata* at Seambiotic, Israel; (b–d) Tubular photobioreactors for *Nannochloropsis* cultivation at JCU-MBD R&D site, Townsville, Australia, (b) and cultivation of *Haematococcus pluvialis* at Ketura, Israel for astaxanthin production (c–d). (c) Biomass and (d) astaxanthin production; (e) Hybrid cultivation system at JCU-MBD R&D site (2009), Townsville, Australia.

Hybrid cultivation either serially combines closed and open cultivation, for example, growing a dense and healthy inoculum in a closed PBR before up-scaling for a short period of time in an open system, or, they represent technically open systems but invasion and environmental exchange potential is limited by a cover and positive air displacement within the system (Figure 18.4e), which allows for temperature control *via* evaporative cooling, but therefore also creates a less water economical system, compared with closed systems. Raceways are cultivation systems most commonly used and are shallow ponds (10–50-cm deep) and either lined with a PVC liner (Figure 18.4a) or painted concrete. The biomass is kept in suspension typically by a paddle wheel (less commonly by air lift, propellers or pumps) creating a flow of about 20–30 cm s^{-1} (Figure 18.4a, far end). The Australian open pond cultivation of *Dunaliella salina* for β-carotene production represents the two largest commercial sites with 740 ha at Hutt Lagoon, Western Australia and Whyalla, South Australia, worldwide (Borowitzka and Moheimani 2013b). Biomass productivities typically in open pond systems average 20–25 g dry weight (DW) m^{-2} day^{-1}, although values of up to 40 g DW m^{-2} day^{-1} have been reported for *Tetraselmis* sp. in Japan (Matsumoto et al. 1995). Generally, a threshold productivity level of 20 g DW m^{-2} day^{-1} has been proposed as the threshold for profitable mass production of microalgae.

TABLE 18.4

Comparison of Open and Closed Systems for Mass Cultivation of Microalgae

Factor	Open Ponds/Raceways	Closed Photobioreactors
Space required	High	Low
Water loss	Very high	Low
CO_2 loss	High	Low
Temperature	Highly variable	Requires cooling
Light utilization efficiency	Low	Moderate–high
Weather dependence	High	Low
Shear	Low	High
Contamination	High	None
Algal species	Restricted	Variable
Biomass quality	Restricted	Flexible
Harvesting efficiency	Low	High
Harvesting cost	High	Lower
Light utilization efficiency	Poor	Good
Capital investments	Low	High

Source: From Singh UB, Ahluwalia A. 2013. *Mitigation and Adaptation Strategies for Global Change* 18(1):73–95.

Note: Hybrid systems (not included) share intermediate characteristics between open ponds and closed PBRs.

Closed systems, usually termed photobioreactors (PBRs), are less commonly used. In contrast to open ponds, PBRs aim to achieve higher biomass productivities by maximizing surface-to-volume ratios. Since their development in the 1950s, they have been tested in different configurations in several pilot studies and the only existing commercial production using tubular PBRs (Biofences) is for astaxanthin using the freshwater green microalga *Haematococcus pluvialis* in Israel (Figure 4c–d, Tables 18.3 and 18.5). Two common PBR platforms exist: tubular and flat panel systems, with a number of variations on the theme. Examples of patented flat panel systems are: Flat-Panel Airlift (FPA, Subitec, Germany), Green Wall Panel (Florence University, Italy), and the ProviAPT (Proviron Holding NV, Belgium). Tubular PBRs typically consist of two compartments, the solar part consisting of glass or plastic tubes and the mixing tank, and the culture is recirculated by pumping between the two compartments (Figure 18.4b far end illustrates the black mixing tanks connected to the solar tubular compartment).

Photosynthesis of the biomass in the solar compartment leads to CO_2 depletion and dissolved oxygen (DO) superconcentrations, which at >150% saturation are known to inhibit photosynthesis. A typical oxygen saturation profile for the tubular PBR, shown in Figure 18.4b, shows that critical oxygen saturation threshold of 150% saturation of pO_2 is exceeded on most of the days of operation (Figure 18.5). Thus the mixing tank receiving this biomass has several functions, gas exchange (alleviate the very high oxygen concentrations and resupply of the biomass with adequate amounts of CO_2), fertilization, and monitoring of system criteria (DO, dissolved CO_2, pH, temperature, nutrients, and turbidity or a similar parameter for monitoring biomass content). Hence degassing requires the rapid circulation of biomass leading to shear forces not tolerated by many microalgae. Solar tubular arrays can be vertical, horizontal, or inclined. Some examples of patented tubular PBRs are the Simgae™ (XL Renewables Inc., Arizona, USA) and the Carbon Capture Recycle Roller-Film PBR (CC&R PBR, A2BE Carbon Capture LCC, Colorado, USA).

In general, biomass productivities in closed systems are higher than those in open systems, but this is not always the case (Table 18.3). Performance comparisons are compromised by effects induced by strains, geography (climate), environment (water source), process control [stocking densities and fertilization (including CO_2 concentrations), cultivation (continuous, semicontinuous,

TABLE 18.5

Commercial Microalgal Cultivation Systems and Bioproducts

Microalgae	Yield (t DW year^{-1})	Metabolites	Price (€)	Commercial Use	Culturing System	Country
Spirulina sp.	3000	Phycocyanin protein	36 kg^{-1}	Human and animal nutrition, phycobiliproteins, cosmetics, food coloring	—	China, India, USA, Myanmar, Japan, Hawaii
Chlorella sp.	2000	Astaxanthin ferredoxin	36 kg^{-1}	Human nutrition, aquaculture, cosmetics, other laboratory uses	Circular ponds with rotating arms, raceway ponds and fermenters	Taiwan, Germany, Japan
Dunaliella sp.	1200	β-Carotene	215–2150 kg^{-1}	Human nutrition, cosmetics, pigmenting agent	Raceway and open ponds	Australia, Israel, USA, China
Aphanizomenon sp.	500	Astaxanthin	—	Human nutrition	—	USA
Haematococcus sp.	300	Aquaculture feed Astaxanthin	50 L^{-1} 7150 kg^{-1}	Aquaculture pigmenting agent	—	USA, India, Israel
Crypthecodinium sp.	240 t DHA oil	Docosahexa-enoic acid (DHA)	4300 kg^{-1}	DHA oil and functional food additives	Fermenter	USA
Schizochytrium sp.	10 t	DHA oil	4300 kg^{-1}	DHA oil	—	USA
Euglena sp.	—	Arachidonic acid	—	Laboratory use	—	USA
Isochrysis sp.	—	DHA	—	Functional food additives	—	USA
Phaeodactylum sp.	—	Eicosapenta-enoic acid (EPA)	—	Medical use	—	USA

Source: Adapted from Carvalho AP, Meireles LA, Malcata FX. 2006. *Biotechnology Progress* 22(6):1490–1506. doi:10.1021/bp060065r; Spolaore P et al. 2006. *Journal of Bioscience and Bioengineering* 101(2):87–96. doi:10.1263/jbb.101.87; Milledge JJ. 2011. *Reviews in Environmental Science and Bio-Technology* 10(1):31–41. doi:10.1007/s11157-010-9214-7; Brennan L, Owende P. 2010. *Renewable and Sustainable Energy Reviews* 14(2):557–577. doi:http://dx.doi.org/10.1016/j.rser.2009.10.009.

Note: DW, dry weight.

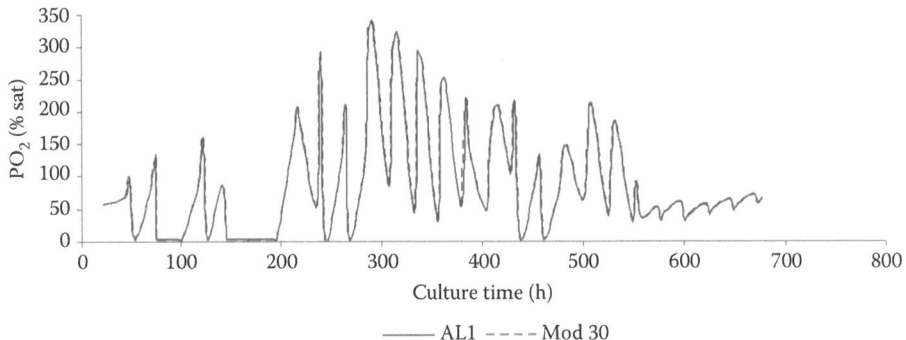

FIGURE 18.5 Twenty-eight day time course of dissolved oxygen concentrations in a tubular PBR operated at James Cook University, Townsville, Australia.

batch, and harvesting regimes)]. The influence of bioreactor type is illustrated in Figure 18.6 using *Nannochloropsis oculata* and *Nannochloris* sp., illustrating that growth performance varies not only with species and system, but also between different runs of the same species in the same system, for example, reactor run 2 and run 3 for *Nannochloris* sp.

In the best case, average productivities of 98 and peaks of 170 g m^{-2} day^{-1} have been claimed from a pilot experiment using a tubular PBR in the Arizona desert (Pulz 2007) (Table 18.3). Despite apparent productivity advantages of closed PBRs, many aspects influence selection of the cultivation platform (Table 18.4). Open systems are generally cheaper ($8–15 kg^{-1} DW in ponds vs. $50 kg^{-1} DW) in closed PBRs (Ozkan et al. 2012), but they are less efficient for CO_2 sequestration owing to effective dissolution of CO_2 in shallow ponds, requiring venturi-mixing of the gas with the broth, which on a large-scale is cost-prohibitive. In comparison, closed systems should better safeguard against contamination and allow for more effective dissolution of CO_2, but the low diameter nature of the tubes or plates leads to high temperatures (unless cooled, which costs again more energy), photosynthesis-inhibiting oxygen buildup requiring fast mixing rates with the mixing tank,

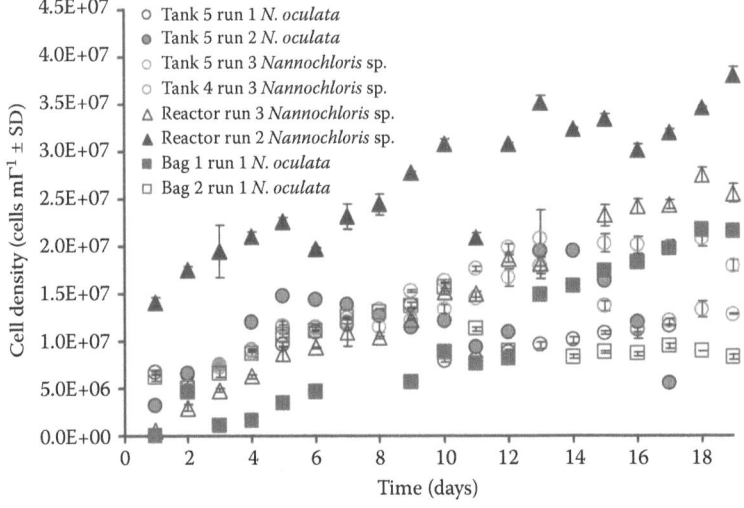

FIGURE 18.6 Growth performance of *Nannochloris* sp. and *Nannochloropsis oculata* in open systems (tank), tubular photobioreactor (reactor) and hybrid system (bag) being a covered system isolated from the environment by positive air displacement.

which leads to excessive shear stress, thereby limiting the system to species with sufficient shear stress tolerance and the high infrastructure and maintenance costs are prohibitive for any large-scale, high-volume applications.

Hybrid systems combine the advantages of open and closed systems, but there are a lower number of published pilot projects for this innovative concept. A good example is the Biological Algal Growth System (BAGS) designed by MBD Energy Ltd. (Melbourne, Australia) comprising a series of horizontal plastic bags, which are inflated by a blower (Figure 18.4e) delivering filtered air to the system with an exit on the other end. This positive air displacement also provides evaporative cooling and safe-guards against invasions of the systems. Owing to the shallow nature of the system, however, other problems associated with shallow tank or raceway cultivation are inherently the same (poor mixing and CO_2 dissolution, light limitation of photosynthesis in high-density cultures), which limit biomass productivities to those comparable to raceways, but at a higher infrastructure cost. The ALDUO™ (HR Petroleum Inc., Hawaii, USA) is a dual cultivation system combining PBR for continuous cultivation and open ponds for batch cultivation. Similar approaches use PBRs for the cultivation of inoculae and raceways for bulk biomass production, which are growing in popularity in recent years (Zittelli et al. 2013).

Produced biomass can be harvested and dewatered using a number of different techniques (Table 18.6), which differ in their degree of efficiency (cells recovered vs. cells delivered), concentration factor (reflecting the degree of dewatering), and the energy inputs and costs. Despite the high energy requirement and costs, centrifugation (e.g., Evodos) is most commonly used for smaller scale operations, as it is less species dependent. Flocculation-coagulation-based harvests either require the use of chemicals (e.g., chitosan, calcium, and aluminum phosphate) or bioflocculation (use of nutrient or pH stress and/or co-flocculating microalgal species) is increasingly being used, but requires fine tuning of conditions depending on the strain cultivated. A new technique called electro-flocculation followed by dissolved air floatation (DAF) is advertised to be cost effective and requiring no chemical input (i.e., the electrodes used are not sacrificial, OriginOil, USA), but effectiveness varies greatly with strain and culture conditions. Ultrafiltration or sand filtration can be used for target products that do not require total dewatering or include a drying step (e.g., aquaculture feed). Sedimentation/decanting is the cheapest and most energy-efficient option, although it is mostly restricted to self-settling strains. In any case, none of the dewatering options are yet workable for high-volume, low-value products (e.g., biofuels) when actually scaled to market size.

TABLE 18.6
Comparison of Techniques for Harvesting and Dewatering Microalgae

Method	Operating Cost (US$)	Removal Efficiency	Concentration Factor	Energy Demand
Magnetic separation	600	95	Unknown	vh
Centrifugation	500	80	40	vh
Coagulation—flotation	450	90	85	h
Coagulation—flocculation sedimentation	400	85	50	h
Ultrafiltration	350	95	50	h
Intermittent sand filtration	250	80	—	l
Micro-strainer	50	50	35	l
Sedimentation	50	80	Unknown	l

Source: With kind permission from Springer Science+Business Media: In: *Algae for Biofuels and Energy*, Harvesting, thickening and dewatering microalgae biomass, 2013, pp. 165–185, Pahl SL et al.

Note: Vh, very high; h, high; l, low.

Biofilm-based PBRs, on the other hand, may offer harvest simplicity (required by such markets), as the produced biomass can be harvested directly by scraping it off the matrix, although development is still at laboratory scale. Harvest-associated energy inputs may be reducible from 21 (PBRs) or 7 MJ kg^{-1} DW (open ponds) to 0.075 MJ kg^{-1} microalgal DW in biofilm-based PBRs (Ozkan et al. 2012).

In addition to biomass production and dewatering/harvesting costs, product-dependent additional downstream processing will ultimately determine the overall cost, energy requirements and carbon footprint of the entire production pathway (Table 18.2). Most studies have focused on life-cycle assessment for the production of biofuels, while other commodities, for example, aquaculture feed has only received recent attention (Taelman et al. 2013). Models for energy input and CO_2 emissions for microalgal biofuel production suggest that 30%–60% are associated with cultivation, with fertilizer demand accounting for 67% of it and harvesting/dewatering for ~30% (Slade and Bauen 2013; Flesch et al. 2013). In general, GHG emissions for open pond cultivation are lower than for closed PBRs, although ranges found in the literature are remarkably broad and overlap for both systems. If assuming a CO_2 sequestration of 1.826 kg CO_2 kg^{-1} microalgal DW produced (Chiang et al. 2011), the carbon footprint for open ponds and biofilm-based PBRs are negative, but generally positive (only negative in some cases) for closed suspension-based PBRs. It has to be emphasized though that the overall GHG balance varies depending on the target product (e.g., biodiesel vs. protein for aquaculture feed) (Table 18.2). Recent studies suggest that laboratory-scale biofilm-based PBRs, with a productivity of 24.9 g DW m^{-2} day^{-1} and virtually no dewatering/harvesting costs, result in a net energy ratio (NER) of 5- and 4–75-fold lower than that of open ponds and suspension-based PBRs, respectively. Therefore, the development of biofilm-based PBRs at large scale arises as a challenging research field aiming at achieving a sustainable microalgal biomass production platform that minimizes energy inputs and carbon footprints particularly of low-value, high-volume bioproducts.

18.5 BIOPRODUCTS FROM MICROALGAE

Microalgae and cyanobacteria, specifically *Spirulina*, have been harvested from natural populations in Mexico, Africa, and Asia as nutritional resources for 400 years (Borowitzka 2013a), whereas commercial interest in using microalgae for bioproduct development started in the 1950s with algae-farming approaches in Germany, Japan, and the United States of America (Tamiya 1957; Krauss 1962). Among potential microalgal bioproducts, microalgae-derived biofuel research has received the greatest attention owing to high biomass productivities and high lipid contents in many species. Life-cycle analyses for microalgal biofuels have yielded contradictory results, with some suggesting that carbon emissions during production may be greater than carbon sequestration achieved through photosynthesis (Gao et al. 2013 and references therein), keeping the question of sustainability under debate. In contrast, recent studies suggest that microalgal production for other commodities, for example, aquaculture feeds, can be accomplished in an environmentally sustainable manner with a reduced carbon footprint (Taelman et al. 2013). In this context, the following section will focus on commercial microalgal bioproducts (other than biofuels) for the agri- and aquaculture industries.

Despite the enormous array of potential microalgal products (pigments, exopolysaccharides, vitamins, protein, lipid, and a range of bioactive compounds), only a few species and limited products are being produced commercially to date. The major genera in commercial production are: the cyanobacterium *Spirulina* (*Arthrospira*) and the green microalgae *Chlorella*, *Dunaliella*, and *Haematococcus*. Estimates of global production volumes are ~10,000, 4000, 1000, and 200 t year^{-1} for *Spirulina*, *Chlorella*, *Dunaliella* and *Haematococcus*, respectively (Benemann 2013).

Examples of commercial production of microalgae (Table 18.5) for specialty products are the pigments β-carotene and astaxanthin, whole algae health supplements (i.e., *Spirulina platensis* and *Chlorella* spp.), and DHA production from *Isochrysis galbana* T-ISO. Commercial production of β-carotene, for use in food coloring to cosmetic applications, uses the hyper-saline-tolerant green marine microalgae *Dunaliella salina* and the industry was already well established using open

ponds in the 1980s with low production costs making the product competitive against synthetically produced or other natural sources of β-carotene (Moulton et al. 1987). Nowadays, β-carotene
is being produced commercially in Western Australia and South Australia, Israel (Figure 18.1),
and Hawai, all using open pond cultivation (Priyadarshani and Rath 2012), as the hyper-salinity
tolerance of the microalgal producer guards against contamination and excessive use of freshwater
because increasing salinity levels due to evaporative water loss can be managed by using brackish
slightly saline water.

In contrast, commercial astaxanthin production uses the freshwater green microalga
Haematococcus pluvialis, as, thus far, it is the only microalga with sufficient astaxanthin production per unit biomass to compete with synthetically produced astaxanthin on the market (Bubrick
1991). However, owing to the essentially freshwater nature of the producer, more sophisticated mass
cultivation systems are required (see Section 18.4) as (a) optimal biomass production requires different cultivation parameters compared with optimization for astaxanthin production (Kobayashi
et al. 1993) and (b) cultures are prone to invasions and dominance by unwanted organisms requiring
the use of closed mass production systems, which are a major capital expenditure (Olaizola 2000).
Commercial production started in the Kibbutz Ketura 50 km inland from the coastal Eilat, Israel
(Figure 18.3b and c) (Lorenz and Cysewski 2000) and Algatech's product AstaPure is still the most
competitive natural source of astaxanthin on the market. It is noteworthy that commercial cultivation of this freshwater microalga is carried out in the desert using slightly saline aquifers since 1998.
The cost for photosynthetic production (mainly due to system and dewatering costs) is, however,
high, resulting in marginal competitiveness with synthetic products. Therefore, mixotrophic and
heterotrophic growth is being investigated for a range of organisms (Han et al. 2013), for example, cane molasses for mixotrophic astaxanthin production by a mutant of *Chlorella zoefingiensis*
appears to be commercially viable (Liu et al. 2013). Mixo- and heterotrophic production pathways
may be viable for high-value products with a comparatively small market size (such as β-carotene
and astaxanthin). Those production pathways are, however, not viable in terms of feedstock costs
and volumes, capital expenditures, and the elimination of the GHG (CO_2) remediation potential
obtainable from mass cultivation for low-value and high-volume markets, such as fuel and potentially protein. As mixo- and heterotrophic cultivation methods have less or no potential for reducing
the carbon footprint in agri- and aquaculture, they are largely excluded from detailed considerations
relating to case studies.

Human brain development relies on an adequate supply of polyunsaturated fatty acids, in particular the long chain ω-6 polyunsaturated fatty acid (LC-ω6-PUFA) arachidonic acid (AA) and
the LC-ω3-PUFA, DHA (Khozin-Goldberg et al. 2011), which cannot be produced in adequate
quantities in nontransgenic plants (Petrie and Singh 2011; Meesapyodsuk and Qiu 2012). Traditional
sources of these fatty acids have been oily fish (Ward and Singh 2005); however, 80% of wild
resources are fully or close to being over-exploited (Frid and Paramor 2012) and, owing to decreasing water qualities, metal accumulation in these wild-caught resources has become a stigma in
recent times (Meujo and Hamann 2013). However, given the well-documented positive effects on
infant cognition, DHA is now being routinely enriched in infant formulae (Khozin-Goldberg et al.
2011; Qawasmi et al. 2012). Some marine algae are natural sources of AA, EPA, and DHA with
naturally low amounts of ω-6 fatty acids, including AA (Khozin-Goldberg et al. 2011). For example,
algae rich in EPA include the eustigmatophyte *Nannochloropsis oculata* (Huerlimann et al. 2010),
the diatom *Skeletonema costatum* (Duong et al. 2012); algae rich in DHA include the haptophytes,
for example, *Isochrysis* aff. *galbana* (T-ISO), while the strain *Isochrysis galbana* (Parke), the cryp-
tophyte *Rhodomonas salina* and the unicellular red alga *Porphyridium cruentum* are rich in both
EPA and DHA (Duong et al. 2012; Fidalgo et al. 1998; Huerlimann et al. 2010). Commercial production of *N. oculata* on coal-fired power station flue gas is being undertaken by Seambiotic in Israel in
raceway cultivation (Figure 18.7), but demands for EPA cannot be met by production of *N. oculata*.

The market for infant formulae DHA is predicted to be USD 10 billion per annum (Spolaore
et al. 2006) and commercial production is presently achieved heterotrophically using the

FIGURE 18.7   Production of *Nannochloropsis oculata* on coal-fired flue gas in raceway ponds by Seambiotic, Israel.

dinoflagellate *Crypthecodinium cohnii* [DHASCA; 240 t in 2003 (Ratledge 2004)] at Martek (Columbia, MD, USA), while DHA derived from heterotrophic cultivation of the thraustochytrid fungus *Schizochytrium* [DHA Gold, 10 t in 2003 (Ratledge 2004)] at OmegaTech, owned by Martek, is being used in adult food and beverage dietary supplements (i.e., yoghurt and cheese), health foods, animal feeds, and mariculture products (Spolaore et al. 2006). Both products meet the US Food and Drug Administration's Good Manufacturing Process and hence it will be hard to compete commercially with photosynthetically-derived DHA, as production costs are typically higher. Commercial reality is that phototrophically grown *Isochrysis* sp. and the thraustochytrid produce around 0.033% to maximally ~0.2% (depending on bioreactor design) of maximal production possible with heterotrophically grown *Schizochytrium* (Fan and Chen 2007). However, while, for example, 12% glucose achieved a 3 g L^{-1} DHA production in *Schizochytrium* (Fan and Chen 2007), feedstock such as glucose for heterotrophic production might become limiting with increasing human population and the predicted scarcity of clean freshwater resources. On the basis of such feedstock used for heterotrophic production of bio-products, such industries will ultimately compete with food production on arable land. However, phototrophically grown DHA producers are also likely to impact negatively on arable land use owing to water and fertilization requirements, but production rates, other than energy requirements for harvesting (dewatering), are of lesser concern for carbon footprint considerations. At present, phototrophic production of DHA is not economically competitive owing to expensive infrastructure requirement for cultivation, harvest, and maintenance.

Given those facts, heterotrophic DHA production appears to be more economically and environmentally sustainable. In a holistic sense, however, dietary requirements, in general, cannot typically be met by producing and extracting a single compound. These lessons are learned in aquaculture. For example, heterotrophically grown *Isochrysis galbana* (fed on glucose or mannose, with a high DHA content) led to higher mortality in larvae of the pacific oyster, *Crassostrea gigas*, as heterotrophic biomass lacked essential sterols, while photosynthetically produced biomass satisfied both nutritional requirements (Park et al. 2002). *Isochrysis galbana* (strain CCMP1324—T-ISO) is produced commercially by Reed Mariculture as *Isochrysis* 1800 for aquaculture feed, but production conditions (photo-, mixo-, or heterotrophic) are not stated. Regardless, it appears that the aim for commercial production should be to produce biomass on limited resources (water, arable land, fertilizers) that closely match the nutritional requirements on the lowest carbon footprint possible, in order to sustain a healthy diet to avoid increasing health costs and loss of human productivity of a growing population. This is a challenging task considering that urban sprawl is likely to further compete with those limiting resources.

As algae respond to changes in light quantity and quality with changes in chlorophyll and accessory pigment content, microalgal industries can exploit this to optimize product range or quality. For example, green light has been shown to significantly increase phycobiliviolin content in the commercially cultivated cyanobacterium *Spirulina plantensis*, while red or white light had no effect (Babu et al. 1991). Phycocyanin and phycoerythrin are used as labeling compounds in biomedical research and are also antioxidants; although *Spirulina platensis* is produced commercially for its high protein content (Chojnacka et al. 2012; Sili et al. 2012; Sotiroudis and Sotiroudis 2013), it is sold as dietary supplements in health food stores. Another interesting compound is scytonemin, which occurs in the sheath of cyanobacteria, which absorbs UV-A and UV-C, protecting these cyanobacteria against UV damage and appears to be far more effective than plant flavonoids and animal melanins (Proteau et al. 1993). Mycosporins also possess UV-protective characteristics similar to scytonemin and are present in several cyanobacterial genera (e.g., *Aphanizonmenon*, *Calothrix*, *Synechococcus*, *Scytonema*). Since many cyanobacteria, including sheath-forming ones, can be grown as biofilms and some are capable of fixing atmospheric nitrogen, sunscreen development could be another profitable product range. Furthermore, nitrogen-fixing cyanobacteria have been proposed as a low-cost source of biofertilizers and soil conditioners (Thajuddin and Subramanian 2005; Benemann 2013), and ω-3/-6/-9 fatty acids for aquaculture feeds and neutraceuticals. This, together with their low nutritional requirements and associated fertilization costs, suggests that cultivation of N_2-fixing cyanobacteria for the production of low-value commodities would be a viable alternative to the rapidly growing microalgae (high nutrient requirements), owing to reduced energy costs and lower carbon footprints.

18.6 CASE STUDY: AQUACULTURE INDUSTRY

According to 2011 data, inland and marine aquaculture production systems reached 44.3 and 19.3 million ton and are continuing to increase steadily (Box 18.1, detailed Australian aquaculture industry). The "fish meal trap" term corresponds to views that aquaculture represents a nonenvironmentally sustainable approach because of the increased demand for feed, which leads to increased fishing efforts for the wild species. Further, it threatens the viability of wild fish stocks, and advocates the view that aquaculture production will be limited by the availability of wild fish (Olsen and Hasan 2012). The volumes of fish meal used in aquaculture have remained steady at around 3.2 million tons per annum (Jackson 2012). Overall feed footprint, that is, the sum of raw materials footprint and energy used to make feed, for salmon farming was calculated as 2.15 t CO_2 t^{-1} feed

---

BOX 18.1 AQUACULTURE INDUSTRY, AUSTRALIA

Aquaculture is the fastest-growing primary industry in Australia; in fact, it is the fastest growing food production sector in the world. More than 40 species are being commercially produced in Australian aquaculture; most of the production comes from high-value species.

The top five species are tuna, pearl oysters, Atlantic salmon, edible oysters, and prawns. Aquaculture is a major industry with a total production of 64,535 t of which Atlantic salmon production accounts for nearly 30,000 t. Aquaculture aims to reach 100,000 t of Atlantic salmon production by end of 2015.

Source: http://www.marevent.com/ASA12australie/Aquaculture%20
in%20the%20States%20of%20Australia.pdf

and a total carbon footprint between 3 and 8 kg CO_2-eq. kg^{-1} for fillet production (Nijdam et al. 2012). Within aquaculture, by far the biggest use of fish meal was for salmon feed (1300 USD t^{-1}) and the most commonly used alternative to fish meal is soy meal (500 USD t^{-1}). Soy meal induces allergic reactions in some carnivorous fish and some essential amino-acid-like compounds (e.g., Taurine) are missing.

The current fish meal (as feed) production is 5 million tons per annum and half of the fish meal produced is used by aquaculture industries. Because of the earlier-stated facts, use of microalgae as fish feed and/or in fish feed formulation is drawing increased attention in recent years owing to potentially high PUFA contents and lower carbon footprints. Also, the required land footprint area is much smaller for microalgae than for soybean cultivation (Priyadarshani and Rath 2012).

A few commonly used genera in aquaculture feed are *Chlorella*, *Nannochloropsis*, and *Thalassiosira* (Hemaiswarya et al. 2011). The major carbon footprint factors are: required large-scale microlagal cultivation systems, energy inputs, fertilizers, and freshwater demand (Clarens et al. 2010b; Handler et al. 2012). To reduce fertilizer- and fresh-water-associated carbon footprints, essentially two approaches are taken: (i) using nitrogen-containing domestic wastewater/aquaculture wastewater (cradle-to-grave cycle) and (ii) recycling of nitrogen contained in the non-TAG portion of microalgae. Additionally, utilizing nitrogen-fixing cyanobacteria for TAG feedstock production instead of eukaryotic microalgae—which cannot biologically fix nitrogen gas—may also circumvent the large external nitrogen requirements for cell growth (Peccia et al. 2013). Further, to reduce energy associated costs, flat panel or attached biofilm photobioreactors, which require less area for mass cultivation, could be effectively considered in place of conventional systems (Draaisma et al. 2013; Medeiros et al. 2013). Moreover, recent studies on biodiesel production from microalgae have shown that these have lower carbon footprints, that is, values range from 0.05 to 0.475 kg CO_2-eq./ MJ bio-jet (Clarens et al. 2010b; Handler et al. 2012). Therefore use of microalgae as aquaculture feed and as alternative nutrient supplements can be expected to also have lower carbon footprints (Draganovic et al. 2013). Compared with raceway pond systems, attached biofilm growth systems for microalgae cultivation are advantageous in terms of reduced (nil) biomass harvesting cost, lower footprint area for cultivation, and 275 time greater (96.4 kg m^{-3}) biomass yield than that obtained using conventional systems (Ozkan et al. 2012).

18.6.1 GOAL AND SCOPE

Considering the growing importance of culturing microalgae as aquaculture feed, the main goal and scope of this work are

 i. to calculate the carbon footprints values for conventional aquaculture meal (i.e., fish and soy meal) and microalgal-based meal production platforms; and
 ii. to consistently compare the systems in terms of net energy ratio (NER), fresh water demand, eutrophication rate, and CO_2 sequestration potentials.

This work assumed that the cultured microalgal biomass is completely used in feed formulation and oil contents; pigments and other coproducts that may be used as additional energy are outside the goal of this work. Also, the study focused on the initial step in microalgal production, that is, the attached biomass system for microalgae is considered in place of conventional suspension systems (e.g., raceway ponds), because the former saves on subsequent harvesting steps. Fertilizers and energy demands for feed formulation were included within the scope of this study. The study did not account for research infrastructure, for example, materials of the different cultivation platforms, as system costs will come down significantly once the nascent microalgal industry matures. As such, incorporation of such costs would skew energy and cost balances to present-day conditions, which we consider inappropriate.

18.6.2 Methodology

Microalgal cultivation system boundaries for conventional and biofilm-based systems are depicted in Figure 18.8. The functional units adopted for the calculation are based on 1 t of aquaculture feed, and all input materials were converted to energy values [mega joules (MJ)] using appropriate conversion factors (detailed in Table 18.7). In addition, a factor of 1.88 CO_2 is assumed for CO_2 sequestration capacity calculations for microalgae, which is 32 times lower compared with soy cultivation as feed for aquaculture industries. No specific microalgal species were considered as case. Instead, generic cultivation data points were assimilated based on our synthesized knowledge as discussed in the earlier sections.

Net energy ratios and carbon balances were calculated as detailed in Sudhakar et al. (2012). In brief, carbon footprint of the aquaculture feed formulation process was calculated from Equation 18.5.

$$CFP = \text{Total } CO_2 \text{ input} - \text{Total } CO_2 \text{ stored} \qquad (18.5)$$

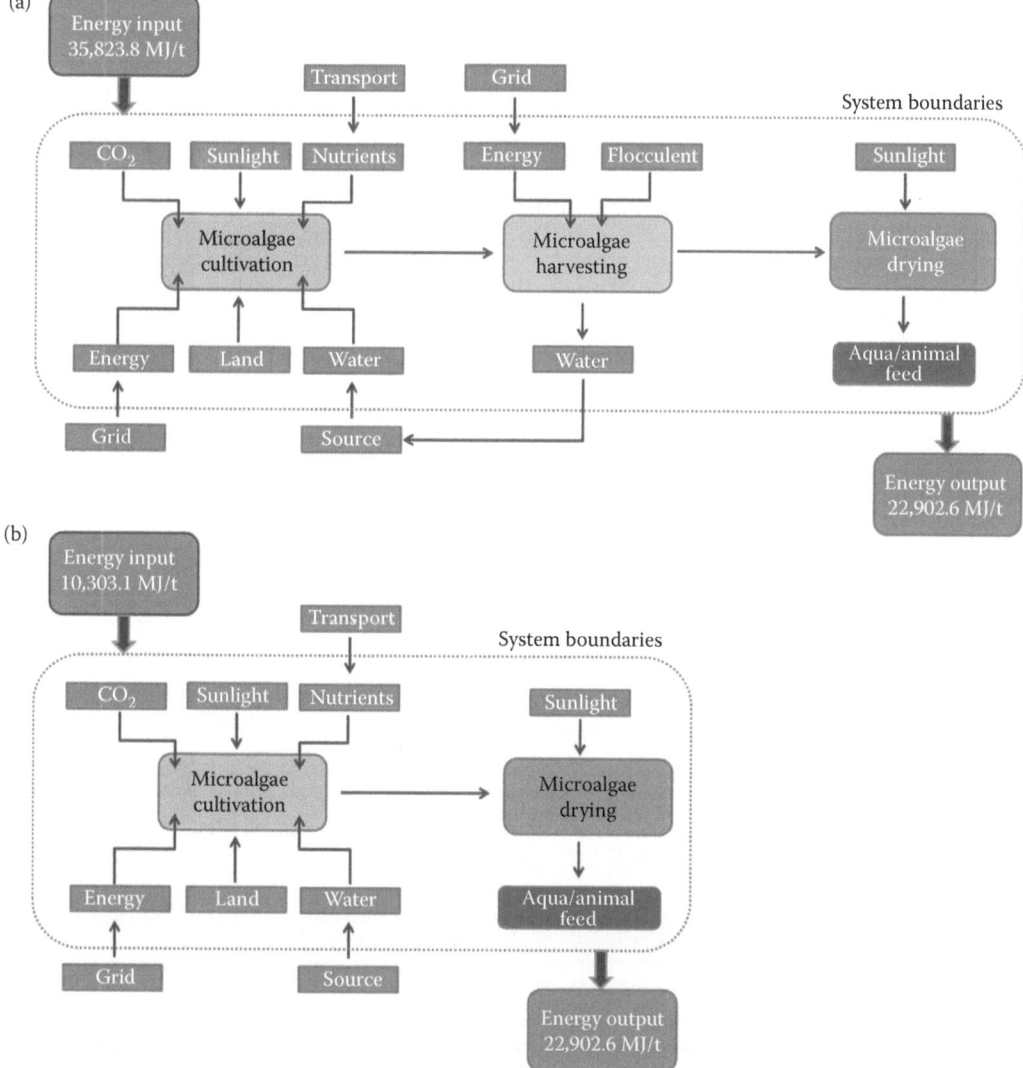

FIGURE 18.8 System boundaries for carbon footprint calculations of microalgae cultivation: (a) conventional systems and (b) attached biomass cultivation systems.

TABLE 18.7
Aquaculture Feed Formulation and Associated Energy Inputs

Particulars	Fishmeal	Soy Meal	Algal Meal (Conventional)	Algal Meal (Attached)
Inputs (MJ/Ton of Feed)				
Herbicides	—	1705.7	—	—
Nitrogen	—	4364.6	5230.1	5230.1
Phosphorous	—	974.5	281.8	281.8
Manure	—	2165.7	—	—
Seeds	—	1067.7	—	—
Elecrticity	912.3	1686.6	9311.9	4716.2
Diesel	8186.2	5845.8	—	—
Natural gas	9776	9651.4	—	—
Harvesting	—	—	21,000	75
Total energy input (MJ/t feed)	18,874.5	27,462.0	35,823.8	10,303.1
Total energy input (kWh/t feed)[a]	5243.3	7628.9	9951.9	2862.2
Carbon input (kg CO_2 eq./t feed)[b]	2335.9	3398.7	4433.6	1275.1
Outputs (MJ/Ton of Feed)				
Protein[c]	16,005.6 (72%)	14,227.2 (64%)	12,004.2 (54%)	12,004.2 (54%)
Carbohydrate[d]	3169.96 (19.4%)	5392.2 (33%)	2287.6 (14%)	2287.6 (14%)
Lipids[e]	3287.8 (8.4%)	1174.2 (3%)	8610.8 (22%)	8610.8 (22%)
CO_2 capture (kg CO_2/t of feed)	—	58.1	1860	1860
Total energy output (MJ/t feed)	22,463.4	20,793.6	22,902.6	22,902.6
Total energy input (kWh/t feed)[a]	6240.3	5776.5	6362.3	6362.3
Carbon output (kg CO_2 eq./t feed)[f]	1953.2	1808.0	1991.4	1991.4
Carbon balance (footprint)	382.7 (−)	1532.6 (−)	582.2 (−)	2576 (+)
NER	1.19	0.76	0.64	2.22
Other Relevant Input Values				
Fresh water demand (L/kg)	—	321	2390 (can be nil if wastewater used)	2390 (can be nil if wastewater used)
Eutrophication (kg NO_3 eq.)	—	5.38×10^7	6.93×10^7 (can be neglected if ·wastewater used)	6.93×10^7 (can be neglected if wastewater used)

Source: Adapted from Ozkan A et al. 2012. *Bioresource Technology* 114:542–548; Draganovic V et al. 2013. *Ecological Indicators* 34:277–289; Draaisma RB et al. 2013. *Current Opinion in Biotechnology* 24:169–177; Medeiros DL, Sales EA, Kiperstok A. 2013. *Paper Presented at the 4th Academic International Workshop Advances in Cleaner Production*, Brazil, May 22–24, 2013.

[a] 0.2778 conversion factor for MJ to kW h.
[b] 0.4455 conversion factor for kW h to CO_2-eq. (electricity delivered).
[c] 22,230 kJ/kg.
[d] 16,340 kJ/kg.
[e] 39,140 kJ/kg.
[f] 0.313 conversion factor for kW h to CO_2-eq. (fossil energy, ex. coal).

Total CO_2 inputs required in each and every stage of biomass cultivation and harvesting (such as energy, fertilizer, etc.) were calculated from material input values. Similarly, CO_2 outputs were calculated from the biomass carbon content and sequestration potentials for different feeds. The NER, that is, the ratio of total energy output to total energy input, was calculated from the following equation:

$$\text{NER} = \sum \text{Energy}_{\text{outputs}} / \sum \text{Energy}_{\text{inputs}} \qquad (18.6)$$

All other conversion factors and assumptions are listed in Table 18.7 (in the note section) along with source references. In addition, fresh water demand and eutrophication potentials, calculated based on literature values for each feed formulation, are discussed.

As calculated in Table 18.7, the inputs and outputs (in terms of NER and CFP) from different feed formulations for aquaculture industries have been modeled for conventional and attached algal cultivation systems. As calculated in Table 18.7, carbon footprint is very low for conventional fish meal preparations (NER of 1.2) compared with soy meal (protein concentrate) preparation (NER of 0.7). Fresh water usage and eutrophication potentials were also minimal in this case. However, considering the rising demand for fish feed in aquaculture production, soy meal is used as alternative feed despite the negative carbon balance and competition for fresh water and arable land use. Considering algal meal production using conventional and attached system concepts, attached systems were found to be more beneficial in terms of lower energy inputs (NER of 2.22) with a great carbon balance to produce 1 ton of feed. Moreover, attached systems require less land compared with conventional algal cultivation systems, which is advantageous.

The carbon footprint could be further reduced, if N and P sources were not supplemented but completely recycled from aquaculture wastewater. However, this calculation is based on literature data sets and not long-term outdoor studies, which may give fluctuating case by case data based on the microalgal species considered, wastewater characteristics, local climatic conditions, and so on, and detailed investigations will be required.

From the real field trial scenario development study, microalgae production (*Nannochloropsis* sp.) in closed photo-bioreactor systems for aquaculture feed formulation appeared to be more sustainable than the production of reference fish meal feed (7.70 MJ, and 0.05 kg CO_2-eq. MJ^{-1} DW) (Taelman et al. 2013). A positive energy balance was attributed to incorporating photo-bioreactor cultivation of microalgae, as a resource footprint, into their calculations. However, closed photo-bioreactors are more energy-intensive than conventional systems that showed a positive energy balance in the study by Taelman et al. (2013). It is also possible to operate attached photobioreactors as closed system with lower energy inputs, but carbon footprints require detailed investigations, as data to date are limiting.

18.7 CASE STUDY: AGRICULTURAL INDUSTRY

Animal husbandry and agriculture industries are closely interlinked with larger land use and potential GHG emissions. Many co-production processes in agricultural industries are being considered for reducing associated environmental impacts and carbon footprints (Jacobsen et al. 2014; Pandey and Agrawa 2014). A detailed review on animal protein production and associated carbon footprints was published recently (Nijdam et al. 2012). Considering different agricultural industries, the pork industry appears to have a medium carbon footprint (i.e., 20–55 kg CO_2-eq. kg^{-1} protein) and energy footprint (i.e., 18–34 MJ) (de Vries and de Boer 2010; Nijdam et al. 2012). Carbon footprints derived from published data attribute 55%–73% to N_2O emissions associated with feed formulation and 24% to manure management, that is, mainly CH_4 emissions (Eriksson et al. 2005; Basset-Mens and van der Werf 2005; Hermansen and Kristensen 2011). Also, the CFP assessment methodologies of different agricultural industry with few case studies and GHG mitigation measures were published elsewhere (Jacobsen et al. 2014; Pandey and Agrawa 2014).

Considering the advantage of growing microalgae in diluted manure systems with the benefits of nutrients (mainly N and P) recycling (Zhu et al. 2013) and using it in feed formulation, carbon footprint in the pork industry might be reducible from an average of 6.5 kg CO_2-eq. kg^{-1} meat to less than 2 kg CO_2-eq. kg^{-1} meat (e.g., Box 18.2). The Australian Pork Industry case.

According to the FAO (Food and Agriculture Organization), pork consumption will reach up to 144 million ton by 2050, which is 40% higher than 2009 production. Therefore, sustainable feed

BOX 18.2   PORK INDUSTRY, AUSTRALIA

According to Australian industry sources, swine production has reached a point of balance between supply and demand. It is expected to increase marginally from 4.6 million to 4.7 million piglets between 2013 and 2014. However, grain prices for feed formulation, which account for 60% of input costs in pork production, have remained relatively high.

The Australian pork industry aims to achieve a carbon footprint of 1 kg CO_{2eq} kg^{-1} HSCW to attain sustainable pork production and to safeguard the environment. Implementation of manure management systems, that is, CH_4 capturing in manure ponds, reduced 50% of the GHG emissions and carbon footprints up to 2.3 kg CO_2-eq. kg^{-1} HSCW in Australia.

Sources:

1. http://gain.fas.usda.gov/Recent%20GAIN%20Publications/Livestock%20and%20 Products%20Annual_Canberra_Australia_9-4-2013.pdf
2. http://www.bpex.org.uk/downloads/302196/301267/Pigs%20and%20the%20 Environment%20%20How%20the%20global%20pork%20business%20is%20 reducing%20its%20impact.pdf

production is important for the pork industry, which requires carbon footprint reduction. Soybean, canola, corn, wheat, pea, rapeseed, and barley meal, and the like are commonly used in feed formulation (Eriksson et al. 2005) to achieve the final feed protein content of 16%–18%. It was reported that 10%–20% of feed formulation could be replaced with microalgae. However, lysine and methionine are two major amino acids very important for growth and need to be considered in feed formulation and selection of microalgae. *Chlorella* sp. had a higher lysine (8.4 g 100 g^{-1} protein) and methionine (2.2 g 100 g^{-1} protein) content than soybean (6.4 and 1.3 g 100 g^{-1} protein, respectively) (Becker 2007). Similarly, *Dunaliella* sp., also had high lysine content (7.0 g 100 g^{-1} protein), but methionine contents were not reported. A few other species have been reported suitable as feed: the kelp *Laminaria digitata* and the dinoflagellate *Schizochytrium* sp. (He et al. 2002; Abril et al. 2003; Sardi et al. 2006). Also, the protein content of the feed, which affects overall *N* emissions, needs to be considered in selection of microalgal species and subsequent feed formulation. Figure 18.9 compares the carbon footprint of different feed formulations with microalgal feed for the pork industry. Algal meal production is negative when using attached biofilm cultivation systems (refer

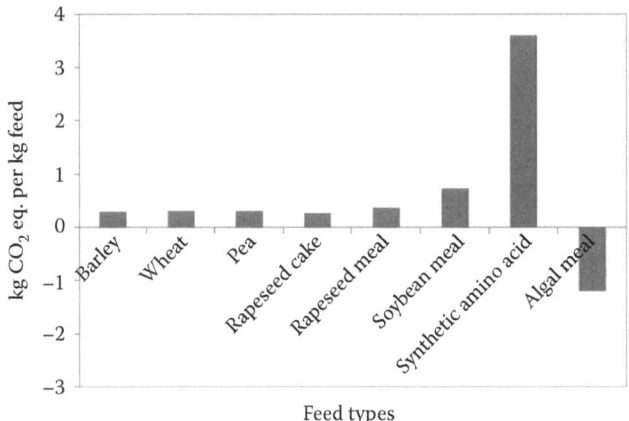

FIGURE 18.9   Carbon footprint of different feed formulations for the pork industry.

TABLE 18.8

Case Scenario for Nutrient Recycling and Associated Carbon Footprints for Piggeries (20,000 Piglets)

Case Scenario	Anaerobic Pond and Biogas Associated Net Emissions (t CO_2-eq.)	Aerobic Treatment and Associated Net Emissions (t CO_2-eq.)	Algal Pond (Operational and Harvesting) Associated Net Emissions (t CO_2-eq.)
AD and effluent discharge	1719.6	—	—
AD and aerobic treatment	1719.6	352.74	—
AD and nutrient recycling with microalgae	1719.6	—	440.9
AD, aerobic treatment and nutrient recycling with microalgae	1719.6	352.74	440.9
Aerobic	—	352.74	—
Aerobic and nutrient recycling with microalgae	—	352.74	440.9

to Table 18.7) compared with that of other feed types. Despite the potential use of microalgae as a source of protein (typically 35%–60% of the biomass) in feed formulation for different animals including pigs, there has been very little published information available on carbon footprint calculations.

According to the Pork Cooperative Research Centre (Pork CRC, Australia), piggeries (Berrybank Farms, Bears Lagoon piggery and Grantham piggery) are well aware of the technology in place for manure digestion and energy recovery systems. For instance, Berrybank Farm installed a seven-stage manure management system producing 1700 m^3 day^{-1} of biogas with an energy potential of 2.6–3.4 MW h day^{-1}. Revenue generated for the AD system accounted for $350,000 p.a., including energy and fertilizer costs (Buchanan and Cheng 2013). However, the feed requirements for 20,000 pigs is approximately 40,000 kg day^{-1}. Further the AD effluents (100,000 L) require treatment before discharge into inland waters, as they contain high nitrogen and phosphorous levels, which are essential nutrients for microalgal growth. These industries are searching for efficient microalgal species for effective recycling of nutrients and subsequently for feed formulation on-site.

Table 18.8 presents different case scenarios for the effective recycling of nutrients recycled from manure management systems at Berrybank Farm (based on values from Grant and Batten (2013)). However, estimates did not include fossil-fuel-derived nitrogen fertilizers used for feed production or water and land footprints. Simple aerobic manure treatment could be effectively coupled with algal ponds, resulting in lower carbon footprints. AD systems, however, could produce more revenue, as discussed earlier, which should be considered along with aerobic treatment options incorporating microalgae as animal feed products, especially for piggeries with 20,000 piglets or more.

Potential cost and energy factors involved in harvesting and drying of microalgae for intracellular water removal are considered by industries to be the main hurdles, hindering technology uptake. However, attached biofilm photobioreactor systems have the potential to reduce capital expenditure, cultivation, and harvesting costs to a greater extent, which could elicit uptake, given the high NER of 2.22. Therefore, the carbon footprint of 440.9 t CO_2-eq. could be significantly reduced, as shown in Table 18.7.

18.8 CONCLUSIONS

On the basis of the aquatic lifestyle, independence of arable land and potable water resources, photosynthesis-based physiology, high biomass productivities, and bioproduct contents, microalgae are ideal candidates for bioproduct development and, at the same time, playing a crucial role in wastewater and waste gas remediation. Yet, the nascent microalgal industry is developing slowly, mainly as a result of investment required into infrastructure, especially for low-value, high-volume bioproducts, such as fuels or protein. In addition, only a handful of the thousands of organisms known are being deployed in commercial bioproduct development. One of the major stumbling blocks for the development of the microalgal industries is the interdependence of bioproduct value and volume chains with choice of cultivation platform and location. This can, however, be overcome if a strategic approach is chosen, that is, select the main outcome of the application first; is it the profit derived from the product or is it the urgency for remediation driving the business case? Once developed, species, cultivation platform and processing options, as well as location and water resource availability can be chosen to fall in line with product volumes and quality requirements.

While microalgae undeniably have the potential for GHG (CO_2) abatement and water remediation, the system boundaries for carbon footprints are more blurred and again highly dependent on the bioproduct market, but in this case more with regard to carbon footprint of bioproduct production compared with existing alternative production pathways. With regard to carbon footprint calculations, the lack of long-term at least demonstration scale outdoor-derived data, agreed measurements and units, as well as stocking densities, fertilization and process parameters makes it difficult to calculate actual or sometimes even likely carbon footprints. Nonetheless, sufficient long-term data exist for the aquaculture and agricultural industries, allowing carbon footprint calculations for using microalgae for fish/soymeal replacement and microalgal feed production from piggery waste. Our analyses show that cultivation systems strongly affect NER for aquaculture fish/soy meal replacement, with attached systems providing better results than conventional open-pond-based suspension cultivation, owing to the removal of required dewatering technology and energy use thereof. Biofilm-based cultivation systems are, however, still at the laboratory investigation scale and therefore up-scaled results shown here should be validated with large demonstration-scale system performance for validation at scale. For, piggeries, microalgal byfeed in their diet could also result in additional carbon emission savings, even when using conventional open pond cultivation. In summary, owing to species-specific cultivation system and process particulars, the nascent microalgal industry should generate and share long-term data from outdoor cultivation at least at demonstration scale and to agree on standards for measurements, units, system, and process [fertilization, harvesting, production (continuous, semi-continuous, batch), species (preferably with genetic strain identity)]. Such data sets would be invaluable for obtaining more reliable mean carbon footprint data, as well as allowing for multifactorial analyses to determine the real drivers for biomass and bioproduct productivities, as well as GHG and wastewater remediation potential.

REFERENCES

Aarab L, Perez-Camacho A, Viera-Toledo MD, de Vicose GC, Fernandez-Palacios H, Molina L. 2013. Embryonic development and influence of egg density on early veliger larvae and effects of dietary microalgae on growth of brown mussel *Perna perna* (L. 1758) larvae under laboratory conditions. *Aquaculture International* 21(5):1065–1076. doi:10.1007/s10499-012-9612-7.

Abril R, Garrett J, Zeller SG, Sander WJ, Mast RW. 2003. Safety Assessment of DHA-rich microalgae from *Schizochytrium* sp. Part V: Target animal safety/toxicity study in growing swine. *Regulatory Toxicology and Pharmacology* 37:73–82.

Adarme-Vega TC, Lim DKY, Timmins M, Vernen F, Li Y, Schenk PM. 2012. Microalgal biofactories: A promising approach towards sustainable omega-3 fatty acid production. *Microbial Cell Factories* 11:96.

Adl SM, Simpson AGB, Farmer MA, Andersen RA, Anderson OR, Barta JR, Bowser SS et al. 2005. The new higher level classification of eukaryotes with emphasis on the taxonomy of protists. *Journal of Eukaryotic Microbiology* 52(5):399–451. doi:10.1111/j.1550-7408.2005.00053.x.

Ahlgren G, Hyenstrand P. 2003. Nitrogen limitation: Effects of different nitrogen sources on nutritional quality of two freshwater organsms *Scenedesmus quadricauda* (Chlorophyceae) and *Synechococcus* sp. (Cyanophyceae). *Journal of Phycology* 39:906–917.

Al-Amoudia OA, Flynn KJ. 1989. Effect of nitrate-N incorporation on the composition of the intracellular amino acid pool of N-deprived *Tetraselmis marina*. *British Phycological Journal* 24:53–61.

Andersen RA. 2005. *Algal Culturing Techniques*. Academic Press, Burlington, San Diego, London.

Andersson I. 2008. Catalysis and regulation in Rubisco. *Journal of Experimental Botany* 59:1555–1568.

ATSE. 2008. *Biofuels for Transport: A Roadmap for Development in Australia*. Melbourne.

AWC. 2009. Vulnarability of arid and semi-arid regions to climate change—Impacts and adaptive strategies. In: *Fifth World Water Forum*, Istanbul. Arab Water Council, p. 16.

Babalola TO, Apata DF. 2012. Effects of dietary lipid source on growth, digestibility and tissue fatty acid composition of *Heterobranchus longifilis* fingerlings. *Journal of Agriculture and Rural Development in the Tropics and Subtropics* 113(1):1–11.

Babu TS, Kumar A, Varma AK. 1991. Effect of light quality on phycobilisome components of the cyanobacterium *Spirulina platensis*. *Plant Physiology* 95(2):492–497. doi:10.1104/pp.95.2.492.

Badger MR, Price GD. 1992. The CO_2 concentrating mechanism in cyanobacteria and microalgae. *Physiologia Plantarum* 84(4):606–615. doi:10.1034/j.1399-3054.1992.840416.x.

Basset-Mens C, van der Werf HMG. 2005. Scenario-based environmental assessment of farming systems: The case of pig production in France. *Agriculture, Ecosystems and Environment* 105:127–144.

Bauer W, Prentice-Hernandez C, Tesser MB, Wasielesky W, Poersch LHS. 2012. Substitution of fishmeal with microbial floc meal and soy protein concentrate in diets for the pacific white shrimp *Litopenaeus vannamei*. *Aquaculture* 342:112–116. doi:10.1016/j.aquaculture.2012.02.023.

Becker EW. 2007. Microalgae as a source of protein. *Biotechnology Advances* 25:207–210.

Belay A. 1997. Mass culture of *Spirulina* outdoors—The earthrise farms experience. In: Vonshak A (ed), *Spirulina platensis (Arthrospira): Physiology, Cell Biology and Biochemistry*. Taylor & Francis, London, pp. 131–158.

Bell RA, Ogden N, Kramer JR. 2002. The biotic ligand model and a cellular approach to class B metal aquatic toxicity. *Comparative Biochemistry and Physiology C-Toxicology & Pharmacology* 133(1–2):175–188. doi:10.1016/s1532-0456(02)00109-6.

Benemann J. 2013. Microalgae for biofuels and animal feeds. *Energies* 6:5869–5886.

Berman-Frank I, Lundgren P, Falkowski P. 2003. Nitrogen fixation and photosynthetic oxygen evolution in cyanobacteria. *Research in Microbiology* 154(3):157–164. doi:10.1016/s0923-2508(03)00029-9.

Bhattacharya D, Schmidt HA. 1997. Division Glaucocystophyta. *Plant Systematics and Evolution* 11:139–148.

Bode A, Botas JA, Fernandez E. 1997. Nitrate storage by phytoplankton in a coastal upwelling environment. *Marine Biology* 129:399–406.

Bohme H. 1998. Regulation of nitrogen fixation in heterocyst-forming cyanobacteria. *Trends in Plant Science* 3(9):346–351. doi:10.1016/s1360-1385(98)01290-4.

Bondioli P, Della Bella L, Rivolta G, Chini Zittelli G, Bassi N, Rodolfi L, Casini D, Prussi M, Chiaramonti D, Tredici MR. 2012. Oil production by the marine microalgae *Nannochloropsis* sp. F&M-M24 and *Tetraselmis suecica* F&M-M33. *Bioresource Technology* 114:567–572.

Boresch K. 1913. Die Färbung von Cyanophyceen und Chlorophyceen in ihrer Abhängigkeit von Stickstoffgehalt des Substrates. *Jahrbuch für wissenschaftliche Botanik* 52:145–185.

Borowitzka MA. 2013a. Energy from microalgae: A short history. In: *Algae for Biofuels and Energy*. Springer, Berlin, pp. 1–15.

Borowitzka MA. 2013b. Species and strain selection. In: *Algae for Biofuels and Energy*. Springer, pp. 77–89.

Borowitzka MA, Moheimani NR. 2013a. Algae culture and applications. In: Buchanan AN, Bolton N, Moheimani N et al. (eds), *Algae for Energy and Feed: A Wastewater Solution. A Review*, Vol. Project 4A-101 112. Co-operative Research Centre for High Integrity, Australian Pork, pp. 79–127.

Borowitzka MA, Moheimani NR. 2013b. Open pond culture systems. In: *Algae for Biofuels and Energy*. Springer, pp. 133–152.

Bothe H, Schmitz O, Yates MG, Newton WE. 2010. Nitrogen fixation and hydrogen metabolism in cyanobacteria. *Microbiology and Molecular Biology Reviews* 74(4):529–551. doi:10.1128/mmbr.00033-10.

Brennan L, Owende P. 2010. Biofuels from microalgae—A review of technologies for production, processing, and extractions of biofuels and co-products. *Renewable and Sustainable Energy Reviews* 14(2):557–577. doi:http://dx.doi.org/10.1016/j.rser.2009.10.009.

Broecker WS, Peng T-H, Beng Z. 1982. *Tracers in the sea*. Lamont-Doherty Geological Observatory, Columbia University.

Brown MR. 2002. Nutritional value of microalgae for aquculture. In: Cruz-Suárez LE, Ricque-Marie D, Tapia-Salazar M, Gaxiola-Cortés MG, Simoes N (eds), *Avances en Nutrición Acuícola VI. Memorias del VI Simposium Internacional de Nutrición Acuícola*, 3 al 6 de Septiembre del 2002, Cancún, Quintana Roo, México,

Brown MR, Blackburn SI. 2013. Live microalgae as feeds in aquaculture hatcheries. In: Allan G, Burnell G (eds), *Advances in Aquaculture Hatchery Technology. Woodhead Publishing in Food Science Technology and Nutrition*. Woodhead Publ Ltd, UK, pp. 117–156. doi:10.1533/9780857097460.1.117.

Brown MR, McCausland MA, Kowalski K. 1998. The nutritional value of four Australian microalgal strains fed to Pacific oyster *Crassostrea gigas* spat. *Aquaculture* 165(3–4):281–293. doi:10.1016/s0044-8486(98)00256-7.

Bubrick P. 1991. Production of astaxanthin from *Haematococcus*. *Bioresource Technology* 38 (2–3):237–239. doi:10.1016/0960-8524(91)90161-c.

Buchanan AN, Cheng R. 2013. Current effluent treatment practice: Anaerobic digetsion. In: *Algae for Energy and Feed: A Wastewater Solution—A Review*, Vol. Project 4A-101 112. Co-operative Research Centre for High Integrity, Australian Pork.

Cai T, Park SY, Li Y. 2013. Nutrient recovery from wastewater streams by microalgae: Status and prospects. *Renewable and Sustainable Energy Reviews* 19:360–369.

Campbell PK, Beer T, Batten D. 2011. Life cycle assessment of biodiesel production from microalgae in ponds. *Bioresource Technology* 102(1):50–56.

Campenni L, Nobre BP, Santos CA, Oliveira AC, Aires-Barros MR, Palavra AMF, Gouveia L. 2013. Carotenoid and lipid production by the autotrophic microalga Chlorella protothecoides under nutritional, salinity, and luminosity stress conditions. *Applied Microbiology and Biotechnology* 97(3):1383–1393. doi:10.1007/s00253-012-4570-6.

Carvalho AP, Meireles LA, Malcata FX. 2006. Microalgal reactors: A review of enclosed system designs and performances. *Biotechnology Progress* 22(6):1490–1506. doi:10.1021/bp060065r.

Cerón MC, García-Malea MC, Rivas J, Acien FG, Fernandez JM, Del Río E, Guerrero MG, Molina E. 2007. Antioxidant activity of *Haematococcus pluvialis* cells grown in continuous culture as a function of their carotenoid and fatty acid content. *Applied Microbiology and Biotechnology* 74:1112–1119.

Chakdar H, Jadhav SD, Dhar DW, Pabbi S. 2012. Potential applications of blue green algae. *Journal of Scientific and Industrial Research* 71(1):13–20.

Chen GQ, Chen F. 2006. Growing phototrophic cells without light. *Biotechnology Letters* 28(9):607–616. doi:10.1007/s10529-006-0025-4.

Chiang CL, Lee CM, Chen PC. 2011. Utilization of the cyanobacteria *Anabaena* sp. CH1 in biological carbon-di-oxide mitigation processes. *Bioresource Technology* 102:5400–5405.

Chojnacka K, Saied A, Michalak I. 2012. The possibilities of the application of algal biomass in the agriculture. *Science* 66(11):1235–1248.

Clarens AF, Resurreccion EP, White MA, Colosi LM. 2010a. Environmental life cycle comparison of algae to other bioenergy feedstocks. *Environmental Science & Technology* 44(5):1813–1819.

Clarens AF, Resurreccion EP, White MA, Colosi LM. 2010b. Environmental life cycle comparison of algae to other bioenergy feedstocks. *Environmental Science and Technology* 44:183–1819.

Coleman JR. 1991. The molecular and biochemical analyses of CO_2-concentrating mechanisms in cyanobacteria and microalgae. *Plant Cell and Environment* 14(8):861–867. doi:10.1111/j.1365-3040.1991.tb01449.x.

Crawford NM, Kahn ML, Leustek T, Long SR. 2000. Nitrogen and sulphur. In: Buchanan BB, Gruissen W, Jones RL (eds), *Biochemistry and Molecular Biology of Plants*, Vol. 16. American Society of Plant Physiologists, Rockville, Maryland, pp. 787–849.

Czeczuga B. 1986. The effect of light on the content of photosynthetically active pigemnts in plants. 4. Chromatic adaptation in blue-green-algae *Anabaena cylindrica* and *Anabaena variabilis*. *Phyton-Annales Rei Botanicae* 26(1):1–9.

Dai YJ, Li J, Wei SM, Chen N, Xiao YP, Tan ZL, Jia SR, Yuan NN, Tan N, Song YJ. 2013. Effect of light with different wavelengths on *Nostoc flagelliforme* cells in liquid culture. *Journal of Microbiology and Biotechnology* 23(4):534–538. doi:10.4014/jmb.1205.05037.

Dawczynski C, Schubert R, Jahreis G. 2007. Amino acids, fatty acids, and dietary fibre in edible seaweed products. *Food Chemistry* 103(3):891–899. doi:10.1016/j.foodchem.2006.09.041.

de Vries M, de Boer IJM. 2010. Comparing environmental impacts for livestock products: A review of life cycle assessments. *Livestock Science* 128(1):1–11.

Dortch Q, Clayton Jr JR, Thoresen SS, Ahmed SI. 1984. Species differences in accumulation of nitrogen pools in phytoplankton. *Marine Biology* 81:237–250.

Doucha J, Livansky K. 2006. Productivity, CO_2 exchange and hydraulics in outdoor open high density microalgal. *Journal of Applied Phycology* 18:811–826.

Dow. 2007. Dow Epoxy advances glycerine-to-epichlorohydrin and liquid epoxy resins projects by choosing Shanghai site. Dow Chemical Company.

Draaisma RB, Wijffels RH, Slegers PM, Brentner LB, Roy A, Barbosa MJ. 2013. Food commodities from microalgae. *Current Opinion in Biotechnology* 24:169–177.

Draganovic V, Jorgensen SE, Boom R, Jonkers J, Riesen G, van der Goot AJ. 2013. Sustainability assessment of salmonid feed using energy, classical exergy and eco-exergy analysis. *Ecological Indicators* 34:277–289.

Ducat DC, Way JC, Silver PA. 2011. Engineering cyanobacteria to generate high-value products. *Trends in Biotechnology* 29(2):95–103.

Duong VT, Li Y, Nowak E, Schenk PM. 2012. Microalgae isolation and selection for prospective biodiesel production. *Energies* 5(6):1835–1849. doi:10.3390/en5061835.

Eriksson IS, Elmguist H, Stern S, Nybrant T. 2005. Environmental systems analysis of pig production: The impact of feed choice. *International Journal of Life Cycle Assessment* 10:143–154.

Eustance E, Gardner RD, Moll KM, Menicucci J, Gerlach R, Peyton BM. 2013. Growth, nitrogen utilization and biodiesel potential for two chlorophytes grown on ammonium, nitrate or urea. *Journal of Applied Phycology* 25(6):1663–1677. doi:10.1007/s10811-013-0008-5.

Evans JR. 2013. Improving photosynthesis. *Plant Physiology* 162(4):1780–1793. doi:10.1104/pp.113.219006.

Falkowski PG, Raven JA. 2007. *Aquatic Photosynthesis*. Chapter 5: *Carbon Acquisition and Assimilation*. Princeton University Press, Princeton, New Jersey.

Falkowski PG, Sukenik A, Herzig R. 1989. Nitrogen limitation in *Isochrysis galbana* (Haptophyceae). II. Relative abundance of chloroplast proteins. *Journal of Phycology* 25(3):471–478.

Fan KW, Chen F. 2007. Chapter 11. Production of high-value products by marine microalgae Thraustochytrids. In: Yang S-T (ed.), *Bio-processing of Value-Added Products from Renewable Resources. New Technologies and Applications*. Elsevier, Amsterdam, pp. 293–324.

Fidalgo JP, Cid A, Torres E, Sukenik A, Herrero C. 1998. Effects of nitrogen source and growth phase on proximate biochemical composition, lipid classes and fatty acid profile of the marine microalga *Isochrysis galbana. Aquaculture* 166 (1–2):105–116. doi:10.1016/s0044-8486(98)00278-6.

Flesch A, Beer T, Campbell PK, Batten D, Grant T. 2013. Greenhouse gas balance and algae-based biodiesel. In: *Algae for Biofuels and Energy*. Springer, Berlin, pp. 233–254.

Flynn KJ, Flynn K. 1992. Nonprotein free amines in microalgae—Consequences for the measurement of intracellular amino-acids and of the glutamine glutamate ratio. *Marine Ecology Progress Series* 89(1):73–79. doi:10.3354/meps089073.

Frid CLJ, Paramor OAL. 2012. Feeding the world: what role for fisheries? *ICES Journal of Marine Science* 69(2):145–150. doi:10.1093/icesjms/fsr207.

Fujita Y. 1996. Flexibility of energy conversion process in thylakoids of cyanobacteria. *Journal of Science and Industrial Research* 55 (8–9):618–629.

Gantar M, Simovic D, Djilas S, Gonzalez WW, Miksovska J. 2012. Isolation, characterization and antioxidative activity of C-phycocyanin from *Limnothrix* sp. strain 37-2-1. *Journal of Biotechnology* 159(1–2):21–26. doi:10.1016/j.jbiotec.2012.02.004.

Gantt E. 1996. Pigment protein complexes and the concept of the photosynthetic unit: chlorophyll complexes and phycobilisomes. *Photosynthesis Research* 48(1–2):47–53. doi:10.1007/bf00040995.

Gao X, Yu Y, Wu H. 2013. Life cycle energy and carbon footprints of microalgal biodiesel production in Western Australia: A comparison of byproducts utilization strategies. *ACS Sustainable Chemistry & Engineering* 1(11):1371–1380.

Garcia MCC, Miron AS, Sevilla JMF, Grima EM, Camacho FG. 2005. Mixotrophic growth of the microalga *Phaeodactylum tricornutum*—Influence of different nitrogen and organic carbon sources on productivity and biomass composition. *Process Biochemistry* 40(1):297–305. doi:10.1016/j.procbio.2004.01.016.

Geider RJ, Macintyre HL, Graziano LM, McKay RML. 1998. Responses of the photosynthetic apparatus of *Dunaliella tertiolecta* (Chlorophyceae) to nitrogen and phosphorus limitation. *European Journal of Phycology* 33(4):315–332.

Grant T, Batten D. 2013. Streamlined LCA of algae for energy for the Pork CRC. In: *Algae for Energy and Feed: A Wastewater Solution—A Review*, Vol. Project 4A-101 112. Co-operative Research Centre for High Integrity, Australian Pork.

Grossman AR, Bhaya D, Apt KE, Kehoe DM. 1995. Light-harvesting complexes in oxygenic photosynthesis: Diversity, control, and evolution. *Annual Review of Genetics* 29:231–288.

Güroy D, Güroy B, Merrifield DL, Ergun S, Tekinay AA, Yigit M. 2011. Effect of dietary *Ulva* and *Spirulina* on weight loss and body composition of rainbow trout, *Oncorhynchus mykiss* (Walbaum), during a starvation period. *Journal of Animal Physiology and Animal Nutrition* 95(3):320–327. doi:10.1111/j.1439-0396.2010.01057.x.

Hagen C, Braune W, Greulich F. 1993. Functional aspects of secondary carotenoids in *Haematococcus lacustris* Girod Rostafinski (Volvocales). 4. Protection from photodynamic damage. *Journal of Photochemistry and Photobiology B-Biology* 20(2–3):153–160. doi:10.1016/1011-1344(93)80145-y.

Han DX, Li YT, Hu Q. 2013. Astaxanthin in microalgae: Pathways, functions and biotechnological implications. *Algae* 28(2):131–147. doi:10.4490/algae.2013.28.2.131.

Handler RM, Canter CE, Kalnes TN, Lupton FS, Kholiqov O, Shonnard DR, Blowers P. 2012. Evaluation of environmental impacts from microalgae cultivation in open-air raceway ponds: Analysis of the prior literature and investigation of wide variane in predicted impacts. *Algal Research* 1:83–92.

Havaux M, Dall'Osto L, Cuine S, Giuliano G, Bassi R. 2004. The effect of zeaxanthin as the only xanthophyll on the structure and function of the photosynthetic apparatus in *Arabidopsis thaliana*. *Journal of Biological Chemistry* 279(14):13878–13888. doi:10.1074/jbc.M311154200.

He ML, Hollwich W, Rambeck WA. 2002. Supplementation of algae to the diet of pigs: A new possibility to improve the iodine content in the meat. *Journal of Animal Physiology and Animal Nutrition* 86:97–104.

Heimann K, Becker B, Harnisch H, Mukherjee KD, Melkonian M. 1997. Biochemical characterization of plasma membrane vesicles of *Cyanophora paradoxa*. *Botanica Acta* 110(5):401–410.

Hemaiswarya S, Raja R, Kumar RR, Ganesan V, Anbazhagan C. 2011. Microalgae: A sustainable feed source for aquaculture. *World Journal of Microbiology and Biotechnology* 27:1737–1746.

Heredia-Arroyo T, Wei W, Ruan R, Hu B. 2011. Mixotrophic cultivation of *Chlorella vulgaris* and its potential application for the oil accumulation from non-sugar materials. *Biomass & Bioenergy* 35(5):2245–2253. doi:10.1016/j.biombioe.2011.02.036.

Hermansen JE, Kristensen T. Management options to reduce the carbon footprint of livestock products. *Animal Frontiers* 1(1):33–39.

Herrero A, Picossi S, Flores E. 2013. Gene Expression during heterocyst differentiation. In: Chauvat F, CassierChauvat C (eds), *Genomics of Cyanobacteria*, Vol. 65. *Advances in Botanical Research*, Elsevier, Amsterdam, pp. 281–329. doi:10.1016/b978-0-12-394313-2.00008-1.

Hoshina R, Fujiwara Y. 2013. Molecular characterization of *Chlorella* cultures of the National Institute for Environmental Studies culture collection with description of *Micractinium inermum* sp nov., *Didymogenes sphaerica* sp nov., and *Didymogenes soliella* sp nov (Chlorellaceae, Trebouxiophyceae). *Phycology Research* 61(2):124–132. doi:10.1111/pre.12010.

Hsieh SI, Castruita M, Malasarn D, Urzica E, Erde J, Page MD, Yamasaki H et al. 2013. The proteome of copper, iron, zinc, and manganese micronutrient deficiency in *Chlamydomonas reinhardtii*. *Molecular & Cellular Proteomics* 12(1):65–86. doi:10.1074/mcp.M112.021840.

Hudgens R, Hercamp R, Francis J, Nyman D, Bartoli Y. 2007. *An Evaluation of Glycerin (Glycerol) as a Heavy Duty Engine Antifreeze/Coolant Base*, SAE International, Warrendale, PA, Vol. 01-4000. doi:10.4271/2007-01-4000.

Huerlimann R, de Nys R, Heimann K. 2010. Growth, Lipid content, productivity, and fatty acid composition of tropical microalgae for scale-up production. *Biotechnology and Bioengineering* 107(2):245–257. doi:10.1002/bit.22809.

Huerlimann R, Heimann K. 2013. Comprehensive guide to acetyl-carboxylases in algae. *Critical Reviews in Biotechnology* 33(1):49–65. doi:10.3109/07388551.2012.668671.

Hughes MN, Poole RK. 1991. Metal speciation and microbial growth—The hard (and soft) facts. *Journal of General Microbiology* 137:725–734.

Huntley ME, Redalje DG. 2007. CO_2 mitigation and renewable oil from photosynthetic microbes: A new appraisal. Mitigation and adaptation strategies for global change 12(4):573–608.

Huo YX, Wernick DG, Liao JC. 2012. Toward nitrogen neutral biofuel production. *Current Opinion in Biotechnology* 23(3):406–413. doi:10.1016/j.copbio.2011.10.005.

IDA. 2007. *IDA and Climate Change*.

IPCC. 2007. Freshwater resources and their management. Climate change 2007: Impacts, adaptation and vulnarability. Contribution of Working Group II to the Fourth Assessment Report of the Intergovernmental Panel on Climate Change. Cambridge, UK.

Islam MA, Magnusson M, Brown RJ, Ayoko GA, Nabi N, Heimann K. 2013. Microalgal species selection for biodiesel production based on fuel properties derived from fatty acid profiles. *Energies* 6:5676–5702. doi:10.3390/en6115676.

Jackson A. 2012. Fishmeal and fish oil and its role in sustainable aquaculture. *International Aquafeed* 15(2):18–21.

Jackson AE, Ayer SW, Laylock MW. 1992. The effect of salinity on growth and amino acid composition in the marine diatom *Nitzschia pungens*. *Canadian Journal of Botany* 70:2198–2201.

Jacobsen R, Vandermeulen V, Venhuylenbroeck G, Gellynck X. 2014. A life cycle assessment application: The carbon footprint of beef in Flanders (Belgium). In: Muthu SS (ed.), *Assessment of Carbon Footprint in Different Industrial Sectors*, Vol. 2. Springer, Singapore, pp. 31–52.

Johnsen G, Sakshaug E. 1996. Light harvesting in bloom-forming marine phytoplankton: Species-specificity and photoacclimation. *Scientifica Marina* 60:47–56.

Jorquera O, Kiperstok A, Sales EA, Embiruçu M, Ghirardi ML. 2010. Comparative energy life-cycle analyses of microalgal biomass production in open ponds and photobioreactors. *Bioresource Technology* 101(4):1406–1413. doi:http://dx.doi.org/10.1016/j.biortech.2009.09.038.

Kanazawa T, Yuhara T, Sasa T. 1958. Mass culture of unicellular algae using the "open circulation method". *Journal of General and Applied Microbiology* 4:135–152.

Keenan JD, Auer MT. 1974. The influence of phosphorous luxury uptake on algal bioassays. *Journal Water Pollution Control Federation* 46(3 Part 1):532–542.

Khoeyi ZA, Seyfabadi J, Ramezanpour Z. 2012. Effect of light intensity and photoperiod on biomass and fatty acid composition of the microalgae, *Chlorella vulgaris*. *Aquaculture International* 20(1):41–49. doi:10.1007/s10499-011-9440-1.

Khozin-Goldberg I, Iskandarov U, Cohen Z. 2011. LC-PUFA from photosynthetic microalgae: Occurrence, biosynthesis, and prospects in biotechnology. *Applied Microbiology and Biotechnology* 91(4):905–915. doi:10.1007/s00253-011-3441-x.

Klein-Marcuschamer D, Turner C, Allen M, Gray P, Dietzgen RG, Gresshoff PM, Hankamer B, Heimann K, Speight R, Nielsen LK. 2013. Technoeconomic analysis of renewable aviation fuel from microalgae, *Pongamia pinnata*, and sugarcane. *Biofuels, Bioproducts and Biorefining* 7(4): 416–428.

Kobayashi M, Kakizono T, Nagai S. 1993. Photo-dependent astaxanthin biosynthesis in a green algal, *Haematococcus pluvialis*. *Journal of Fermentation and Bioengineering* 71(4):233–237.

Kotzabasis K, Dornemann D. 1998. Differential changes in the photosynthetic pigments and polyamine content during photoadaptation and photoinhibition in the unicellular green alga *Scenedesmus obliquus*. *Zeitschrift Fur Naturforschung C-Journal of Biosciences* 53(9–10):833–840.

Krauss R. 1962. Mass culture of algae for food and other organic compounds. *American Journal of Botany* 49:425–435.

Kropat J, Hong-Hermesdorf A, Casero D, Ent P, Castruita M, Pellegrini M, Merchant SS, Malasarn D. 2011. A revised mineral nutrient supplement increases biomass and growth rate in Chlamydomonas reinhardtii. *Plant Journal* 66(5):770–780. doi:10.1111/j.1365-313X.2011.04537.x.

Lardon L, He´lias A, Sialve B, Steyer J-P, Bernard O. 2009. Life-cycle assessment of biodiesel production from microalgae. *Environmental Science & Technology* 43(17):6475–6481.

Li Y, Horsman M, Wang B, Wu N. 2008. Effects of nitrogen sources on cell growth and lipid accumulation of the green alga *Neochloris oleoabundans*. *Applied Microbiology and Biotechnology* 81:629–636.

Liu J, Sun Z, Zhong YJ, Gerken H, Huang JC, Chen F. 2013. Utilization of cane molasses towards cost-saving astaxanthin production by a *Chlorella zofingiensis* mutant. *Journal of Applied Phycology* 25(5):1447–1456. doi:10.1007/s10811-013-9974-x.

Lomas MW, Glibert PM. 2000. Comparisons of nitrate uptake, storage, and reduction in marine diatoms and flagellates. *Journal of Phycology* 36:903–913.

Lopezfigueroa F. 1991. Control by light quality of chlorophyll synthesis in the brown alga *Desmarestia aculeata*. *Zeitschrift Fur Naturforschung C-Journal of Biosciences* 46(7–8):542–548.

Lorenz RT, Cysewski GR. 2000. Commercial potential for *Haematococcus* microalgae as a natural source of astaxanthin. *Trends in Biotechnology* 18(4):160–167. doi:10.1016/s0167-7799(00)01433-5.

Lutz C, Seidlitz HK, Meindl U. 1997. Physiological and structural changes in the chloroplast of the green alga *Micrasterias denticulata* induced by UV-B simulation. *Plant Ecology* 128(1–2):54–64.

Magnus W, Schindler B. 1912. Über den Einfluss der Nährsalze auf die Färbung der Oscilarien. *Berichte der Deutschen Botanischen Gesellschaft* 30:314–320.

Malasarn D, Kropat J, Hsieh SI, Finazzi G, Casero D, Loo JA, Pellegrini M, Wollman FA, Merchant SS. 2013. Zinc deficiency impacts CO_2 assimilation and disrupts copper homeostasis in *Chlamydomonas reinhardtii*. *Journal of Biological Chemistry* 288(15):10672–10683. doi:10.1074/jbc.M113.455105.

Malerba ME, Connolly SR, Heimann K. 2012. Nitrate–nitrite dynamics and phytoplankton growth: Formulation and experimental evaluation of a dynamic model. *Limnology and Oceanography* 57(5):1555–1571. doi:10.4319/lo.2012.57.5.1555.

Marchetti J, Bougaran G, Jauffrais T, Lefebvre S, Rouxel C, Saint-Jean B, Lukomska E, Robert R, Cadoret JP. 2013. Effects of blue light on the biochemical composition and photosynthetic activity of *Isochrysis* sp (T-iso). *Journal of Applied Phycology* 25(1):109–119. doi:10.1007/s10811-012-9844-y.

Margulis L, Corliss JO, Melkonian M, Chapman DJ. 1989. *Handbook of Protoctista*. Jones and Bartlett, Boston, MA.

Marshall AT, Haverkamp RG. 2008. Production of hydrogen by the electrochemical reforming of glycerol–water solutions in a PEM electrolysis cell. *International Journal of Hydrogen Energy* 33(17):4649–4654.

Martinez ME, Jimenez JM, El Yousfi F. 1999. Influence of phosphorus concentration and temperature on growth and phosphorus uptake by the microalga *Scenedesmus obliquus*. *Bioresource Technology* 67:233–240.

Matsumoto H, Shioji N, Hamasaki A, Ikuta Y, Fukuda Y, Sato M, Endo N, Tsukamoto T. 1995. Carbon dioxide fixation by microalgae photosynthesis using actual flue gas discharged from a boiler. *Applied Biochemistry and Biotechnology* 51–52(1):681–692. doi:10.1007/BF02933469.

Maxwell DP, Falk S, Trick CG, Huner NPA. 1994. Growth at low temperature mimics high-light acclimation in *Chlorella vulgaris*. *Plant Physiology* 105(2):535–543.

McGinn PJ, Dickinson KE, Bhatti S, Frigon J-C, Guiot SR, O'Leary SJB. 2011. Integration of microalgae cultivation with industrial waste remediation for biofuel and bioenergy production: opportunities and limitations. *Photosynthesis Research* 109 (1–3):231–247.

Medeiros DL, Sales EA, Kiperstok A. 2013. Energy production from microalgae biomass: the carbon foot print and energy balance. *Paper Presented at the 4th Academic International Workshop Advances in Cleaner Production*, Brazil, May 22–24, 2013.

Meesapyodsuk D, Qiu X. 2012. The front-end desaturase: Structure, function, evolution and biotechnological use. *Lipids* 47(3):227–237. doi:10.1007/s11745-011-3617-2.

Melis A, Neidhardt J, Baroli I, Benemann JR. 1998. Maximizing photosynthetic productivity and light utilization in microalgae by minimizing the light-harvesting chlorophyll antenna size of the photosystems. *Biohydrogen*. Plenum Press, New York.

Meujo DF, Hamann M. 2013. Science, policy, and risk management: Case of seafood safety. In: Laws EA (ed.), *Environmenal Toxicology*. Springer, New York, pp. 461–501.

Meyer M, Griffiths H. 2013. Origins and diversity of eukaryotic CO_2-concentrating mechanisms: Lessons for the future. *Journal of Experimental Botany* 64(3):769–786. doi:10.1093/jxb/ers390.

Milledge JJ. 2011. Commercial application of microalgae other than as biofuels: A brief review. *Reviews in Environmental Science and Bio-Technology* 10(1):31–41. doi:10.1007/s11157-010-9214-7.

Moheimani NR, Borowitzka MA. 2006. The long-term culture of the coccolithophore *Pleurochrysis carterae* (Haptophyta) in outdoor raceway ponds. *Journal of Applied Phycology* 18(6):703–712.

Moheimani NR, Borowitzka MA, Isdepsky A, Sing SF. 2013. Standard methods for measuring growth of algae and their composition. In: *Algae for Biofuels and Energy*. Springer, Berlin, pp. 265–284.

Moreno J, Vargas M, Rodríguez H, Rivas J, Guerrero MG. 2003. Outdoor cultivation of a nitrogen-fixing marine cyanobacterium, *Anabaena sp.* ATCC 33047. *Biomolecular Engineering* 20(4):191–197.

Moroney JV, Bartlett SG, Samuelsson G. 2001. Carbonic anhydrases in plants and algae. *Plant Cell and Environment* 24(2):141–153. doi:10.1046/j.1365-3040.2001.00669.x.

Moroney JV, Jungnick N, DiMario RJ, Longstreth DJ. 2013. Photorespiration and carbon concentrating mechanisms: Two adaptations to high O-2, low CO_2 conditions. *Photosynthesis Research* 117(1–3):121–131. doi:10.1007/s11120-013-9865-7.

Morschel E. 1991. The light-harvesting antennae of cyanobacteria and red algae. *Photosynthetica* 25(1):137–144.

Moulton TP, Borowitzka LJ, Vincent DJ. 1987. The mass culture of *Dunaliella salina* for beta-carotene—From pilot-plant to production-plant. *Hydrobiologia* 151:99–105. doi:10.1007/bf00046114.

Murali P. 1992. Regulation of Rubisco activity through phosphorylation of the small subunit in dwarf pea. *Photosynthesis Research* 34(1):196–196.

Muro-Pastor AM, Hess WR. 2012. Heterocyst differentiation: From single mutants to global approaches. *Trends in Microbiology* 20(11):548–557. doi:10.1016/j.tim.2012.07.005.

Nath PR, Khozin-Goldberg I, Cohen Z, Boussiba S, Zilberg D. 2012. Dietary supplementation with the microalgae *Parietochloris incisa* increases survival and stress resistance in guppy (*Poecilia reticulata*) fry. *Aquaculture Nutrition* 18(2):167–180. doi:10.1111/j.1365-2095.2011.00885.x.

Neff JM. 1997. Ecotoxicology of arsenic in the marine environment. *Environmental Toxicology and Chemistry* 16(5):917–927. doi:10.1897/1551-5028(1997)016<0917:eoaitm>2.3.co;2.

Nijdam D, Rood T, Westhoek H. 2012. The price of protein: Review of land use and carbon footprints from life cycle assessments of animal food products and their substitutes. *Food Policy* 37(6):760–770. doi:10.1016/j.foodpol.2012.08.002.

Niyogi KK, Bjorkman O, Grossman AR. 1997. *Chlamydomonas* xanthophyll cycle mutants identified by video imaging of chlorophyll fluorescence quenching. *Plant Cell* 9(8):1369–1380.

Nobel PS. 1999. *Plant Physiology. 5. Photochemistry of photosynthesis*. 2nd edn. Academic Press, San Diego.

O'Connor S, Moltschaniwskyj N, Bolch CJS, O'Connor W. 2012. Dietary influence on growth and development of flat oyster, *Ostrea angasi* (Sowerby, 1871), larvae. *Aquaculture Research* 43(9):1317–1327. doi:10.1111/j.1365-2109.2011.02935.x.

Olaizola M. 2000. Commercial production of astaxanthin from *Haematococcus pluvialis* using 25,000-liter outdoor photobioreactors. *Journal of Applied Phycology* 12(3–5):499–506. doi:10.1023/a:1008159127672.

Oliva D, Abarca A, Gutierrez R, Celis A, Herrera L, Pizarro V. 2013. Effect of stocking density and diet on growth and survival of post-larvae of the taquilla clam *Mulinia edulis* cultivated in sand in a hatchery. *Revista de Biologia Marina y Oceanografia* 48(1):37–44.

Olsen RL, Hasan MR. 2012. A limited supply of fishmeal: Impact on future increases in global aquaculture production. *Trends in Food Science & Technology* 27(2):120–128. doi:10.1016/j.tifs.2012.06.003.

Orosa M, Franqueira D, Cid A, Abalde J. 2001. Carotenoid accumulation in *Haematococcus pluvialis* in mixotrophic growth. *Biotechnology Letters* 23(5):373–378. doi:10.1023/a:1005624005229.

Osmond CB. 1981. Photorespiration and photoinhibition: Some implications for the energetics of photosynthesis. *Biochimica et Biophysica Acta* 639:77–98.

Ott L, Bicker M, Vogel H. 2006. The catalytic dehydration of glycerol in sub- and supercritical water: A new chemical process for acrolein production. *Green Chemistry* 8(2):214–220. doi:10.1039/b506285c.

Ozkan A, Kinney K, Katz L, Berberoglu H. 2012. Reduction of water and energy requirement of algae cultivation using an algal biofilm photobioreactor. *Bioresource Technology* 114:542–548.

Pahl SL, Lee AK, Kalaitzidis T, Ashman PJ, Sathe S, Lewis DM. 2013. Harvesting, thickening and dewatering microalgae biomass. In: *Algae for Biofuels and Energy*. Springer, Berlin, pp. 165–185.

Pal D, Khozin-Goldberg I, Didi-Cohen S, Solovchenko A, Batushansky A, Kaye Y, Sikron N, Samani T, Fait A, Boussiba S. 2013. Growth, lipid production and metabolic adjustments in the euryhaline eustigmatophyte Nannochloropsis oceanica CCALA 804 in response to osmotic downshift. *Applied Microbiology and Biotechnology* 97(18):8291–8306. doi:10.1007/s00253-013-5092-6.

Pandey D, Agrawa M. 2014. Carbon footprint estimation of agriculture sector. In: Muthu SS (ed.), *Assessment of Carbon Footprint in Different Industrial Sectors, Vol. 1. Ecoproduction Series—Environmental Issues in Logistics and Manufacturing*. Springer, Singapore, pp. 25–48.

Park DW, Jo QT, Lim HJ, Veron B. 2002. Sterol composition of dark-grown *Isochrysis galbana* and its implication in the seed production of Pacific oyster, *Crassostrea gigas*. *Journal of Applied Phycology* 14(5):351–355.

Patil V, Kallqvist T, Olsen E, Vogt G, Gislerod HR. 2007. Fatty acid composition of 12 microalgae for possible use in aquaculture feed. *Aquaculture International* 15(1):1–9. doi:10.1007/s10499-006-9060-3.

Patterson D, Gatlin DM. 2013. Evaluation of whole and lipid-extracted algae meals in the diets of juvenile red drum (*Sciaenops ocellatus*). *Aquaculture* 416:92–98. doi:10.1016/j.aquaculture.2013.08.033.

Peccia J, Haznedaroglu B, Gutierrez J, Zimmerman JB. 2013. Nitrogen supply is an important driver of sustainable microalgae biofuel production. *Trends in Biotechnology* 31(3):134–138.

Pedroni P, Lamenti G, Prosperi G, Ritorto L, Scolla G, Capuano F, Valdiserri M. 2004. Enitecnologie R&D project on microalgae biofixation of CO_2: Outdoor comparative tests of biomass productivity using flue gas CO_2 from a NGCC power plant. In: *Proceedings of the Seventh International Conference on Greenhouse Gas Control Technologies (GHGT-7)*, Vancouver, Canada.

Peltier G, Tolleter D, Billon E, Cournac L. 2010. Auxiliary electron transport pathways in chloroplasts of microalgae. *Photosynthesis Research* 106(1–2):19–31. doi:10.1007/s11120-010-9575-3.

Perez-Garcia O, Escalante FME, de-Bashan LE, Bashan Y. 2011. Heterotrophic cultures of microalgae: Metabolism and potential products. *Water Research* 45(1):11–36. doi:10.1016/j.watres.2010.08.037.

Petrie JR, Singh SP. 2011. Expanding the docosahexaenoic acid food web for sustainable production: Engineering lower plant pathways into higher plants. *AoB Plants*. doi:10.1093/aobpla/plr011.

Poot-Lopez GR, Gasca-Leyva E, Olvera-Novoa MA. 2012. Nile tilapia production (*Oreochromis niloticus* L.) using tree spinach leaves (*Cnidoscolus chayamansa* McVaugh) as a partial substitute for balanced feed. *Latin American Journal of Aquatic Research* 40(4):835–846. doi:10.3856/vol40-issue4-fulltext-2.

Powell N, Shilton AN, Pratt S, Chisti Y. 2008. Factors influencing luxury uptake of phosphorus by microalgae in waste stabilization ponds. *Environmental Science and Technology* 42(16):5958–5962.

Praveenkumar R, Shameera K, Mahalakshmi G, Akbarsha MA, Thajuddin N. 2012. Influence of nutrient deprivations on lipid accumulation in a dominant indigenous microalga *Chlorella* sp., BUM11008: Evaluation for biodiesel production. *Biomass & Bioenergy* 37:60–66. doi:10.1016/j.biombioe.2011.12.035.

Priyadarshani I, Rath B. 2012. Commercial and industrial applications of micro algae—A review. *Journal of Algal Biomass and Utilisation* 3(4):89–100.

Proteau PJ, Gerwick WH, Garciapichel F, Castenholz R. 1993. The structure of Syconemin, an ultraviolet sunscreen pigment from the sheaths of cyanobacteria. *Experientia* 49(9):825–829. doi:10.1007/bf01923559.

Pulz O. 2007. Performance summary report. Evaluation of GreenFuel's 3D matrix algae growth engineering scale unit Performance summary report. IGV Institut für Getreideverarbeitung GmbH, Germany

Qawasmi A, Landeros-Weisenberger A, Leckman JF, Bloch MH. 2012. Meta-analysis of long-chain polyunsaturated fatty acid supplementation of formula and infant cognition. *Pediatrics* 129(6):1141–1149. doi:10.1542/peds.2011-2127.

Raghothama KG. 1999. Phosphate acquisition. *Annual Review of Plant Physiology and Plant Molecular Biology* 50:665–693.

Raposo MFD, de Morais R, de Morais A. 2013. Health applications of bioactive compounds from marine microalgae. *Life Sciences* 93(15):479–486. doi:10.1016/j.lfs.2013.08.002.

Ratledge C. 2004. Fatty acid biosynthesis in microorganisms being used for single cell oil production. *Biochimie* 86(11):807–815. doi:10.1016/j.biochi.2004.09.017.

Raven JA. 1995. Phycological reviews: 15. Photosynthetic and nonphotosynthetic roles of carbonic anhydrase in algae and cyanobacteria. *Phycologia* 34(2):93–101. doi:10.2216/i0031-8884-34-2-93.1.

Raven JA, Girard-Bascou J. 2001. Algal model systems and the elucidation of photosynthetic metabolism. *Journal of Phycology* 37(6):943–950. doi:10.1046/j.1529-8817.2001.01079.x.

Redfield AC. 1934. *On the Proportions of Organic Derivations in Sea Water and Their Relation to the Composition of Plankton.* James Johnstone Memorial Volume. University Press of Liverpool, Liverpool.

Renaud SM, Thinh LV, Lambrinidis G, Parry DL. 2002. Effect of temperature on growth, chemical composition and fatty acid composition of tropical Australian microalgae grown in batch cultures. *Aquaculture* 211:195–214.

Renuka N, Sood A, Ratha SK, Prasanna R, Ahluwalia AS. 2013. Evaluation of microalgal consortia for treatment of primary treated sewage effluent and biomass production. *Journal of Applied Phycology* 25(5):1529–1537.

Richmond A, Vonshak A, Arad SM. 1980. Environmental limitations in outdoor production of algal biomass. In: Shelef G, Soeder C (eds), *Algae Biomass: Production and Use.* Elsevier/North Holland Biomedical Press, Amsterdam, pp. 65–72.

Roessler PG. 1990. Environmental control of glycerolipid metabolism in microalgae: Commercial implications and future research directions. *Journal of Phycology* 26(3):393–399.

Ronquillo JD, Fraser J, McConkey AJ. 2012. Effect of mixed microalgal diets on growth and polyunsaturated fatty acid profile of European oyster (*Ostrea edulis*) juveniles. *Aquaculture* 360:64–68. doi:10.1016/j.aquaculture.2012.07.018.

Round FE. 1981. *The Ecology of Algae. Chapter 2. Habitats and Communities of Algae.* Cambridge University Press, Cambridge.

Sardi L, Martelli G, Lambertini L, Parisini P, Mordenti A. 2006. Effects of a dietary supplement of DHA rich marine algae on Italian heavy pig production parameters. *Livestock Science* 1–2:95–103.

Saucedo PE, Gonzalez-Jimenez A, Acosta-Salmon H, Mazon-Suastegui JM, Ronson-Paulin JA. 2013. Nutritional value of microalgae-based diets for lions-paw scallop (*Nodipecten subnodosus*) juveniles reared at different temperatures. *Aquaculture* 392:113–119. doi:10.1016/j.aquaculture.2013.02.001.

Schindler B. 1913. Über den Farbenwechsel der Oscillarien. *Zeitschrift für Botanik* 5:497–575.

Schubert H, Kroon BMA, Matthijs HCP. 1994. *In vivo* manipulation of the xanthophyll cycle and the role of zeaxanthin in the protection against photodamage in the green alga *Chlorella pyrenoidosa*. *Journal of Biological Chemistry* 269(10):7267–7272.

Schubert N, Garcia-Mendoza E, Pacheco-Ruiz I. 2006. Carotenoid composition of marine red algae. *Journal of Phycology* 42(6):1208–1216. doi:10.1111/j.1529-8817.2006.00274.x.

Shea R, Chopin T. 2007. Effects of germanium dioxide, an inhibitor of diatom growth, on the microscopic laboratory cultivation stage of the kelp, *Laminaria saccharina*. *Journal of Applied Phycology* 19(1):27–32.

Sheikhzadeh N, Tayefi-Nasrabadi H, Oushani AK, Enferadi MHN. 2012. Effects of *Haematococcus pluvialis* supplementation on antioxidant system and metabolism in rainbow trout (*Oncorhynchus mykiss*). *Fish Physiology and Biochemistry* 38(2):413–419. doi:10.1007/s10695-011-9519-7.

Shi DL, Kranz SA, Kim JM, Morel FMM. 2012. Ocean acidification slows nitrogen fixation and growth in the dominant diazotroph Trichodesmium under low-iron conditions. *Proceedings of the National Academy of Sciences of the United States of America* 109(45):E3094-E3100. doi:10.1073/pnas.1216012109.

Sili C, Torzillo G, Vonshak A. 2012. *Arthrospira* (*Spirulina*). In: *Ecology of Cyanobacteria II*. Springer, Berlin, pp. 677–705.

Simon A, Glöckner G, Felder M, Melkonian M, Becker B. 2006. ST analysis of the scaly green flagellate *Mesostigma viride* (Streptophyta): Implications for the evolution of green plants (Viridiplantae). *BMC Plant Biology* 6(2). doi:10.1186/1471-2229-6-2.

Singh UB, Ahluwalia A. 2013. Microalgae: A promising tool for carbon sequestration. *Mitigation and Adaptation Strategies for Global Change* 18(1):73–95.

Slaby E. 2013. Going green for protein: Algae for profit. *Paper Presented at the International Conference on Algae Biomass, Biofuels and Bioproducts*, Toronto, Canada, June 16–19, 2013.

Slade R, Bauen A. 2013. Micro-algae cultivation for biofuels: Cost, energy balance, environmental impacts and future prospects. *Biomass and Bioenergy* 53:29–38.

Sotiroudis TG, Sotiroudis GT. 2013. Health aspects of *Spirulina (Arthrospira)* microalga food supplement. *Journal of Serbian Chemical Society* 78(3):395–405. doi:10.2298/jsc121020152s.

Spiegel AJ, Noseworthy MM. 1963. Use of nonaqueous solvents in parenteral products. *Journal of Pharmaceutical Sciences* 52(10):917–927.

Spolaore P, Joannis-Cassan C, Duran E, Isambert A. 2006. Commercial applications of microalgae. *Journal of Bioscience and Bioengineering* 101(2):87–96. doi:10.1263/jbb.101.87.

Spreitzer RJ. 2003. Role of the small subunit in ribulose-1,5-bisphosphate carboxylase/oxygenase. *Archives of Biochemistry and Biophysics* 414(2):141–149. doi:10.1016/s0003-9861(03)00171-1.

Spreitzer RJ, Salvucci ME. 2002. Rubisco: Structure, regulatory interactions, and possibilities for a better enzyme. *Annual Review of Plant Biology* 53:449–475.

Stammbach MR, de Nys R, Heimann K, Rogers A. 2010. *Method of Culturing Photosynthetic Organisms.* PCT Patent Application, WO 2010/132917 A1 (25 November, 2010).

Stephenson AL, Kazamia E, Dennis JS, Howe CJ, Scott SA, Smith AG. 2010. Life-cycle assessment of potential algal biodiesel production in the United Kingdom: A comparison of raceways and air-lift tubular bioreactors. *Energy & Fuels* 24(7):4062–4077.

Stevens A. 2000. Preserving flowers and decorative foliages with glycerine and dye. Kansas State University MF-2446.

Strain HH, Sherma J. 1972. Investigations of chloroplast pigments of higher plants, green-algae and brown algae and their influence upon invention, modifications and applications of Tswetts chromatographic method. *Journal of Chromatography* 73(2):371. doi:10.1016/s0021-9673(01)91216-6.

Sudhakar K, Premalatha M, Sudharshan K. 2012. Energy balance and exergy analysis of large scale algal biomass production. *Paper Presented at the Second Korea–Indonesia Workshop & International Symposium on Bioenergy from Biomass*, Indonesia, June 13–15, 2012.

Taelman SE, Meester SD, Roef L, Michiels M, Dewulf J. 2013. The environmental sustainability of microalgae as feed for aquaculture: A life cycle perspective. *Bioresource Technology* 150:513–522.

Tamiya H. 1957. Mass culture of algae. *Annual Review in Plant Physiology* 8:309–344.

Tcherkez G. 2013. Modelling the reaction mechanism of ribulose-1,5-bisphosphate carboxylase/oxygenase and consequences for kinetic parameters. *Plant Cell and Environment* 36(9):1586–1596. doi:10.1111/pce.12066.

Thajuddin N, Subramanian G. 2005. Cyanobacterial biodiversity and potential applications in biotechnology. *Current Science* 89(1):47–57.

Thompson PA, Guo M, Harrison PJ. 1993. The influence of irradiance on the biochemical composition of 3 phytoplankton species and their nutritional value for larvae of the pacific oyster (*Crassostrea gigas*). *Marine Biology* 117:259–268.

Titlyanov EA, Lee BD. 1978. Adaptation of benthic plants to light. 4. Variations of photosynthetic pigment content and ratio in green-algae under changing light conditions. *Biologiya Morya-Marine Biology* (4):36–41.

Tiwari A, Pandey A. 2012. Cyanobacterial hydrogen production—A step towards clean environment. *International Journal of Hydrogen Energy* 37(1):139–150. doi:10.1016/j.ijhydene.2011.09.100.

Ugwu CU, Ogbonna JC, Tanaka H. 2005. Light/dark cyclic movement of algal culture (*Synechocystis aquatilis*) in outdoor inclined tubular photobioreactor equipped with static mixers for efficient production of biomass. *Biotechnology Letters* 27(2):75–78. doi:10.1007/s10529-004-6931-4.

Ulloa G, Otero A, Sanchez M, Sineiro J, Nunez MJ, Fabregas J. 2012. Effect of Mg, Si, and Sr on growth and antioxidant activity of the marine microalga *Tetraselmis suecica*. *Journal of Applied Phycology* 24(5):1229–1236. doi:10.1007/s10811-011-9764-2.

UN. 2004. World Population to 2300. United Nations, New York.

van den Berg H, Faulks R, Granado HF, Hirschberg J, Olmedilla B, Sandmann G, Southon S, Stahl W. 2000. The potential for the improvement of carotenoid levels in foods and the likely systemic effects. *Journal of the Science of Food and Agriculture* 80(7):880–912. doi:10.1002/(sici)1097-0010(20000515)80:7<880::aid-jsfa646>3.3.co;2-t.

Van den Dorpel P. 2010. AlgaeLink®. *Paper Presented at the Fourth Annual Algae Biomass Summit*, Phoenix, AZ, 28–30 September.

Veith T, Buchel C. 2007. The monomeric photosystem I-complex of the diatom *Phaeodactylum tricornutum* binds specific fucoxanthin chlorophyll proteins (FCPs) as light-harvesting complexes. *Biochimica Et Biophysica Acta-Bioenergetics* 1767 (12):1428–1435. doi:10.1016/j.bbabio.2007.09.004.

Vershinin A. 1999. Biological functions of carotenoids—Diversity and evolution. *Biofactors* 10(2–3):99–104.

Vincent WF. 1992. The daily pattern of nitrogen uptake by phytoplankton in dynamic mixed layer environments. *Hydrobiologia* 238:37–52.

von Alvensleben N, Stookey K, Magnusson M, Heimann K. 2013. Salinity tolerance of *Picochlorum atomus* and the use of salinity for contamination control by the freshwater cyanobacterium *Pseudanabaena limnetica*. *PLoS ONE* 8(5). doi:10.1371/journal.pone.0063569.

Ward OP, Singh A. 2005. Omega-3/6 fatty acids: Alternative sources of production. *Process Biochemistry* 40(12):3627–3652. doi:10.1016/j.procbio.2005.02.020.

Wassef EA, Saleh NE, El-Abd El-Hady HA. 2009. Vegetable oil blend as alternative lipid resources in diets for gilthead seabream, *Sparus aurata*. *Aquaculture International* 17(5):421–435. doi:10.1007/s10499-008-9213-7.

Watanabe M, Iida T, Aizawa Y, Aida TM, Inomata H. 2007. Acrolein synthesis from glycerol in hot-compressed water. *Bioresource Technology* 98(6):1285–1290.

Wen ZY, Chen F. 2000. Production potential of eicosapentaenoic acid by the diatom *Nitzschia laevis*. *Biotechnology Letters* 22(9):727–733. doi:10.1023/a:1005666219163.

Wikfors GH, Patterson GW, Ghosh P, Lewin RA, Smith BC, Alix JH. 1996. Growth of post-set oysters, *Crassostrea virginica*, on high-lipid strains of algal flagellates *Tetraselmis* spp. *Aquaculture* 143(3–4):411–419. doi:10.1016/0044-8486(96)01265-3.

Wildman RB, Bowen CC. 1969. Phycobilisomes in blue-green algae. *Journal of Cell Biology* 43(2P2):A173.

Wood BJB, Grimson PHK, German JB, Turner M. 1999. Photoheterotrophy in the production of phytoplankton organisms. *Journal of Biotechnology* 70(1–3):175–183. doi:10.1016/s0168-1656(99)00070-x.

Xu F, Hu HH, Cong W, Cai ZL, Ouyang F. 2004. Growth characteristics and eicosapentaenoic acid production by *Nannochloropsis* sp in mixotrophic conditions. *Biotechnology Letters* 26(1):51–53. doi:10.1023/B:BILE.0000009460.81267.cc.

Xu ZB, Yan XJ, Pei LQ, Luo QJ, Xu JL. 2008. Changes in fatty acids and sterols during batch growth of *Pavlova viridis* in photobioreactor. *Journal of Applied Phycology* 20:237–243.

Yadavalli V, Jolley CC, Malleda C, Thangaraj B, Fromme P, Subramanyam R. 2012. Alteration of proteins and pigments Influence the function of photosystem I under iron deficiency from *Chlamydomonas reinhardtii*. *PLoS ONE* 7(4). doi:10.1371/journal.pone.0035084.

Yazdani SS, Gonzalez R. 2007. Anaerobic fermentation of glycerol: A path to economic viability for the biofuels industry. *Current Opinion in Biotechnology* 18(3):213–219.

Yoshimura T, Okada S, Honda M. 2013. Culture of the hydrocarbon producing microalga *Botryococcus braunii* strain Showa: Optimal CO_2, salinity, temperature, and irradiance conditions. *Bioresource Technology* 133:232–239. doi:10.1016/j.biortech.2013.01.095.

Zhang K, Kurano N, Miyachi S. 1999. Outdoor culture of a cyanobacterium with a vertical flat-plate photobioreactor: Effects on productivity of the reactor orientation, distance setting between the plates, and culture temperature. *Applied Microbiology and Biotechnology* 52(6):781–786.

Zhao B, Su Y. 2014. Process effect of microalgal-carbon dioxide fixation and biomass production: A review. *Renewable and Sustainable Energy Reviews* 31:121–132.

Zhekisheva M, Boussiba S, Khozin-Goldberg I, Zarka A, Cohen Z. 2002. Accumulation of oleic acid in *Haematococcus pluvialis* (Chlorophyceae) under nitrogen starvation or high light is correlated with that of astaxanthin esters. *Journal of Phycology* 38:325–331.

Zhila NO, Galina S, Kalacheva GS, Volova TG. 2005. Influence of nitrogen deficiency on biochemical composition of the green alga *Botryococcus*. *Journal of Applied Phycology* 17:309–315.

Zhu L, Wang L, Takala J, Hiltunen E, Qin L, Xu Z, Qin X, Yuan Z. 2013. Scale-up potential of cultivating *Chlorella zofingiensis* in piggery wastewater for biodiesel production. *Bioresource Technology* 137:318–325.

Zittelli GC, Rodolfi L, Bassi N, Biondi N, Tredici MR. 2013. Photobioreactors for microalgal biofuel production. In: *Algae for Biofuels and Energy*. Springer, Berlin, pp. 115–131.

19 Carbon Footprint of Agricultural Products

Bidisha Chakrabarti, S. Naresh Kumar, and H. Pathak

CONTENTS

19.1 Introduction..431
19.2 Carbon Footprint..432
19.3 Why Estimate the Carbon Footprint of Agricultural Products?..432
19.4 How to Estimate the Carbon Footprint of Agricultural Products?......................................433
19.5 Carbon Footprint of Agricultural Products 434
19.6 Carbon Footprint of Livestock Sector 440
19.7 Carbon Footprint of Piggery and Poultry 441
19.8 Carbon Footprint of Fisheries..441
19.9 Carbon Footprint of Food Habits...441
19.10 Strategies to Reduce the Carbon Footprint of Agricultural Produce 442
    19.10.1 Growing Leguminous Crops..442
    19.10.2 Fertilizer Management..442
    19.10.3 Water Management 443
    19.10.4 Crop Cultivars 443
    19.10.5 Tillage Operation 443
    19.10.6 Use of N-Transformation Inhibitors..443
    19.10.7 Organic Manure..444
    19.10.8 Diversification of Cropping System 444
    19.10.9 Livestock Management 444
    19.10.10 Management Options for Fisheries 445
    19.10.11 Dietary Pattern...445
19.11 Limitations and Scope for Future Research 445
19.12 Conclusion..445
References..446

19.1 INTRODUCTION

The carbon cycle holds center stage in global change studies owing to its role in influencing weather and determining food, fiber, and wood supply for human use through plant productivity. In recent times, a continuous rise in the concentration of greenhouse gases (GHGs) in the atmosphere has led to global warming and associated climate change. Increased emissions of GHGs such as carbon dioxide (CO_2), methane (CH_4), and nitrous oxide (N_2O) have led to temperature rise, change in precipitation pattern, melting of glaciers, rise in sea level, and increased frequency of extreme events. Anthropogenic emissions of GHGs increased by 70% between 1970 and 2004, and these are estimated to increase further by 25 to 95% by 2030 (Rose and McCarl, 2008). The Intergovernmental Panel on Climate Change (IPCC), in its 5th Assessment Report (AR5) published in 2013–2014, warned about dire consequences of climate change on agriculture, human health, settlements, and natural resources, if no measures are taken to curb the ill effects of global warming. Baseline scenarios, those without additional mitigation, result in global mean surface temperature increases in

2100 from 3.7 to 4.8°C when compared with pre-industrial levels. On the basis of the longest global surface temperature data set available, the observed change between the average of the period 1850–1900 and of the Assessment Report 5 reference period (1986–2005) is 0.61°C (IPCC 2014).

World Agriculture sector emits GHGs such as methane (CH_4) and nitrous oxide (N_2O). Agriculture contributes about 13.5% of global GHG emission (IPCC, 2007). The latest UK Greenhouse Gas Inventory estimates the proportion of the nation's overall carbon footprint (CFP) due to agriculture to be around 8%, of which 75% is directly related to fertilizer use (Choudrie et al., 2008). In India, the agriculture sector contributes 18% of the total GHG emission, of which 63.4% is from enteric fermentation, 20.9% from rice cultivation, 13% from agricultural soils, 2.4% from manure management, and 2% from burning of crop residues (INCCA, 2010). The increase in nitrous oxide (N_2O) emission is primarily due to fertilizer use in agriculture (IPCC, 2007). Nearly two-thirds of agricultural emissions occur as N_2O, which has 300 times the global warming potential of CO_2 (Forster et al., 2007). New technologies need to be developed to reduce GHG emission from the agriculture sector. This chapter is aimed to review the CFP, that is, the total set of GHG emission from agricultural products. The necessity to address this issue and the methods to quantify CFP in agriculture is also elaborated. Scientific literature available on CFP, its assessment in the agricultural sector, and different case studies from various countries was presented. Case studies on CFP of different cereals, legumes, vegetables, fruits, and the livestock sector (including dairy, poultry, piggery, and fisheries) were included in the review. This was followed by a discussion on the methods to reduce CFP from the agricultural sector. Various strategies by which we can reduce CFP were also mentioned with suitable examples. The ultimate goal of the chapter was to illustrate the need of carbon foot printing in agriculture sector, as well as development and implementation of low-carbon technologies at various levels to make world agriculture more sustainable.

19.2 CARBON FOOTPRINT

Wiedmann and Minx (2008) introduced CFP for quantifying the impact of GHG emissions on environmental sustainability. The term originally stemmed from the first academic publication discussing "ecological foot printing" by Rees (1992). In the present situation, CFP becomes a widely used term and concept dealing on the responsibility and abatement actions against the threat of global climate change. CFP is the total set of GHG emissions caused by a product (Pathak et al., 2010) and is expressed in terms of carbon dioxide (CO_2) equivalent. It is basically accepted as a measure of an individual's contribution to global warming in terms of the amount of GHGs produced (Lynas, 2007). Assessment of CFP has been widely accepted for identifying and developing low-carbon options in different sectors.

CFP of food is an estimate of all the emissions caused by the production, manufacture, and delivery to the consumer and the disposal of packaging (Bokern, 2010). Recently there is a growing interest in reducing the CFP of agricultural products, that is, the total GHG emissions associated with the amount of grain produced (Williams and Wikstrom, 2011). Producer, consumer, and even policy makers want to know how farming practices could be improved so as to produce high-quality food while ensuring its low CFP.

19.3 WHY ESTIMATE THE CARBON FOOTPRINT OF AGRICULTURAL PRODUCTS?

Carbon emission occurs during different stages in the life cycle of a product. Carbon foot printing can be used to assess and compare the GHG emission of different products along with identification of strategies to reduce CFP, and thereby improve the environmental quality and improve resource use efficiency. According to Wiedmann and Minx (2007), CFP has become a widely used term and concept in the public debate on responsibility and abatement action against the threat of global climate change. Assessing CFP from an activity helps in emission management and evaluation of mitigation measures (Carbon Trust, 2007). In recent times, there is a growing awareness toward climate change and global warming. This has led to the development of personal CFP facilities, particularly in developed countries (Padgett et al., 2008). UNDP (2007) published country-wise per capita CFP, which clearly showed that developed countries leave the biggest footprint while CFP of developing nations is significantly lower. CFPs of products can guide consumers in choosing more climate-friendly products and also identify the emission 'hotspots' in the processing and delivery process, helping manufacturers identify ways to save energy (Bokern, 2010). As such, there is a need to quantify typical emissions for the cultivation of common food crops, which will help in exploring the potential for mitigation of the C emissions from food crops (Hilier et al., 2009). Effect of each individual activity on the overall footprint helps in determining how reduction of the overall footprint can be achieved. There are opportunities for reduction in GHG emission from cultivated soils if better management practices are identified and adopted (Hutchinson et al., 2007). GHG emissions from crop production can be reduced by adopting improved farming practices having lower CFP. Estimates of GHGs emission due to the production of crops are necessary for evaluating the environmental impact of current agricultural management practices and to develop strategies for lowering the CFP of existing cultivation methods by formulating suitable mitigation strategies. Once the CFP is known, a strategy can be devised to reduce it by technological developments, better process and product management, and alternate consumption strategies (Pathak et al., 2010). Environmental impact of food consumption has also received political and social attention in recent years. The Intergovernmental Panel on Climate Change (IPCC) in its Fourth Assessment Report (AR4) pointed out that lifestyle changes can contribute to climate change mitigation and that reducing animal protein consumption can lower GHG emission. According to Pathak et al. (2010), in order to efficiently work out "climate smart diets," more knowledge is needed about life-cycle impact of single products and about connections between diets and how the food chain is affected by changed diets. Pandey et al. (2011) has stressed the need that carbon foot printing must be promoted as a strong tool for GHG emission reductions and should be included as an indicator of sustainable development.

19.4   HOW TO ESTIMATE THE CARBON FOOTPRINT OF AGRICULTURAL PRODUCTS?

The carbon footprint of a product can be quantified by assessing GHG emissions at all stages during the life cycle of a product. Life-cycle analysis is also called "cradle-to-grave analysis." Quantification of CFP is done during all the stages in the life cycle of a product, including procurement of raw material, its production, processing, value addition, packaging, storage, transport, and finally consumption. The methodologies for quantification of carbon footprint of products are still evolving and it is emerging as an important tool for GHG management (Pandey et al., 2011). CFP of crop production was assessed by taking into account the total GHGs emission in terms of carbon equivalent (CE) through material added, and from mechanical operati on performed in a single whole cycle of crop production (Hillier et al., 2009). In order to estimate CFP of agricultural products, GHG emission data is collected for ploughing the field, applying agrochemicals, harvesting, storage, processing, packaging, transport, and consumption. GHG emission from production of agri-inputs such as fertilizers, pesticides also needs to be estimated to obtain CFP of the product. In CFP estimation, all major GHGs are taken into account. However, the most significant GHGs in agriculture are methane (CH_4), nitrous oxide (N_2O), and carbon dioxide (CO_2). In order to allow comparison across different GHGs, non-CO_2 GHGs are converted to the common unit of carbon dioxide "equivalent" (CO_2-eq.) based on their global warming potential, relative to that of CO_2 (ITC, 2012). Methodologies and standards are available for estimating GHG emission at different stages of the life cycle of crops. GHG accounting methods are given by IPCC 2006 guidelines, GHG protocol of World Resource Institute (WRI), ISO 14064 (parts 1 and 2), Publicly Available Specifications-2050 (PAS 2050) of British Standard Institution (BSI), ISO 14025, and ISO 14067 (Pandey et al., 2011). IPCC (2006) guidelines are available for quantification of GHGs emitted from different sectors such as energy, industrial processes, agriculture, forestry, and waste. Data on GHG emission can be obtained by direct measurement in the field, or by estimation based on emission factors. Default emission factors for different climate, land use, soil type, crops, and management practices are available in IPCC (2006) guidelines. Country-specific emission factors are also generated by conducting field experiments, collecting GHG samples, and analyzing them. CFP is generally expressed as per unit of product. In case of agricultural products, it is expressed as kg CO_2 emitted per kg of produce. So yield of crops obtained has a major role on the CFP of agricultural products. Increasing yield along with judicial use of agri-inputs can help in lowering the carbon footprint of agriculture. According to IPCC guidelines, GHG emission from cultivated land can be calculated following Tier 1, Tier 2, or Tier 3 approach. Tier 1 is the simplest among all for which equations and default parameter values (e.g., emission and stock change factors) are provided. Tier 2 uses the same methodological approach as Tier 1 does, but applies emission and stock change factors that are based on country- or region-specific data for important land-use and livestock categories. In Tier 3 approach, models and inventory measurement systems are used to estimate GHG emission. Repeated field sampling at regular time intervals is done, and simulation models are used to estimate emission of GHGs under different land use, soil type, and crop management practices. Moving to higher tiers improves the accuracy of the inventory and reduces uncertainty. But the complexity and data requirement increases for higher tiers. According to IPCC, sources of uncertainties need to be mentioned while reporting CFP of agri-products. In case of agriculture, uncertainties are related to non-availability of country- and location-specific emission factors. Emission factors for different GHGs are influenced by climate, soil type, crops as well as management practices. Measured data on GHG emission for every location is not available, which leads to uncertainties in estimation of CFP using Tier 1 and Tier 2 approach. Besides this, there is an immediate need for uniformities in GHG estimation techniques in agriculture to make more accurate comparisons (Pandey and Agrawal, 2014).

19.5   CARBON FOOTPRINT OF AGRICULTURAL PRODUCTS

Anthropogenic impact on global warming is through our consumption of various products, and attempts need to be made to quantify the relationship between our consumption of different products and its impact on global warming. In order to understand the mitigation potential and develop low-carbon practices in world agriculture, great efforts have been given worldwide in quantifying the CFP of agricultural production and the effects by agricultural management activities (Perry et al., 2008; Hillier et al., 2009). Tukker et al. (2006) identified the products or services with the greatest environmental impact throughout their life cycle. Food and drink, private transport, and housing were found to be responsible for 20%–30% of the total environmental impact of the overall consumption of citizens in the European Union.

Agriculture contributes to GHG emission owing to both on-farm and off-farm activities. Crop production, food processing, and product marketing all these stages generate GHG, contributing to global climate change (Dyer et al., 2010). The CFP for production varied among regions even for a specific crop (Table 19.1). Agricultural soils contribute to the greenhouse effect primarily through the emission of GHGs such as methane, nitrous oxide, and carbon dioxide. Submerged rice fields are the major source of methane. Application of nitrogenous fertilizers to soil leads to emission of N_2O. Carbon dioxide is emitted as a result of burning fossil fuel in various farm machineries and during burning of crop residues in the field. Soil management practices such as tillage operation also lead to emission of carbon dioxide through biological decomposition of soil organic matter. Off-farm sources include production of carbon dioxide for manufacturing agro-chemicals such as fertilizers and pesticides.

An important contributor to the C footprint of rice crop is methane (CH_4). Paddy fields are important sources of methane. Indian rice fields covering an area of 43.86 Mha (MoA, 2008) emitted 3.37 Mt of CH_4 in 2007. Nitrous oxide emission occurs mainly as a result of the application of nitrogenous fertilizers. The total N_2O–N emission from India was estimated to be 0.14 Mt in

TABLE 19.1

Carbon Footprint (CFP) for Production of Various Field Crops

Crop	CFP (kg CO_2-eq. kg^{-1} Produce)	Country	Reference
Rice	1.34–2.5	China	Xu et al. (2013)
	0.65–0.66	India	Pathak et al. (2012a)
	2.9	Italy	Blengini and Busto (2009)
Wheat	0.37–0.40	India	Pathak et al. (2012a)
	0.343	Canada	Gan et al. (2012)
	0.383–0.56	Canada	Gan et al. (2011b)
	0.34	New Zealand	Barber et al. (2011)
Durum-wheat	0.27–0.42	Canada	Gan et al. (2011a)
Maize	0.1–0.583	Canada	Ma et al. (2012)
	0.19	New Zealand	Barber et al. (2011)
Mustard	0.601–1.56	Canada	Gan et al. (2011b)
Chickpea	0.254–0.456		
Lentil	0.164–0.237		
Barley	0.28–0.32	Canada	Gan et al. (2012)
Maize silage	0.13		
Ryegrass seed	1.3		
Potato	0.1–0.16	Sweden	Roos et al. (2010)
	0.2	UK	Williams et al. (2010)

2007 (Bhatia et al., 2012). GHG emission from agriculture in different states of India has shown that Punjab, Haryana, Uttar Pradesh, and Andhra Pradesh emit higher amount of N_2O emissions because of the application of higher doses of nitrogenous fertilizer.

Pathak et al. (2012a) assessed the potential of various technologies to mitigate GHGs emission in rice and wheat crop at site as well as at regional scales in the Indo-Gangetic Plains (IGP) and evaluated the economic viability of those technologies. Results showed that compared with the current cultivation practices of farmers, 15 technologies in upper-IGP and 14 technologies in lower-IGP have the potential to reduce the global warming potential (GWP). But adoption of some technologies involves some extra cost, which makes them not feasible for adoption by the farmers. Technologies such as sprinkler irrigation, direct seeded rice, nitrification inhibitor, urea super granules, leaf-color-chart-based nitrogen application, site-specific nutrient management, and crop diversification were able to reduce GWP without any extra cost in the upper IGP. On the other hand, in the lower-IGP, use of nitrification inhibitor, leaf-color-chart-based nitrogen application, site-specific nutrient management, and crop diversification reduced GWP with no additional cost. A comparative study on GHG emission was done by Pathak et al. (2012b) at Jalandhar, Punjab with transplanted as well as direct seeded rice (DSR). Growing DSR reduced GHG emission and saved irrigation water, human labor, and machine labor without any yield penalty when compared with conventional transplanted rice.

Lal (2004) synthesized the available information on energy use in farm operations, and its conversion into carbon equivalent (CE). Synthesis of the data showed that several agricultural practices such as ploughing, fertilizers and pesticides application, and irrigation are C-intensive because of the fossil fuel and energy involved in their use. Conversion of conventional tillage to no-tillage, using integrated nutrient and pest management practices, and enhancing water use efficiency by adopting drip irrigation and sub-irrigation practices can save C emission and at the same time increase soil C pool. Sustainability of agricultural systems depends on their CFP, and the $C_{output}:C_{input}$ ratio (Dubey and Lal, 2009). The C-based sustainability index was found to be much higher in Ohio, USA than in Punjab, India. But results of the study indicated that management practices such as mulching, use of high analysis fertilizers, contour farming, no-till farming, manuring, retention of crop residues such as mulch, judicious use of chemical fertilizers, crop rotation using legume crops, and integrated water management techniques will improve soil quality in both the regions.

Hilier et al. (2009) used farm survey data from the east of Scotland combined with published GHG emission data to quantify the carbon footprint of different crops. Across all farm and crop types, the average estimated carbon footprint was 351.7 kg CE ha^{-1} year^{-1} ranging from just 18 to 903 in different crops. Over all crops and farm types, 75% of the carbon footprint resulted from use of nitrogenous fertilizer and direct nitrous oxide emissions from the soil. The contributions of the other management practices such as ground preparation, sowing, crop protection, fertilizer (P and K), and harvesting operations contributed less than 8% of the average carbon footprint across all sites (Figure 19.1). The carbon footprint for leguminous crops was found to be lower than for cereal crops. Organic farms had a significantly lower carbon footprint when compared with conventional and integrated farm types.

CFP of wheat crop grown in four different rotation systems involving fallow sequences was determined by Gan et al. (2012a). Average CFP of wheat crop was quantified as 0.343 kg CO_2-eq. kg^{-1} of grain irrespective of crop rotations. The inverse relationship between fallow frequency and GHG emissions largely reflects that cropping the land more intensively requires a greater input of N fertilizer, the main contributor to GHG emission. When soil C changes were included in the calculations, the CFP values became negative and differed significantly among cropping systems. Continuous wheat system produced maximum grain yield and gained highest soil organic carbon over the years, leading to the smallest CFP value of −0.441 kg CO_2-eq. kg^{-1} of grain, which was significantly lower than the CFP values from the three other systems. CFP of wheat crop was also calculated by Ho (2011) in Ontario, Canada by conducting the life-cycle analysis of crop. In the study, GHG emission was quantified during production, transportation, use of machinery, and application of agricultural chemicals in wheat crop.

CFP of spring barley crop was quantified in relation to various preceding oilseed crops that were fertilized at various rates of inorganic N the previous years (Gan et al., 2012b). CFP of barley crop

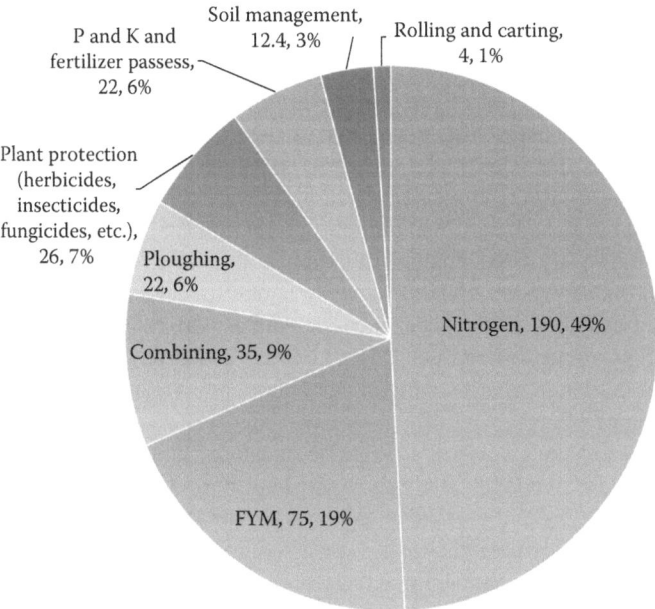

FIGURE 19.1 Relative contributions of different farm management practices to the total carbon footprint (kg C-eq. ha^{-1} year^{-1}), as well as percentage of total) for all crops surveyed. (Modified from Hilier, J. et al. 2009. *Int. J. Agric. Sustain.* 7(2):107–118.)

ranged from 0.28 to 0.32 kg CO_2-eq. kg^{-1} of grain in two different growing locations. Emissions due to N fertilizers were 26.6 times the emission from the use of phosphorous, 5.2 times the emission from pesticides, and 4.2 times the emission from various farming operations. Decomposition of crop residues contributed 19% of the total emission. GHG emission had a positive linear correlation with rate of nitrogenous fertilizer applied to the preceding oilseed crops.

Carbon footprint of China's crop production was assessed by Cheng et al. (2011) using national statistical data available for the period of 1993 to 2007. The mean CFP of China's crop production was estimated to be 0.78 ± 0.08 tCE ha^{-1} $year^{-1}$ and 0.11 ± 0.01 tCE t^{-1} $year^{-1}$, for land use and bulk production, respectively. Use of fertilizer, electricity for irrigation, and mechanical operation accounted for 89% to the total CFP while plastic film accounted for only 6%, on average (Figure 19.2). Fertilizer-use-induced emissions accounted for 57% of the total emission from chemical inputs. Pesticide-induced emissions contributed only 5% to the total CFP.

Rice has the largest yield among all food crops in China, and rice cultivation emits GHGs such as CH_4 and N_2O. Complete life-cycle assessment (LCA) of rice cultivation was done by Xu et al. (2013) to quantify CFP of rice production in selected five provinces of China. CFPs of rice production were 2504.2, 2326.5, 1890, 1538.9, and 1344.92 kg CO_2-eq. ton^{-1} rice production in Guangdong, Hunan, Heilongjiang, Sichuan, and Jiangsu province, respectively. GHG emission was mostly from cultural and irrigating machinery, and emission from soil during the crop growth period. More irrigation resulted in higher CH_4 emission while N_2O emission was directly attributed to the rate of urea application. The study concluded that reducing the amount of urea applied would decrease the N_2O emission and intermittent irrigation system would decrease CH_4 emission, thereby reducing C footprint of rice cultivation in China.

One of the effective management options to reduce GHG emissions from farming could be diversification of cropping systems. A study conducted by Gan et al. (2011a) aimed to determine CFP of durum wheat (*Triticum turgidum* L.) produced in diverse cropping systems in Canada. Carbon footprint of durum wheat was lowered by an average of 22% owing to diversification of cropping systems. Including biological N-fixers in the systems lowered CFP of durum wheat by as much as 34% over monoculture wheat systems. Reduction in use of fertilizer N contributed greatly to the lowered carbon footprint of durum grain in diversified cropping systems.

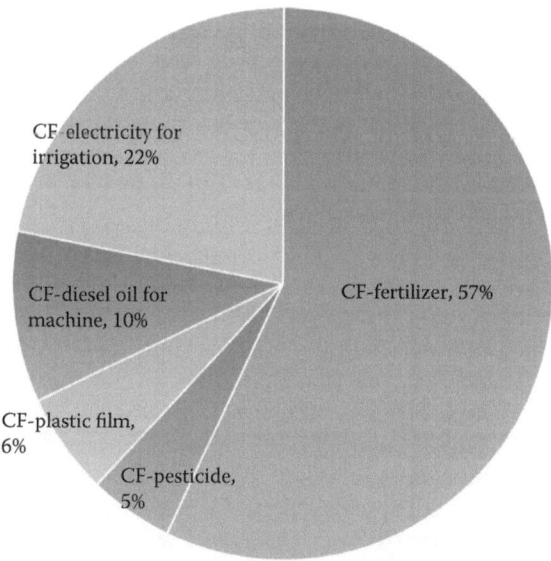

FIGURE 19.2 Mean proportions of different inputs to total CFP of crop production in China. (Reprinted from *Agric. Ecosyst. Environ.*, 142, Cheng, K. et al. Carbon footprint of China's crop production—An estimation using agro-statistics data over 1993–2007, 231–237, Copyright 2011, with permission from Elsevier.)

The carbon footprint of maize monoculture as well as maize-legume rotation was assessed by Ma et al. (2012) in a 19-year-old field experiment in Ontario, Canada. Although GHG emission was higher in maize-forage and maize-soybean rotation when compared with maize monoculture but C footprint of maize-legume rotations was lower than sole maize because of increased productivity of maize-legume system. In this study, higher doses of N fertilizer increased maize grain yield, but also increased the C footprint of maize production across all the rotation systems. Results suggested that increased grain yield and reduced CFPs in maize production can be achieved through appropriate N application and rotation with a forage legume or soybean.

A study was done by Gan et al. (2011b) to assess CFP of field crops grown in different soil types of the Canadian prairie. The CFPs were, in decreasing order, 0.80 kg CO_2-eq. for canola, 0.59 kg CO_2-eq. for mustard and flaxseed, 0.46 kg CO_2-eq. for spring wheat, and 0.20–0.33 kg CO_2-eq. kg^{-1} of product for pulses grown in the brown and the dark brown soil zones. The carbon footprint of crops grown in the black soil zone had greater magnitude. The higher emissions associated with crop residue decomposition for the black soil zone are mainly due to higher crop yield and due to higher emission factor because of more favorable climatic conditions during the growing season.

For quantification of CFP of agricultural produce GHG emission at different stages in the life cycle of the product needs to be quantified. In the whole life-cycle assessment, GHG emission should be quantified in production, processing, marketing, and consumption stage (Pathak et al., 2012c). GHG emission was quantified during the whole life cycle of rice of wheat crop. In case of wheat, GWP of emissions was found to be higher in the upper-IGP than in the lower-IGP. Emissions during production contributed maximum (38%–47%) to the total GWP in both the IGP regions, followed by GWP of emissions in the marketing (30%–32%) stage. In case of rice, GHG emissions during the production contributed more than 50% to the total GWP, followed by marketing (17%–23%) in both the IGP regions. During the processing stage, parboiling of rice was found to be the most energy-intensive process having GWP of 198.4 CO_2-eq. in the lower-IGP.

The CFP varied greatly for various steps of production, processing, transport, and storage of horticultural crops (Table 19.2). However, it is difficult to compare the CFP across the regions since the estimates included different combinations of those steps. Stoessel et al. (2012) used life-cycle assessment methodology to study the environmental impact of 34 fruits and vegetables of a large Swiss retailer, with the aim of providing environmental decision-support to the retailer, and establishing lifecycle inventories (LCI). Various stages in the LCA analysis include seedling production, use of farm machinery, use of fuels for heating of greenhouses, irrigation, fertilizers, pesticides, storage, and transport to and within Switzerland. Asparagus, lettuce, banana, tomato, and pear were the top scorers in having maximum CFP. The results showed that reduction in CFP can be achieved by consuming seasonal fruits and vegetables, followed by reduction in transport by airplane. Sourcing fruits and vegetables locally is only a good strategy to reduce the CFP if no greenhouse heating with fossil fuels is involved. Rab et al. (2008) made the preliminary estimate of the CFP of total vegetable industry of Australia as 1,047,008 t CO_2-eq. $year^{-1}$.

Life-cycle analysis of arable crop production of New Zealand was done and report was prepared for "Foundation for Arable Research, Ministry of Agriculture and Forestry" (Barber et al., 2011). GHG emissions of each arable crop were broadly similar when compared on a per hectare basis, ranging from 2,190 to 2,820 kg CO_2-eq. ha^{-1} (Figure 19.3). Field emission following nitrogen fertilizer application is the single largest source of GHG emission. Energy (fuel and electricity) contributed less than 20% of total emissions.

A study by Tzilivakis et al. (2005) assessed the CFP of sugar beet crop grown in the United Kingdom. The crop had an average CFP of 0.024 eq. t CO_2 per one of clean beet harvested. The high-yielding peat and silt soils had the lowest CFP while sandy soils that used farmyard manure or mineral fertilizer in addition to irrigation had the largest CFP. It was concluded that in the United Kingdom, beet production on peat, silt, or clay loam soils that are more water retentive tend to produce larger yields and minimizes the global warming potential (GWP) from the inputs.

TABLE 19.2

Carbon Footprint (CFP) during Production, Processing, Packaging, Transport, and Storage of Various Horticultural Crops

Crop	Country	CFP (kg CO_2-eq. kg^{-1} Produce)	Stage in Life Cycle	Reference
Apple	UK	0.27	Production and storage	Saunders et al. (2006)
Onion		0.17	Production, packaging, and storage	
Sugarbeet		0.024	Production	Tzilivakis et al. (2005)
Lettuce		0.13	Production	Koerbar et al. (2009)
Broccoli		0.14	Production	
Brinjal	India	0.045	Production, processing, and transport	Pathak et al. (2010)
Cauliflower		0.042	Production, processing, and transport	
Banana		0.1	Production, processing, and transport	
Carrot	Italy	0.63	Production, storage, and transport	Carlsson-Kanyama (1998)
Tomato	Sweden	0.43	Production, storage, and transport	Carlsson-Kanyama (1998)
Carrot	Denmark	0.21	Production	Halberg (2008)
Onion		0.15	Production, packaging, and transport	Fogelberg and Carlsson-Kanyama (2006)
Carrot	Netherlands	0.16	Production, packaging, and transport	Fogelberg and Carlsson-Kanyama (2006)
Cauliflower	Australia	2.28	Production and transport	O'Halloran et al. (2008)
Tomato		0.17	Production and transport	
Cucumber		0.09	Production and transport	
Pumpkin		0.46	Production and transport	

Source: Adapted from Caracciolo, F. et al. 2010. CO_2 emission in the fresh vegetables chains: A meta-analysis. Department of Agricultural Economics and Policy, University of Naples Federico II, Italy.

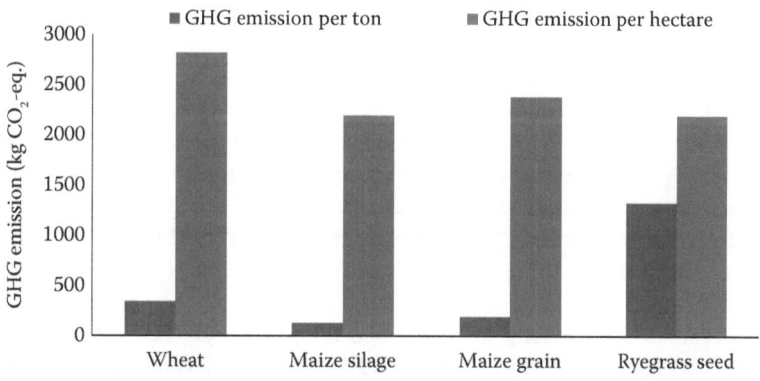

FIGURE 19.3 GHG emission from arable crop production in New Zealand. (Data from Barber, A., Pellow, G., and M. Barber. 2011. Carbon footprint of New Zealand arable production—Wheat, maize silage, maize grain and ryegrass seed. MAF Technical Paper No. 2011/97. Foundation for Arable Research Ministry of Agriculture and Forestry, Agri LINK New Zealand Ltd., pp. 1–58.)

19.6   CARBON FOOTPRINT OF LIVESTOCK SECTOR

Livestock is another sector mainly dependent on natural resources. According to Bokern (2010), more than half of the total GHG emissions from food come from farming and fishing. Of these, nearly two-thirds come from animal products, which provide only about one-third of the food energy in UK food. Livestock products have nearly twice the CFP than of plant-based foods. Livestock products carry a large CFP compared with other foods (Hermansen and Kristensen, 2011). The major contributing factors are emissions related to feed use and manure handling. Methane is produced during enteric fermentation and during storage of manure, while emission of N_2O is related to application of manure to the field. Weidema et al. (2008) estimated that the consumption of meat and dairy products was responsible for 14% of the total global warming resulting from the consumption of all kinds of products and services (e.g., food, transportation) within the 27 European Union member countries. GHG emission from food consumption in United States also revealed a similar trend. Higher meat consumption resulted in increased GHG emissions (Kim and Neff, 2009).

De Vries and de Boer (2010) compared different livestock products in terms of their CFP generation. The study showed that beef had the largest CFP (15–32 kg CO_2-eq. kg^{-1}) followed by pork (4–11 kg CO_2-eq. kg^{-1}) and chicken (4–6 kg CO_2-eq. kg^{-1}). FAO recently reported that global average carbon footprint of milk is 2.4 kg CO_2-eq. kg^{-1} (FAO, 2010). Emissions vary greatly among different regions and countries.

Livestock production contributes GHGs to the atmosphere both directly and indirectly (IPCC, 2006). Methane from enteric fermentation and manure management are the main sources of CH_4 emissions from livestock. In 2011, an estimated 137 million metric tons of methane was produced in the United States from enteric fermentation, comprising about 70% of the total agricultural methane emissions (Smith, 2014). In United States, beef cattle are the largest contributor of enteric methane, nearly three times that of dairy cattle and 50 times greater than that of swine. Stored manures are also an important source of GHGs. According to IPCC report, majority of the N_2O emission from animal agriculture is from manure management, which is the second largest N_2O emitter in the agricultural sector. In 2011, an estimated 52 million tons of CO_2 equivalents of CH_4 and 18 metric tons of CO_2 equivalents of N_2O were produced from animal manure deposition, storage, and application (Smith, 2014). According to Dunkley and Dunkley (2013), based on the life-cycle assessment of beef cattle, 86.15% of the GHGs are emitted during the production stage, while 68.51% of emissions take place during the production of pork, and 47.82% of GHG emissions occur during the production stage of broiler chickens.

Livestock constitutes an integral component of Indian agriculture sector and also a major source of GHGs emissions. Enteric fermentation in livestock released 212.10 million tons of CO_2-eq. (INCCA, 2010). This constituted 63.4% of the total GHG emissions from agriculture sector in India. Chhabra et al. (2013) prepared a detailed inventory of GHG emission from Indian livestock. According to the study, total methane emission (including enteric fermentation and manure management of livestock) was estimated to be 11.75 Tg $year^{-1}$ for the year 2003, while nitrous oxide emission from Indian livestock was estimated at 1.42 Gg $year^{-1}$. Enteric fermentation constitutes 91% of the total methane emission and 86.1% of N_2O emission occurs from the poultry sector.

CFP of dairy products of Canada was estimated by Verge et al. (2013). CFP estimates of dairy products ranged between 1 and 3 kg of CO_2-eq. kg^{-1} of product. Average CFP of milk was 1.07 kg of CO_2-eq. L^{-1}, ranging from 0.86 to 1.14 kg of CO_2-eq. L^{-1}. Higher CFP was observed for products like cheese (5.3 kg of CO_2-eq. kg^{-1}), butter (7.3 kg of CO_2-eq. kg^{-1}), and milk powder (10.1 kg of CO_2-eq. kg^{-1}). The CFP of Canadian beef production increased from 25 to 32 Tg of CO_2-eq. during the period of 1981 and 2001 owing to the expansion of beef industry (Verge et al., 2008). But the emission intensity per kg of live animal weight produced decreased from 16.4 to 10.4 kg of CO_2 equivalent.

19.7   CARBON FOOTPRINT OF PIGGERY AND POULTRY

Globally, pig and chicken supply chains are estimated to produce GHG emissions of 668 and 606 million tons of CO_2-eq., respectively (MacLeod et al., 2013). The average emission intensity of chicken is 5.4 kg CO_2-eq. kg^{-1} meat. Feed production is the main source of emissions, but emissions from manure are also significant. The average emission intensity of pig meat is 6.1 kg CO_2-eq. kg^{-1}, with feed production and manure management representing the main categories of emission. According to Jacob (2009), CFP of the U.S. broiler industry is 5.77 tons CO_2 equivalents per 1000 broilers marketed, which is about 0.6% of the GHG emission attributed to agricultural GHGs. In United States, ruminants, such as beef and sheep, have the highest CFP because of their release of methane, while poultry has the lowest CFP, followed by pigs. The largest contributor to GHG emissions from poultry production is feed production (Dunkley and Dunkley, 2013).

19.8   CARBON FOOTPRINT OF FISHERIES

Fishing activities are also source of GHG emission. Fishing is considered as the most energy-intensive food production method in the world (Wilson, 1999). Use of large amounts of fuel in activities such as fish catch, onboard processing, refrigeration, and freezing of fishes results in considerable emissions of GHGs. According to an estimate made in year 2000, fisheries account for 1.2% of global oil consumption and fishing boats released 134 million tons of CO_2 into the atmosphere (Tyedmers et al., 2005). On average, 1.7 tons of CO_2 was released per ton of live weight landed product. Carbon footprint of tuna fisheries in Philippines was estimated by Tan and Culaba (2009). The carbon footprint was estimated at 0.25–0.30 kg per kg of tuna, of which about half comes directly from fishing fleet operations. In India, CO_2 emission for every ton of marine fish caught has increased from 0.50 to 1.02 t during the period of 1961–2000 (Vivekanandan et al., 2013). At Visakhapatnam, India, 0.382 kg C and 1.404 kg CO_2 was emitted per kilogram of marine fish over all its life-cycle stages (Ghosh et al., 2014). Maximum energy requirement and highest CO_2 emission (90.2%) occurred at the harvest phase, while emissions at the pre and post-harvest phases were found to be trivial. Hospido and Tyedmers (2005) examined the "life cycle environmental impacts" of the Spanish purse seine tuna fleets. In terms of total carbon dioxide output, marine transport accounted for 30%, while pretransport diesel use accounted for 60%.

19.9   CARBON FOOTPRINT OF FOOD HABITS

The Intergovernmental Panel on Climate Change (IPCC) has pointed out that lifestyle changes such as reducing animal protein consumption patterns can contribute to climate change mitigation by lowering GHG emission. Quantification of carbon footprints of Indian food items revealed that animal food products such as meat mostly contributed to methane (CH_4) emission (Pathak et al., 2010). GHG emission from mutton is 11.9 times more than that of milk, 12.1 times fish, 12.9 times rice, and 36.5 times chapatti. GHG emission during the life cycle of cooked rice was 2.8 times the GHG emission during the life cycle of chapatti, a product of wheat flour. For a balanced vegetarian diet, CFP of an adult Indian male is 723.7 g CO_2 eq. day^{-1}, while CFP of nonvegetarian meal with mutton is 1.8 times that of a vegetarian meal, 1.5 times of a nonvegetarian meal with chicken and an ovo-vegetarian meal, and 1.4 times a lacto-vegetarian meal (Figure 19.4).

Reduced meat consumption is becoming increasingly popular owing to concerns about personal health and also because of its environmental impact. A United Nations report explains that globally the livestock sector is "one of the two or three most significant contributors to the most serious environmental problems," including climate change, air and water pollution, land degradation, and biodiversity loss (Steinfeld et al., 2011). Beef generates 27.1 kg of CO_2-eq. per kg consumed (Hamerschlag, 2011). According to Kim and Neff (2009), meat and dairy consumption in USA was the primary diet-related contributor to GHG emission.

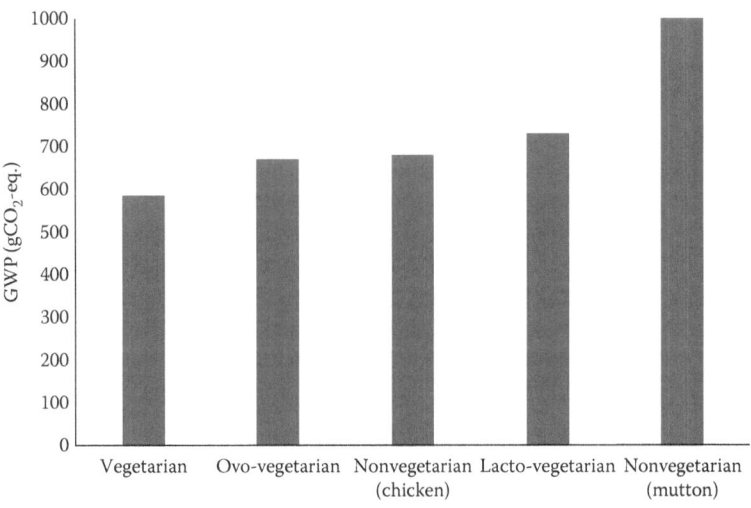

FIGURE 19.4 Global warming potential of various vegetarian and nonvegetarian meals in India. (Reprinted from *Agric. Ecosyst. Environ.*, 139, Pathak, H. et al. Carbon footprints of Indian food items, 66–73, Copyright 2010, with permission from Elsevier.)

19.10 STRATEGIES TO REDUCE THE CARBON FOOTPRINT OF AGRICULTURAL PRODUCE

The CFP of a product varies with crop type, climatic condition, soil type, and management options adopted by the farmers.

Some of the methods to reduce the CFP of our food have been suggested by Bokern (2010). More efficient farming practices using less fertilizer and improving the productivity of animals can help reduce the CFP from farm products. Emissions can also be curtailed by changing the storage process and also the process of food preparation. Switching to plant-based diets generally reduces the CFP, particularly. Removing meat completely from the diet reduces the food carbon footprint by 20%, whereas reducing all livestock products by 50% can reduce the food CFP by 13%.

19.10.1 Growing Leguminous Crops

Improving cropping systems may help lower the emission of GHGs. Using biological N-fixation through the inclusion of legume crops helps reduce the CFP by fixing atmospheric nitrogen via symbiosis, resulting in lower N_2O fluxes compared with cropping systems that are fertilized with a fertilizer N. In addition, they also play a key role in sequestering soil C. Inclusion of herbaceous legumes in pastures, either as sole crops, green or brown manures, cover-crops, or intercrops in reduced tillage cropping systems, has been shown to enhance soil C accumulation (Jensen et al., 2011). Pulse crops in crop rotations can reduce the dependence of agriculture on synthetic N fertilizers (Crews and Peoples, 2004) and thus reduce agricultures CFP. Study by Gan et al. (2011b) also showed that nitrogen-fixing crops such as chickpea, lentil, and dry pea had an average emission of 65% lower than the emissions of canola and spring wheat. Intercropping with legumes and certain permutations of crop rotations are also found to be greatly effective for carbon sequestration in cultivated soil (Nishimura et al., 2008).

19.10.2 Fertilizer Management

Appropriate fertilizer management practices that lead to increased N-use efficiency and higher yield while reducing N losses can help in mitigating nitrous oxide emission, thereby lowering CFP of agricultural crops. Improved N fertilizer application technology such as side-banding, timely application, and site-specific approaches (Malhi et al., 2001; Peng et al., 2010; Sieling and Kage, 2010) can help in reducing N losses and improving nitrogen use efficiency. Site-specific nutrient management (SSNM) includes site-specific quantitative knowledge of crop nutrient requirements, indigenous nutrient supply, and recovery efficiency of applied fertilizer N. Chlorophyll meter and leaf-color-chart-based N application can help in increasing the fertilizer use efficiency because of less fertilizer application and loss of nitrous oxide. Gan et al. (2012b) also reported that CFP in barley crop can be lowered by increasing grain yield, making a wise choice of crop types, reducing nitrogen inputs to previous and current crops, and improving N use efficiency.

Application of urea in plough layers gives less emission of nitrous oxide than band application of urea fertilizer. Besides this, foliar fertilization is one way of supplying N during periods of rapid plant growth and N demand. Direct measurement of gaseous emission in such systems showed that very little N was lost from foliar applied urea unless rainfall washed unassimilated urea from the plant onto the soil (Pathak et al., 2011).

19.10.3 WATER MANAGEMENT

Soil water regime is an important factor for the gas exchange between soil and atmosphere and has a direct impact on the processes involved in methane emission from rice field. A single mid-season drainage may reduce seasonal methane emission. This emission could be reduced further by intermittent irrigation, yielding a 30% reduction when compared with mid-season drainage (Lu et al., 2000). Applying irrigation as and when required increases N use efficiency and reduces the emission of nitrous oxide. Moreover, applying water as needed saves both water and energy for pumping, thereby reducing CFP generated from those activities.

19.10.4 CROP CULTIVARS

Plants influence the methane efflux by (i) providing channels (aerenchyma) for the transport of methane from soil to the atmosphere, (ii) releasing root exudates or root autolysis products to methanogenic bacteria, and (iii) creating oxic environment in the anoxic soil through the transport of O_2 into the rhizosphere, stimulating the oxidation of methane and inhibiting methanogenesis (Wassmann et al., 2000). There are significant variations in the quantities of methane emitted from soils growing different cultivars. Early-maturing cultivars emit less methane compared to that from late-maturing cultivars. Crop cultivars with (a) low emission, (b) high N-use efficiency, (c) faster decomposition of residues, and (d) less water-demanding have potential to reduce nitrous oxide emission from soil.

19.10.5 TILLAGE OPERATION

During a tillage event, soil aggregates are broken, increasing oxygen supply and surface area exposure of organic material. This promotes the decomposition of organic matter. Thus, more CO_2 emission can occur from a tilled than from an undisturbed soil (Doran, 1980). In contrast, conservation tillage favored organic carbon enrichment of soils (Lal, 1999). Optimizing tillage practices including (a) no-tillage or minimum tillage, (b) bed planting, (c) reducing compaction, and (d) improving drainage also help in mitigating CO_2 and N_2O emission from soil by less soil disturbance. Carbon can be sequestered in soils by minimizing soil disturbance and increasing inputs of organic matter. Some of the recommended management practices to sequester carbon in soils include mulching, no tillage, and organic manure application (Lal, 2004).

19.10.6 USE OF N-TRANSFORMATION INHIBITORS

Nitrification inhibitors such as DCD (dicyandiamide), nitrapyrin, sodium thiosulphate, 2-amino-4-chloro-6-methyl-peyremidine (AM), and acetylene inhibit the nitrification process (Prasad and

Power, 1995). These chemicals are recommended to increase nitrogen-use efficiency and reduce N-loss through leaching and denitrification. Urease inhibitors retard the hydrolysis of urea to ammonia. This results in the reduction of direct nitrous oxide emission from nitrifiers.

19.10.7 Organic Manure

Soil organic carbon levels were either maintained or increased with an adequate manure treatment (Katyal, 2001). Long-term nutrient management of crops, including organic nutrient sources, can help in sequestering carbon in soil, thereby mitigating the negative impact of climate change. Incorporation of crop residues in soil also leads to increased soil organic matter levels. Long-term studies have shown that organic manure application increases the carbon sequestration capacity of soil in the range of 70–551 kg C ha^{-1} when compared with mineral fertilizer use (Mandal et al., 2007, 2008).

19.10.8 Diversification of Cropping System

Diversified cropping systems can reduce the CFP of crop products. Gan et al. (2003) found that diversified cropping systems compared with monoculture systems significantly reduced the production inputs and increased yields of durum wheat, thereby lowering carbon footprint when compared with monoculture systems. A study by Gan et al. (2011a) showed that diversifying cropping systems with oilseeds and biological N-fixers significantly lowered CFP of durum wheat. Durum wheat preceded by an oilseed crop the previous year had 7% lower carbon footprint than durum in cereal–cereal–durum system, while durum wheat preceded by a biological N-fixing crop the previous year lowered its carbon footprint by 17%. Lowest CFP was observed in durum wheat produced in a pulse–pulse–durum system, which had 34% lower carbon footprint than durum grown in cereal–cereal–durum systems.

19.10.9 Livestock Management

Level of feed intake and its composition are important factors influencing CH_4 and N_2O emissions from livestock sector. According to Hermansen and Kristensen (2011), mitigation options to reduce CFP from livestock include better feed conversion; use of feeds that increase soil carbon sequestration, which ensures that the manure produced substitutes for synthetic fertilizer; and use of manure for bio-energy production. Increasing forage digestibility and digestible forage intake will generally reduce GHG emissions from rumen fermentation from livestock and is a highly recommended mitigation practice (Gerber et al., 2013). This will also reduce GHG emission from animal manure since diet has a significant impact on manure quality. Providing balanced diets to the animals can also help in lowering the emission. Regular analysis of the nutrient content of forage can help producers to alter their grain and supplements to meet the nutritional needs of cattle without overloading on certain nutrients (Beef Sector Report, 2003). Similarly making wise choices about species combinations in pastures will increase the nutritional value of the forage and have a net reducing effect on the emissions.

Proper manure management by reducing the time of manure storage, providing proper aeration, and stacking are recommended for reducing GHG emission from manure. Anaerobic digesters are a recommended mitigation strategy for CH_4 generate renewable energy, and provide sanitation opportunities for developing countries (Gerber et al., 2013). A variety of livestock production systems operates in the United States, and different manure management systems are utilized depending on the type of livestock or poultry produced (Del Grosso et al., 2008). When cattle are raised under conditions where the manure is collected and spread daily and there is no storage before it is spread onto the soil, there is low CH_4 emissions and no N_2O emissions (Dunkley and Dunkley, 2013).

19.10.10 MANAGEMENT OPTIONS FOR FISHERIES

GHG emission from the fisheries sector can be reduced with technological improvements. By implementing fuel efficiency norms for fishing vessels, there is scope for reducing CO_2 emission in India (Vivekanandan et al., 2013). A shift from fuel-intensive active fishing methods such as trawling to passive methods such as seining, lining, and gillnetting may provide a sustainable long-term solution. Eliminating inefficient fleet structures, improving fisheries management, reducing post-harvest losses, and increasing waste recycling will help in reducing CO_2 emissions from this sector (FAO, 2008).

19.10.11 DIETARY PATTERN

Animals require food over time, which means more N is used to produce a steak than a vegetarian meal. Producing meat also tends to increase the chances for N to be lost as pollution from animal wastes into the atmosphere and waterways (Pathak et al., 2011). Reduction in CFP also requires changes in behavioral pattern, including measures to change the dietary habit of people. Selection of food products having lesser CFP can help in minimizing the environmental impact. Some options to reduce the GHG emissions of food consumptions are consumption of locally produced foods, less mutton consumption, and substitution of meat and milk by other vegetable protein (Pathak et al., 2010). Many highly processed snack foods require GHG-intensive production with few beneficial returns on nutritional value (Carlsson-Kanyama et al., 2003). Reducing consumption of these foods will help in reducing GHG emission and will have many health cobenefits.

19.11 LIMITATIONS AND SCOPE FOR FUTURE RESEARCH

Although many researchers have estimated CFP of different agricultural products, further work is still needed now to refine some calculations taking into account different cropping systems. In some studies, quantification of CFP of crops is done until harvesting of the crop. The studies can be further expanded to estimate GHG emission during various post-harvest activities. In many cases, limitation of data has not allowed researchers to do the whole life-cycle analysis. An attempt has been made in this chapter to synthesize CFP of various crops and vegetables. But the values could not be compared because there is difference in agricultural management, soil, and weather parameters, and the methodology adopted for calculation of CFP is also varying. There is a need to have uniform guidelines regarding calculation of CFP of agricultural products. The crop management practices followed by different farmers are also very much diverse. Hence CFP, of agricultural products are different for different regions. So region-specific emission factors are required to quantify the CFP. Hence more research is needed in this area to reduce the uncertainties and develop low C technologies for agriculture. In many studies, Tier 1 or Tier 2 methodology has been used to calculate the CFP. In order to get more precise estimate of the CFP, IPCC Tier 3 methodology should be used. This will lead to a better understanding of the processes, leading to higher footprint and will help in formulating suitable options to reduce the CFP in the agriculture sector. More studies are required on quantification of CFP of agri-products under alternative management practices, which will help in assessing the environmental impact of those alternative management options in terms of GHG emission. This might help in development of low C technologies for agriculture.

19.12 CONCLUSION

Carbon foot printing is becoming an important approach for determining the developmental pathways that also determine the CFP. In pursuit of sustainable agricultural systems, it is essential that the researchers and developmental agencies are equally aware of, as well as appreciative of, the importance of developing and implementing the low-carbon technologies in agricultural activities at various levels. It is not far away that any technology developed with a potential for large-scale implementation will have to undergo CFP estimates before final judgment of its economic and ecological implications. Although awareness is on rise with regard to footprint estimates, acceptance in a wide range of stakeholder spectrum is poor. There is a need for developing universal protocols for estimating footprints of various aspects of agricultural activities.

REFERENCES

Barber, A., Pellow, G., and M. Barber. 2011. Carbon footprint of New Zealand arable production—Wheat, maize silage, maize grain and ryegrass seed. MAF Technical Paper No. 2011/97. Foundation for Arable Research Ministry of Agriculture and Forestry, Agri LINK New Zealand Ltd., pp. 1–58.

Beef Sector Report. 2003. Reducing greenhouse gas emissions through feeding and breeding. Canadian Cattlemen's Association, Calgary, Alberta.

Bhatia, A., Jain, N., and H. Pathak. 2012. Greenhouse gas emissions from Indian agriculture. In: Pathak, H. and P.K. Aggarwal (Eds.), *Low Carbon Technologies for Agriculture: A Study on Rice and Wheat Systems in the Indo-Gangetic Plains*, pp. 1–11.

Blengini, G.A. and M. Busto. 2009. The life cycle of rice: LCA of alternative agri-food chain management systems in Vercelli (Italy). *J. Environ. Manage.* 90:1512–1522.

Bokern, D.A. 2010. Understanding the carbon footprint of our food. *Complete Nutr.* 10(5):61–63.

Caracciolo, F., Cicia, G., Giudice, T.D., Menna, I., and L. Cembalo. 2010. CO_2 emission in the fresh vegetables chains: A meta-analysis. Department of Agricultural Economics and Policy, University of Naples Federico II, Italy.

Carbon Trust. 2007. Carbon footprinting. An introduction for organizations. http://www.carbontrust.co.uk/publications/publicationdetail

Carlsson-Kanyama, A. 1998. Food consumption patterns and their influence on climate change: Greenhouse gas emissions in the life-cycle of tomatoes and carrots consumed in Sweden. *Ambio* 27: 528–534.

Cheng, K., Pan, G., Smith, P., Luo, T., Li, L., Zheng, J., Zhang, X., Han, X., and M. Yan. 2011. Carbon footprint of China's crop production—An estimation using agro-statistics data over 1993–2007. *Agric. Ecosyst. Environ.* 142:231–237.

Chhabra, A., Manjunath, K.R., Panigrahy, S., and J.S. Parihar. 2013. Greenhouse gas emissions from Indian livestock. *Clim. Change* 117:329–344.

Choudrie, S.L., Jackson, J., Watterson, J.D., Murrells, T., Passant, N., Thompson, A., Cardenas, L. et al. 2008. UK greenhouse gas inventory, 1990 to 2006. Annual Report for submission under the Framework Convention on Climate Change. ISBN 0-9554823-4-2. http://www.airquality.co.uk/archive/reports/cat07/0804161424_ukghgi-90-06_main_chapters_UNFCCCsubmission_150408.pdf (accessed March 24, 2009).

Crews, T.E. and M.B. Peoples. 2004. Legume versus fertilizer sources of nitrogen: Ecological tradeoffs and human needs. *Agric. Ecosyst. Environ.* 102:279–297.

De Vries, M. and I.J.M. de Boer. 2010. Comparing environmental impacts for livestock products: A review of life cycle assessments. *Livest. Sci.* 128:1–11.

Del Grosso, S.J., Ogle, S., Wirth, J., and S. Skiles. 2008. U.S. agriculture and forestry greenhouse gas inventory: 1990–2005. United States Department of Agriculture Technical Bulletin 1921.

Doran, J.W. 1980. Soil microbial and biochemical changes associated with reduced tillage. *Soil Sci. Soc. Am. J.* 47:102–107.

Dubey, A. and R. Lal. 2009. Carbon footprint and sustainability of agricultural production systems in Punjab, India, and Ohio, USA. *J. Crop Improvement* 23:332–350.

Dunkley, C.S. and K.D. Dunkley. 2013. Greenhouse gas emissions from livestock and poultry. *Agric. Food Anal. Bacteriol.* 3(1):17–29.

Espinoza-Orias, N., Stichnothe, H., and A. Azapagic. 2011. The carbon footprint of bread. *Int. J. Life Cycl. Assess.* 16(4):351–365.

FAO. 2008. Climate change for fisheries and aquaculture. Technical Background Document on Climate Change, Energy and Food, FAO, Rome, HLC/08/BAK/6, p. 18.

FAO. 2010. Greenhouse gas emissions from the dairy sector: A life cycle assessment. Food and Agriculture Organization, Rome, Italy.

Fogelberg, C. and A. Carlsson-Kanyama. 2006. Environmental assessment of foods—An LCA inspired approach. Swedish Defence Agency (FOI).

Forster, P., Ramaswamy, V., Artaxo, P., Berntsen, T., Betts, R., Fahey, D.W., Haywood, J., Lean, J., Lowe, D.C., and G. Myhre. 2007. Changes in atmospheric constituents and in radiative forcing. In: Solomon, S.,

Qin, D., Manning, M., Chen, Z., Marquis, M., Averyt, K.B., Tignor, M. and M.L. Miller (Eds.), *Climate Change 2007: The Physical Science Basis. Contribution of Working Group I.* Fourth Assessment Report of the Intergovernmental Panel on Climate Change. Cambridge University Press, New York, pp. 129–234.

Gan, Y., Liang, C., Campbell, C.A., Zentner, R.P., Lemke, R.L., Wanga, H., and C. Yang. 2012a. Carbon footprint of spring wheat in response to fallow frequency and soil carbon changes over 25 years on the semiarid Canadian prairie. *Eur. J. Agron.* 43:175–184.

Gan, Y., Liang, C., Hamel, C., Cutforth, H., and H. Wang. 2011b. Strategies for reducing the carbon footprint of field crops for semiarid areas. A review. *Agron. Sust. Dev.* 31:643–656.

Gan, Y., Liang, C., Wang, X., and B. McConkey. 2011a. Lowering carbon footprint of durum wheat by diversifying cropping systems. *Field Crops Res.* 122:199–206.

Gan, Y.T., Liang, B.C., May, W., Malhi, S.S., Niu, J., and X. Wang. 2012b. Carbon footprint of spring barley in relation to preceding oilseeds and N fertilization. *Int. J. Life Cycl. Assess.* 17:635–645.

Gan, Y.T., Miller, P.R., McConkey, B.G., Zentner, R.P., Stevenson, F.C., and C.L. McDonald. 2003. Influence of diverse cropping sequences on durum wheat yield and protein in the semiarid northern Great Plains. *Agron. J.* 95:245–252.

Ghosh, S., Hanumantha Rao, M.V., Satish Kumar, M., Mahesh, V.U., Muktha, M., and P.U. Zacharia. 2014. Carbon footprint of marine fisheries: Life cycle analysis from Visakhapatnam. *Curr. Sci.* 107(3):515–521.

Hamerschlag, K. 2011. *A Meat Eater's Guide to Climate Change + Health: What You Eat Matters.* Environmental Working Group, 2011.

Hermansen, J.E. and T. Kristensen. 2011. Management options to reduce the carbon footprint of livestock products. *Anim. Frontiers* 1(1):33–39.

Hilier, J., Hawes, C., Squire, G., Hilton, A., Wale, S., and P. Smith. 2009. The carbon footprints of food crop production. *Int. J. Agric. Sustain.* 7(2):107–118.

Ho, J.A. 2011. Calculation of the carbon footprint of Ontario wheat. *Stud. Undergraduate Res. Guelph* 4(2):49–55.

Hospido, A. and P. Tydemers. 2005. Life cycle environmental impacts of Spanish tuna fisheries. *Fish. Res.* 76:174–186.

Hutchinson, J.J., Campbell, C.A., and R.L. Desjardins. 2007. Some perspectives on carbon sequestration in agriculture. *Agric. For. Meteorol.* 142:288–302.

INCCA (Indian Network for Climate Change Assessment). 2010. Assessment of the greenhouse gas emission: 2007. The Ministry of Environment & Forests, Govt. of India, New Delhi.

IPCC. 2006. Guidelines for national greenhouse gas inventories. In: Eggleston, H.S., Buendia, L., Miwa, K., Ngara, T., and K. Tanabe (Eds.), *National Greenhouse Gas Inventories Programme.* IGES, Japan.

IPCC. 2007. The physical science basis. In: Solomon, S., Qin, D., Manning, M., Chen, Z., Marquis, M., Averyt, K.B., Tignor, M., and H.L. Miller (Eds.), *Climate Change (2007) Contribution of Working Group I.* Fourth Assessment Report of the Intergovernmental Panel on Climate Change, Cambridge University Press, Cambridge, UK and New York, NY, USA.

IPCC. 2014. Summary for policymakers. In: Edenhofer, O., Pichs-Madruga, R., Sokona, Y., Farahani, E., Kadner, S., Seyboth, K., Adler, A., Baum, I., Brunner, S., Eickemeier, P., Kriemann, B., Savolainen, J., Schlomer, S., von Stechow, C., Zwickel, T., and J.C. Minx (Eds.), *Climate Change 2014, Mitigation of Climate Change.* Contribution of Working Group III to the Fifth Assessment Report of the Intergovernmental Panel on Climate Change. Cambridge University Press, Cambridge, UK and New York, NY, USA.

ITC. 2012. Product carbon footprinting standards in the agri-food sector. Technical paper. International Trade Centre, Geneva, xiii, 46 pp.

Jacob, J. 2009. Poultry's carbon footprint. *Cheeps Chirps* 2(4):5–8.

Jensen, E.S., Peoples, M.B., Boddey, R.M., Gresshoff, P.M., Haugga ard-Nielsen, H., Alves, B.J.R., and M.J. Morrison. 2011. Legumes for mitigation of climate change and the provision of feedstock for biofuels and bio-refineries. A review. *Agron. Sustain. Dev.* 32(2):329–364.

Katyal, J.C., Rao, N.H., and M.N. Reddy. 2001. Critical aspects of organic matter management in the tropics - Example India. *Nutr. Cycl. Agroecosyst.* 93:77–88.

Kim, B. and R. Neff. 2009. Measurement and communication of greenhouse gas emissions from U.S. food consumption via carbon calculators. *Ecol. Econ.* 69:186–196.

Koerber, G.R., Jones, G.E., Hill, P.W., Milài Canals, L., Nyeko, P., York, E.H., and D.L. Jones. 2009. Geographical variation in carbon dioxide fluxes from soils in agro-ecosystems and its implications for life-cycle assessment. *J. Appl. Ecol.* 46(2):306–314.

Lal, R. 1999. Soil management and restoration for carbon sequestration to mitigate the accelerated greenhouse effect. *Prog. Environ. Sci.* 1:307–326.

Lal, R. 2004. Carbon emission from farm operations. *Environ. Int.* 30:981–990.

Lu, W.F., Chen, W., Duan, B.W., Guo, W.M., Lu, Y., Lantin, R.S., Wassmann, R., and H.U. Neue. 2000. Methane emission and mitigation options in irrigated rice fields in southeast China. *Nutr. Cycl. Agroecosyst.* 58:65–74.

Lynas, M. 2007. *Carbon Counter.* Harper Collins Publishers, Glasgow, UK.

Ma, B.L., Liang, B.C., Biswas, D.K., Morrison, M.J., and N.B. McLaughlin. 2012. The carbon footprint of maize production as affected by nitrogen fertilizer and maize-legume rotations. *Nutr. Cycl. Agroecosyst.* 94:15–31.

MacLeod, M., Gerber, P., Mottet, A., Tempio, G., Falcucci, A., Opio, C., Vellinga, T., Henderson, B., and H. Steinfeld. 2013. Greenhouse gas emissions from pig and chicken supply chains—A global life cycle assessment. Food and Agriculture Organization of the United Nations (FAO), Rome.

Malhi, S.S., Grant, C.A., Johnston, A.M., and K.S. Gill. 2001. Nitrogen fertilization management for no-till cereal production in the Canadian Great Plains: A review. *Soil Till. Res.* 60:101–122.

Mandal, B., Majumder, B., Adhya, T.K., Bandyopadhyay, P.K., Gangopadhyay, A., Sarkar, D., Kundu, M.C. et al. 2008. Potential of double-cropped rice ecology to conserve organic carbon under subtropical climate. *Glob. Change Biol.* 14:2139–2151.

Mandal, B., Majumder, B., Bandyopadhyay, P.K., Hazra, G.C., Gangopadhyay, A., Samantarayz, R.N., Mishraz, A., Chaudhury, J., Saha, M.N., and S. Kundu. 2007. The potential of cropping systems and soil amendments for carbon sequestration in soils under long-term experiments in subtropical India. *Glob. Change Biol.* 13:357–369.

MoA. 2008. Agricultural statistics at a glance 2008. Directorate of Economics and Statistics, Department of Agriculture and Cooperation (DAC), Ministry of Agriculture, Government of India, New Delhi.

Nishimura, S., Yonemura, S., Sawamoto, S., Shirato, Y., Akiyama, H., Sudo, S., and K. Yagi. 2008. Effect of land use change from paddy rice cultivation to upland crop cultivation on soil carbon budget of a cropland in Japan. *Agric. Ecosyst. Environ.* 125:9–20.

Padgett, J.P., Steinemann, A.C., Clarke, J.H., and M.P. Vandenbergh. 2008. A comparison of carbon calculators. *Environ. Impact Assess. Rev.* 28:106–115.

Pandey, D. and M. Agrawal. 2014. Carbon footprint estimation in the agriculture sector. In: Muthu, S.S. (Ed.), *Assessment of Carbon Footprint in Different Industrial Sectors*, Vol. 1. Springer, Singapore, pp. 25–47.

Pathak, H., Agarwal, T., and N. Jain. 2012c. Greenhouse gas emission from rice and wheat systems: A life-cycle assessment. In: Pathak, H. and P.K. Aggarwal (Eds.), *Low Carbon Technologies for Agriculture: A Study on Rice and Wheat Systems in the Indo-Gangetic Plains*, Venus printers & publishers, New Delhi, pp. 41–53.

Pathak, H., Bhatia, A., and B. Chakrabarti. 2011. Nitrogen management in agriculture for climate change mitigation. In: Jain, V. and Ananda Kumar (Eds.), *Nitrogen Use Efficiency in Plants*. New India Publishing Agency, New Delhi.

Pathak, H., Chakrabarti, B., Bhatia, A., Jain, N., and P.K. Aggarwal. 2012a. Potential and cost of low carbon technologies in rice and wheat systems: A case study for the Indo-Gangetic Plains. In: Pathak, H. and P.K. Aggarwal (Eds.), *Low Carbon Technologies for Agriculture: A Study on Rice and Wheat Systems in the Indo-Gangetic Plains*, Venus printers & publishers, New Delhi, pp. 12–40.

Pathak, H., Jain, N., Bhatia, A., Patel, J., and P.K. Aggarwal. 2010. Carbon footprints of Indian food items. *Agric. Ecosyst. Environ.* 139:66–73.

Pathak, H., Sankhyan, S., Dubey, D.S., Harit, R.C., Tomar, R., Jain, N., and A. Bhatia. 2012b. Low carbon technologies in agriculture: On-farm and simulation studies on direct-seeded rice. In: Pathak, H. and P.K. Aggarwal (Eds.), *Low Carbon Technologies for Agriculture: A Study on Rice and Wheat Systems in the Indo-Gangetic Plains*, Venus printers & publishers, New Delhi, pp. 41–53.

Peng, S., Buresh, R.J., Huang, J., Zhong, X., Zou, Y., Yang, J., Wang, G. et al. 2010. Improving nitrogen fertilization in rice by site-specific N management—A review. *Agron. Sustain. Dev.* 30, 649–656. doi:10.1051/agro/2010002.

Prasad, R. and J.F. Power. 1995. Nitrification inhibitors for agriculture, health and the environment. *Adv. Agron.* 54:233–281.

Rab, M.A., Fisher, P.D., and N.J. O'Halloran. 2008. Preliminary estimation of the carbon footprint of the Australian vegetable industry. Horticulture Australia Limited (HAL).

Rees, W.E. 1992. Ecological footprints and appropriated carrying capacity: What urban economics leaves out? *Environ. Urbanis.* 4:121–130.

Roos, E., Sundberg, C., and P.A. Hansson. 2010. Uncertainties in the carbon footprint of food products: A case study on table potatoes. *Int. J. Life Cycl. Assess.* 15:478–488.

Rose, S.K. and B.A. McCarl. 2008. Greenhouse gas emissions, stabilization and the inevitability of adaptation: Challenges for U.S. agriculture. *Choices: The Magazine of Food, Farm & Resource Issues* 23:23–18.

Saunders, C., Barber, A., and G. Taylor. 2006. Food Miles—Comparative energy/emissions performance of New Zealand's agriculture industry. The Agribusiness and Economics Research Unit (AERU), Lincoln University.

Sieling, K. and H. Kage. 2010. Efficient N management using winter oilseed rape—A review. *Agron. Sustain. Dev.* 30:271–279. doi:10.1051/agro/2009036.

Smith, D.W. 2014. Contribution of greenhouse gas emissions: Animal agriculture in perspective. Animal Agriculture & Climate Change. Department of Biological & Agricultural Engineering, Texas A&M University.

Steinfeld, H., Gerber, P., Wassenaar, T., Castel, V., Rosales, M., and C. Haan. 2011. Livestock's long shadow: Environmental issues and options. Food and Agriculture Organization of the United Nations.

Stoessel, F., Juraske, R., Pfister, S., and S. Hellweg. 2012. Life cycle inventory and carbon and water foot print of fruits and vegetables: Application to a Swiss retailer. *Environ. Sci. Technol.* 46:3253–3262.

Tan, R.R. and A.B. Culaba. 2009. Estimating the carbon footprint of tuna fisheries. Center for Engineering and Sustainable Development Research De La Salle University, Manila, Philippines.

Tukker, A., Huppes, G., Jeroen Guinée, Heijungs, R., Koning, A.D., Oers, L.V., and S. Suh. 2006. Environmental Impacts of Products (EIPRO): Analysis of the life cycle environmental impacts related to the final consumption of the EU-25. Institute for Prospective Technological Studies, European Science and Technology Observatory, Spain.

Tyedmers, P.H., Watson, R., and D. Pauly. 2005. Fueling global fishing fleets. *Ambio* 34(8):635–638.

Tzilivakis, J., Warner, D.J., May, M., Lewis, K.A., and K. Jaggard. 2005. An assessment of the energy inputs and greenhouse gas emissions in sugar beet (*Beta vulgaris*) production in the UK. *Agric. Syst.* 85:101–119.

UNDP. 2007. Human development report 2007/2008: Fighting climate change: Human solidarity in a divided world. Palgrave Macmillan, New York.

Vergé, X.P.C., Dyer, J.A., Desjardins, R.L., and D. Worth. 2008. Greenhouse gas emissions from the Canadian beef industry. *Agric. Syst.* 98:126–134.

Vergé, X.P.C., Maxime, D., Dyer, J.A., Desjardins, R.L., Arcand, Y., and A. Vanderzaag. 2013. Carbon footprint of Canadian dairy products: Calculations and issues. *J. Dairy Sci.* 96:6091–6104.

Vivekanandan, E., Singh, V.V., and J.K. Kizhakudan. 2013. Carbon footprint by marine fishing boats of India. *Curr. Sci.* 105(3):361–366.

Wassmann, R., Buendia, L.V., Lantin, R.S., Bueno, C., Lubigan, L.A., Umali, A., Nocon, N.N., Javellana, A.M., and H.U. Neue. 2000. Mechanism of crop management on methane emissions from rice fields in Los Banos, Philippines. *Nutr. Cycl. Agroecosyst.* 58:107–119.

Weidema, B.P., Wesnæs, M., Hermansen, J., Kristensen, T., Halberg, N., Eder, P., and Delgado, L. 2008. Environmental improvement potentials of meat and dairy products. Institute for Prospective Technological Studies, Sevilla, Spain (EUR23491 EN).

Wiedmann, T. and J. Minx. 2008. A definition of 'carbon footprint'. In: Pertsova, C.C. (Ed.), *Ecological Economics Research Trends*, Chapter 1. Nova Science Publishers, Hauppauge, NY, USA, pp. 1–11.

Williams, A.G., Audsley, E., and D.L. Sandars. 2010. Environmental burdens of producing bread wheat, oilseed rape and potatoes in England and Wales using simulation and system modeling. *Int. J. Life Cycl. Assess.* 15:855–868.

Williams, H. and F. Wikstrom. 2011. Environmental impact of packaging and food losses in a life cycle perspective: A comparative analysis of five food items. *J. Cleaner Prod.* 19:43–48.

Wilson, J.D.K. 1999. Fuel and financial savings for operators of small fishing vessels. *FAO Fish. Tech. Pap.* 383:46.

Xu, X., Zhang, B., Liu, Y., Xue, Y., and B. Di. 2013. Carbon footprints of rice production in five typical rice districts in China. *Acta Ecol. Sin.* 33:227–232.

20 The Carbon Footprint of Sugar Production in Eastern Batangas, Philippines

Teodoro C. Mendoza, Rex B. Demafelis, Anna Elaine
D. Matanguihan, Justine Allen S. Malabuyoc, Richard
V. Magadia Jr., Amabelle A. Pector, Klarenz A. Hourani,
Lavinia Marie A. Manaig, and Jovita L. Movillon

CONTENTS

20.1 Introduction 452
20.2 Methodology 452
    20.2.1 Goal, Scope, and Systems Boundary 452
    20.2.2 Source of Data 453
    20.2.3 Calculating the Carbon Footprint of Sugar Production 455
        20.2.3.1 Sugarcane Production at the Farm Level 455
        20.2.3.2 Sugarcane Milling 456
        20.2.3.3 Carbon Footprint of Factory Construction 457
        20.2.3.4 Carbon Emission from Hauling 457
        20.2.3.5 Carbon Emission from Factory Operation and Products 458
        20.2.3.6 Carbon Emission from Material Inputs 459
        20.2.3.7 Carbon Emission from Wastewater Treatment Facility 459
    20.2.4 Calculating Sequestered CO_2 in Sugarcane Production 460
        20.2.4.1 Net CO_2 Sequestration/Emissions 461
20.3 Results 461
    20.3.1 The Carbon Footprint of Sugarcane Production at the Farm Level 461
        20.3.1.1 The Associated GHG Emission (Expressed in CO_2 Equivalence)
            in Cane Burning 462
    20.3.2 Carbon Sequestered 463
    20.3.3 The Carbon Footprint of Sugarcane Milling 464
        20.3.3.1 Carbon Emission of Factory Construction 464
        20.3.3.2 Carbon Emission from Hauling 464
        20.3.3.3 Carbon Emission from Milling Operation 464
        20.3.3.4 Carbon Emission from Material Inputs 466
        20.3.3.5 Carbon Emission from Wastewater Treatment Facility 466
    20.3.4 Total Carbon Footprint of Raw Sugar Production (Sugarcane Production and
        Sugarcane Milling) 467
    20.3.5 Payback Period 467
20.4 Discussions, Conclusion, and Recommendations 468
    20.4.1 Field Level 468
    20.4.2 Factory Level 469
References 470

20.1   INTRODUCTION

Sugar production (cane production in the field and sugar processing in the factory) is an intensive energy-requiring process (Corpuz and Aguilar 1992; Mendoza and Samson 2000; Mendoza et al. 2004). All of the operations, processes, and inputs that are used, directly and indirectly, involve burning of a large amount of fossil fuel. Growing and hauling canes to the mill use machines (tractors, trucks) that burn oil. The manufacture and transport of agrochemical inputs (fertilizer, herbicide) also use fossil fuel (natural gas). Certain field practices such as cane burning directly emit CO_2 and other greenhouse gases (GHG) (e.g., methane, carbon monoxide, nitrous oxide), which have powerful global warming potential (GWP) relative to CO_2 (Weier 1998; Mendoza and Samson 2000). Factory operations for converting sugarcane to raw sugar and its co-products consume electricity that is sourced from burning certain fuels, which, in turn, contribute to carbon emission. More so, the construction of the whole factory has equivalent carbon emission. Collectively, when all these GHG are added together, they represent the carbon footprint (CF) of sugar. Does sugar production from sugarcane have a net carbon emission?

These GHGs cause climate change. There is now overwhelming evidence that human activities emit GHGs that cause climate change (IPCC 2007). Although the Philippines contributes only a small fraction (0.27%) of the global GHG emissions (Godilano 2009), it is still important to find ways to reduce emissions. Since the Philippines is a major producer of sugar from sugarcane, it is important to determine whether sugar production is carbon neutral or net contributory to carbon emission. If it is contributory, then proper measures should be implemented to address the generation of the carbon emissions in sugar production. As defined, carbon footprint (CF) is the total amount of carbon dioxide (CO_2) and other greenhouse gas (GHG) emissions (e.g., methane, CO, N_2O) associated with a product or activity causing climate change (Wiedmann and Minx 2008; Walser et al. 2010). The cost of CF in society is called the social cost of carbon (SCC). SCC includes the net effects or impacts on the economies and societies of the long-term trends in climate conditions, including the extreme events, related to these anthropogenic emissions of GHGs (Downing et al. 2005; Parry et al. 2007). SCC is typically expressed in US dollar per metric ton of carbon (Tol and Richard 2008).

Quantifying the carbon footprint of a product such as sugar can be used as the basis for reducing its GHG emission (Wiedmann and Minx 2008). The so-called hot spots in terms of energy consumption and the associated CO_2 emission in the production cycle can be identified. The recent study by Mendoza (2014) calculated the carbon footprint of sugar production at the farm level only. There was no detailed calculation from the processing or sugar manufacture in the factory. Hence, this paper introduced the procedure in calculating the carbon footprint for each operation and inputs used in sugar production starting from the cane production in the farm up to processing in the factory to produce raw sugar. Carbon footprint is expressed in terms of per ha, per ton cane, and per kg sugar.

20.2   METHODOLOGY

20.2.1   GOAL, SCOPE, AND SYSTEMS BOUNDARY

The goal of the paper is to calculate the carbon footprint of raw sugar production starting from sugarcane production to the manufacture and use of products (cradle-to-grave). In the farm level cane production (as summarized in Figure 20.1), two crop types were involved: plant cane and ratoon crop. The associated operations for plant crop were delineated as follows: (1) land preparation–plowing, harrowing, furrowing; (2) planting–cane point preparation, hauling, distribution, planting; (3) cultivation–ridge busting, off-barring, hilling-up; (4) application of fertilizer and other chemicals; and (5) harvesting and hauling of canes. For ratoon crop, since ratoon crop starts with what is left in the field after the harvest of a plant crop, only the data in numbers 3, 4, and 5 were considered.

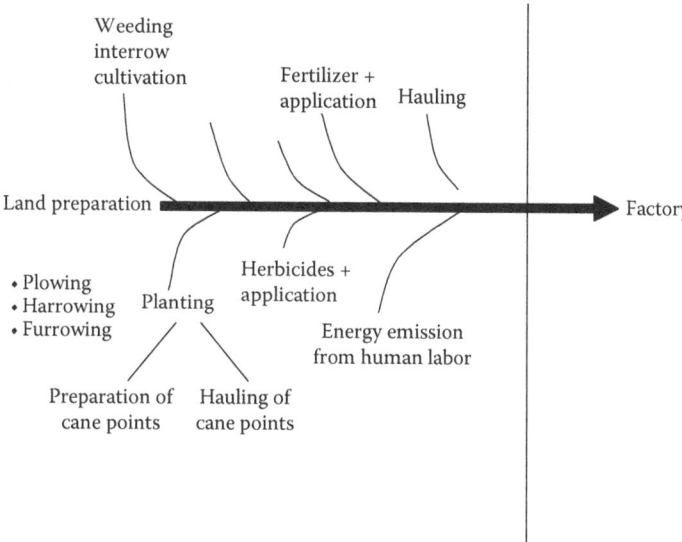

FIGURE 20.1 The different operations involved in sugarcane production that contribute to the total carbon footprint.

For the manufacture of the final product, the main areas of raw sugar life cycle were considered for carbon inventory, namely: factory construction, operations, and processes including power generation and wastewater treatment. Figure 20.2 shows the sources of carbon emissions, represented by the dark gray circles; the light gray circles denote the supposed emission savings due to the capacity of sugarcane plant to sequester carbon dioxide from the atmosphere during photosynthesis and the utilization of bagasse as fuel instead of fossil fuel for power generation. On top of the required supply of electricity for the factory, the study also included savings that could be realized by co-generation (i.e., excess bagasse is used to produce electricity that could be sold to the grid). The savings were used in calculating the payback period or the length of time it will take to recover the construction emissions of the factory. The biogas recovered from the wastewater treatment plant was not utilized in this assessment. It was assumed that this could be sold to other industries. It could replace natural gas, which basically contains the same chemicals, thus, no comparative savings was calculated. Also, the amount of commercial fertilizer that could be replaced by using the mud and mill ash produced in the milling process was not considered, as most sugarcane planters do not use these on a commercial scale.

20.2.2 SOURCE OF DATA

For sugarcane production, the primary field survey data obtained by Mendoza et al. (2007) from Eastern Batangas Province, Philippines were used. These data were collected by interviewing the sugarcane planter-members of the Batangas Integrated Sugarcane Planters Multipurpose Cooperative (BISPMC).

For the factory, operations, processes, and ranges of values established in the *Handbook of Cane Sugar Engineering by Hugot* (1986) served as this work's primary guide. The chief data were sourced from practicum reports (Busco Sugar Milling Co., Inc., Cagayan Robina Sugar Milling Company, and Davao Sugar Central Co., Inc.), online journals, chemical engineering books (Foust 1960; Peters and Timmerhaus 1991; Perry and Green 2008), and actual field experience. Calculations were based on an existing operational sugar factory, San Carlos Bioenergy, Inc. (SCBI), Negros Occidental, Philippines.

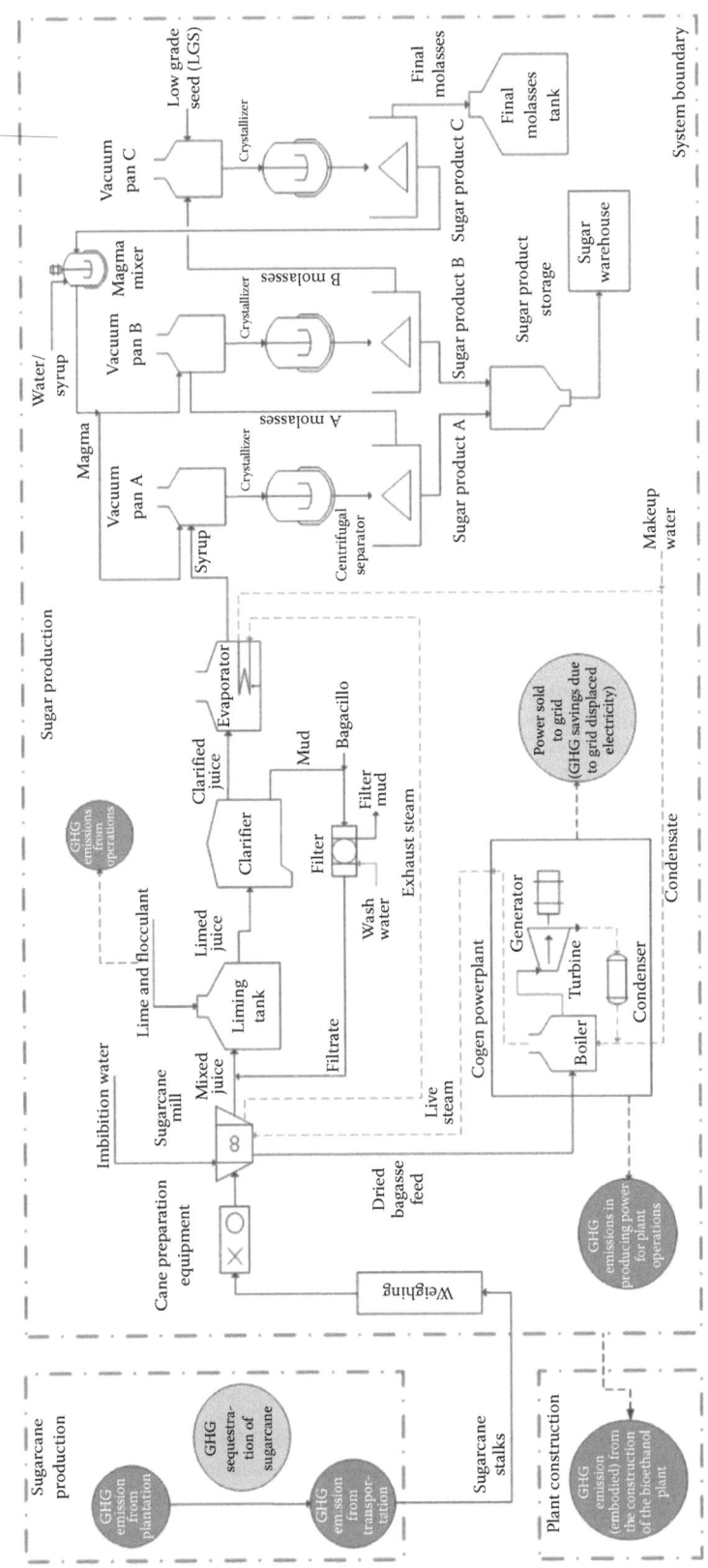

FIGURE 20.2   Carbon sources for raw sugar production.

20.2.3 Calculating the Carbon Footprint of Sugar Production

The GHG emissions accumulated in the entire process were standardized to equivalent carbon dioxide (CO_2-eq.) both in sugarcane production at the farm level and sugar manufacture.

20.2.3.1 Sugarcane Production at the Farm Level

The calculations of the carbon footprint (CF) of sugarcane production at the farm level were adopted from Mendoza (2014) as follows: First, the energy used in cane production (land preparation to harvesting), including the inputs—embedded energy of machines up to transport of millable stalks to the factory (Figure 20.1)—was estimated and converted into common energy expression as liter diesel oil equivalent (LDOE) (Corpuz and Aguilar 1992). Appropriate energy values for the various field operations and inputs (fertilizer, NPK) were obtained from Pimentel (1980), Pimentel et al. (1983), and Panesar and Fluck (1993) as cited by Mendoza et al. (2004) and Mudahar and Hignett (1985). The energy usage was converted into liter diesel oil equivalent (LDOE) as calculated by Mendoza et al. (2007). The carbon foot print (CF) was computed by multiplying the LDOE/unit input or operation × 3.96 g CO_2/LDOE. The CO_2 emission from 1 L diesel oil = 3.96 kg CO_2/L. It is the sum of the direct (3.332 kg CO_2/L) (IPCC 2001) and the indirect emission (0.63 kg CO_2/L) (Pimentel 1980).

Nitrogen fertilizer consumed a high amount of energy to manufacture. Application in the field emits CO_2. A separate estimate for the CF of N was done as follows:

$$N_e \ ha^{-1} = N_e \times N_{ha}$$

where $N_e \ ha^{-1}$ is CO_2 emission for N applied per ha, N_e the CO_2 emission from 1 kg N applied, N_{ha} the nitrogen applied per ha, and $N_e = (E_N \times C_e do) + N_2 O_e$, where N_e is the CO_2 emission from 1 kg N, E_N the Li diesel oil equivalent in the manufacture of N+ transport to the Philippines (2.04 LDOE + 0.11 LDOE, 2.15 LDOE/kg N^{-1}, Mendoza et al. 2007); $C_e do$ is carbon dioxide emission from diesel oil; $N_2 O_e$ is $N_2 O$ emission from applied nitrogen, where the $N_2 O$ emission from applied N is 1.5% (IPCC 2001).

The $N_2 O$ emission was adjusted to its global warming potential (GWP-100 years) relative to carbon dioxide. The global warming potential (GWP) of $N_2 O = 298$ (IPCC 2007).

Then, the carbon footprint of sugarcane production (C_e scp) was estimated using the following formula:

$$C_e scp - C_e ce + C_e cc + C_e i + C_e m$$

where $C_e scp$ is the carbon emission from sugarcane production; $C_e ce$, the carbon emission from crop establishment; $C_e cc$, carbon emission from crop care to harvesting and transport of canes to the factory; $C_e i$, carbon emission from inputs i.e., fertilizer, herbicides, cane points; $C_e m$, carbon emission from the manufacture of machines (tractors and implements: hauling trucks).

Sugarcane production involved pre- and postharvest burning of canes. Preharvest burning is done to facilitate cutting and loading of stalk for hauling to the sugarcane mills. Postharvest burning is done to clean the fields and facilitate ratooning operations. Crop residue burning leads to direct and indirect CO_2 emissions (Mendoza 2014). The CO_2 emission from sugarcane crop residues (dried leaves + tops) included: (1) direct CO_2 emission from residues burning and (2) emission from other gases (CH_4, CO, $N_2 O$):

1. Direct CO_2 emission from residues burning was estimated as follows:

$$CO_{2escrb} = Q_{scrb} * [F_b \times F_f \times C_f] \times CO_2\text{-}C$$

where CO_{2esrb} denotes direct CO_2 emission from sugarcane residue burning, Q_{srb} the quantity of sugarcane residues burned (tops and leaves), F_b the fraction of residues burned in the field, 0.65 (Mendoza and Samson 2000), F_f the fraction fully oxidized, 0.90 (IPCC 1996), C_f the carbon fraction in residues, 0.4235 (IPCC 2000), CO_2–C 3.7 (44/12).

2. Emission from other gases (CH_4, CO, N_2O)
    a. Methane

$$CO_2\text{–}CH_4 = CH_4C \times GWP\ CH_4\ (t\ CO_2/ha)$$

where CO_2–CH_4 is the carbon dioxide equivalent from methane emission from burning of sugarcane residues, CH_4C the methane emission from burning sugarcane trash in ton C/ha * CR, emissions in ton C/ha, ER * Total C Released (tC/ha year), ER the emission ratio, 0.005 (IPCC 1996), CR the conversion ratio, 1.33 (IPCC 1996), and GWP CH_4 (100 years), 25 (IPCC 2007).

    b. Carbon monoxide

$$CO_2\text{–}CO = CO_e \times GWP\ CO\ (t\ CO_2/ha)$$

where CO_2–CO stands for carbon dioxide emission equivalent from carbon monoxide emission from burning sugarcane residues (t/ha), CO_e is the carbon monoxide emission from burning (t/ha), emission in t C/ha × conversion ratio.

$$Emission\ in\ t\ C/ha = ER \times Total\ C\ released\ (tC/ha)$$

where ER (= 0.06) is the emission ratio (IPCC 1996), CR (= 2.3) is the conversion ratio (IPCC 1996), and GWP-CO (= 3.2) is the global warming potential of carbon monoxide (100 years) (IPCC 2001).

    c. Nitrous oxide

$$N_2O\text{–}CO_2 = Ec \times GWP\ N_2O\ (t\ CO_2/ha)$$

where Ec (= Ne × CR) is emissions in t C, CR (= 1.6) is the conversion ratio (IPCC 1996), Ne (= Tb × Nf × ER) is the emission in N/ha, Tb (= 6.65) is the total biomass burned (t/ha), Nf (= 0.004) is the N fraction in biomass (Yadav 1995), and ER (= 0.007) is the emission ratio (IPCC 1996).

20.2.3.2 Sugarcane Milling

The computation of the total carbon emission in milling ($C_e$scf) involved a comprehensive design of the processing plant and computations of material and energy balance, based on a capacity of 4000 tons per day sugarcane stalks, operating for 270 days per year (off-milling season already reflected), 24 hours per day. The balances determined the requirements on input materials and flow rates for each unit operation for equipment sizing and plant layout. The raw sugar production was divided into several distinct operations and processes, namely: cane preparation and milling, heating and clarification, evaporation, crystallization, centrifugation, and steam generation. Four-mill tandem was used in the design. Pol extraction was 92.5%. The extracted juice was clarified using the hot liming method to produce a clarified juice of 83% apparent purity. Afterwards, the clarified juice was fed to the quadruple-effect evaporator to produce syrup with 65°Brix. The design yields 2.05 L-kg raw sugar per ton cane (1 L-kg = 50 kg).

The wastewater generated from the process was treated before discharge; some was reused. Treatment was done first in an aerobic digester, then followed by a facultative lagoon. The plant generated 2960 L of wastewater per ton cane crushed (Patwardhan 2008). Initially, the wastewater was treated in the aerobic digester for 10 days that removed 92% of COD. Until such time, it was released to the 30-acre facultative lagoon with COD and BOD loading of 234 and 117 kg ha^{-1} day^{-1}, respectively (0.5 COD/BOD). At 5 days detention time, a 95% BOD conversion was attained and produced a final BOD output of 6 mg/L, qualifying for a Class C water quality (Philippine standard) (DENR 2008) or that which can be used for irrigation. The conversion produced an equivalent amount of 400 L methane per kg COD removed (92% efficiency).

The size and specifications of the equipment were based on the supposed capacity established in the material balance. Final report on the total carbon emission for the production of raw sugar included the emissions from transportation routines, power consumption of the plant during operation, water treatment facility, and input chemicals and other supplies (Figure 20.2). The CO_2 emitted during the construction of the factory was considered as debt.

20.2.3.3 Carbon Footprint of Factory Construction

The carbon footprint computations from construction of the factory were based on the equipment and structural designs, and the embodied CO_2 emissions of the materials used. It involved weight calculations of the equipment, facility, and additional footing.

$$C_e \text{ construction} = C_e (\text{Facility, equipment, additional footing})$$
$$= \text{Weight of the materials} \times \text{Carbon emission factor}$$

The carbon emission factors (mentioned in the above equation) for construction materials used to build the plants are summarized in Table 20.1. These data were obtained from the Inventory of Carbon and Energy (ICE) by Hammond and Jones (2011) as cited by Greenspec UK. Essentially, the term "embodied carbon" of a material translates to the total CO_2 released over its life cycle.

The sugar factory was divided into several divisions, namely: cane supply and transport; milling area; boiling house; power plant; wastewater treatment facility; and other facilities. The carbon emission from construction relied on the life-cycle analysis of the materials used for building the facilities, manufacture of equipment, and assembly, including base support, pipes and pumps, and additional footing. The bill of materials for the construction of the facility included the major components: roofing system; wall and framings; flooring; beams and girders; staircases, hand railings, and bracings.

20.2.3.4 Carbon Emission from Hauling

Calculating GHG emission from carrying the sugarcane stalks from the site entrance up to the unloading area considered that the truck has a hauling capacity of 10 tons with fuel economy of

TABLE 20.1
Embodied Carbon Dioxide Emission of Materials for Plant Construction

Material	Embodied CO_2 Emission (kg-CO_2 kg-Material^{-1})
Concrete	0.16
Stainless steel	6.15
Steel	1.37
Cast iron	1.91

Source: Adapted from Hammond, G. and C. Jones (2011), cited by Greenspec UK.

TABLE 20.2
GHG Emission Factors for Overall Plant Operation

Chemicals or Reagents	CO_2 Emission (kg-CO_2 kg-material^{-1})
Lime (CaO)[a]	1.0302
Fuels	**CO_2 Emission (kg-CO_2 L^{-1})**
Diesel[b]	3.96
Gasoline[c]	2.35
Ethanol[d]	1.51
Power Generating Feedstocks	**CO_2 Emission**
Coal[e]	0.534 kg-CO_2 kWh^{-1}
Biogas[f]	0.25 kg-CO_2 kWh^{-1}
Bagasse[g]	0.522 kg-CO_2 kWh^{-1}
Bunker oil[h]	0.778 kg-CO_2 kWh^{-1}

[a] Biograce GHG Calculation Tool Standard Values.
[b] Mendoza (2014).
[c] US EPA (2013) United States Environmental Protection Agency.
[d] Derived value.
[e] IPCC Carbon Dioxide Intensity of Electricity—Philippines.
[f] Clark (2013).
[g] Derived (US EPA, 1993).
[h] BHP Billington (2011).

2.126 km per L diesel (Greenhouse gases, Regulated Emissions, and Energy use in Transportation (GREET), 2013). From the gate to the cane unloading area, the distance to be travelled by the truck was assumed to be 60 m.

$$C_e \text{ hauling} = \text{Weight of usage of fuel} \times \text{Carbon emission factor}$$
$$= \frac{(\text{Distance traveled})(\text{Number of travel})(\text{Emission factor})}{(\text{Fuel consumption})}$$

Carbon emission factors (mentioned in the above equation) for fuel consumption in hauling are summarized in Table 20.2.

20.2.3.5 Carbon Emission from Factory Operation and Products

The carbon footprint from operation and production was estimated from the emissions of fuels in power generation and in the end use of products of the factory (raw sugar and molasses). The factory has its own facility that generates power from the bagasse, thereby accounting most of the emissions in bagasse fueling. However, additional fuel combustion from bunker oil was added to start the milling operation since bagasse was not yet available during the start up.

$$\text{Ce plant operation} = (\text{Power requirement of equipment} + \text{Excess power generated})$$
$$\times \text{Carbon emission factor} + \Sigma(X \times Cfx \times CF)$$

where x is the amount of raw sugar or molasses (raw sugar = 10.25% of cane; molasses = 3.3%); carbon fraction (Cf) in raw sugar = 0.4105; Cf in molasses = 0.2315; carbon emission factors for power depends on the feedstock used to produce the power, as summarized in Table 20.2.

20.2.3.6   Carbon Emission from Material Inputs

Like fertilizer for the plantation, some unit operations require chemical inputs to proceed with the reaction. In raw sugar production, the mixed juice extracted from the sugarcane stalks are clarified using hot lime (CaO) solution ($Ca(OH)_2$).

$$C_e \text{ input materials} = \text{Weight of usage of chemicals} * \text{Carbon emission factor}$$

Carbon emission factors (mentioned in the above equation) for the use of chemicals/reagents are summarized in Table 20.2.

20.2.3.7   Carbon Emission from Wastewater Treatment Facility

Carbon emission from the wastewater treatment facility comes from the aerobic facultative lagoon, which is an open area. The biogas produced consists of 65% (v/v) methane, 5% hydrogen sulfide (H_2S), and 30% CO_2. The values are then converted to equivalent carbon dioxide.

$$C_e \text{ wastewater treatment} = \Sigma(\text{Weight of pollutant produced} * \text{Carbon emission factor})$$

To sum it all up, the carbon dioxide emission from sugarcane milling is given in the following equation:

$$C_e \text{ scf} = C_e(\text{contruction} + \text{hauling} + \text{plant operation} + \text{material inputs} + \text{wastewater treatment facility})$$

where C_e denotes potential CO_2 emission.

For reference, shown in Table 20.3 are the flows of materials during milling; moreover, shown in Table 20.4 are the dimensions of some equipment, the material used to construct the equipment, and the equivalent CO_2 emission.

TABLE 20.3
Material Balance for Milling Section

Streams	Description	Components	%, w/w	kg/Day
IN	Cane stalks	Fiber	13.97	558,800.00
		Ash	0.90	36,000.00
		Water	70.48	2,819,200.00
		Sugar	12.00	480,000.00
		Nonsugar	2.65	106,000.00
		Total		**4,000,000.00**
IN	Maceration	Water	100	**1,229,360.00**
OUT	Final bagasse	Fiber	46.15	553,800.00
		Ash	0.800	9,600.00
		Water	49.00	588,000.00
		Sugar	3.00	36,000.00
		Nonsugar	1.050	12,600.00
		Total		**1,200,000.00**
OUT	Raw mixed juice	Fiber	0.12	5,000.00
		Ash	0.66	26,400.00
		Water	85.88	3,460,560.00
		Sugar	11.02	444,000.00
		Nonsugar	2.32	93,400.00
		Total		**4,029,360.00**

TABLE 20.4
Equipment Size, Total Weight, and Equivalent Carbon Dioxide Emission

Equipment for Clarification

Equipment	Capacity m³ h⁻¹	Dimension Diameter (m)	Height/ Length (m)	Construction Material	# of Units	Weight (Tons)	Potential CO₂ Emission (Tons)
MOL prep stirrer	0.89	1.50	1.67	c	1	0.464	0.886
MOL tank	0.89	2.78	9.71	c	2	4.928	18.824
Juice heater	157.95	1.03	5.59	g	1	7.146	13.648
Flash tank	175.86	5.00	5.50	g	1	7.024	13.416
Liming tank (clarifier)	176.85	5.85	6.44	g	1	8.237	15.732
Rotary vacuum filter	6.77	4.27	5.5	f	1	4.283	26.341
Heat exchanger	30.26	0.90	4.77	g	1	4.623	8.830
Cooling tower	71.44	1.8	6.2	g	1	6.864	13.110

a Cast Iron, flake graphite, plain, or low alloy.
b Ductile iron (higher strength and hardness may be attained by composition and heat treatment or both).
c Ni-resist corrosion-resistant cast irons.
d Mild steel, also low alloy irons and steels.
e Stainless steel, ferritic 17% Cr type.
f Stainless steel, austenitic 18% Cr; 8% Ni type.
g 14% silicon iron.

20.2.4 CALCULATING SEQUESTERED CO₂ IN SUGARCANE PRODUCTION

While sugarcane production is emitting CO_2 in the atmosphere, it is also fixing CO_2 in the plantation on a yearly basis represented by the stalks harvested and milled plus the carbon sequestered by the trash, roots, and stumps. The carbon content from the stalks was estimated from the total carbon in bagasse biomass and the carbon in raw sugar and final molasses, measured from sucrose content. The fixed CO_2 in the stalks harvested and milled was calculated as follows:

$$CO_2 - \text{fxd. (stalks)} = \text{Mass of bagasse} * C_f * CF + \Sigma(\text{sucrose}_x * C_{suc} * CF)$$

where x is the raw sugar (rs), molasses (mol); bagasse percent in cane = 30%; carbon fraction (Cf) in bagasse = 0.4235 (IPCC, 2000); sucrose % in rs = 97.5%, mol = 55%; $C_{suc} = 0.421$.

The CO_2 sequestered in the roots and stumps were calculated using the data of Rosario and Mendoza (1977), as follows:

$$CO_2 - \text{seq. } (t + r + s) = B * C_f * CF$$

where CO_2 seq. (r + s) is CO_2 sequestered as computed from the carbon content of the roots, stumps, and trash; $B_{t/r/s}$, the amount of biomass (11.36 tons/ha trash, 1.2 tons/ha root biomass, 3.6 tons/ha stump); C_f, carbon fraction = 0.4235 (IPCC 2000); CF, conversion factor of C to CO_2 = 3.7.

The sequestered CO_2 in the plantation or the total carbon dioxide sequestered was obtained by simply adding the sequestered CO_2 in the stalks harvested and milled, and the CO_2 sequestered in the trash, roots, and stumps that stayed in the field. The sequestered CO_2 in the plantation was used to compute for the total amount of CO_2 sequestered by multiplying it to the total plantation area, which is 13,090 hectares. This was needed in calculating the number of years to offset the CO_2 emission in the mill construction, termed as payback period, which was estimated simply as follows:

$$\text{Payback period (years)} = \frac{\text{Total } CO_2 \text{ emitted during construction}}{\text{Total annual carbon savings}}$$

where total annual carbon savings = net CO_2 sequestration/emission.

20.2.4.1 Net CO_2 Sequestration/Emissions

The net CO_2 sequestration/emission was estimated by subtracting the total emissions from the total carbon dioxide fixed in the biomass: Total CO_2 fxd.–CO_2-eq. (plantation + milling).

20.3 RESULTS

20.3.1 THE CARBON FOOTPRINT OF SUGARCANE PRODUCTION AT THE FARM LEVEL

The average carbon foot print (CF) of sugarcane production (average yield of 1 plant crop and 3 ratoons = 82.51 TC/ha) was estimated at 5563 kg CO_2/ha, which is the average of 1 plant crop and 3 ratoons (Table 20.5). Slightly higher CF was calculated in the plant cane (6415 kg/ha) than in the ratoon cane (5279 kg CO_2/ha) owing to the energy bill (CO_2 emissions) incurred during land preparation and planting (crop establishment). Of the various operations and inputs used in cane production, fertilizer had the highest emission at 3927 and 3834 kg CO_2/ha for plant and ratoon

TABLE 20.5
Carbon Footprint (Equivalent CO_2 Emission) of Sugarcane Production (by Stages of Production)

Items	CO_2 Emission
A. Plant Crop	kg/ha
Land preparation	902.88
Cane points and planting	42.69
Fertilizer (fertilizer + hauling) (2)	3927.66
Cultivation + weeding	118.80
Hauling of canes (from field to factory gate)	990.00
Agricultural machines + implement (3)	120.78
Emissions in man labor (4)	222.24
Equivalent in cane points	90.41
Total (per/ha)	6415.45
B. Ratoon Cane	
1. Fertilizing (fertilizer + hauling)	3834.04
2. Cultivation + weeding	237.60
3. Hauling of canes to the mill	990.00
4. Agricultural machines + implement	79.20
5. Emissions in man labor	138.60
Total	5279.44
Average (1 plant cane + 3 ratoons)	5563.44
kg CO_2 per TC	67.44

cane, respectively. This was due to the high energy expended to manufacture and transport the fertilizer in the Philippines and to apply it in the field at 2.17 LDOE/ha (see CF of N fertilizer). N-fertilizer was applied at 300 kg/ha in Eastern Batangas, Philippines (Table 20.5).

20.3.1.1 The Associated GHG Emission (Expressed in CO_2 Equivalence) in Cane Burning

Conventional sugarcane production entails burning of canes (Figure 20.3). Burning is being done before or after cutting and loading the cane stalks. From the point of view of the sugarcane planters and the harvesters, burning is necessary. For the harvesters, burning is essential to facilitate cutting the sugarcane stalks. Sugarcane is trashy. Removing the trashes (dried leaves of tops) consumes time for the cane cutters, and when the field is weed-infested, they hinder cutting the stalks close to the ground. The sugarcane planters burn the trashes to facilitate farm operations. Hauling them in one corner in the field or mulching them between the cane rows is difficult and too risky. The trash might be accidentally burned during the height of summer months, which could also burn the growing ratoon canes. Burning canes liberates a considerable amount of CO_2 and other GHGs. The estimated direct CO_2 emission from cane burning was 10,410 kg/ha (Table 20.6). An additional 1791 kg CO_2/ha was estimated from the other gases (CH_4 = 467 kg CO_2, CO = 1241 kg CO_2,

FIGURE 20.3 Cane burning scenario.

TABLE 20.6
Carbon Footprint (CO_2 Emission) from Burning Canes at Harvest Time (Other GHG, N_2O, CH_4, CO)

	tons CO_2/ha
1. Direct CO_2 emission from burning canes at harvest time	
CO_2 eq. tCO_2 ha^{-1}	10.41
2. Emission from other GHG (CH_4, CO, N_2O) due to burning SC residues	
a. CH_4	
CH_4 emission in ton C	0.47
b. CO	
CO emission in ton C	1.24
c. N_2O	
N_2O emission in ton C	0.083
Total CO_2 emission eq. from other GHG (N_2O, CH_4, CO)	1.79
Total GHG emission from cane residues burning per ha	12.20
kg CO_2 per ton cane	147.93

and $N_2O = 830$ kg CO_2). This summed up to 12,204 kg CO_2/ha, which translates to about 42.69% of the total greenhouse gas emission at 28,587.08 kg CO_2/ha (see Table 20.6).

Treating the direct CO_2 emission from cane trash burning is complex. The effect of cane burning on soil organic matter is to decrease its contribution to soil functions. Soil Organic-C affects water retention, soil aggregate formation, energy supply, and carbon for the myriad forms of microorganism involved in N-fixation, P-solubilization, and bacteria synthesizing plant growth hormone, and so on. This leads to decreased soil fertility, requiring higher fertilizer application. The added fertilizer finds its ways through run-off in bodies of water causing eutrophication (pollution, algal bloom). Altogether, this leads to higher emission (direct and indirect) of CO_2-eq. These were not calculated being outside the boundary set in this study.

20.3.2 CARBON SEQUESTERED

While sugar manufacture emits a considerable amount of CO_2-GHG, it also sequesters CO_2 through photosynthesis, stored in various plant parts (leaves, stems, roots). The team has computed 60.11 tons per hectare CO_2 sequestration capacity (Table 20.7) of the sugar plantation. The computed value lies between the values generated by previous local reports by Guingab (2009) with a value of 39.26 tons CO_2 per hectare and Patricio (2014) at 72.56 tons/ha.

TABLE 20.7
Indicative Carbon Fixed Sequestered in the Various Parts of Sugarcane Crop

	CO_2-eq., Ton/ha
1. By the stalk:	
Bagasse	19.615
Raw-sugar	12.844
Molasses	2.330
2. By the biomass left in the field:	
Trash	17.801
Roots	1.880
Stumps	5.641
Total (1 + 2)	**60.111**

TABLE 20.8

Carbon Footprint of Sugarcane Production (in kilograms)

A. Cane production (1 plant cane + 3 ratoons) (average per ha)	5563.44
per TC	67.44
per kg sugar	0.67
B. CO_2 emission from sugarcane crop residue burning (per ha)	12,204.00
CO_2 emission from sugarcane crop residue burning per kg sugar (Table 20.6)	1.46
C. Total CO_2 emission per ha (A + B)	**17,767.44**
CO_2 emission/TC	**215.36**
Total kg CO_2 emission/kg sugar	**2.13**
D. Equivalent CO_2 fixed per ha	(60,111.00)
CO_2 fixed/TC	(728.62)
Total kg CO_2 fixed/kg sugar	(7.11)
E. Net CO_2 emission per ha (C – D)	**(42,343.56)**
CO_2 emission/TC	**(513.26)**
Total kg CO_2 emission/kg sugar	**(4.98)**

The calculated net carbon foot print after deducting carbon sequestered is –42,343.56 kg CO_2 ha^{-1} or about –513.26 kg CO_2 TC^{-1}. The total emission per kg sugar is estimated at –4.98 kg (Table 20.8). The system at the farm level emits less carbon dioxide compared with the amount of carbon that the crops can sequester, which means that cane production at the farm level is carbon-negative, or has an equivalent GHG savings.

20.3.3 The Carbon Footprint of Sugarcane Milling

20.3.3.1 Carbon Emission of Factory Construction

Prior to operation, construction of the factory already emits greenhouse gases. Table 20.9 summarizes the carbon emissions from each division and shows the distributions of the source.

The total carbon debt accumulated in the construction of the factory amounted to 32,164.88 tons CO_2-eq. (Table 20.7). Construction of the foundation and structure of buildings contributed the most to the carbon inventory constituting to about 86% of the total (32,164.88 tons). Equipment, base support, pipes, and pumps constituted to only 14%, which is contributed mostly by the equipment in the boiling house (46% of 4530.17 tons) and the power plant (45%). However, the construction of areas other than processing area contributes the largest at 44% of the total.

20.3.3.2 Carbon Emission from Hauling

The computation started at the gate of the factory and the total carbon emission calculated was 24,139.98 kg CO_2-eq. year^{-1}, equivalent to only 0.01% (Table 20.10) of the total factory emission.

20.3.3.3 Carbon Emission from Milling Operation

On the basis of the total amount of bagasse left after extraction, the amount of power that can be produced by the plant was 32.26 MW. The total power requirement of the milling plant, however, was 5.30 MW only. With these figures, it followed that the total electricity demand for the overall plant operation could be generated entirely from bagasse as fuel for boiler. However, for operation start-up when bagasse is not yet available, one day of the 270-day operation was assumed to consume electricity generated from banker oil. Therefore, using the conversions from Table 20.7, the carbon emission from milling operation amounted to 457,590,857.29 tons CO_2-eq. year^{-1} (Table 20.10), constituting 88.40% of the total carbon emission.

TABLE 20.9

Carbon Emission Breakdown for the Construction of the Sugar Factory

Plant Division	Structural		Foundation		Major Equipment, Base Support, Pipes, and Pumps		Combined (Overall)	
	CO$_2$ Emissions (tons)	% CO$_2$ Distribution	CO$_2$ Emissions (tons)	% CO$_2$ Distribution	CO$_2$ Emissions (tons)	% CO$_2$ Distribution	CO$_2$ Emissions (tons)	% CO$_2$ Distribution
Cane Supply and Transport	**490.30**	**5**	**1021.06**	**6**	**300.29**	**7**	**1811.65**	**6**
Cane supply and transport area	490.30	5	75.27	0	300.29	7	865.86	3
Unloading areas	0.00	0	945.79	5	0.00	0	945.79	3
Milling Area	**867.97**	**10**	**292.20**	**2**	**141.84**	**3**	**1302.01**	**4**
Boiling House	**3135.21**	**34**	**895.30**	**5**	**2062.54**	**46**	**6093.05**	**19**
Clarification	997.67	11	235.74	1	1033.94	23	2267.34	7
Evaporation	437.50	5	83.60	0	357.27	8	878.37	3
Pan house	448.24	5	153.67	1	625.28	14	1227.18	4
Tank farm	1192.12	13	293.51	2	46.06	1	1531.69	5
Sugar storage and bagging	59.69	1	128.78	1	0.00	0	188.47	1
Wastewater Treatment Facility	**0.00**	**0**	**2.50**	**0**	**7.85**	**0**	**10.35**	**0**
Power Plant	**4375.21**	**48**	**2425.33**	**13**	**2017.65**	**45**	**8818.19**	**27**
Power plant area	4291.86	47	2228.61	12	2017.65	45	8538.13	27
Bagasse storage	83.35	1	194.22	1	0.00	0	277.56	1
Switchyard	0.00	0	2.50	0	0.00	0	2.50	0
Miscellaneous Area	**238.76**	**3**	**13,890.88**	**75**	**0.00**	**0**	**14,129.65**	**44**
Warehouse/workshop area	66.83	1	136.20	1	0.00	0	203.02	1
Fire station area	21.19	0	33.09	0	0.00	0	54.28	0
Clinic	15.34	0	19.39	0	0.00	0	34.72	0
Laboratory	14.60	0	17.53	0	0.00	0	32.14	0
Canteen	36.25	0	66.30	0	0.00	0	102.55	0
Administration building	84.56	1	193.85	1	0.00	0	278.41	1
Mud bin	0.00	0	155.46	1	0.00	0	155.46	0
Parking area	0.00	0	2.50	0	0.00	0	2.50	0
Others	0.00	0	13,266.56	72	0.00	0	13,266.56	41
Total	**9107.45**	**28.31**	**18,527.26**	**57.60**	**4530.17**	**14.09**	**32,164.88**	**100**

TABLE 20.10
Summary of Total Annual Carbon Inventory of the Raw Sugar Factory

	Carbon Inventory (kg CO_2 year^{-1})	% Contribution
Hauling	24,139.98	0.01
Factory operation	457,590,857.29	98.93
Material input	556,308.00	0.12
Wastewater treatment	4,358,634.55	0.94
Total	**462,529,939.82**	**100**

The cogeneration facility produces excess electricity that could be sold to the grid. Electricity from bagasse has less carbon dioxide intensity than the average in the generation of electricity in the Philippines, thus, saving an annual GHG emission of about 2,089,109.95 kg CO_2.

20.3.3.4 Carbon Emission from Material Inputs

For the given amount of input, the total amount of lime needed in the operation was 2000 kg/day, which is equivalent to 556,308.00 kg CO_2-eq. year^{-1} (Table 20.10).

20.3.3.5 Carbon Emission from Wastewater Treatment Facility

The 30-acre facultative lagoon that is used to further treat the wastewater from the anaerobic digester produces 660.58 kg/day methane and 949.81 kg/day CO_2. Methane has global warming potential equivalent to 23 (Biograce 2013). The total CO_2-eq. then amounts to 4358.63 tons per year, which is 0.94% of the total annual carbon emission.

It can be observed from Table 20.10 that factory operations and products, which includes power generation and products end-use, is the major contributor of emission in the factory (98.93%). The breakdown of the power consumption per division is presented in Table 20.11.

Because of the pieces of equipment for cane preparation, which consume a large amount of electricity, the cane supply and transport division contributes the largest amount to power consumption at 70.41% (Table 20.11).

The factory was assumed to operate for 270 days per year. For the given daily capacity (4000 TCD), the plant can produce 2.05 L-kg sugar ton cane^{-1} (110,700 tons year^{-1}). With this amount, emissions are translated to 4.18 kg carbon dioxide per kg raw sugar produced, 428.27 kg CO_2-eq. per ton cane, and 35,332.15 kg CO_2 per hectare.

TABLE 20.11
Summary of the Total Power Requirement per Division

Plant Division	Power Rating, kW	% Contribution
Cane supply and transport	3727.38	70.41
Mills	894.98	16.91
Boiling house	534.51	10.10
Wastewater treatment facility	22.07	0.42
Power plant	88.41	1.67
Miscellaneous buildings	26.61	0.50
Total	**5293.96**	**100**

20.3.4 TOTAL CARBON FOOTPRINT OF RAW SUGAR PRODUCTION (SUGARCANE PRODUCTION AND SUGARCANE MILLING)

The total carbon footprint for sugarcane production and sugar manufacture without carbon sequestration is 6.31 kg CO_2-eq. per kg raw sugar or 643.62 kg CO_2-eq. per ton cane as summarized in Table 20.12.

Cane burning contributed 22.98% of the total carbon emission at 12,204 kg CO_2-eq./ha. This is about 69% of the CO_2-eq. emitted at the farm level, which is 33.46% of the total emission. Milling contributes to a greater fraction with 15.47% of the total at 35,332.15 kg CO_2-eq./ha.

Accounting for the carbon fixed in the crops, the entire process becomes a carbon sink. The system saves 7,011.41 kg CO_2-eq. per hectare (84.99 kgCO_2 per ton cane; 0.83 kg CO_2 per kg raw sugar). Adding the savings from the grid, which amounts to 2,089,109.95 kg CO_2 per year, the whole system saves 7,170.97 kg CO_2 ha^{-1} (86.92 kg CO_2 tc^{-1} or 0.85 kg CO_2 kg-raw sugar^{-1}).

20.3.5 PAYBACK PERIOD

The payback period is the time when the carbon emitted during preoperation is compensated. The payback period was computed because carbon savings were realized from the inventory. The analysis showed that the carbon emitted during construction could be paid back or offset immediately in a matter of 0.25 years.

TABLE 20.12
Summary of the Total Carbon Footprint (Cane Production and Milling) of Raw Sugar Production in the Philippines

Unit	kg CO_2-eq.	% Emission Contribution at Farm Level	Total% Emission Contribution
A. Farm Level			
Cane Production		31.31	10.48
Per ha	5563.44		
Per ton cane	67.44		
Per kg raw sugar	0.67		
Cane Burning		68.69	22.98
Per ha	12,204.00		
Per ton cane	147.92		
Per kg raw sugar	1.46		
B. Milling		0	66.54
Per ha	35,332.15		
Per ton cane	428.27		
Per kg raw sugar	4.18		
Total (without Sequestration)		100	100
Per ha	53,099.59		
Per ton cane	643.63		
Per kg raw sugar	6.31		
C. Sequestration			
Per ha	(60,111.00)		
Per ton cane	(728.62)		
Per kg raw sugar	(7.11)		
Net Total (with Sequestration)			
Per ha	(7011.41)		
Per ton cane	(84.99)		
Per kg raw sugar	(0.83)		

20.4   DISCUSSIONS, CONCLUSION, AND RECOMMENDATIONS

20.4.1   FIELD LEVEL

In the field level of producing cane at Eastern Batangas, Philippines, two practices (N-fertilizer application and cane residue burning) were identified as major hotspots of sugarcane production. But the high N-fertilizer application is the consequence of the several decades of widespread practice of cane burning. This is true for the entire Philippines as reported in previous studies (Mendoza and Samson 2002). As a result, cane burning accounted for 22.98% of the total carbon footprint. This century-old practice of burning the trash could no longer find merit at this point. The explanation is simple. Trash-burning deprives the soil from the much needed soil organic matter (SOM). At least 15% of organic C in the trash is converted into Humus-C (Batjes 1999). SOM had decreased by almost 50% in Philippine sugarcane soils (Rosario et al. 1992). Low SOM leads to low fertilizer use efficiency. This means more fertilizer should be applied to obtain the same yield or to prevent yield decline. In Eastern Batangas, since soils are high in P and K, only N-fertilizer is added but in a huge amount, which is already at 300 kg ha^{-1}. Should trash burning, the conventional practice in the past century, be sustained, then SOM will decrease further. This will happen at a time when oil prices will increase (Schlesinger 2010; Smith 2010), consequently, the price of N-fertilizer will follow (Pfeiffer 2003; Rodolfo 2008). Should the application of N-fertilizer increase further, then accordingly, the carbon footprint of sugar production will be intensified even more. As of the moment, oil price increase had ceased owing to the new hydraulic fracking technology in oil extraction. However, this does not mean that there will be no 'peak oil' anymore to trigger another round of oil price increase.

On top of this, trash burning is hazardous to human health for the farm workers and the residents in the immediate vicinity where sugarcane is grown as synthesized by Mendoza et al. (2003). Sugarcane workers have been observed to have significantly high rates of mortality owing to illnesses originating from agricultural operations. A case-control study in the USA suggests that people engaged in sugarcane farm-related occupations have significantly higher rates of lung cancer. Sugarcane workers also have an increased risk of lung cancer. This condition may be related to the practice of burning in sugarcane fields that releases fly soot, containing polycyclic aromatic hydrocarbons (PAH) that have mutagenic and carcinogenic properties, into the atmosphere (Zamperlini et al. 1997). In India, it was also found out that there was an increased risk of lung cancer for workers employed in a sugarcane farm (Amre et al. 1999). Work involving burning after harvesting and exposure to fibers of biogenic amorphous silica may increase the risks of lung cancer and possibly mesothelioma among sugarcane farmers (Poolchund 1991).

Eliminating the field burning of residues through trash farming reduces the health hazards associated with exposure to airborne particulate matter (fly soot and biogenic amorphous silica). Moreover, no trash burning means that the trash left in the field will decompose upon the onset of rains. It is true that leaving the field with thick trashes is too risky because accidental or intentional fire that will damage the growing canes may occur. Examining the benefits may offset the risks. The fertilizing value of sugarcane trash was about 54 kg N ha^{-1} year^{-1} (Boddey et al. 1995). Nitrogen fixation levels of 50–200 kg N ha^{-1} occur in trash-farmed sugarcane fields, with the higher range associated with higher trash levels. A mean value of 125 kg N ha^{-1} was recorded when trash farming was established as a practice (Patriquin 2001). Burned cane fields lose an average of 44 kg N ha^{-1} year^{-1} and some of the P and K are also lost through burning. In trash farming, P uptake appears more efficient as the mulch protects the soil from desiccation and permits root proliferation in the soil surface, where P levels are high. Mulching permits a greater recycling of P from residues than burning and that P fertilization rates is lower where burning is stopped (Ball-Coelho et al. 1993). Trash farming helps increase organic matter in the soil since 15% of the C in the trash is converted into humus-C (Batjes 1999).

Trash burning is quick and easy but soil impoverishing and heavy GHG-emitting. N-fertilizer is now PhP 53.3 per kg (Urea = PhP 1200/50 kg-bag; 1$ = PhP 44). At 125 kg N, the gain in N fixation alone is worth US$151 (PhP 6667/ha/year). The cost of managing the trash without burning was estimated at US$122/ha. Trash farming involves detrashing at 2 × PhP 1500/ha = PhP 3000, trash shredding at PhP 2000/ha, roving guard at PhP 400 for 3 months at 1 roving guard/5 ha; a total of PhP 5400/ha. The monetary cost of managing the trash could be more than compensated by the N-gain alone (excluding P and K-fertilizer).

20.4.2 Factory Level

Meanwhile, for the factory, it was identified that at the construction level, the mere assembly of the buildings contributes the most (in comparison to equipment manufacture and set-up) to the total emission from the preoperational period. The way to go around this is to ensure that the buildings be constructed in strict accordance to the guidelines for infrastructure projects set by the government so as to be sure of the quality and durability of the frames and foresee possible occurrences of unwanted conditions. In this case, risks that may destroy the buildings could be minimized.

During the operations, it was observed that factory operations and products were the major source of carbon emission at 98.93% of the total. Burning bagasse to power the mill is practical because it is a by-product and it saves the mill from huge amount of coal or bunker oil to start the mill. There are possible options as earlier suggested (Mendoza et al. 2004; Mendoza and Samson 2006) and they are as follows: (1) to increase the boiler efficiency to reduce the amount of bagasse to be burned (Doon and Thompson 1998) and produce excess bagasse for other uses (i.e., co-generation; particle board manufacture) (ESMAP 1993; EDUFI 1994); and (2) to efficiently use the heat produced to generate electricity which can be sold to the grid. This offsets the grid-sourced electricity, in turn, reducing the electricity generated by burning fossil oil, thereby reducing the CO_2-GHG loading (Brown 2008). If compared to coal, with the utilization of bagasse, the system is able to accumulate savings that amounts to 2,089.11 tons CO_2 per year. Also, the use of high-pressure boilers will lessen carbon emission. High-pressure boilers will require smaller equipment size than their low-pressure counterpart. With the more efficient high-pressure boilers and the resolution to utilize the heat of the exhaust steam, more excess power will be provided by the cleaner biomass, leading to bigger carbon savings.

Additionally, breaking down the contributors to the total power requirement, it was found that about 87% comes from cane preparation and milling. This hotspot is best handled through proper maintenance. To prevent inefficiencies and losses due to possible emergency operation shutdown, regular maintenance and replacement of worn parts should be practiced. Moreover, the consumption of electricity is greatly affected by the motion of the objects, which in this case could be hindered by friction (if parts are not properly greased) that affects the ease of the rolling equipment to rotate around its shaft. Less-resisted motion requires less power; thus, regular lubrication of the equipment would help reduce the power consumption.

The sugar industry is one of the premiere industries in the Philippines. It is a major source of employment because it directly employs about 0.5 million people and an additional 5 million employees and family members indirectly dependent on the sugar industry. It must lead in re-greening and in pursuing green labor (discontinue burning canes) in the industry. There is a need to fund the needed R/D, infrastructure, equipment, and extension program to assist the farmers and millers in shifting to the alternative system of sugar production that reduce burning in sugarcane fields (for the planters), and encourage investment to more efficient boilers and more recycling of heat for power co-generation (for the millers).

Mendoza (2014) had estimated the social costs of carbon in sugar production at P2.29/kg. In Eastern Batangas, the total sugar yield is about 182 million kg sugar. The SCC of sugar amounted to PhP 417 million or US$ 9.47 million. If the estimates will be extended to the whole sugar industry (at 2.25 million metric ton sugar produced for crop year 2010–2011), then, the amount is enormous at PhP 5.15 billion or US4117 million (1US$ = PhP 44). If this amount can be collected yearly, the sugar industry can self-finance the greening initiatives while generating additional green labor benefiting the rural economy, the environment, and the society at large. An incentive scheme must be devised. For those who will adopt the alternative system of cane production systems of not burning canes, no sugar levy will be charged because the system will serve as the government counterpart in the shifting or conversion process.

In summary, although the whole system for raw sugar manufacture is a carbon sink, there are still the "hotspots" that can be managed to further lessen carbon emission. The alternative system of sugarcane production (No cane burning + reduced N-fertilizer application), improving the efficiency of boilers, and increasing heat recycling to generate more electricity for the grid could reduce significantly the carbon foot print of sugar. The sugar industry, the premiere industry in the country, must lead in re-greening and in pursuing green labor in the industry consistent to the global thrusts of pursuing low-carbon or green economy.

REFERENCES

Amre, D.K, C. Infante-Rivard, A. Dufresne, P.M. Durgawale and P. Ernst. 1999. Case-control study of lung cancer among sugar cane farmers in India. *Occupational and Environmental Medicine* 56(8): 548–552.

Brown, L.R.2008. PLAN B 3.0. Mobilizing to Save Civilization. EARTH POLICY INSTITUTE. W.W. Norton & Company, Inc., 500 Fifth Avenue, New York, N.Y. 10110 www.wwnorton.com

Ball-Coelho, B., H. Tiessen, J. Stewart, I. Salcedo, and E. Sampaio, 1993. Residue management. Effects on sugarcane yield and soil properties in northeast Brazil. *Agronomy Journal* 85: 1004–1008.

Batjes, N.H., 1999. Management options for reducing CO_2 concentrations in the atmosphere by increasing carbon sequestration in the soil. *Report 410-200-031*, Dutch National Research Programme on Global Air Pollution and Climate Change & Technical Paper 30, International Soil and Reference and Information.

BHP Billington. 2011. Greenhouse Gas. *Olympic Dam Expansion Supplementary Environmental Impact Statement 2011*. Retrieved from http://www.bhpbilliton.com/home/society/regulatory/Documents/Olympic%20Dam%20Supplementary%20EIS/Documents/Chapter%2013%20Greenhouse%20Gas.pdf

Biograce GHG Calculation Tool. 2013. Retrieved from http://www.biograce.net/content/ghgcalculationtools/overview.

Boddey, R.M., O.C. de Olivera, S. Urquiaga, et al., 1995. Biological nitrogen fixation associated with sugarcane and rice: Contributions and prospects for improvement. *Plant and Soil* 174: 195–209.

Bruce, J., M. Frome, E. Haites, H. Janzen, R. Lal, and K. Paustian. 1999. Carbon sequestration in soils. *Journal of Soil and Water Conservation* 54: 382–389.

Clark, D. 2013. CO_2 emissions from biomass and biofuels. *Cundall Johnston & Partners LLP.*

Corpuz, F.H. and P.S. Aguilar. 1992. Specific energy consumption of Philippine sugar mills. Energy consumption of Philippine sugar mills. *Proceedings of the PHILSUTECH 39th Annual Convention*, Bacolod City, pp. 410–421.

Department of Environment and Natural Resources [DENR]. 2008. DENR Administrative Order 2008–XX: Water Quality Guidelines and General Effluent Standards. Retrieved December 3, 2014 from http://www.emb.gov.ph/wqms/Draft%20DAO%20on%20the%20Revised%20WQG%20and%20GES%20rev%20121807.pdf

Doon, R. and J. Thompson. 1998. Powering the Philippine sugar industry into the next millennium. *Proceedings of the 45th PHILSUTECH Annual National Convention*, Cebu City, Philippines, pp. 243–252.

Downing et al. 2005. Social Cost of Carbon: A Closer Look at Uncertainty. Retrieved May 13, 2011 from http://www.decc.gov.uk/assets/decc/what%20we%20do/a%20low%20carbon%20uk/carbon%20valuation/social_cost/sei-scc-report.pdf

EDUFI (Energy Development and Utilization Foundation Inc). 1994. Biomass Co-generation Potential in the Philippines Sugar Milling Industry. A Joint report of EDUFI, Office of Energy Affairs, Sugar Regulatory Administration and the Philippine Sugar Milling Assoc. Inc. (PSMAI), Manila, April 1994.

ESMAP. 1993. Philippines: Commercial Potential for Power Production from Agricultural Residues. Joint study of Energy Sector Management Assistance Program (ESMAP, UNDP) and the Philippines Department of Energy.

Foust, A.S. 1960. *Principles of Unit Operations.* New York, Wiley.

Godilano, E.C. 2009. *Global Climate Change and its Impacts on Agriculture and Fishery Production in the Philippines.* Department of Agriculture. Information Technology Center for Agriculture and Fishery

(ITCAF) Enterprise. Geospatial Information Systems for Analysis and Learning Laboratory, Quezon City, Philippines.

GREET 2013. A software of Argonne National Laboratory.

Guingab, M.V.D. 2009. Carbon Sequestration of Selected C4 Plants in Isabela, Philippines. Retrieved December 3, 2014 from http://ejournals.ph/index.php?journal=ISUCJR&page=article&op=view&path%5B%5D=5191&path%5B%5D=5379.

Hammond, G. and C. Jones. 2011. Embodied Energy. Greenspec. Retrieved from http://www.greenspec.co.uk/embodied-energy.php

Hansen J. et al. 2010. Target Atmospheric CO_2: Where Should Humanity Aim? Retrieved May 13, 2011. DOI: 10.2174/1874282300802010217.

Hugot, E. 1986. *Handbook of Cane Sugar Engineering*. Elsevier Science Publishers, The Netherlands.

IPCC. 2000. Intergovernmental Panel on Climate Change. Good Practice Guidance and Uncertainty Management in National Greenhouse Gas Inventories. Retrieved March 19, 2011 from http://www.ipcc-nggip.iges.or.jp/public/gp/english/4_Agriculture.pdf.

IPCC. 2006. Intergovernmental Panel on Climate Change Guidelines for National Greenhouse Gas Inventories. Retrieved March 30, 2011 from http://www.ipcc-nggip.iges.or.jp/public/2006gl/index.html.

IPCC. 2007. Climate Change 2007: Mitigation. Contribution of Working Group III to the Fourth Assessment Report of the Intergovernmental Panel on Climate Change. Cambridge University Press, Cambridge, UK and New York. Retrieved March 11, 2010 from http://www.ipcc.ch/publications_and_data/ar4/wg3/en/ch3s3-5-3-3.html.

Lal, R., 1997. Residue management, conservation tillage and soil restoration for mitigating greenhouse effect by CO_2 enrichment. *Soil and Tillage Research* 43: 81–107.

Mendoza, T.C. 2003. CO_2-greenhouse gas-reducing potentials of some ecological agriculture practices in the Philippine landscape. *Philippine Journal of Crop Science* 26(3): 31–44.

Mendoza, T.C. 2014. Reducing the carbon footprint of sugar production in the Philippines. *Journal of Agricultural Technology* 10(1): 289–308. Retrieved September 15, 2014 from the world wide web: http://www.ijat-aatsea.com.

Mendoza, T.C. and R. Samson. 2000. Estimates of CO_2 production from the burning of crop residues. *Journal of Environmental Science and Management* 3(1): 25–33.

Mendoza, T.C. and R. Samson. 2002. Energy costs of sugar production in the Philippine context. *Philippine Journal of Crop Science*. 27(2): 17–26.

Mendoza, T.C. and R. Samson. 2006. Relative bioenergy potential of major agricultural crop residues in the Philippines. *Philippine Journal of Crop Science* 31(1): 11–28.

Mendoza, T. C., E.T. Castillo, and R. Demafelis. 2007. The costs of ethanol production from sugarcane under Eastern Batangas conditions. *Philippine Journal of Crop Science* 32(3): 25–48.

Mendoza, T.C., R. Samson, and A.R. Elepano. 2004. Renewable biomass fuel as 'Green Power' alternative for sugarcane milling in the Philippines. *Philippine Journal of Crop Science* 27(3): 23–39.

Mendoza, T.C., R. Samson, and T. Helwig. 2003. Evaluating the many benefits of sugarcane trash farming systems. *Philippine Journal of Crop Science* 27(1): 43–51.

Mudahar, M.S. and T.P. Hignett. 1985. Energy efficiency in nitrogen fertilizer production. *Energy in Agriculture* 4, 159–177.

Parry, M.L., O.F. Canziani, J.P. Palutikof, P.J. van der Linden, and C.E. Hanson (Eds.). 2007. *Climate Change 2007: Impacts, Adaptation and Vulnerability*. Contribution of Working Group II to the Fourth Assessment Report of the Intergovernmental Panel on Climate Change. Cambridge University Press, Cambridge, UK and New York, NY, USA. Retrieved February 13, 2011 from http://www.ipcc.ch/publications_and_data/ar4/wg2/en/spmsspm-c-15- magnitudes-of.html.

Patricio, J.H.P. 2014. How Much Soil Organic Carbon is there in Agricultural Lands? A Case Study of a Prime Agricultural Province in Southern Philippines. AES Bioflux. Retrieved December 3, 2014 from: http://www.aes.bioflux.com.ro/docs/2014.194-208.pdf.

Patriquin D. 2000. Overview of N_2 fixation in sugarcane residues: Levels and effects on decomposition. In: *Strategies for Enhancing Biomass Energy Utilization in the Philippines. Resource Efficient Agricultural Production*. Zoology Department, Dalhousie University, Halifax, Nova Scotia. Retrieved January 15 from http://www.p2pays.org/ref/19/18956.pdf.

Patwardhan AD. 2008. Industrial Waste Water Treatment. PHI Learning Pvt. Ltd. Retrieved December 3, 2014 from http://books.google.com.ph/books?id=psf56CPmZsYC&dq=raw+sugar+factory+wastewater+treatment&source=gbs_navlinks_s.

Perry, R.H. and D.W. Green. 2008. *Perry's Chemical Engineers' Handbook*, 7th ed. McGraw-Hill, New York: Print.

Peters, M.S. and K.D. Timmerhaus. 1991. *Plant Design and Economics for Chemical Engineers*. McGraw-Hill, New York.

Pfeiffer, D.A. 2003. Eating Fossil Fuels. Retrieved March 13, 2011 from http://www.fromthewilderness.com/free/ww3/100303_eating_oil.html.

Pimentel, D. (editor). 1980. *Handbook of Energy Utilization in Agriculture*. CRC Press, Boca Raton, FL. 430 p.

Pimentel, D., G. Berardi, and S. Fast. 1983. Energy efficiency of farming systems: Organic and conventional Agriculture. *Agriculture, Ecosystem and Environment* 9: 359–372.

Poolchund, H.N. 1991. Aspects of occupational health in the sugarcane industry. *Journal of the Society of Occupational Medicine* 41(3): 133–136.

Rodolfo, K. 2008. Peak oil: The global crisis of diminishing petroleum supply and its implications for the Philippines. *Asian Studies Journal* 41(1): 41–101.

Rosario, E.L. and T.C. Mendoza. 1977. Root and shoot growth pattern of six commercial sugarcane varieties as influenced by population density and nitrogen fertilization. *Philippine Journal of Crop Science* 2(3): 163.

Rosario E.L, E.P. Paningbatan and A.G.P. Dionora. 1992. Study on the causal factors of declining sugarcane quality in the La Carlota Mill District. *Proceedings of the 39th Annual Convention PHILSUTECH*. Assoc. Inc., Bacolod City, Philippines. pp. 169–184.

Schlesinger, J. 2010. The Peak Oil Debate Is Over. Retrieved November 01, 2010 from http://www.countercurrents.org/schlesinger011110.htm.

Smith, C. 2010. New Zealand Parliament Peak Oil Report: The Next Oil Shock? Energy Bulletin, Retrieved October 1, 2010 from http://www.energybulletin.net/stories/2010-10-14/next-oil-shock.

Tol, S. and J. Richard. 2008. *Social Cost of Carbon*. Economic and Social Research Institute, Dublin.

US EPA. 1993. Emission Factor Documentation for AP-42 Section 18. Office of Air Quality Planning and Standards. Retrieved from http://www.epa.gov/ttnchie1/ap42/ch01/bgdocs/b01s08.pdf.

US EPA. 2011. *Greenhouse Gas Emissions from a Typical Passenger Vehicle*. Office of Transportation and Air Quality Office. EPA-420-F-11-041.

Walser, M.L., S.C. Nodvin, and S. Draggan. 2010. Carbon footprint. In: Cutler J. Cleveland (Ed.), *Encyclopedia of Earth*. Environmental Information Coalition, National Council for Science and the Environment, Washington, DC: Retrieved June 27, 2011 http://www.eoearth.org/article/Carbon_footprint.

Weier, K.L. 1998. Sugarcane fields; sources or sinks for greenhouse gas emissions? *Australian Journal of Agricultural Research* 49: 1–9.

Wiedmann, T. and J. Minx. 2008. A definition of 'Carbon Footprint'. In: C.C. Pertsova (Ed.), *Ecological Economics Research Trends*, Chapter 1, pp. 1–11. Nova Science Publishers, Inc, Hauppauge NY, USA. https://www.novapublishers.com/catalog/product_info.php?products_id=5999.

Yadav, R.L., 1995. Soil organic matter and NPK status as influenced by integrated use of green manure, crop residues, cane trash and urea in sugarcane based crop sequences. *Bioresource Technology* 54: 93–98.

Zamperlini, G.C.M., M.R.S. Silva and W. Villegas. 1997. Identification of polycyclic aromatic hydrocarbons in sugar cane soot by gas chromatography-mass spectroscopy. Chromatographia 46(11/12): 655–666.

21 A Two-Phase Carbon Footprint Management Framework

A Case Study on the Rockwool Supply Chain

Eirini Aivazidou, Christos Keramydas, Agorasti Toka,
Dimitrios Vlachos, and Eleftherios Iakovou

CONTENTS

21.1 Introduction 474
21.2 A Carbon Footprint Management Framework for Supply Chains 475
    21.2.1 Systematic Recording of the International Legislative Frameworks and Trends
    in Carbon Footprint Management 476
    21.2.2 Identification of the Best Practices in Supply Chain Carbon Footprint
    Management 477
    21.2.3 Setting of the Strategic Goals for Supply Chain Carbon Footprint Management 477
    21.2.4 Ranking of the Critical Operations and Identification of the Supply Chain
    Partners' Role 477
    21.2.5 Measurement of the Total Supply Chain's Carbon Footprint 478
    21.2.6 Evaluation of the Outcomes and Reconsideration of the Goals 478
    21.2.7 Centralized and Decentralized Decision Making for Supply Chain Carbon
    Footprint Management 478
    21.2.8 Monitoring and Reporting of the Supply Chain's Carbon Footprint 479
    21.2.9 Reevaluation and Update of the Decisions 479
21.3 Case Study: A Real-World Rockwool Supply Chain 479
    21.3.1 The Rockwool Supply Chain 479
    21.3.2 Implementation of the Carbon Footprint Management Framework 481
        21.3.2.1 Recording of the European Regulatory Framework 481
        21.3.2.2 Identifying the Best Practices of the Top Global Competitors ... 481
        21.3.2.3 Setting Goals for Carbon Footprint Management in the Rockwool
        Supply Chain 481
        21.3.2.4 Identifying the Critical Rockwool Supply Chain Operations and the
        Stakeholders' Role 481
        21.3.2.5 Quantifying the Rockwool Supply Chain Carbon Footprint 481
        21.3.2.6 Evaluating Carbon Footprint Findings for Reconsidering the Goals 483
        21.3.2.7 Decision Making for Carbon Footprint Management in the Rockwool
        Supply Chain 483
        21.3.2.8 Monitoring and Reporting the Rockwool Supply Chain Carbon
        Footprint 484
        21.3.2.9 Reevaluating Outcomes for Updating the Carbon Footprint
        Management Decisions 484

21.4 Conclusions and Recommendations 484
Acknowledgments...485
References...485

21.1 INTRODUCTION

Climate change along with global warming are two of the greatest environmental challenges humanity has encountered during the last years, thus the efforts to tackle them have emerged as a crucial issue of international debate (Scipioni et al. 2010). According to Eurostat, 83% of the European Union citizens consider ecological footprint as a critical factor when deciding to purchase a product (European Commission 2009). As a consequence of this rising environmental awareness, governments have already been focusing on sustainability policies through enacting regulations for mitigating greenhouse gas (GHG) emissions, such as the Kyoto Protocol and the European Union Emissions Trading Scheme (EU ETS). At the same time, the unique characteristics of a product's life cycle along with the entire supply chain activities, namely procurement, manufacturing, warehousing, distribution, use and waste disposal, are responsible for multiple environmental impacts, such as natural resource depletion, GHG emissions, pollution due to toxic materials' use and waste production (Lenzen et al. 2007). In this context, it is imperative that corporations should not only comply with environmental legislation, but also exceed customers' requirements for green products and services through launching corporate social responsibility (CSR) programs for protecting both environment and society.

Over the last decade, green business mentality has emerged as a profitability center and a core competency for companies (Kumar et al. 2012). To that effect, a plethora of firms are gradually adopting sustainable practices into their corporate strategy, such as the management of the carbon footprint (CF) across their wide supply chain networks. Particularly, CF constitutes a key performance indicator of the climate change impact caused by human activities (Clark 2012). CF can be considered as a part of Life-Cycle Analysis (LCA) and defined as a measure of the total amount of CO_2 emitted directly and indirectly during a single activity or over a product's life stages (Wiedmann and Minx 2008). Nevertheless, it is evident that the majority of the definitions regarding CF also include other noncarbon GHGs. According to a more elaborate definition, CF methodology estimates the total GHG emissions, expressed in CO_2 equivalent units (CO_2-eq.), across a full product's life cycle, from the stage of raw materials' extraction, via manufacturing, packaging, logistics and use, to the recycling and waste disposal (Carbon Trust 2012).

Several scientific publications provide case studies that deal with the assessment of the CF across global supply chains of real-world manufacturing firms. Indicatively, in the agrifood sector, farming and overseas transportation constitute the crucial contributors to the total GHG emissions of the Ecuadorian banana supply chain (Iriarte et al. 2014), while raw materials' extraction and processing, as well as distribution, are the main CF hotspots in the Danish rye bread supply chain (Jensen and Arlbjørn 2014). Moreover, in the US dairy feed industry, raw materials for the feed production are responsible roughly for 80% of the total carbon emissions, while both energy utilization and transportation contribute approximately by 10% to the whole supply chain's CF (Adom et al. 2013). In the biofuel industry, the CF evaluation of a soy-to-biodiesel supply chain and a corn-to-ethanol supply chain indicates that the former generates lower GHG emissions than the latter (Young et al. 2012). At the same time, in the automobile industry, the operations of Hyundai Motor Company's suppliers are considered as a key factor that influences the CF of a car's supply chain (Lee 2011). Concerning the textile industry, the recycling of the preconsumer process waste appears to contribute more effectively to the reduction of the CF across the clothing supply chain than the recycling of the post-consumer end-of-life waste (Muthu et al. 2012). As far as the building sector is concerned, fossil fuel utilization is considered as the critical factor of various environmental impacts, including CF, of the brick supply chain (Koroneos and Dompros 2007).

Finally, with respect to the insulating materials industry, few research efforts have been made for addressing the evaluation of the CF of various products. More specifically, according to

Li et al. (2014), rockwool consumes twice as much electrical energy per functional unit as expanded polystyrene leading to a more severe environmental footprint. Likewise, Proietti et al. (2013) argue that reflective foil panels have a lower total life-cycle CF compared to the expanded polystyrene and rockwool products. The limited research on the CF assessment of insulating materials addresses the identified gap in the related scientific literature. Thus, the need for conducting further research in order to foster the environmental sustainability of the insulating materials sector is highlighted.

The purpose of this chapter is to provide a new two-phase methodological approach for assessing and holistically managing the total GHG emissions across insulating materials supply chains. More specifically, the conceptual framework includes a strategic and a tactical phase, following the natural hierarchy of the corporate decision-making process. In order to validate the proposed methodology, we apply the related framework in a real-world supply chain of rockwool, which is a widely used insulating material for buildings produced from mineral wool. Since the existing literature on sustainable rockwool supply chain management focuses on the application of the LCA techniques for rockwool products (Proietti et al. 2013; Li et al. 2014), this case study is considered of significant added-value providing discrete steps for managing efficiently the CF of rockwool supply chains.

The remainder of the chapter is structured as follows. In Section 21.2, the two-phase methodological framework for CF management in supply chains is provided. In Section 21.3, the case study on CF management in the rockwool supply chain of a Greek insulating materials manufacturer is presented. More specifically, the basic echelons of the rockwool supply chain are described, while the implementation of the proposed CF methodology across the entire supply chain of the rockwool manufacturer is demonstrated. Finally, in Section 21.4, conclusions and recommendations for further research are discussed.

21.2 A CARBON FOOTPRINT MANAGEMENT FRAMEWORK FOR SUPPLY CHAINS

Several state-of-the-art methodologies for CF assessment during a product's life cycle or across its entire supply chain have been proposed in the literature (Sundarakani et al. 2008; Rigot-Muller et al. 2013; Montoya-Torres et al. 2014). In 2013, the International Standardization Organization (ISO) published ISO/TS 14067 that specifies the requirements and guidelines for quantifying and reporting the CF of a product either across its full life cycle or only for a set of specific processes (ISO 2013). Nonetheless, CF offsetting and mitigation activities are excluded from the ISO/TS 14067.

To the best of our knowledge, although there are several methodologies for CF quantification at a product's life-cycle level, there is a lack of systemic efforts that incorporate both strategic and tactical CF management into supply chain management within the highly dynamically moving international field. To that effect, Aivazidou et al. (2013) propose a holistic methodological framework for CF management across entire supply chains. The novelty of the framework lies in the fact that it follows the natural hierarchy of the decision-making process and is comprised of two phases, namely a strategic and a tactical one, as illustrated in Figure 21.1. More specifically, the strategic phase involves the mapping of the company's internal and external environment, the setting of the strategic goals, as well as the measurement and evaluation of the current status of the supply chain CF for reviewing the relevant strategic goals. Then, the tactical phase follows in order to implement the insights gained from the strategic phase. Particularly, the tactical phase includes the decision-making process for the management of the supply chain CF, the monitoring and reporting of the total GHG emissions, as well as the re-evaluation of the CF for updating the decisions concerning the carbon emissions mitigation. Finally, the operational phase (day-to-day decisions) is omitted from the proposed CF management decision-making framework.

Below, a brief description of each of the nine involved steps of the framework proposed for rockwool manufacturers is provided. Specifically, the strategic phase of the framework includes six (6) steps, presented in Subsections 21.2.1–21.2.6, while the tactical phase is comprised of three (3) steps, presented in Subsections 21.2.7–21.2.9 accordingly.

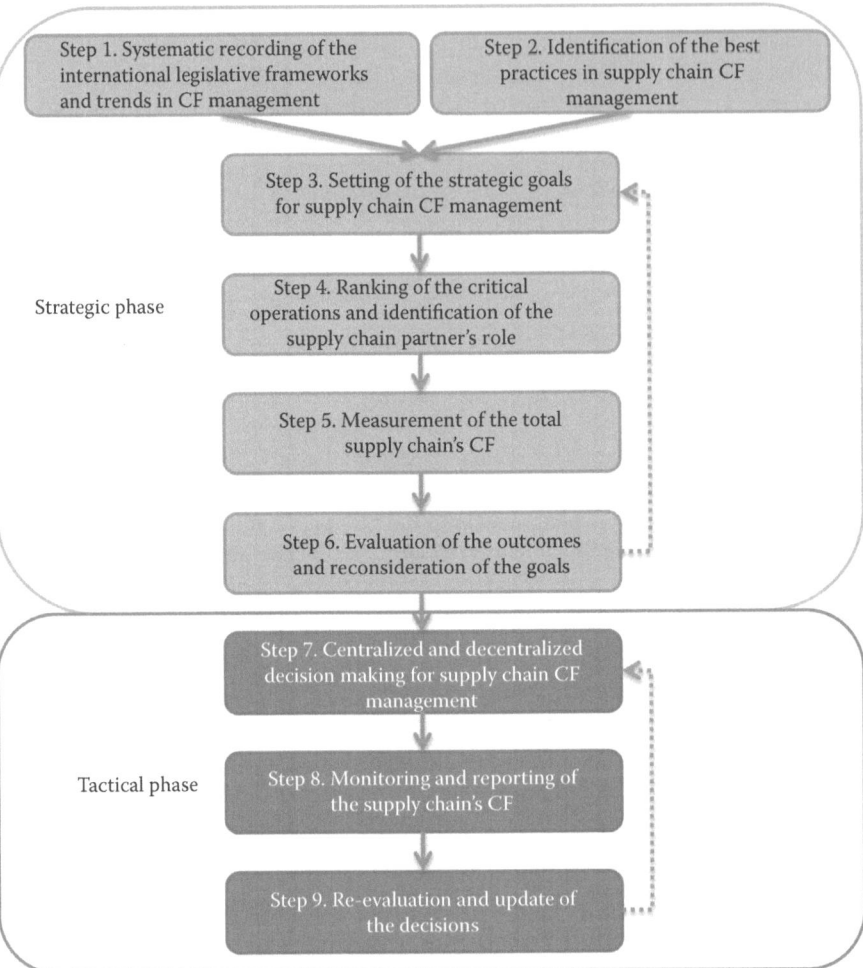

FIGURE 21.1 The supply chain CF management framework. (From Aivazidou, E. et al. 2013. *Chemical Engineering Transactions* 35: 313–318. With permission.)

21.2.1 SYSTEMATIC RECORDING OF THE INTERNATIONAL LEGISLATIVE FRAMEWORKS AND TRENDS IN CARBON FOOTPRINT MANAGEMENT

Companies should systematically follow the enacted international environmental regulatory interventions and trends regarding the mitigation of the industrial CF for effectively adapting their goals, processes, and strategic decision making to their requirements. Indicatively, power stations, industrial plants, including insulating material manufacturers, as well as commercial airlines operating in Europe, have already entered the EU ETS, which is the largest cap-and-trade system for GHG emission allowances worldwide (European Commission 2013). In particular, rockwool manufacturers should apply the sectoral methodology for the allocation of the emission allowances within the EU ETS, specially designed for the mineral wool industry (Ecofys 2009).

Further, as globalization gave a new impetus for the establishment of global supply chain networks, the European Union aims at creating other transportation-related emission schemes for achieving substantial reductions in emissions by 2050 (European Commission 2012). Since sourcing and procurement of the rockwool's raw materials from globally dispersed suppliers require lengthy transportation, rockwool manufacturers and their supply chain stakeholders should be continually aware of any novel CF schemes regarding transportation.

21.2.2 IDENTIFICATION OF THE BEST PRACTICES IN SUPPLY CHAIN CARBON FOOTPRINT MANAGEMENT

Companies should take into consideration all the environmental policies that top organizations have already implemented in their supply chains and proved to be efficient in practice. In particular, a plethora of multinational organizations, including top rockwool manufacturing industries, have embraced ambitious green practices for reducing their supply chain's CF and thus improving their sustainability indices. Indicatively, Rockwool Group (2013) reduced its direct emissions by 6% in 2012 compared to the base line year 2009 through investing in capital-intensive production processes. At the same time period, the company's electrical energy emissions decreased by 5% through deploying a monitoring system for avoiding unnecessary electricity use, as well as utilizing renewable energy resources. Moreover, Paroc (2013) reduced its total CO_2 emissions approximately by 120,000 tons in 2013 compared to the year 2011 through: (i) implementing green practices in production processes for reducing heat losses, (ii) utilizing energy-efficient machines in manufacturing and applying energy savings in all working environments, (iii) auditing its external suppliers' environmental efficiency, (iv) optimizing its global transportation network (e.g., factories located close to market areas, maximization of track capacity utilization), and (v) developing recycling and reuse of waste systems.

21.2.3 SETTING OF THE STRATEGIC GOALS FOR SUPPLY CHAIN CARBON FOOTPRINT MANAGEMENT

Corporations should set the goals and objectives for managing the CF of their supply chains in the context of the triple helix (economy, environment, and society) of sustainable development. To that end, industrial manufacturers have to consider: (i) the manner in which CF management affects their traditional business goals, such as operations' efficiency and profitability, and (ii) the new goals that emerge, such as the development of green supply chains and the establishment of a green image factor. In order to communicate their new environmental goals, such as the mitigation of the supply chain's carbon emissions, companies launch CSR programs. Nonetheless, apart from the environmental perspective of sustainability, a corporate key objective is to achieve economic sustainability through creating a green competitive advantage in the marketplace that leads to improved competitiveness and profitability.

21.2.4 RANKING OF THE CRITICAL OPERATIONS AND IDENTIFICATION OF THE SUPPLY CHAIN PARTNERS' ROLE

When considering CF management from a product's life-cycle perspective, companies should take into account the emissions of all supply chain stages, namely from the raw material extraction to the recycling or disposal of waste. Depending on the industry sector in which the company operates, the ranking of the supply chain activities according to their CF contribution varies considerably. Indicatively, the production phase is responsible for the majority of carbon emissions due to high energy requirements in various manufacturing plants (US Department of Energy 2010). In general, transportation generates significant amounts of CF in a product's supply chain. According to the Stern Review (2006), transportation accounts for the 14% of the total GHG emissions, with three-quarters of them emitted due to commercial road transport.

At the same time, since CF management is considered in a supply chain network context, the need for a holistic approach involving all stakeholders is recognized. Companies should extend their point of view toward the upstream and downstream levels of the supply chain, and identify the role of each partner (e.g., suppliers, distributors, etc.) in the overall CF chain, their strengths and weaknesses, as well as potential opportunities for improvement.

21.2.5 MEASUREMENT OF THE TOTAL SUPPLY CHAIN'S CARBON FOOTPRINT

Companies should calculate the CF of each supply chain activity, as well as of the total. Given that the legislative CF measures are exogenously defined, businesses should focus on: (i) the implementation of accurate data-gathering mechanisms, (ii) the allocation of the relevant roles and responsibilities to the company's departments, (iii) the acquisition of the necessary measurement tools and software, and (iv) the establishment of reliable procedures for measuring the CF and sharing the results to the rest supply chain partners.

With regard to the calculation techniques, CF measurement (carbon accounting) is performed with life-cycle-oriented tools and software that quantify the GHG emissions. The most widely used international tools are ISO 14064, ISO/TS 14067, and the GHG Protocol. Specifically, ISO 14064 provides guidance for: (i) the quantification and reporting of the GHG emissions at the company level, (ii) the design of project-based activities for reducing emissions, as well as (iii) the validation and verification of the GHG projects (ISO, 2006a,b,c). At the same time, ISO/TS 14067 specifies the principles, requirements, and guidelines for the quantification and communication of the total life-cycle CF at the product level (ISO 2013). Finally, the GHG Protocol provides accounting and reporting standards for companies in order to calculate, manage, and report the GHG emissions at the corporate or product level (WRI and WBCSD 2012a). Particularly, the GHG Protocol classifies the GHG emissions in three scopes, those of: (i) direct emissions obtained from sources owned and controlled by the company (Scope 1), (ii) indirect emissions of purchased electricity or heat (Scope 2), as well as (iii) indirect emissions, such as the raw materials' extraction or transportation (Scope 3) (WRI and WBCSD 2012b).

21.2.6 EVALUATION OF THE OUTCOMES AND RECONSIDERATION OF THE GOALS

It is essential that companies evaluate whether the outcomes verify that the supply chain activities, previously considered as critical (e.g., the manufacturing stage in a rockwool supply chain due to high energy requirements), actually generate high levels of carbon emissions. In general, a company's decision makers may take actions when results indicate a high concentration of direct emissions, as they are exclusively responsible for their controlling. On the other hand, if the outcomes reveal an excess in the indirect emissions (e.g., emissions due to the utilization of electrical energy or transportation), this may be a sign that manufacturers should scrutinize their relationship with the rest supply chain stakeholders, such as suppliers or third party logistics (3PL) providers. Furthermore, the comparison of the company's supply chain CF metrics with the CF indicators of other global competitors (benchmarking) could also provide valuable findings. After the evaluation process, the initial company's goals should be reconsidered according to the gained up-to-date information and insights.

21.2.7 CENTRALIZED AND DECENTRALIZED DECISION MAKING
FOR SUPPLY CHAIN CARBON FOOTPRINT MANAGEMENT

Taking into account the managerial insights obtained from the previous steps, the insulating materials manufacturer should firstly decide on the CF management strategies, as well as on the relevant supporting procedures that should be implemented. Secondly, the company should define which department will be in charge of deciding the methodology to be employed for reducing the direct emissions during the internal supply chain activities, such as manufacturing, packaging, or even warehousing. Finally, the company should appropriately select, negotiate, and contract with the supply chain partners who are responsible for producing all the indirect emissions related to the sourcing and procurement of the raw materials, fuels, and energy, as well as the distribution and transportation of the end products to the retailers. Consequently, the implementation of the final decisions requires the coordination among all supply chain stakeholders. This cooperative scheme adds value in the long-run concept of stable and harmonized relationships and common strategic planning regarding CF management across supply chains.

21.2.8 Monitoring and Reporting of the Supply Chain's Carbon Footprint

Apart from the officially existing provided tools (i.e., the reporting tool for National Implementation Measures under the EU ETS), insulating material manufacturers should be capable of monitoring their internal procedures on a more frequent basis, rather than the one-year period during which the firm is usually obliged to report its CF performance to the related authorities. To that end, managers should agree on a solid information reporting scheme within the company's borders, as well as on a communication scheme between the company and its supply chain partners. Stakeholders have a crucial contribution to the overall environmental footprint of a product and thus cooperation in terms of reporting and meetings is a prerequisite for the efficiency of the entire venture.

21.2.9 Reevaluation and Update of the Decisions

In light of a dynamically changing global environment, manufacturers need to reevaluate and periodically update their adopted CF management decisions. In the case of GHG emissions, for example, manufacturers should take into account potentially lower regulatory boundaries in order to reassess their decisions and apply new green technological advancements for further reducing the supply chain CF. Finally, apart from the regulation-driven changes and in terms of the economic sustainability, companies along with their stakeholders should develop any profitability-driven action (e.g., contracts with environmentally friendly partners, implementation of green production technologies, recycling policies, etc.) targeted to improve their performance through the efficient supply chain CF management.

21.3 CASE STUDY: A REAL-WORLD ROCKWOOL SUPPLY CHAIN

In this section, we present the applicability of the proposed methodology to the rockwool supply chain of a real-world company that belongs to the Greek insulating materials industry. The manufacturer counts 40 years of experience in producing and distributing several types of insulating materials, such as extruded polystyrene, expanded polystyrene, and rockwool. Notably, the company ranks third in terms of extruded polystyrene productivity in Europe. Thus, with regard to environmental sustainability, the company has already evaluated the life-cycle GHG emissions of its extruded polystyrene products. However, given that the rockwool supply chain requires both a significant amount of energy resources and a wide distribution network, the firm now intends to follow the aforementioned framework in order not only to quantify, but also to holistically manage the total supply chain CF of rockwool.

In the following subsections, first the definition and the boundaries of the system under study are presented. Particularly, the main activities of the rockwool supply chain are described in brief. Then, the two-phase CF management framework is implemented in the real-world rockwool supply chain.

21.3.1 The Rockwool Supply Chain

Global rockwool supply chains exhibit certain unique characteristics. Specifically, in this case study, the sourcing and procurement of the raw materials (mineral rocks) and other auxiliary materials for the rockwool manufacturing process indicate high spatial variability that necessitates the design of wide and well-organized supply networks. At the same time, rockwool manufacturing requires the consumption of high volumes of electrical energy, notably for the melting of the mineral rocks and the production of the final product. Finally, the end products are transported and distributed through an extensive network of customers worldwide.

Virtually, the rockwool supply chain under study is comprised of seven (7) discrete activities, namely procurement and transportation of raw materials, melting of raw materials in an electrical furnace, fiberization in a spinner, heating in a polymerization furnace, cutting and final facing, packaging and warehousing as well as transportation of final products. These operations, along with the associated materials and resources, are depicted in Figure 21.2.

More specifically, the rockwool supply chain echelons of the system are described in brief as follows:

- Procurement of raw materials, including the extraction of mineral rocks for the rockwool formation, as well as the production of secondary raw materials for the manufacturing and packaging processes.
- Transportation of necessary resources from worldwide dispersed suppliers to the manufacturing plant, either by trucks or vessels for longer distances.
- Manufacturing: The mineral rocks enter the electrical furnace for melting. Afterwards, the molten material flows into the spinner, where the rockwool fibers are produced. Then, rockwool enters the polymerization furnace for achieving the fibers' bonding and the desired thickness of the end product. Then, rockwool goes through the cutting machines, where the final dimensions are configured. Finally, after the final facing, the end product of rockwool is formed.
- Packaging and warehousing of final products within the borders of the manufacturing plant.
- Transportation and distribution of final products from the plant to worldwide-dispersed customers mainly by trucks or else by vessels.

In the supply chain under study, the phases of the recycling of defective products and the use of final products are omitted from the CF quantification. Moreover, since the rockwool is used for internal insulation in buildings for a long time span, the stage of the recycling or waste disposal of used products has not been taken into account.

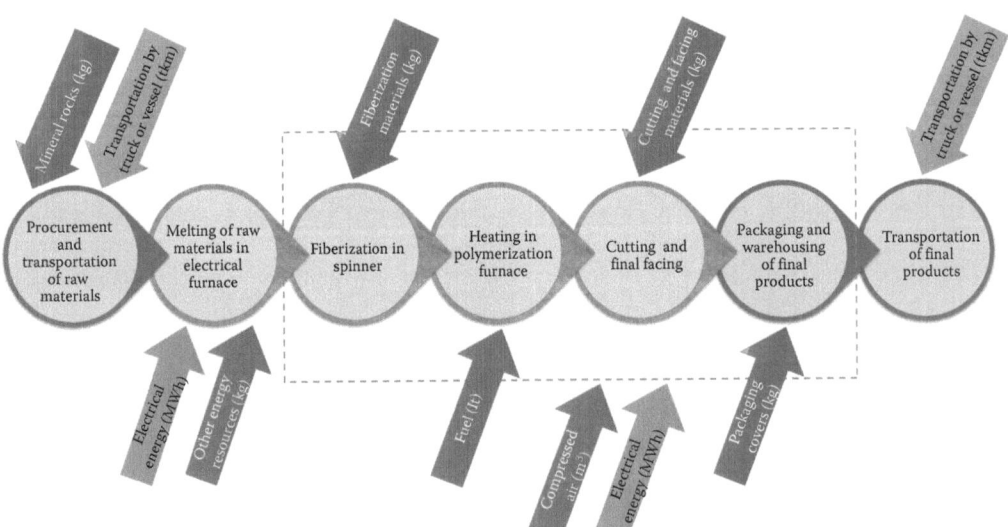

FIGURE 21.2 The rockwool supply chain echelons. (From Aivazidou, E. et al. 2014. *Proceedings of the 5th International Conference on Manufacturing Engineering (5th ICMEN)*, Thessaloniki, Greece. With permission.)

21.3.2 Implementation of the Carbon Footprint Management Framework

The implementation of the CF management methodological framework in the supply chain of the rockwool manufacturer is provided through the description of each specific step as described below.

21.3.2.1 Recording the European Regulatory Framework

The rockwool manufacturer entered the EU ETS in 2013 as an industrial plant operating in Europe. According to the EU ETS, the company is responsible for its direct GHG emissions produced during the manufacturing processes, as well as the indirect emissions due to the utilization of electrical energy during all internal operations (e.g., manufacturing, packaging, and offices). To date, there is a lack of an integrated emissions trading system in Europe referring to the entire supply chain of a product. Hence, currently the company is not legally liable for the indirect CF associated with the procurement and transportation stages.

21.3.2.2 Identifying the Best Practices of the Top Global Competitors

The company under study took into consideration the recorded best practices that top foreign competitors (e.g., Rockwool Group, Paroc, and Knauf Insulation) have already deployed across their rockwool supply chains, succeeding in mitigating the total CF. Particularly, the manufacturer's Research and Development (R&D) department studied thoroughly the competitors' sustainability reports in order to map the carbon emissions produced along the rival supply chains, as well as to identify and adopt green polices that have proven to be efficient in practice.

21.3.2.3 Setting Goals for Carbon Footprint Management in the Rockwool Supply Chain

Concerning the legal responsibilities, the manufacturer's primary objective is twofold: (i) first not to exceed its emission's cap determined by the EU ETS for avoiding penalties, and (ii) then to trade its surplus emission permits with relevant companies in the trading system. At the same time, the key strategic goal of the rockwool manufacturer is the reduction of the total CF for fostering the company's CSR, as well as ensuring economic sustainability for all supply chain stakeholders. Finally, the company aims at outperforming its global competitors through: (i) benchmarking its supply chain's CF performance indicators compared to the competitors, and (ii) improving its green image and competitiveness through using environmentally friendly practices across the rockwool supply chain.

21.3.2.4 Identifying the Critical Rockwool Supply Chain Operations and the Stakeholders' Role

Due to the fact that rockwool products require significant amounts of energy resources during their production, the company under study is concerned about the CF generated during the rockwool manufacturing. Given that electrical energy in Greece is produced mainly from lignite extraction and burning, which are environmentally harmful processes, the manufacturing stage is considered to be the critical operation in the rockwool supply chain when considering CF production. With respect to the external supply chain activities, the company has already taken into account the significant role of its suppliers and 3PL providers, since the company's wide supplier and customer network requires lengthy road transportation that implies significantly high GHG emissions.

21.3.2.5 Quantifying the Rockwool Supply Chain Carbon Footprint

After mapping the internal and external environment of the company, the R&D department was in charge of quantifying the CF of the entire rockwool supply chain, namely from the procurement and transportation of the raw materials, via manufacturing, packaging, and warehousing, to the distribution of the final products. In this study, the CF is calculated in CO_2-eq. units per functional unit, which represents 1 ton of produced rockwool.

At the same time, there are two types of data for the CF assessment across the rockwool supply chain: (i) the input quantities for each of the system's operations required for the production of 1 ton of rockwool, and (ii) the emissions factors (in CO_2-eq.) per input unit. The data for the first category (quantities of mineral rocks, secondary raw materials, packaging materials, electrical energy, fuels, and transport distances) were directly provided by the manufacturer's R&D department. The majority of data for the second category were obtained from the Ecoinvent Center, which is a top Swiss provider for LCA data (Ecoinvent 2013). In case the coefficients for certain processes were not available at the employed database, emission factors of similar processes were used. Concerning the utilization of fuels during the manufacturing process, the relevant emissions factor was obtained from the UK Department of Energy and Climate Change. Regarding the decomposition of carbonates during rockwool manufacturing, the emission factors were calculated based on the relevant chemical reactions that take place.

Finally, the calculation of the total CF (in CO_2-eq.) of 1 ton of produced rockwool was conducted on Excel spreadsheets, while the evaluation of the rockwool supply chain CF was based upon the specifications of the GHG Protocol. The obtained results, according to the rockwool supply chain activities and the GHG Protocol scopes, are illustrated in Figures 21.3 and 21.4, respectively. Quantitative estimates are presented in percentile form for reasons of confidentiality.

According to the numerical analysis, the procurement of the raw and auxiliary materials contributes by 21% to the GHG emissions of the rockwool supply chain, while the transportation of the necessary resources and final rockwool products is responsible for 12% of the total CF. Notably, the manufacturing process constitutes the main CF hotspot in the rockwool supply chain accounting for 2/3 of the carbon emissions. Within the context of the GHG Protocol, the indirect emissions of the procurement and transportation stages (Scope 3) are responsible approximately for 1/3 of the total CF, while the indirect emissions derived from the electricity use (Scope 2) constitute 64% of the GHG emissions in the rockwool supply chain. Finally, the direct emissions during the production of rockwool (Scope 1), mostly generated due to relevant chemical reactions, contribute by less than 3% in the total supply chain CF.

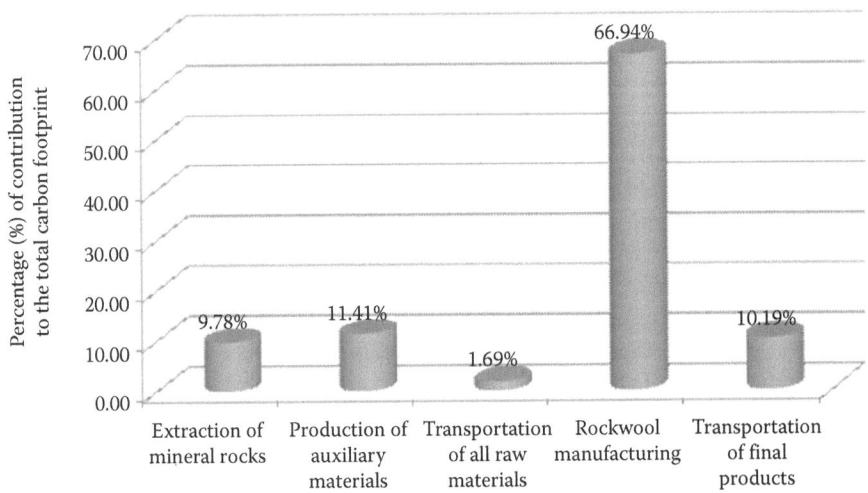

FIGURE 21.3 CF allocation in the rockwool supply chain operations. (From Aivazidou, E. et al. 2014. *Proceedings of the 5th International Conference on Manufacturing Engineering (5th ICMEN)*, Thessaloniki, Greece. With permission.)

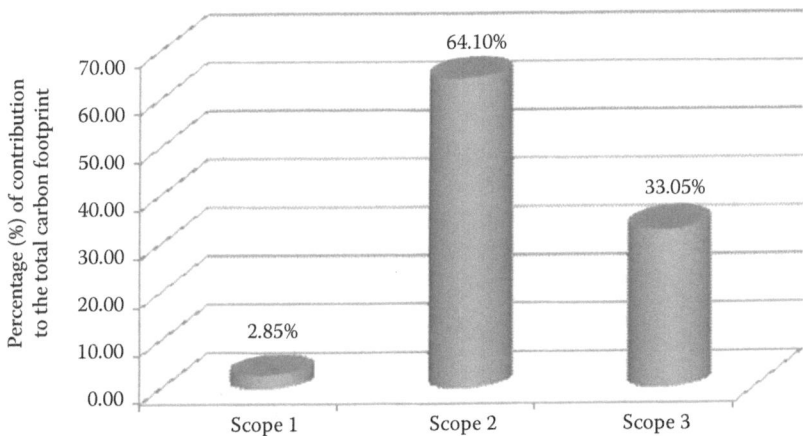

FIGURE 21.4 CF allocation in the GHG Protocol scopes. (From Aivazidou, E. et al. 2014. *Proceedings of the 5th International Conference on Manufacturing Engineering (5th ICMEN)*, Thessaloniki, Greece. With permission.)

21.3.2.6 Evaluating Carbon Footprint Findings for Reconsidering the Goals

It is evident that the stage of rockwool manufacturing is the key contributor to the total CF, particularly due to the utilization of a significant amount of electrical energy that is mainly produced from lignite burning. Specifically, after conducting various sensitivity analyses concerning the electrical energy's source, the hypothetical utilization of environmentally friendly or renewable energy resources (i.e., natural gas burning, photovoltaic panels, and hydroelectric power) proved to reduce the rockwool supply chain CF by 16% up to 64% (Aivazidou et al. 2014). On the other hand, the quantity of the direct emissions during rockwool manufacturing, for which the company is solely responsible, is sufficiently low. Furthermore, the extraction of the mineral rocks and the production of several auxiliary materials for production of the end product are considered to be environmentally unsound thus augmenting the emissions of the procurement stage. Finally, because the distribution of the raw materials and final products is mainly performed through the road transport, indirect GHG emissions from transportation are considerably increased.

Moreover, the company compared its supply chain CF performance metrics to the best values recorded for the industry, which were found in the CSR reports of the top global insulating material manufacturers. The benchmarking results reveal that, except for the CF generated from the use of electrical energy, the rest of the supply chain GHG emissions per unit of produced rockwool are considered approximately equal to the emissions of the company's competitors. Finally, when reconsidering the initial goals, the manufacturer is specially focused on: (i) the potential minimization of CF produced from electricity use, and (ii) the negotiation with the transportation-related stakeholders in order to optimize its global distribution network to further reduce GHG emissions across the rockwool life cycle.

21.3.2.7 Decision Making for Carbon Footprint Management in the Rockwool Supply Chain

After the strategic evaluation of the rockwool supply chain's carbon emissions, the decision-making process follows in order to define the actions for effectively managing the total CF. These actions should be taken within the company's borders as well as in collaboration with the rest of the supply chain partners. With regard to the mitigation of the electricity emissions, the company's manufacturing department recommends: (i) partially up-grading the manufacturing procedures to environmentally friendly, for example, through replacing any outdated manufacturing equipment with high-tech and energy-efficient machinery, and (ii) utilizing company-owned photovoltaic panels in order to reduce the use of electrical energy purchased from the Greek electricity provider and mainly produced from lignite power plants. At the same time, efforts are made through negotiating with 3PL providers regarding the distribution routes in order to optimize the GHG emissions in the rockwool global supply chain network. Particularly, the rockwool manufacturer in cooperation with its stakeholders aims at increasing the deployment of sea or rail transport modes where feasible, since these options are not only low-carbon but also cost-effective. Finally, in the long-run, in case the R&D department detects improved formulas for rockwool products, the company's target is to negotiate with its existing suppliers or contract with new suppliers in order to purchase environmentally safe materials for decreasing the CF of procurement.

21.3.2.8 Monitoring and Reporting the Rockwool Supply Chain Carbon Footprint

To date, the manufacturer is obliged to annually monitor and report only the direct and indirect emissions produced during its internal processes within the context of the EU ETS. However, the company's managers have already raised the issue of creating an integrated CF reporting scheme (i.e., based on the ISO/TS 14067 specifications for CF communication), which will include all supply chain stakeholders, in order to assess the impact of the implementation of the green policies. Moreover, the outcomes of the GHG emissions reporting across the rockwool life cycle are going to be published in the annual sustainability reports for promoting CSR and enhancing the company's environmental image.

21.3.2.9 Reevaluating Outcomes for Updating the Carbon Footprint Management Decisions

Finally, after monitoring the improvement in the CF of the rockwool supply chain for a long time span, the company will able to evaluate the effectiveness of the actions already taken in order to reduce the GHG emissions. In case the outcomes are not satisfactory, the company, along with its stakeholders, should reconsider its decisions and perform further rigorous actions, such as implementing recycling of defective products or other waste management techniques, which may ensure environmental and economic sustainability for all partners across the rockwool supply chain.

21.4 CONCLUSIONS AND RECOMMENDATIONS

The management of the environmental impacts, and particularly GHG emissions, of global supply chain networks has emerged as a key priority for corporations in order not only to comply with the strict environmental regulations, but also to exhibit a sustainable supply chain performance and gain a green competitive advantage. In this chapter, a holistic two-phase methodological framework that incorporates the natural decision-making hierarchy (strategic and tactical levels) for supply chain CF management is presented. Specifically, the proposed framework could serve as a comprehensive guide for a company to progressively map its external and internal environment, set its goals regarding CF mitigation, and assess its supply chain GHG emissions, in order to decide the appropriate actions that should be taken for managing the total CF efficiently during a product's life cycle.

Particularly, the framework is implemented in the case of a real-world rockwool supply chain of a Greek insulating materials manufacturer. The results indicate that the company should take drastic actions toward the reduction of the high manufacturing indirect emissions due to the utilization of electrical energy produced from lignite, such as utilization of energy-efficient equipment or consumption of low-carbon renewable energy resources. At the same time, the company is currently negotiating contracts with the rest of the supply chain partners in order to minimize the CF produced during the procurement of raw materials and the transportation of final products. Accordingly, potential collaboration with environmentally friendly 3PL providers that deploy sea or rail transport modes, where feasible, could significantly decrease the transportation emissions.

Furthermore, the benchmarking of the rockwool supply chain CF performance reveals that the manufacturer is competitive to other foreign rival companies, when considering the amount of the direct GHG emissions produced during the internal processes, as well as indirect emissions of procurement and transportation. Notably, the electricity emissions are considerably higher due to the fact that the Greek energy production sector is less environmentally friendly compared to other countries.

Virtually, the implementation of the tactical phase of the framework, and particularly the decision-making process, is still in progress. Thus, the evaluation of the decisions will be conducted after monitoring and assessing the reductions in the total CF when the green policies will have been employed across the rockwool supply chain.

Concerning future research, the case study of the CF management in the rockwool supply chain can act as an example for multiple supply chains that display similar attributes, especially that of the mineral wool and metal industry. Particularly, it is recommended that the decision making framework be implemented in multiple supply chains of other insulating materials and similar products, in order to: (i) identify similarities and discrepancies in the CF quantification, management, and mitigation for various products, and (ii) foster environmental sustainability and profitability for the sectoral manufacturing companies.

ACKNOWLEDGMENTS

This research has been conducted in the context of the GREEN-AgriChains project that has received funding from the European Union's Seventh Framework Programme (FP7-REGPOT-2012-2013-1) under Grant Agreement No. 316167. All the above reflect only the authors' views; the European Union is not liable for any use that may be made of the information contained herein.

In addition, E. Aivazidou is grateful to the Alexander S. Onassis Public Benefit Foundation for awarding her a fellowship to conduct this research work as a part of her doctoral studies.

REFERENCES

Adom F, Workman C, Thoma G, Shonnard D. 2013. Carbon footprint analysis of dairy feed from a mill in Michigan, USA. *International Dairy Journal* 31:S21–S28.

Aivazidou E, Iakovou E, Vlachos D, Keramydas C. 2013. A methodological framework for supply chain carbon footprint management. *Chemical Engineering Transactions* 35: 313–318.

Aivazidou E, Toka A, Iakovou E, Vlachos D. 2014. Assessing the carbon footprint of the rockwool supply chain: A real-world case study. *Proceedings of the 5th International Conference on Manufacturing Engineering (5th ICMEN)*, Thessaloniki, Greece.

Carbon Trust. 2012. Carbon footprinting. http://www.carbontrust.com/media/44869/j7912_ctv043_carbon_footprinting_aw_interactive.pdf (accessed July 23, 2014).

Clark D. 2012. What's a carbon footprint and how is it worked out? http://www.guardian.co.uk/environment/2012/apr/04/carbon-footprint-calculated (accessed July 23, 2014).

Ecofys. 2009. Methodology for the free allocation of emission allowances in the EU ETS post 2012. Sector report for the mineral wool industry. http://ec.europa.eu/clima/policies/ets/cap/allocation/docs/bm_study-mineral_wool_en.pdf (accessed July 23, 2014).

Ecoinvent. 2013. The ecoinvent Database. http://www.ecoinvent.org/database/database.html (accessed May 6, 2015).

European Commission. 2009. Europeans' attitudes towards the issue of sustainable consumption and production. http://ec.europa.eu/public_opinion/flash/fl_256_en.pdf (accessed July 23, 2014).

European Commission. 2012. EU transport GHG: Routes to 2050 project. http://www.eutransportghg2050.eu/cms/?flush=1 (accessed July 23, 2014).

European Commission. 2013. The EU emissions trading system (EU ETS). http://ec.europa.eu/clima/policies/ets/index_en.htm (accessed July 23, 2014).

International Standardization Organization (ISO). 2006a. ISO 14061-1:2006 Greenhouse gases—Part 1: Specification with guidance at the organization level for quantification and reporting of greenhouse gas emissions and removals. https://www.iso.org/obp/ui/#iso:std:iso:14064:-1:ed-1:v1:en (accessed November 12, 2014).

International Standardization Organization (ISO). 2006b. ISO 14061-2:2006 Greenhouse gases—Part 2: Specification with guidance at the project level for quantification, monitoring and reporting of greenhouse gas emission reductions or removal enhancements. https://www.iso.org/obp/ui/#iso:std:iso:14064:-2:ed-1:v1:en (accessed November 12, 2014).

International Standardization Organization (ISO). 2006c. ISO 14061-3:2006 Greenhouse gases—Part 3: Specification with guidance for the validation and verification of greenhouse gas assertions. https://www.iso.org/obp/ui/#iso:std:iso:14064:-3:ed-1:v1:en (accessed November 12, 2014).

International Standardization Organization (ISO). 2013. ISO/TS 14067:2013 Greenhouse gases—Carbon footprint of products—Requirements and guidelines for quantification and communication. https://www.iso.org/obp/ui/#iso:std:59521:en (accessed November 12, 2014).

Iriarte A, Almeida MG, Villalobos P. 2014. Carbon footprint of premium quality export bananas: Case study in Ecuador, the world's largest exporter. *Science of the Total Environment* 472:1082–1088.

Jensen JK, Arlbjørn JS. 2014. Product carbon footprint of rye bread. *Journal of Cleaner Production* 82:45–47.

Koroneos C, Dompros A. 2007. Environmental assessment of brick production in Greece. *Building and Environment* 42:2114–2123.

Kumar S, Teichman S, Timpernagel T. 2012. A green supply chain is a requirement for profitability. *International Journal of Production Research* 50(5):1278–1296.

Lee K. 2011. Integrating carbon footprint into supply chain management: The case of Hyundai Motor Company in the automobile industry. *Journal of Cleaner Production* 19(11):1216–1223.

Lenzen M, Murray J, Sack F, Wiedmann T. 2007. Shared producer and consumer responsibility—theory and practice. *Ecological Economics* 61:27–42.

Li Z, Gong X, Wang Z, Liu Y, Liping M, Wang S, Guo J. 2014. Life cycle assessment of rock wool board and EPS board. *Materials Science Forum* 787:106–110.

Montoya-Torres JR, Gutierrez-Franco E, Blanco EE. 2014. Conceptual framework for measuring carbon footprint in supply chains. *Production Planning and Control: The Management of Operations* (in press).

Muthu SS, Li Y, Hu JY, Ze L. 2012. Carbon footprint reduction in the textile process chain: Recycling of textile materials. *Fibers and Polymers* 13(8):1065–1070.

Paroc. 2013. Book of sustainability. http://www.paroc.com/about-paroc/focus-on-sustainability/~/media/Files/Brochures/COM/Paroc-Book-of-sustainability_2013-INT.ashx (accessed July 23, 2014).

Proietti S, Desideri U, Sdringola P, Zepparelli F. 2013. Carbon footprint of a reflective foil and comparison with other solutions for thermal insulation in building envelope. *Applied Energy* 112:843–855.

Rigot-Muller P, Lalwani C, Mangan J, Gregory O, and Gibbs D. 2013. Optimising end-to-end maritime supply chains: A carbon footprint perspective. *The International Journal of Logistics Management* 24(3):407–425.

Rockwool Group. 2013. Sustainability report for the reporting year 2012. http://www.rockwool.com/files/COM2011/CSR/Committed-to-society/Sustainability-Report/Reporting-year-2012_Published-2013/SR-2012_2013_EN.pdf (accessed July 23, 2014).

Scipioni A, Mastrobuono M, Mazzi A, Manzardo A. 2010. Voluntary GHG management using a life cycle approach: A case study. *Journal of Cleaner Production* 18(4):299–306.

Stern N. 2006. Stern review: The economics of climate change. http://webarchive.nationalarchives.gov.uk/20130129110402/http://www.hm-treasury.gov.uk/d/Executive_Summary.pdf (accessed July 23, 2014).

Sundarakani B, Goh M, de Souza R, Shun C. 2008. Measuring carbon footprints across the supply chain. *Proceedings of the 13th International Symposium on Logistics (ISL2008): Integrating the Global Supply Chain*, Bangkok, Thailand, pp. 555–562.

US Department of Energy. 2010. Understanding manufacturing energy and carbon footprint. http://www1.eere.energy.gov/manufacturing/pdfs/understanding_energy_footprints_2012.pdf (accessed July 23, 2014).

Wiedmann T, Minx J. 2008. A definition of 'carbon footprint'. In: Pitsova CC (ed.), *Ecological Economics Trends*. Hauppauge, NY: Nova Science Publishers, pp. 1–11.

World Resources Institute (WRI), World Business Council for Sustainable Development (WBCSD). 2012a. About the GHG protocol. http://www.ghgprotocol.org/about-ghgp (accessed November 12, 2014).

World Resources Institute (WRI), World Business Council for Sustainable Development (WBCSD). 2012b. FAQ. http://www.ghgprotocol.org/calculation-tools/faq (accessed July 23, 2014).

Young DM, Hawkins T, Ingwersen W, Lee S, Ruiz-Mercado G, Sengupta D, Smith RL. 2012. Designing sustainable supply chains. *Chemical Engineering Transactions* 29:253–258.

22 Product Carbon Footprint Estimation of a Ton of Paper

Case Study of a Paper Production Unit in West Bengal, India

Debrupa Chakraborty

CONTENTS

22.1 Introduction...487
    22.1.1 Background..487
22.2 Methodology for Estimating PCF..490
    22.2.1 Process Description 490
        22.2.1.1 Paper Manufacturing Process 490
        22.2.1.2 Sheet Forming Section...490
        22.2.1.3 Press Section..490
        22.2.1.4 Drying..491
        22.2.1.5 Calendaring..491
        22.2.1.6 Reeling...491
    22.2.2 Study Boundary and Data Collection 491
        22.2.2.1 Study Boundary 491
        22.2.2.2 Data Collection 493
        22.2.2.3 Scope, Research Gaps, Assumptions, and Limitations of Study 493
22.3 Results and Discussions...497
22.4 Conclusion 499
Acknowledgment 499
Appendix...500
References..500

22.1 INTRODUCTION

22.1.1 BACKGROUND

The growing importance of low-carbon pledges and policies has been conceptualized at the global landscape through binding commitments and voluntary agreements. This includes ratification of global covenants (such as Kyoto protocol, Copenhagen Accord) that have created impetus for the Indian industries to undertake the bottom-up approach leading to low-carbon innovations. Low-carbon innovations in turn have led to produce low-carbon products with reduced emissions, definitely helping industries to remain competitive in the market. To achieve this, a consumption scrutiny of natural resources and its associated emissions would help businesses and consumers to assess their roles toward contributing to climate change and on the impact their purchase-related decisions have on emissions. The consumption-based approach to evaluate resource use and its associated emissions is becoming all the more important in equity debate in the climate change context.

The carbon footprint (CF) concept emerged as a measure to estimate the total amount of greenhouse gases (GHGs) emitted during the life cycle of the goods and services, that is, from extraction of raw materials, production, transportation, storage, and use to waste disposal. They are calculated by businesses, governments, and other stakeholders in order to understand the emissions of GHGs from different products (Bolwig and Gibbon, 2009). Businesses can use the result of a CF to achieve emissions reductions throughout their operations, to influence the use of different elements of their supply chains and to communicate their CF to their customers (Plassmann et al., 2010). In this context, product carbon footprint (PCF) addresses businesses' need to better understand the extent to which their products and activities along the supply chains participate in carbon emissions. It also attempts to respond to stakeholders' growing demand for low-carbon products with associated carbon information. It again provides an opportunity for the businesses to identify the drivers behind emissions, scope to reduce emissions across the product life cycle leading to cost savings, supporting the development of future low-carbon products, and thereby finally helping mitigate the climate change. PCF is one of the evolving environmental performance indicators that work as a part of larger contributory effort to expedite the move toward a low-carbon economy.

Applications of the footprint method to assess industrial systems are sparse. The industry sector plays an important role with 27.3 GDP share (2012–2013) in the fast economic growth of India (http://indiabudget.nic.in/es2013-14/echap-01.pdf). Given that the industrial sector plays an important role in the fast economic growth of India and with further expected economic development, it may be interesting to know the impact that the products of these industries are creating on ecological resources and the associated energy consumed by them. Further, the emissions that are being generated in their production processes necessitate an in-depth analysis.

Estimation of PCF for an energy-intensive industry such as pulp and paper is necessary, particularly for a developing economy such as India. At present in India, there are in total 759 pulp and paper mills with an installed capacity of 12.7 million tons producing around 10.11 million tons/annum of paper/paper board and newsprint out of the total world production of around 402 million tons (Kulkarni, 2013). The Indian paper industry structure consists of small-, medium-, and large-sized paper mills having production capacities ranging from 10 to 1150 tons per day (Kulkarni, 2013). These industries employ wood, agro residues, and recycled/wastepaper as the major raw material for manufacturing different varieties of paper, paper board, and newsprint. In 2011, the share in production of paper from wood-based raw materials, agro residues, and recycled/wastepaper had been 31%, 22%, and 47%, respectively (Kulkarni, 2013). Paper mills in India continue to face challenges with wastepaper-based raw materials. Of the 4.5 million tons of wastepaper required for recycling, the indigenous wastepaper available for recycling is 2.0 million and 2.5 million tons depending on import (Kulkarni, 2013). This occurs because agro-based industries are closing down as a result of pollution-related problems and wastepaper quality, thereby putting pressure on the paper industry.

In this context also the PCF estimation will definitely help the industries to cope up with the challenge of creating impetus for the Indian industries to undertake bottom-up response leading to low-carbon innovations. This will lead toward production of low-carbon products with reduced emissions and retain competitiveness in the global market. To achieve this, a consumption scrutiny of natural resources and its associated emissions would help businesses and its consumers to access their role contributing to climate change and on the impact their purchase-related decisions have on emissions. The consumption-based approach to evaluate resource use and its associated emissions is becoming all the more important in equity debate in the climate change context.

In this study, PCF for one ton of paper, produced by a paper mill manufacturing "newsprint" varieties in West Bengal, India since 1994, is considered. PCF of a ton of paper has been calculated for the year 2013–2014.

The reason for measuring the PCF of one ton of paper is that production capacity or selling price of the product is usually based on 1 ton in India. More precisely the objectives of this study are to

1. Estimate the amount of GHG emission of a ton of paper product and its allocation in the production and distribution process.
2. Access the ability of the unit to reduce emissions and evaluate the impacts the methodological choices have on the CF of product.
3. Improve the environmental performance of the case study unit and to reduce the resultant PCF.

In this context, PCF of one ton of paper available for consumption has been calculated by taking into consideration the emission associated with production process and activities along the supply chains to produce and deliver 1 ton of paper to customers. The case study unit manufactures more or less 18,000 tons of paper per annum, primarily from wastepaper. The unit is certified by International agency, Ms. DNV, the Netherlands for ISO 9001:2000 for Quality Management System, ISO 14001:2004 for Environmental Management System & OHSAS 18001: 1996. The case study unit chosen for the present study is a good representation of an important segment of Indian paper manufacturing units and belongs to a company that is a forerunner in terms of modernization and follower of sustainable development goals.

CF estimation of products such as bananas across the supply chain in the United States (Craig et al., 2012) shows the impact of transportation on the structure of a supply chain introducing significant variability in the CF required to serve different customers; CF of Indian food items revealed (Pathak et al., 2010) that animal food products (meat and milk) and rice cultivation mostly contributed to methane (CH_4) emission, while food products from crops contributed to emission of nitrous oxide (N_2O) and emission of CO_2 occurred during farm operations, production of farm inputs, transport, processing, and preparation of food; a study on global warming potential of biogas plants in India (Pathak et al., 2009a) suggests that policy makers should promote biogas technology to combat climate change, and integration of carbon revenues will help farmers to develop biogas as a profitable activity exists in literature. In Indian context, CF calculation of Indian household reveals that when households become more affluent, the demand for energy-intensive goods and services increases disproportionately in comparison with less emission-intensive consumption categories (Grunewald et al., 2013). CF of road transport in Kolkata City, India shows that per capita footprint from transport use increases with income and further income, vehicle ownership, and per capita transport expenditure has significant and positive impact on per capita footprint of households from road transport use (Dasgupta, 2014). Literature survey in the Indian context reveals that a considerable gap in knowledge domain exists in the area of performance evaluation of the industries through resource use assessments and PCF calculations.

This study demonstrates the ways in which PCF can be estimated using unit level data collected from the specific industry. Final choice of the study unit has been determined by willingness of the unit to cooperate in data sharing and time commitment for a face-to-face interview.

Several methodologies for calculating PCF developed by government or businesses varying in their approach and mode of functioning are in operation or in the course of being developed. Of the various models for calculating the CF of a product, the life-cycle assessment (LCA) model framed by GHG Protocol, ISO 14067 and British PAS 2050 framework are the most significant ones. ISO 14067 is the most robust and the more general standard, with some specific requirements, for example, requirements on the use of green electricity. On the other hand, PAS 2050 and GHG Protocol provide more detailed requirements and guidance with less space for interpretation.

PAS 2050 is the first internationally applicable standard method for calculating life-cycle GHG emissions of products. PAS 2050 provides a framework for business organizations to understand challenges of climate change across supply chain. It has been adopted by a large number of companies for a wide range of products (The Carbon Trust, 2008). Publicly Available Specification (PAS) 2050 [developed by British Standard Institute (BSI), co-sponsored by the Carbon Trust and Department of Environment, Forest and Rural Affairs (DEFRA)] is based on process life-cycle assessment (LCA) approach. Following the footsteps of PAS 2050 and adopting it as the key document, World

Resource Institute (WRI) and World Business Council on Sustainable Development (WBCSD) has developed a Product GHG protocol and International Organization for Standardization (ISO) has also developed and published a standard for CF of products, ISO 14067 (WRI and WBCSD, 2011a). Again, economic input–output analysis (IOA) model has been used for tracing all transactions throughout the supply chain leading up to the final demand of produced commodities by consumers (Huang et al., 2009). However the needs for PCF are not fully met by some of the existing standards for LCA or company GHG accounting such as GHG protocol developed by WRI and WBCSD. Both these standards allow considerable flexibility, thereby providing varying decisions that are arrived at length (Plasmann et al., 2010).

The British PAS 2050 and ISO 14067 are the only finalized PCF methodology that has detailed calculation methods in the public domain since 2009 (Plasmann et al., 2010). Although ISO 14067 is more robust and recent (published in 2013), PAS 2050 is the first attempt to provide a consistent and well-integrated approach that assesses the role of consumption of goods and services at the product level that contributes toward GHG emissions. PAS refines, clarifies, and simplifies the existing LCA methods. Also PAS 2050 is considered to be the seed document for ISO 14067 and, if implemented, can make it easier for an organization to adopt to a robust standard similar to that of ISO 14067. Also the robust nature of ISO14067 calls for detailed primary and secondary data requirements, which the case study unit found difficult to share with. For this reason, for PCF calculation, Publicly Available Specification (PAS 2050) method is applied.

22.2 METHODOLOGY FOR ESTIMATING PCF

The PAS 2050 method has been used to calculate PCF of the case study product (a ton of paper) as other available standards measure emissions at the company level rather than focusing at identifying the system boundary and data quality rules for secondary data. GHG emissions from energy use, combustion processes, chemical reactions, refrigerant losses and other fugitive gases, operations, service provisions and delivery, land use change, livestock, other agricultural processes, and wastes have to be included in the assessment (Plassmann et al., 2010). Capital goods are not taken into consideration in the PAS 2050. Emissions related to human energy inputs, transport of consumers to and from shops, and transport of employees to and from work are excluded from CF calculations.

22.2.1 Process Description

The manufacture of paper is a continuous process. A brief description of the technological process, major equipment, and facilities of the production unit are described as follows:

22.2.1.1 Paper Manufacturing Process

The paper is manufactured in the "Fourdrinier Machine" having different subsections. The necessary amount of fiber, fillers, and other substances are supplied to the machine in the form of fluid suspension to moving woven wire screen for making paper.

22.2.1.2 Sheet Forming Section

The stock is discharged on the moving wire and stock is deposited on the wire. This wire section consists of elements such as rolls, foils for drainage of water, and finally vacuum is applied to improve dryness up to 20%.

22.2.1.3 Press Section

At the end of the Fourdrinier table, the free water contained in the wet sheet is removed so as to improve the sheet compactness.

22.2.1.4 Drying

The sheet is passed through the drying cylinder by means of drying screen. Steam is supplied to heat the cylinder.

22.2.1.5 Calendaring

The function of calendaring is to produce the desired finish and thickness of the sheet.

22.2.1.6 Reeling

Finally the paper is reeled on a moving drum for conversion.

22.2.2 STUDY BOUNDARY AND DATA COLLECTION

22.2.2.1 Study Boundary

The responsibility of the study unit for GHG emitted by the facilities while producing the selected product defines the organizational boundary of this study. The operational boundary includes cradle-to-gate approach.

This approach takes into account all life-cycle stages from raw material extraction up to the point at which it leaves the organization undertaking the assessment. It enables the CFs to be calculated and reported across the supply chain. Once the boundaries are selected, necessary primary data are collected from a face-to-face interview of factory managers based on a preset questionnaire and CO_2 emissions are quantified. This study uses activity data based on the primary data collected from the factory managers of the case study unit and the emission factors provided by a number of secondary sources. Emission factors have mainly been taken from secondary sources such as Transport (Department for Environment, Food and Rural Affairs [DEFRA]/Department of Energy and Climate Change [DECC] GHG reporting factors); Fuel, Electricity (Indian Network for Climate Change Assessment, India (INCCA: 2010), and CO_2 Baseline Database for the Indian Power Sector: User Guide, Version 8.0. January 2013 published by Government of India, Ministry of Power, Central Electricity Authority); wastewater, freshwater, waste (emission factors published by the PAS 2050: 2011 guidelines developed by British Standard Institute (BSI), co-sponsored by the Carbon Trust and DEFRA).

Emission factors quantify the emissions resulting from individual processes (as CO_2-eq.). GHG emission for each processes is calculated from the product of the organization's activity data and the corresponding and relevant emission factors (expressed in units of measurement) are used. At this stage, all the data having relevance to this case study are collected. The data collected are according to PAS 2050 analysis requirements. Using the following formula, the PCF has been calculated:

$$\text{Activity data (kg/L/kW h/t km etc.)} \times \text{emission factor (kg } CO_2\text{-eq. per kg/L/kW h/t km etc.)}$$
$$(22.1)$$

All these data are summed up to provide the total CF for each life-cycle stage and for the total system.

Organizational Boundary   In this case study, organizational boundaries encompass the operations over which the organization can exercise its control and also those such as transportation of raw materials, wastes, and finished goods.

Operational Boundary   The greenhouse-gas (GHG) emitted for producing a ton of paper measured in terms of kg CO_2-eq. has been considered as the main operation in this case study. The PAS 2050 method has been used to calculate the PCF. Figure 22.1 (process map of one ton of paper production) shows the mapping of activities involved at the various stages of production, namely pulping stage,

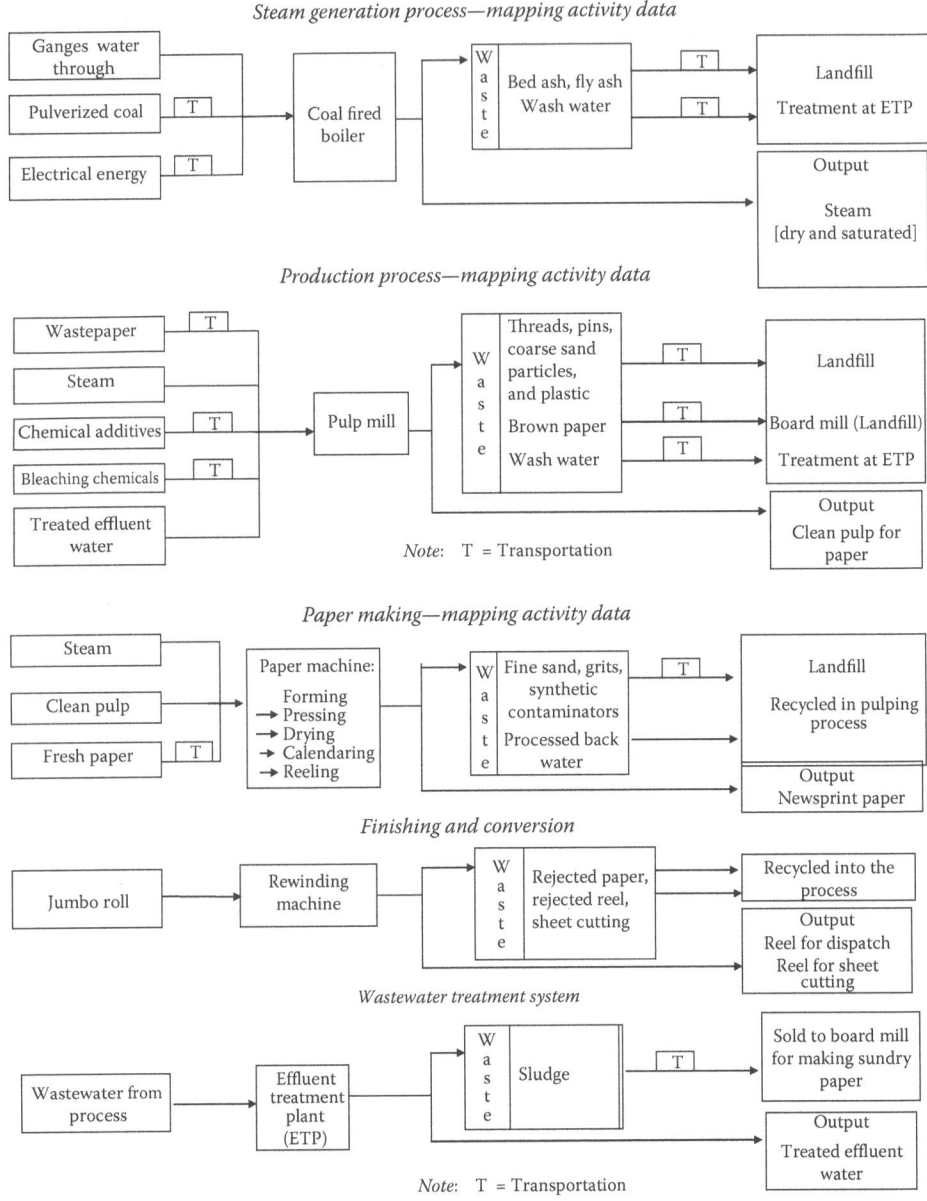

FIGURE 22.1 Process map of one ton of paper production.

paper-making stage, finishing and conversion stage, steam generation, and wastewater treatment. The product-related GHG emissions have been classified under six stages or heads, taking into consideration the consumption of electricity, freshwater, wastewater, and generation of wastes:

- Pulping stage
- Paper-making stage
- Finishing and conversion stage
- Steam generation
- Water treatment and
- Distribution of finished products

22.2.2.2 Data Collection

PCF analysis of the selected paper production unit has been studied on the basis of the primary data collected from a factory located in the state of West Bengal, India. Data were primarily collected from a face-to-face interview of factory managers based on a preset questionnaire. The required data, their sources of collection, and sources from which emission factors have been derived are provided in Table 22.1. The detailed calculation of PCF encompasses the components of production with the corresponding units, emission factors used, sources of the used emission factors, and the total emission in each stage of production for producing a ton of paper is given in Table 22.2.

22.2.2.3 Scope, Research Gaps, Assumptions, and Limitations of Study

The study shall help the paper production unit to better understand the extent to which their products and activities along the supply chains participate in carbon emissions. It also attempts to respond to stakeholders' growing demand for low-carbon products with associated carbon information. Also providing an opportunity for the unit to identify the drivers behind emissions, scope to reduce emissions across the product life cycle leading to cost savings, supporting the development of future low-carbon products and thereby finally helping toward mitigating climate change. Lack of specific emission factors for inputs and processes of the case study country (India) has necessitated the use of emission factors from databases derived mostly from European systems. This major assumption may have led to an underestimation or overestimation of resultant emission for the case study. This is a major research gap and also a limitation necessitating guidelines for developing countries. For the majority of the components of production, emission factors published by the PAS 2050: 2011 guidelines [developed by British Standard Institute (BSI), co-sponsored by the Carbon Trust and DEFRA] have been used in this study (Table 22.2). However, some available country-specific emission factors developed and published by Indian Network for Climate Change Assessment, India (INCCA: 2010) and CO_2 Baseline Database for the Indian Power Sector: User Guide, Version 8.0.January 2013 published by Government of India, Ministry of Power, Central Electricity Authority have also been

TABLE 22.1
Required Data, Sources of Collection, and Emission Factors for Estimating PCF

Data Required	Sources of Collection	Sources of Emission Factor
	Primary Data	
Energy, water, and material use in the manufacture of pulp	Primary data from the case study unit	Secondary data source
Energy, water, and material use in the manufacture of paper	Primary data from the case study unit	Secondary data source
Fuel, electricity, water and materials used in steam generation	Primary data from the case study unit	Secondary data source
Transport of raw materials, such as wastepaper, chemicals, coal, finished products, and wastes generated	Primary data from the case study unit	Secondary data source
Waste generated	Primary data from the unit	Secondary data source

Note: Emission factors have mainly been taken from secondary sources such as transport (Department for Environment, Food and Rural Affairs [DEFRA]/Department of Energy and Climate Change [DECC] GHG reporting factors); fuel, electricity (Indian Network for Climate Change Assessment, India (INCCA: 2010), and CO_2 Baseline Database for the Indian Power Sector: User Guide, Version 8.0. January 2013 published by Government of India, Ministry of Power, Central Electricity Authority); wastewater, freshwater, waste (emission factors published by the PAS 2050: 2011 guidelines developed by British Standard Institute (BSI), co-sponsored by the Carbon Trust and DEFRA).

TABLE 22.2

Detailed Calculation of CF of 1 ton of Paper

Components of Production	Quantity (Total: 18,000 MT)	Unit	Emission Factor	Reference/Source	GHG Emission (kg CO$_2$-eq.) for 18,000 t of Paper	GHG Emission (kg CO$_2$-eq.) per Ton of Paper
Production Process: Pulping						
Wastewater	483,450,000	liter	0.0008 kg CO$_2$-eq./L	PAS 2050:2011 Guidelines	386,760	21.48
Energy (grid electricity)	3,968,653	kW h	1.29 kg CO$_2$-eq./kW h	CO$_2$ Baseline Database for the Indian Power Sector:User Guide, Version 8.0.January 2013	5,119,562	284.42
Transportation of wastepaper and pulping chemicals and wastes (4,373,900 + 216,004 + 10,000)	4,599,904	t km	0.10 kg CO$_2$-eq./t km	PAS 2050:2011 Guidelines	459,990	25.55
Waste:						
i. Kachra (plastic, thread, brown paper) landfill waste	300,000	kg	1.59 kg CO$_2$-eq./kg	PAS 2050:2011 Guidelines	477,000	26.5
ii. Sludge (sold to board mill and recycled)	900,000	kg	0	PAS 2050:2011 Guidelines	0	0
Subtotal (A)						357.95
Production Process: Paper Making						
Freshwater (for paper making)	153,000,000	L	0.0003 kg CO$_2$-eq./L	PAS 2050:2011 Guidelines	459,000	25.50
Freshwater (sealing and cooling)	132,000,000	L	0.0003 kg CO$_2$-eq./L	PAS 2050:2011 Guidelines	39,600	2.20
Energy (grid electricity)	6,975,209	kW h	1.29 kg CO$_2$-eq./kW h	CO$_2$ Baseline Database for the Indian Power Sector:User Guide, Version 8.0.January 2013	8,998,019	499.8
Freshwater (chemical preparation)	19,800,000	L	0.0003 kg CO$_2$-eq./L	PAS 2050:2011 Guidelines	5940	0.33
Freshwater (Evaporation loss)	28,900,000	L	0.0003 kg CO$_2$-eq./L	PAS 2050:2011 Guidelines	86,700	4.81
Transportation of paper-making chemicals	22,161	t km	0.10/kg CO$_2$-eq./t km	PAS 2050:2011 Guidelines	2216	0.123

(Continued)

TABLE 22.2 (Continued)
Detailed Calculation of CF of 1 ton of Paper

Components of Production	Quantity (Total): 18,000 MT	Unit	Emission Factor	Reference/Source	GHG Emission (kg CO$_2$-eq.) for 18,000 t of Paper	GHG Emission (kg CO$_2$-eq.) per Ton of Paper
Wastes:						
Fine sand and grits, synthetic contaminates (landfill wastes)	200,000	kg	1.59 kg CO$_2$-eq./kg		318,000	17.66
Transportation of wastes	10,000	t km	0.10/0.018 kg CO$_2$-eq./t km	PAS 2050:2011 Guidelines	1000	0.05
Subtotal (B)						550.47
Production Process: Finishing and Conversion						
Wastes:						
Rejected paper and reels (recycled)	556,000	Kg	0	PAS 2050:2011 Guidelines	00.00	00.00
Energy	340,000	kW h	1.29 kg CO$_2$-eq./kW h	CO$_2$ Baseline Database for the Indian Power Sector:User Guide, Version 8.0.January 2013	438,600	24.37
Subtotal (C)						24.37
Production Process: Steam Generation						
Pulverized coal	6379	MT		INCCA : 2010	11,996,000	666.44 [WN.1]
Energy (grid electricity)	408,000	kW h	1.29 kg CO$_2$-eq./kW h	CO$_2$ Baseline Database for the Indian Power Sector:User Guide, Version 8.0.January 2013	526,320	29.24
Freshwater in boiler	26,400,000	L	0.0003 kg CO$_2$-eq./L	PAS 2050:2011 Guidelines	7920	0.44
Waste:						
Landfill (boiler ash and fly ash)	1,600,000	kg	1.59 kg CO$_2$-eq./kg	PAS 2050:2011 Guidelines	2,544,000	141.33
Transportation of boiler ash and fly ash	384,000	t km	0.10/kg CO$_2$-eq./t km	PAS 2050:2011 Guidelines	38,400	2.13

(Continued)

TABLE 22.2 (Continued)
Detailed Calculation of CF of 1 ton of Paper

Components of Production	Quantity (Total: 18,000 MT)	Unit	Emission Factor	Reference/Source	GHG Emission (kg CO$_2$-eq.) for 18,000 t of Paper	GHG Emission (kg CO$_2$-eq.) per Ton of Paper
Transportation of coal	1,333,211	t km	0.10/kg CO$_2$-eq./t km	PAS 2050:2011 Guidelines	133,321	7.40
Transportation of chemicals::effluent treatment	14,502	t km	0.10/kg CO$_2$-eq./t km	PAS 2050:2011 Guidelines	1450	0.08
Boiler water treatment	1876	t km			188	0.10
Subtotal (D)						847.16
Water Treatment						
Energy (grid electricity)	646,000	kW h	1.29 kg CO$_2$-eq./kW h	CO$_2$ Baseline Database for the Indian Power Sector:User Guide, Version 8.0.January 2013	833,340	46.29
Subtotal (E)						**46.29**
Distribution of Finished Products						
Energy (lighting and air-conditioning of office, warehouse)	190,000	kW h	1.29 kg CO$_2$-eq./kW h	CO$_2$ Baseline Database for the Indian Power Sector:User Guide, Version 8.0.January 2013	245,100	13.61
Transportation of finished goods	6,717,375	t km	0.10/0.018 kg CO$_2$-eq./t km	PAS 2050:2011 Guidelines	6,717,375	37.31
Subtotal (F)						**50.92**
Grand total (A + B + C + D + E + F)					**33,788,800**	**1877.16**

adopted for CF calculation of coal and energy, respectively (Table 22.2). Another major limitation is the cradle-to-gate PCF assessment adopted in this study instead of cradle-to-grave assessment. This is due to the uncertainty in the availability of data from end users (i.e., consumers) and disposal at the end of life. Cradle-to-gate assessment lacks the completeness of a total cradle-to-grave assessment, thereby overlooking a large proportion of the impact of the product under consideration.

22.3 RESULTS AND DISCUSSIONS

Total emissions produced by the production unit for the year 2013–2014 is 33,788,800 kg CO_2 eq. for manufacturing 18,000 tons of paper. This result relates to one year (2013–2014), with calculations being based on formula (1) given in Section 22.2.2.1. PCF of a ton of paper is 1877.16 kg CO_2-eq. Table 22.3 shows the results of the calculations of annual CO_2-eq. emissions inventory summary of the production unit. Breakdown of the unit's CF (in kg of CO_2 eq.) has been shown under the various stages of production: pulping papermaking, finishing and conversion, steam generation, water treatment, and distribution of final products. The results of the product CF reveal that in 2013–2014, the largest source of emission for producing a ton of paper resulted from steam generation process (45.12%), followed by papermaking (29.36%) and pulping (19.06%), together these three sources form the lion's share (93.54%) of the total emission per ton of production. The remaining emission results from distribution of finished goods (2.71%) occupying the fourth largest position in the matter of emission followed by water treatment process (2.46%), with finishing and conversion taking the last two positions (1.29%) is being the least contributor. The reason behind steam generation being the largest emission contributor is the pulverized coal used in the process (78.66%) and 141.33 kg CO_2-eq. landfill waste generated per ton of paper produced in the form of fly ash and boiler ash (16.62%). Energy (purchased from grid in the eastern India) was the third reason for highest emission resulting from steam generation (3.45%). Transportation of coal also generated considerable amount of emission.

In papermaking stage, energy consumption (constituting of both grid energy and energy from coal) of 499.80 kg CO_2-eq. per ton of paper was the largest emitter (90.79%) followed by freshwater used (4.62%) and generated landfill waste such as fine sand and grits, as well as synthetic contaminates (3.20%).

Energy consumption of 284.95 kg CO_2-eq. per ton of paper production formed the largest share of emission (79.45%) in the pulping process, whereas landfill wastes (7.40%); transportation of raw materials and wastes—plastic, thread, brown paper (7.13%); wastewater used in the pulping process (6.0%) comes second, third, and fourth in this ranking. Consumption of energy is the primary component at the finishing and conversion stage and also in water treatment process. In the finishing stage, the generated wastes in the form of rejected reels are recycled and used as raw material in production of paper.

TABLE 22.3
Results of CF of a Ton of Paper

Stages	GHG Emissions (in kg CO_2-eq. per Process Stage)	% of Total
Pulping process	357.95	19.06
Paper-making process	550.47	29.36
Finishing and conversion process	24.37	1.29
Steam generation process	847.16	45.12
Water treatment process	46.29	2.46
Distribution process	50.92	2.71
Total	1877.16	100.00

Emissions resulting from transportation of finished goods to different destinations are the major contributor (73.27%) in the last stage. This is followed by the energy consumed (26.7%) for lighting and air conditioning in the warehouse and office of the paper production unit.

Analysis of percentagewise contribution of production and distribution components toward GHG emission (in kg CO_2-eq.) per ton of paper production (Table 22.4) reveals that grid energy comes first in order of ranking with 47.82%, followed by fuel (non-coking coal grade C & D) consumption (35.50%), and waste (9.88%). Transportation of raw materials, wastes, and finished goods (3·87%), freshwater (1.75%), and wastewater (1.18%) hold the fourth, fifth, and sixth position, respectively.

The objective of this study is to estimate the amount of GHG emission of a ton of paper and its allocation in the production and distribution process, and to highlight some possible ways of reducing emissions so as to guide the paper mill to achieve zero CO_2 emission. Some of the possible emission reduction strategies to achieve a GHG reduction program can be:

- To reduce emission at the origin
- To reduce emission in the process
- To promote renewable sources of energy
- To introduce CO_2 labeling at the product level (BSI, 2011. PAS 2050:2011)

Results of percentagewise contribution of production and distribution components toward GHG emission of a ton of paper produced by the case study unit (Table 22.4) reveal that the most important source of emission is the grid energy consumption followed by consumption of fossil fuel (coal) used for generation of steam used in the pulping and paper making process. The proportion of purchased electricity from grid in eastern India is derived from coal, diesel, and gas, the proportion of contribution of coal is 83.72% and that of gas (CNG) and diesel are 1.12% and 0.10%, respectively (Central Electricity Authority, 2013). Reduction in this hot spot area can be achieved by replacing fossil fuel by fuel with fewer impacts or by replacing it with a biomass source. Introduction of photovoltaic panels can also help toward achieving this target of reducing GHG emission in origin.

Reduction in the process level can be achieved by adopting heat recovery in the drying section and compressed air system revision, both of which the plant has already undertaken. Adoption of high-energy efficiency motor and vacuum pumps, introduction of lighting system efficiency, and technologically advanced processes are other alternatives for achieving energy efficiency. So far as energy and efficiency management are concerned, the plant is an ISO 14001:2004 certified unit for environmental management system and ISO 14001:2000 for quality management system.

The distribution of the finished product is also a major contribution (fourth largest) to the total footprint. This is because large distances are travelled for delivery of the finished products or

TABLE 22.4
Share of Production Component Used in the Production and Distribution Process

Production Components	GHG Emissions (in kg CO_2-eq. per Ton of Production and Distribution)	% of Total
Energy (grid)	897.73	47.82
Fuel (coal)	666.44	35.50
Wastes	185.49	9.88
Transportation of raw materials, wastes and finished goods	72.69	3.87
Freshwater	32.87	1.75
Wastewater	21.48	1.18
Total	**1877.16**	**100.00**

material purchase. This can be controlled through use of low-emission lorries (lorries confirming to stringent emission standards), including advanced technology components and producing less emissions than average vehicles do (e.g., Bharat Stage IV or Euro 4) and improving or reducing the delivery routes to the maximum possible extent.

22.4 CONCLUSION

PCF of a ton of newsprint paper manufactured from wastepaper by a paper production unit situated in West Bengal, India has been estimated and analyzed. The PAS 2050 method (the first internally applicable standard on product carbon footprinting) has been used to calculate PCF as these method measure emissions at a product level, providing the specific rules for identifying the system boundary and data quality rules for secondary data. GHG emissions for each process (measured in terms of kg CO_2-eq.) are calculated from the product of the organization's activity data and the corresponding and relevant emission factors (expressed in units of measurement) are used. At this stage, all the data having relevance to this case study are collected. The data collected are in accord with PAS 2050 analysis requirements. On the basis of the activities involved at the various stages of production, the product-related GHG emissions have been classified under six stages or heads, namely pulping stage, papermaking stage, finishing and conversion stage, steam generation and wastewater treatment, and distribution of finished products. Primary data based on a preset questionnaire were collected from a face-to-face interview of the environmental head of the case study unit under consideration. For the majority of the components of production and distribution, emission factors published by the PAS 2050: 2011 guidelines are considered. However some country-specific available emission factors developed and published by INCCA:2010 and Government of India, Ministry of Power have also been adopted.

Results revealed that total emissions produced by the production unit for the year 2013–2014 is 33,788,800 kg CO_2 eq. for producing 18,000 tons of paper. PCF of a ton of paper is 1877.16 kg CO_2-eq. The breakdown of the unit's CF (in kg of CO_2 eq.) shown under the various stages of production reveals that the largest source of emission for producing a ton of paper resulted from steam generation process, followed by papermaking and pulping taking the third position. The remaining emission results from distribution of finished goods (occupying the fourth largest segment) and emission from finishing and conversion being the least contributor.

Again, analysis of percentagewise contributions of production and distribution components toward GHG emission (in kg CO_2-eq.) per ton of paper production reveals that grid energy comes first in order of ranking, followed by fuel (non-coking coal grade C & D) consumption and waste in the second and third place, respectively. Transportation of raw materials, wastes and finished goods; freshwater; and wastewater hold the fourth, fifth, and sixth position in the contribution league.

On the basis of these results, the hotspots are identified and some recommendations have been provided for the manufacturing unit toward GHG emission reductions.

The emissions that are being generated in the production processes provide an in-depth analysis of the impact this product has created on ecological resources and the associated energy consumed by them. From the calculation of the value and analysis of results of PCF, it may be inferred that this footprint calculation results in communicating information on the environmental impacts to the business, to its stakeholders, and to the public at large. The results obtained by using this new performance footprint indicator provide an assessment of the production unit's environmental impacts over time and help in developing an assessment of sustainable limits for the product manufactured by the case study unit.

ACKNOWLEDGMENT

I extend my heartfelt thanks to Dr. Niladri Chakraborty, Professor, Department of Power Engineering, Jadavpur University, Kolkata, India for providing useful guidance and suggestions.

I also acknowledge with thanks the anonymous referee(s) for their encouraging and valuable observations on the earlier versions of the paper.

APPENDIX

WORKING NOTE 1 [W.N.1]

CF Calculation for Coal

Scope 1

The case study unit uses 6379 metric tons (MT) or tons (t) of (non-coking) coal for steam generation required in the production process for producing the annual output of 180,000 MT of paper.

6379 t of non-coking coal is = 6.379 kt (1 kt = 1000 t)

Net calorific value of (NCV) of non-coking coal = 6.379 kt × 19.63 Tj/kt (INCCA: 2010, pp. 13)
$$= 125.21 \text{ Tj}$$

Emission factor for coal (non-coking) = 95.81 t/Tj (INCCA: 2010, pp. 13)
$$= 125.21 \text{ Tj} \times 95.81 \text{ t/Tj (INCCA: 2010, pp. 13)}$$
$$= 11,996 \text{ t } CO_2 \text{ (1 ton = 1000 kg)}$$
$$= 11,996,000 \text{ kg } CO_2$$

Annual emissions from for steam generation for 18,000 tons of paper
$$= 11,996,000 \text{ kg } CO_2$$

REFERENCES

Bolwig, S., Gibbon, P., 2009. Emerging product carbon footprint standards and schemes and their possible trade impacts. Riso-R Report 1719 (EN). Riso National Laboratory for Sustainable Energy. Technical University of Denmark (DTU). ISBN: 978- 87-55or-r3796-0.

BSI, 2011. PAS 2050:2011. The Guide to PAS 2050:2011. Specification for the assessment of the life cycle greenhouse gas emissions of goods and services. British Standards Institution, London, UK.

Central Electricity Authority. 2013. *CO₂ Baseline Database for the Indian Power Sector: User Guide*, Version 8.0. Government of India, Ministry of Power, Central Electricity Authority.

Craig, J.A., Blanco, E.E., Sheffi, Y., 2012. A supply Chain View of Product Carbon Footprints. Results from the Banana Supply Chain. MIT working paper, Cambridge.

Dasgupta, M., 2014. Carbon footprint from road transport use in Kolkata city. *Transport and Environment* 32, 397–410.

Grunewald, N., Harteisen, M., Lay, J., Minx, J., Renner, S., 2012. The Carbon Footprint of Indian Household. Paper Prepared for the 32nd General Conference of The International Association for Research in Income and Wealth, Boston, USA.

http://www.rbi.org.in/Scripts/PublicationsView.aspx?id=151230.

http://indiabudget.nic.in/es2013-14/echap-01.pdf

Huang, A.Y., Manfred, L., Weber, C.L., Murray, J., Mathews, H.S., 2009. The Role of Input—Output analysis for the screening of corporate carbon footprints. *Economic Systems Research* 21, 217–242.

INCCA. 2010. Indian Network for Climate Change Assessment, India: Greenhouse Gas Emission, 2007. Ministry of Environment & Forest, Government of India, May 2010, pp. 13–27.

Kulkarni, H.D., 2013. Pulp and Paper Industry Raw Material Scenario—ITC Plantation a Case Study. *IIPTA* 25(1), 79–90.

Pathak, H., Jain, N., Bhatia, A., Mohanty, S., Gupta, N., 2009a. Global warming mitigation potential of biogas plants in India. *Environ. Monit. Assess.* 157, 407–418.

Pathak, H., Jain, N., Bhatia, A., Mohanty Patel, J., Agarwal, P.K., 2010. Carbon footprints of Indian food items. *Agric. Ecosyst. Environ.* 139, 66–73.

Plassmann, K., Norton, A., Attarzadeh, N., Jensen, M.P., Brenton, P., Edward-Jones, G. 2010. Methodological complexities of product carbon foot printing: A senility analysis of key variables in a developing country context. *Environ. Sci. Policy.* doi: 10:1016/j.envsci.2010.03.013.

The Carbon Trust. 2008. Product carbon foot printing: The new business opportunity, Experience from leading companies. The Carbon Trust, UK.

WRI (World Resource Institute) and WBCSD (World Business Council on Sustainable Development). 2011a. Green house gas protocol. Product Life Cycle Accounting and Reporting Standard. Earth Print Limited, USA.

23 Product Carbon Footprint Assessment of a Personal Electronic Product

Case Study of an Electronic Scale

Winco K.C. Yung and Subramanian Senthilkannan Muthu

CONTENTS

23.1 Introduction 504
23.2 Methodology...504
    23.2.1 Scope..504
        23.2.1.1 Product System and Its Function(s)...504
        23.2.1.2 Functional Unit 504
        23.2.1.3 Data and Data Quality 505
        23.2.1.4 Cut-Off Criteria and Cut-Offs 506
        23.2.1.5 Allocation Procedures...506
        23.2.1.6 Geographical and Time Boundary of Data...506
        23.2.1.7 System Boundary..506
        23.2.1.8 Relevant Assumptions in This Study 507
        23.2.1.9 Treatment of Electricity 509
23.3 Carbon Footprint Analysis 509
    23.3.1 Carbon Emission Calculation 509
23.4 Life-Cycle Inventory for PCF...511
    23.4.1 Life-Cycle Inventory of "Raw Material" Stage..511
        23.4.1.1 Sources of Data...511
    23.4.2 Life-Cycle Inventory of "Manufacturing" Stage..511
        23.4.2.1 Process Flow in "Manufacturing" Stage 511
    23.4.3 Life-Cycle Inventory of "Transportation and Storage" Stage 511
    23.4.4 Life-Cycle Inventory of "Use" Stage...512
    23.4.5 Life-Cycle Inventory of "End-of-Life" Stage 512
23.5 Life-Cycle Impact Assessment of PCF Analysis..512
23.6 Life-Cycle Interpretation 512
    23.6.1 Results of Life-Cycle Interpretation 512
    23.6.2 Specific GHG Emissions and Removals...512
    23.6.3 Limitations...512
        23.6.3.1 Focus on a Single Environmental Issue 512
        23.6.3.2 Limitations Related to the Assumptions...514
        23.6.3.3 Limitations Related to the Methodology 514
    23.6.4 Disclosure and Justification of Value Choices..514
    23.6.5 Sensitivity Analysis 514

          23.6.5.1 Sensitivity Analysis of Significant Input Data in Material Stage 514
          23.6.5.2 Sensitivity Analysis of Significant Input Data in Manufacturing Stage 518
          23.6.5.3 Sensitivity Analysis of Use Profile 518
          23.6.5.4 Sensitivity Analysis of End-of-Life Scenarios.. 518
      23.6.6 Uncertainty Analysis 520
23.7 Conclusion 520
Acknowledgment 521
References..521

23.1 INTRODUCTION

Carbon footprint is becoming more and more popular nowadays due to increasing awareness among consumers to use low-carbon and ecofriendly products. And one more dimension is the pressure on industries to design and manufacture low-carbon products. Integrating sustainable measures into their production phase will definitely help their product to get an edge over other products of the same category in the market. This is also a value-added component to their product and simultaneously it also helps in decision making. Product carbon footprint (PCF) analysis of an electronic scale performed and reported here is the first of its kind in this field. The results will help the manufacturer identify the hotspots and reduce carbon emissions accordingly in all the life-cycle phases.

    The main objective in carrying out this PCF study is to calculate the potential contribution of the chosen product, Color Coach Electronic Scale (hereafter named as "the scale"), to global warming expressed as CO_2-eq., by quantifying all significant GHG emissions and removals over the product's life cycle. The type and format of this report follows the format of the PCF study report in ISO/TS 14067 [1]. The results, data, methods, assumptions, and life-cycle interpretations are transparent and presented in sufficient detail to allow the reader to comprehend the complexities and trade-offs inherent in the PCF study. The PCF value is measured by quantifying the carbon emission for the entire product life cycle expressed in carbon equivalent value (kg CO_2-eq.), as per ISO/TS 14067.

$$CF_{total} = CF_{RM} + CF_M + CF_D + CF_U + CF_{EOL} \tag{23.1}$$

where CF_{RM} is the carbon footprint of the raw material (kg CO_2-eq.), CF_M is the carbon footprint in manufacturing (kg CO_2-eq.), CF_D is the carbon footprint in distribution (kg CO_2-eq.), CF_U is the carbon footprint during use (kg CO_2-eq.), and CF_{EOL} is the carbon footprint at the end of life (kg CO_2-eq.).

23.2 METHODOLOGY

23.2.1 Scope

23.2.1.1 Product System and Its Function(s)

The product selected for this study is an electronic scale, which is a home appliance for measuring body weight. The scale is mainly made from glass and stainless steel. It also provides the weight memory-saving history for up to four users. The scale is powered by four AAA batteries. Figure 23.1 shows the outlook of the electronic scale.

23.2.1.2 Functional Unit

As there is no product category rule (PCR) for an electronic scale, the functional unit (f.u.) of the scale is referenced to the PCRs of electronic products in Korea, Sweden, and Taiwan. The functional unit of this study is defined as one unit of electronic scale with primary packaging (PE bag) and secondary packaging (carton box). The reference flow is defined as one unit of electronic scale in this study.

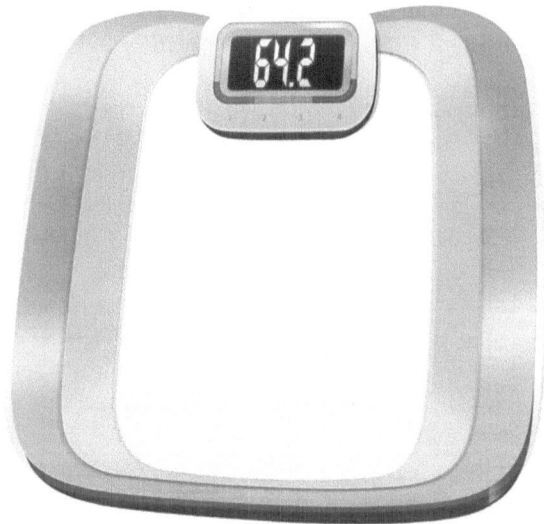

FIGURE 23.1   Electronic scale.

23.2.1.3   Data and Data Quality

23.2.1.3.1   Description of Data

Data on the unit process collected from the raw material, manufacturing, and transportation were primarily collected by Terraillon Asia Pacific Limited, which had financial and operational control of these unit processes. The data for the use stage and end-of-life stage were based on the scenario established by Terraillon Asia Pacific Limited. The emission factors used were as local as possible, with geographic, technology, and time-related considerations. Table 23.1 shows details of the data.

23.2.1.3.2   Characterization of Data Quality

The best quality data were used to reduce bias and uncertainty as far as practically possible. Primary and secondary data were selected to enable the goal and scope of the CFP study to be met.

TABLE 23.1
Description of Data

Life-Cycle Stage	Type of Data	Data
Raw material stage	Primary data (site-specific)	Amount of materials and components used
	Secondary data	Emission factors of material and components
Manufacturing stage	Primary data (site-specific)	Electricity consumption of manufacturing process
	Secondary data	Emission factors of electricity
Transportation stage	Primary data	Distance of transportation routes
		Transportation tools used
	Secondary data	Emission factors of transportation route
Use stage	Scenario	Resource consumption in use
	Secondary data	Emission factors of resource used
End-of-life stage	Scenario	Distance of transportation routes
		Transportation tools used
	Secondary data	Emission factors of disposal activities

23.2.1.4   Cut-Off Criteria and Cut-Offs

No cut-off was applied in this study, so no cut-off criteria were established.

23.2.1.5   Allocation Procedures

Allocation of this study was based on the ratio of electronic scales manufactured and total number of products manufactured in the company's factory in Shenzhen, from October 2012 to March 2013. The allocation was applied for the electricity, water, and consumables used in the manufacturing stage (Section 23.4.2).

23.2.1.6   Geographical and Time Boundary of Data

The electronic scale was manufactured in Shenzhen in China from October 2012 to March 2013 and was to be sold, used, and disposed of in France. According to the test report of the scale provided by Terraillon Asia Pacific limited, the scale passed an endurance test that suggested the scale can be used for over 10,000 times.

23.2.1.7   System Boundary

As a product manufacturer, Terraillon provides service and products to customers (B2C business model). The system boundary for the study covered the whole life cycle (cradle-to-grave) of the electronic scale, from raw material extraction to product end-of-life. Figure 23.2 shows the system boundary of this study.

All unit processes included in this study directly contributed to the end result (GHG emission). The overheads of the factory and equipment maintenance in all life-cycle stages were excluded in this study as they were not significant and not directly related to the GHG emission and removal of the scale.

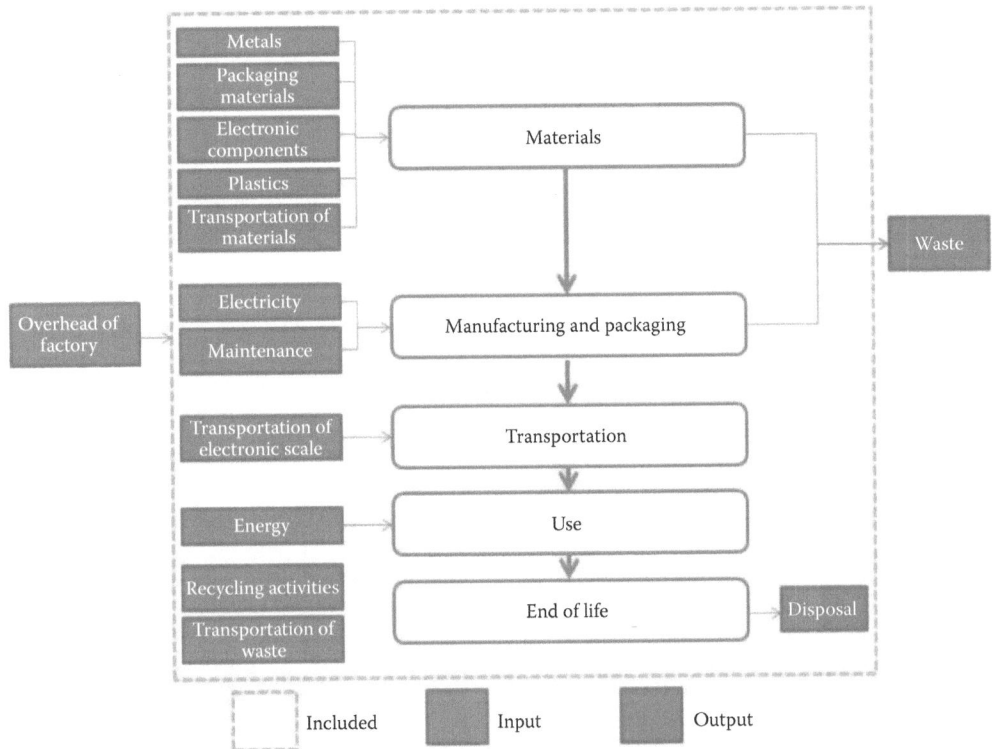

FIGURE 23.2   System boundary of carbon footprint analysis of "the scale."

23.2.1.8   Relevant Assumptions in This Study

23.2.1.8.1   Assumptions in "Material Stage"

- Packaging of material and packaging of packaging material were randomly sampled. The contribution to the carbon footprint from the packaging of material was proportionally increased according to the ratio of the selected packaging material weight and the total packaging weight.
- The transportation of material and packaging material was randomly sampled. The contribution to the carbon footprint of the material transportation was proportionally increased according to the ratio of the selected material weight and the total product weight.
- The QC label was made using 0.0006 g glue (33.33% of total weight), 0.0006 g LDPE (33.33% of total weight), and 0.0006 g paper (33.33% of total weight).
- The transportation of the packaging material "printed box" was not measured as the location of the printed box supplier was just next to the factory. Transportation of the printed box was excluded in this study.

23.2.1.8.2   Assumptions in the "Manufacturing Stage"

- The energy consumption of each scale manufactured in the factory was assumed to be the same.
- The consumables used for each scale manufactured in the factory were assumed to be the same.
- The annual energy consumption was averaged over 6 months of data as the scale was only manufactured from October 2012 to March 2013. No seasonal production factors were included in this study.

23.2.1.8.3   Assumptions in "Use Stage"

As there is no PCR for electronic scales, the use stage scenario for the scale was created by Terraillon. The scale passed an endurance test for 80 kg loading over 10,000 cycles. The scale could be used three times per day for 10 years under normal conditions.

- The useful life of "the scale" is 10 years.
- The scale is powered by four AAA batteries which provide 1200 mA h battery capacity.
- The energy consumption of "the scale" provided by Terraillon Asia Pacific is shown in Table 23.2.
- The emission factor of alkaline AAA battery is obtained from Ecoinvent 2.2. The emission factor for AAA battery is 5.37 kg CO_2-eq./kg.
- The operation duration for each use is 8 s, three times per day. Thus, the annual operation duration is 2.41 h. The annual standby duration is 8757.57 h.
- The annual energy consumption of "the scale" is shown in Table 23.3.
- Number of changes of batteries in 10 years = (annual energy consumption × 10 years)/battery capacity of four AAA batteries = 2.36, say three times.

TABLE 23.2
Energy Consumption of "the Scale"

Mode	Voltage (V)	Power Consumption (mA)	Power Consumption (W)
On mode	6	45	0.2700
Standby mode	6	0.0200	0.0001

TABLE 23.3
Annual Energy Consumption of "the Scale"

Mode	Power Consumption (mA)	Duration (h)	Annual Energy Consumption (mA h)
On mode	45	2.4300	109.3500
Standby mode	0.0200	8757.5700	175.1500
Total annual energy consumption			**283.6**

23.2.1.8.4 Assumptions in "End-of-Life" Stage

As there is no PCR for electronic scales, the end-of-life stage scenario of the scale follows the WEEE (waste electrical and electronic equipment) disposal scenario in EU. According to the WEEE directive (2012/19/EU) recast [2], electronic scales (category 9, monitoring and control instrument) should meet the minimum recovery targets applicable from August 13, 2013 to August 14, 2015:

- At least 70% shall be recovered
- At least 50% shall be recycled

According to the recycling percentage estimated by engineers in Terraillon Asia Pacific limited, the electronic scale can be disposed of with the recycling details shown in Table 23.4.

The electronic scale studied in this project was assumed to be recycled as detailed in Table 23.5. The used scales were sent to a WEEE disassembly facility in the municipal waste collection facility. It is assumed that recyclable materials for the used scale are transferred to the recyclers. Materials

TABLE 23.4
Recycling Details of the Scale

Parts	Recycling Percentage
Plastic parts	100
Glass	100
Metal parts	100
Load cells	100
PCB	0
LCD	0
UV glue	0
Electrical wires	100
Packaging	100

TABLE 23.5
Treatment Methods for the Disassambled Parts of "the Scale"

Treatment Method	Percentage	Weight (g)	Emission Factor (kg CO_2-eq./kg)	Database Source
Dismantling	100	2591.4582	0.27	Ecoinvent 2.2
Recycling	98.09	2541.9222	0.29	Ecoinvent 2.2
Landfill	1.91	49.5360	0.56	Ecoinvent 2.2

TABLE 23.6

Assumption of Traveling Distance and Tools for Transportation of Used Scale

Journey	Transportation Tool	Distance (km)	Weight (kg)	Emission Factor (kg CO_2-eq./t km)	Database Source
Municipal waste collection point to WEEE treatment plant	Euro4 16 t truck	60	2.2	0.17	Ecoinvent 2.2
WEEE treatment plant to recycler		120	2.11	0.17	Ecoinvent 2.2
WEEE treatment plant to landfill		120	0.09	0.17	Ecoinvent 2.2

in the used "scale" that cannot be recycled were sent to landfill for disposal. Tools for transportation and the treatment methods for used scale are shown in Table 23.6.

23.2.1.9 Treatment of Electricity

The electricity emission factor for the manufacturing stage was obtained from the Chinese Government National Development and Reform Commission (NDGC). The emission factor for southern China in 2012 was 0.9344 kg CO_2-eq./kW h [3]. The emission factor of alkaline AAA batteries is from Ecoinvent 2.2 and is given as 5.37 kg CO_2-eq./kg.

23.3 CARBON FOOTPRINT ANALYSIS

23.3.1 CARBON EMISSION CALCULATION

According to the data provided by Terraillon Asia Pacific Limited, calculation of the PCF is based on the product life cycle, which includes the raw material stage, the manufacturing stage, the distribution stage, the use stage, and the end-of-life stage. The analysis is based on the concept of life-cycle assessment and ISO/TS 14067: 2013 [1] and allows for computation of the total amount of carbon emission within the system boundary. This PCF study, according to ISO/TS 14067: 2013, includes four phases of life-cycle assessment (LCA), that is, goal and scope definition, life-cycle inventory (LCI), life-cycle impact assessment (LCIA), and life-cycle interpretation (LCI). Figure 23.3 shows the framework of life-cycle assessment from ISO 14040 [4].

The unit processes comprising the product system are grouped into life-cycle stages, that is, raw material acquisition, manufacturing, transportation, use, and end-of-life. GHG emissions and removals from the product's life cycle are assigned to the life-cycle stage in which the GHG emissions and removals occur. Partial PCFs may be added together to quantify the PCF, provided that they are performed according to the same methodology and that no gaps and overlaps exist. The PCF is calculated and presented as a carbon dioxide equivalent (CO_2-eq.) using the relevant 100-year global warming potential (GWP 100). Findings in the impact assessment and the sensitivity and contribution analysis which are performed as part of the interpretation help identify the most relevant and contributing ("key") processes and basic flows of the system.

$$\text{Carbon emission} = \text{Activity data} \times \text{Emission factor} \times \text{Global warming potential} \quad (23.2)$$

The "global warming potential" should be excluded, where the unit of "emission factor" was "kg CO_2-eq./unit." The activity data collected from each unit process was given as an input to an online PCF analyzer (G-BOM Analyzer provided by the Hong Kong Polytechnic University) and product carbon footprint assessment was done. The working principle, the methodology behind the software, and access can be found in the following link: http://www.pctech.ise.polyu.edu.hk/ecodesign/gbom_analyzer.html

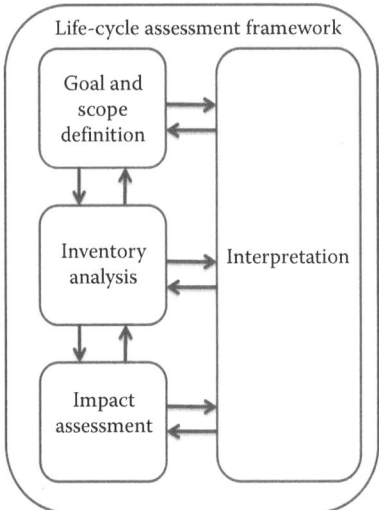

FIGURE 23.3 Framework of life-cycle assessment. (Adapted from International Organization for Standardization (ISO). 2006. ISO 14040:2006 Environmental management—Life-cycle assessment—Principles and Framework.)

Green Bill of Materials (G-BOM) analyzer is an on-line software, which is based on the structure of the bill of material. The G-BOM analyzer allows manufacturers to input their product information/data according to the product life cycle, which includes the raw material stage, the manufacturing stage (as shown in Figure 23.4), the distribution stage, the use stage and the end-of-life stage, and simply calculates the product carbon footprint using the greenhouse gas (GHG) emissions database provided. Users can define the material names, emission factors, and quantities according to their life-cycle inventory. The GHG emission database is pregrouped into five life-cycle stages. The emission factor can be selected manually in life-cycle groups or by keyword search engine.

During the quantification of this study, values were corrected to four decimal places in the life-cycle inventory data input. The results of the PCF in the impact assessment were corrected to two significant figures [5].

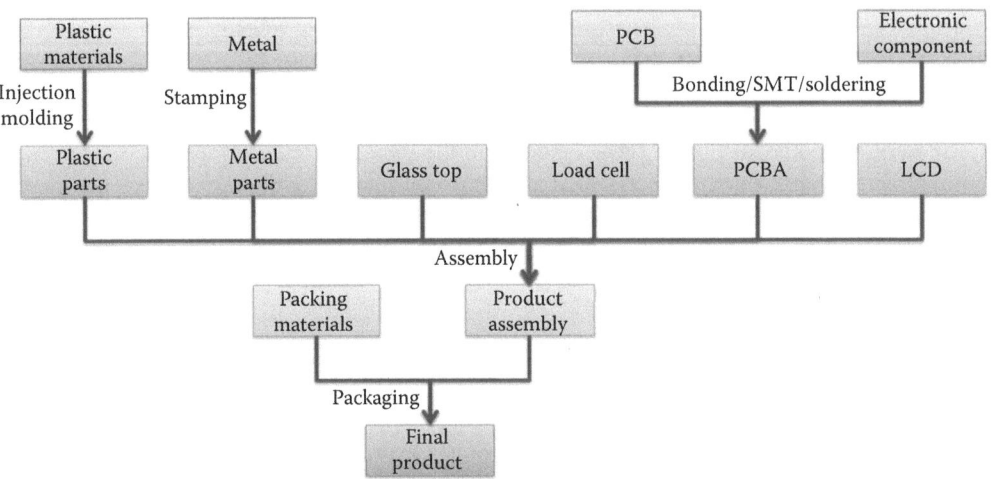

FIGURE 23.4 Process flow in "manufacturing" stage.

23.4 LIFE-CYCLE INVENTORY FOR PCF

23.4.1 Life-Cycle Inventory of "Raw Material" Stage

23.4.1.1 Sources of Data

1. NDRC—Chinese Government National Development and Reform Commission
2. JEMAI—Japan Environmental Management Association for Industry 2012
3. Korean LCI—Korean LCI database 2008
4. Ecoinvent—Swiss Ecoinvent data version 2.2

Database selection was based on credibility and geographical, technological, and time-related considerations, in order of priority. As such, the first priority is NDRC, followed by Ecoinvent, JEMAI, and Korean LCI. The data used in this study are mainly from Ecoinvent and JEMAI, when the emission factor of material is not available from NDRC.

- Contribution of randomly sampled packaging of material was proportionally increased by the following ratio of material weight: (335 g + 1600 g + 87 g)/2207.6582 = 91.6%.
- Contribution of randomly sampled packaging of packaging material was proportionally increased by the ratio of material weight: 184 g/383.8 g = 47.9%.
- The contribution of transportation of randomly sampled material was proportionally increased by the ratio of material weight: (335 g + 1600 g + 87 g)/2207.6582 = 91.6%.
- The transportation of the printed box cannot be measured as the location of the printed box supplier is right next to the factory.

23.4.2 Life-Cycle Inventory of "Manufacturing" Stage

23.4.2.1 Process Flow in "Manufacturing" Stage

- All activity data of the resources and energy consumption in the "manufacturing" stage were measured onsite and provided by Terraillon Asia Pacific Ltd. in January 2013.
- The emission factor of the electricity used was obtained from the Chinese Government National Development and Reform Commission (NDGC). The emission factor of southern China for 2012 was 0.9344 kg CO_2-eq./kW h [3].
- The energy consumption in the manufacturing stage was provided by Terraillon Asia Pacific. It was based on the allocation of total energy consumption of the factory from October 2012 to March 2013.
- The energy consumption for manufacturing each electronic scale was 2.2158 kW h.
- The emission factor of the tap water used was 0.168 CO_2-eq./m^3 (Source: Ecoinvent 2.2).
- The consumables used in the manufacturing stage were provided by Terraillon Asia Pacific. It was based on the allocation of the consumables purchase records of the factory from October 2012 to March 2013.
- The defect rate of the product within the factory was 0.5%. The contribution from the manufacturing stage was proportionally increased by 0.5%.

23.4.3 Life-Cycle Inventory of "Transportation and Storage" Stage

- Transportation routes and tools for distributing the finished product were provided by Terraillon Asia Pacific.
- The distance in each route taken was based on electronic mapping.
- The AAA batteries were not installed in the scale during transportation. "The scale" had no specific storage conditions. There was no energy consumption for storage.

23.4.4   Life-Cycle Inventory of "Use" Stage

- The useful life of "the scale" is 10 years.
- The scale was powered by four AAA batteries which provided 1200 mA h capacity.
- The energy consumption of "the scale" was provided by Terraillon Asia Pacific.
- The operation duration of each use was 8 s, for three times per day. Thus, the annual operation duration was 2.41 h. The annual standby duration was 8757.57 h.
- Annual energy consumption of "the scale" is shown in Table 23.3.
- Number of battery changes in 10 years = (Annual energy consumption × 10 years)/battery capacity of four AAA batteries = 2.36 = say three times = 12 AAA batteries.
- The weight of each AAA battery used in the scale was 11.3 g.
- The emission factor of a battery was 5.37 kg CO_2-eq./kg (Source: Ecoinvent 2.2).

23.4.5   Life-Cycle Inventory of "End-of-Life" Stage

According to the data provided by Terraillon, the electronic scale is disposed of in the scenarios shown in Tables 23.4–23.7.

The electronic scale could be recycled according to the details mentioned in Table 23.5. The used scales were sent to the WEEE disassembly facility from the municipal waste collection facility. It was assumed that recyclable materials in the used scale were transferred to the recyclers. Materials of the used "scale" that cannot be recycled were sent to landfill for disposal. Distance, tools for transportation, and treatment methods of used products were assumed as per the details in Table 23.6.

23.5   LIFE-CYCLE IMPACT ASSESSMENT OF PCF ANALYSIS

The calculation of the carbon footprint of the scale was based on a comprehensive coverage of raw material, manufacturing, transport and storage, use, and end-of-life stages. By assigning quantities for input and output in each process, the carbon footprint of each input, output, and process were obtained and are shown in Table 23.8.

23.6   LIFE-CYCLE INTERPRETATION

23.6.1   Results of Life-Cycle Interpretation

1. The carbon footprint per functional unit (one unit of electronic scale with packaging) is 13 kg CO_2-eq. The summary of the scale's carbon footprint for each life cycle is presented in Table 23.9.
2. The "raw materials" stage is the most significant life-cycle stage and contributes to a significant amount of carbon emission (60%) in the whole life cycle. In the "raw materials" stage, the PCB-1 (15.04%) has a significant contribution, as shown in Table 23.9 and Figure 23.5.
3. The "manufacturing" stage contributes the second most carbon emission (16.97%) in the whole life cycle.

23.6.2   Specific GHG Emissions and Removals

Table 23.10 separately shows the specific GHG emissions and removals arising from this study.

23.6.3   Limitations

23.6.3.1   Focus on a Single Environmental Issue

The PCF only reflects GHG emissions and removals of a product system, expressed as CO_2-eq., associated with raw material acquisition, manufacturing, transportation and storage, use, and end-of-life

TABLE 23.7

Disposal Scenarios of the Scale in End-of-Life Stage

No.	Name of Component	Material	Amount (g)	Treatment Methods
1	Screen printing glass plate	Reinforced glass	1600	Recycling
2	Bottom case (including battery house)	HIPS	147.3	
3	Upper case	HIPS	140	
4	FU100 weighing sensor	Stainless steel (including chip and wire)	71.5	
5	Stainless steel sheet (left and right)	HL304 stainless steel sheet	65	
6	Supportive plate	Stainless steel sheet	28.1	
7	LCD	LCD	26	Landfill
8	Backlight film	PMMA	23	Recycling
9	Display housing (bottom)	HIPS	20	
10	UV glue	UV glue for glass	11.5	Landfill
11	Display stand	HIPS	11	Recycling
12	Stainless steel sheet (bottom)	HL304 stainless steel sheet	9	
13	Stainless steel sheet (upper)	HL304 stainless steel sheet	8.6	
14	Round pin button	HIPS	7.7	
15	PCB-1	Printed circuit board (PCB)	7.3	Landfill
16	Display housing (upper)	PMMA	7	Recycling
17	Stainless steel display ring	HL304 stainless steel sheet	4.4	
18	PCB stand	HIPS	3.9	
19	Wire AWG30 (3.86 M)	Wire	3.7442	
20	M2.6X8 PA screws	Screws	2.4	
21	Paint for silk printing	Paint	2.399	
22	PCB-2	Printed circuit board (PCB)	2.3	Landfill
23	M2X6 PA screws	Screws	1.6	Recycling
24	Lead free tin wire	Lead free solder	1.477	Landfill
25	Metal film capacitor 0.47 uF/63 V DC	SMD type capacitor	0.425	
26	Bonding agent 8004 HH	Epoxy adhesive	0.207	
27	Electrolytic capacitor 10 uF/25 V	SMD type capacitor	0.2	
28	Lead free tin paste FLY-905 (RoHS)	Lead free solder	0.185	
29	Electrolytic capacitor 1 uF/50 V	SMD type capacitor	0.18	
30	Inductor	Inductor	0.15	
31	Microchip	Logic integrated circuit	0.15	
32	Parallel pin connector	Pin connector	0.1	
33	Hot press flat cable	Flat cable	0.1	
34	IC (OTP S 3P72N4xZZ -C0C4)	Logic integrated circuit	0.1	
35	Touch sensor HT BS804B NSOP 16 PIN	Touch sensor	0.1	
36	Hot melt glue stick (RoHS)	Adhesive	0.095	
37	Diode 1N5817	SMD type diode	0.071	
38	ATMEL EEPROM 1K AT93C46DN-SH-T SOP 1.5V	Logic integrated circuit	0.07	
39	Voltage stabilizer	Voltage stabilizer	0.054	
40	Lead free tin bar	Lead free solder	0.035	
41	Diode	SMD type diode	0.026	
42	SMD transistor	SMD type transistor	0.007	
43	Micro triode PNP	Triode	0.007	
44	Ceramic capacitor	SMD type capacitor	0.042	
45	Chip resistors	SMD type resistor	0.115	

(*Continued*)

TABLE 23.7 (*Continued*)
Disposal Scenarios of the Scale in End-of-Life Stage

No.	Name of Component	Material	Amount (g)	Treatment Methods
46	Lead free tin bar Sn 0.7 Cu	Lead free solder	0.017	
47	QC label	Label	0.002	Recycling
	Total weight		2207.6582	
		Packaging Materials		
1	Printed box	Cardboard	184	Recycling
2	Carton box	Carton box	104.8	
3	Protective cardboard	Cardboard	58	
4	User manual	Paper	25	
5	PE bag	PE film	12	
	Total weight		383.8	

treatment of a product. While the PCF can be an important environmental aspect of the life cycle of a product, affecting "climate," a product's life cycle may have other environmental impacts which are also of concern (e.g., resource depletion, air, water, soil, and ecosystems).

23.6.3.2 Limitations Related to the Assumptions

Assumptions were made in each stage of this study which may cause uncertainty and may affect the accuracy of the final result.

23.6.3.3 Limitations Related to the Methodology

The PCF was calculated based on LCA methodology. ISO 14040 and ISO 14044 address the inherent limitations and trade-offs in LCA methodology. These include the establishment of a functional unit and the system boundary, the availability and selection of appropriate data sources, allocation rules and assumptions regarding the transport, user behavior, and end-of-life scenarios. Some of the chosen data may be limited to a specific geographical area (e.g., national electricity grid) and/or may vary in time (e.g., seasonal variations). The choice of the functional unit and allocation rules may also be important in more accurately modeling a life cycle.

23.6.4 Disclosure and Justification of Value Choices

One of the important value choices is the database selection. The priority for the database selection was based on geographical, technological, and time-related considerations. The first priority was NDRC, followed by Ecoinvent, JEMAI, and Korean LCI.

23.6.5 Sensitivity Analysis

This section describes a sensitivity check regarding the influence of input data, electricity, alternative use profiles, and end-of-life scenarios on the final result.

23.6.5.1 Sensitivity Analysis of Significant Input Data in Material Stage

As shown in Table 23.8, the printed circuit board used in the material stage and the electricity in manufacturing stage contribute large amounts of carbon emissions, and their emission factors are secondary data, which are not localized. Sensitivity analysis was conducted for these two inventories. Different emission factors were used to analyze the sensitivity. For PCB emission factors,

TABLE 23.8
Carbon Footprint Results in Different Stages

No.	Item	Carbon Footprint (kg CO_2-eq.)	Percentage of Total PCF
	Raw Material		
	Basic Material and Components		
1	Screen printing glass plate	1.5758	12.82
2	Bottom case (including battery house)	0.5181	4.22
3	Upper case	0.4925	4.01
4	FU100 weighing sensor	0.1523	1.24
5	Stainless steel sheet (left and right)	0.1385	1.13
6	Supportive plate	0.0599	0.49
7	LCD	0.2744	2.23
8	Backlight film	0.1942	1.58
9	Display housing (bottom)	0.0704	0.57
10	UV glue	0.0332	0.27
11	Display stand	0.0387	0.31
12	Stainless steel sheet (bottom)	0.0192	0.16
13	Stainless steel sheet (upper)	0.0183	0.15
14	Round pin button	0.0271	0.22
15	PCB-1	1.8487	15.04
16	Display housing (upper)	0.0591	0.48
17	Stainless steel display ring	0.0094	0.08
18	PCB stand	0.0137	0.11
19	Wire AWG30 (3.86M)	0.0338	0.27
20	M2.6X8 PA screws	0.0086	0.07
21	Paint for silk printing	0.0069	0.06
22	PCB-2	0.5825	4.74
23	M2X6 PA screws	0.0057	0.05
24	Lead free tin wire	0.0088	0.07
25	Metal film capacitor 0.47 uF/63 V DC	0.0246	0.20
26	Bonding agent 8004 HH	0.0014	0.01
27	Electrolytic capacitor 10 uF/25 V	0.0116	0.09
28	Lead free tin paste FLY-905 (RoHS)	0.0011	0.01
29	Electrolytic capacitor 1 uF/50 V	0.0104	0.08
30	Inductor	0.0063	0.05
31	Microchip	0.1528	1.24
32	Parallel pin connector	0.0007	0.01
33	Hot press flat cable	0.0009	0.01
34	IC (OTP S 3P72N4xZZ -C0C4)	0.1018	0.83
35	Touch sensor HT BS804B NSOP 16 PIN	0.0049	0.04
36	Hot melt glue stick (RoHS)	0.0001	0.00
37	Diode 1N5817	0.0164	0.13
38	ATMEL EEPROM 1K AT93C46DN-SH-T SOP 1.5V	0.0713	0.58
39	Voltage stabilizer	0.0027	0.02
40	Lead free tin bar	0.0002	0.00
41	Diode	0.0060	0.05
42	SMD transistor	0.0010	0.01
43	Micro triode PNP	0.0003	0.00

(Continued)

TABLE 23.8 (*Continued*)
Carbon Footprint Results in Different Stages

No.	Item	Carbon Footprint (kg CO_2-eq.)	Percentage of Total PCF
44	Ceramic capacitor 0603 33 pF/50 V	0.0024	0.02
45	Chip resistors 0603 0.1 W 5% 10 K ohm (PHPS CHIP RES)	0.0150	0.12
46	Lead free tin bar Sn 0.7 Cu	0.0004	0.00
47	QC label	0.0000	0.00
Product Packaging			
55	Printed box	0.2663	2.17
56	Carton box	0.1517	1.23
57	Protective cardboard	0.0839	0.68
58	User manual	0.0807	0.66
59	PE bag	0.0534	0.43
Packaging of Basic Material and Components			
60	Paper bag (for HIPS)	0.0017	0.01
61	Foam (for glass)	0.0038	0.03
62	Paper (for glass)	0.0004	0.00
63	Cable tie (for glass)	0.0147	0.12
64	LDPE film (for stainless steel)	0.0063	0.05
Packaging of Packaging Materials			
65	Paper (for printed box)	0.0044	0.04
Transportation of Basic Materials, Components, and Packaging Materials			
66	HIPS (road)	0.0029	0.02
67	HIPS (ship)	0.0001	0.00
68	Glass (road)	0.0318	0.26
69	Stainless steel (road)	0.0012	0.01
Subtotal of material stage		7.3254	59.59
Manufacturing Process			
70	Energy consumption for production of one unit of studied product	2.0808	16.93
71	Water consumption for production of one unit of studied product	0.0031	0.03
72	Ethanol consumption (L)	0.0012	0.01
73	Hydraulic fluids consumption (L)	0.0010	0.01
74	Grease consumption (L)	0.0001	0.00
Subtotal of manufacturing stage		2.0862	16.97
Transportation			
75	Factory to Yantian port	0.0066	0.05
76	Yantian port to Le Havre port	0.5051	4.11
77	Le Havre port to retail shop in Paris	0.0968	0.79
Subtotal of transportation stage		0.6084	4.95
Use			
78	AAA batteries (g)	0.7282	5.92
Subtotal of use stage		0.7282	5.92
End of Life			
79	Dismantling	0.6997	5.69
80	Recycling	0.7372	6.00

(Continued)

TABLE 23.8 (*Continued*)
Carbon Footprint Results in Different Stages

No.	Item	Carbon Footprint (kg CO$_2$-eq.)	Percentage of Total PCF
81	Landfill	0.0277	0.23
82	Transportation for municipal waste collection point to WEEE treatment plant	0.0264	0.22
83	Transportation for WEEE treatment plant to recycler	0.0519	0.42
84	Transportation for WEEE treatment plant to landfill	0.0010	0.01
	Subtotal of end of life stage	1.5439	12.56
	Total carbon footprint	12.2921	

TABLE 23.9
Contribution of Carbon Footprint for Each Product Life-Cycle Stage

Life-Cycle Stage		Carbon Footprint (kg CO$_2$-eq.)		Contribution (%)	
Raw materials	Basic materials and components	6.62		53.87	
	Product packaging	0.03	7.33	5.17	59.59
	Packaging of basic materials and components	0.00		0.04	
	Transportation of basic materials, components and packaging materials	0.04		0.29	
	Manufacturing process		2.09		16.97
	Transportation		0.61		4.95
	Use		0.73		5.92
	End-of-life		1.54		12.56
	Total		12.29		100.00

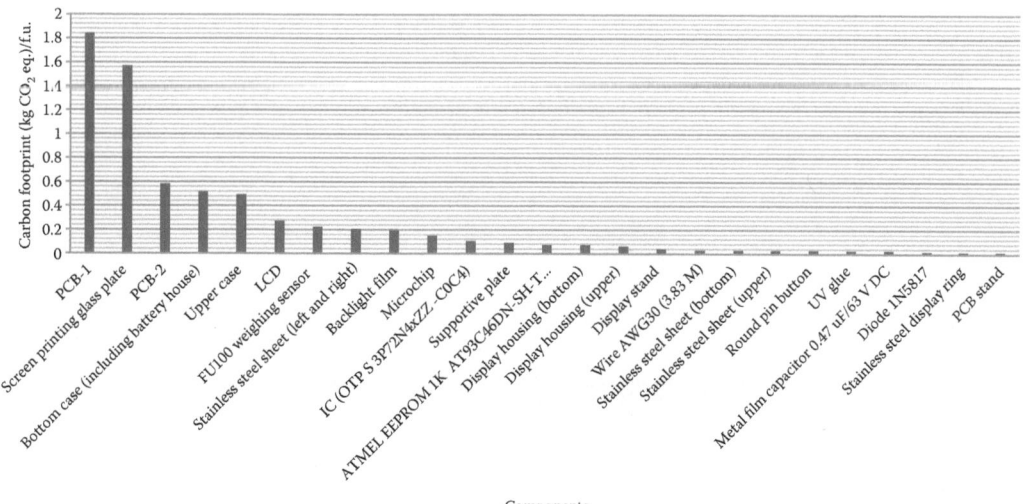

FIGURE 23.5　Contribution of carbon footprint by each part/component in the raw material stage.

TABLE 23.10

Specific GHG Emissions and Removals Arising from This Study

Source	Emission and Removals
Fossil carbon sources and sinks	N.A.
Biogenic carbon sources and sinks	N.A.
Direct land use change (dLUC)	N.A.
Aircraft transportation	N.A.

Table 23.11 and Figure 23.6 show the different carbon footprint results when changing the emission factors of the PCB from Eco-invent to Korean LCI.

23.6.5.2 Sensitivity Analysis of Significant Input Data in Manufacturing Stage

Table 23.12 and Figure 23.7 show different carbon footprint results in changing the emission factor of electricity provided by NDRC, Eco-invent. The emission factor obtained in a *Chinese Research Journal* [6] was used in the manufacturing stage.

23.6.5.3 Sensitivity Analysis of Use Profile

Different usage profiles were used to check the sensitivity regarding the effect of significant input(s) in the usage stage on the result. Table 23.13 and Figure 23.8 show the results of using different usage durations, that is, 5, 10, 15, and 20 years.

23.6.5.4 Sensitivity Analysis of End-of-Life Scenarios

Table 23.14 and Figure 23.9 show the different results from changing the end-of-life scenarios, that is, WEEE recast, and Hong Kong Environmental Protection Department.

TABLE 23.11

Result of Changing Emission Factor of PCB

Material	Weight (Included Defect Rate)	Database	Emission Factor (kg CO_2-eq./kg)	Carbon Emission (kg CO_2-eq.)/f.u.	Final Carbon Footprint Result (kg CO_2-eq./f.u.)
PCB	7.3365 g	Eco-invent 2.2	252	1.85	12.11
		Korean LCI 2008	10.2	0.07	10.34

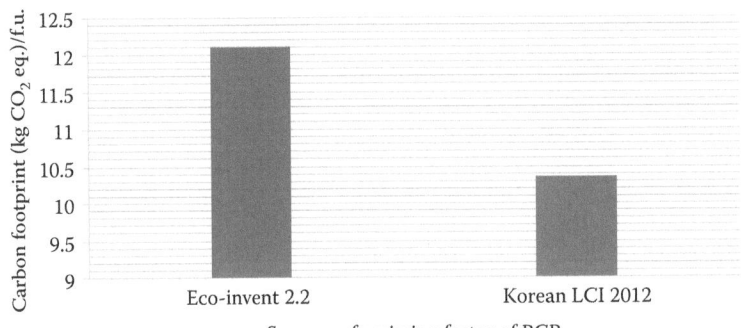

FIGURE 23.6   Result of changing emission factor of PCB.

TABLE 23.12

Result of Changing Emission Factor of Electricity Used in Injection Moulding

Electricity Consumption (Included Defect Rate)	Database	Emission Factor (kg CO₂-eq./kW h)	Carbon Emission (kg CO₂-eq./f.u.)	Final Carbon Footprint Result (kg CO₂-eq./f.u.)
2.226879	NDRC	0.9344	2.0808	11.00
	Eco-invent 2.2	1.51	3.3626	12.28
	HOU Ping et al., 2012	0.853	1.8995	10.82

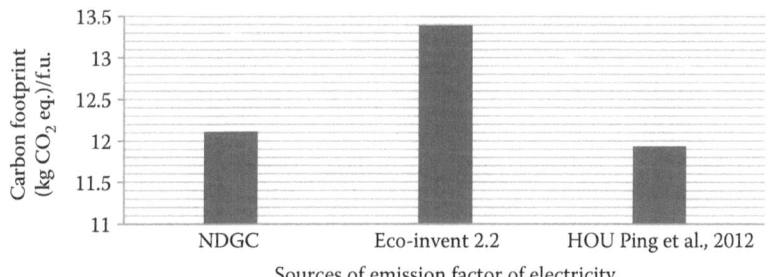

FIGURE 23.7   Result of changing emission factor of electricity used in manufacturing.

TABLE 23.13

Result of Using Different Usage Duration of 5, 10, 15, and 20 Years

Use Duration (Year)	Set of Batteries Required	Carbon Emission (kg CO₂-eq./f.u.)	Final Carbon Footprint Result (kg CO₂-eq./f.u.)
5	2	0.47	11.87
10	3	0.71	12.11
15	4	0.95	12.35
20	5	1.18	12.58

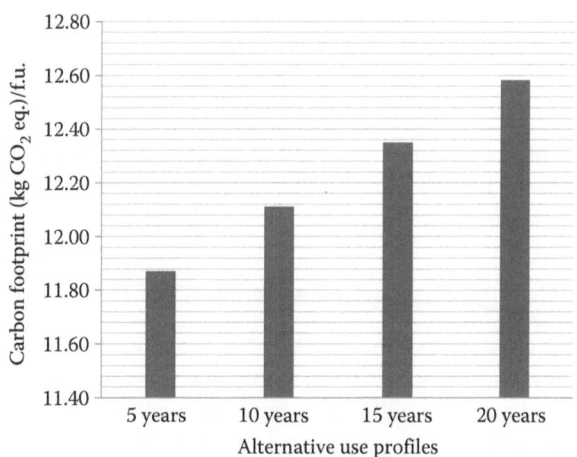

FIGURE 23.8   Result of using different usage duration for 5, 10, 15, and 20 years.

TABLE 23.14
Results by Changing End-of-Life Scenario

End-of-Life Scenario	Treatment Method	Percentage (%)	Weight (g)	Emission Factor (kg CO_2-eq./kg)	Database Source	Carbon Emission (kg CO_2-eq./f.u.)	Final Carbon Footprint Result (kg CO_2-eq./f.u.)
Data provided by	Dismantling	100.00	2591.46	0.27	Ecoinvent 2.2	0.70	12.11
Terraillon	Recycling	98	2541.92	0.29		0.74	
	Landfill	2	49.54	0.56		0.03	
Subtotal						1.46	
WEEE recast [2]	Dismantling	100.00	2591.34	0.27	Ecoinvent 2.2	0.70	12.42
	Recycling	50.00	1295.67	0.29		0.38	
	Recovery	20.00	518.27	0.50		0.26	
	Landfill	30.00	777.40	0.56		0.44	
Subtotal						1.77	
Waste disposal	Dismantling	100.00	2591.34	0.27	Ecoinvent 2.2	0.70	12.46
statistics of 2011	Recycling	48.00	1243.84	0.29		0.36	
by HKEPD [7]	Landfill	52.00	1347.50	0.56		0.75	
Subtotal						1.79	

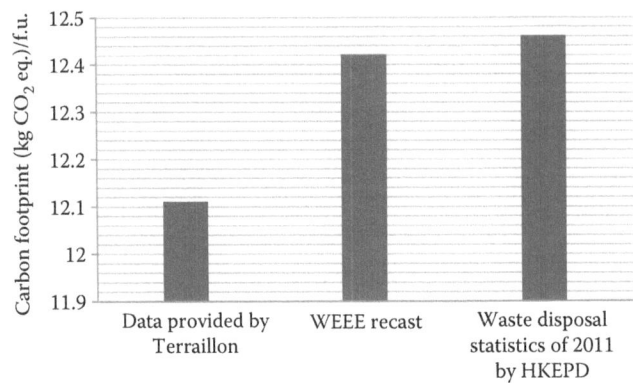

FIGURE 23.9   Results by changing end-of-life scenario.

23.6.6 UNCERTAINTY ANALYSIS

In this study, the data include primary data and secondary data. Some uncertainties may exist. Table 23.15 summarizes the type of data and the possible errors that may contribute to the uncertainty of the results.

23.7 CONCLUSION

According to the results of interpretation, the PCB (15%) and glass top (13%) have large contributions to the total GHG emission. Alternative materials can be employed in substitution. Also reducing the usage of PCB and glass top will reduce the GHG emissions to certain extent. The carbon footprint result also reflects that any improvement in the raw material stage (60%) and the manufacturing stage (17%) may effectively reduce the carbon footprint of the product. The design and the usage of the PCB in the material stage are critical in eco product design. Integrating an eco product

TABLE 23.15

Summary of Type of Data and Possible Error That May Contribute to Uncertainty of the Results

Type of Data	Type of Error Contributing Uncertainty to Results
Process data	Allocation of electricity for manufacturing one unit of scale
Process data	Seasonal factor of production
Process data	Scenarios adopted in use stage and end-of-life stage may be different with the reality
Emission factor	Localized emission factor is the first choice and only adopts other emission factors if localized emission factor is not applicable or available

design with low-carbon emissions is the future. Hence, we hope that this case study will point in the right direction toward future PCF analysis of personal electronic products for many manufacturers who are keen in delivering green products to their consumers.

ACKNOWLEDGMENT

We would like to express our thanks to SGS Hong Kong Limited for their support of the Project "Development of an embedded Greenhouse Gas (GHG) emissions database with a G-BOM analyzer and an SME advisory kit for electrical and electronic industries to respond to the implementation and compliance of ISO 14067 (carbon footprint of products)."

This case study report (Brand: Terraillon; Product model: Color Coach) of an electronic scale's PCF was made possible by the technical and professional contribution of our collaborating organization: SGS Hong Kong Limited.

We would like to thank the above organization which has given valuable technical support and professional advice and has participated on site assessment.

REFERENCES

1. International Organization for Standardization (ISO). 2013. ISO/TS 14067:2013 Greenhouse gases—Carbon footprint of products—Requirements and guidelines for quantification and communication.
2. European Commission. 2012. Directive 2012/19/EU of the European parliament and of the council on waste electrical and electronic equipment (WEEE) recast.
3. National Development and Reform Commission. 2012. 2012 Baseline Emission Factors for Regional Power Grids in China. cdm.ccchina.gov.cn/WebSite/CDM/UpFile/.
4. International Organization for Standardization (ISO). 2006. ISO 14040:2006 Environmental management—Life cycle assessment—Principles and framework.
5. British Standards Institution (BSI). 2011. PAS 2050:2011 Specification for the assessment of the life cycle greenhouse gas emission of goods and services.
6. Hou P, Wang H-T, Zhang H, Fan C-D, Huang N. 2012. Greenhouse gas emission factors of Chinese power grids for organization and product carbon footprint. *China Environmental Science* 32:961–967.
7. Hong Kong Waste Treatment and Disposal Statistics of Environmental Protection Department. 2014. http://www.epd.gov.hk/epd/english/environmentinhk/waste/data/stat_treat.html.

Index

A

Abiotic depletion potential (ADP), 119
ACCD, *see* Andalusia Construction Cost Database (ACCD)
Accoya® Wood Cladding, 89; *see also* Wood cladding CF
Acidification potential (AP), 119
ACQ, *see* Alkaline copper quaternary (ACQ)
ADP, *see* Abiotic depletion potential (ADP)
AEC (Architecture, Engineering, and Construction), 303; *see also* Building information modeling (BIM)
AFOLU, *see* Agriculture, Forestry and Other Land Uses (AFOLU)
Agricultural CF, 231, 389, 417–420, 431, 432, 434, 445–446; *see also* Aquaculture CF; Carbon footprint (CF); Grain crop production CF; Livestock CF; Microalgae
    of agricultural systems and cultivation practices, 135–136
    C-based sustainability index, 435–436
    of China's crop production, 437
    crop cultivars efficiency, 443
    crop diversification effect, 437–438, 444
    dietary management effect, 445
    estimation, 433–434
    farm management practices on, 436
    fertilizer effect, 442–443
    of field crops, 435, 438
    future research, 445
    GHG emission, 434, 437, 438, 439
    global warming potential reduction, 435, 438, 439
    of horticultural crops, 436, 439
    of land use, 227–228, 232
    LCA analysis, 438
    leguminous crop effect, 442
    livestock management effect, 444
    nitrogenous fertilizer effect, 436
    N-transformation inhibitors effect, 443–444
    organic manure effect, 444
    reason for, 432–433
    reduction strategies, 442
    soil management effect, 434
    of sugar beet crop, 438
    tillage effect, 443
    water management effect, 443
Agriculture, Forestry and Other Land Uses (AFOLU), 126
Agrochemical inputs, 452
Algae, 33
Algal biofuels, 16; *see also* Biofuel; Carbon sequestration
    as energy source, 27–32
    high lipid strains, 16
    lipid potential, 31
    potential analysis, 29
Alkaline copper quaternary (ACQ), 93
2-Amino-4-chloro-6-methyl-peyremidine (AM), 443
Andalusia Construction Cost Database (ACCD), 261
Anderson–Schulz–Flory distribution (ASF distribution), 112

Anthropocentric carbon footprint, 224
Anthropogenic-changed landscapes, 209; *see also* Urban soil carbon stocks and fluxes
Antioxidants, 390
AP, *see* Acidification potential (AP)
Aquaculture CF, 413; *see also* Carbon footprint (CF); Microalgae
    CO_2 outputs, 417
    feed formulation and energy inputs, 416
    goal, 414
    methodology, 415
    system boundaries, 415
AR, *see* Autotrophic respiration (AR)
AR4, *see* Fourth Assessment Report (AR4)
AR5, *see* 5th Assessment Report (AR5)
Arachidonic acid (AA), 411
ASF distribution, *see* Anderson–Schulz–Flory distribution (ASF distribution)
ATR, *see* Attenuated total reflectance (ATR)
Attenuated total reflectance (ATR), 19
Autotrophic respiration (AR), 187

B

BAGS, *see* Biological Algal Growth System (BAGS)
Basal respiration (BR), 187
Batangas Integrated Sugarcane Planters Multipurpose Cooperative (BISPMC), 453
BC, *see* Black carbon (BC)
B2C business model, 506
BEES, *see* Building for Environmental and Economic Sustainability (BEES)
β-carotene, 392
BIM, *see* Building information modeling (BIM)
Biochemical oxygen demand (BOD), 11, 35, 229
Biofuel; *see also* Algal biofuels; Carbon sequestration
    as energy source, 20
    for GHG estimation studies, 133–134
Biological Algal Growth System (BAGS), 409
Biomass; *see also* Synthetic biofuels
    gasification, 104
    pyrolysis, 103
Bio-oil, 103; *see also* Synthetic biofuels
    hydrotreating, 104
Biorefinery plant, 104, 107; *see also* Synthetic biofuels
Biosyngas, *see* Syngas
BISPMC, *see* Batangas Integrated Sugarcane Planters Multipurpose Cooperative (BISPMC)
Black carbon (BC), 187; *see also* Urban soil carbon stocks and fluxes
BLCC, *see* Building Life-Cycle Cost (BLCC)
BR, *see* Basal respiration (BR)
British PAS 2050 framework, 489
British Standard Institute (BSI), 9, 46, 372, 433, 489
BSI, *see* British Standard Institute (BSI)
Building for Environmental and Economic Sustainability (BEES), 242

Building information modeling (BIM), 303, 326–327;
    see also Construction industry and MS Excel;
    Embodied energy; Revit stands (REVise
    InsTantly)
    analysis, 322–323, 326
    BEAM Plus sustainable building rating, 307
    challenges, 304–306
    maturity levels, 305
    sustainability appraisal, 306–308
Building Life-Cycle Cost (BLCC), 242
Business-based CF, 380–382

C

Carbon, 187; *see also* Urban soil carbon stocks and fluxes
    cycle, 431
    turnover, 186
Carbonation, 152
Carbon-concentrating mechanisms (CCMs), 397
Carbon dioxide
    emissions per day of electricity and gas, 380
    sequestration in algae, 10, 16
Carbon dioxide equivalents (CO_2-eq.), 6, 509
    contributor, 382
Carbon Dioxide Information Analysis Center
        (CDIAC), 177
Carbon equivalent (CE), 433, 435
    value, 504
Carbon fluxes, 197; *see also* Urban soil carbon stocks and
        fluxes
    CH_4 flux, 197
    laboratory measurements of, 198
    measurement in soil, 197–198
    microbial respiration, 198
    spatial analysis and modeling of soil, 198–199
    from urban soils, 190–191
Carbon footprint (CF), 4, 45, 62, 282, 299–300, 432, 488;
        see also Agricultural CF; Carbon sequestration;
        Construction CF of India; Food production;
        Green house gas emission; Life-cycle impact
        assessment (LCIA); Life-cycle interpretation
        (LCI); Life-cycle inventory (LCI); Livestock
        CF; Paper PCF; Restaurant CFP; Sensitivity
        analysis; Sugar production CF; Urban CF;
        Urban planning and CF; Wetland ecosystem;
        Wood-based products CF
    allocation method, 285, 286
    analysis, 145, 509
    anthropocentric, 224
    assessment, 4–6
    background, 283
    British PAS 2050 framework, 489
    business-based, 380–382
    CO_2 emissions, 240
    data sources, 286
    emission factor, 518
    of energy activities, 226, 229–230
    of food, 432
    on food production, 131–132
    G-BOM analyzer, 510
    GHG flows in lifecycle, 147
    indirect emissions, 7, 11
    of industrial production, 227, 230–231
    ISO 14067, 489

land use, 285–286
literature on, 371–372
low-carbon products, 504
manufacturing process flow, 510
methods, 284
mitigation measures, 299
on municipal wastewater, 11
at product life-cycle stages, 517
quantification issues, 7–10
results, 295, 296, 515–517
scope and system boundaries, 284
sensitivity, 296–299
specifications, 102
standards, 282, 284
uncertainty analysis, 520
of wastes, 228–229, 231–232
Carbon footprint calculation, 62, 284, 509; *see also*
        Emission factors; Grain crop production CF;
        Greenhouse gas emissions (GHG emissions);
        Milk production CF
    activity data, 74–75
    case studies, 75
    CO_2 emissions, 66
    crop production, 65–67
    data sources, 69
    energy use in farm management, 68
    field survey data, 74
    GHG emissions, 66, 68
    livestock production, 67–69
    methane emissions, 65–66, 67
    methods, 65
    nitrous oxide emissions, 65, 67–68
    statistical data, 74
Carbon footprint of product (CFP), 10
Carbon labeling (C labeling), 136–137; *see also* Food
        production
Carbon sequestration, 16, 186; *see also* Algal biofuels;
        Biofuel; Carbon footprint; Greenhouse gas
        emissions (GHG emissions); *Lepocinclis ovum*;
        Urban soil carbon stocks and fluxes; Wetland
        ecosystem
    algal growth, 18–19, 22–24
    band assignments for infrared spectra, 25
    comparative assessment of cell densities, 23
    fatty acid composition, 19
    lipid composition, 24–26
    lipid extraction, 19
    materials and methods, 17
    microalgae, 16
    sampling, 17, 18
    spectral signature, 19
    study area, 17
    in wastewater algae, 26
Carbon stock, 187, 191; *see also* Urban soil carbon stocks
        and fluxes
    calculation, 195
    chemical analysis, 193
    digging soil pits, 192
    DSM, 196–197
    functional zoning, 192
    microbial biomass carbon, 195
    microbiological analysis, 193–195
    soil carbon analysis, 193
    soil sampling, 191–193

spatial modeling and mapping, 209–213
spatial variability, 196
total soil organic, 196
Carotenes, 395
C-based sustainability index, 435
CBS, *see* Construction-work breakdown system (CBS)
CCMs, *see* Carbon-concentrating mechanisms (CCMs)
CDIAC, *see* Carbon Dioxide Information Analysis Center (CDIAC)
CDM, *see* Clean development mechanism (CDM)
CDW, *see* Construction and demolition waste (CDW)
CE, *see* Carbon equivalent (CE)
Center for Sustainable Economy (CSE), 167
CF, *see* Carbon footprint (CF)
CFP, *see* Carbon footprint of product (CFP)
Chemical oxidation method (CO method), 193
Chemical oxygen demand (COD), 11, 35, 229
Chinese Government National Development and Reform Commission (NDGC), 509, 511
Chlorella, 392
CHP, *see* Combined heat and power (CHP)
CIC, *see* Construction Industry Council (CIC)
C labeling, *see* Carbon labeling (C labeling)
Clean development mechanism (CDM), 358
Climate
    and emissions data, 381
    footprint, 146
C_{mic}, *see* Microbial carbon (C_{mic})
CNA, *see* National Water Commission (CNA)
Combined heat and power (CHP), 156
CO method, *see* Chemical oxidation method (CO method)
CONAPO, *see* National Population Council (CONAPO)
Construction and demolition waste (CDW), 264
Construction CF of India, 239, 255; *see also* Carbon footprint (CF); Construction CF of Spain
    case studies, 247–252
    CF estimation factors, 243
    CF methodologies, 241–243
    clean development mechanism, 240
    CO_2 emission comparison, 251
    componentwise energy saving, 254
    direct energy, 241
    energy conservation, 252–254
    energy consumption and GHG emissions, 254
    GHG emission, 241, 244, 249, 250
    green buildings, 252
    LCA framework, 242
    recommendations, 254–255
    studies on, 244–247
    use of low-cost materials, 251
Construction CF of Spain, 259, 277–278; *see also* Carbon footprint (CF); Construction CF of India
    advantages, 278
    budget and resources, 272–273
    buildings, 260
    construction characteristics, 261
    electricity consumption, 265–266, 275
    emission factors, 264–265, 266, 270
    energy intensity, 267
    food consumption, 267, 272
    food footprint parameters, 268
    fuel consumption and emission coefficients, 266
    limitations, 278
    machinery, 271, 274

manpower CF, 277
material CFs, 269–271, 274, 277
methodology, 263, 265
MSW, 269, 273
project characteristics, 273
project selection, 260
quantities in concrete slabs per ACCD, 264
research potential, 278
residential building CF, 276
resource consumption quantification, 261–263
results, 272, 276–277
total CF of project number 25, 275
waste CF, 271–272, 274
water consumption, 266–267, 275
workers mobility, 268, 273
Construction industry and MS Excel, 319; *see also* Building information modeling (BIM)
    choosing output options, 321
    final export stages of QTO to MS Excel, 322
    modified output in MS Excel, 324–325
    output from MS Excel, 323
    preliminary export process to, 320
    QTO schedule, 319
    results and analysis, 322–323, 326
    saving QTO as delimited text file, 320
    specification of file type, 321
Construction Industry Council (CIC), 305
Construction-work breakdown system (CBS), 261
Corporate social responsibility (CSR), 474
Cradle-to-grave, 506, *see* Life-cycle assessment (LCA)
CRF, *see* Cumulative radiative forcing (CRF)
CSE, *see* Center for Sustainable Economy (CSE)
CSR, *see* Corporate social responsibility (CSR)
Cumulative radiative forcing (CRF), 147
Cyanobacteria, 395, 397, 410; *see also* Microalgae

D

DAF, *see* Dissolved air floatation (DAF)
Data
    collection questions, 375, 377
    loss, 304
DCD, *see* Dicyandiamide (DCD)
DE, *see* Digestible energy (DE)
DEFRA, *see* Department of Environment, Forest and Rural Affairs (DEFRA)
Deionized (DI), 336
Department of Environment, Forest and Rural Affairs (DEFRA), 489
Development and Reform Commission (NDRC), 337
DHA, *see* Docosahexaenoic acid (DHA)
DI, *see* Deionized (DI)
Dicyandiamide (DCD), 443
Digestible energy (DE), 292
Digital soil mapping (DSM), 192
Direct energy, 241; *see also* Construction CF of India
Direct land use change (dLUC), 518
Direct seeded rice (DSR), 435
Dissolved air floatation (DAF), 409
Dissolved organic matter (DOM), 21
Dissolved oxygen (DO), 18
dLUC, *see* Direct land use change (dLUC)
Docosahexaenoic acid (DHA), 390
DOM, *see* Dissolved organic matter (DOM)

Dry weight (DW), 395
DSM, *see* Digital soil mapping (DSM)
DSR, *see* Direct seeded rice (DSR)
DW, *see* Dry weight (DW)

E

Ecological footprint (EF), 4, 163 164; *see also* Urban
          planning and CF
    and biocapacity, 168
    online carbon/EF calculators, 167
    of residential site and consumption, 165
Economic input–output (EIO), 245
Ecosystem carbon payback time (ECPT), 134
ECPT, *see* Ecosystem carbon payback time (ECPT)
EF, *see* Ecological footprint (EF)
EGIS, *see* Environmental Geographic Information System
          (EGIS)
Eicosapentaenoic acid (EPA), 390
EIO, *see* Economic input–output (EIO); Environmental
          input–output (EIO)
Ekranic Technosols, 191
Ekranozems, *see* Ekranic Technosols
Electronic PCF, 504, 520; *see also* Carbon footprint (CF);
          Product carbon footprint (PCF)
Embodied energy, 303, 326; *see also* Building information
          modeling (BIM); Revit stands (REVise
          InsTantly)
    analysis, 322–323, 326
    and CO_2 analysis, 308–310
    computation of, 311
    forms of, 308
Emission factors, 69; *see also* Carbon footprint
          calculation; Grain crop production CF;
          Greenhouse gas emissions (GHG emissions);
          Milk production CF
    CO_2, 71
    conversion factor, 70
    for electricity generation, 74
    for enteric fermentation, 72, 73
    for irrigation, 72
    for manufacturing process, 72
    methane emission, 69, 70
    of N_2O emissions, 69
    net calorific values, 71
Emissions associated with electricity, 383
ENEHOPE, *see* Energy efficient housing options
          evaluation model (ENEHOPE)
Energy; *see also* Building information modeling (BIM);
          Construction CF of India; Construction CF of
          Spain
    analysis software, 303
    intensity, 267
    -intensive industry, 488
    mix, 382
    -saving buildings, 253
Energy efficient housing options evaluation model
          (ENEHOPE), 250
Environmental Geographic Information System
          (EGIS), 170
Environmental input–output (EIO), 9
Environmental products declarations (EPDs), 89
    European, 90
Environmental scanning electron microscope (ESEM), 19

EP, *see* Eutrophication potential (EP)
EPDs, *see* Environmental products declarations (EPDs)
ESEM, *see* Environmental scanning electron microscope
          (ESEM)
EU ETS, *see* European Union Emissions Trading Scheme
          (EU ETS)
European Union (EU), 85; *see also* Wood cladding CF
    forest strategy, 86
European Union Emissions Trading Scheme (EU ETS),
          474, 476
Eutrophication potential (EP), 119
Exergetic efficiency, 117; *see also* Synthetic biofuels
Exergy, 116; *see also* Synthetic biofuels
    efficiency, 117
    of product, 118
    of PYR-HT and GAS-FT Processes, 118

F

FAME, *see* Fatty acid methyl esters (FAME)
Fatty acid methyl esters (FAME), 24
FCP, *see* Fucoxanthinchlorophyll protein (FCP)
Fischer–Tropsch (FT), 103, 106; *see also* Synthetic
          biofuels
5th Assessment Report (AR5), 431
Food and Agriculture Organization (FAO), 62, 417
Food miles, 127–128
Food production, 125, 137; *see also* Carbon footprint (CF);
          Greenhouse gas emissions (GHG emissions)
    biofuels, 133–134
    carbon labeling and policy implications, 136–137
    case studies, 134
    CF application to, 131–132
    crop cultivation, 126–127
    C sequestration, 132
    efficiency improvisation, 129–130
    emission reduction, 128, 129
    food miles, 127–128
    and GHG emissions, 126
    GWPbio indicator, 133
    land-use change, 133
    livestock rearing and dairy, 127
    milk and beef production CF, 128
    packaging and transport, 130
    research, 137
    stages and supply chain, 126
Footprint, 4
Forest-based sector, 86; *see also* Wood cladding CF
Fossil; *see also* Wood cladding CF; Wood-based
          products CF
    carbon, 91, 94
    -fuel combustion, 143
Fourier transform-infrared (FTIR), 19
Fourth Assessment Report (AR4), 433
French Retification process, 88; *see also* Wood
          cladding CF
FT, *see* Fischer–Tropsch (FT)
FTIR, *see* Fourier transform-infrared (FTIR)
FU, *see* Functional unit (FU)
Fucoxanthinchlorophyll protein (FCP), 395
Functional unit (FU), 45, 58, 108, 146
    case study, 47–51
    consumer products, 53
    definitions, 46

design and industrial operations, 56
durable products, 55–56
impacts of, 47
inappropriate assignation of, 57–58
in industry, 56
of LCD panel, 334
level of inducements, 52
literature review, 46–47
multi-item consumable products, 54–55
at organizational level, 52
policy making, 51
single-item consumable products, 54
specification, 57
weighing scale, 504
Functional zoning, 192; *see also* Carbon stock

G

Gas chromatography (GC), 19, 26
GAS-FT, *see* Gasification-based system (GAS-FT)
Gasification-based system (GAS-FT), 108, 109; *see also*
    Synthetic biofuels
    Aspen Plus® flowsheet, 113
    exergy results of, 118
    inventory data of, 114
    production process, 111–112
Gasifier, 106; *see also* Synthetic biofuels
G-BOM, *see* Green Bill of Materials (G-BOM)
GE, *see* Gross energy (GE)
General linear model (GLM), 196
Genetically modified organisms (GMOs), 403
GFN, *see* Global Footprint Network (GFN)
GLM, *see* General linear model (GLM)
Global covenants on low-carbon emission, 487
Global Footprint Network (GFN), 167, 171
Global Reporting Initiative (GRI), 52
Global warming, 239; *see also* Greenhouse gas emissions
    (GHG emissions)
Global warming potential (GWP), 4, 63, 435, 509
    of Indian meals, 442
Glyceraldehyde-3-phosphate (G-3-P), 394
G-3-P, *see* Glyceraldehyde-3-phosphate (G-3-P)
Grain crop production CF, 75; *see also* Agricultural CF;
    Carbon footprint calculation
    calculation, 76–77
    data source, 76
    of maize, 438
    original data for CF calculation, 76
    results, 77
    scope and objective, 75–76
    sensitivity analysis, 78–79
    of spring barley crop, 436
    of wheat crop, 436
Green Bill of Materials (G-BOM), 338
    analyzer, 510
Green buildings, 252; *see also* Construction CF of India;
    Construction CF of Spain
Greenhouse gas (GHG), 4, 452, 488
Greenhouse gas emission in LAC, 351, 352; *see also*
    Greenhouse gas emissions (GHG emissions)
    biogenic origin, 353
    CF of representative technology, 356–358, 359–360
    CF reduction, 360–363
    CO_2 emissions, 352–353

electricity consumption, 359
emission factor, 366–367
methane emissions, 353–354
nitrous oxide emissions, 354
of representative technologies, 358–359
Greenhouse gas emission in Mexico, 362; *see also*
    Greenhouse gas emissions (GHG emissions)
    economic aspects of, 364–366
    emission factor for electricity, 366–367
    parameters for CO_2e emission, 365
    projected, 364
    technologies used in wastewater treatment, 364
    wastewater treatment in Mexico, 363
Greenhouse gas emissions (GHG emissions), 62, 239; *see*
    also Carbon footprint (CF); Carbon footprint
    calculation; Carbon sequestration; Emission
    factors; Food production; Grain crop production
    CF; Greenhouse gas emission in LAC;
    Greenhouse gas emission in Mexico; *Lepocinclis
    ovum*; Milk production CF; Wetland ecosystem
    anthropogenic, 5
    from arable crop production, 439
    from buildings, 244, 249, 250
    CO_2 emissions, 64, 65
    control, 241
    in crop, 63
    direct emission, 63, 64
    electricity consumption, 15
    emission factor estimation, 15
    emissions from diesel generator, 15
    emissions of treatment plants, 15
    enteric fermentation, 64
    farm management, 64
    indirect emission, 15, 64, 65
    by irrigation, 64
    in livestock, 64
    from manufacturing of agricultural inputs, 64
    manure treatment, 64
    methane emissions, 14, 63
    nitrogen estimation, 14
    nitrous oxide emissions, 63
    organically degradable material estimation, 13
    organic carbon stock changes in soil, 63–64
    quantification, 13–15, 438
    by sector, 5
    system boundaries, 63
    from wastewater sector, 10–13
Greenhouse gases, Regulated Emissions, and Energy use
    in Transportation (GREET), 458
Green Rating for Integrated Habitat Assessment
    (GRIHA), 253
GREET, *see* Greenhouse gases, Regulated Emissions, and
    Energy use in Transportation (GREET)
GRI, *see* Global Reporting Initiative (GRI)
GRIHA, *see* Green Rating for Integrated Habitat
    Assessment (GRIHA)
Gross energy (GE), 292
GWP, *see* Global warming potential (GWP)
GWPbio indicator, 133

H

HCF, *see* Housing CF (HCF)
Heating, ventilation, and air-conditioning (HVAC), 239

Heterotrophic respiration (HR), 187
Hill reaction, 394
Hot-watersoluble organic C (HWSOC), 193
Housing CF (HCF), 169
HR, *see* Heterotrophic respiration (HR)
HWSOC, *see* Hot-watersoluble organic C (HWSOC)
Hydrothermolysis, 89; *see also* Wood cladding CF
Hydrotreating, 104

I

ICE, *see* Inventory of Carbon and Energy (ICE)
IDF, *see* International Dairy Federation (IDF)
IES, *see* Integrated Environmental Solutions (IES)
IGP, *see* Indo-Gangetic Plains (IGP)
ILCD, *see* International Reference Life Cycle Data (ILCD)
INCCA, *see* Indian Network for Climate Change Assessment, India (INCCA)
Indian Network for Climate Change Assessment, India (INCCA), 493
Indo-Gangetic Plains (IGP), 435
Industrial process and product use (IPPU), 131
INEGEI, *see* National Emissions Inventory of Greenhouse Gases (INEGEI)
Infra-red gas analyzers (IRGA), 197
Input–output analysis (IOA), 490
Integrated Environmental Solutions (IES), 304
Intergovernmental Panel on Climate Change (IPCC), 5, 86, 282, 431
International Dairy Federation (IDF), 134, 282
International Organization for Standardization (ISO), 46, 475
International Organization for Standardization's technical specification (ISO/TS), 145
International Reference Life Cycle Data (ILCD), 91, 282
Inventory of Carbon and Energy (ICE), 457
IPCC, *see* Intergovernmental Panel on Climate Change (IPCC)
IPPU, *see* Industrial process and product use (IPPU)
IRGA, *see* Infra-red gas analyzers (IRGA)
ISO/TS, *see* International Organization for Standardization's technical specification (ISO/TS)

K

Kebony Cladding, 89; *see also* Wood cladding CF
KICSD, *see* Korean Institute Center for Sustainable Development (KICSD)
Korean Institute Center for Sustainable Development (KICSD), 169, 171

L

LAC, *see* Latin America and the Caribbean (LAC)
Land competition (LC), 119
Land-use change (LUC), 285
Latin America and the Caribbean (LAC), 351; *see also* Wastewater management in LAC
LC, *see* Land competition (LC)
LCA, *see* Life-cycle assessment (LCA)
LCI, *see* Life-cycle interpretation (LCI); Life-cycle inventory (LCI)

LCIA, *see* Life-cycle impact assessment (LCIA)
LC-ω-3-PUFA, *see* Long chain polyunsaturated omega-3 fatty acids (LC-ω-3-PUFA)
LDOE, *see* Liter diesel oil equivalent (LDOE)
Leadership in Energy and Environmental Design (LEED), 307
LEED, *see* Leadership in Energy and Environmental Design (LEED)
Lepocinclis ovum, 16; *see also* Carbon footprint (CF); Carbon sequestration; Greenhouse gas emissions (GHG emissions); Wetland ecosystem
    cellular characteristics of, 20
    chromatogram of fatty acids in, 28
    densities of, 22
    distribution, 20
    fatty acid profile of, 27
    FTIR spectra of, 23
    growth environment parameters, 21
    morphological features, 20
    nutrient concentrations and growth, 20–22
    proliferation, 22
    SEM image of, 26
LHCs, *see* Light-harvesting centers (LHCs)
Life-cycle assessment (LCA), 6, 433, 45, 67, 89, 509; *see also* Carbon footprint (CF); Product carbon footprint (PCF); Wood cladding CF
    of arable crop production, 438
    carbon storage in wood products, 93–94
    cascade use of wood, 95–96
    CF calculation, 91–92
    CO_2 uptake, 91
    fossil carbon, 91, 94
    framework, 242, 338, 510
    GHG Protocol, 489
    of kiln-dried softwood, 91
    PAS 2050, 91
    purpose of, 90
    steps, 90
    of wood products, 92–93
Life-cycle impact assessment (LCIA), 90, 242, 337, 509, 512; *see also* Carbon footprint (CF); Life-cycle assessment (LCA); Product carbon footprint (PCF)
Life-cycle interpretation (LCI), 337, 346, 509, 512; *see also* Carbon footprint (CF); Product carbon footprint (PCF); Sensitivity analysis
    changing emission factor, 348, 349
    contribution to carbon footprint, 346, 347
    disposal, 513–514
    GHG emissions and removals, 346, 347, 512
    limitations, 346, 347, 514
    results of, 346, 512
    sensitivity analysis, 348–349
    single environmental issue, 512
    uncertainty analysis, 349
    value choice justification, 347, 514
Life-cycle inventory (LCI), 90, 337, 438, 509; *see also* Carbon footprint (CF); Life-cycle assessment (LCA); Product carbon footprint (PCF)
    of end-of-life stage, 512
    of manufacturing stage, 511
    of raw material stage, 511
    sources of data, 511

of transportation and storage Stage, 511
of usage stage, 512
Light-harvesting centers (LHCs), 395
Lignocellulosic energy crops, 102
Lipid-soluble accessory pigments, 395
Liquid crystal display (LCD), 334; *see also* Product carbon
        footprint (PCF)
    module, 335
Liter diesel oil equivalent (LDOE), 455
Livestock CF, 440; *see also* Agricultural CF; Milk
        production CF
    dairy processing, 288–289
    DE values for fodder, 293
    emissions from cattle breeding, 292
    emissions from fodder production, 289–292
    emissions from transport, 294
    emission sources within covered system
        boundaries, 286
    feed consumption, 287
    fisheries management effect, 441, 445
    food habit effect on, 441
    GHG emissions, 440
    global warming potential of indian food, 441, 442
    manure storage and usage, 292–294
    methane conversion factors and manure storage
        systems, 293
    methodology of studies regarding CF of milk, 283
    of milk and beef production, 128
    nitrogen losses from manure, 294
    on piggery and poultry, 441
    raw materials and farm level, 286–288
    in supply chain, 489
Long chain polyunsaturated omega-3 fatty acids (LC-ω-3-
        PUFA), 390
LUC, *see* Land-use change (LUC)

M

Macronutrients, 396
Mass spectroscopy (MS), 19, 26
MBC, *see* Microbial biomass carbon (MBC)
MC, *see* Moisture content (MC)
Methane flux, 197; *see also* Carbon fluxes
Microalgae, 16, 390, 419–420; *see also* Agricultural CF;
        Aquaculture CF
    astaxanthin, 410, 411
    β-carotene, 410
    biofilm-based PBRs, 410
    biofuels, 402
    bioproducts from, 410
    Calvin cycle, 394
    carbon-concentrating mechanisms, 397
    carbonic anhydrases, 397
    CO_2 emissions for cultivation, 403
    comparison of cultivation system, 406, 409
    cultivation systems and bioproducts, 407
    dissolved oxygen concentrations, 408
    energy requirements, 410
    feed production, 402
    fertilization, 396–398, 400–401
    food-supplement organisms, 390
    GHG remediation potential, 411
    green algae, 410
    growth performance, 408

harvesting and dewatering techniques, 409
heterotrophic DHA production, 412
Hill reaction, 394
hybrid cultivation, 405, 409
inorganic carbon, 396
light and temperature, 398–390
light-harvesting centers, 395
light harvesting pigments, 395–396
lipid-soluble accessory pigments, 395
mass cultivation systems, 401, 405
mixotroph, 393
nitrogen assimilation, 397
as nutrient supplements, 390–391
oxygen saturation effect, 406
photosynthesis, 393–395
physiology, 392–393
productivity of outdoor cultivation, 404
raceway ponds, 405, 412
Redfield ratio, 396
response to light, 413
scytonemin, 413
temperature effect, 395–396
trace element, 398, 401
Microbial biomass carbon (MBC), 193, 195; *see also*
        Carbon stock
Microbial carbon (C_{mic}), 187
Microbial respiration (MR), 198, 201; *see also* Carbon
        fluxes
Micronutrients, 396
Milk production CF, 79; *see also* Carbon footprint (CF);
        Livestock CF
    CF calculation, 79–81
    data for, 80
    data source, 79
    emission factors for enteric fermentation, 80
    emissions from milk processing, 294–295
    functional unit, 285
    milk CF, 295–296
    milk processing, 296
    results, 81
    scope, 79
    for semi-skimmed UHT milk, 295
    sensitivity analysis, 81–82
Million liters per day (MLD), 10
Ministry of New and Renewable Energy (MNRE), 253
Mixotrophic algae, 10, 16
MLD, *see* Million liters per day (MLD)
MNRE, *see* Ministry of New and Renewable Energy
        (MNRE)
Modeling software, 303; *see also* Building information
        modeling (BIM)
Moisture content (MC), 88
Monounsaturated (MUFA), 398
MR, *see* Microbial respiration (MR)
MSW, *see* Municipal solid waste (MSW)
MUFA, *see* Monounsaturated (MUFA)
Municipal solid waste (MSW), 26

N

National Emissions Inventory of Greenhouse Gases
        (INEGEI), 363
National Population Council (CONAPO), 362
National Water Commission (CNA), 362

NCVs, *see* Net Calorific Values (NCVs)
NDGC, *see* Chinese Government National Development and Reform Commission (NDGC)
NDRC, *see* Development and Reform Commission (NDRC)
NDVI, *see* Normalized difference vegetation index (NDVI)
NEP, *see* Net ecosystem production (NEP)
NER, *see* Net energy ratio (NER)
Net Calorific Values (NCVs), 71
Net ecosystem production (NEP), 225
Net energy ratio (NER), 410
Nitrate, 33
Nitrification inhibitors, 443; *see also* Agricultural CF
Normalized difference vegetation index (NDVI), 196
Norwegian University of Technology and Science (NTNU), 247
NTNU, *see* Norwegian University of Technology and Science (NTNU)

O

Off-farm emissions, 178; *see also* Urban planning and CF
OHT-Process, *see* Oil-heat treatment process (OHT-Process)
Oil-heat treatment process (OHT-Process), 88; *see also* Wood cladding CF
On-farm emissions, 178; *see also* Urban planning and CF
Online carbon/EF calculators, 167
OPC, *see* Ordinary Portland cement (OPC)
Optical density (OD), 16
Ordinary Portland cement (OPC), 47

P

PA, *see* Process analysis (PA)
Paper manufacturing process, 490; *see also* Paper PCF
    calendaring, 491
    drying, 491
    press section, 490
    sheet forming section, 490
Paper PCF, 487, 499; *see also* Carbon footprint (CF); Paper manufacturing process
    assumption, 493
    calculation of CF, 494–496
    data collection, 493
    Indian paper industry, 488
    objectives of, 489
    results, 497–499
    scope, 493
    study boundary, 491
Parliamentary Office of Science and Technology (POST), 241
Parts per million by volume (ppmv), 4
PAS, *see* Publically Available Specification (PAS)
Path analysis, 170; *see also* Urban planning and CF
Payback period, 467
PBRs, *see* Photobioreactors (PBRs)
PCF, *see* Product carbon footprint (PCF)
PCR, *see* Product category rule (PCR)
Per-purchasable-product (PPP), 57
Personal footprint calculator, 167
PGA, *see* Phosphoglycerate (PGA)
Philippine sugar industry, 469; *see also* Sugar production CF
Phosphoglycerate (PGA), 394
Photobioreactors (PBRs), 401, 406

Phycocyanin, 413
Phycoerythrin, 413
PLATO technology, *see* Providing Lasting Advanced Timber Option technology (PLATO technology)
PMF, *see* Proton-motive force (PMF)
Polycyclic aromatic hydrocarbons (PAH), 468
Polyethylene Terephthalate (PET), 334
POST, *see* Parliamentary Office of Science and Technology (POST)
ppmv, *see* Parts per million by volume (ppmv)
PPP, *see* Per-purchasable-product (PPP)
Pressure-swing-adsorption (PSA), 104
Process analysis (PA), 9
Product-based CF, 378–380
Product carbon footprint (PCF), 46, 350, 488, 504; *see also* Carbon footprint (CF); Life-cycle interpretation (LCI); Life-cycle inventory (LCI)
    allocation procedures, 335, 506
    assumptions, 336, 507–509
    CF calculation, 337, 341–345
    cut-off criteria, 335, 506
    data, 334–335, 505, 506
    functional unit, 334
    G-BOM analyzer, 338
    LCA, 338
    LCI of manufacturing, 339–341
    life-cycle impact assessment, 512
    product system and function, 334
    product unit, 504
    system boundary, 335–336, 506
    treatment of electricity, 337, 509
    waste disposal, 341
Product category rule (PCR), 45, 90, 334, 504
Proton-motive force (PMF), 394
Providing Lasting Advanced Timber Option technology (PLATO technology), 88; *see also* Wood cladding CF
PSA, *see* Pressure-swing-adsorption (PSA)
Publically Available Specification (PAS), 46, 91, 372, 489, 490
Pyrolysis-based system, 108; *see also* Synthetic biofuels
    Aspen Plus® flowsheet of PYR-HT biorefinery, 110
    exergy results of, 118
    inventory data of, 112
    kinetic reaction model, 111
    production process, 109–111
Pyrolysis products, 104; *see also* Synthetic biofuels

Q

QTO, *see* Quantity takeoff (QTO)
Quantity takeoff (QTO), 311; *see also* Revit stands (REVise InsTantly)
    component selection for, 313
    field selection for, 314
    output, 314
    preliminary, 313
    project parameter creation, 315

R

RBCs, *see* Red blood cells (RBCs)
Readily oxidizable C (ROC), 193

RED, *see* Renewable Energy Directive (RED)
Red blood cells (RBCs), 34
Renewable Energy Directive (RED), 101; *see also* Synthetic biofuels
Renewable energy sources, 252
Research and Development (R&D), 481
Resource quantification database (RQDB), 263
Restaurant CFP, 369, 385
    for alternative menu items, 379
    business-based CF, 380–382
    calculations for transportation, 385–387
    climate and emissions data, 381
    CO_2 emissions, 380
    CO_2 equivalent contributor, 382
    comparison, 384
    complexity in, 370
    contributions to life-cycle CF, 380
    data collection, 374, 375, 377
    emission due to electricity, 383
    emission factors, 378
    energy consumption by equipments, 382–383
    energy mix effect, 382
    functional unit, 373
    initial data collection plan, 374–376
    interview responses, 376
    inventory, 376
    limitations, 384–385
    literature on CF, 371–372
    methodology, 372, 373
    product-based CF, 378–380
    project description, 370
    sensitivity analysis, 383
    study scope, 370, 372
    system boundaries, 373–374
    uncertainty analysis, 384
    for utilities, 381
REVise InsTantly, *see* Revit stands (REVise InsTantly)
Revit stands (REVise InsTantly), 310; *see also* Building information modeling (BIM); Embodied energy; Quantity takeoff (QTO)
    analysis, 322–323, 326
    changing quantities, 318
    components, 310, 317
    exploiting model in, 315
    material properties, 318
    model description, 311, 312
    modeling formulae in, 316
    output of project parameters, 316
Ribulose-1,5-bisphosphate (RuBP), 394
ROC, *see* Readily oxidizable C (ROC)
Rockwool supply chain, 475, 479; *see also* Supply chain CF; Two-phase CF framework
    best practice identification, 481
    CF allocation, 482, 483
    CF implementation, 481
    CF monitoring, 484
    CF quantification, 481–482
    decision making for CF, 483–484
    echelons of system, 480
    European regulatory framework recording, 481
    evaluation of CF findings, 483
    goal setting for CF management, 481
    re-evaluation of outcomes, 484
    role identification, 481

RQDB, *see* Resource quantification database (RQDB)
Rs, *see* Soil respiration (Rs)
RuBP, *see* Ribulose-1,5-bisphosphate (RuBP)

S

San Carlos Bioenergy, Inc. (SCBI), 453
Saturated fatty acid (SFA), 398
SCBI, *see* San Carlos Bioenergy, Inc. (SCBI)
SCC, *see* Social cost of carbon (SCC)
SCMs, *see* Supplementary cementitious materials (SCMs)
Scytonemin, 414
Second-generation biofuels, 102; *see also* Synthetic biofuels
Selective inhibition (SI), 201
Sensitivity analysis, 514; *see also* Life cycle interpretation (LCI); Product carbon footprint (PCF)
    of end-of-life scenarios, 518
    manufacturing stage input data, 518
    material stage input data, 514
    of usage profile, 518–520
Sewage treatment plant (STP), 17, 37
SFA, *see* Saturated fatty acid (SFA)
SI, *see* Selective inhibition (SI)
SIC, *see* Soil inorganic carbon (SIC)
SIR, *see* Substrate-induced respiration (SIR)
Site-specific nutrient management (SSNM), 443
Small and medium enterprises (SMEs), 305
SMEs, *see* Small and medium enterprises (SMEs)
SOC, *see* Soil organic carbon (SOC)
Social cost of carbon (SCC), 452
Soil, 187; *see also* Agricultural CF; Carbon stock; Urban soil carbon stocks and fluxes
    anthropogenic effects on, 186
    augering, 192
    for carbon cycle, 187–188
    carbon efflux, 188
    chemical features, 191
    microbial community, 200
    respiration, 187
    sealing, 191
    SOC distribution in, 190
    water regime, 443
Soil inorganic carbon (SIC), 187; *see also* Urban soil carbon stocks and fluxes
Soil organic carbon (SOC), 186
    distribution in urban and natural soils, 190
Soil organic matter (SOM), 468
Soil respiration (Rs), 187; *see also* Urban soil carbon stocks and fluxes
SOM, *see* Soil organic matter (SOM)
SSNM, *see* Site-specific nutrient management (SSNM)
STP, *see* Sewage treatment plant (STP)
Substrate-induced respiration (SIR), 194
Sugar factory, 457; *see also* Sugar production CF
Sugar production CF, 452; *see also* Carbon footprint (CF)
    calculation, 455, 461–462, 464, 467
    cane burning, 462
    carbon emission, 457, 460, 462–463, 464–466
    carbon sources for sugar, 454
    co-generation, 453
    crop types, 452
    data source, 453
    factory construction CF, 457

Sugar production CF (*Continued*)
  factory level, 469–470
  at farm level, 455–456
  field level, 468–469
  GHG emission factors, 458
  material balance for milling section, 459
  operations involved in, 453
  payback period, 467
  Philippine sugar industry, 469
  results, 461–467
  sequestered carbon dioxide, 460–461, 463–464
  sugarcane milling, 456–457
Supplementary cementitious materials (SCMs), 49
Supply chain CF; *see also* Rockwool supply chain;
        Two-phase CF framework
  best practices in, 477
  decision making for, 478–479
  measurement, 478
  monitoring and reporting, 479
  strategic goals for, 477
Sustainability, 303; *see also* Building information
        modeling (BIM)
  analysis software systems, 304
  assessment concept, 307
  environmental, 306
Syngas, 104; *see also* Synthetic biofuels
  energy products from, 106
Synthetic biofuels, 101, 120; *see also* Gasification-
        based system (GAS-FT); Pyrolysis-based
        system
  agriculture, 109
  biofuel use, 112
  biorefinery plant, 104, 107
  CF and GHG emission savings, 114–116
  CF as indicator, 116
  CF framework, 106
  data acquisition, 109
  environmental perspective, 119–120
  exergy of system, 117
  gasification and FT synthesis, 104–106
  production, 102–103
  pyrolysis and hydroupgrading, 103–104
  results, 114
  thermochemical routes, 103
  thermodynamic perspective, 116–118
System boundary, 147, 506; *see also* Wood-based products
        CF
  carbon dynamics, 151
  end-of-life stage, 150–152
  energy for alternative production, 149
  GHG dynamics quantification, 151
  maintenance, 150
  operation stage and service life, 149–150
  production stage, 148–149
  system boundaries, 148, 152–154

T

TAG, *see* Triacylglyceride (TAG)
Takeoff, 311; *see also* Quantity takeoff (QTO)
TCF, *see* Travel CF (TCF)
tCO_2, *see* Tones of carbon dioxide (tCO_2)
TDSs, *see* Total dissolved solids (TDSs)
TERI, *see* The Energy and Resources Institute (TERI)

The Energy and Resources Institute (TERI), 253
Thin-layer chromatography (TLC), 19
Third party logistics (3PL), 478
3PL, *see* Third party logistics (3PL)
TLC, *see* Thin-layer chromatography (TLC)
TOC, *see* Total organic carbon (TOC)
Tones of carbon dioxide (tCO_2), 169
Total dissolved solids (TDSs), 18
Total organic carbon (TOC), 15
Total primary energy supply (TPES), 143; *see also*
        Wood-based products CF
  breakdown of, 144
Total production cost (TPC), 266
TPC, *see* Total production cost (TPC)
TPES, *see* Total primary energy supply (TPES)
Transportation CF calculations, 385–387
Travel CF (TCF), 169
Triacylglyceride (TAG), 19
Two-phase CF framework, 474; *see also* Rockwool supply
        chain; Supply chain CF
  case study, 479–484
  decision re-evaluation, 479
  EU ETS, 476
  evaluation of goals, 478
  recommendations, 484–485
  supply chain CF, 475, 476
  supply chain partners' role, 477–478
  systematic recording of international legislative
        frameworks, 476
  trends in CF management, 476

U

UASB, *see* Upflow Anaerobic Sludge Blanket Reactor
        (UASB)
UGD, *see* Underground drainage (UGD)
UHT, *see* Ultrahigh temperature (UHT)
Ultrahigh temperature (UHT), 285
Underground drainage (UGD), 30
UNFCCC, *see* United Nations Framework Convention on
        Climate Change (UNFCCC)
United Nations Framework Convention on Climate
        Change (UNFCCC), 358
Upflow Anaerobic Sludge Blanket Reactor (UASB), 35
Urban CF, 224, 235–236
  accounting categories, 226
  of agricultural activities, 227, 231
  carbon absorption, 228
  carbon budget, 233
  carbon emissions, 224, 227, 228
  data sources, 225
  decoupling analysis of economic growth and CF,
        234–235
  of energy activities, 226, 229–230
  evaluation methods, 226
  of industrial production, 227, 230–231
  intensity and productivity, 233–234
  of land use, 227–228, 232
  literature review, 224
  policy implications, 236
  prediction of, 235
  pressure analysis, 233
  results, 229
  of wastes, 228–229, 231–232

Urban planning and CF, 163, 164, 179–180; *see also*
    Ecological footprint (EF)
  agricultural CF, 177–179
  case studies, 168
  emissions, 178
  energy consumption in demand sectors, 172
  energy use determinants, 165
  findings on spatial-energy relationships, 166
  G statistics and *z*-scores of footprint variables, 175
  household CF, 168–171
  housing footprint components, 172
  individual CF, 171–177
  measures of variables, 169
  path analysis, 170
  path diagram, 176
  personal footprint calculator, 167
  spatial patterns, 173, 174
  travel footprint components, 173
Urban soil carbon stocks and fluxes, 186, 213; *see also*
    Carbon fluxes; Carbon stock
  anthropogenic effects, 186, 188, 189
  case studies, 199
  SOC distribution in, 190
  soil microbial community, 200
  soil sealing, 191
  spatial modeling and mapping, 209
  spatial–temporal variability of, 191
  spatial–temporal variability of respiration, 204–208
  spatial variability of, 200–204
  urban heat island effect, 188

W

Wastewater management in LAC, 354–355; *see also*
    Wastewater treatment GHG emission
  accumulated treated flow, 355
  challenges in, 352
  representative treatment systems, 357
  technology selection criteria, 355–356
  treatment technologies, 354
  water and sanitation program, 351
Wastewater treatment GHG emission, *see* Greenhouse gas
    emission in LAC; Greenhouse gas emission in
    Mexico
Wastewater treatment plants (WWTPs), 353
  in Bengaluru, 12
  mechanical, 13
Wastewater treatment systems, 33
Water-soluble organic C (WSOC), 193
WBCSD, *see* World Business Council on Sustainable
    Development (WBCSD)
Wetland ecosystem, 32; *see also* Carbon footprint (CF);
    Carbon sequestration; Green house gas
    emission; *Lepocinclis ovum*
  algal composition, 36
  BOD and COD, 35
  functional aspects of, 37–39

GHG emission mitigation, 37
  macrophyte composition, 37
  nutrients, 33
  sampling locations, 34
  wastewater management, 34, 35
  as wastewater treatment systems, 32
  for water management, 38
Wood-based products CF, 143, 157; *see also* System
    boundary
  analysis, 145
  CF characterization, 146–147
  climate implications, 145–146
  electricity production, 154–155
  functional unit, 146
  GHG assessment standards, 145
  GHG flows, 147
  low-carbon materials, 144
  recommendations, 156–157
  system boundaries, 147
  system-wide material flow of wood products, 144
  treatment of allocation, 155–156
Wood cladding CF, 87, 97; *see also* Life-cycle assessment
    (LCA); Wood-based products CF
  Accoya® Wood Cladding, 89
  biopolymers, 94
  as building material, 87
  external wooden claddings, 87
  French retification process, 88
  heat-treatment processes, 88
  hydrothermolysis, 89
  Kebony Cladding, 89
  OHT-Process, 88
  PLATO technology, 88
  products and climate change, 86
  in sustainable buildings, 96–97
  ThermoWood, 89
Wood plastic composite (WPC), 93
World Business Council on Sustainable Development
    (WBCSD), 46, 131, 490
World population, 389
World Reference Base (WRB), 188
World Resource Institute (WRI), 131, 245, 433, 490
World Wildlife Fund (WWF), 167
WPC, *see* Wood plastic composite (WPC)
WRB, *see* World Reference Base (WRB)
WRI, *see* World Resource Institute (WRI)
WSOC, *see* Water-soluble organic C (WSOC)
WWF, *see* World Wildlife Fund (WWF)
WWTPs, *see* Wastewater treatment plants (WWTPs)

X

Xanthophylls, 392, 395

Z

Z-scheme of photosynthesis, 394